The Physical Geography of Western Europe

THE OXFORD REGIONAL ENVIRONMENTS SERIES

PUBLISHED

The Physical Geography of Africa
edited by William M. Adams, Andrew S. Goudie, and Antony R. Orme

The Physical Geography of North America
edited by Antony R. Orme

The Physical Geography of Northern Eurasia
edited by Maria Shahgedanova

The Physical Geography of Southeast Asia
edited by Avijit Gupta

The Physical Geography of Fennoscandia
edited by Matti Seppälä

FORTHCOMING

The Physical Geography of South America
edited by Tom Veblen, Kenneth Young, and Antony R. Orme

The Physical Geography of the Mediterranean Basin
edited by Jamie Woodward

The Physical Geography of Western Europe

edited by

Eduard A. Koster

OXFORD
UNIVERSITY PRESS

OXFORD
UNIVERSITY PRESS

Great Clarendon Street, Oxford OX2 6DP

Oxford University Press is a department of the University of Oxford.
It furthers the University's objective of excellence in research, scholarship,
and education by publishing worldwide in

Oxford New York

Auckland Cape Town Dar es Salaam Hong Kong Karachi
Kuala Lumpur Madrid Melbourne Mexico City Nairobi
New Delhi Shanghai Taipei Toronto

With offices in

Argentina Austria Brazil Chile Czech Republic France Greece
Guatemala Hungary Italy Japan South Korea Poland Portugal
Singapore Switzerland Thailand Turkey Ukraine Vietnam

Oxford is a registered trade mark of Oxford University Press
in the UK and in certain other countries

Published in the United States
by Oxford University Press Inc., New York

First published 2005

British Library Cataloguing in Publication Data
Data available

Library of Congress Cataloging in Publication Data
Data available

ISBN 0-19-927775-3

10 9 8 7 6 5 4 3 2 1

Typeset by Graphicraft Limited, Hong Kong
Printed in Great Britain
on acid-free paper by Antony Rowe Ltd., Chippenham, Wilts

Foreword

The Physical Geography of Western Europe is the sixth in a series of advanced books that is being published by Oxford University Press under the rubric of Oxford Regional Environments.

The aim of the series is to provide a durable statement of physical conditions on each of the continents, or major regions within those continents. Each volume includes a discussion of the systematic framework of the region (for instance, tectonism, climate, biogeography), followed by an evaluation of dominant environments (such as mountains, forests, and deserts) and their linkages, and concludes with a consideration of the main environmental issues related to the human use and misuse of the land (such as resource exploitation, agricultural and urban impacts, pollution, and nature conservation). While books in the series are framed within an agreed context, individual books seek to emphasize the distinctive qualities of each region. We hope that this approach will provide a coherent and informative basis for physical geography and related sciences, and that each volume will be an important and useful reference source for those concerned with understanding the varied environments of the continents.

Andrew Goudie, University of Oxford
Antony Orme, University of California, Los Angeles

Preface

European landscapes are to a large extent 'cultural landscapes'. Millennia of human activities have reshaped the surface of Europe intensively and irreversibly. Due to the complex geological and geomorphological history and the resulting intricate landscape patterns as well as the exceptionally strong human impact on the landscape a highly diverse physical geographical region has developed in western Europe.

Although landscape typologies, textbooks, and maps discussing and depicting physical geographical regions have been published at a national level, they are still lacking on a European scale. In contrast to Physical Geography there are several excellent overviews on the Geology of Europe, including extensive information on stratigraphy, geophysics, natural resources, geological hazards, and environmental issues (Rutten 1969; Ager 1980; Ziegler 1991; Blundell *et al.* 1992; Lumsden *et al.* 1994).

This regional physical geography text is the sixth in a series that is produced by Oxford University Press. The series involves a finite number of volumes devoted to major regions of the world. Each volume presents a detailed and current statement of knowledge written by specialists in the various research fields of physical geography. With this book, we aspire to fill a void in geographical and earth science literature, namely, the lack of a comprehensive, high-quality systematic and regional overview on the Physical Geography of Western Europe.

The twenty-one chapters are broadly divided into three groups: systematic framework, regional environments, and environment and human impact. The volume includes an overview of the main natural environmental factors (structural geology, tectonic evolution, landscape evolution and Quaternary climate changes, glacial geomorphology, periglacial processes, river, marine, aeolian, and peatland environments), a discussion on major physical geographical regions (Danish–German–Dutch Wadden region, German Uplands and Alpine Forelands, French–Belgian Uplands, Parisian Basin, French Alps and Alpine Forelands), and a consideration of environmental issues partly related to human impact (climate, soils, forestry, geomorphic hazards, air-soil-water pollution, urban geoscience, and geoconservation).

Although the chapters are framed within a more or less agreed context, the contributions by the individual authors inevitably involve much synthesis and some subjectivity. All chapters have been subject to an elaborate reviewing and editing procedure in order to promote some measure of uniformity in format, consistency, and depth of the discussions. To the same end all figures have been redrawn. To assist the reader in pursuing certain subjects, each chapter contains a selective bibliography, including a limited number of classic works and more or less recent, generally available, scientific publications.

Embleton's reference book, the *Geomorphology of Europe* (1984), was the first comprehensive survey on the geomorphological regions of Europe. It contains a wealth of information on the geomorphological evolution and related issues in Earth Sciences, such as (Quaternary) Geology, Physical Geography, and Soils. The present volume on the Physical Geography of Western Europe might be seen as the successor of Embleton's work. Seen in this perspective I now realize how difficult it is to 'stand on the shoulders of a giant'.

Eduard A. Koster
Utrecht, Huizen, Bédoin 2004

References

Ager, D. V. (1980), *The geology of Europe*. McGraw-Hill, New York.
Blundell, D., Freeman, R., and Müller, S. (eds.) (1992), *A continent revealed. The European Geotraverse.* Cambridge University Press, Cambridge.
Embleton, C. (ed.) (1984), *Geomorphology of Europe*. Macmillan, London.
Lumsden, G. I., and the Editorial Board of the Directors of the Western European Geological Surveys (eds.) (1994), *Geology and the Environment in Western Europe*. A co-ordinated statement by the Western European Geological Surveys. Clarendon Press, Oxford.
Rutten, M. G. (1969), *The geology of Western Europe*. Elsevier, Amsterdam.
Ziegler, P. A. (1991), *Geological Atlas of Western and Central Europe*. Elsevier, Amsterdam.

Acknowledgements

The authors, editor, and publishers wish to thank the following, those who helped with the preparation of this volume, and those who have kindly given permission for the use of copyright material.

Franca Geerdes (Utrecht University) has been a great help during all stages of preparation of this volume. She conducted a lively correspondence with (potential) authors, reviewers, and publishers, and assisted in the preparation of the manuscripts in many ways. Nigel Pears (formerly University of Leicester) has been very helpful in correcting the English texts prepared by German, French, Dutch, Danish, and Italian authors. Special thanks are due to Joke Koster-van Dijk (Huizen) for thoroughly checking manuscripts during various stages of preparation.

We are grateful to the members of the Cartographic Department of the Faculty of Geosciences (Utrecht University), in particular to Margriet Ganzeveld, Margot Stoete, and Gerard van Bethlehem, who have prepared all artwork, including the reproduction of the photographic plates. Their graphic and photographic expertise is gratefully acknowledged and their patience with a demanding editor is very much appreciated.

Unless otherwise indicated all photographs are made by the authors. The following persons have kindly given permission for the use of additional photographic material: Yvette Dewolf (Paris) for Fig. 6.4; Leo Tebbens (Utrecht) for Fig. 6.7; Henk Berendsen (Utrecht) for Figs. 6.14 and 6.15; Philippe Larroudé (Grenoble) for Figs. 7.11 and 7.12; Simon Smit—Fotografie (Den Burg) for Fig. 8.13; Bernd Probst (Melsdorf) for Fig. 10.5; Pim Beukenkamp (Utrecht) for Figs. 11.6, 11.7, 11.8, 11.13, 11.15, 19.2, 19.3, and 20.3; Hans Middelkoop (Utrecht) for Fig. 11.12; Lothar Eissmann (Leipzig) for Figs. 11.14 and 21.6; Jean-François Billet (Lille) for Fig. 14.10; Michèle Evin (Amiens) for Fig. 14.16; Roger Langohr (Ghent) for Fig. 16.14; Jan van Mourik (Universiteit van Amsterdam) for Fig. 16.15; Stiboka Collection (Wageningen) for Fig. 16.21; Jean Trautmann (Strasbourg) for Fig. 18.2; Daniel Delahaye (Université de Rouen) for Fig. 18.4; Jean Philippe Malet (Université de Strasbourg) for Fig. 18.16; Monsieur Casabonne (Maire d'Arette, France) for Fig. 18.8; François Valla (Cemagref Grenoble) for Fig. 18.19; François Rapin (Cemagref Grenoble) for Fig. 18.20. Eduard Koster has provided additional photographs for Figs. 3.9B, 3.9C, 3.12A, 3.12B, 4.4, 4.6, 4.7, 4.8, 4.11, 4.12, 21.4, 21.7, and 21.12.

Our sincere thanks go to the following reviewers, whose comments helped to improve earlier drafts of the chapters: Theo van Asch, Pieter Augustinus, Henk Berendsen, Pim Beukenkamp, Jan Boerma, Wim Hoek, Hans Middelkoop, Peter de Ruiter, Henk van Steijn, Leo Tebbens, Joost Terwindt, Jos Verhoeven, and Stan White (Utrecht University), Yvonne Battiau-Queney (University of Science and Technology of Lille), Tom Bloemers, Bas van Geel, and Joke Koster-van Dijk (University of Amsterdam), Margot Böse (Free University of Berlin), Peter Buurman (Wageningen University), Alain Demoulin and Albert Pissart (University of Liège), Jos Dijkmans (Netherlands Institute of Applied Geoscience [TNO]), Richard Dikau (University of Bonn), Jürgen Ehlers (Hamburg Geological Survey), Bernard Etlicher (University of Saint-Étienne), Clare Goodess (University of East Anglia), Leendert Louwe Kooijmans (University of Leiden), Koen Kramer (Alterra Wageningen), Jean-Pierre Lautridou (Centre de Géomorphologie [CNRS], Caen), Ernst Look (Niedersächsisches Landesambt für Bodenforschung, Hannover), Jaap van der Meer (Queen Mary, University of London), Hans-Heinrich Meyer (University of Applied Sciences, Erfurt), Maurizio

Parotto (University of Rome), Ken Pye (Royal Holloway, University of London), Hans Renssen (University of Louvain), Jeff Vandenberghe (Free University of Amsterdam).

The following colleagues and institutions have kindly given permission for the use of copyright material: Derek Blundell (Royal Holloway, University of London) for Figs. 1.2 and 1.5; Lothar Lippstreu (Kleinmachnow, Germany) and Herbert Liedtke (Bochum, Germany) for Figs. 4.9 and 4.10; Common Wadden Sea Secretariat (Wilhelmshaven) for Fig. 10.1 and Hansjörg Streif (Hannover) for Fig. 10.4; Société Géologique de France (Paris) for Figs. 13.4, 13.6, and 13.11; Springer-Verlag (New York) for Figs. 15.2 and 15.3; Kluwer Academic Publishers (Dordrecht) and James Hurrell (Boulder, Colo.) for Figs. 15.6, 15.7, and 15.8; Jean-Noël Salomon (Presses Universitaire de Bordeaux) for Fig. 18.1; Daniel Delahaye (Université de Rouen) for Fig. 18.3; Jérôme Lambert (BRGM, Orléans) for Fig. 18.7; Gottfried Grünthal (Annali di Geofisica, Rome) for Fig. 18.9; Dominique Weber (Université de Strasbourg) for Fig. 18.18. We are also grateful to individual authors and other colleagues for freely providing many other figures in the book. Although every effort has been made to trace and contact copyright owners, we apologize for any apparent negligence.

'Que faites-vous ici?' dit le petit prince.
'Je suis géographe,' dit le vieux Monsieur.
'Qu'est-ce qu'un géographe?'
'C'est un savant qui connaît où se trouvent les mers, les fleuves, les villes, les montagnes et les déserts.'
'Ça c'est bien intéressant,' dit le petit prince. 'Ça c'est enfin un véritable métier!'

Antoine de Saint-Exupéry, *Le Petit Prince*

Contents

List of Figures

List of Tables

List of Contributors

Yvonne Battiau-Queney is Professor in Geomorphology and Physical Geography at the University of Science and Technology of Lille, France. Her research focuses on long-term landform development in relationship with plate tectonics, especially in Great Britain, eastern North America, and France. She is also concerned with coastal geomorphology, dune dynamics, and shoreline change in northern France.

Margot Böse is Professor of Physical Geography and Quaternary Research and Dean of the Department of Earth Sciences of the Free University Berlin, Germany. Her fields of research include Holocene environmental changes caused by man-induced geomorphologic processes, Quaternary glacial geomorphology, stratigraphy and periglacial landscape development, in particular in northern Germany, western Poland, and high mountain areas in East Asia.

Chris Bremmer is Head of the Geomechanical Research Section of the Netherlands Institute of Applied Geosciences (TNO), National Geological Survey. He studied hydrogeology at the Free University of Amsterdam. He is currently involved in national and European research projects on engineering geology and geo-hazards in deltaic environments.

Alain Demoulin is a Research Associate of the Belgian National Fund for Scientific Research. He is working at the University of Liège, Department of Physical Geography. His research interests include the long-term landform development and the neotectonic evolution of north-western Europe. His current work focuses on the use of geodetic techniques in the study of present-day crustal motion in intraplate regions.

Yvette Dewolf was Professor in Physical Geography at the University Paris VII Denis Diderot, France. She is ex-president of the Association of Geologists of the Paris Basin. Her research interests include present (Spitsbergen, Northern Canada, Siberia) and ancient (France, Maghreb) periglacial environment, and arid geodynamic processes (Sahara, Niger, Burkina Faso, Australia). At present she is editing a book (fifty-one contributors) on 'Les Formations Superficielles—Genèse, Typologie, Classification—Paysages et environnements—Ressources et Risques'.

Francesco Dramis is Professor of Geomorphology and Dean of Geology at the Department of Geological Sciences, Roma Tre University, Italy. He is the chairman of the Italian Association of Physical Geography and Geomorphology (AIGEO). His research focuses on Quaternary Geology, including morphotectonics, geological hazards, and periglacial studies. His current scientific interests concentrate on the morphotectonic evolution of the Apennines, with special reference to the geomorphological effects of the Quaternary uplift, and geomorphological-stratigraphic indicators of Holocene climate changes in Eastern Africa and the Mediterranean area.

Jürgen Ehlers is Head of the Mapping Department of the Geological Survey of Hamburg and Lecturer at the Institute for Geography at Bremen University, Germany. His research interests include Quaternary and coastal geology and geomorphology. He is author and editor of many papers and monographs. His current work focuses on an INQUA project concerning the extent and chronology of Quaternary glaciations.

Bernard Etlicher is Professor in Physical Geography at the University Jean Monnet, Saint-Étienne, France. His research focuses on glacial and periglacial deposits of hercynian uplands in France, mainly in the Massif Central. He performed experimental freeze-thaw tests on igneous material. His current projects analyse relations between surficial deposits and landscapes.

Josef Fanta was Professor in Landscape Ecology at the University of Amsterdam, and in Forest Ecology at the Wageningen Agricultural University, The Netherlands. His research concerns primary forest succession in blown sand landscapes, restoration of forested landscapes damaged by environmental pollution in Central European mountains, and inventory and strategy for sustainable protection and management of natural forests in south-east European countries.

Andrew Farmer is a Senior Fellow at The Institute for European Environmental Policy in London, UK. He is a biologist by training and specializes in EU legislation in relation to air and water pollution and wider issues of pollution control. Before joining IEEP in 1997, he worked for English Nature as its atmospheric pollution specialist and sustainable development co-ordinator; as a Research Assistant at both the university of Wisconsin, USA, and Imperial College, University of London; and as a Research Associate at the University of Florida, USA.

Tim Favier studied Physical Geography at Utrecht University. Since 2003 he has worked as a Quaternary geologist at the Netherlands Institute of Applied Geosciences (TNO)—National Geological Survey in Utrecht, The Netherlands.

Gerard Gonggrijp† was trained as a Physical Geographer and Geomorphologist at the University of Amsterdam, The Netherlands. He devoted his scientific career—at the Netherlands State Institute for Nature Research—to the conservation, preservation, and rehabilitation of geological, geomorphological, and pedological important sites in The Netherlands and in a European context; he was the first president of ProGEO (the European Association for the Conservation of the Geological Heritage). To our great sorrow Gerard Gonggrijp died shortly after he had finished his contribution to this book, on 13 March 2002.

Jacobus Hofstede is Senior Coastal Defence Manager at the Schleswig-Holstein State Ministry of the Interior, Germany. Further, he is vice-president of the German branch of the European Union on Coastal Conservation. He has studied coastal geomorphology, and published about forty articles. His current work concentrates on coastal defence and integrated coastal zone management (ICZM).

Else Kolstrup is Professor and Chair of Physical Geography, especially Geomorphology, at the Department of Earth Sciences, Uppsala University, Sweden. Her research interests include the development of past and present cold-climate areas in relation to changing environmental conditions. She has extensive fieldwork experience in temperate and Subarctic regions.

Eduard Koster is Professor in Physical Geography and Geomorphology at the Department of Physical Geography, Faculty of Geosciences, Utrecht University, The Netherlands. His research interests concern Quaternary geology of north-western Europe, periglacial and cold-climate aeolian studies (Fennoscandia, Greenland, northern Canada, Alaska), and impact studies of global climatic change. He has supervised a very large number of graduate and Ph.D. students. Until recently he was vice-chairman of the Netherlands Division for Earth and Life Sciences of the Netherlands Science Foundation.

Aart Kroon is Lecturer in Physical Geography in the Department of Physical Geography at the Faculty of Geosciences, Utrecht University, The Netherlands. His research mainly concerns coastal, fluvial and aeolian morphodynamics. Sandy beaches and tidal inlets are the main topics in the national and international research projects he's working on. Video-remote sensing systems and field experiments often play a key role in his research approach.

Olivier Maquaire is Associate Professor of Geomorphology and Engineering Geology at the Faculty of Geography and at the École et Observatoire des Sciences de la Terre, University Louis Pasteur in Strasbourg, France. Since September 2004, he has been Professor at the University of Caen, Basse Normandie. His research focuses on geomorphological hazard mapping, risk analysis, and temporal evolution of landslides in mountainous areas. Since 2000, he is the Director of the European Centre on Geomorphological Hazards (CERG) of the EUR-APO Major Hazard Agreement. He is author of more than fifty publications.

Eduardo de Mulder is the President of the International Union of Geological Sciences (IUGS) and senior adviser of the Netherlands Institute of Applied Geosciences (TNO), National Geological Survey, The Netherlands. He is co-founder of the International Working Group on Urban Geology and co-author of the Balkema book on Urban Geosciences.

Charles Pomerol, emeritus Professor of Geology at the University Paris VI Pierre et Marie Curie (France), is a specialist on Tertiary stratigraphy. He is (former) president of the Geological Society of France and of the international subcommission on Palaeogene Stratigraphy. He is supervisor of the series of regional geological guides (Masson editions). He is also (co-)author of a handbook on stratigraphy and palaeogeography in four volumes (*Elements de Géologie*).

Cor Schuurmans is a retired Professor of Climate Dynamics at Utrecht University, The Netherlands. Also through his affiliation with the Royal Netherlands Meteorological Institute (KNMI), he was involved in climate research. His principal research interest focuses on the general circulation of the atmosphere and the response of it to solar forcing.

Jan Sevink holds a Chair in Soil Science at the University of Amsterdam (The Netherlands), additionally being director of the Institute for Biodiversity and Ecosystem Dynamics (IBED) within the Faculty of Science. His main fields of interests are soil geography and ecopedology. He has extensive experience with field studies and surveys of western and southern European soils.

Otto Spaargaren has been working for the past thirty years in soil survey and tropical soil management, in worldwide soil correlation and classification, and in teaching at universities and post-graduate courses in The Netherlands and abroad. He recently authored several chapters on soil classification in soil science handbooks and published several papers on the World Reference Base for Soil Resources. He obtained a Ph.D. in Mathematics and Physics from the University of Amsterdam on a study of weathering and soil formation on limestone in the Mediterranean environment of Italy.

Emanuele Tondi is a research scientist, specializing in Structural Geology, at the Department of Earth Sciences, University of Camerino, Italy. He has carried out extensive field studies in Central and Southern Italy concerning the evaluation of the seismogenic potential associated with different fault segments, and the understanding of geodynamic processes and fault zone evolution in the Apennines.

I Systematic Framework

1 Tectonic Evolution, Geology, and Geomorphology

Alain Demoulin

Introduction

The present-day major relief features of western Europe are to a great extent determined by the underlying geological structures, either passively or actively. To get a comprehensive picture of their morphological evolution and interrelations, this chapter provides an overview of the spatial and temporal characteristics of the large-scale tectonic framework of the continent. After having described the west European landscape at the end of the Palaeozoic, to which time the oldest preserved landforms date back, an outline of the Mesozoic and Cenozoic history of the major tectonic domains follows. Finally, some denudation estimates highlighting the relationship between tectonics, erosion, and the resulting relief, will be discussed.

The three main influences on the present-day topographic patterns are those of the Alpine orogeny, the Cenozoic West European rifting, and the imprint of Variscan structures. They combine within a regional stress field determined by the Africa–Eurasia collision and the Alpine push as well as the mid-Atlantic ridge push. Since the end of the Miocene, this stress field is characterized by a fan-shaped distribution of S_{Hmax} along the northern border of the Alpine arc. This gives way to a more consistent NW–SE to NNW–SSE direction of compression further from the chain (Bergerat 1987; Müller *et al.* 1992).

Topographically, western Europe may be roughly divided into a series of belts parallel to the Alpine chain (Fig. 1.1). The Alpine chain culminates in a number of peaks exceeding 4,000 m in elevation (4,810 m at Mont Blanc) but the average altitude is in the order of 2,000 m. To the north, the mountainous Alps are bordered by the Molasse foredeep basin whose surface makes an inclined plane descending northwards from *c.*1,000 m to *c.*300 m near the Donau River in the Regensburg-Passau area. To the north-west, the Molasse basin narrows between the Alps and the Jura Mountains and is occupied by several extended lakes inherited from Quaternary glacial activity. Next to the Molasse basin in the north and west is a wide belt of recently more or less uplifted areas between 200 and 1,000 m in elevation (and locally in excess of 1,000 m in the French Massif Central and the Bohemian massif). It includes most Palaeozoic massifs, in which the Variscan structural setting strongly influences the medium-scale relief. This belt also encompasses all or part of the intervening Mesozoic basins. The presently uplifting eastern parts of the Aquitaine and Paris basins pertain to this belt which also comprises the south German basin, clearly separated from the Molasse basin by the highs of the Swabian and Franconian Alps. The relief of all three Mesozoic basins is characterized by the development of a number of more or less parallel, arcuate cuestas. Lastly, a belt of low-elevation areas runs from the Atlantic coast of France in the west to the Baltic Sea in the north-east. Although it also includes the slightly uplifted Armorican massif, it is mainly comprised of areas of Tertiary subsidence: the western parts of the Aquitaine and Paris basins, the Tertiary Belgian basin and the north German basin. Extended zones within the latter (e.g. the northern Netherlands) are still subsiding today. This continental-scale alternation of belts of higher and lower terrains fits remarkably well into the lithospheric buckling hypothesis of Nikishin *et al.* (1997). Finally, cutting more or less north–south across this belt pattern, a series of aligned troughs extending from the Rhône area in southern France to the Roer graben in The Netherlands mark the trace of the Cenozoic rift system of western Europe.

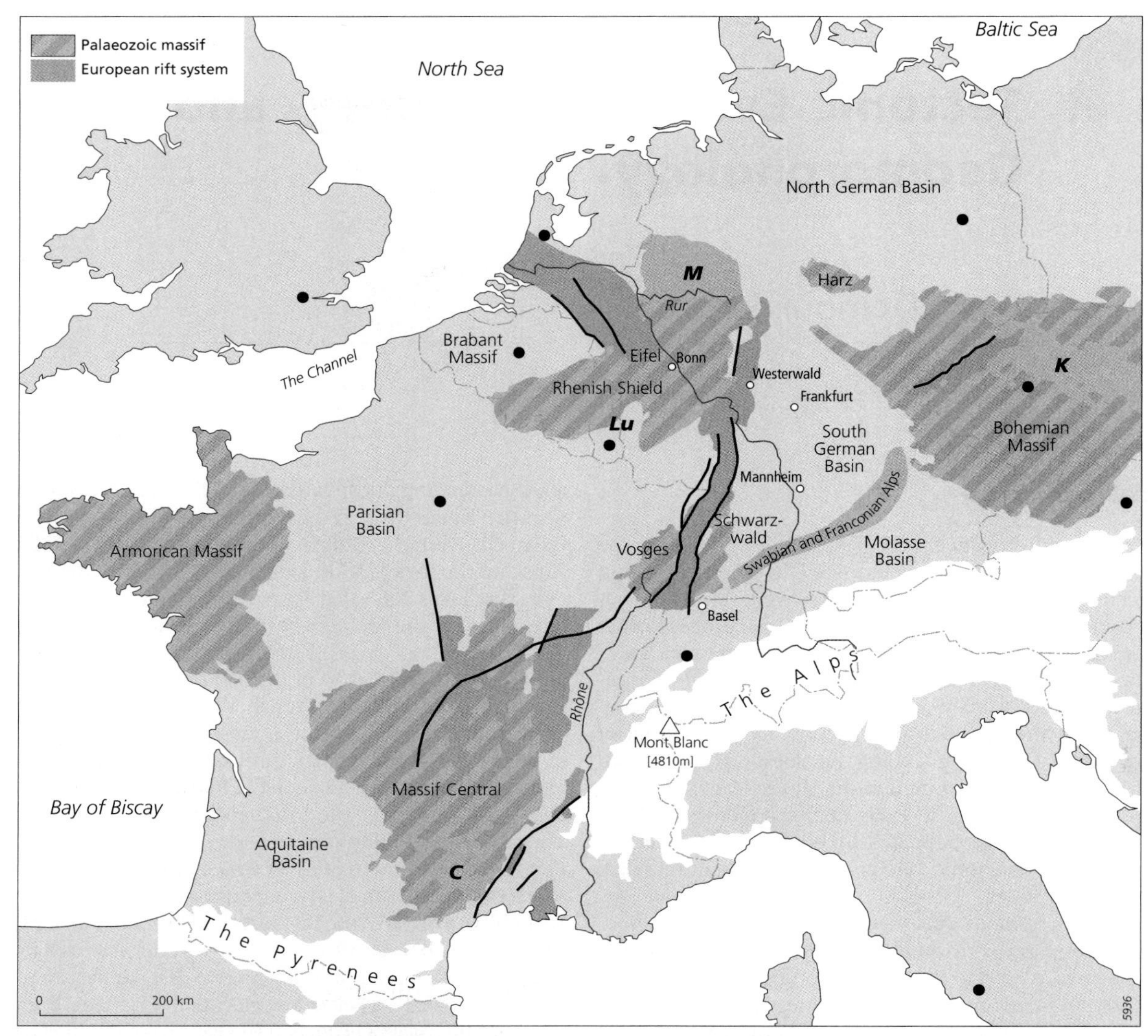

Fig. 1.1. The main relief features of western Europe. Hatched areas correspond to the Palaeozoic massifs. In dark grey, the European rift system. **C**. Causses (Jurassic). **K**. Cretaceous basin of Bohemia. **M**. Münster basin (Cretaceous). **Lu**. Luxemburg Embayment.

The Variscan Landscape

Plate Tectonic Setting of the Variscan Orogeny

Different models have been proposed for the plate tectonic history leading to the Variscan orogeny of Europe (Berthelsen 1992; Franke 1992). Although still speculative with respect to several important points, they agree on a broad frame in which, at the dawn of the Devonian, two probably narrow oceanic domains separated the Laurussia plate to the north, the Gondwana plate to the south, and the intervening Armorica microplate, which includes the north Armorican and Bohemian massifs and parts of the Vosges, Schwarzwald, and Massif Central (Fig. 1.2A). Laurussia itself had resulted from the collision, causing the Caledonian orogeny (*c*.430–400 Ma), of three tectonic plates: Baltica (approximately Proterozoic north-east Europe), Laurentia (including North America, Greenland, and north-west Scotland), and eastern Avalonia (a microplate detached

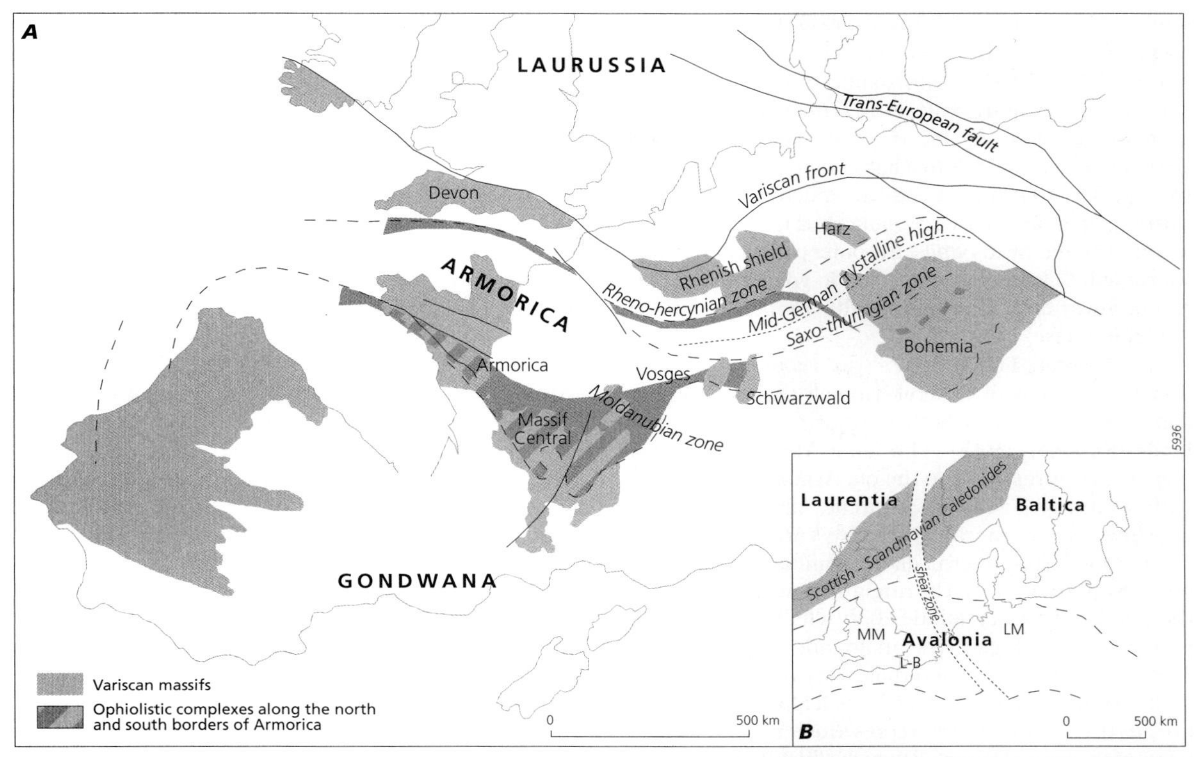

Fig. 1.2. A. Main structural zones of the Variscan orogen. The currently outcropping Variscan massifs are outlined in light grey, the zones of ophiolitic complexes along the north and south borders of Armorica are indicated in dark grey and dark/light grey hatching. B. Plates involved in the Caledonian welding giving birth to the Laurussia megaplate. **MM**. Midlands massif. **L–B.** London–Brabant platform. **LM**. Lüneburg massif (modified after Berthelsen 1992).

from Gondwana in earlier time and to which the Midland massif, the London–Brabant platform and the buried north German Lüneburg massif pertain) (Fig. 1.2B).

Owing to the general plate convergence that started during the Silurian in the eastern Variscan domain (Bohemian massif) and that would eventually lead to the Pangea supercontinent, the oceanic crust of both the Rheic (to the north) and the Mesogean (to the south) oceans was subducted beneath Armorica. Obducted ophiolitic complexes and high pressure metamorphic rocks still bear witness to this stage (Debelmas and Mascle 1993). According to Matte (1986), the ocean closure and continent collision took place earlier (*c.*380 Ma) on the Gondwanian margin, causing the acceleration of subduction in its northern counterpart. After collision had occurred in both basins (*c.*350 Ma), post-collisional plate convergence went on for almost another 100 Ma producing large-scale crustal stacking which progressively migrated from the centre towards the marginal parts of the belt.

Main Orogenic Phases and Tectonic Zones

The tectonic phases identified by earlier authors within the Variscan orogeny actually are only paroxystic episodes of an otherwise continuous process successively marked by rifting, subduction, and subsequent continent collision. Whilst transient extensional episodes occurred up to the upper Devonian (Franke 1989; Roig and Faure 2000), the collisional stage was fully active from the basal Carboniferous onwards. A first compressional phase (Acadian phase) has, however, been described in the middle Devonian (Walter 1992). Although occurring on both sides of the axial zone of the orogen, the Bretonian phase (360–350 Ma) especially affected the meridional Gondwanian margin where it is marked by increased thrusting, metamorphism, and granitoid generation (Debelmas and Mascle 1993). The Sudetic phase (350–320 Ma) is characterized by the outward propagation of thrusting preceded by flysch sedimentation in flexured basins and spilitic to rhyo-dacitic volcanism. The resulting crustal thickening also determined granite intrusion. As for the Asturian phase (320–280 Ma), it corresponds to a last compression episode. In this episode, the inner Variscan structures underwent isostatic uplift and active erosion, thus feeding sedimentation in limnic basins whilst the outer paralic basins were shortened. Calc-alkaline to rhyolitic volcanism and the rise of lower crustal granitic melts are ascribed to the Asturian phase.

The Variscan 'final product' was a roughly east–west trending orogen with bilateral symmetry, with thrusts and folds verging towards both the north and south forelands (Kossmat 1927). The axial Moldanubian zone (Fig. 1.2A) corresponds to the old Armorica microplate, with a strong Cadomian tectonic and metamorphic imprint that was mostly spared by Variscan thrusting. To the north, the intermediate, granitized, but weakly metamorphosed Saxo-Thuringian zone is followed by the mid-German Crystalline High which is assumed to represent an Andean-type magmatic arc (Berthelsen 1992). Its southern counterpart may be found notably in the crystalline nappes of the French Massif Central. The northern externides are represented by the Rhenohercynian zone (Devon, Rhenish shield, Harz), made of piled north-vergent nappes overriding the late Carboniferous coal-bearing foreland basins. The northern Variscan front is clearly marked from Ireland to Bohemia. To the south of the belt, the frontal thrust is clearly identified only at the south-eastern border of the Bohemian massif and in north Spain.

Permian Tectonics and Palaeogeography

While the Variscan orogeny was in its waning phase during the late Carboniferous/early Permian, continued convergence in the Uralides and the Appalachians, respectively to the east and to the west of the Variscides, led to a last episode of compression marked by escape tectonics associated with major transverse shear faulting in the front of the thickened orogen (Burg *et al.* 1994). Not only was the Variscan thrust-and-fold belt cut by great strike-slip faults (e.g. the 'Sillon Houiller', the Upper Rhine graben border faults, the Hunsrück border fault) but its foreland too was fractured (Fig. 1.3). Beyond pull-apart basins associated with these faults, a later Basin-and-Range-type extension determined the development of important intramontane basins following the Variscan structural trend. This more intense extensional phase was caused by gravitational collapse of the thickened crust due either to a change in plate kinematics during the Stephano-Permian (ibid.) or even to tensile plate boundary forces (Henk 1997). Such Permian basins as the Saar-Nahe and the Kraichgau basins (Fig. 1.3) were filled by continental deposits up to several kilometres thick.

At the same time, more or less north–south trending basins formed in the Variscan foreland of north Germany as a branching-off from the south Permian basin towards the Hessian depression. Ziegler (1978) suggested that these basins could have resulted from post-orogenic foreland collapse. Culminating in the late Carboniferous/early Permian, extended volcanism arose from combined high heat flow and the presence of crustal faults. Within the orogen, the Permian volcanics mainly occurred in the same places as the earlier granites.

During the late Permian, due to cooling of the heated lithosphere, extension and basin subsidence increased in the North Sea (and in north Germany). As the sedimentation rate could not keep pace with subsidence, this led to the development of large topographic depressions, north and south Permian basins, which finally were invaded by the hypersaline Zechstein sea when a seaway was opened through the northern North Sea. Up to 1,500 m of evaporites and carbonates were deposited over the early Permian red beds and evaporites. The southward transgression of the Zechstein sea marked the end of the Variscan and post-Variscan era. Meanwhile, post-orogenic uplift of the belt was not only accompanied by basin formation but also allowed extremely active denudation largely to level the Variscan mountains. Consequently, by the end of the Permian, the so-called post-Hercynian peneplain was well developed through much of France, Belgium, and south Germany. The Mesozoic era, starting with the Pangea break-up then began.

Mesozoic and Cenozoic History of the Major Tectonic Domains

Implications of the Atlantic Rifting: the Aquitaine Basin and the Pyrenees

For the sake of consistency, the evolution of the Aquitaine basin and that of the Pyrenees fold belt will be jointly presented, and also linked to the development of the small oceanic basin of the Bay of Biscay (Fig. 1.4). As a possible consequence of the Pangea break-up, a first east–west trending basin developed during the Triassic and Jurassic to the south of the Arcachon–Toulouse flexure (Debelmas and Mascle 1993). During the Dogger, succeeding extended evaporite deposition, marine sedimentation overlapped the deep basin's northern edge and

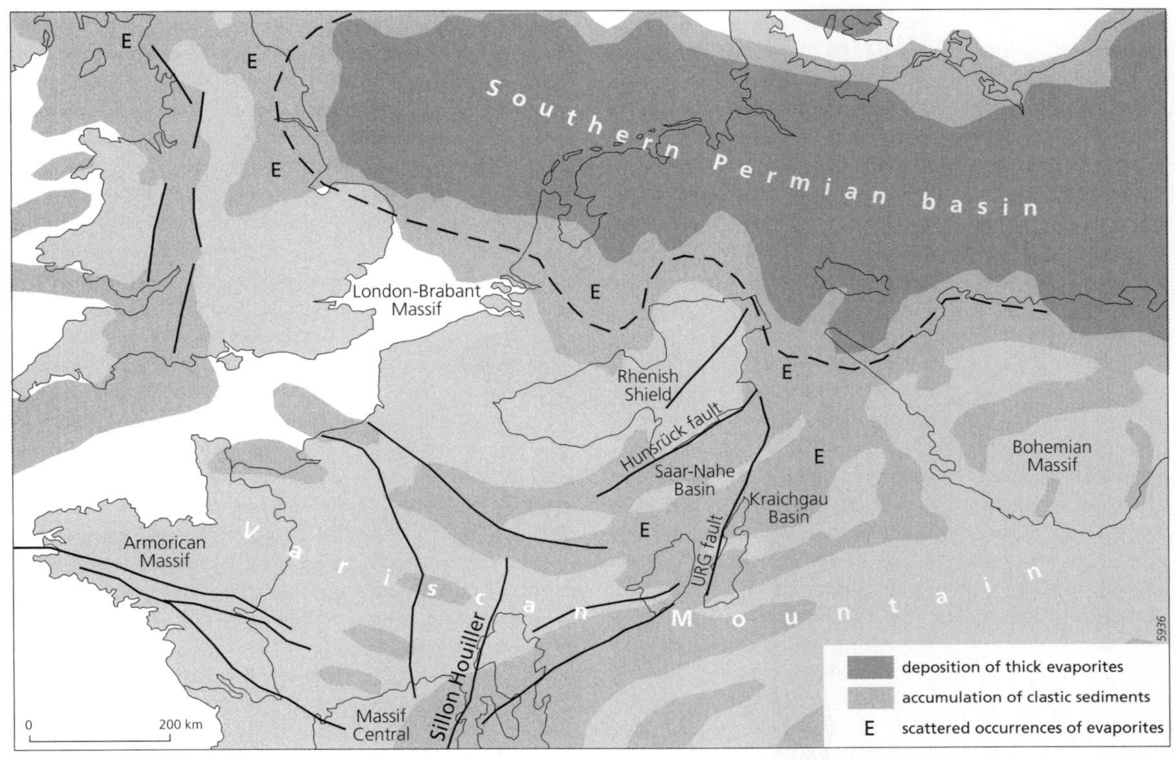

Fig. 1.3. Palaeogeography of north-western Europe during the upper Permian (modified after Walter 1992). The zone of deposition of thick evaporites is indicated. The accumulation of clastic sediments occurred on the margins of the South Permian basin and in a series of basins resulting from the collapse of the Variscan orogen.

a carbonate platform extended northwards towards the embryonic north Aquitaine basin. Afterwards the upper Jurassic regression allowed denudation to remove a *c.*200–250-m thick layer of sediments in a 40-Ma-long period of time (Simon-Coinçon and Astruc 1994). Then the north Atlantic opening caused renewed extension and subsidence of the north Pyrenean basin during the Aptian and Albian and a marine transgression proceeding from the west invaded it during the Cenomanian. In this time, the basin became a true rift.

From the upper Albian onwards a left-lateral movement took place along a transform zone between the Iberian and European cratons. Superimposed on this 200-km eastward drift of Iberia with respect to Europe (Boillot 1984), rotation of the Iberian block took place (Fig. 1.4). This induced the opening of the Bay of Biscay as well as incipient shortening of the eastern end of the future Pyrenees and first uplift of their axial zone. The subsidence axis was displaced northwards as the upper Cretaceous transgression reached a maximum and covered the whole Aquitaine during the Santonian.

From the Campanian onwards, as a result of the south Atlantic opening, the Iberian craton moved northwards for *c.*100 km. This led not only to the subduction of the southern part of the Bay of Biscay basin beneath the Cantabrian margin but also to the definitive closure of the north Pyrenean rift. While the Aquitaine basin became a failed rift arm (or *aulacogen*), compression and folding started in the Pyrenees, which finally would display bivergent, asymmetric thrusting. In the same time, after the upper Cretaceous regression, the Aquitaine basin was again drowned by an Eocene sea with the exception of its north-west margin which was covered by the so-called 'siderolithic' continental deposits. However, during the middle Eocene, shortening and uplift of the Pyrenees, which had started at their eastern end, had reached the whole domain. The Bartonian shoreline was thus displaced westward and most of the Aquitaine underwent continental evolution during the Oligocene. This was marked by intense chemical weathering, karstification, and removal of part of the Mesozoic cover, 150–200 m of Palaeogene denudation being assumed

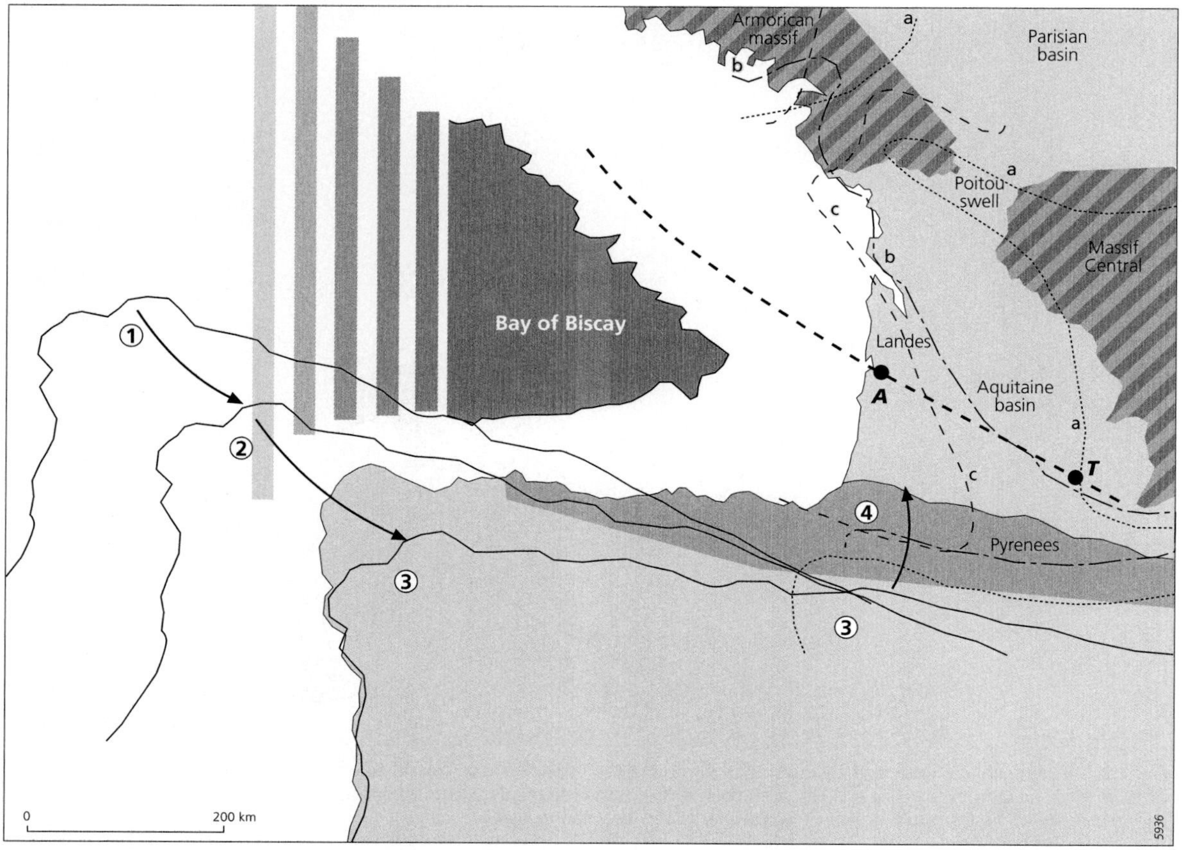

Fig. 1.4. The Aquitaine basin and its structural environment. The oceanic crust of the Bay of Biscay is indicated in dark grey. In bold dashed line, the Arcachon (A)–Toulouse (T) flexure limiting the basin to the north during the lower Mesozoic. The numbers 1 to 4 refer to the location of Spain with respect to Europe respectively at 110, 100, 75, and 0 Ma (Debelmas and Mascle 1993). The shorelines are mapped for the Cenomanian (a), the middle Eocene (b), and the middle Miocene (c).

by Simon-Coinçon and Astruc (1994). During the Aquitanian, a last epicontinental sea still encroached on the western Aquitaine while continental deposits covered the eastern part of the basin. From the middle Miocene onwards, fluvio-deltaic deposits then spread over the whole Aquitaine, alternately supplied by the Pyrenees and the Massif Central (Dubreuilh *et al.* 1995). The axis of the Plio-Pleistocene delta of Les Landes progressively moved northwards, suggesting a recent north-tilting of the whole area at the foot of the Pyrenees.

As for the Tertiary evolution of the Pyrenees, Calvet (1999) shows that an active denudation almost kept pace with the Eocene uplift so that the belt was narrow and of rather low elevation (< 2,000 m) when the Oligocene erosion started to reduce it, leaving a flight of stepped planation surfaces. Moreover, in the east block tectonics during the lower Miocene caused accelerated denudation by offering multiple scarps to the attack of erosion. The present-day mountainous topography of the belt thus results from a renewed, multistage uplift starting in the upper Miocene and showing paroxystic phases at the Mio-Pliocene boundary and in the Pleistocene.

The Alpine Domain

Tectonic Evolution

The Alpine chain results from the collision between the Eurasian and African plates, with participation of the intervening Adriatic microplate. In relation with the global-scale left-lateral strike-slip system which led to the Pangea break-up and opening of the Central Atlantic in Triassic times, the shallow shelf seas that occupied the future Alpine domain in that time underwent extension

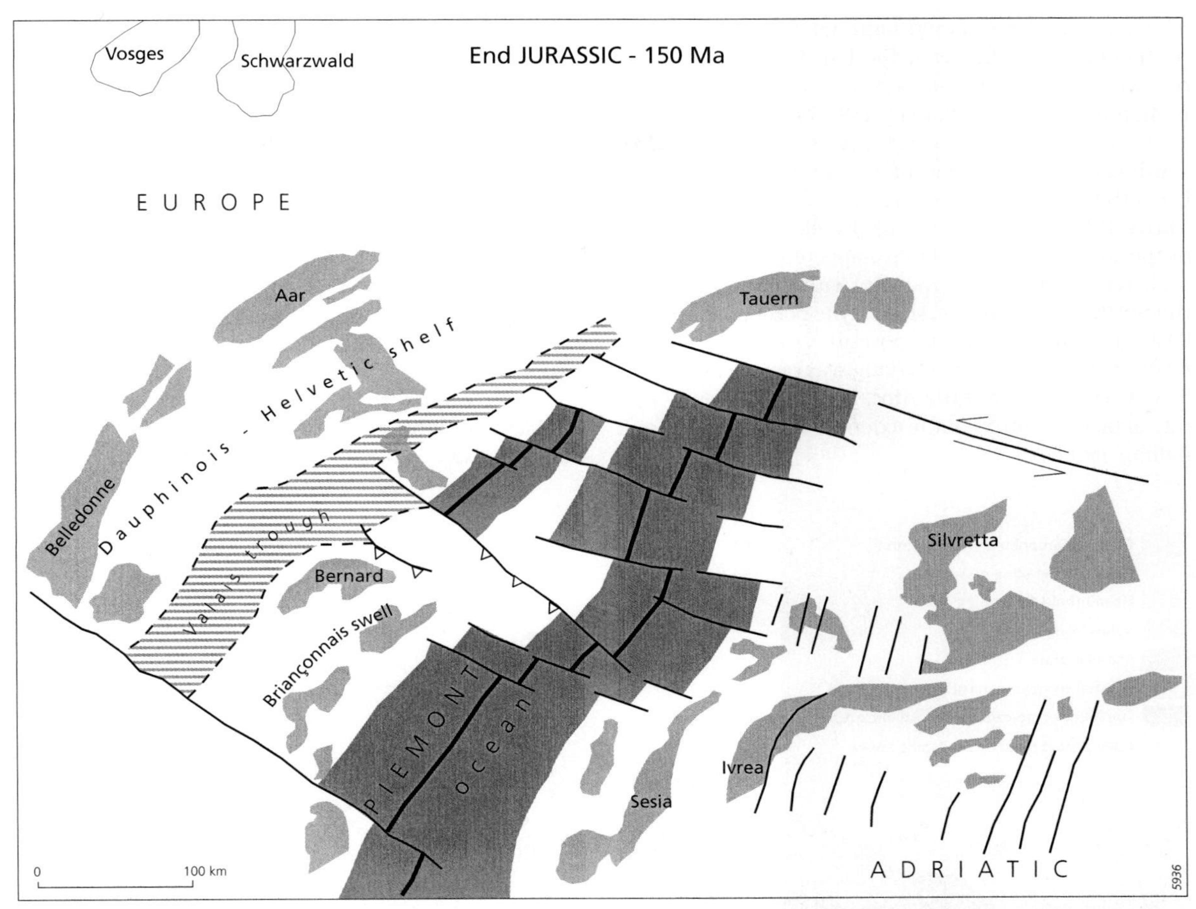

Fig. 1.5. Plate tectonic context of the Alpine domain at the end of the Jurassic. In dark grey, oceanic crust on both sides of the Piémont mid-oceanic ridge (bold line). The Valais area of strongly thinned continental crust is horizontally hatched. The present crystalline massifs appear in light grey (after Pfiffner 1992).

from the Jurassic onwards. Associated with left-lateral movement between Europe and Africa, a SW–NE trending rift developed, followed by the opening of the so-called Tethys, or Piémont ocean during the late middle Jurassic (Dewey *et al.* 1989; Pfiffner 1992) (Fig. 1.5). The upper Jurassic and lower Cretaceous were marked by continuous pelagic sedimentation, mainly of carbonates, in a wide area corresponding to the thermally subsiding European passive margin.

Regional extension came to an end during the Aptian when the Tethysian oceanic domain started to subduct south-eastwards beneath the Adriatic microplate. This caused an accretionary wedge to accumulate along the active Adriatic margin (Eoalpine phase of Hunziker *et al.* 1989). From the end of the upper Cretaceous, this NW–SE oblique plate convergence was progressively replaced by a north–south movement of Africa relative to Europe which soon induced plate margin collision. The normal faults of the rifted European margin were reactivated in inverse faults rooted in flat crustal-scale thrusts delaminating the thinned continental crust. This Mesoalpine phase of intense folding and thrusting culminated in the Eocene and could have also involved north–south left-lateral strike-slip tectonics (Ricou and Siddans 1986). Then, the Neoalpine phase, starting in the Oligocene, corresponds to the hypercollision stage involving deformation of the European continental crust of normal thickness beyond the margin. Shortening of the basement and subsequent decollement of its cover in north to north-west vergent nappes prevailed, accompanied by dextral shearing (ibid.; Steck and Hunziker 1994). The Eocene-Miocene phases of Alpine collision were a major phase of mountain uplift in the inner part of the chain. Finally, from the late Miocene onwards, renewed shortening was associated with a change in motion of the African plate from S–N to SE–NW (Dewey *et al.* 1989). This shortening was realized through southward underplating of the foreland's basement beneath the interior zones, creating the Frontal Pennine Thrust (FPT) and inducing the decollement and folding of the foreland's cover. Final nappe emplacement also occurred in this time, namely with the thrusting of the northern calcareous Alps on the perialpine Molasse basin in Austria. The important crustal thickening resulting from the nappe stacking was responsible for the recent strong uplift of the Alps, going on presently at a rate of up to 2 mm/yr in the external massifs (Banda and Balling 1992).

Present-Day Structural Pattern of the Western Alps

The main structural units of the Alps have an east–west orientation in the central part of the chain turning to a north–south trend in its western part. These units are described going from the foreland towards the core of the Alps (Fig. 1.6), especially those belonging to the European plate and exposed in the area with which this book is concerned.

The external (Dauphinois and Helvetic) zones correspond to that part of the European plate where the crust displays a normal thickness of *c.* 30 km. The outwardmost Dauphinois zone is primarily comprised of the French Subalpine chains. Most Subalpine massifs are made of the Mesozoic limestones of the cover which were folded during the late Miocene above a decollement in Triassic

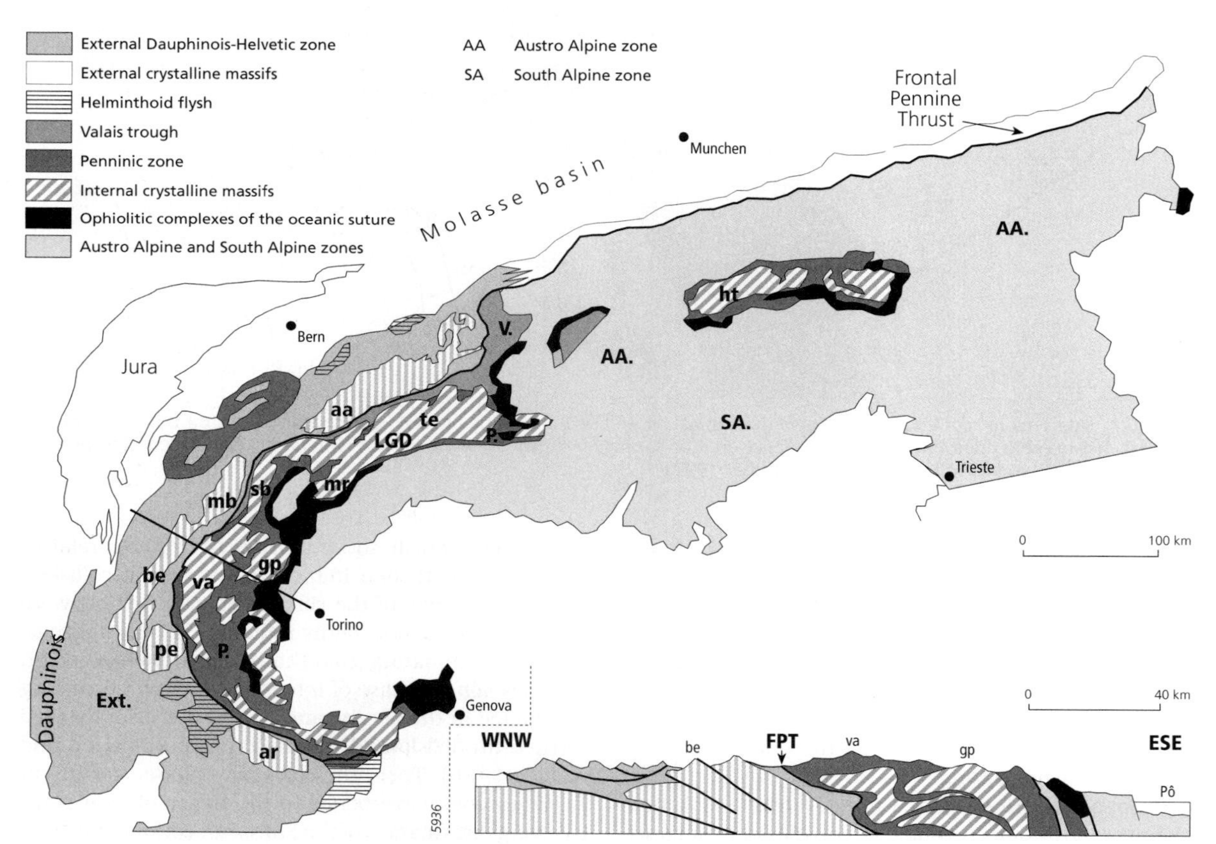

Fig. 1.6. Map of, and cross-section through, the main structural units of the Alpine chain. The cross section is indicated by a straight bold line on the map. The external Dauphinois–Helvetic zone (**Ext.**) comprises the crystalline massifs (**aa**: Aar; **ar**: Argentera; **be**: Belledonne; **mb**: Mont Blanc; **pe**: Pelvoux). The Penninic zone (**P.**) comprises the internal crystalline massifs (**gp**: Gran Paradiso; **ht**: Hohe Tauern; **mr**: Monte Rosa; **sb**: Grand Saint Bernard; **te**: Tessin; **va**: Vannoise). The Austro Alpine (**AA.**) and South Alpine (**SA.**) zones are indicated in light grey, whereas the ophiolitic complexes of the oceanic suture are horizontally hatched. **LGD** shows the outline of the Lepontine Gneiss Dome (modified after Debelmas and Mascle 1993).

evaporites. The longitudinal *Sillon subalpin* separates the Subalps from the external crystalline massifs of the Helvetic zone (from south to north: Argentera, Ecrins-Pelvoux, Belledonne, Mont Blanc, and Aar massifs). These granitic and gneissic massifs originated in the late Oligocene phase of intense shortening which brought them up to the surface and expelled northwards their sedimentary cover, now appearing as the north to north-west vergent Helvetic nappes of Switzerland.

The ancient European passive margin proper with its thinned and faulted continental crust constitutes the internal, or Penninic zones of the western Alps. Separated from the externides by the FPT and the Valais trough, where small amounts of ophiolites crop out (Schmid *et al.* 1989), the Penninic domain is divided into a number of zones inherited from the faulted and more or less tilted blocks of the Jurassic rift. The Briançonnais and Piémont zones correspond to north-west to west vergent cover nappes, whereas the granitic/gneissic basement of the Penninic domain is exposed in the extended internal crystalline massifs (Vannoise, Gran Paradiso, Grand Saint-Bernard, Monte Rosa, Simplon) (Fig. 1.6). To the east and south of the Penninic zones, the suture between the European and African/Adriatic plates is marked by a number of ophiolitic complexes (e.g. Ivrea). More eastwards, this suture is buried under the northwards thrusted Austroalpine nappes, belonging to the Adriatic plate margin. The sedimentary cover of the suture zone ('Liguro-Piémontais') was folded and metamorphosed during the upper Cretaceous eoalpine phase of plate convergence and subduction. Finally, the northernmost part of the Austroalpine nappes, i.e. the northern calcareous Alps, should be mentioned. They correspond to the Mesozoic cover of the Adriatic margin, separated from its basement by a decollement and independently thrusted northwards to such an extent that it partly covered the Molasse basin in north-west Austria and south Bavaria.

The Molasse Basin in Southernmost Germany

The stacked nappes in the Alps caused a strong loading of the cold and thick, thus rigid European lithosphere and induced its flexure, the process migrating northwards with the development of the chain. Crustal flexure along the northern Alpine front resulted in the subsidence of the Molasse foreland basin (Sinclair *et al.* 1991). This basin, fed since the Eocene by clastic sediments supplied by the intense erosion of the uplifting Alps, extends in Switzerland and south Germany (Fig. 1.6). It shows a clear transverse asymmetry, with the greatest sediment thickness (up to 4 km) closest to the Alpine chain, and a subsidence axis progressively displaced northwards. Phases of increased subsidence accompanied the Oligocene and Miocene periods of nappe emplacement (Walter 1992). Two main marine/continental cycles are recognized within the late Eocene to late Miocene basin sedimentation. In the east however, the first continental episode is lacking, with the sea retreating only at the end of the lower Miocene. Today, the extensive Miocene molassic deposits are covered by Quaternary glacial deposits.

The Jura Mountains

The Jura Mountains are an arcuate, up to 1,723 m high, mountain range resulting from thin-skinned tectonics deforming a mainly calcareous Mesozoic cover. The Mesozoic Jura shelf emerged at the end of the Cretaceous and underwent Palaeogene subtropical weathering and erosion as well as tectonic extension in relation with the development of the West European rift. A last marine transgression of the perialpine sea invaded it during the Miocene. During the late Miocene/lower Pliocene, the internal part of the Jura was reached by the Alpine compression. Simultaneously and in spatial continuation with the basement of the Subalpine chains, into whose folds the folded Jura also merges, its basement was underthrusted beneath the Alpine external crystalline massifs. At the same time its cover (again above a decollement located within the Triassic evaporites) was folded and displaced north-west and westwards, notably onto the lacustrine Miocene of the Bresse graben. Strike-slip faults oblique to the fold axes, and partly inherited from the southern continuation of the Oligocene Upper Rhine graben into the Jura, acted as regional stops for the nappe displacement. The more intense folding of the internal Jura was responsible for a thickened cover which in turn induced the uplift of this zone. Transition towards the unfolded Triassic and Jurassic of the high Saône area is realized through the French Jura plateau, which are alternating subhorizontal or south-tilted monoclinal blocks and strongly folded and faulted narrow zones (Walter 1992). Relying upon the present-day stress field pattern in the Jura and its surroundings, Becker (1999) has suggested that most of the fold belt is no longer active today.

The West European Rift System

General Configuration

The European Cenozoic rift system (ECRS) certainly is one of the most prominent present-day megastructures of western Europe and, although each of its constituting segments has its own history, it must be presented as a whole. It extends in an almost north–south direction over 1,500 km from the western Mediterranean to the North

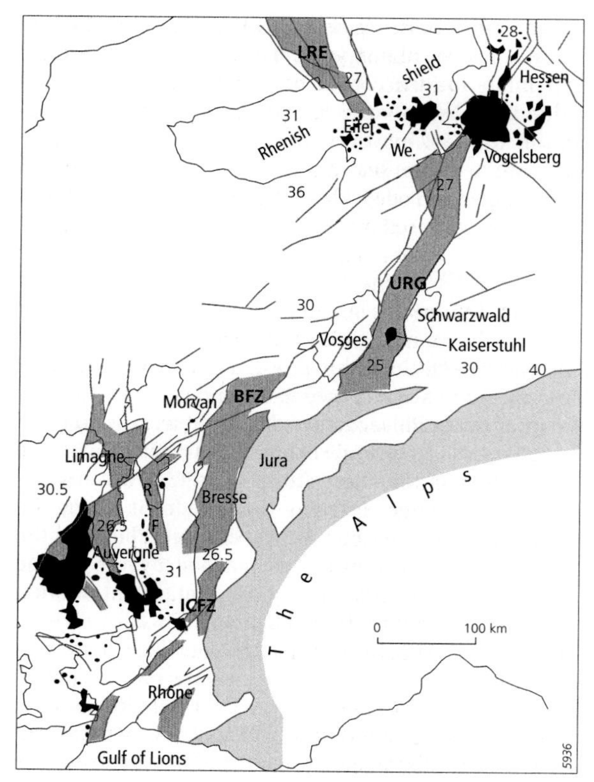

Fig. 1.7. The European Cenozoic rift system (ECRS). The main grabens are in dark grey, the volcanic areas in black. Numbers indicate the Moho depth in kilometres and are taken from Mechie *et al.* (1983) and Prodehl *et al.* (1995). **LRE**: Lower Rhine Embayment, **URG**: Upper Rhine graben, **R-F**: Roanne and Forez basins, **We**: Westerwald, **BFZ**: Burgundy transform fault zone, **ICFZ**: Isère–Cévennes fault zone.

Sea (Fig. 1.7). To the south, it is continued offshore by the Gulf of Lions and the Valencia trough in direction of the Alboran Sea (Ziegler 1992). In its onshore part, it is comprised, from south to north, of the Rhône depression, the Limagne (*s.l.*), Bresse and Saône grabens, the Upper Rhine graben, the Leine graben *s.l.*, and the Lower Rhine Embayment. Moreover, the Eger graben of the Bohemian massif is considered to belong to the ECRS. Whereas the Rhône, Saône, and Bresse grabens are aligned to separate the French Massif Central from the Jura and the Alps, the Limagne branched grabens are developed within the Massif Central. Except in its central part, the ECRS cuts the Variscan structures at a steep angle whereas it is parallel to the deformation front of the Western Alps in south-east France. The location of most individual grabens of the ECRS was determined by the presence of late Variscan fracture systems. The north-east striking Burgundy sinistral transform fault zone, on which the Limagne and Bresse grabens as well as the Upper Rhine graben are abutting, corresponds to a lateral shift of the rift axis of *c.* 100 km. The Upper and Lower Rhine and the Leine grabens meet at a triple junction near Frankfurt, close to the southern border of the Rhenish shield (Prodehl *et al.* 1995). From there, the presently extinct Leine graben prolongs the Upper Rhine graben in the same NNE direction towards the north German basin. The north-west striking rift branch, however, cuts the Variscan structures of the Rhenish shield at a right-angle before being continued by the Lower Rhine Embayment, which merges north-westwards in the West Netherlands basin and dies out by entering the North Sea (Zijerveld *et al.* 1992).

The ECRS started to develop during the middle Eocene in its central part (southern Upper Rhine graben and Saône graben) and rapidly propagated to the south and the north, locally accompanied by strong volcanic activity. It thus evolved in the Alpine foreland contemporaneously with the main and late phases of the Alpine orogeny, its main centre of subsidence apparently migrating northwards with time.

The Main Rift Segments

Located to the south of the Isère–Cévennes fault zone (Fig. 1.7), the *Rhône area* is a wide north–south trending depression that underwent strong Mesozoic subsidence and is filled with 8–10-km thick marls and limestones of mainly Jurassic-Cretaceous age. In the Oligocene to early Miocene times, renewed subsidence was limited to small basins. The southern offshore continuation of the Rhône depression bears witness to a much more intense subsidence, and oceanic crust has even developed in the Gulf of Lions (Burrus *et al.* 1987). The southern Rhône valley accordingly is one of the ECRS onshore zones with greatest Moho uplift. Below its thick sedimentary cover, the depressed crystalline crust is only *c.* 15 km thick, giving a depth to Moho of 24–25 km (Sapin and Hirn 1974).

Terminating to the north against the Burgundy rift–rift transform, the *Limagne subsidence area* is a complex zone of north–south trending grabens penetrating deeply into the French Massif Central. The main subsident units, the Limagne graben to the west and the aligned Roanne and Forez basins to the east, join together northwards when coming out of the massif (Prodehl *et al.* 1995). At the southern end of the Limagne graben, several smaller basins are still present before the basement rapidly rises to the surface. In the Limagne grabens, lacustrine sedimentation started during the middle Eocene, with main subsidence phases occurring during the late Eocene–early Oligocene and late Oligocene (Blès *et al.* 1989). At the beginning of the Miocene, the subsidence was interrupted, probably in relation with thermal doming of the Massif Central and the associated volcanic activity. The Limagne graben is presently being uplifted (Ziegler 1992). Maximum thickness of *c.*2,500 m of Cenozoic is observed in the western part of the strongly asymmetric Limagne graben. The Limagne subsidence is also an area of marked crustal thinning, with 5–6 km of Moho uplift (Zeyen *et al.* 1997*a*).

Running to the east of the Massif Central parallel to the Limagne and Forez grabens, the *Bresse-Saône graben* forms the northern prolongation of the Rhône depression and abuts on the Burgundy transform zone. Its Cenozoic evolution is largely similar to that of the Limagne grabens except that it is devoid of accompanying volcanic activity. Palaeogene sedimentation also started during the middle Eocene and culminated in the Oligocene. However, after an early/middle Miocene interruption, subsidence resumed during the late Miocene. During the latest Miocene, the eastern margin of the graben was overriden by the frontal thrusts of the Jura Mountains before Pliocene lacustrine sedimentation finally sealed the thrusted elements (Debelmas and Mascle 1993). The highly asymmetric sedimentary fill of the now uplifting Bresse graben reaches a maximum thickness of 5 km (Bergerat *et al.* 1990), of which about 2,000 m correspond to the Oligocene to Pliocene deposits (Ziegler 1992). The crustal thickness seems to be reduced by *c.*10% below the Bresse graben.

The *c.*310 km long *Upper Rhine graben* (URG) constitutes the central segment of the ECRS. It extends in a NNE direction from Basle, located at the eastern tip of the Burgundy transform zone, to Frankfurt near the southern border of the Rhenish shield. The graben is 35 km wide on average and is flanked by uplifted shoulders, especially in its southern half where the Vosges and the Schwarzwald dominate it by more than 1,000 m. The regional Mesozoic cover shows no structural relationship with the future graben (Prodehl *et al.* 1995). The graben fill proper spans the period from middle Eocene to Quaternary. Heralded by an early Eocene volcanic activity, subsidence started first in the southern part of the URG with middle/late Eocene brackish sedimentation and advanced northwards during the early Oligocene. At this time, the graben was invaded by the Alpine sea, and marls and evaporites were deposited. At the beginning of the Miocene, the sea retreated, owing to the thermal uplift of the Vosges–southern URG–Schwarzwald rift dome, and the Kaiserstuhl volcanic complex developed. The southern URG ceased to subside and its sedimentary fill was even partly eroded during the late Miocene whereas the northern part of the graben underwent increased subsidence (Ziegler 1992). During the Plio-Pleistocene, due to the reorientation of the regional maximum horizontal stress S_{Hmax} to N145°E (Müller *et al.* 1992), the graben subsidence was reactivated in a transtensional mode. The main centre of subsidence remained in the northern URG, but slow, intermittent subsidence occurred again in its southern half, paralleled by renewed uplift of the Vosges and Schwarzwald. The maximum thickness of Cenozoic sediments is recorded in the Mannheim area, where it reaches 3,400 m. Not only does the URG show a strong transverse asymmetry with regard to sedimentary infill and crustal structure, like most other rift segments, but its polarity is reversed along the transverse Erstein swell (Chorowicz *et al.* 1989). Crustal stretching in the URG has been estimated in the range 5–7 km (Doebl and Teichmüller 1979; Meier and Eisbacher 1991), corresponding to maximum crustal thinning and a Moho depth of 24–25 km in the southern URG. The Moho discontinuity, however, dips slightly towards the north and reaches a depth of 27–28 km near Frankfurt (Meissner and Vetter 1974).

Although it is located in the right prolongation of the URG, the *Hessen depression*, including the Leine graben to the north, displayed limited subsidence during the Cenozoic and may thus be considered as a rapidly failing rift arm to the north of the Frankfurt triple junction. Northward propagation of the ECRS in Hessen occurred in the early Oligocene, reopening for a time a narrow seaway between the North Sea and the Alpine foreland basin (Walter 1992). However, the sea had already retreated during the upper Oligocene. The middle Eocene to middle Miocene sedimentary fill of the depression does not exceed 300 m. At the beginning of the Miocene, the depression underwent regional uplift accompanied by increased faulting and extended volcanic activity in northern Hessen and especially in the Vogelsberg area, where it culminated between 17 and 15 Ma.

Extending in the direction of the current S_{Hmax}, the wedge-shaped *Lower Rhine Embayment* (LRE) is the presently most active segment of the ECRS. To the north-west, the embayment widens out to merge into a subsidence area encompassing the western Netherlands and Central basins. At its south-eastern end, it is linked across the Rhenish shield to the Frankfurt triple junction by a zone of diffuse faulting and seismicity followed by the Rhine river since the Pliocene. According to Illies *et al.* (1979), this hiatus in the rift surface expression results from the tectonic incompetence of the shales making the bulk of the massif. The LRE is comprised of a number of tectonic blocks, of which the westernmost Roer graben underwent maximum subsidence (Ahorner 1962). The embayment as a whole thus appears asymmetric. Moreover, its main elements are clearly tilted to the north-east. The Roer graben is a late Permian structure that had already undergone late Permian/early Triassic strong subsidence as well as a shorter subsidence episode during the late Jurassic. Despite upper Cretaceous basin inversion, Mesozoic sediments in the graben still reach a thickness of *c.*2 km. Rifting propagation within the LRE dates back to the late middle Oligocene. In this time, the area began to subside and was drowned by a sea coming from the north-west up to Bonn, a marine connection even existing for a short time with the URG. During the early Miocene, coinciding with the onset of the thermal uplift of the Rhenish shield, the sea retreated and the sedimentation was temporarily interrupted. During the middle Miocene, subsidence started again, continuing up to the present, accumulating a maximum 1,500 m of sediments. Widespread volcanic activity marked by a number of eruptions from the upper Eocene to the upper Pleistocene in the east and west Eifel, in the Westerwald, and in the Siebengebirge accompanied the LRE opening and the Rhenish shield uplift. Cenozoic crustal extension within the Roer graben has been estimated at 1–2 km and crustal thinning is moderate, the Moho rising from 30–31 km depth under the graben flanks to 28 km beneath the graben (Zijerveld *et al.* 1992).

The ENE–WSW trending *Eger graben* is located in the north-western Bohemian massif. Comprising a number of small basins, it started to subside during the upper Oligocene, the main subsidence phase spanning the upper Oligocene and the early/middle Miocene. Polyphase intense volcanic activity within the graben and its Bohemian surroundings developed from the upper Oligocene to the Pleistocene (Walter 1992), each volcanic cycle being generally followed by basin subsidence (Malkowsky 1987). The Eger graben area is presently uplifting (Ziegler 1992).

Associated Volcanism

The main volcanic fields associated with the development of the ECRS are located within the Rhenish shield and the French Massif Central. In the Massif Central, the volcanic activity is especially concentrated in the Auvergne, west and south-west of the Limagne graben (Chaîne des Puys, Mont Dore, etc.). The volcanics derive from primitive mafic alkaline magmas and show wide variations in composition (Wilson and Downes 1992). Another important volcanic district is located near the south-east border of the Limagne. The Massif Central volcanism started during the lower Miocene. After a paroxystic phase of widespread activity during the Pliocene, it was limited to the Chaîne des Puys and the Vivarais in recent times. No systematic spatial or temporal connection exists between volcanism and the development of the Limagne graben (Prodehl *et al.* 1995).

In the Rhenish shield area, rifting-related volcanism started in the Eifel, west of the Rhine valley, during the middle Eocene. On the eastern side of the rift, first volcanic eruptions are recorded during the upper Oligocene. With the exception of a limited Pliocene reactivation in the Westerwald, this activity came to rest during the middle Miocene, in a time when volcanism began farther east in the north Hessen and when the Vogelsberg stratovolcano was under construction (Lippolt 1983; Seck 1983). To the east of the Vogelsberg, the Rhön volcanism is dated between 22 and 11 Ma (Walter 1992). The volcanic activity of the Rhenish shield and its surroundings thus appears to have shifted from west to east during the Tertiary. Except for a few eruptions in the Westerwald close to the Rhine valley, Quaternary volcanism is restricted to the western part of the Rhenish massif, where two distinct NW–SE trending provinces are recognized in the Eifel. Tertiary volcanics are mainly derived from alkaline basaltic magmas originating in homogeneous upper mantle sources.

In the URG, first volcanic manifestations date from the upper Cretaceous. Eocene activity is then recorded in the high bordering areas of the northern URG, but the main volcanic activity is located in the southern URG and its shoulders during the upper Miocene. In this time, the Kaiserstuhl complex was built within the graben at the intersection of the eastern graben border fault and a WNW–ESE fault system. On the other side of the Schwarzwald (i.e. beyond the graben shoulder proper), the Hegau and Urach volcanic fields also developed during the upper Miocene. The Kaiserstuhl volcanics mainly derive from primitive olivine nephelinitic magmas.

In all ECRS-related volcanic fields, from Cantal to north Hessen and the Eger graben, the magmatic melts come from the upper mantle (lower lithosphere and upper asthenosphere) without a deep mantle contribution (Ziegler 1992). In the Rhenish shield and the Massif Central, they originated in 60–80-km deep sources corresponding to asthenospheric bulges, whereas in the URG their composition rather suggests an origin in the asthenospheric mantle at depths of *c.*100 km (Prodehl *et al.* 1995). No clear connection exists between volcanic activity and the phases of increased rift subsidence. In general, the former show closer association with the episodes of uplift, either regional or limited to the graben shoulders. Spatially, the volcanic provinces are correlated with the zones of lithospheric thinning (Zeyen *et al.* 1997*b*).

Rifting Dynamics

In the last three decades, numerous large-scale seismic and teleseismic studies have provided a lot of information on the structure of the European lithosphere (Ansorge *et al.* 1992; Prodehl *et al.* 1995; Ritter *et al.* 2000). They clearly revealed that the crust is thinned beneath most parts of the ECRS. A maximum Moho uplift of 5–6 km has been observed below the southern URG, the Limagne graben, and the Rhône depression, where crustal thickness (including the basin fill) amounts to *c.*25–26 km (Zeyen *et al.* 1997*b*). In contrast, Moho depths of 27–28 km are characteristic of the LRE and the Bresse and Eger grabens. Whereas crustal thinning correlates well with thickness of the basins' sedimentary fill, lithospheric thickness is much more correlated with the surrounding volcanic activity. The lithosphere is indeed *c.*80 km thick below the French Massif Central (Sobolev *et al.* 1996) and *c.*60 km below the Eifel-Westerwald area (Babuska and Plomerová 1992). Moreover, in the topography, the uplifted shoulders of the southern URG may be opposed to more regional uplifts affecting the whole Rhenish shield and the Massif Central. On the contrary, the LRE does not display any uplifted flanks. Regional or more strictly rift-related uplifts appear anyway to have a thermal origin that Ziegler (1992) explains by two different models. In the URG, where shoulder uplift follows the episodes of graben subsidence, the tensional failure model proposes that crustal extension has caused an asthenolith to rise and to flatten at the crust-mantle limit. Concerning the Rhenish shield, it is believed that the thermally driven regional updoming and the associated widespread volcanic activity are better explained by a model involving a mantle plume below a thinned lithosphere. However, the extent of the identified Eifel plume is better correlated with the sole area of Quaternary volcanism in the Eifel (Ritter *et al.* 2000), and it seems not able to explain the Tertiary uplift and volcanism of the whole Rhenish shield.

The ECRS may thus be considered as a 'passive' rift in that its evolution is primarily governed by far-field lithospheric stresses. Sengör *et al.* (1978) saw it as a result of foreland splitting in response to continent–continent Pyrenean and Alpine collisions. However, the southward extension of the rift towards the Valencia trough is not accounted for by this model. Moreover, despite the generally compressional stress regime resulting in west Europe from plate collision, rifting occurred during phases of predominating far-field extensional stresses (Bergerat 1987; Lacombe *et al.* 1993). These phases are integrated by the plate kinematics reconstruction of Le Pichon *et al.* (1988), showing divergent motion between westernmost Europe and the rest of Eurasia during the Oligocene. Consequently, Ziegler (1992) argued that repeated changes in the Cenozoic stress field of north-west Europe not only depend on changing convergence patterns between Africa and Europe and varying mid-Atlantic sea-floor spreading rates but also result from a global plate reorganization, leading notably to the west European rifting.

Zeyen *et al.* (1997*b*) furthermore observe that the rift opening occurs obliquely with respect to the lithospheric mantle displacement defined by the direction of Europe–Africa collision and the opening of the mid-Atlantic ridge. According to them, the low level of volcanism and insignificant lithospheric thinning under the main part of the ECRS imply that the mantle is hardly affected by extension and is decoupled from the independently moving and deforming crust. They calculate that such a detachment of the crust from the mantle is possible along the whole rift length due to the low strength of the European lower crust. This different movement of crust and lithospheric mantle explains why the whole European lithosphere did not break apart. In this scheme, differential velocity between crust and mantle of the western European lithosphere, which cannot propagate into oceanic lithosphere, would be accommodated by compressional crustal tectonics giving rise to basin inversion in south England, the North Sea, and the north German basin. As for the Massif Central and Eifel mantle plumes, Zeyen *et al.* (1997*b*) suggest that, hitting a thinned lithosphere, they are able rapidly to penetrate levels where melting occurs. Therefore, they induce neither lithospheric bending nor large-scale rifting but when the plume material enters the brittle part of the lithosphere, deformation is controlled by the crustal stress field.

The Paris Basin

The Paris basin is the archetype of an intracratonic basin. It finds its origin in the Permian rifting that affected the Cadomian-Variscan basement of western Europe during the initial stages of the Pangea break-up, creating several east–west trending basins (Mascle 1990). Filled with 500–1,000 m of Permian sediments, these early grabens still correspond to the places of maximum later Mesozoic/Cenozoic sedimentation (Fig. 1.8). After the Permian rifting phase, the broader depressed area of the Paris basin would subside for as long as 230 Ma. The mechanism of this long-lasting subsidence still remains unclear. Post-rift thermal relaxation of the

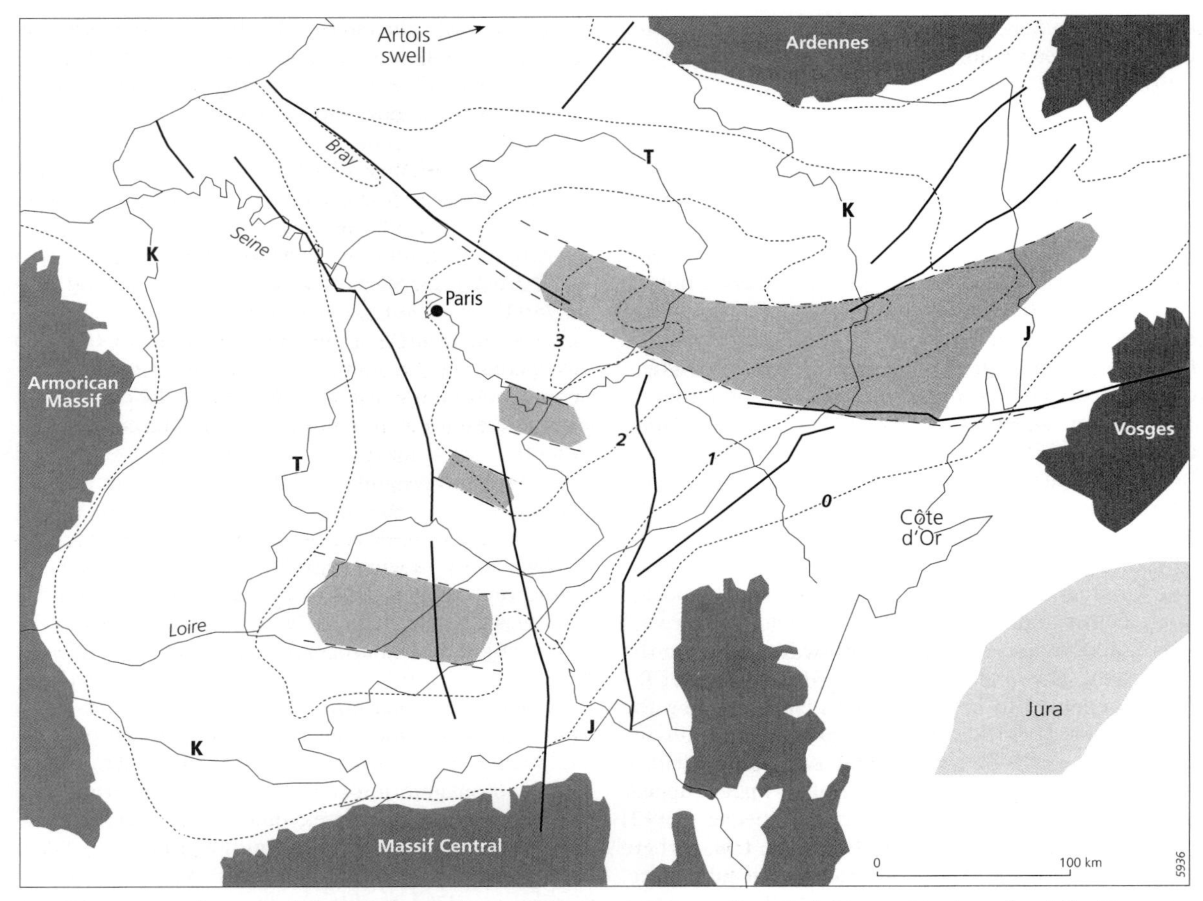

Fig. 1.8. Structural setting of the Paris basin. The initial Permian basins are shown in light grey. The isopachs of the Mesozoic-Cenozoic basin fill (thin dashed lines) are given in kilometres. **T** gives the outward limit of the current continuous outcrop area of the Tertiary deposits (**K** and **J**: same for the Cretaceous and Jurassic outcrop areas; modified after Debelmas and Mascle 1993).

lithosphere certainly played a role but does not explain the duration of the subsidence, all the more as the Moho is encountered at normal depths of 35–37 km over an upper mantle devoid of any dense body (Loup and Wildi 1994). Robin *et al.* (2000) show that most subsidence rate variations (or even transient positive movements) leading to first-order transgression/regression alternations with a 10–40 Ma cyclicity are related to major tectonic events involving the whole west European domain. As for the eustatic control, it was especially responsible for the upper Cretaceous and Stampian transgressions. In the central part of the Paris basin, up to 3,000-m thick sediments were accumulated during the Mesozoic and the Cenozoic.

During the Mesozoic, nine major stratigraphic cycles may be recognized in the period extending from the lower Triassic to the Cenomanian (Robin *et al.* 2000). During the Triassic/early Jurassic, sandstones and marls representing three cycles were deposited mainly in the eastern half of the basin, in a gulf proceeding from the east and originating in the German Sea. Then, the Toarcian–Aalenian major discontinuity corresponds to the mid-Kimmerian tectonic phase of Ziegler (1990) and could represent the basin response to the pre-rift updoming of the North Sea (Underhill and Partington 1993). During the middle and upper Jurassic, three further cycles recorded the tectonically induced fore and back movements of epicontinental seas coming from the Alpine domain through the Côte d'Or strait (Fig. 1.8) and drowning the whole basin. Marked by two successive unconformities, the following upper Jurassic main regressive phase is associated with the neo-Kimmerian tectonic phase (Ziegler 1990). It caused the upper part of the Jurassic carbonates to be widely karstified and partly eroded in many parts of the basin. During the early Cretaceous, Weald continental deposits accumulated in the north-west of the basin, whereas in the east marine clays and sands were deposited in a shallow sea coming from the Alpine domain. The upper Cretaceous transgression then covered the whole basin, with for the first time a direct link to the opening Atlantic ocean via the Channel and the Poitou swell, where a connection was also established with the Aquitaine basin. At the end of the Campanian, another major discontinuity developed in relation with a sharp change towards continental conditions prevailing from the Maastrichtian to the Thanetian in some places (Dewolf 1982; Thiry and Simon-Coinçon 1996).

Three major stratigraphic cycles may be distinguished in the Tertiary evolution of the Paris basin (Robin *et al.* 2000). They have developed in progressively more and more continental conditions, so that marine transgressions were often restricted to the central and north-western parts of the basin, whilst highly variable shorelines were found in the south-west and contemporaneous lagoonal to lacustrine environments were present in the south-east (Dewolf 1982). The north-eastern borders of the basin no longer experienced marine conditions. During the Palaeocene and the Eocene, several transgressions invaded the basin and accumulated limestones, sands, and clays. In these times, owing to the Alpine orogeny, the depocentre had shifted towards the north-west and the seas came from north-west or north, in a gulf originating in the North Sea. Moreover, the basin was open towards the Atlantic ocean through the Loire valley and the Channel. During the Lutetian, the Paris basin was definitively separated from the Belgian basin by the upwarping of the Artois axis (Fig. 1.8). At the end of the Eocene, as a consequence of the Mesoalpine tectonic phase, the Bartonian emersion marks another major discontinuity. It was followed by a mostly detritic sedimentation during the Oligocene/lower Miocene cycle which began with endoreic lagoonal conditions, continued with the last major marine invasion during the Stampian, and ended with a new continental episode (Dewolf 1982). At the beginning of the Miocene, the centre of deposition had again shifted, this time by more than 150 km towards the south-west (Debelmas and Mascle 1993). During the Burdigalian, while previous endoreic conditions had given way to the basin reopening towards the Channel, sediments chiefly originating from the Massif Central were spread in alluvial nappes on a glacis corresponding to its south-western part. After a last limited Helvetian transgression in the north-west, Pliocene continental erosion of the basin and its borders superficially reshaped a pre-Pliocene polygenic erosion/accumulation surface (Dewolf 1982).

The North German Basin

The North German basin, opening northwards in the present-day North Sea basin, is part of the South Permian basin, developed on varying crustal domains including Variscan, Caledonian, and even pre-Caledonian. It originates from the Permo-Carboniferous Variscan foreland collapse and crustal fragmentation along the south-west border of the Baltic shield, which developed an east–west trending rift occupied by mainly rhyolitic volcanics (Benek *et al.* 1996). After abortion of this rift, subsequent thermal relaxation led to the subsidence of a broader Permian-early Triassic intraplate basin (Fig. 1.9). As noted by Scheck and Bayer (1999), this basin displays a strong north–south asymmetry opposing a steep southern border and a gently dipping northern slope. It underwent a long-lasting (> 250 Ma) polyphase evolution marked by phases of subsidence, with repeated shifts of centre of deposition and intervening deformation phases.

During the Permian and the early Triassic, thermal subsidence was responsible for the accumulation of up to 5,000 m of clastics and evaporites in the North German basin. From the middle Triassic, however, the subsidence rate decreased exponentially and gave way to basin differentiation. In that time, a number of NNE striking troughs appeared along the southern flank of the basin, among which the Hessian depression constituted a seaway linking the Triassic basins of north-west Europe and the Tethys (Ziegler 1978). Strong post-depositional salt movements also started during the middle Triassic (Pompeckj block). During the upper Triassic, a short subsidence episode caused the deposition of up to 1,200-m thick continental deposits into subbasins. In contrast, the Jurassic and lower Cretaceous recorded limited sedimentation, moreover mostly removed as a consequence of a middle Jurassic regional uplift related to thermal doming in the North Sea (Ziegler 1990). At the same time, seemingly in relation with wrench movements between the Danish–North German block and the Variscan massifs to the south, WNW–ESE trending marginal troughs (Lower Saxony basin in the south-west, Altmark and Brandenburg basins in the south-east) appeared along the southern border of the North German basin (Ziegler 1978; Betz *et al.* 1987). The main subsidence of these basins occurred in the late Jurassic and early Cretaceous when up to 3,000-m thick deposits accumulated, accompanied

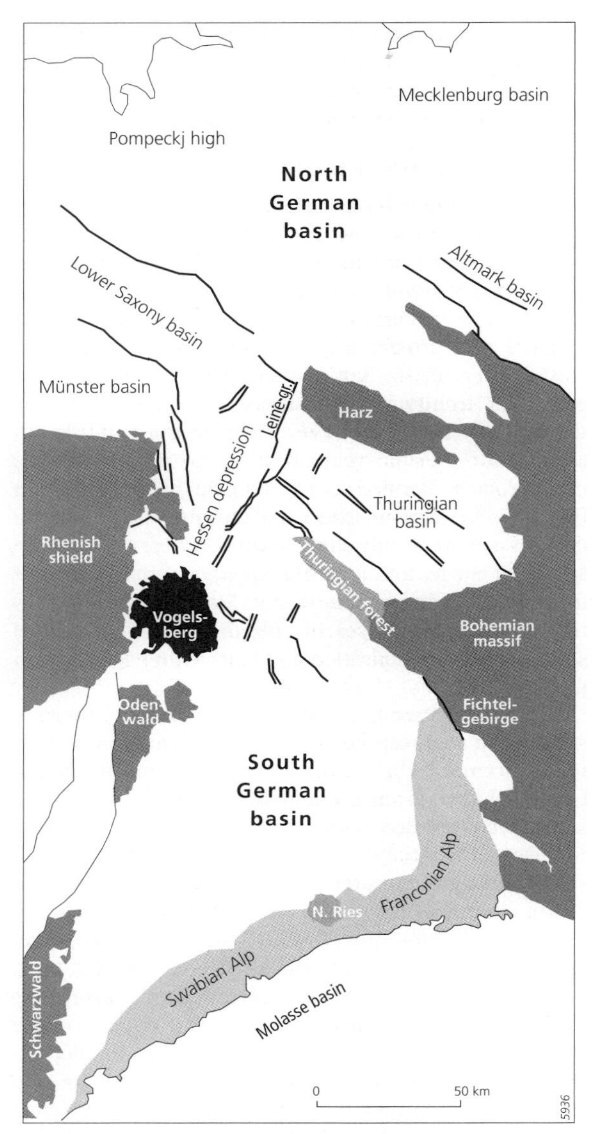

Fig. 1.9. The North and South German basins and the surrounding structural units. See text for explanation.

by strong diapirism of the underlying Zechstein evaporites (Walter 1992).

Then, during the late Cretaceous, a regional tectonic inversion was caused in north-western Europe by the build-up of intraplate compressional stresses in relation with the north Atlantic opening and the first Alpine phases. The sedimentary fill of the strongly inverted marginal troughs was folded and uplifted (with the Lower Saxony basin notably giving rise to the present-day Lower Saxony tectogen) but surprisingly, in contrast with its southern border, the North German basin continued to subside at least locally (e.g. Mecklenburg basin) (Scheck and Bayer 1999). Nevertheless, during the Cenozoic, the influence of the North Sea thermal subsidence became predominant in northern Germany and the North German basin underwent renewed subsidence, allowing a further *c.*1,000 m of Tertiary brackish-marine clastics to be deposited. Backstripping calculations by Scheck and Bayer (1999) indicate that around 30% of the total basin subsidence (*c.*8-km thick basin infill) was tectonically induced, the remainder being isostatically induced by the sediment load.

Finally, a veil of Quaternary glacial sediments, 50–100 m thick on average in north Germany and up to 500 m in the adjacent northern part of The Netherlands (Walter 1992), covered and shaped the present-day landscape. Holocene and current subsidence of the coastal areas of north Germany results from the resorption of the elastic forebulge after the removal of the Scandinavian ice sheet (Franke 1992).

The Mesozoic South German Basin and the Thuringian Basin

The Mesozoic South German basin is located between the Rhenish shield and the URG to the west, the Bohemian massif to the east, and the Miocene Molasse basin to the south (Fig. 1.9). The Swabian and Franconian Alps respectively form its southern and south-eastern/eastern limits. Present-day depth of its Palaeozoic basement may reach 1,500 m in limited lower Permian troughs but does not exceed a few hundred metres in the south. A first marine incursion coming from the north invaded the basin during the upper Permian. The succeeding continuous Triassic sedimentation accumulated freshwater sands delivered by the Rhenish and Bohemian massifs, followed by shallow marine carbonates deposited in an epicontinental northern extension of the Tethys, and lagoonal deposits. During the Triassic, the South German basin thus corresponds to a tectonically quiet platform (Walter 1992). The Jurassic is characterized by the transgression of several shallow seas, not only connected to the seas drowning the Paris basin but also to the North Sea domain through the Hessen depression. Dark-coloured clays, marls, sands, and carbonates were successively deposited. At the end of the Jurassic, the sea retreated and the long continental evolution of the basin began. It was first characterized by a weak south tilting as a consequence of the Rhenish shield uplift (Ziegler 1978). The erosion therefore started in the highest places of the time, i.e. in the Odenwald area (Fig. 1.9). It removed the Jurassic strata from the main part of the basin, whereas the remaining upper Jurassic carbonates of the Swabian and Franconian Alps were strongly karstified. It seems that the resulting typical landscape of cuestas may be at least partly dated as far back as the lower Tertiary (Walter 1992). At present, the southern part of the basin, the Swabian and south Franconian Alps, is most uplifted.

Due to its proximity to the Bohemian massif, the eastern border of the South German basin underwent a different evolution. During the early Cretaceous, reactivation of upper Carboniferous basement-involving faults individualized a number of tectonic blocks (Schröder *et al.* 1997) and the relative vertical motion of the basin with respect to the uplifted western border of the Bohemian massif amounted to *c.*1,500 m. Consequently, during the upper Cretaceous, a last marine invasion drowned the depressed eastern part of the basin, corresponding to the subsiding northern Franconian Alp. Then, Senonian to Palaeocene transpressional deformation led to the upthrusting of major basement blocks. These blocks were mildly reactivated during the Mio-Pliocene in relation to the development of the volcano-tectonic Eger zone and the associated uplift of the Bohemian massif.

Beyond the ECRS-related Miocene volcanic fields of the Hegau and Urach, another unusual feature of the Swabian Alp is the 15-Ma-old round meteoritic crater of the Nördlinger Ries. Located at the junction between the Swabian and Franconian Alps and accompanied by the much smaller Steinheim impact crater, the Nördlinger Ries is about 26 km in diameter and is encompassed with a *c.*100 m high wall.

As for the Thuringian basin, it extends north of the Thuringian Forest (Fig. 1.9), and mainly received upper Permian and Triassic sediments with a similar succession of facies as in the North and South German basins. Maximum infill thickness is observed in the north-east where it reaches *c.*2,200 m. Sparse remainders of lower Jurassic and marine Cenomanian covers are also still present, as well as Palaeogene lacustrine sediments. These lacustrine sediments include lignites that accumulated in karstic depressions due to solution of the underlying Zechstein evaporites. First tectonic movements probably started in relation to the mid-Kimmerian phase and continued during the Tertiary (Walter 1992).

The Palaeozoic Massifs (Rhenish Shield–Ardennes, French Massif Central, Brittany, Harz, Bohemia)

These Palaeozoic massifs remained emerged during most of the Mesozoic and the Cenozoic. During these times, their geological evolution amounted to alternations of tectonic stability and uplift. This evolution can be unravelled by using the last sedimentary remnants of the few seas that encroached on their borders and the stepwise arrangement of morphological erosion surfaces within the massifs.

The Mesozoic history of the Rhenish shield and its Ardennian annexe, of Brittany, the French Massif Central, and the Bohemian massif and its Harz northern outpost began with the continuation of the Permian levelling of the Variscan mountains. These mountains certainly never reached extreme elevations since active erosion more or less kept pace with uplift. The subdued topography of the massifs allowed the Zechstein sea to drown the Harz and the eastern Rhenish shield completely (Walter 1992). During the Triassic and the first half of the Jurassic, extended seas still invaded subsiding areas within these massifs, like the Luxemburg Embayment or the Causses, with up to 1,500-m thick Jurassic limestones to the south of the Massif Central (Fig. 1.1). The Vosges and Schwarzwald were also totally covered by the Liassic seas, appearing as individualized massifs only during the middle Jurassic. Then, from the upper Jurassic onwards, a large continental area encompassed from west to east the Brabant massif, the Ardennes, the Rhenish shield, the Hessen area, the Harz, and the Bohemian massif. During the lower Cretaceous, this continental zone extended southwards to include the Palatinate Mountains, the Vosges, and the Schwarzwald (Ziegler 1978) and its low topography underwent generalized deep ferrallitic weathering and erosion under humid tropical conditions. The upper Cretaceous transgression invaded the emerged lands of north-west Europe from the north and covered large parts of the Ardennes, the Rhenish shield, the Harz, and the Bohemian massif, with thick marine Cretaceous deposits accumulating in peripheral basins (Münster basin, with a maximum thickness of 2,000 m of sands, marls, and carbonates) and inner basins (upper Cretaceous basin of north Bohemia with an up to 1,000-m thick infill).

During the Tertiary, all massifs underwent an almost uninterrupted continental evolution. Under generally hot, but alternately wetter and drier climates, their rejuvenated relief was sculptured in a morphological succession of stepped erosion surfaces, in many cases of progressively lesser extent with younger age. This bears witness to the Tertiary uplift which brought the so-called Palaeozoic massifs to their present elevation. Authors dealing with the morphological evolution of different massifs generally agree that, after as early as Eocene first signs of motion, the main uplift phase started during the Miocene (Meyer *et al.* 1983; Klein 1990;

Walter 1992; Battiau-Queney 1993; Peterek and Schröder 1997). Moreover, the Plio-Pleistocene incision of the valleys within the older surfaces highlights the acceleration of uplift in recent times (Meyer and Stets 1998; Bonnet *et al.* 2000). Demoulin (1995) states for instance that the Neogene and Quaternary uplift of the Ardennes amounted to 300–350 m in average, of which 130–175 m are ascribed to the Quaternary. Meyer and Stets (1998) also conclude that the Eifel underwent *c.*150-m uplift in the last 0.8 Ma, whereas east of the Rhine River the Westerwald was uplifted by 60 m in the same time (Meyer *et al.* 1983) and by 200 m since the Miocene. In the Massif Central, Battiau-Queney (1993) mentions uplift values of several hundred metres. Similar values are suggested by Peterek and Schröder (1997) for the Fichtelgebirge area (W. Bohemia). The recent uplift of ancient massifs acting as shoulders of the central segments of the European rift (Vosges, Schwarzwald, Morvan) may also be compared: Vogt (1992) calculates a 700-m Pliocene uplift for the southern Vosges and a 100–300-m Pleistocene uplift for the central and northern Vosges. Moreover, reactivation of major Variscan faults was associated with the Tertiary updoming of the massifs.

Because it did not undergo as marked a Tertiary uplift as the other massifs of western Europe, the Armorican massif should be treated separately. Indeed, in the centre of an otherwise emerged massif since 250 Ma, the extended Rennes basin was drowned by several Miocene and Pliocene seas (Battiau-Queney 1993). Although both Klein (1990) and Wyns (1991) conclude that the Armorican massif underwent an Eocene updoming of 100–150 m, no clear regional uplift may be evidenced. Rather, Neogene and Quaternary movements are observed along major fault zones, with the subsidence of numerous small basins and narrow grabens, whose present-day elevation of the Pliocene deposits suggests marked recent local uplift. According to Bonnet *et al.* (2000), the Quaternary fault reactivation and associated differential uplift might be responsible for the present-day large-scale variations in elevation and for the locally marked Quaternary incision of the hydrographic network. Moreover, at present, the east–west trending main drainage divide separating Atlantic and Channel rivers has clearly shifted northwards, suggesting a recent south tilting of the whole massif (Battiau-Queney 1993).

It seems that a number of causes may be held responsible for the Neogene and Quaternary generalized uplift of the massifs throughout north-west Europe. Some authors ascribe a main role to lithospheric buckling induced by the Africa–Eurasia collision and the push exerted by the Alpine belt on its foreland (Ziegler *et al.* 1995; Nikishin *et al.* 1997). This push would have determined long-wavelength (400–600-km) deformation of the topography. Beyond a first belt of molasse basins trending along the northern margin of the collisional belt, a line of neotectonic highs extends from the French Massif Central in the west through Bohemia to the south-eastern Turan platform in the east and is interpreted as a chain of peripheral lithospheric flexural bulges (Nikishin *et al.* 1997). Northwards, these highs are bordered by a second belt of intracratonic basins (North Sea basin, North German basin) whose strong syncollisional reactivation since the Oligocene and especially during the Plio-Pleistocene is well documented.

Other, more regional, factors of uplift have already been invoked. The development of the ECRS and related shoulder uplift determined episodes of vertical motion in the Vosges and Schwarzwald, in the Morvan, along the eastern border of the Massif Central, in the Bergisches Land (Rhenish shield), and around the Eger graben. Mantle plumes, resulting in lithospheric thinning and thermal anomaly, have been identified beneath the Massif Central and the north-west Rhenish shield (Eifel), where they help to explain regionally higher amounts of uplift. Finally, Ziegler (1992) suggests that uplift of the Ardennes–Artois–Weald axis since the Oligocene could be related to wrench deformation due to clockwise rotation of the Armorican-Paris basin block with respect to the Franconian-Bohemian block in a frame of global plate reorganization.

Long-Term Denudation Rates in Land Areas

It seems appropriate to address here the questions of long-term denudation rates in land areas and of the relationship of denudation with the tectonic uplift of these areas. Clearly, the debate is concerned with geomorphological (surface processes) as well as tectonic aspects. In the last two decades, significant advances were made in this field of research. A fast-growing number of thermochronological studies benefiting from the development of radiometric dating methods (apatite and zircon fission track, $^{40}Ar/^{39}Ar$, K/Ar, Rb/Sr) have yielded cooling and thus denudation histories (Gleadow and Brown 2000). Information on denudation in inverted basins has, moreover, been supplied by compaction studies (Bulat and Stoker 1987). However, many studies concentrated on specific tectonic environments (orogens, passive margins) and only scattered data are presently available for north-western Europe.

An important point is to clearly define what is meant by uplift and denudation and not to misuse denudation as surface or rock uplift. England and Molnar (1990) have presented the equation:

surface uplift = uplift of rock—exhumation

where surface uplift represents the change in elevation of the mean topographic surface with respect to a reference level (generally the geoid or its approximation by an assumed fixed mean sea-level); uplift of rocks equals the change in elevation of a crustal parcel with respect to the same reference level, and exhumation (or denudation) refers to vertical displacement of that crustal parcel with respect to the evolving topographic surface. With the exception of the uninteresting case of no change at all, all three variables can never be equal. A parallel statement is that due to the highly variable interplay between uplift and erosion, the rate and amount of surface uplift will generally display only a remote link to the rate and amount of exhumation.

It is further necessary to consider that the surface uplift of a massif results not only from tectonic causes but also from the isostatic response to erosional unloading. Denudation is most often initiated by an increase in relief energy which in many cases results from tectonic (surface and rock) uplift. However, there is a non-negligible positive feedback of enhanced rock uplift due to the isostatic adjustment to unloading. In this, rock uplift cannot be compared with surface uplift which, moreover, integrates lowering by erosion and thus is often much smaller. Gilchrist *et al.* (1994) calculate that the isostatic uplift amounts to *c.*82% of denudation and would therefore account for a maximum one-fourth (*c.*1,000 of 3,500 m) of the elevation of the highest peaks of the Alpine orogen in eastern Switzerland. However, such a partition between tectonic and isostatic uplift of rock relies on the unrealistic assumption that no denudation occurred above the summit peaks. This has been demonstrated not to be true by thermochronological data. Introducing this extra denudation amount dramatically increases the percentage of isostatically induced rock uplift (with respect to actual peak elevation) whereas taking also into account the flexural rigidity of the lithosphere has an opposite effect.

Steck and Hunziker (1994) have reviewed the cooling history of the Central Alps on the basis of numerous thermochronological studies. A first denudation-related cooling episode is identified from 38 Ma onwards, in relation with regional updoming of the Lepontine Gneiss Dome (LGD) (Fig. 1.6). A second episode took place from 20 Ma. After cooling had slowed down between 10 and 3 Ma, renewed very rapid cooling characterized the last 3 Ma. Soom (1990) calculates an exhumation rate of 0.5–0.7 mm/yr for the LGD north of the Simplon low-angle normal fault during the last 5 Ma. Since 38 Ma, up to 24 km of overburden (to be compared with the *c.*2 km mean surface elevation) would have been eroded in the LGD. This corresponds to an average denudation rate of 0.63 mm/yr (but spatially and temporally varying between 0.2 and 2.2 mm/yr). Such denudation values are confirmed by Grundmann and Morteani (1985) for the eastern Alps (*c.*0.5 mm/yr between 10 and 5 Ma) and by Pettke *et al.* (1999) for the north-western Alps (0.2–1 mm/yr).

In the north-western Pyrenees, Vanara *et al.* (1997) demonstrate a Plio-Pleistocene uplift of 1,000 m. Based on the reconstruction of stepped morphologies, this should be interpreted as rock uplift. According to their calculations, the concomitant denudation should not have exceeded 200–300 m in limestones and only somewhat more in marls, because of the time-lag between the triggering uplift and the subsequent denudation, at least at this scale of moderate uplift and final elevation.

Outside orogens, denudation values are generally much smaller. Illies (1974) calculated that the Cenozoic crustal exhumation of the Vosges-Schwarzwald updoming could have locally reached 2,500 m, but did not exceed 1,000 m in the shoulders of the northern part of the graben. By comparison, Japsen (1997) reviewed published exhumation estimates of Britain and the western North Sea and proposed that the Tertiary exhumation of these areas occurred in two episodes, each of *c.*1 km amplitude. The first one started towards 60 Ma and affected only onshore Britain during the Laramide tectonic phase. This episode also caused basin inversion and important exhumation in the Roer graben, with the erosion of *c.*1,500-m thick Cretaceous strata (Zijerveld *et al.* 1992). By contrast, the Neogene episode of exhumation beginning towards 15 Ma is recognized in both onshore and offshore areas of Britain, and is paralleled by renewed intense subsidence of the central North Sea. This British case study especially highlights the difference between significant denudation and limited actual surface uplift.

Other denudation data in north-western Europe are scarce. In the Bohemian massif, Hejl *et al.* (1997) conclude, based on fission-track dating, that the Fichtelgebirge would have undergone an average 3-km denudation since the upper Cretaceous, most of it occurring during the Laramide phase. In the nearby basin, about 2 km of late Cretaceous-Palaeogene sediments would have been eroded during the last 30 Ma. In the Belgian Brabant massif, van den Haute and

Vercoutere (1989) found apatite fission-track evidence for a 3-km denudation which culminated towards 150–140 Ma and probably removed an upper Carboniferous cover during an uplift episode related to the Mid-Kimmerian tectonic phase.

In other areas, denudation appears much more limited or even almost absent. In spite of a mean Cenozoic rock uplift of *c.* 500 m, the Ardennes and the western Rhenish shield display very small amounts of Tertiary denudation since their summits still preserve remnants of an upper Cretaceous cover. It seems, however, that the main part of the uplift of the north-western European Palaeozoic massifs is recent, dating back to the Plio-Pleistocene, and that the subsequent dissection continues to the present. Moreover, in the absence of thermochronological data, the proposed patterns of Tertiary uplift of these massifs strongly rely on the identification and reconstruction of stepped ancient surfaces. Due to the frequent absence or scarcity of dated correlative deposits, it unfortunately often proves difficult to recognize the polycyclic, tectonic, or lithological nature of such staircases of erosion surfaces and therefore to estimate timing and amount of uplift and denudation.

References

Ahorner, L. (1962), Untersuchungen zur quartären Bruchtektonik der Niederrheinischen Bucht. *Eiszeitalter und Gegenwart* 13: 24–105.

Ansorge, J., Blundell, D., and Mueller, S. (1992), Europe's lithosphere—seismic structure. In: D. Blundell, R. Freeman, and S. Mueller (eds.) *A continent revealed. The European geotraverse.* Cambridge University Press, Cambridge, 33–69.

Babuska, V., and Plomerová, J. (1992), The lithosphere in Central Europe—seismological and petrological aspects. *Tectonophysics* 207: 141–63.

Banda, E., and Balling, N. (1992), Europe's lithosphere—Recent activity. In: D. Blundell, R. Freeman, and S. Mueller (eds.) *A continent revealed. The European geotraverse.* Cambridge University Press, Cambridge, 111–37.

Battiau-Queney, Y. (1993), *Le Relief de la France. Coupes et croquis.* Masson, Paris.

Becker, A. (1999), *In situ* stress data from the Jura Mountains—new results and interpretation. *Terra Nova* 11: 9–15.

Benek, R., Kramer, W., McCann, T., Scheck, M., Negendank, J., Korich, D., Hübscher, H., and Bayer, U. (1996), Permo-Carboniferous magmatism and related subsidence of the northeast German Basin. *Tectonophysics* 266: 379–404.

Bergerat, F. (1987), Stress fields in the European platform at the time of Africa–Eurasian collision. *Tectonics* 6: 99–132.

—— Mugnier, J., Guellec, S., Truffert, C., Cazes, M., Damotte, B., and Roure, F. (1990), Extensional tectonics and subsidence of the Bresse basin: An interpretation from ECORS data. *Mémoires Société Géologique France*, NS 156: 145–56.

Berthelsen, A. (1992), Mobile Europe. In: D. Blundell, R. Freeman, and S. Mueller (eds.) *A continent revealed. The European geotraverse.* Cambridge University Press, Cambridge, 11–32.

Betz, D., Führer, F., and Plein, E. (1987), Evolution of the Lower Saxony Basin. *Tectonophysics* 137: 127–70.

Blès, J., Bonijoly, D., Castaing, C., and Gros, Y. (1989), Successive post-Variscan stress fields in the European plate (French Massif Central and its borders): Comparison with geodynamic data. *Tectonophysics* 169: 79–111.

Boillot, G. (1984), Le Golfe de Gascogne et les Pyrénées. In: G. Boillot, L. Montadert, M. Lemoine, and B. Biju-Duval (eds.) *Les Marges continentales actuelles et fossiles autour de la France.* Masson, Paris, 7–73.

Bonnet, S., Guillocheau, F., Brun, J., and van den Driessche, J. (2000), Large-scale relief development related to Quaternary tectonic uplift of a Proterozoic-Paleozoic basement: The Armorican massif, north-west France. *Journal of Geophysical Research* 105: 19273–88.

Bulat, J., and Stoker, S. (1987), Uplift determination from interval velocity studies, United Kingdom, southern North Sea. In: J. Brooks and K. Glennie (eds.) *Petroleum geology of northwest Europe.* Graham & Trotman, London, 293–305.

Burg, J. P., van den Driessche, J., and Brun, J. P. (1994), Syn- to post-thickening extension in the Variscan belt of Western Europe: modes and structural consequences. *Géologie de la France* 3: 33–51.

Burrus, J., Bessis, F., and Doligez, B. (1987), Heat flow, subsidence and crustal structure of the Gulf of Lions (north-west Mediterranean): A quantitative discussion of the classic passive margin model. In: C. Beaumont and A. Tankard (eds.) *Sedimentary basins and basin-forming mechanisms.* Memoirs Canadian Society Petroleum Geologists 12: 1–15.

Calvet, M. (1999), Rythmes et vitesses d'évolution morphogénétique dans un orogène alpin. Le Cas des Pyrénées orientales franco-espagnoles. *Zeitschrift für Geomorphologie*, Suppl. 118: 91–105.

Chorowicz, J., Deffontaines, B., and Villemin, T. (1989), Interprétation des structures transverses NE–SW du fossé rhénan en termes de failles de transfert. Apport de données multisources. *Comptes Rendu Académie Scientifique Paris* 309, ser. 2, 1067–73.

Debelmas, J., and Mascle, G. (1993), Les Grandes Structures géologiques. Masson, Paris, 2nd edn.

Demoulin, A. (1995), L'Ardenne bouge toujours. Néotectonique du massif ardennais. In: A. Demoulin (ed.) *L'Ardenne. Essai de géographie physique.* Département Géographique Physique de Université de Liège, 110–35.

Dewey, J., Helman, M., Turco, E., Hutton, D., and Knott, S. (1989), Kinematics of the western Mediterranean. In: M. Coward, D. Dietrich, and R. Park (eds.) *Alpine tectonics.* Geological Society, London, Special Publication 45: 265–83.

Dewolf, Y. (1982), Le Contact Île de France—basse Normandie. Évolution géodynamique. *Mémoires et Doctorat de Géographie*, CNRS, Paris.

Doebl, F., and Teichmüller, R. (1979), Zur Geologie und heutigen Geothermik im mittleren Oberrhein-Graben. *Fortschritt Geologie Rheinland Westphalen* 27: 1–17.

Dubreuilh, J., Capdeville, J. P., Farjanel, G., Karnay, G., Platel, J. P., and Simon-Coinçon, R. (1995), Dynamique d'un comblement continental néogène et quaternaire: l'exemple du bassin d'Aquitaine. *Géologie de la France* 4: 3–26.

England, P., and Molnar, P. (1990), Surface uplift, uplift of rocks, and exhumation of rocks. *Geology* 18: 1173–7.

Franke, W. (1989), Variscan plate tectonics in Central Europe—current ideas and open questions. *Tectonophysics* 169: 221–8.

—— (1992), Phanerozoic structures and events in central Europe. In: D. Blundell, R. Freeman, and S. Mueller (eds.) *A continent revealed. The European geotraverse.* Cambridge University Press, Cambridge, 164–80.

Gilchrist, A., Summerfield, M., and Cockburn, H. (1994), Landscape dissection, isostatic uplift, and the morphologic development of orogens. *Geology* 22: 963–6.

Gleadow, A., and Brown, R. (2000), Fission-track thermochronology and the long-term denudational response to tectonics. In: M. Summerfield (ed.) *Geomorphology and global tectonics*. Wiley, Chichester, 57–75.

Grundmann, G., and Morteani, G. (1985), The young uplift and thermal history of the central Eastern Alps (Austria/Italy), Evidence from apatite fission track ages. *Jahrbuch Geologische Bundesanstalt Austria* 128: 197–216.

Hejl, E., Coyle, D., Lal, N., van den Haute, P., and Wagner, G. (1997), Fission-track dating of the western border of the Bohemian massif: thermochronology and tectonic implications. *Geologische Rundschau* 86: 210–19.

Henk, A. (1997), Gravitational orogenic collapse vs plate-boundary stresses: a numerical modelling approach to the Permo-Carboniferous evolution of Central Europe. *Geologische Rundschau* 86: 39–55.

Hunziker, J., Desmons, J., and Martinotti, G. (1989), Alpine thermal evolution in the central and western Alps. In: M. Coward, D. Dietrich, and R. Park (eds.) *Alpine tectonics*. Geological Society of London, Special Publication 45: 353–67.

Illies, J. (1974), Taphrogenesis and plate tectonics. In: J. Illies and K. Fuchs (eds.) *Approaches to taphrogenesis*. Schweizerbart, Stuttgart, 433–60.

—— Prodehl, C., Schmincke, H., and Semmel, A. (1979), The Quaternary uplift of the Rhenish shield in Germany. *Tectonophysics* 61: 197–225.

Japsen, P. (1997), Regional Neogene exhumation of Britain and the western North Sea. *Journal of the Geological Society, London* 154: 239–47.

Klein, C. (1990), L'Évolution géomorphologique de l'Europe hercynienne occidentale et centrale. Aspects régionaux et essai de synthèse. *Mémoires et Doctorat de Géographie*, CNRS, Paris.

Kossmat, F. (1927), Gliederung des varistischen Gebirgsbaues. *Abhandlung Sächsiche Geologische Landes-Amt* 1: 1–39.

Lacombe, O., Angelier, J., Byrne, D., and Dupin, J. (1993), Eocene-Oligocene tectonics and kinematics of the Rhine-Saône continental transform zone (eastern France). *Tectonics* 12: 874–88.

Le Pichon, X., Bergerat, F., and Roulet, M. (1988), Plate kinematics and tectonics leading to Alpine belt formation. A new analysis. In: *Continental lithospheric deformation*, Geological Society of America, Special Paper 218: 111–31.

Lippolt, H. (1983), Distribution of volcanic activity in space and time. In: K. Fuchs, K. Von Gehlen, H. Mälzer, H. Murawski, and A. Semmel (eds.) *Plateau uplift. The Rhenish shield—a case history*. Springer, Berlin, 112–20.

Loup, B., and Wildi, W. (1994), Subsidence analysis in the Parisian Basin: a key to Northwest-European intracontinental basins? *Basin Research* 6: 159–77.

Malkowsky, M. (1987), The Mesozoic and Tertiary basins of the Bohemian Massif and their evolution. *Tectonophysics* 137: 31–42.

Mascle, A. (1990), Géologie pétrolière des bassins permiens français. Comparaison avec les bassins permiens du Nord de l'Europe. *Chron. Rech. Min.* 499: 69–87.

Matte, P. (1986), Tectonics and plate tectonics model for the Variscan belt of Europe. *Tectonophysics* 126: 329–74.

Mechie, J., Prodehl, C., and Fuchs, K. (1983), The long-range seismic refraction experiment in the Rhenish massif. In: K. Fuchs, K. Von Gehlen, H. Mälzer, H. Murawski, and A. Semmel (eds.) *Plateau uplift. The Rhenish shield—a case history*. Springer, Berlin, 260–75.

Meier, L., and Eisbacher, G. (1991), Crustal kinematics and deep structure of the Northern Rhine graben, Germany. *Tectonics* 10: 621–30.

Meissner, R., and Vetter, U. (1974), The northern end of the Rhinegraben due to some geophysical measurements. In: J. Illies and K. Fuchs (eds.) *Approaches to taphrogenesis*. Schweizerbart, Stuttgart, 236–43.

Meyer, W., Albers, H., Berners, H., von Gehlen, K., Glatthaar, D., Löhnertz, W., Pfeffer, K., Schnütgen, A., Wienecke, K., and Zakosek, H. (1983), Pre-Quaternary uplift in the central part of the Rhenish massif. In: K. Fuchs, K. von Gehlen, H. M. Mälzer, H. Murawski, and A. Semmel (eds.) *Plateau uplift. The Rhenish shield—a case history*. Springer, Berlin, 39–46.

—— and Stets, J. (1998), Junge Tektonik im Rheinischen Schiefergebirge und ihre Quantifizierung. *Zeitschrift der Deutschen Geologischen Gesellschaft* 149: 359–79.

Müller, B., Zoback, M. L., Fuchs, K., Mastin, L., Gregersen, S., Pavoni, N., Stephansson, O., and Ljunggren, C. (1992), Regional patterns of tectonic stress in Europe. *Journal Geophysical Research* 97: 11783–803.

Nikishin, A., Brunet, M., Cloetingh, S., and Ershov, A. (1997), Northern Peri-Tethyan Cenozoic intraplate deformations: influence of the Tethyan collision belt on the Eurasian continent from Paris to Tian-Shan. *Comptes Rendus Académie Scientifique Paris* 324: 49–57.

Peterek, A., and Schröder, B. (1997), Neogene fault activity and morphogenesis in the basement area north of the KTB drill site (Fichtelgebirge and Steinwald). *Geologische Rundschau* 86: 185–90.

Pettke, T., Diamond, L., and Villa, I. (1999), Mesothermal gold veins and metamorphic devolatilization in the northwestern Alps: The temporal link. *Geology* 27: 641–4.

Pfiffner, A. (1992), Alpine orogeny. In: D. Blundell, R. Freeman, and S. Mueller (eds.) *A continent revealed. The European geotraverse*. Cambridge University Press, Cambridge, 180–90.

Prodehl, C., Mueller, S., and Haak, V. (1995), The European Cenozoic rift system. In: K. Olsen (ed.) *Continental rifts: evolution, structure, tectonics*. Developments in Geotectonics 25, Elsevier, Amsterdam, 133–212.

Ricou, L., and Siddans, A. (1986), Collision tectonics in the western Alps. In: M. Coward and A. Ries (eds.) *Collision tectonics*. Geological Society London, Special Publication 19: 229–44.

Ritter, J., Achauer, U., Christensen U., and the Eifel Plume Team (2000), The teleseismic tomography experiment in the Eifel region, Central Europe: Design and first results. *Seismic Research Letters* 71: 437–43.

Robin, C., Guillocheau, F., Allemand, P., Bourquin, S., Dromart, G., Gaulier, J., and Prijac, C. (2000), Echelles de temps et d'espace du contrôle tectonique d'un bassin flexural intracratonique: le bassin de Paris. *Bulletin Société Géologique de France* 171: 181–96.

Roig, J., and Faure, M. (2000), La Tectonique cisaillante polyphasée du Sud Limousin (Massif central français) et son interprétation dans un modèle d'évolution polycyclique de la chaîne hercynienne. *Bulletin Société Géologique de France* 171: 295–307.

Sapin, M., and Hirn, A. (1974), Results of explosion seismology in the southern Rhône valley. *Ann. Geophys.* 30: 181–202.

Scheck, M., and Bayer, U. (1999), Evolution of the Northeast German Basin—inferences from a 3D structural model and subsidence analysis. *Tectonophysics* 313: 145–69.

Schmid, S., Aebli, H., Heller, F., and Zingg, A. (1989), The role of the Periadriatic Line in the tectonic evolution of the Alps. In: M. Coward, D. Dietrich, and R. Park (eds.) *Alpine tectonics*. Geological Society London, Special Publication 45: 153–71.

Schröder, B., Ahrendt, H., Peterek, A., and Wemmer, K. (1997), Post-Variscan sedimentary record of the SW margin of the Bohemian massif: a review. *Geologische Rundschau*. 86: 178–84.

Seck, H. (1983), Eocene to recent volcanism within the Rhenish massif and the northern Hessian depression—summary. In: K. Fuchs, K. von Gehlen, H. Mälzer, H. Murawski, and A. Semmel (eds.) *Plateau uplift. The Rhenish shield—a case history*. Springer, Berlin, 153–62.

Sengör, A., Burke, K., and Dewey, J. (1978), Rifts at high angles to orogenic belts: test for their origin and the Upper Rhine graben as an example. *American Journal of Sciences* 278: 219–42.

Simon-Coinçon, R., and Astruc, J. (1994), Des paléotopographies continentales au remplissage des bassins sédimentaires: l'exemple du sud-ouest du Massif central et de ses bordures au Cénozoïques. *Proceedings Special Meeting SGF/GFG at Rennes, September 1994*, 40.

Sinclair, H., Coakley, B., Allen, P., and Watts, A. (1991), Simulation of foreland basin stratigraphy using a diffusion model of mountain uplift and erosion: an example from the Central Alps, Switzerland. *Tectonics* 10: 599–620.

Sobolev, S., Zeyen, H., Stoll, G., Werling, F., Altherr, R., and Fuchs, K. (1996), Upper mantle temperatures from teleseismic tomography of French Massif Central including effects of composition, mineral reactions, anharmonicity, anelasticity and partial melt. *Earth Planetary Science Letters* 139: 147–63.

Soom, M. (1990), Abkühlungs- und Hebungsgeschichte der Externmassive und der Penninischen Decken beidseits der Simplon-Rhône-Linie seit dem Oligozän: Spaltspurdatierungen an Apatit/Zirkon und K-Ar Datierungen an Biotit/Muskowit (Westliche Zentralalpen). Ph.D. Thesis, University of Berne.

Steck, A., and Hunziker, J. (1994), The Tertiary structural and thermal evolution of the Central Alps—compressional and extensional structures in an orogenic belt. *Tectonophysics* 238: 229–54.

Thiry, M., and Simon-Coinçon, R. (1996), Tertiary paleoweatherings and silcretes in the southern Parisian Basin. *Catena* 26: 1–26.

Underhill, J., and Partington, M. (1993), Jurassic thermal doming and deflation in the North Sea: implications of the sequence stratigraphic evidence. In: J. R. Parker and I. D. Bartholomew (eds.) *Petroleum geology of the Northwest Europe*, Proceedings 4th Conference, London, 337–45.

Vanara, N., Maire, R., and Lacroix, J. (1997), La Surface carbonatée du massif des Arbailles (Pyrénées Atlantiques): un exemple de paléoréseau hydrographique néogène déconnecté par la surrection. *Bulletin Société Géologique de France* 168: 255–65.

van den Haute, P., and Vercoutere, C. (1989), Apatite fission-track evidence for a Mesozoic uplift of the Brabant massif: preliminary results. *Ann. Soc. Géol. Belg.* 112: 443–52.

Vogt, H. (1992), Le Relief en Alsace. Étude géomorphologique du rebord sud-occidental du fossé rhénan. Oberlin, Strasburg.

Walter, R. (1992), *Geologie von Mitteleuropa*. Schweizerbart, Stuttgart.

Wilson, M., and Downes, H. (1992), Mafic alkaline magmatism associated with the European Cenozoic rift system. *Tectonophysics* 208: 173–82.

Wyns, R. (1991), Evolution tectonique du bâti armoricain oriental au Cénozoïque d'après l'analyse des paléosurfaces continentales et des formations géologiques associées. *Géologie de la France* 3: 11–42.

Zeyen, H., Novak, O., Landes, M., Prodehl, C., Driad, L., and Hirn, A. (1997*a*), Refraction seismic investigations of the northern Massif Central, France. *Tectonophysics* 275: 99–118.

—— Volker, F., Wehrle, V., Fuchs, K., Sobolev, F., and Altherr, R. (1997*b*), Styles of continental rifting: crust-mantle detachment and mantle plumes. *Tectonophysics* 278: 329–52.

Ziegler, P. (1978), North-western Europe: tectonics and basin development. *Geologie en Mijnbouw* 57: 589–626.

—— (1990), Geological atlas of western and central Europe (2nd edn.). *Shell Internationale Petroleum Maatschappij B.V.*, The Hague.

—— (1992), European Cenozoic rift system. *Tectonophysics* 208: 91–111.

—— Cloetingh, S., and van Wees, J. (1995), Dynamics of intra-plate compressional deformation: the Alpine foreland and other examples. *Tectonophysics* 252: 7–59.

Zijerveld, L., Stephenson, R., Cloetingh, S., Duin, E., and van den Berg, M. (1992), Subsidence analysis and modelling of the Roer Valley graben (SE Netherlands). *Tectonophysics* 208: 159–71.

2 Neotectonics

Francesco Dramis and Emanuele Tondi

Introduction

Debate in neotectonics mainly hinges on how far back in time the prefix 'neo' should be taken. The term 'neotectonics' means, in a first approximation, geologically young, recent or living (active) crustal structures and processes. Some of the many definitions (Angelier 1976; Mercier 1976; Beloussov 1978; Hancock and Williams 1986; Vita-Finzi 1986; Winslow 1986) focus neotectonic studies only on active deformation (late Quaternary–Present) and accept *neotectonics* as more or less synonymous to *active tectonics*, while others trace the neotectonic period mainly from the Middle Miocene. It is very difficult to identify a standard time period for defining the beginning of neotectonics, but the present-day opinion is that it depends on the individual characteristics of each geological environment. According to Fourniguet (1987), no time limit is fixed and the field of investigation extends from the present as far back into the past as necessary to understand present or active deformation. The INQUA (International Union for Quaternary Research) Tectonic Commission has accepted the definition of Mörner (1978): 'Neotectonics is defined as any earth movements or deformations of the geodetic reference level, their mechanisms, their geological origin, their implications for various practical purposes and their future extrapolations.' Pavlides (1989) proposed a definition along the following lines: 'Neotectonics is the study of young tectonic events (deformation of upper crust), which have occurred or are still occurring in a given region after its final orogeny (at least for recent orogenies) or more precisely after its last significant reorganization.'

When western Europe is considered, a major change in boundary conditions occurred in the Upper Miocene (7 Ma) when the motion of Africa became directed to the north-west (Dewey *et al.* 1989). Geological, seismological, and geodetic data in the Mediterranean region and in continental Europe show that the relative motion of Africa and Europe is still in this direction. For this reason we think that for the neotectonics of western Europe one cannot go far back in time beyond the Upper Miocene.

State of Stress

The study of the state of stress of the lithosphere around the world has recently been attempted within the World Stress Map Project of the International Lithosphere Programme (Zoback 1992). A compilation of new and existing data has led to a large database which includes results from a variety of geological and geophysical techniques: earthquake focal mechanisms, hydraulic fracturing, borehole break-outs, overcoring, and fault-slip orientations. Some of these results and their interpretation were published in a special issue of the *Journal of Geophysical Research* in which one paper is specifically devoted to Europe (Müller *et al.* 1992).

Measurements of tectonic stress in western Europe take into account regional stress patterns as well as more local perturbations. The only way to distinguish between these is that of defining a large-scale stress (long wavelength) pattern with a large number of consistent and regionally distributed observations. This can then be interpreted as either the result of large-scale tectonic forces due to movements of the plate, or the effects of large-scale flexural loading or unloading, or in homogeneous density contrasts within the lithosphere–asthenosphere system, or else as some other large-scale phenomena. Local factors such as topography, erosion,

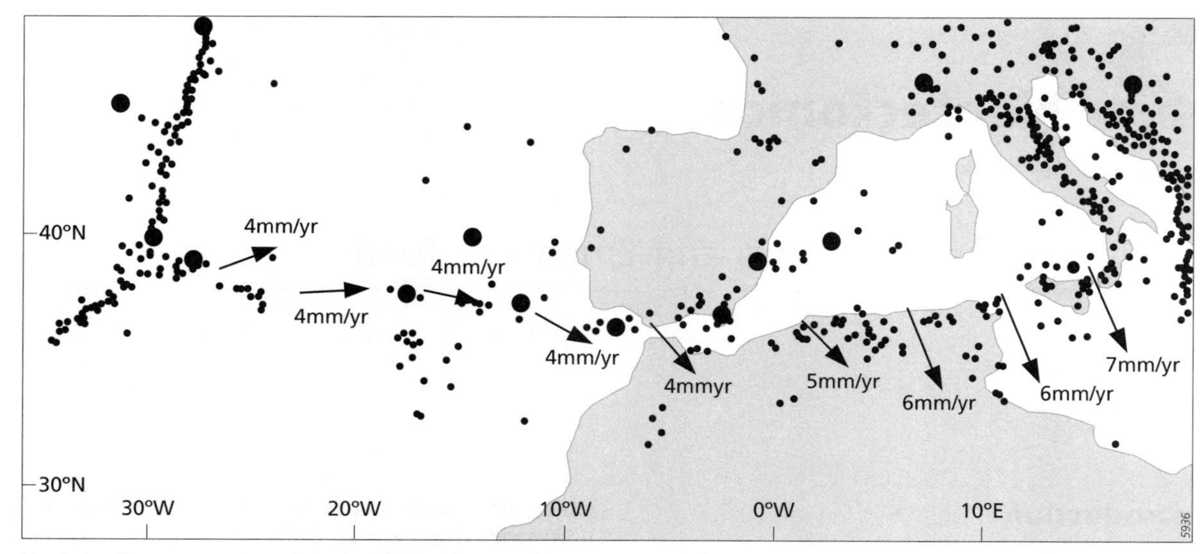

Fig. 2.1. Convergence rates along the African–Eurasian plate boundary (taken from Argus *et al.* 1989; Dewey *et al.* 1989). Directions and velocities along the plate boundary are calculated from major earthquake slip vectors and global modelling of plate tectonics (extensional rates at the medio-Atlantic ridge) where Africa is rotated counter-clockwise and the pole has Lat. 18.8° N and Long. 20.3° W.

and other local or induced modest-wavelength stresses can then be considered responsible for the lesser regional pattern perturbations. In western Europe a well-constrained maximum principal stress orientation trending NW–SE to NNW–SSE has been determined.

In the Swabian Jura and in the Pyrenees almost all geological data and all fault plane solutions show dominant sinistral (left-lateral) and dextral (right-lateral) strike-slip movements along respectively, NNE–SSW and E–W striking planes. This suggests that in a large area of western Europe the intermediate principal stress is dominantly vertical. However, in the Upper Rhine graben recent movements include strike-slip, together with normal faulting, which seem to be due to the interaction of the regional stress field with graben features. In contrast current seismic activity shows predominantly normal faulting in the Lower Rhine graben, the Rhenish massif, and the Cologne basin (north-western part of the area).

Focal mechanisms determined for nearly two hundred events from the northern and western Alpine arc in general show P-axis (maximum principal stress) orientation radially perpendicular to its trend, and thus T-axis (minimum principal stress) generally parallel to the strike (Eva *et al.* 1990; Pavoni 1990). The P-trajectories are aligned with the maximum horizontal crustal shortening derived from kinematic analysis of neotectonic structural features. This indicates that the stress field and the mode of deformation have not changed much over the past few Ma (Pavoni 1990). Local variations occur in the south-western part of the western Alps towards the Ligurian Sea where a complex pattern of compression and extension is observed.

Finally, available stress studies in Europe demonstrate a fairly consistent orientation of the maximum horizontal stress. Such a consistency, coinciding with the direction of ridge push from the North Atlantic and the relative motion of the African and Eurasian plates, suggests that the stresses are actually controlled by forces generated by plate tectonics (Fig. 2.1). The maximum principal stress has been found to be horizontal except in the Lower Rhine Embayment. The NW–SE to NNW–SSE maximum principal stress in western Europe is affected only locally by major geological structures such as the Alps, confirming the suggestion that the stress is largely controlled by plate driving forces acting on its boundaries.

Seismicity

For most tectonic provinces in western Europe, macroseismic observations, including historical seismicity and detailed instrumental observations covering the past 20–30 years, are available. The combined use of information from regional and local networks often gives detailed information about the activity of an area. Epicentres, hypocentral depths, and focal mechanisms (fault plane solutions) can be determined accurately and these allow seismotectonic information to be extracted.

The distribution of large and intermediate-size earthquakes in Europe (the western part of the Eurasian plate) is clearly controlled by plate tectonic processes. Present-day activity is concentrated (Fig. 2.2) along the African–European plate boundary zone (Apennines, western Alps, and the Pyrenees). However, several large historical and instrumentally recorded earthquakes ($M > 6$) have occurred in some intraplate areas of western Europe. Surface ruptures associated with large, prehistorical earthquakes have been recently identified in the south of France (Provence) and in the Upper Rhine graben, which represent the most active present-day tectonic elements of western Europe.

In the northern and western Alps, as also in the intraplate areas in western Europe, seismicity is limited to depth levels corresponding to isotherms of 450–600 °C (Fig. 2.3; Cloetingh and Banda 1992). The maximum depth of seismicity is in the range of 20–30 km, being essentially restricted to crustal levels. A comparison of strength profiles and seismicity shows a minimum in seismicity coinciding with a minimum in crustal strength (cf. Fig. 2.3). Marked lateral variations of the mechanical properties of the lithosphere can be deduced from the rheological profiles and seismicity distribution. For instance, inferred spatial variation in the mechanical properties of the lithosphere is important since it controls the magnitude of the stresses within the lithosphere. Whereas a thick, strong lithosphere will decrease the stress level, stress magnitudes will be amplified when the lithosphere is thin and weak.

Recent Crustal Deformation

Western Europe, in terms of plate motions, is affected by the spreading of the North Atlantic and by the interaction of the African and Eurasian plates. Space geodesy techniques, such as Very Long Baseline Interferometry (VLBI), Satellite Laser Ranging (SLR), and the Global Positioning System (GPS), have recently begun to yield independent information for tests of plate tectonic motion models. In the North Atlantic, the current rate of spreading has been confirmed by such techniques (Smith *et al.* 1990), although it is slightly slower than that suggested on geological grounds (de Mets *et al.* 1990). Geodetic data are still scanty in the Eurasia–Africa interaction area, although those available point to the fact that Africa and Europe are converging, as geological models appear to suggest. Along the western part of the southern Eurasian Plate boundary, interplate tectonic effects attributable to the convergence of Africa and Europe are evident. However, intraplate motions, revealed by use of geodetic techniques, allow no significant relative tectonic motion to be inferred in central and western Europe, but rather suggest that the block is a single one.

The theory of plate tectonics assumes that plate interiors behave rigidly while deformation is confined along their boundaries. In western Europe most of the deformation is concentrated in the Alps and the Pyrenees, two mountain ranges forming part of the African–European plate boundary zone (e.g. Philip 1987). Space geodetic measurements during the last two decades have confirmed the validity of a global plate tectonic theory, while the increasing accuracy and density of space measurements in the last few years have allowed the testing of the plate rigidity hypothesis at a level of a few millimetres per year. Nocquet *et al.* (2001) found residual velocities for western Europe that are consistent with known active tectonic features, identifying active deformation < 1 mm/yr in the eastern Alps and western Carpathians. In the Alpine range their results indicate east–west extension across the western Alps and north–south compression across the central and eastern Alps. This is in agreement with the strain regime deduced from seismotectonic observations. In Belgium and The Netherlands they found residual velocities of 1–1.5 mm/yr to the north-west, most likely accommodated along the Upper and Lower Rhine graben structures. An important outcome of their study was the identification of internal deformation of the order of 1–2 mm/yr in an area usually interpreted as 'stable' Europe.

In the following sections, neotectonics will be analysed on the basis of the kind of deformation. Those areas will be distinguished where the present-day dominant active tectonic events are due to faulting processes and where several observations evidence a dominance of vertical crustal movements.

Faulting Processes

Faulting processes and associated earthquakes have been recorded in different parts of western Europe, in particular in the Rhine graben area, in the south of France (Provence), in the south of the Armorican massif, and in the Pyrenees (cf. Fig. 2.2). However, we will focus our attention on the areas where major crustal-scale faults are known and where evidence for active faulting or seismic activity has been reported.

The Provence Area

In the Provence, NE–SW trending reverse faults show recent activity, and historical strong earthquakes were

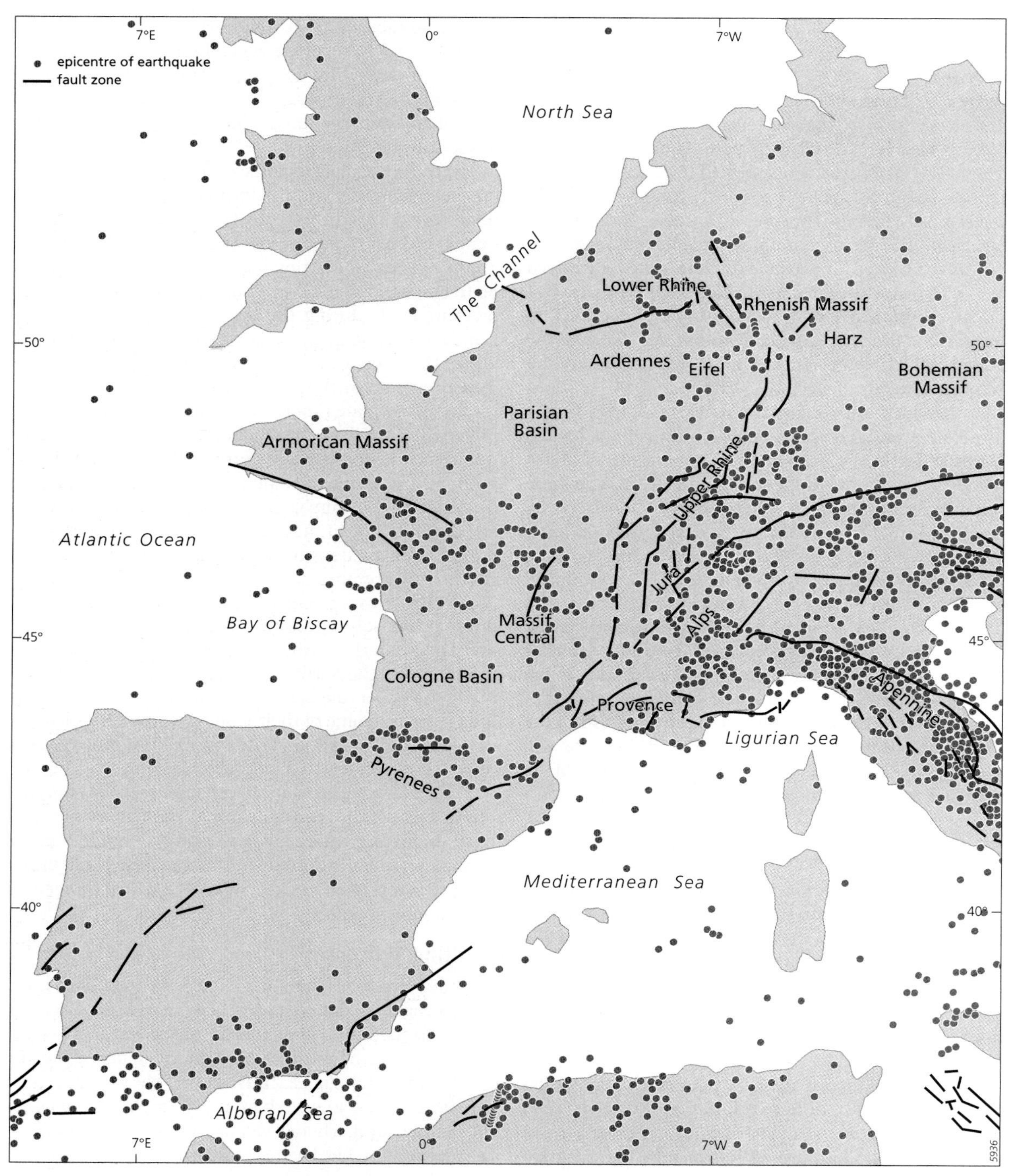

Fig. 2.2. Seismotectonic map of western Europe indicating epicentres of earthquakes and fault zones (after Armijo *et al.* 1986 and Simkin *et al.* 1989).

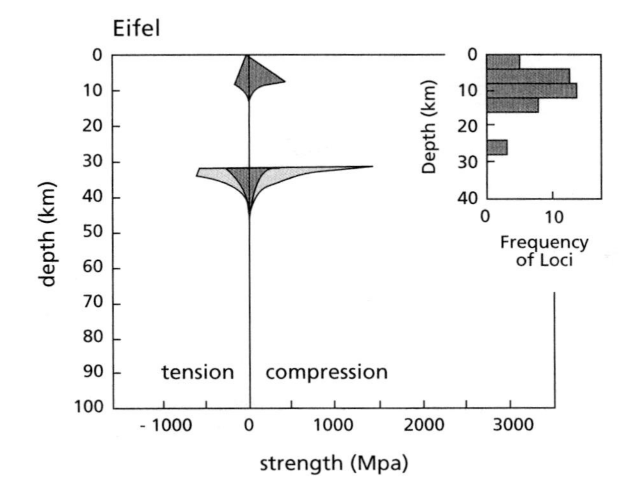

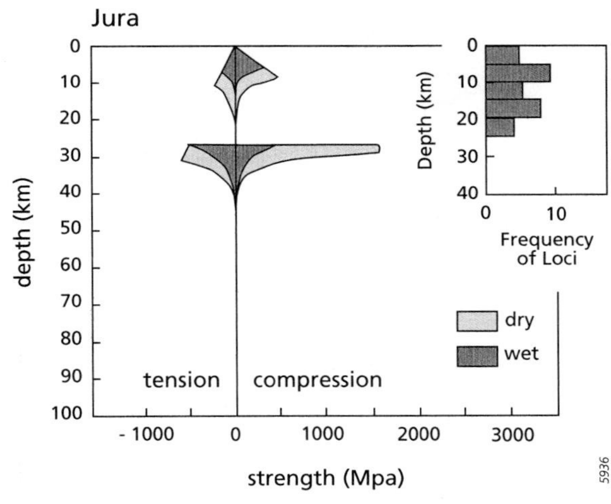

Fig. 2.3. Seismicity distribution and strength profiles for intraplate area of western Europe (Eifel) and for an area in the northern part of the Alpine belt (Jura) (after Cloetingh and Banda 1992).

identified by palaeoseismological studies. Sebrier *et al.* (1993) studied in detail two main fault zones in the Provence: the Nimes and the Middle Durance zones (Fig. 2.4). The most reliable information was provided by trenching observations on the Middle Durance fault, where a N 35° E-striking knee fold resulted from a single coseismic slip event associated with a strong earthquake (Mw = 6.4–6.9). Different palaeoseismic data from several fault zones in the south of France indicate long recurrence intervals, at least of the order of 10 ka, for earthquakes with a magnitude around 7.

The Rhine Graben

The Rhine graben system forms the most active present-day tectonic province in western Europe. In this area, the neotectonic deformation is dominated by faulting processes: normal faulting and subordinate strike-slip faulting. Present-day seismicity shows several earthquakes of intermediate magnitude distributed over a large area in a north–south direction (cf. Fig. 2.2). The Rhine graben system extends over a distance of 600 km from the North Sea to the northern margin of the Alps. It traverses the Caledonian and Variscan tectonic units of western Germany, which are covered in the north by the sediments of the North German basin and in the south by those of the Tertiary Molasse basin (Fig. 2.5A).

The Early Tertiary graben structures trending NNE–SSW (Upper Rhine graben) probably mark the main horizontal compressive stress axis in the Alpine foreland (Sengor *et al.* 1978; Dewey *et al.* 1989). The NW–SE trending grabens in the Middle and Lower Rhine areas were determined by a later, anticlockwise rotation of the stress field, while the Upper Rhine graben became a sinistral wrench-zone (Ahorner *et al.* 1983). The Alpine orogeny, for which geochronological investigations have shown the earliest events to be approximately 110 Ma (Lippoldt 1983), is partly responsible for the Late Cretaceous to Recent volcanism in the Rhenish massifs. Volcanic activity attained its widest extent during the Oligocene to Miocene and the youngest eruptions are only 11,000 years old. Frequent emanations of CO_2 and helium isotopes in source waters (Oxburgh and O'Nions 1987) are proof of the fact that volcanism must not be regarded as extinct. The geochemistry of the Cenozoic volcanic rocks is typical of an intraplate setting (Fuchs *et al.* 1983; Wedepohl 1987).

The Tertiary rifting and magmatism brought about, in some areas, a profound re-equilibration of the crust. The Upper Rhine graben is marked by a pronounced Moho high and correlated effects such as a gravity anomaly, a fairly shallow brittle/ductile transition at approximately 15 km deduced from seismicity, and high heat-flow (Zeis *et al.* 1990). The Meier and Eisbacher (1991) crustal model for the Lower Rhine graben points to the fact that the upper and lower crust are thinned by separate systems of similarly oriented shear zones, with mid-crustal detachment. A low-angle shear zone dipping from the Moho directly beneath the graben transfers the extension into the upper mantle.

The existence of a low-velocity zone extending from Lake Constance in northern Switzerland to the northern margin of the Rhine–Hercynian belt, at about 10–15 km depth, might also explain Tertiary to Recent thermal activity. As the DEKORP Research Group (1990) suggested,

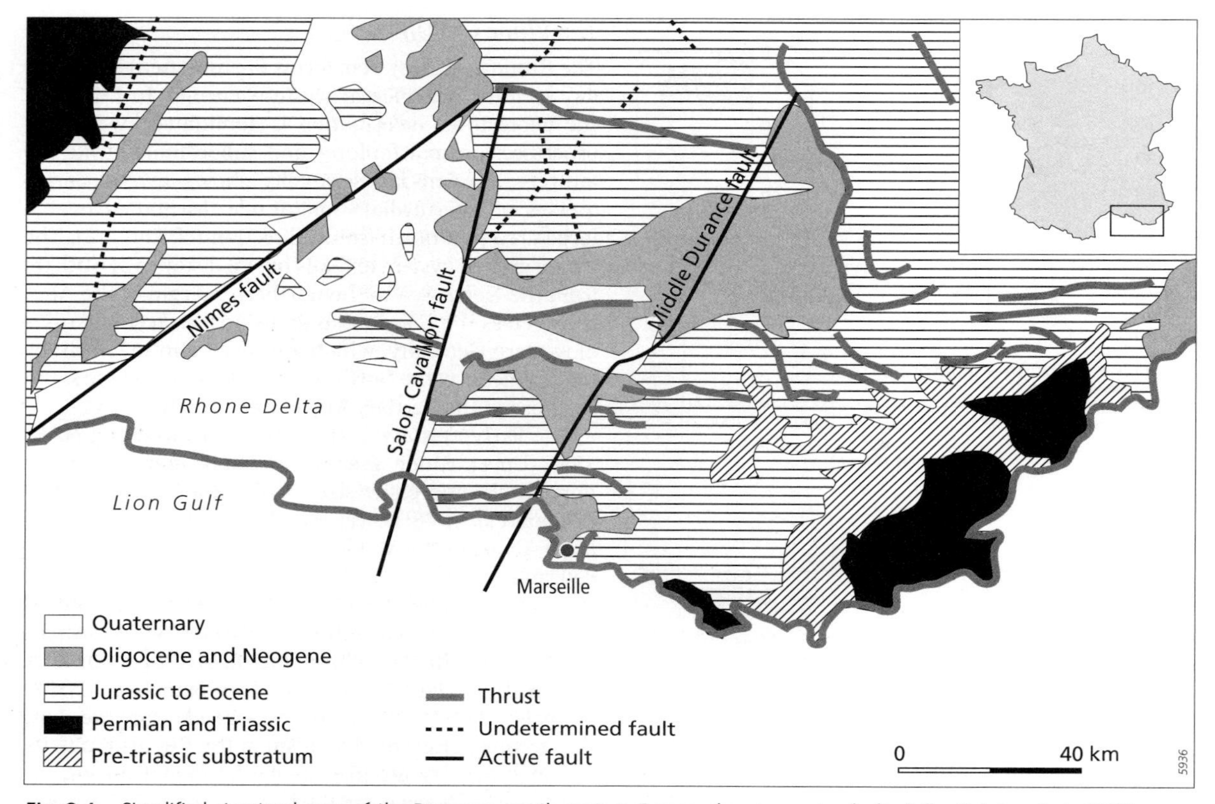

Fig. 2.4. Simplified structural map of the Provence, south-eastern France, showing major faults (after Sebrier *et al.* 1993).

this low-velocity zone might be ascribable to the presence of fluids in the crust. These fluids would have derived from the mantle and/or been expelled from the lower crust, as a consequence of magmatism in the Cenozoic. Meteorite impacts approximately 15 Ma ago (Gall *et al.* 1977) were responsible for the large crater of the Nordlinger Ries and a smaller one near Steinheim in southern Germany (locations see Fig. 2.5A).

The Rhine graben system forms the most active present-day elements north of the Alps (cf. Fig. 2.5A), the seismicity being characterized by low- and intermediate-size earthquakes. Since the occurrence of 13 April 1992 Roermond earthquake (Ms = 5.3) in the south-eastern Netherlands, the largest instrumentally recorded event in the region, the question was raised among seismologists as to whether larger earthquakes rupturing the whole seismogenic layer could occur in the Lower Rhine graben area as well, and if so, what the return period of such an event would be. Considering the relatively low rates of deformation characterizing the area (0.06 mm/yr during the Pleistocene, 0.2–0.3 mm/a during the last 150 ka, Geluk *et al.* 1994), it is not possible to address this question when evidence is based on only about one century of instrumental recordings and *c.*1,000 years of historical data.

Recent palaeoseismological studies were carried out along the Bree Fault scarp in northern Belgium (Camelbeeck and Meghraoui 1996; Vanneste *et al.* 1999; Fig. 2.5B). This fault scarp corresponds to the geomorphologically most prominent segment of the Belgian portion of the Feldbiss Fault, a normal fault forming the south-western border of the Roer Valley graben, the central graben of the Lower Rhine graben system; the Feldbiss is the counterpart of the Peel Fault, which was the probable source of the Roermond event in 1992 (Ahorner 1994; Camelbeeck *et al.* 1994). The Bree fault scarp is between 10 and 17.5 m high, and trends NW–SE over a distance of *c.*9 km.

The first results of these seismotectonic and palaeoseismological investigations were reported by Camelbeeck and Meghraoui (1998) and revealed the existence of fault displacements at ground surface, indicating severe

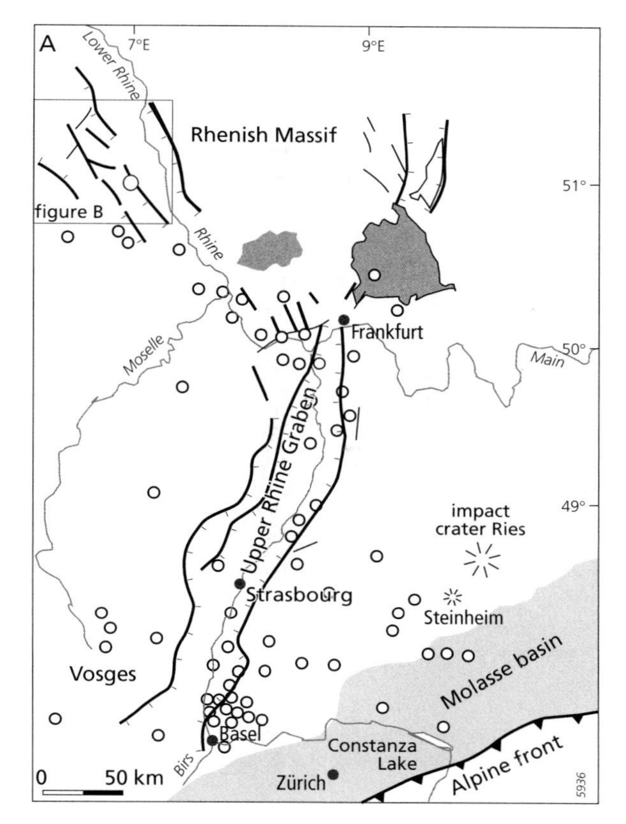

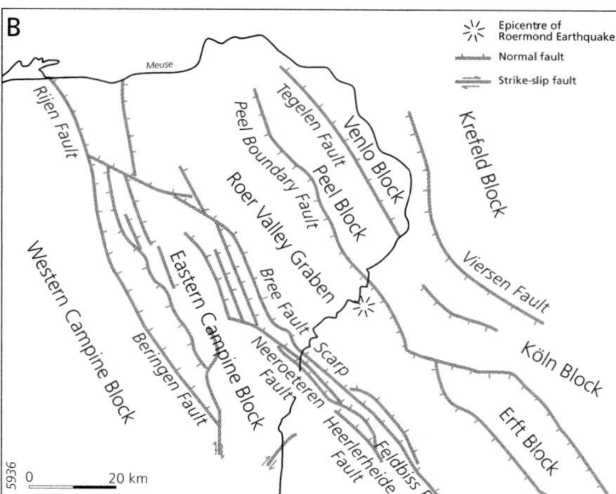

Fig. 2.5. A. Main faults and seismicity for the Rhine graben and adjacent areas (after Balling and Banda 1992); B. Structural sketch of the epicentral area of 13 April 1992, Ms = 5.3, Roermond earthquake (after Vanneste *et al.* 1999).

past seismic shaking of at least Mw = 6.1–6.2. At least three separate surface rupturing earthquakes are inferred, with an average return period of 12 ± 5 ka, and the last event possibly occurring between 1000 and 1350 BP.

Delouis *et al.* (2001) presented new data on the late Quaternary normal faulting of the Birs valley (Fig. 2.5A) western margin (the epicentral area of the 1356 Basle earthquake; intensity IX to X on the MSK-scale). The major fault branches are located at the foot of a 30–40-m high NNE–SSW trending cumulative scarp, that can be followed over a distance of 7 km or more. Several secondary faults, sometimes affecting the present-day topsoil, are also observed in trenches. Preliminary ^{14}C dating of stratigraphic layers from new trenches, as well as those obtained in previous excavations, indicate that at least two or three seismic events must have occurred in this fault zone within the last 4,500 years.

Vertical Crustal Movements

The occurrence of Neogene-Quaternary vertical crustal movements in the Alpine belt as well as in the intraplate domains of western Europe has been mainly inferred from geomorphological observations (planation surfaces, drainage network, topography) and geodetic levelling on a timescale of around 10^2 years (Demoulin *et al.* 1995; Kooi *et al.* 1998; Lenotre *et al.* 1999; Schenk *et al.* 2000; Dirkzwager *et al.* 2001). These domains are characterized by a moderate seismicity, with low- and intermediate-size earthquakes, distributed over large areas without any defined orientation. No direct relationships between the early structures of the movement masses and the main uplift or subsidence pattern can be observed (Rutten 1969).

Understanding the causes of the vertical crustal movements is still problematic (Ollier and Pain 2000), even if different mechanisms have been proposed, such as: subduction; underplating by lighter material; lithospheric buckling induced by the Africa–Europe collision; pushing by the Alpine belt on its foreland; granite intrusion; incidence of mantle plumes resulting in lithospheric thinning and thermal anomaly; phase boundary displacements and/or adjustment within the lower lithosphere (Mörner 1990; Ziegler *et al.* 1995; Burg and Ford 1997). For the areas previously covered by the late Pleistocene ice sheet, the isostatic response to deglaciation must be taken into account as a triggering factor of vertical crustal movements (Mörner 1990; Lambeck 1997). An additional factor for all the areas is the isostatic response to erosional unloading (Gilchrist *et al.* 1994).

In western Europe, uplift processes have been described in the western Alps and in the Pyrenees, along the

African–European plate boundary zone, and in the following intraplate areas: Armorican massif, Ardennes, Bohemian massif, Central massif, Harz, Rhine shield. After the post-Hercynian planation, these Palaeozoic massifs were essentially stable during the Mesozoic. Then, starting from the Eocene, they were uplifted to their present elevation, with a major uplifting phase during the Late Tertiary and an uplift acceleration in the Pleistocene (Andres 1989; Demoulin 1995; Bonnet *et al.* 2000).

Crustal movements may trigger salt flow (*halokinesis*; Trusheim 1960) in evaporite deposits, producing uprising salt domes, salt walls, and elongated diapirs (Rutten 1969). Widespread deformation of the Zechstein (Upper Permian) evaporites, at the base of the sedimentary sequence of northern Germany, started during the Mesozoic. This was mostly as a consequence of sedimentary loading in subsiding basins, while acceleration of halokinetic salt flow, with associated warping of the overlying strata, was induced by Tertiary-Quaternary tectonics (Kallenbach 1995). A striking morpho-structural effect of halokinetic warping is the cuesta arrangement of the Mesozoic layers on the northern side of the central Uplands of Germany.

The main subsidence area in western Europe is The Netherlands (Fig. 2.6) where, according to Kooi *et al.* (1998) subsidence is due to the combined influence of glacioisostasy, compaction, and tectonics. Moreover, land subsidence not related to tectonic activity may have been caused by solutional removal of salt deposits underground (Andres 1989; Semmel 1989). In the following sections we will describe the areas where tectonic mechanisms have been documented as the major cause of vertical crustal movements.

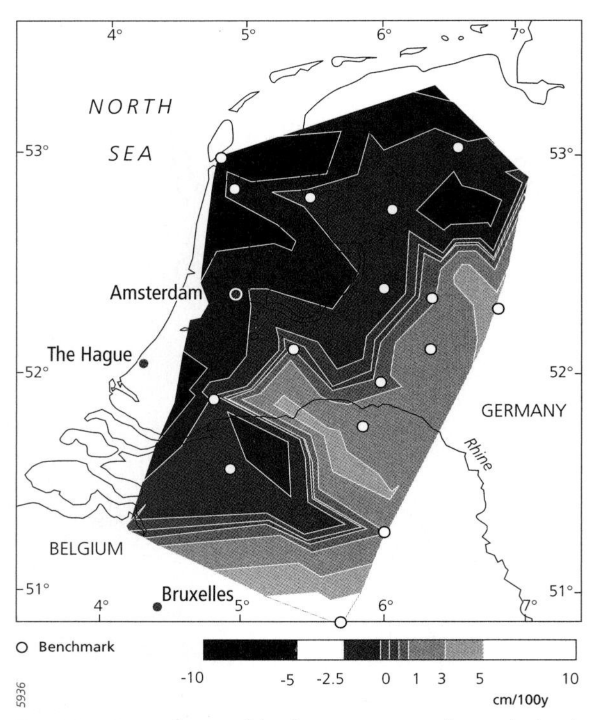

Fig. 2.6. Rate of vertical land movement in The Netherlands inferred from repeated levelling of underground benchmarks (after Kooi 1997).

The Alps

The tectonic evolution of the Alps started in the Early Mesozoic with the deposition of thick sedimentary sequences in a Tethys Sea geosyncline. The main tectogenetic phase occurred in the Early and Middle Tertiary (Rutten 1969). In the later stages, fault-bounded basins, filled with several thousand metres of Plio-Pleistocene marine or lacustrine sediments, were formed. Recent faulting processes can also be recognized in the Jura Mountains (Baize *et al.* 2001). The youngest thrust-and-fold belt is related to a French Alpine orogeny. Several sources of evidence (geology, seismicity) suggest that transverse strike-slip faults, which characterize the most recent deformations, are still active there.

After a first uplifting phase, the whole Alpine region was planated or even deeply lowered to small hills during the Pliocene (Trümpy 1980) and, subsequently, strongly uplifted again and eroded to the present rough topography (Ollier and Pain 2000). The old uplifted landscape is still testified by the summit-level (*Gipfelflur*) of peaks (Rutten 1969).

On the basis of apatite fission-tracks data, Steck and Hunziker (1994) outlined the main cooling episodes of the Alps: a first episode around 38 Ma, in relation to the updoming of the Lepontine Gneiss Dome in the central Alps; a second episode around 20 Ma, connected with normal faulting and tectonic unroofing of the Rhône–Simplon Line in the western Alps; a very rapid cooling event during the last 3 Ma. Exhumation rates of 0.5–0.7 mm/yr were calculated for the central-eastern Alps during the last 5 Ma. Similar values were obtained by Grundmann and Morteani (1985) for the eastern Alps and by Pettke *et al.* (1999) for the north-western Alps.

Also the distribution of Alpine metamorphism (Frey *et al.* 1980) proves that the crust has been affected by an asymmetric uplift, the maximum of which lies north of the Insubric Line, which divides the central and southern Alps.

According to Mueller (1984), the excessive depth of the M-discontinuity in the Alps was the primary cause

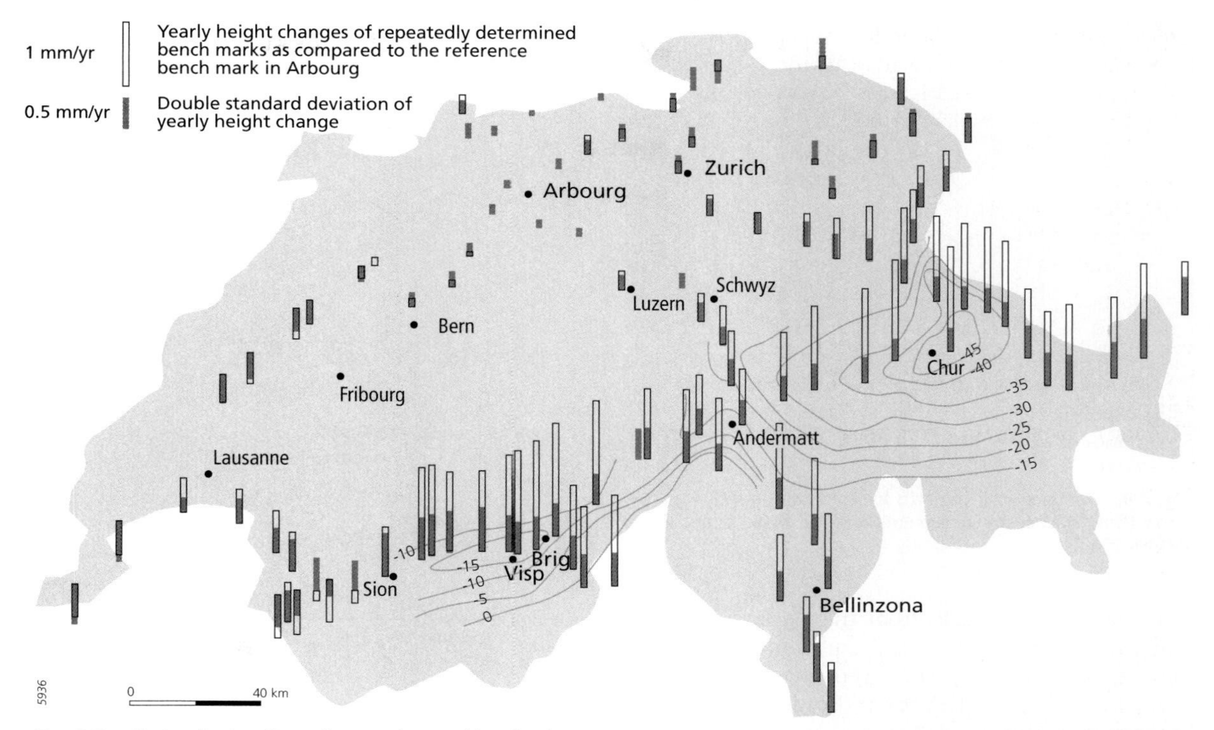

Fig. 2.7. First-order levelling of net and annual height changes in Switzerland with reference to an Arbourg bench mark (taken from Gubler *et al.* 1981). The regions with highest uplift are in the vicinity of Chur (Canton Grisons) and Brig/Visp (Canton Valais). A contour map of regional isostatic gravity anomalies (mgal) is also shown (taken from Kahle *et al.* 1980). A perfect correlation between maximum uplift and the minima of isostatic anomalies may be observed.

for vertical isostatic movements. The area of highest uplift (*c.*1.7 mm/yr) near Chur in the central Alps is clearly associated with negative isostatic anomalies reaching a minimum of −48 mgal (Fig. 2.7). Moreover, it is interesting to note that the uplift rate decreases almost linearly from Chur to Andermatt, whereas the isostatic anomalies increase proportionately. A comprehensive picture of the present-day dynamics in the Alps can be obtained by analysing all the available measurements of recent vertical crustal movements in Switzerland (Gubler *et al.* 1981). Figure 2.7 shows the uplift rates determined from the first repetition of the first-order levellings (between 1943 and 1978) as compared with the initial measurements (between 1903 and 1925) carried out by the Swiss Federal Office of Topography.

The Pyrenees

The Pyrenees are a symmetric tectonic belt with south-dipping faults in the north and north-dipping faults in the south. The chain tectogenetic phase occurred in the Early Tertiary, rejuvenating previous Hercynian structures (Rutten 1969; Soquet *et al.* 1975). In subsequent times the belt underwent planation and uplift (Peña Monné 1994; Calvet 1999; Ollier and Pain 2000). On the basis of erosion surface analyses, 2,000 m of uplift can be inferred since the Early Pliocene. Vanara *et al.* (1997) calculated a Plio-Pleistocene uplift of 1,000 m in the north-western Pyrenees. According to Sala (1984), the planation surfaces developed during Pliocene and Early Quaternary, when the denudation rate exceeded that of uplift.

The Armorican Massif

The Armorican massif is an Upper Proterozoic to Palaeozoic basement cropping out in north-western France between the Bay of Biscay and the English Channel (Fig. 2.8). It is surrounded by three Meso-Cenozoic sedimentary basins: the Western Approaches Trough and the South Armorican Margin, which are presently located offshore, and the Parisian basin, which is presently located onshore. While the northern portion of the Armorican basement can be considered part of the

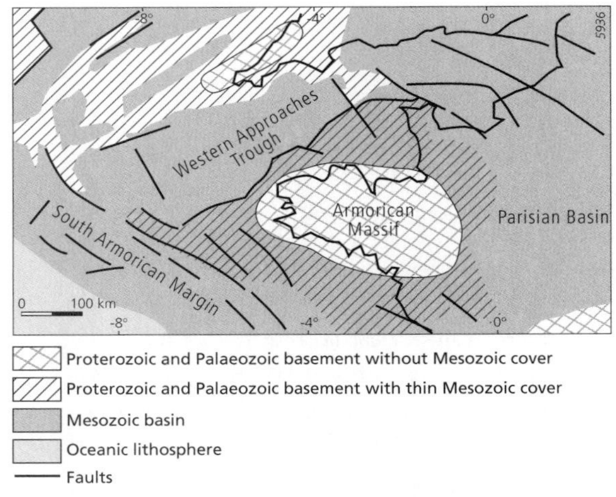

Fig. 2.8. Tectonic map of north-western Europe showing Proterozoic and Palaeozoic basements and Mesozoic basins (after Ziegler 1987).

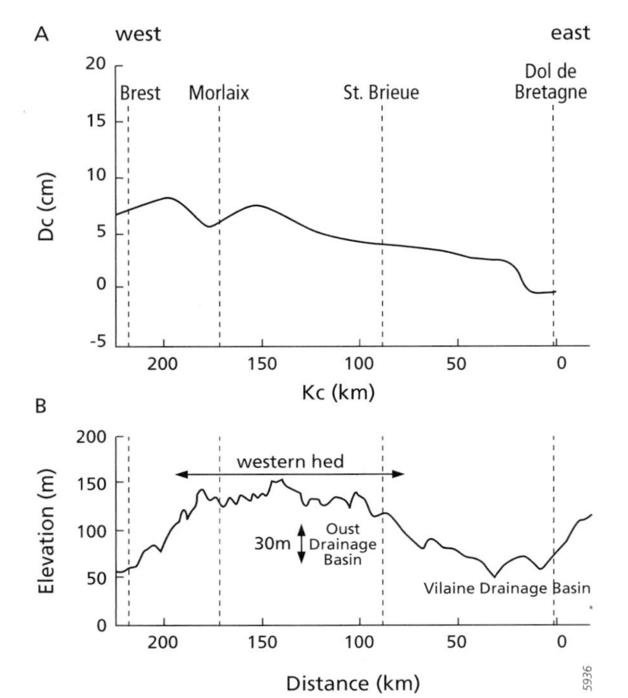

Fig. 2.9. Comparison between levelling and average topographic profiles of the western part of the Armorican massif (after Bonnet *et al.* 2000). A. Comparative levelling profile of Brittany, calculated from levelling data measured in 1889 and 1962 (taken from Lenotre *et al.* 1999); B. Average east–west topographic profile of Brittany (taken from Bonnet *et al.* 2000). A discrepancy exists between the two data. Uplift rates measured by levelling range between 0.2 and 0.6 mm/yr, whereas those estimated by Bonnet *et al.* (2000) are about 0.04–0.06 mm/yr.

Cadomian Orogenic belt, from the Upper Proterozoic (Rabu *et al.* 1990), its central and southern portions were the southern branch of the European Variscan belt dating from the Upper Palaeozoic (Brun and Burg 1982; Le Corre *et al.* 1991).

The Alpine compressive deformation (Hayward and Graham 1989) is deducible from all the basins around the Armorican massif. To the north of this massif, seismic studies have suggested a kilometre-scale tectonic inversion of the Western Approaches Trough occurring during Late Oligocene/Early Miocene (Hayward and Graham 1989). At present, the Armorican massif has become a western European lithosphere intraplate domain corresponding to the extended foreland of the Pyrenees and the Alps (Ziegler *et al.* 1995). Instrumental seismicity shows a concentration of earthquakes, which indicates that the Armorican massif is still an active domain.

Uplift in the Armorican massif was documented by Bonnet *et al.* (2000) on a timescale of 10^5 to 10^6 years, using a detailed study of multiple drainage basins, and by Lenotre *et al.* (1999) on a timescale of 10^2 years, using levelling data (Fig. 2.9). A comparison between the east–west levelling profile and the average topography of the same zone shows a good agreement in the profile shapes, which suggests that uplift measured by levelling is the result of uplift on a longer timescale, as reflected by the topography. The pattern and timescale of this uplift are clearly inconsistent with a glacioisostatic origin, and thus a tectonic one can be ascribed to it. The uplift pattern can be referred to the stress field of the western European lithosphere, which is still today under compression. The influence of this compressional stress field is documented in the Armorican massif by, amongst other things, focal mechanisms (Nicolas *et al.* 1990) which display pure reverse movements along N70 nodal planes. According to Bonnet *et al.* (2000), this low-amplitude and long-wavelength deformation suggests that the present-day mechanism of deformation in north-west France mainly involves lithospheric buckling.

The Bohemian Massif

In the last decade geodynamical investigations of the Bohemian massif based on seismotectonic analyses of selected regions (Schenk *et al.* 1989; Grunthal *et al.* 1990) gave a basic idea on the possible geodynamical behaviour of individual geological blocks of the Bohemian massif. Preliminary results of a GPS network confirm the geological and geophysical evidence for recent movements of the Earth's surface in the Bohemian massif. Models have assumed an active pushing of the Alps to the north that has caused the recent mobility among individual structural blocks of the Bohemian massif.

On the basis of the new available data, a preliminary geodynamical model was introduced by Schenk *et al.* (2000), in which the Bohemian massif is situated between the Russian (east-European) Platform and the Alps. Preliminary GPS data detected two pronounced movement trends in the northern area of the Moravo-Silesian part of the Bohemian massif.

From thermochronological data collected in the Fichtelgebirge (north-western Bohemian massif) about 2-km denudation of Cretaceous-Palaeogene sediments during the last 30 Ma was calculated. Then, on the basis of the study of planation surfaces, Migon and Lach (1999) stated that the tectonic uplift of the Sudeten Mountains, a part of the Bohemian massif, mainly occurred in Pliocene–Early Quaternary. The recent crustal movements of the Bohemian massif can be due to different causes, among which are postglacial rebound and a present-day mechanism of lithospheric buckling.

References

Ahorner, L. (1994), Fault-plane solutions and source parameters of the 1992 Roermond, The Netherlands, main shock and its stronger aftershocks from regional seismic data. *Geologie and Mijnbouw* 73: 199–214.

—— Baier, B., and Bonjer, K. P. (1983), General pattern of seismotectonic dislocation and the earthquake generating stress field in Central Europe between the Alps and the North Sea. In: K. Fuchs, K. von Gehlen, H. Mälzer, H. Murawski, and A. Semmel (eds.) *Plateau uplift. The Rhenish Shield—a case history*. Springer, Berlin, 187–97.

Andres, W. (1989), The Central German uplands. In: F. Anhert (ed.) *Landforms and Landform Evolution in West Germany*. Catena Suppl. 15: 25–44.

Angelier, J. (1976), La Néotectonique cassante et sa place dans un arc insulaire: l'Arc Égéen méridional. *Revue Géographie Physique Géologie Dynamique* 18: 1257–65.

Argus, D. F., Gordon, R. G., Demets, C., and Stein, S. (1989). Closure of the Africa–Eurasia–North America plate motion circuit and tectonics of the Gloria fault. *Journal of Geophysical Research* 94: 5585–602.

Armijo, R., Deschamps, A., and Poirier, J. P. (1986), *Carte seismotectonique Europe et bassin mediterraneen*. Inst. Phys. du Globe, Paris.

Baize, S., Mathieu, F., Calais, E., Scotti, O., Jouanne, F., Marc Cushing, E., and Peyridieu, G. (2001), Semi-permanent GPS network for active faults survey and for seismic hazard assessment in France. Technical aspects, objectives and first installation in Jura. European Union of Geosciences, EUGXI, *Journal of Conference Abstracts* 6/1.

Balling, N., and Banda, E. (1992), Europe's lithosphere. In: D. Blundell, R. Freeman, and St. Mueller (eds.) *A continent revealed: the European Geotraverse*. Cambridge University Press, Cambridge, 111–23.

Beloussov, V. V. (1978), *Géologie structurale*. Edit. Mir, Moscow.

Bonnet, S., Guillocheau, F., Brun, J., and Driessche, J. (2000), Large-scale relief development related to Quaternary tectonic uplift of a Proterozoic-Paleozoic basement. The Armorican massif, NW France. *Journal of Geophysical Research* 105: 19273–88.

Brun, J. P., and Burg, J. P. (1982), Combined thrusting and wrenching in the Ibero-Armorican arc: a corner effect during continental collision. *Earth and Planetary Science Letters* 61: 319–32.

Burg, J. P., and Ford, M. (eds.) (1997), *Orogeny through time*. Geological Society, London, Special Publication 121.

Calvet, M. (1999), Rythmes et vitesses d'évolution morphogénètique dans un orogèn alpin. Le cas des Pyrénées orientales franco-espagnoles. *Zeischrift für Geomorphologie*, Suppl. 118: 91–105.

Camelbeeck, T., and Meghraoui, M. (1996), Large earthquake in northern Europe more likely than once thought. *Eos* 77: 405–9.

—— Meghraoui, M. (1998), Geological and geomorphological evidence for large paleoearthquakes with surface faulting in the Roer Graben (northwest Europe). *Geophysical Journal International* 12: 347–62.

—— van Eck, T., Pelzing, R., Ahorner, L., Looh, J., Haak, H. W., Hoang-Trong, P., and Hollnack, D. (1994), The 1992 Roermond earthquake, The Netherlands, and its aftershocks. *Geologie and Mijnbouw* 73: 181–97.

Cloetingh, S., and Banda, E. (1992), Europe's lithosphere-physical properties. In: D. Blundell, R. Freeman, and St. Mueller (eds.) *A continent revealed: the European Geotraverse*. Cambridge University Press, Cambridge, 71–91.

DEKORP Research Group (1990), Crustal structure of the Rhenish Massif: results of deep seismic reflection lines. DEKORP 2-N and 2-N-Q. *Geologische Rundschau* 79: 523–66.

Delouis, B., Meghraoui, M., Ferry, M., Huggenberger, P., Spottke, I., Giardini, D., and Granet, M. (2001), Paleoseismic investigations in the Basel Region (Switzerland): new evidence for Holocene normal faulting, European Geophysical Society, 26th General Assembly. *Geophysical Research Abstracts* 3.

de Mets, C., Gordon, R. G., Argus, D. F., and Stein, S. (1990), Current plate motion. *Geophysical Journal International* 101: 425–78.

Demoulin, A. (1995), L'Ardenne bouge toujours. Néotectonique du massif ardennais. In: A. Demoulin (ed.) *L'Ardenne. Essai de Géographie Physique*. Department of Physical Geography, Liège University, 110–35.

—— Pissart, A., and Zippelt, K. (1995), Neotectonic activity in and around the southwestern Rhenish shield (west Germany): indications of a leveling comparison, *Tectonophysics* 249: 203–16.

Dewey, J. F., Helman, M. L., Turco, E., Hutton, D. H. W., and Knott, S. D. (1989), Kinematics of the western Mediterranean. In: M. P. Coward, D. Dietrich, and R. G. Park (eds.) *Alpine tectonics*. Geological Society, London, Special Publication 45: 265–83.

Dirkzwager, J. B., Connolly, P. T., and Nieuwland, D. A. (2001), 3D finite element modelling of upper crustal faults and their near surface expression for the west Netherlands basin, European Geophysical Society, 26th General Assembly. *Geophysical Research Abstracts* 3.

Eva, C., Augliera, P., Cattaneo, M., and Giglia, G. (1990), Some considerations on seismotectonics of northwestern Italy. In: R. Freeman, P. Giese, and St. Mueller (eds.) *The European geotraverse: integrative studies*. European Science Foundation, Strasburg, 389–96.

Fourniguet, J. (1987), Néotectonique. In: J.-C. Miskovsky (ed.) *Géologie de la Préhistoire*. GEOPRE, Paris, 281–92.

Frey, M., Bucher, K., Frank, E., and Mullis, J. (1980), Alpine metamorphism along the Geotraverse Basel-Chiasso—a review. *Ecologae Geologica Helvetica* 73: 527–46.

Fuchs, K., von Gehlen, K., Mälzer, H., Murawski, H., and Semmel, A. (eds.) (1983), *Plateau uplift. The Rhenish Shield—a case history*. Springer, Berlin.

Gall, H., Huttner, R., and Müller, D. (1977), Erlauterungen zur Geologischen Karte des Rieses 1:50.000. *Geol. Bavaric* 76: 1–171.

Geluk, M. C., Duin, E. J. Th., Dusar, M., Rijkers, M. H. B., van den Berg, M. W., and van Rooijen, P. (1994), Stratigraphy and tectonics of the Roer Valley Graben. *Geologie and Mijnbouw* 73: 129–41.

Gilchrist, A., Summerfield, M., and Cockburn, H. (1994), Landscape dissection, isostatic uplift, and the morphologic development of orogens. *Geology* 22: 963–6.

Grundmann, G., and Morteani, G. (1985), The young uplift and thermal history of the central eastern Alps (Austria/Italy), Evidence from apatite fission track ages. *Jahrbuch Geologische Bundesanstalt Austria* 128: 197–216.

Grunthal, G., Schenk, V., Zeman, A., and Schenkova, Z. (1990), Seismotectonic model for the earthquake swarm 1985/86 in the Vogtland/West Bohemia focal area. *Tectonophysics* 174: 369–83.

Gubler, E., Kahle, H. G., Klingelé, E., Mueller, St., and Olivier, R. (1981), Recent crustal movements in Switzerland and their geophysical interpretation. *Tectonophysics* 71: 125–52.

Hancock, P., and Williams, G. (1986), Neotectonics. *Journal Geological Society London* 143: 325–6.

Hayward, A. B., and Graham, R. H. (1989), Some geometrical characteristics of inversion. In: M. A. Cooper and G. D. Williams (eds.) *Inversion tectonics*. Geological Society, London, Special Publication 44: 17–39.

Kahle, H. G., Mueller, St., Klingelé, E., Egloff, R., and Kissling, E. (1980), Recent dynamics, crustal structure and gravity in the Alps. In: N. A. Mörner (ed.) *Earth rheology, isostasy and eustasy*. Wiley, Chichester, 377–88.

Kallenbach, H. (1995), Quaternary morphology of the Berlin landscape. In: W. Schirmer (ed.) *Quaternary field trips in central Europe*. INQUA Berlin Excursion Guide 3, Dr Friedrich Pfeil, Munich, 1142–6.

Kooi, H. (1997), Contribution to tectonics, isostasy and natural compaction to vertical land movement in The Netherlands. *Report Meetkundige Dienst* MDGAP-9770, Rijkswaterstaat, The Netherlands.

—— Johnston, P., Lambeck, K., Smither, C., and Molendijk, A. (1998), Geological causes of recent (ca. 100 yr) vertical land movements in The Netherlands. *Tectonophysics* 299: 297–316.

Lambeck, K. (1997), Sea-level change along the French Atlantic and Channel coasts since the time of the Last Glacial Maximum. *Palaeogeography, Palaeoclimatology, Palaeoecology* 129: 1–22.

Le Corre, C., Auvray, B., Ballevre, M., and Robardet, M. (1991), Le Massif Armoricain. *Sci. Geological Bulletin* 44: 31–103.

Lenotre, N., Thierry, P., Blanchin, R., and Brochard, C. (1999), Current vertical movement demonstrated by comparative leveling in Brittany (France). *Tectonophysics* 301: 333–44.

Lippoldt, H. J. (1983), Distribution of volcanic activity in space and time. In: K. Fuchs, K. von Gehlen, H. Mälzer, H. Murawski, and A. Semmel (eds.) *Plateau uplift. The Rhenish Shield—a case history*. Springer, Berlin, 112–20.

Meier, L., and Eisbacher, G. H. (1991), Crustal kinematics and deep structure of the northern Rhine Graben, Germany. *Tectonics* 10: 621–30.

Mercier, J. L. (1976), La Néotectonique, ses méthodes et ses buts. Un example: l'Arc Égéen (Méditerranée orientale). *Revue Géographie Physique Géologie Dynamique* 18: 323–46.

Migon, P., and Lach, J. (1999), Geomorphological evidence of neotectonics in the Kaczawa sector of the Sudetic marginal fault, southwestern Poland. In: D. Krzyszkowski (ed.) *The Late Cainozoic Evolution of the Sudeten and its Foreland*. Geologia Sudetica 32/2: 307–16.

Mörner, N. A. (1978), Faulting, fracturing and seismic activity as a function of glacialism in Scandinavia. *Geology* 6: 41–5.

—— (1990), Glacial isostasy and long-term crustal movements in Fennoscandia with respect to lithospheric and asthenospheric processes and properties, *Tectonophysics* 176: 13–24.

Mueller, St. (1984), Dynamic processes in the Alpine arc. *Annales Geophysicae* 2: 161–4.

Müller, B., Zoback, M. L., Fuchs, K., Mastin, L., Gregersen, S., Pavoni, N., Stephansson, O., and Ljunggren, Ch. (1992), Regional patterns of tectonic stress in Europe. *Journal of Geophysical Research* 97: 11783–803.

Nicolas, M., Santoire, J. P., and Delpech, P. T. (1990), Intraplate seismicity: new seismotectonic data in western Europe. *Tectonophysics* 179: 27–53.

Nocquet, J. M., Calais, E., Altamini, Z., Sillard, P., and Boucher, C. (2001) Intraplate deformation in western Europe deduced from an analysis of the International Terrestrial Reference Frame 1997 (ITRF97) velocity field. *Journal of Geophysical Research* 106: 11239–57.

Ollier, C., and Pain, C. (2000), *The origin of mountains*. Routledge, London.

Oxburgh, E. R., and O'Nions, R. K. (1987), Helium loss, tectonics and the terrestrial heat budget. *Science* 237: 1583–8.

Pavlides, B. S. (1989), Looking for a definition of neotectonics. *Terra Nova* 1/3: 233–5.

Pavoni, N. (1990), Seismicity and fault plane solutions along the EGT: data selection and representation as illustrated by the seismicity of Switzerland. In: R. Freeman and St. Mueller (eds.) *Sixth EGT Workshop: Data Compilations and Synoptic Interpretation*. European Science Foundation, Strasburg, 341–8.

Peña Monné J. L. (1994), Cordillera Pirenaica. In: M. Gutierrez Elorza (ed.) *Geomorfología de España*. Editorial Rueda, Madrid, 159–225.

Pettke, T., Diamond, L., and Villa, I. (1999), Mesothermal gold veins and metamorphic devolatilization in the northwestern Alps: the temporal link. *Geology* 27: 641–4.

Philip, H. (1987), Plio-Quaternary evolution of the stress field in Mediterranean zones of subduction and collision. *Annales Geophysicae* 3: 301–20.

Rabu, D., Chantraine, J., Chauvel, J.-J., Denis, E., Bale, P., and Bardy, P. (1990), The Brioverian (Upper Proterozoic) and the Cadomian orogeny in the Armorican Massif. In: R. S. D'Lemos, R. A. Strachan, and C. G. Topley (eds.) *Cadomian orogeny*. Geological Society, London, Special Publication 51: 81–94.

Rutten, M. G. (1969), *The geology of Western Europe*. Elsevier, Amsterdam.

Sala, M. (1984), Pyrenees and Ebro basin complex. In C. Embleton (ed.) *Geomorphology of Europe*. Macmillan, London, 269–93.

Schenk, V., Cacon, S., Bosy, J., Kontny, B., Kottnauer, P., and Schenkova, Z. (2000), GPS network 'Sudeten'-preliminary results of the campaigns 1998–1999. In: J. Sledzinski (ed.) *Proceedings of the 2nd Czech–Polish Workshop on Recent Geodynamics of the East Sudety Mts. and Adjacent Areas, Boleslawow, Poland, 6–8 April 2000*. Reports on Geodesy, Warsaw University of Technology 7/53: 25–33.

—— Schenkova, Z., and Pospisil L. (1989), Fault system dynamics and seismic activity—two examples from the Bohemian Massif and the Western Carpathians. *Geophysical Transactions* 35: 101–16.

Sebrier, M., Ghafiri, A., and Blès, J. L. (1993), Paleoseismicity in France: Fault trench studies in a region of moderate seismicity. *Journal Geodynamics* 24: 207–17.

Semmel, A. (1989), The Central German uplands. In: F. A. Anhert (ed.) *Landforms and Landform Evolution in West Germany*. Catena Suppl. 15: 25–44.

Sengor, A. M. C., Burke, K., and Dewey, J. F. (1978), Rifts at high angles to orogenic belts: test for their origin and the Upper Rhine graben as an example. *American Journal Science* 278: 219–42.

Simkin, T., Tilling, R., Taggart, J. J. W., and Spall, H. (1989), *This dynamic planet: world map of volcanoes, earthquakes, and plate tectonics*. US Geological Survey/Smithsonian Institute, USA.

Smith, D. E., Kolenkiewicz, R., Dunn, P. J., Robbins, J. W., Torrence, M. H., Klosko, S. M., Williamson, R. G., Pavlis, E. C., Douglas, N. B., and Fricke, S. K. (1990), Tectonic motion and deformation from Satellite Laser ranging to LAGEOS. *Journal of Geophysical Research* 95: 22013–41.

Soquet, P., Billotte, M., Canerot, J., Debroas, E. J., and Peybernes, B. (1975), Nouvelle interpretation de la structure des Pyrénées. *Comptes Rendus Académie Scientifique Paris* 281: 609–12.

Steck, A., and Hunziker, J. (1994), The Tertiary structural and thermal evolution of the Central Alps—compressional and extensional structures in an orogenic belt. *Tectonophysics* 238: 229–54.

Trümpy, R. (1980), *An outline of the geology of Switzerland*. Wepf, Basle.

Trusheim, F. (1960), Mechanism of salt migration in northern Germany. *Bulletin American Association Petroleum Geologists* 44: 1519–40.

Vanara, N., Maire, R., and Lacroix, J. (1997), La Surface carbonatée du massif des Arbailles (Pyrénées Atlantiques): un example de paléoréseau hydrographique néogène déconnecté par la surrection. *Bulletin Société Géologique France* 168: 255–65.

Vanneste, K., Meghraoui, M., and Camelbeeck, T. (1999), Late Quaternary earthquake-related soft-sediment deformation along the Belgian portion of the Feldbiss Fault, Lower Rhine Graben system. *Tectonophysics* 309: 57–79.

Vita Finzi, C. (1986), *Recent earth movements: an introduction to neotectonics*. Academic Press, London.

Wedepohl, K. H. (1987), Kontinentaler Intraplatten-Vulkanismus am Beispiel der tertiaren Basalte der Hessischen Senke. *Fortschr. Miner.* 65: 19–47.

Winslow, M. A. (1986), Neotectonics: concepts, definitions and significance. *Neotectonics* 1: 1–5.

Zeis, S., Gajewski, D., and Prodehl C. (1990), Crustal structure of southern Germany from seismic refraction data. *Tectonophysics* 176: 59–86.

Ziegler, P. A. (1987), Celtic Sea-Western Approaches area: an overview. *Tectonophysics* 137: 341–6.

—— Cloething, S., and van Wees, J. (1995), Dynamics of intra-plate compressional deformation: the Alpine foreland and other examples. *Tectonophysics* 252: 7–59.

Zoback, M. L. (1992), First and second order patterns of stress in the lithosphere: the World Stress Map Project. *Journal of Geophysical Research* 97B/11: 11703–28.

Quaternary Climatic Changes and Landscape Evolution

Jürgen Ehlers

Introduction

The last 2–3 Ma have witnessed climatic changes of a scale unknown to the preceding 300 Ma. In the cold periods vegetation was reduced to a steppe, giving rise to large-scale aeolian deposition of sand and loess and river sands and gravels. In the warm stages, flora and fauna recolonized the region. Parts of Europe were repeatedly covered by mountain glaciers or continental ice sheets which brought along huge amounts of unweathered rock debris from their source areas. The ice sheets dammed rivers and redirected drainage towards the North Sea. They created a new, glacial landscape. This chapter presents an outline of the climatic history, and in particular the glacial processes involved in shaping the landscapes of western Europe.

Definition of the Quaternary

By convention, geologists generally tend to draw stratigraphical boundaries in marine deposits because they are more likely to represent continuous sedimentation and relatively consistent environments in comparison to terrestrial sediments. However, marine deposits from the period in question are relatively rarely exposed at the surface. According to a conclusion of the International Geological Congress 1948 the Tertiary/Quaternary boundary was defined as the base of the marine deposits of the Calabrian in southern Italy. In the Calabrian sediments fossils are found that reflect a very distinct climatic cooling (amongst others the foraminifer *Hyalinea baltica*). This climatic change roughly coincides with a reversal of the earth's magnetic field; it is situated at the upper boundary of what is called the Olduvai Event. Consequently, it is relatively easy to identify; its age is today estimated at 1.77 Ma (Shackleton *et al.* 1990).

However, in contrast to the older parts of the earth's history, the significant changes within the Quaternary are not changes in faunal composition but changes in climate. For reasons of long-term climatic evolution the base of the Calabrian is not a very suitable global boundary. Its adoption excludes some of the major glaciations from the Quaternary. Therefore, in major parts of Europe another Tertiary/Quaternary boundary is in use, based on the stratigraphy of the Lower Rhine area (Fig. 3.1) (e.g. Zagwijn 1989). Here the most significant climatic change is already recorded as far back as the Gauss/Matuyama magnetic reversal (some 2.6 Ma ago). At this time the catchment area of the Rhine extended for the first time into the Alpine foreland, resulting in a dramatic change in heavy-mineral composition of the sediments (Boenigk 1982). At the same time the typical Pliocene pollen spectra changed to those of a cooler, characteristic Quaternary aspect. Similarly, important changes are found in gravel composition and in the mollusc and vertebrate assemblages. These changes are also used to mark the Tertiary/Quaternary boundary in the British Isles and the northern Alpine area (e.g. Gibbard *et al.* 1991; Schreiner 1992; Jerz 1993). It can be easily identified worldwide, both biostratigraphically and magnetostratigraphically. Moreover, the Chinese loess record began at about that time (Ding *et al.* 2000).

The Quaternary is traditionally subdivided into the Pleistocene (the ice-age period proper) and the Holocene (the present interglacial). A single period of ice-sheet expansion is called a 'glaciation'. The adjective 'glacial' is used whenever the temporal aspect of an event or deposit is intended; where genesis by a glacier is to be expressed, the term 'glacigenic' is used. Stratigraphic subdivision of the Quaternary is largely based on climate. Cold periods are referred to as 'cold stages'

Fig. 3.1. Location map.

(chronostratigraphy), 'glacials', or 'glaciations' (climatostratigraphy). The intervening temperate periods are 'warm stages', 'interglacials', or 'interglaciations'. Minor climatic oscillations within these stages are called 'stadials' or 'stades' (cold) and 'interstadials' or 'interstades' (temperate).

Subdivision of the Quaternary

The first attempt to subdivide the 'Ice Age' into a number of individual events, i.e. glaciations and intervening interglacials was undertaken in the Alps. The classic Alpine Quaternary stratigraphy is largely based on morphostratigraphic criteria (Penck and Brückner 1901/9). It is based on the concept that the so-called 'glacial series' consists of a sequence of tongue basins with drumlins, moraine belts, and gravel spreads supposedly of the same age. Comprehensive morphological sequences of this type were initially found for only three glaciations; for the fourth this evidence was produced much later through investigations in upper Austria. The glaciations were named after rivers Günz, Mindel, Riss, and Würm, from oldest to youngest respectively.

The morphostratigraphical method was developed in the northern Alpine foothills; it is there where it is best applied. It is much more difficult to use in the southern and western Alps. In the south-German Alpine foreland later observations indicated the involvement of more than the four glaciations postulated by Penck and Brückner (1901/9). Here Eberl (1930) identified older, highly elevated gravels in the Iller-Lech region that he regarded as deposits of a 'Donau Ice Age', that predated the Günz Glaciation of Penck and Brückner. Schaefer (1953) confirmed the existence of this Donau Ice Age and later added another, earlier 'Biber Ice Age', represented by gravels at Staufenberg and in the Aindlingen flight of terraces east of the Lech River. Those additional cold stages have been generally accepted. It is implied that the gravels are of glaciofluvial origin, although the equivalent till sheets and end moraines have not yet been identified.

Originally it was assumed that north Germany, like the Alps, had been affected by three glaciations. This concept, however, only found general acceptance after a third (oldest) till unit had been identified in the Berlin and Hamburg areas. From 1911 onwards these glaciations were termed Elsterian, Saalian, and Weichselian. After it had been discovered that the Alps had been glaciated (at least) four times instead of three (Penck and Brückner 1901/9), there has been no lack of attempts to identify corresponding numbers of Nordic glaciations. However, this has not been possible.

Whilst originally Quaternary stratigraphical work was almost exclusively concerned with the glaciations, the advent of pollen analysis in the mid-twentieth century, interest increased in the intervening warm interglacials. In The Netherlands and in the Dutch–German border area a number of sites with Early Pleistocene interglacial deposits have been found. However, palynological interpretation of these sediments is often problematic. Many of the pollen analyses were undertaken on clastic deposits, or in thin peat or mud beds intercalated in gravelly deposits. In clastic deposits, however, a certain degree of reworking must always be considered. This is often indicated by pollen and spores derived from Tertiary or Mesozoic rocks. Furthermore, mechanical abrasion of the pollen grains can influence the composition, and some oscillations in pollen composition may reflect changes in depositional environment rather than vegetational changes (see e.g. Gibbard *et al.* 1991).

More favourable conditions for the preservation of good pollen sequences are locally found in north Germany. For example, a karstic depression on the Lieth salt dome in Schleswig-Holstein contains a sedimentary sequence with five Early Pleistocene warm stages that has been preserved. According to Menke (1975) these events represent the period from the Tiglian to beyond the Menapian Stages (Fig. 3.2). The sediments are autochthonous peats and muds, which in contrast to their Dutch equivalents show a clear vegetational development for each stage. Moreover, the younger part of the Early Pleistocene (probably from the end of the Menapian to the early Elsterian) has been encountered in boreholes in a karstic depression on top of the Gorleben salt dome, only a brief account of which has so far been published (Müller 1992). However, the discrepancy between the number of events identified in the terrestrial records and those in the ocean sediment marine isotope record remains to be resolved.

It was always suspected that the continental glacial stratigraphy of north-west Europe was incomplete. However, it was not before the advent of deep-sea research that the extent of the gaps became known. The marine isotope record from the deep sea floor sediments suggests that the last 2.5 Ma were characterized by about fifty climatic cycles (Shackleton *et al.* 1990); so far not even half these cycles have been identified on land, not even in the supposedly rather complete Dutch–German border area stratigraphic sequences. Although the climatic changes of the Late Pleistocene are relatively well known, the events before the last, Eemian Interglacial can be reconstructed only in broad outline. Thus correlation of the terrestrial stratigraphy with the

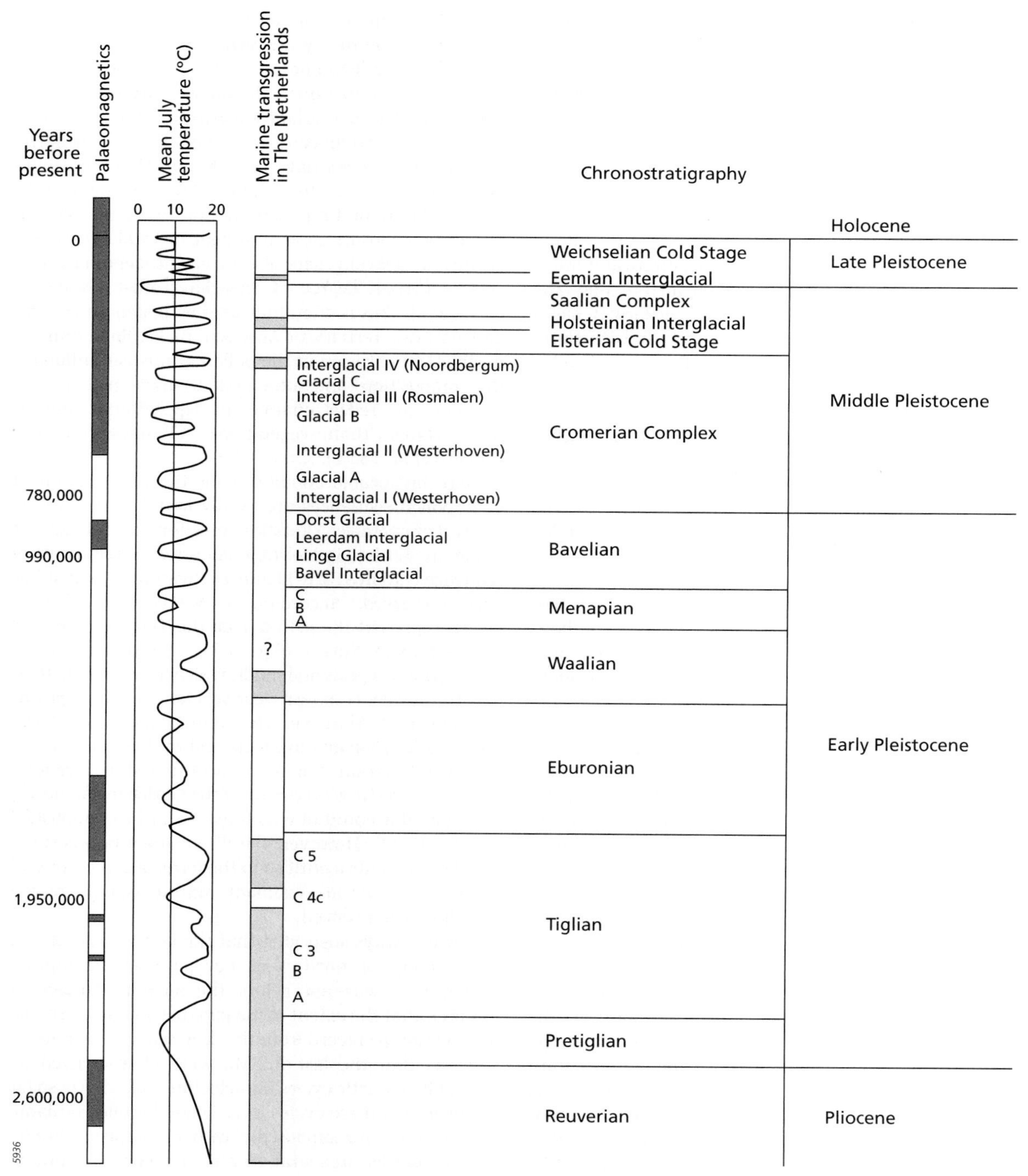

Fig. 3.2. Climatic curve and chronostratigraphy of the Pleistocene in The Netherlands and adjoining regions (after Gibbard *et al.* 1991).

marine isotope record is impeded. However, some fixed points have been established through palaeomagnetic analyses.

The negatively magnetized Late Early Pleistocene 'Cromerian Interglacial I' probably corresponds to Marine Isotope Stage 21. Consequently, 'Cromerian Interglacials II to IV' might be equivalent to Stages 19, 17, and 15. Yet assuming that traces of extensive glaciation in 'Cromerian Glacial A' should be clearly reflected in the marine isotope sequence, it should probably be correlated with Stage 16. In addition, investigations of Massif Central craters (Reille and De Beaulieu 1995) and in Poland (Vandenberghe 2000) could indicate correlation of the Holsteinian with Marine Isotope Stage 11, whereas the Elsterian is assigned to the pronounced Marine Isotope Stage 12. However, many open questions remain. There is as yet no obvious correlation for the Alpine Donau and Biber cold Stages.

The Glaciations

The Günz Glaciation

The oldest Alpine tills belong to the Günz glaciation. Originally it was assumed that the Günz was younger than the Matuyama/Brunhes palaeomagnetic boundary. However, at the Höchsten and Heiligenberg sites near the German–Swiss border (north of Lake Constance~Bodensee) the tills and gravels of the Günz Glaciation are reversely magnetized. Consequently, those tills should be assigned to the Matuyama Epoch (Schreiner 1992). It is not yet clear, however, if all deposits of the Alpine region formerly assigned to the Günz really form part of the same single glaciation. The revision of the Alpine Quaternary stratigraphy has begun, yet it is far from being completed.

Evidence of Pre-Elsterian Glaciation

In the area of the Nordic glaciations, investigations in the North Sea have produced evidence of ancient ice sheets. According to recent evaluations, the first major continental glaciation that reached the shelf edge off Norway occurred at about 1,100,000 BP. Its ice sheet may have reached into northern Germany (Sejrup *et al.* 2000). Possible corresponding sediments in the Dutch–German border region may be the so-called 'Hattem Beds' which provide strong indirect evidence of glaciation. These gravel-rich beds contain strongly weathered Nordic material (Lüttig and Maarleveld 1961) and are assigned chronostratigraphically to the Menapian Stage. The Hattem Beds contain Scandinavian rocks as well as material from the Weser catchment area and the Thuringian Forest in Central Germany. It is possible that this material was, in part, transported by drift ice.

In other North Sea drillings, Stoker and Bent (1985) detected till in the normally magnetized part of the Aberdeen Ground Formation which they correlated to a 'Cromerian'-aged glaciation. This may correspond to a second bed with Scandinavian indicator erratics found in The Netherlands in the so-called 'mixed zone' at the base of the fluvial Urk Formation (e.g. Zandstra 1983). It seems that an ice sheet was close to The Netherlands during deposition of this 'Weerdinge Member' which is correlated with the 'Cromerian Glacial C' event.

From the continental section of north-western Europe there is little concrete evidence of glaciation prior to the Elsterian. However, near Harreskov, in Denmark (Jutland), Cromerian interglacial lacustrine sediments, found in the mid-1960s, are underlain by meltwater sands 5 m thick. These in turn overlie a till-like deposit about 5 m thick (Andersen 1967). Almost the same sequence was encountered at another Danish site called Ølgod, where it was described by Jessen and Milthers (1928). At neither of these sites have the potential glacial deposits been studied in detail. However, pre-Elsterian glacigenic deposits occur much more widely and are well expressed in eastern Europe (Ehlers *et al.* 1995).

The Elsterian

The oldest glaciation that has been proved throughout the North-west European Lowlands is the Elsterian Glaciation (Fig. 3.3). During the advance of the Elsterian ice the drainage system was completely rearranged, rivers that hitherto mostly drained towards the Baltic Sea were partially dammed by the advancing ice sheet and forced to flow west or east. At the southern margin of the glaciated area, in the central German Uplands, deposits of ice-dammed lakes have been found. They are more extensive in eastern than in western Germany. Here, for example, the Elbe River was dammed and a large lake formed south of Dresden, which probably drained westward.

The major ice sheets of the Pleistocene cold periods advanced very rapidly. This is best documented for the Elsterian Glaciation, an average advance velocity of *c.*750 m/year was reconstructed from overridden varved lake deposits by Junge (1998). This is in good agreement with the advance rates postulated by Clayton *et al.* (1985) for the last ice sheet in the Great Lakes region of North America. Whereas such rapid ice advances seem plausible for the Alpine glaciers because of the short distances and the special character of the glaciation (ice stream network), the rapid advances of the north European and North American ice sheets are

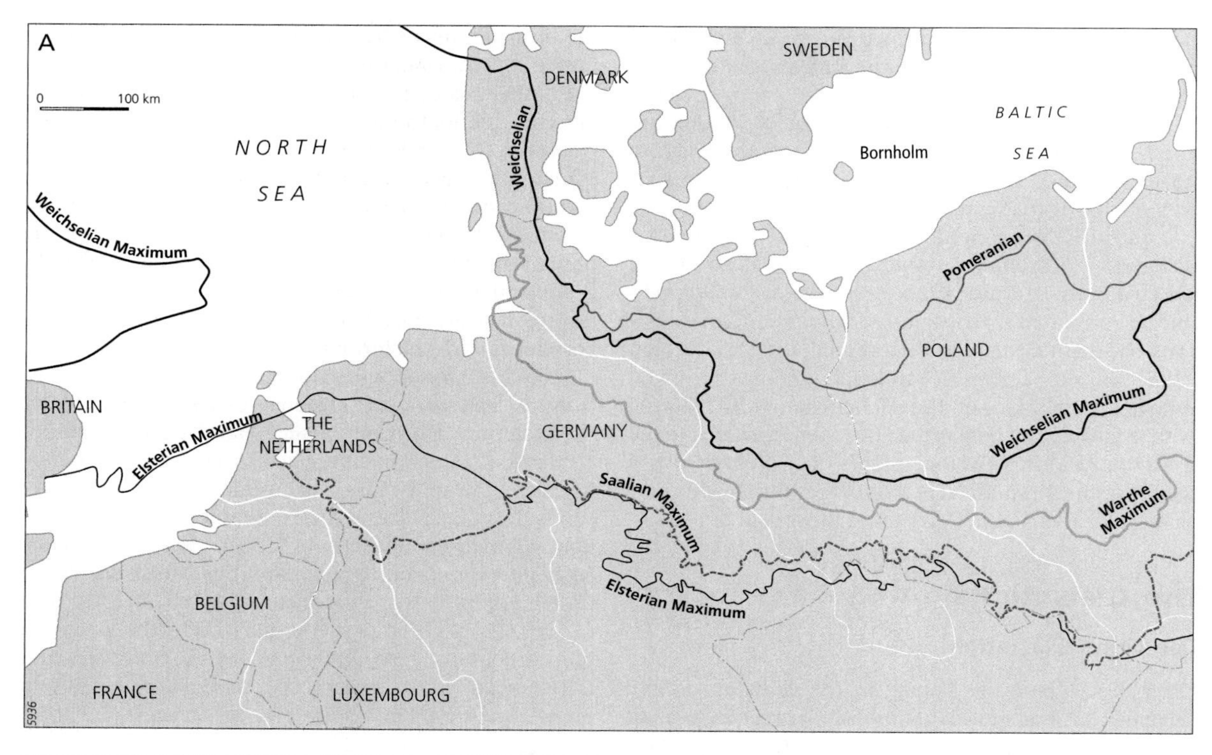

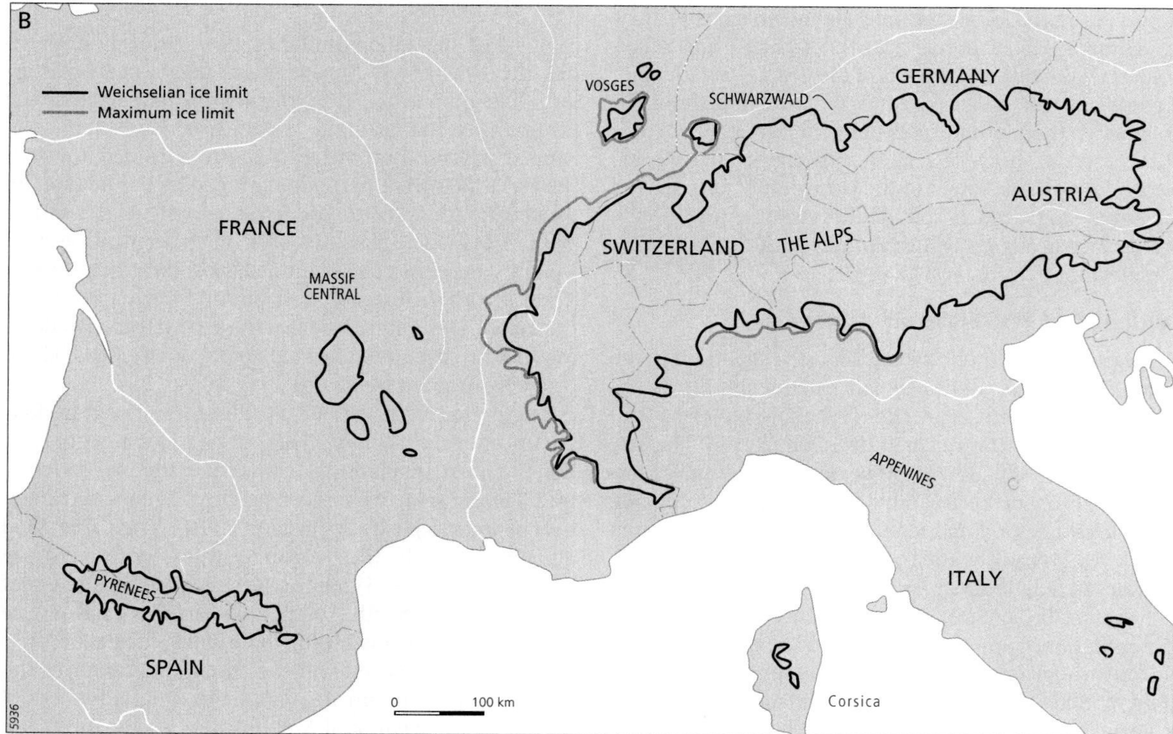

Fig. 3.3. Glacial limits, A. Nordic glaciations; B. Alpine glaciations.

hard to comprehend using traditional explanations. Other possibilities must be taken into consideration.

By investigating glacial processes in a tunnel under an Icelandic glacier, Boulton and Jones (1979) showed that not only the glacier was moving but also the sediment at its base. The high flow rates of certain recent and ancient glaciers had found a simple explanation. The glacier dynamics of the Scandinavian ice sheet suggests that at least temporarily a deformable bed may have played a part in glacier movement (Boulton *et al.* 2001). On the other hand, investigations in the vast opencast lignite mines of Sachsen and Sachsen-Anhalt suggest that basal sliding may have been more important than the 'deformable bed' (Junge and Eissmann 2001).

The extent of the Elsterian glaciation is not clear in all regions. Thrust moraines at the bottom of the North Sea are attributed to the Elsterian, but in The Netherlands, occurrences of Elsterian till are so far restricted to a few thin layers (Zandstra 1983). It must be borne in mind that major parts of the Elsterian tills have been eroded during subsequent cold periods. Further east, a more complete till sheet can be found. For example, tills and meltwater sands of Elsterian age are accessible in several exposures in the area north of Bremen (Höfle 1983).

The southern peripheral zone of the North European ice sheets is characterized by a network-like system of channels (Fig. 3.4), running mostly in a radial pattern from the centre of glaciation to the former ice margin. In contrast to the most recent channels formed during

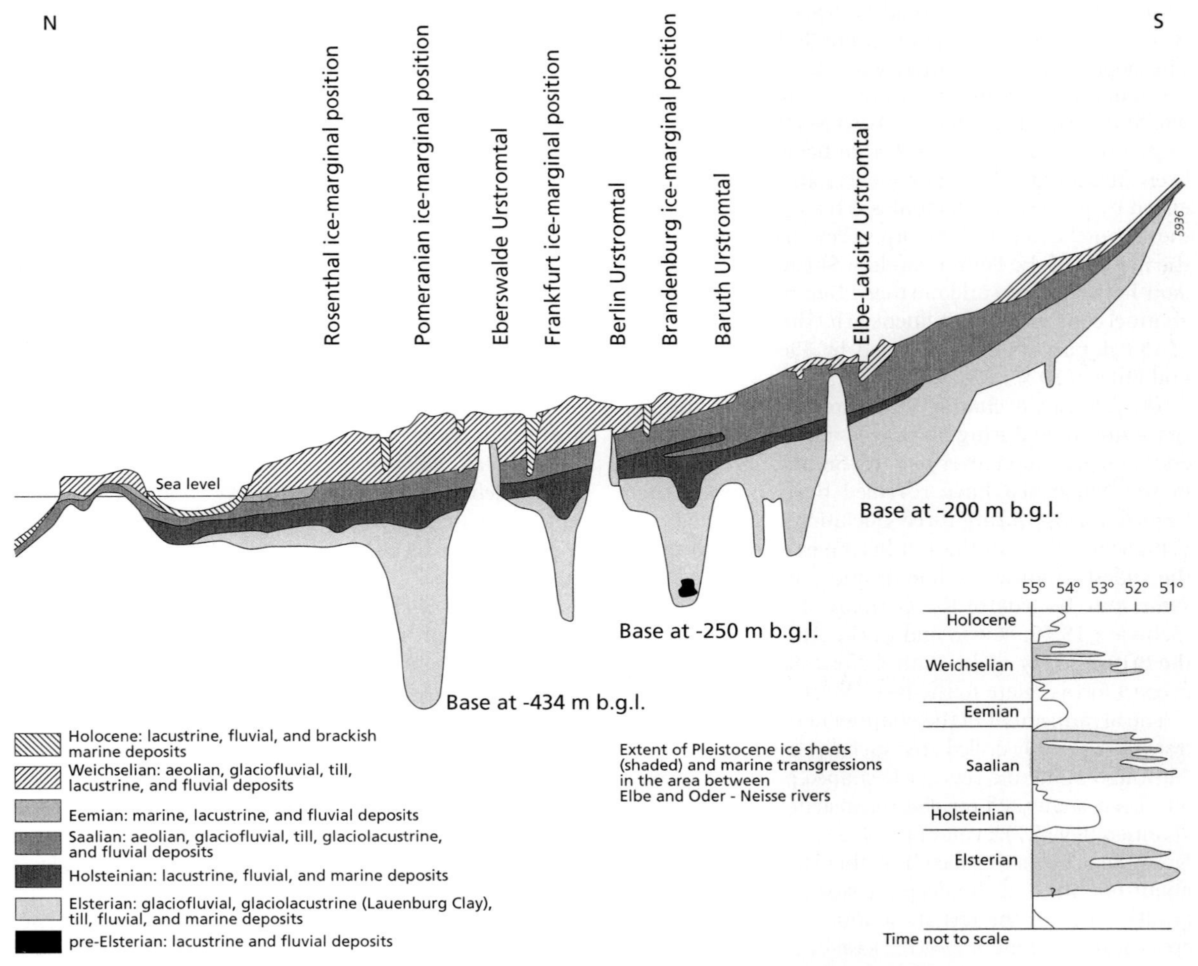

Fig. 3.4. N–S profile through the formerly glaciated part of north-eastern Germany (simplified after Eissmann *et al.* 1995).

the last ice age, which are still clearly visible at the land surface, the Elsterian channels have been completely filled. These forms are often referred to as tunnel valleys, because they were formed by meltwater flowing through tunnels underneath the ice. Tunnel channels might be a more appropriate term, because in contrast to normal valleys their thalweg has no continuous unidirectional gradient.

Modern geological investigation of the tunnel channels in mainland Europe was initiated in the former East Germany (Eissmann 1967). Recently, Eissmann (2002) reviewed the huge body of information on the Elsterian, Saalian, and Weichselian glacial stages as well as the intervening Holsteinian and Eemian sequences, which have been extensively investigated in central Germany. The first detailed investigations of ice age channel patterns were made by Eissmann (1975). In some cases the course of the channels could be traced for many kilometres in the opencast lignite mines in the Leipzig area. The depths of the channels vary greatly. In northern Germany, a maximum depth of 434 m below sea-level was found. Equally, in the southern North Sea, maximum depths of channels of over 400 m below sea-level have been measured. Channel courses can be locally influenced by deep-seated tectonics. This is most obvious in the case of the so-called 'Alnarp Valley', a trough following the margin of the Fennoscandian Shield through Skåne (southern Sweden) and northern Denmark. In addition, channel courses can be influenced by the strike directions of salt domes, as demonstrated by Piotrowski (1994) and Ehlers (1995; Fig. 3.5).

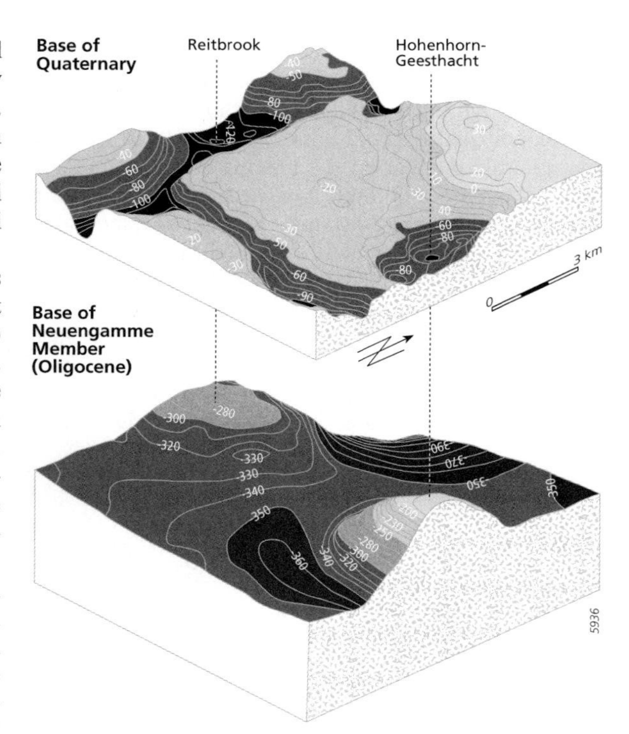

Fig. 3.5. Buried channel on top of the Reitbrook salt dome near Hamburg (after Ehlers 1995).

The intensity of channel formation has varied through space and time during all the ice ages. Yet these differences do not apply everywhere. Seismic investigations in the North Sea have revealed that channels there formed at least during three glaciations (e.g. Ehlers and Wingfield 1991). In the southern part of the North Sea the oldest channel system is most pronounced. The same may be true of the German and Danish sectors (Schwarz 1996; Huuse and Lykke-Andersen 2000). In the British sector of the central North Sea, however, the deepest forms relate to the last, Weichselian Stage.

The arrangement of the channel network is best illustrated in a densely drilled area such as Hamburg (Fig. 3.6). Although on land a regular U-shaped profile has generally been suggested for the reconstruction of Elsterian channels (Fig. 3.7), evaluation of seismic profiles from the North Sea shows this form is rather the exception. Only about one-third of the deep incisions have a simple U-profile, whereas the rest show much more complicated cross-sections (Ehlers 1990). Detailed seismic investigations have shown that channels often consist of a system of parallel incisions. This indicates that the channels were not generated in a single process, but through a sequence of several similar events. Occasional reactivations of old, abandoned channels during later glaciations are known (Piotrowski 1994). The shape of the channels, with their local overdeepenings and irregularities of profile, definitely excludes extra-glacial formation, such as river valleys. In most examples subglacial meltwater erosion was the main force behind the channel formation.

When the Elsterian ice melted, large ice-dammed lakes formed in which silt and clay were deposited. In northern Germany these glaciolacustrine deposits are called *Lauenburger Ton* (Lauenburg Clay). Their distribution reaches west into The Netherlands, where this sediment is referred to as *potklei* (pottery clay) in the provinces of Friesland, Groningen, and Drente. It consists of a complex of glaciolacustrine clays, silts, and fine sands; in the channels it can reach a thickness of over 150 m. The composition of the Lauenburg Clay reflects the progressive decay of the Elsterian ice. Whereas the older layers are rich in dropstones and mixed with sand, the sorting increases towards the top where a sequence of laminated sediments is found. If they represent

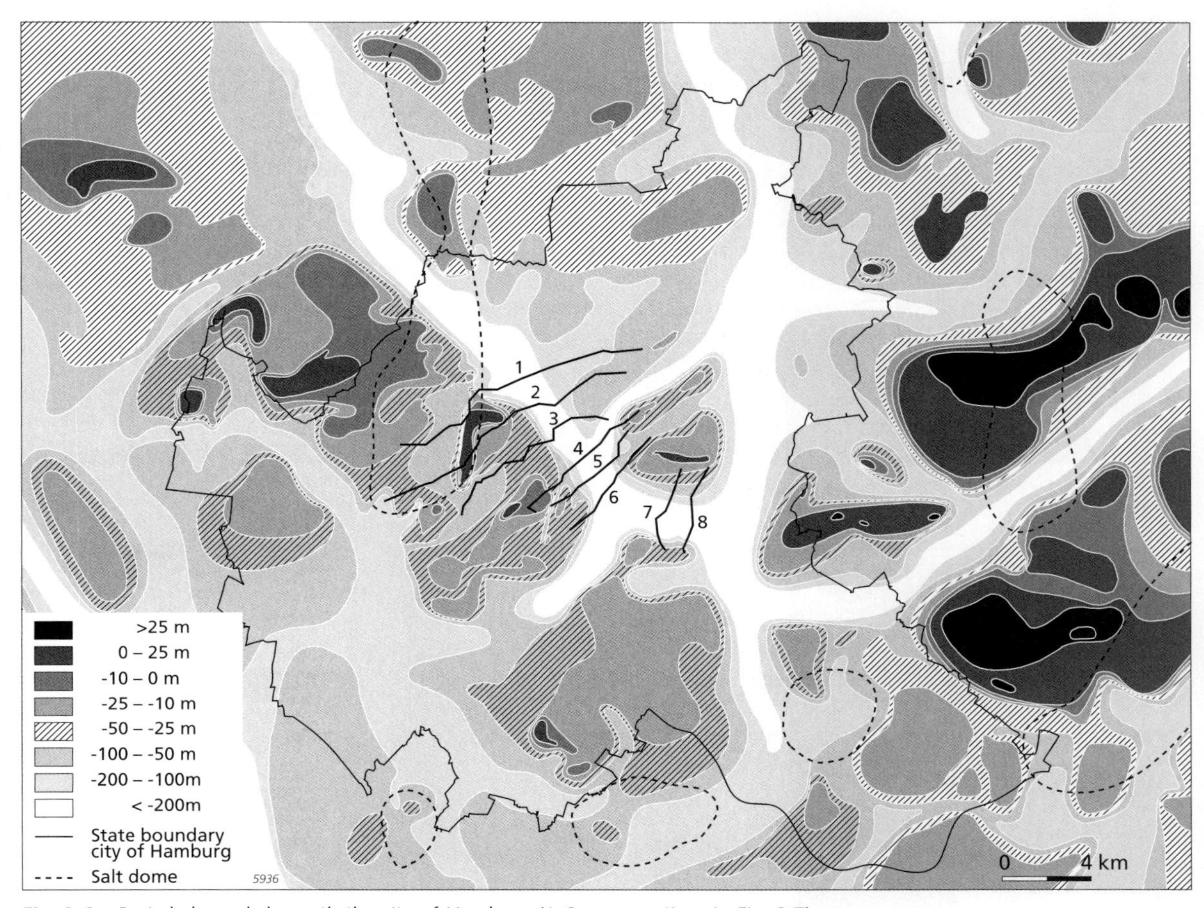

Fig. 3.6. Buried channels beneath the city of Hamburg (1–8 cross-sections in Fig. 3.7).

annual layering, deposition in the Hamburg region seems to have lasted for more than 2,000 years.

Mindel Glaciation

Moraines in southern Germany associated with the 'Jüngere Deckenschotter' of the Mindel valley caused Penck and Bruckner (1901/9) to recognize the second oldest Alpine glaciation as the 'Mindel Cold Stage'. Since then, there has been no end of attempts to reinterpret layers that had originally been classified as Mindel in age. After comprehensive geomorphological mapping Habbe (1986) failed to confirm any of these suggestions. The outcrop conditions do not allow any far-reaching subdivision. The homogeneity of the sediments and the overall palaeogeographical situation speak in favour of a single body of gravel that was formed during one glacial period. However, it is unclear whether or not this can be correlated with the Elsterian of north-west Europe, as suggested in the past.

Holsteinian Interglacial

In the Late Tertiary the sea had withdrawn from the North West European basin, and subsequently fluvial deposition prevailed (Ziegler 1982). Only the region of The Netherlands remained under a strong marine influence well into the Early Pleistocene. It was not before the Holsteinian Warm Stage that north-west Europe experienced another major transgression, that affected the southern North Sea basin.

The early (pre-Weichselian) development of the Baltic Sea is largely unknown. At the time of the Baltic river system (from the Late Tertiary to the Early Pleistocene Waalian Stage) fluvial sediments were transported from Fennoscandian source areas into north Germany and

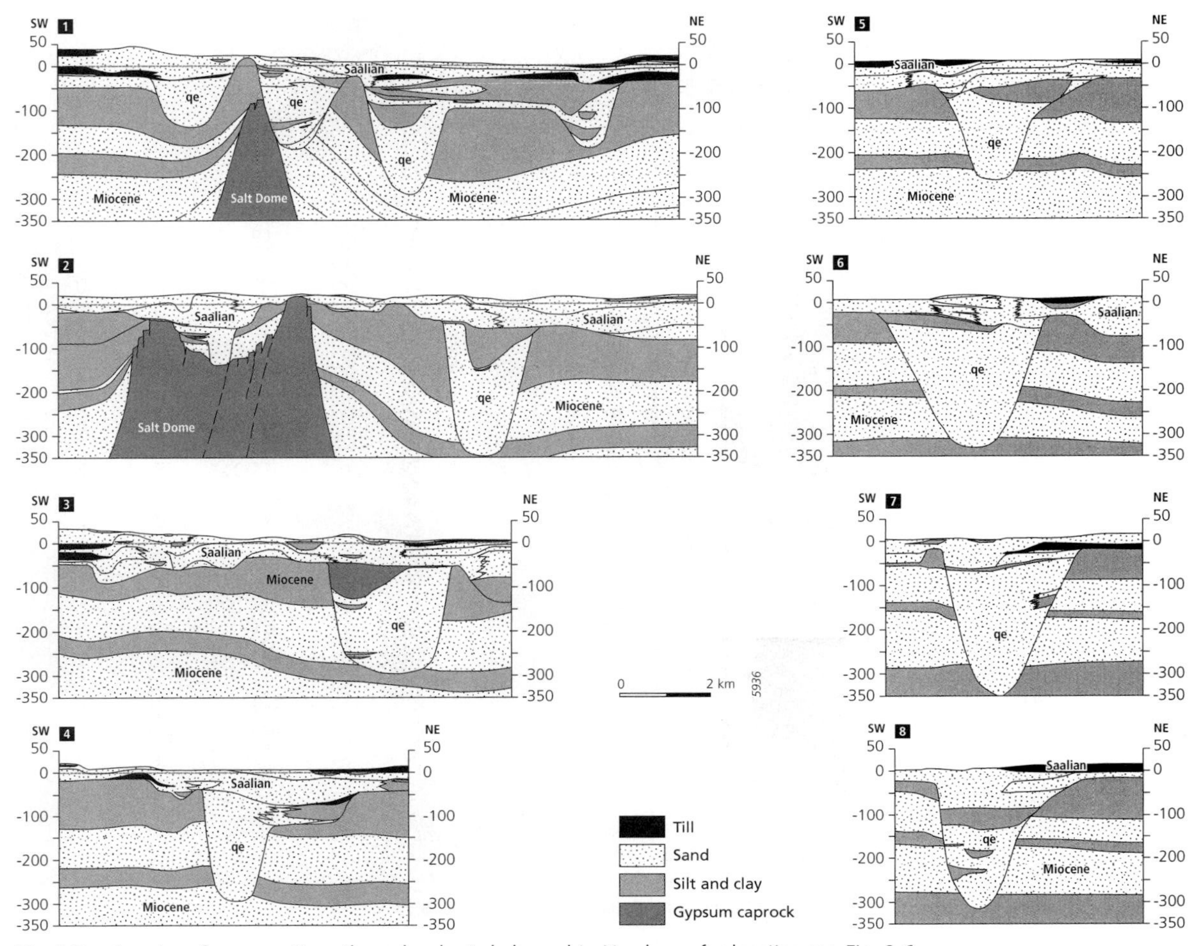

Fig. 3.7. A series of cross-sections through a buried channel in Hamburg; for location see Fig. 3.6.

The Netherlands. These include the Loosen Gravels in Mecklenburg and the Kaolin Sand on the Isle of Sylt. During their deposition the Baltic Sea depression cannot have existed. The further development from the Menapian to the Elsterian Cold Stage is unknown. If and how far earlier glaciations in this period were able to form a proto-Baltic Sea cannot be said at present. There can be no doubt, however, that glacial erosion played a prominent part in forming the Baltic Sea basin, and that the basin has increased in size and depth through the course of the Pleistocene. The relatively low percentage of Cretaceous chalk and flint in the Elsterian tills in north Germany may be an indication that at that time the Tertiary cover overlying the Cretaceous in the western Baltic Sea region was still largely intact (K.-D. Meyer 1991). The first unequivocal evidence of an initial Baltic Sea is found in the Holsteinian Interglacial. A Holsteinian marine transgression into the Baltic Sea area can be followed across Schleswig-Holstein into eastern Mecklenburg-Vorpommern. Several occurrences of marine Holsteinian deposits in western Mecklenburg and north-eastern Niedersachsen seem to indicate a marine connection between the North and Baltic Seas via the Lower Elbe region (Woldstedt and Duphorn 1974).

Beyond the reaches of the marine transgression, the interglacial is represented by fossil soils and lacustrine and organic strata. In particular, the diatomite of the northern Lüneburger Heide area has provided valuable information on the evolution of the interglacial. K.-J. Meyer (1974) proved that the fine rhythmic lamination of the diatomite represents annual layers. By extrapolation from the annually-layered part of the sequence it was

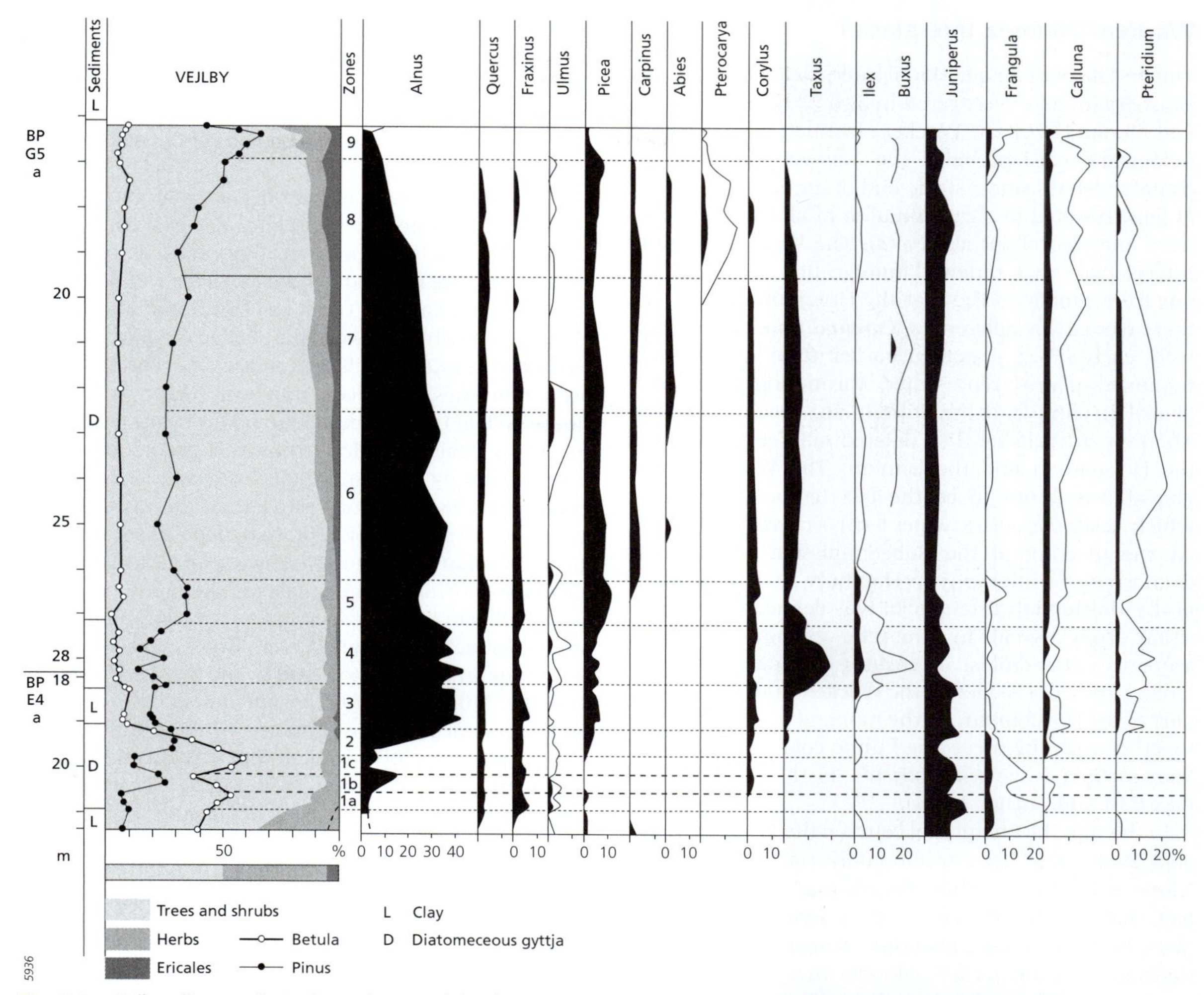

Fig. 3.8. Pollen diagram from the Holsteinian lake deposits at Vejlby, Denmark (after Andersen 1965 in Sjørring 1983).

possible to calculate the overall duration of the interglacial to about 15,000–16,000 years Müller (1974*b*).

The vegetational development of the Holsteinian has been described by numerous authors (e.g. Andersen 1965; Fig. 3.8; Erd 1970; Müller 1974*b*; Reille and de Beaulieu 1995). In contrast to the Eemian (last interglacial), the thermophilous trees in north Germany appeared almost simultaneously during the Holsteinian. An early appearance of fir is characteristic for the north German Holsteinian. Hackberry (*Celtis*) and the water fern (*Azolla*) also occur. Towards the end of the interglacial, during the period of the oak and elm decline, wing nut (*Pterocarya*) occurs regularly; a phenomenon also observed in the Alpine region, in France and in the Hoxnian deposits of Britain, which are the equivalent of the continental Holsteinian.

Holsteinian mammal faunas are relatively rare in central Europe. Van Kolfschoten (1990) has described mammal remains from Neede in The Netherlands. Apart from horse, forest rhinoceros, and red deer (*Cervus elaphus*), mainly rodents such as the red-backed vole (*Clethrionomys*) and the water vole (*Arvicola terrestris cantiana*) have been found. Shrew (*Sorex*) and the extinct vole *Pliomys* are absent. The faunal assemblage indicates an open forest landscape. That the climate must have been relatively warm is supported by finds of water buffalo, forest elephant, aurochs, giant deer, and forest rhinoceros in south-west Germany (Czarnetzki 1983).

Wacken/Dömnitz Interglacial

The first hints of an additional interglacial following the Holsteinian were discovered by Erd (1973) at Dömnitz and Menke (1968) at Wacken. Evaluation of the palynological record provided the following picture. The climatic deterioration at the end of the Holsteinian had at least resulted in an elimination of all thermophilous trees and also of spruce (*Picea*). The Wacken/Dömnitz Interglacial saw a renewed immigration of thermophilous trees similar to those at the Holsteinian. However, there were a few differences. *Carpinus* immigrated relatively early. *Picea* appeared earlier than *Carpinus* but remained poorly represented throughout the interglacial. In contrast to the Holsteinian, *Abies* was lacking. The new interglacial thus differed markedly from both the Holsteinian and the Eemian. The Wacken Interglacial was found to be the last temperate event in which *Azolla filiculoides* (water fern) occurred (Erd 1978). At the meeting of the Subcommission of European Quaternary Stratigraphy in Hamburg in 1986 the end of the Holsteinian Interglacial was defined at the transition from boreal to subarctic climate in specific sequences (boreholes) in eastern Germany (Jerz and Linke 1987). Consequently the Wacken Interglacial is not part of the Holsteinian. At the moment it is combined—together with the preceding Fuhne cold period and the Saalian Cold Stage *sensu stricto*—in the Saalian Complex (Litt and Turner 1993).

In the Alps the interglacial between the Mindel and Riss cold stages has been traditionally referred to as the 'Great Interglacial'. This concept may result from the fact that it actually represents a series of two interglacials. In the research borehole Samerberg 2, near the German-Austrian border, a double interglacial has been discovered under a thin Riss Stage till. The interglacial has been correlated with the north German Holsteinian and Wacken/Dömitz interglacials (Grüger 1983). At Schöningen in Niedersachsen, Urban *et al.* (1991) and Urban (1995), claim to have found an additional 'intra-Saalian' interglacial, post-dating the Wacken/Dömnitz event.

Saalian Cold Stage

The Saalian Cold Stage in Germany is traditionally subdivided into two major ice advances, the Drenthe and the Warthe advance. Whilst on land the glacial limits are relatively well known, the offshore limits of the Saalian ice sheet are still open to question. In the Dutch sector of the North Sea, Saalian till can only be traced some 40 km offshore (Laban 1995). Saalian till is missing in the central and northern North Sea. This either indicates that there was no contact between British and continental ice sheets at the time (e.g. Long *et al.* 1988) or that the till has been removed subsequently. On the basis of ice movement directions in The Netherlands, Rappol *et al.* (1989) concluded that the British and continental ice must have been in contact.

Within the Saalian Complex in north Germany, three till sheets separated by meltwater deposits can be distinguished. As the local stratigraphies still do not match, they are referred to here as 'Older', 'Middle', and 'Younger' Saalian till. Ter Wee (1962) and Maarleveld (1981) originally subdivided the Saalian Glaciation in The Netherlands into five different phases, all represented by push moraines. However, nowhere has more than a single Saalian till sheet been found. This is correlated with the 'Older Saalian' till in Germany. It occurs in a number of strikingly different lithofacies types. Nevertheless, Zandstra (1983) demonstrated that the clast associations involved do not represent deposits of different glaciations but were deposited by a sequence of different ice masses within a single major ice advance.

The Older Saalian ice sheet covered almost all of Niedersachsen. The Weser River was dammed (Winsemann and Asprion 2001), the ice sheet intruded into the Münsterland Bight and advanced up to its southern margin (Klostermann 1992; Skupin *et al.* 1993). It also entered the Leipzig Bight in Saxony (Eissmann 1987; Eissmann *et al.* 1995). These authors showed and discussed a variety of glaciotectonic disturbances, which have been exposed in the huge browncoal quarries in central Germany. Eissmann (1987) has also discussed the formation of glacigenic disturbances using the example of the Schmiedeberg end moraine in eastern Germany. He assumes that the disturbances are mainly caused by an upward-directed vertical sediment injection during the thrusting. On the contrary, van der Wateren (1992) emphasized the horizontal component of glacial thrusting. Field evidence seems to indicate that both processes can play a decisive role depending on local circumstances. Different examples of glaciotectonic deformations are illustrated in Fig. 3.9A, B and C).

The large thrust moraines in the marginal areas of the Nordic glaciations represent major sediment accumulations (Fig. 3.10). As a rule, they must be related to equivalent adjacent zones of glacial erosion ('exaration'). In the Dutch literature the depressions are mostly referred to neutrally as glacial basins (e.g. de Gans *et al.* 1986), a typical example of which is the Amsterdam basin (Fig. 3.11). In the German literature the basins are called *Zungenbecken* (tongue basins). Such basins, particularly in the area of the older glaciations, have

Fig. 3.9. Glaciotectonic deformations shown as A. large-scale folding of clayey and fine sandy ice-pushed deposits in central Germany, the Rettmer brickyard near Lünenburg (photo by Hinrich Prigge, published in 1937 Naturwiss. Monatschrift aus der Heimat 50/2); B. almost vertical thrust planes in a cliff section in north-western Jutland; C. small-scale folding of finely laminated sands in ice-pushed deposits in the Veluwe area (central Netherlands) (photos B and C: Eduard Koster).

Fig. 3.9. (cont.)

subsequently been infilled with sediments and are therefore no longer visible at the surface.

Investigations in the Ülsen and Dammer Berge end moraines in Germany have revealed that partially subhorizontal nappe-like thrust sheets are the dominant features, which in small exposures are difficult to identify as thrust phenomena (van der Wateren 1992; Kluiving 1994). As in orogeny, in glaciotectonics it is not easy to explain the mechanism behind the formation of large nappes moved over long distances. A prerequisite is the formation of a glide plane or zone of *décollement* on which the nappes can move with minimal friction. Fine-grained sediments (silts and clays) found at the base of many thrust zones, are ideal for this purpose. The Dammer Berge and the Ülsen end moraines lie a short distance from the central German uplands. Here Tertiary and older clays come sufficiently close to the land surface to have been susceptible to glaciotectonics (van der Wateren 1992).

The most prominent thrust moraine of the Older Saalian Glaciation is the Rehburg end moraine. Previously interpreted as recessional end moraines (of the so-called 'Rehburger Phase'), its formation during the advance phase and subsequent overriding by the ice is documented by the occurrence of the same sandy basal till both in the foreland and on top of the thrust ridges (K.-D. Meyer 1987). Later, ice masses from the eastern part of the Baltic Sea overrode the area and deposited a red-brown till in north-west Germany and The Netherlands. This till is characterized by east Baltic indicators, i.e. rich in Palaeozoic limestone, with little or no flint and usually some dolostones.

The warmer intervals between the individual ice advances of the Saalian in north Germany and adjoining areas are characterized by a lack of organic deposits. There are strong indications that in protected positions dead ice from the Older Saalian advance even survived until the very end of the Saalian Stage (e.g. Wansa and Wimmer 1990). This clearly speaks against the existence of any additional temperate stages intervening between the Older Saalian and the Eemian interglacial.

The subsequent 'Middle Saalian' ice advance began in north Germany with the deposition of meltwater sands. Extensive meltwater deposits, often in the form of outwash plains or sandrs, kames, kame terraces, and other slope deposits cover large areas in northern Germany and adjacent countries. Sedimentary structures sometimes reveal transport during cold-climate conditions (Fig. 3.12A, B). In contrast to the glaciofluvial deposits of the Elsterian, which are largely concentrated in buried channels in north Germany, the meltwaters of the Middle Saalian accumulated vast outwash fans (Fig. 3.13), the deposits of which can be several tens of metres thick. The vast sand bodies of the Lüneburger Heide largely formed in this period. The Warthe ice sheet terminated more than 100 km behind the Drenthe maximum in western Europe. During the maximum phase of this advance, when the ice lay south of the

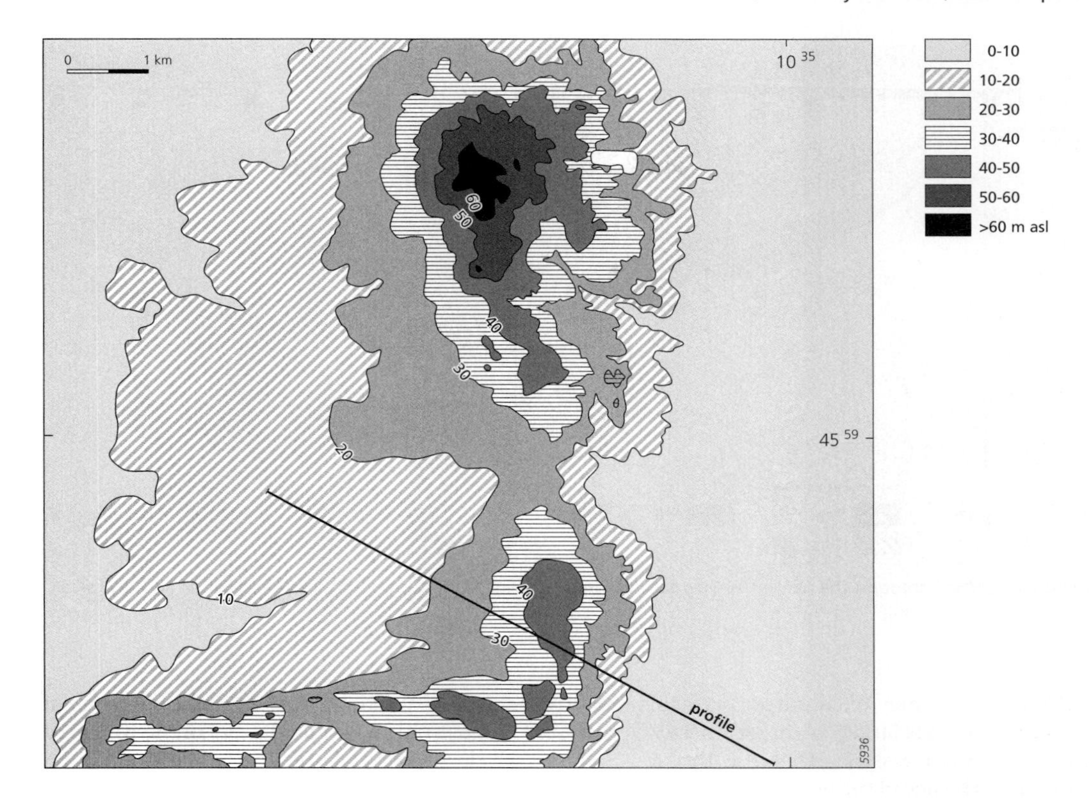

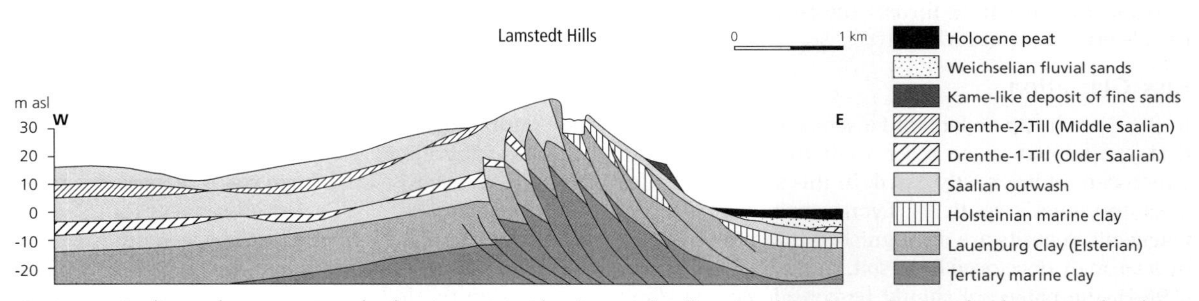

Fig. 3.10. Outline and cross-section of a thrust morainic ridge (Lamstedt Hills, west of Hamburg) showing the shearing and upthrust of older strata by the advancing Middle Saalian ice sheet. To the west a sequence of tills and meltwater deposits occurs (after Höfle and Lade 1983).

present Elbe valley, the meltwater drained southwards and then via the Aller-Weser ice-marginal valley (*Urstromtal*) towards the North Sea (K.-D. Meyer 1983). It is thought that the Middle Saalian ice advanced at least as far west as the Altenwalder Geest hills south of Cuxhaven (Höfle and Lade 1983; van Gijssel 1987). After the Middle Saalian glaciation the ice margin retreated far to the north-east. Parts of the ice sheet stagnated and melted *in situ*. The active ice margin was probably situated in the present Baltic Sea area. In the proglacial zone, widespread dead-ice masses were covered by outwash during the subsequent readvance.

A line of marked end moraines in the northern Lüneburger Heide area has traditionally been associated with the Younger Saalian 'Warthe' ice advance. However, more recent investigations show that many assumed Younger Saalian end moraines in Niedersachsen really have a much older core (Höfle 1991). The

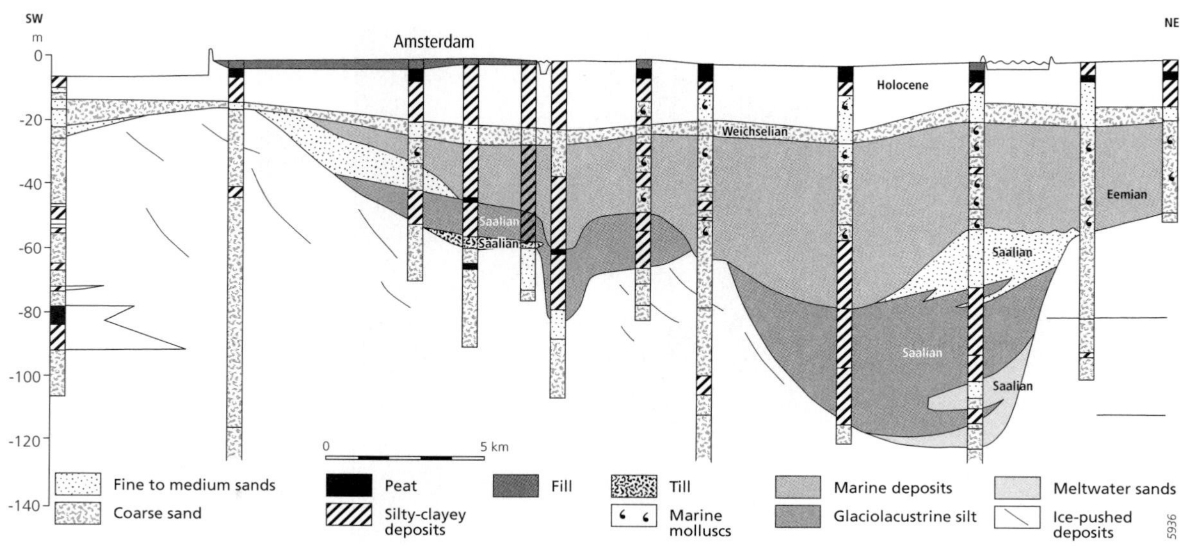

Fig. 3.11. Longitudinal section through the Saalian Amsterdam glacigenic basin, × 100 vertical exaggeration (after de Gans *et al.* 1986).

Garlstorfer Wald–Toppenstedter Wald ridge south of Hamburg, for instance, consists largely of thrust Tertiary strata. The internal structure shows that the thrust occurred from the east. The age of the push moraine can be deduced from the stratigraphy, as Elsterian deposits were incorporated in the thrusts, the ridge must post-date the Elsterian Glaciation.

Riss Glaciation

In the French Alpine foreland a series of moraines exist that are known to be older than the Würmian but which cannot be exactly dated. In this zone of so-called 'Riss moraines', south of Lyons, two complexes of glacigenic deposits are distinguished that are separated by a strongly developed fossil soil. According to Mandier (1984), the palaeosol should be considered the result of an interglacial, which might be the Mindel/Riss Interglacial (~Holsteinian).

In southern Germany conceptions of what was to be defined as the 'Riss Cold Stage' have undergone a marked change through the course of time. When they originally established their Alpine Quaternary chronology, Penck and Brückner (1901/9) assumed that their four cold stages represented single, uniform glacial events. Eberl (1930) later subdivided a Riss I, II, and III Glaciation, which he interpreted as individual glacier advances within one and the same glacial period. Schaefer (1973), on the other hand, assumed that the original Rissian Stage had been bipartite. He introduced an independent older, 'Paar Cold Stage' from the original Riss Cold Stage. In his opinion, glaciers of the Riss Cold Stage overrode deposits of the Paar Cold Stage, and locally advanced furthest north in southern Germany. Conversely, Schreiber and Müller (1991) came to the conclusion that it had not been the penultimate (Riss) Glaciation that had advanced furthest into the Alpine foreland, but an older ice advance that they assigned provisionally to Schaefer's Paar Cold Stage. Lack of precise dating makes it impossible currently to clarify the chronostratigraphic situation. It cannot be excluded that the classic Riss Cold Stage really comprises a number of glaciations.

Neither the exact extent of the glaciers, nor the number of ice advances or their age, can yet be determined with certainty. Under these conditions a direct correlation of the classic Riss or parts of it with the Saalian Glaciation of northern Europe is not yet possible.

Eemian Interglacial

The Eemian or last interglacial has been identified at numerous sites throughout western Europe. Its duration is usually estimated at about 11,000 years (Müller 1974*a*), but there is still argument about the precise ages, within 2,000–3,000 years, of the upper and lower boundaries of MIS 5e (Turner 2000). Its vegetational development can be summarized as beginning with an early phase characterized by reimmigration of birch and pine (zones I and II) (Andersen 1965; Fig. 3.14;

Fig. 3.12. Coarse-grained fluvioglacial sands: A. containing blocks of finely laminated sands which must have been transported downslope in a frozen condition (location Veluwe, central Netherlands); B. with fine-grained channel fills (southern Germany) (photos: Eduard Koster).

Litt 1994). In contrast to the Holocene this was followed—prior to the appearance of hazel (*Corylus*)—by immigration of oak (*Quercus*) and elm (*Ulmus*) (zone III). This light pine-mixed-oak forest was rapidly joined by hazel. Hazel became the dominant tree type even in the upland sites (oak-hazel period, zone IVa, and hazel-*Taxus*-lime period, zone IVb). A characteristic of the Eemian vegetational development is also the strong *Carpinus* phase, succeeded by an equally strong spruce (*Picea*) peak. Subsequently fir (*Abies*) and pine (*Pinus*) immigrated (pine-spruce-fir period, zone VI). Finally, pine became dominant (pine period, zone VII), until towards the end of the interglacial birch began to re-occupy the region once again.

The vegetational history of the Eemian is known in such detail that Menke (1984) could undertake a close comparison with the Holocene for Schleswig-Holstein. Both temperate stages show strong resemblances. Thus in both interglacials the mistletoe (*Viscum*) immigrated about 2,000–5,000 years after the beginning of the interglacial and disappeared about 7,000–8,000 years after afforestation from the landscape of northern

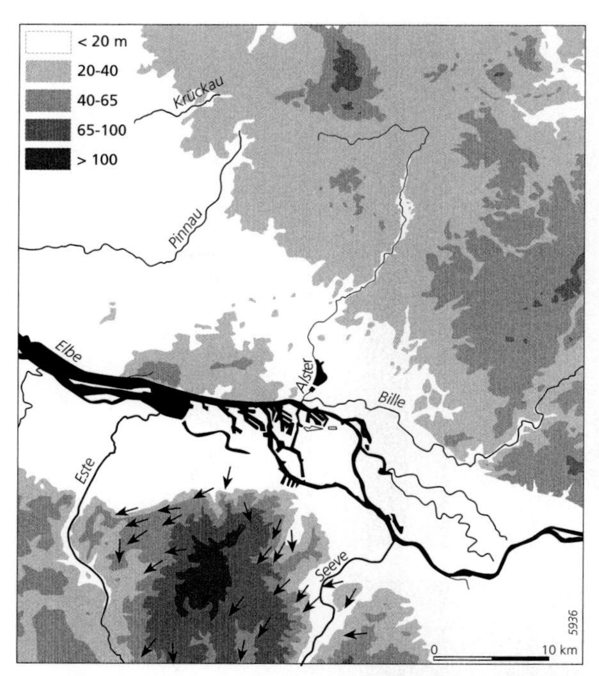

Fig. 3.13. Harburger Berge, south of Hamburg; Middle Saalian drainage directions based on palaeocurrent measurements. When the Elbe valley was blocked by the Middle Saalian ice, a large outwash fan accumulated in the Harburger Berge area. Drainage was directed south-westwards towards the Weser valley (after Ehlers 1990).

Holstein. Mistletoe is an indicator of warm summers. During their climatic optima both interglacials had July temperatures about 2° warmer than at present (Menke 1984).

In contrast, the Eemian mammalian fauna of western and central Europe differed markedly from that of the Holocene. *Hippopotamus amphibius incognitus* had spread into central England, and the Upper Rhine Graben was inhabited by water buffalo (*Bubalus murrensis*). Forest elephant and forest rhinoceros, as well as giant deer (*Megaloceros giganteus*), occurred together with deer, roe, dama, and boar (von Koenigswald 1988; van Kolfschoten 2000). Eemian faunal assemblages within a firm stratigraphical record have mainly been collected from sites in eastern Germany, whereas most of the so-called Eemian fossils in the Rhine valley have been dredged up from the river.

The best-investigated pollen section of the Alpine region comprising the last interglacial and three Early Würmian age interstadials, however, is the Samerberg profile in Bavaria, south-east of Rosenheim (Grüger 1979). A comparable Riss/Würm Interglacial site has been identified in the Wurzacher Becken, a tongue basin of the Rissian Stage Rhine Glacier. The vegetational succession corresponds largely with that of the north German Eemian. The montane character of the site is expressed through high values of pine pollen. *Picea* appears early and soon reaches a dominance among the tree pollen. After a short, pronounced *Taxus* maximum, the immigration of *Abies* and *Carpinus* follows. At the end of the interglacial all three species vanish and are replaced by *Pinus*. In the lower-altitude Riss/Würm Interglacial sequence of Zeifen, the montane species *Picea* and *Abies* appear in smaller numbers and *Carpinus* dominates persistently (Grüger 1989).

The climatic development of the Eemian very much resembled that of the Holocene. Overall, however, the Eemian seems to have been slightly warmer. Therefore it is no surprise that the Eemian sea-level globally rose slightly higher than that of the Holocene.

Sea-level depends to a large degree on how much water is bound up worldwide in glacier ice. The formation of extensive terrestrial ice sheets resulted in a eustatic lowering of sea-level. Recent evidence suggests that for the last glacial maximum a lowering of global sea-level for about 130 m must be assumed (Shackleton 1987). Simultaneously, the loading of glaciated areas by ice up to several thousand metres thick results in a gradual sinking of the earth's crust. The balance between the loaded, sinking part of the crust and its surroundings occurs in the asthenosphere. In Scandinavia and North America postglacial uplift has been reconstructed by evaluation of raised shorelines. This method, however, can only provide information about the period after the onset of the postglacial transgression, not about the period when the area was still ice-covered. More recent investigations in Scandinavia have shown that about 10,300 BP central Scandinavia was depressed by about 450 m (Svendsen and Mangerud 1987). By extrapolating this value back into the period of maximum depression about 13,000 BP considerably higher values have been estimated. Mörner (1980) considers the maximum depression of Scandinavia to be about 800 m.

Whereas glaciated areas were depressed under the overburden of ice, the adjoining unglaciated areas were simultaneously slightly uplifted. That the build-up of the north European ice sheet resulted in the formation of such a marginal forebulge is indicated by investigations in western Norway. Here, Svendsen and Mangerud (1987) demonstrated a postglacial downwarping of about 20 m, which is interpreted as forebulge collapse after Weichselian uplift. Boulton (1990) and Fjeldskaar *et al.* (2000) consider maximum forebulges of about

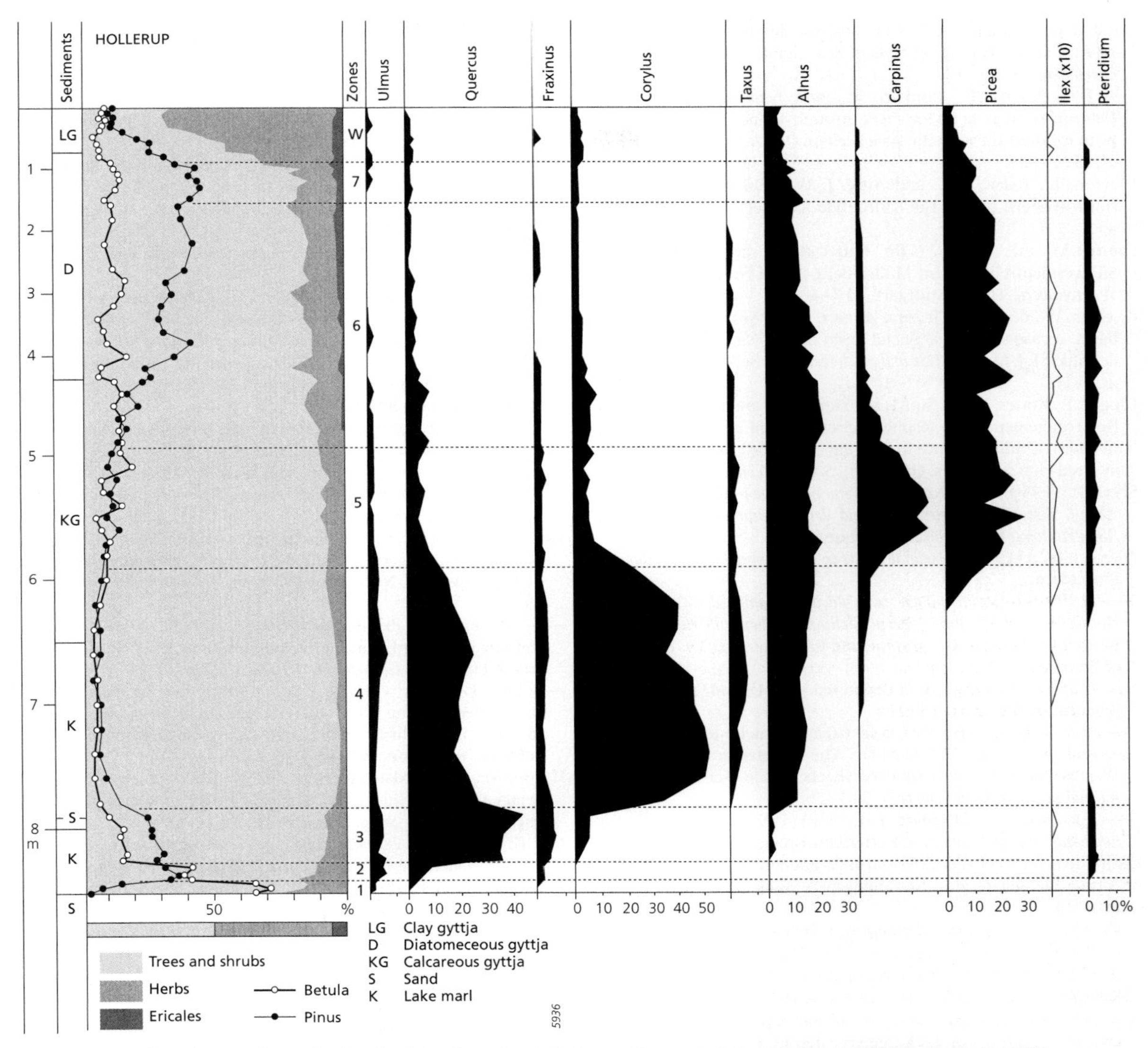

Fig. 3.14. Pollen diagram from the Eemian lake deposits at Hollerup, Denmark (after Andersen 1965 in Sjørring 1983).

30 m for the Weichselian. However, as the ice-load history and the viscosity-depth structure of the Earth's mantle are not sufficiently well known, geophysical modelling of isostasy and related sea-level changes is still in its infancy. Isostatic deformation of the earth's crust must have influenced ice movement and the development of drainage patterns during all major glaciations. Because of the lack of shoreline evidence, the extent of isostatic movements in continental areas is very difficult to assess. It would be a mistake, however, to assume that no such adjustments have taken place.

References

Andersen, S. Th. (1965), Interglacialer og interstadialer i Danmarks kvartær. *Meddelelser fra Dansk Geologisk Forening* 15: 486–506.

—— (1967), Istider og mellemistider. In: A. Nørrevang and T. J. Meyer (eds.) *Danmarks Natur*, i. 199–250.

Boenigk, W. (1982), Der Einfluß des Rheingraben-Systems auf die Flußgeschichte des Rheins. *Zeitschrift für Geomorphologie* 42: 167–75.

Boulton, G. S. (1990), Sedimentary and sea level changes during glacial cycles and their control on glacimarine facies architecture. In: J. A. Dowdeswell and J. D. Scourse (eds.) *Glacimarine Environments, Processes and Sediments*. Geological Society Special Publication 53: 15–52.

Boulton, G. S., and Jones, A. S. (1979), Stability of temperate ice caps and ice sheets resting on beds of deformable sediment. *Journal of Glaciology* 24: 29–43.

—— Dongelmans, P., Punkari, M., and Broadgate, M. (2001), Palaeoglaciology of an ice sheet through a glacial cycle, the European ice sheet through the Weichselian. *Quaternary Science Reviews* 20: 591–625.

Clayton, L., Teller, J. T., and Attig, J. W. (1985), Surging of the south-western part of the Laurentide Ice Sheet. *Boreas* 14: 235–41.

Czarnetzki, A. (1983), Zur Entwicklung des Menschen in Südwestdeutschland. In: Müller-Beck (ed.) *Urgeschichte in Baden Württemberg*. Theiss, Stuttgart, 217–40.

de Gans, W., de Groot, Th., and Zwaan, H. (1986), The Amsterdam basin, a case study of a glacial basin in The Netherlands. In: van der Meer, J. J. M. (ed.) *Tills and glaciotectonics*. Balkema, Rotterdam, 205–16.

Ding, Z. L., Rutter, N. W., Sun, J. M., Yang, S. L., and Liu, T. S. (2000), Re-arrangement of atmospheric circulation at about 2.6 Ma over northern China: evidence from grain size records of loess-palaeosol and red clay sequences. *Quaternary Science Reviews* 19: 547–58.

Eberl, B. (1930), *Die Eiszeitenfolge im nördlichen Alpenvorlande. Ihr Ablauf, ihre Chronologie auf Grund der Aufnahmen des Lech- und Illergletschers*. Benno Filser, Augsburg.

Ehlers, J. (ed.) (1983), *Glacial deposits in north-west Europe*. Balkema, Rotterdam.

—— (1990), *Untersuchungen zur Morphodynamik der Vereisungen Norddeutschlands unter Berücksichtigung benachbarter Gebiete*. Bremer Beiträge zur Geographie und Raumplanung 19, University of Bremen.

—— (1995), Hamburg. In: L. Benda (ed.) *Das Quartär Deutschlands*. Bornträger, Stuttgart, 14–22.

—— (1996), *Quaternary and Glacial Geology*. Chichester, Wiley.

—— and Wingfield, R. (1991), The extension of the Late Weichselian/Late Devensian ice sheets in the North Sea Basin. *Journal of Quaternary Science* 6: 313–26.

—— Kozarski, S., and Gibbard, P. L. (eds.) (1995), *Glacial deposits in north-east Europe*. Balkema, Rotterdam/Brookfield, 439–64.

Eissmann, L. (1967), Glaziäre Destruktionszonen (Rinnen, Becken) im Altmoränengebiet des Norddeutschen Tieflandes. *Geologie* 16: 804–33.

—— (1975), Das Quartär der Leipziger Tieflandsbucht und angrenzender Gebiete um Saale und Elbe. Modell einer Landschaftsentwicklung am Rand der europäischen Kontinentalvereisung. *Schriftenreihe für Geologische Wissenschaften* 2: tables.

—— (1987), Lagerungsstörungen im Lockergebirge. Exogene und endogene Tektonik im Lockergebirge des nördlichen Mitteleuropa. *Geophysik und Geologie, Geophysikalische Veröffentlichungen der Karl-Marx-Universität Leipzig III* (4): 7–77.

—— (2002), Quaternary geology of eastern Germany (Saxony, Saxon-Anhalt, South Brandenburg, Thüringia), type area of the Elsterian and Saalian Stages in Europe. *Quaternary Science Reviews* 21: 1275–346.

—— Litt, Th., and Wansa, S. (1995), Elsterian and Saalian deposits in their type area in central Germany. In: J. Ehlers, S. Kozarski, and P. L. Gibbard (eds.) *Glacial deposits in north-east Europe*. Balkema, Rotterdam/Brookfield, 439–64.

Erd, K. (1970), Pollenanalytical classification of the Middle Pleistocene in the German Democratic Republic, *Palaeogeography, Palaeoclimatology, Palaeoecology* 8: 129–45.

—— (1973), Vegetationsentwicklung und Biostratigraphie der Dömnitz-Warmzeit (Fuhne/Saale 1) im Profil von Pritzwalk/Prignitz. *Abhandlungen aus dem Zentralen Geologischen Institut* 18: 9–48.

—— (1978), Pollenstratigraphie im Gebiet der skandinavischen Vereisungen. *Schriftenreihe für Geologische Wissenschaften* 9: 99–119.

Fjeldskaar, W., Lindholm, C., Dehls, J. F., and Fjeldskaar, I. (2000), Postglacial uplift, neotectonics and seismicity in Fennoscandia. *Quaternary Science Reviews* 19: 1413–22.

Gibbard, P. L., West, R. G., Zagwijn, W. H., Balson, P. S., Burger, A. W., Funnell, B. M., Jeffery, D. H., de Jong, J., van Kolfschoten, T., Lister, A. M., Meijer, T., Norton, P. E. P., Preece, R. C., Rose, J., Stuart, A. J., Whiteman, C. A., and Zalasiewicz, J. A. (1991), Early and Middle Pleistocene correlations in the Southern North Sea Basin. *Quaternary Science Reviews* 10: 23–52.

Grüger, E. (1979), Spätriß, Riß/Würm und Frühwürm am Samerberg in Oberbayern—ein vegetationsgeschichtlicher Beitrag zur Gliederung des Jungpleistozäns. *Geologica Bavarica* 80: 5–64.

—— (1983), Untersuchungen zur Gliederung und Vegetationsgeschichte des Mittelpleistozäns am Samerberg in Oberbayern. *Geologica Bavarica* 84: 21–40.

—— (1989), Palynostratigraphy of the last interglacial/glacial cycle in Germany. *Quaternary International* 3/4: 69–70.

Habbe, K. A. (1986), *Zur geomorphologischen Kartierung von Blatt Grönenbach (I)—Probleme, Beobachtungen, Schlußfolgerungen*. Erlanger Geographische Arbeiten 47.

Höfle, H.-Ch. (1983), Strukturmessungen und Geschiebeanalysen an eiszeitlichen Ablagerungen auf der Osterholz-Scharmbecker Geest. *Abhandlungen des Naturwissenschaftlichen Vereins zu Bremen* 40: 39–53.

—— (1991), Über die innere Struktur und die stratigraphische Stellung mehrerer Endmoränenwälle im Bereich der Nordheide bis östlich Lüneburg. *Geologisches Jahrbuch A* 126: 151–69.

—— and Lade, U. (1983), The stratigraphic position of the Lamstedter Moraine within the Younger Drenthe substage (Middle Saalian). In: J. Ehlers (ed.) *Glacial deposits in north-west Europe*. Balkema, Rotterdam, 343–6.

Huuse, M., and Lykke-Andersen, H. (2000), Overdeepened Quaternary valleys in the eastern Danish North Sea, morphology and origin. *Quaternary Science Reviews* 19: 1233–53.

Jerz, H. (1993), Das Eiszeitalter in Bayern—Erdgeschichte, Gesteine, Wasser, Boden. *Geologie von Bayern*, ii. Schweizerbart, Stuttgart.

—— and Linke, G. (1987), Arbeitsergebnisse der Subkommission für Europäische Quartärstratigraphie, Typusregion des Holstein-Interglazials (Berichte der SEQS 8). *Eiszeitalter und Gegenwart* 37: 145–8.

Jessen, K., and Milthers, V. (1928), Stratigraphical and Paleontological Studies of Interglacial Fresh-Water Deposits in Jutland and Northwest Germany. *Danmarks Geologiske Undersøgelse*, 2. Ser. 48 + Atlas.

Junge, F. W. (1998), *Die Bändertone Mitteldeutschlands und angrenzender Gebiete*. Altenburger Naturwissenschaftliche Forschungen 9. Mauritianum, Altenburg.

—— and Eissmann, L. (2001), Postsedimentäre Deformationsbilder in mitteldeutschen Vorstoßbändertonen—Hinweise auf den Bewegungsmechanismus des quartären Inlandeises, *Brandenburgische Geowissenschaftliche Beiträge* 7 (1/2): 21–8.

Klostermann, J. (1992), *Das Quartär der Niederrheinischen Bucht. Ablagerungen der letzten Eiszeit am Niederrhein*. Geologisches Landesamt Nordrhein-Westfalen, Krefeld.

Kluiving, S. J. (1994), Glaciotectonics in the Itterbeck-Uelsen push moraines, Germany, *Journal of Quaternary Science* 9: 235–44.

Laban, C. (1995), *The Pleistocene glaciations in the Dutch sector of the North Sea. A synthesis of sedimentary and seismic data*. Ph.D. Thesis, University of Amsterdam.

Litt, Th. (1994), *Paläoökologie, Paläobotanik und Stratigraphie des Jungquartärs im nordmitteleuropäischen Tiefland unter besonderer*

Berücksichtigung des Elbe-Saale-Gebietes. Dissertationes Botanicae 227. Bornträger, Stuttgart.
—— and Turner, Ch. (1993), Arbeitsergebnisse der Subkommission für Europäische Quartärstratigraphie, Die Saalesequenz in der Typusregion. *Eiszeitalter und Gegenwart* 43: 125–8.
Long, D., Laban, C., Streif, H., Cameron, T. D. J., and Schüttenhelm, R. T. E. (1988), The sedimentary record of climatic variation in the southern North Sea. *Philosophical Transactions of the Royal Society of London B* 318: 523–37.
Lüttig, G., and Maarleveld, G. C. (1961), Nordische Geschiebe in Ablagerungen prä Holstein in den Niederlanden (Komplex von Hattem). *Geologie en Mijnbouw* 40: 163–74.
Maarleveld, G. C. (1981), The sequence of ice-pushing in the Central Netherlands. *Mededelingen Rijks Geologische Dienst* 34/1: 2–6.
Mandier, P. (1984), Le Relief de la moyenne vallée du Rhône au Tertiaire et au Quaternaire, 3 vols. University of Lyons, ii.
Menke, B. (1968), Beiträge zur Biostratigraphie des Mittelpleistozäns in Norddeutschland. *Meyniana* 18: 35–42.
—— (1975), Vegetationsgeschichte und Florenstratigraphie Nordwestdeutschlands im Pliozän und Frühquartär. Mit einem Beitrag zur Biostratigraphie des Weichsel-Frühglazials. *Geologisches Jahrbuch A* 26: 3–151.
—— (1984), Wie stabil ist das Ökosystem 'Wald'? *Allgemeine Forst Zeitschrift 1984 (6)*, 122–6.
Meyer, K.-D. (1983), Zur Anlage der Urstromtäler in Niedersachsen. *Zeitschrift für Geomorphologie* 27: 147–60.
—— (1987), Ground and end moraines in Lower Saxony. In: J. J. M. van der Meer (ed.) *Tills and Glaciotectonics*. Balkema, Rotterdam, 197–204.
—— (1991), Zur Entstehung der westlichen Ostsee. *Geologisches Jahrbuch A* 127: 429–46.
Meyer, K.-J. (1974), Pollenanalytische Untersuchungen und Jahresschichtenzählungen an der holstein-zeitlichen Kieselgur von Hetendorf. *Geologisches Jahrbuch A* 21: 87–105.
Mörner, N.-A. (1980), The Fennoscandian Uplift, Geological data and their geodynamical implication. In: N.-A. Mörner (ed.) *Earth rheology, isostasy and eustasy*. Wiley, Chichester, 251–84.
Müller, H. (1974*a*), Pollenanalytische Untersuchungen und Jahresschichtenzählungen an der eemzeitlichen Kieselgur von Bispingen/Luhe. *Geologisches Jahrbuch A* 21: 149–69.
—— (1974*b*), Pollenanalytische Untersuchungen und Jahresschichtenzählungen an der holsteinzeitlichen Kieselgur von Munster-Breloh. *Geologisches Jahrbuch A* 21: 107–40.
—— (1992), Climate changes during and at the end of the interglacials of the Cromerian Complex. In: G. H. Kukla and E. Went (eds.) *Start of a glacial*. Springer, Berlin, 51–69.
Penck, A., and Brückner, E. (1901/9), *Die Alpen im Eiszeitalter*, 3 vols. Tauchnitz, Leipzig.
Piotrowski, J. A. (1994), Tunnel-valley formation in north-west Germany—geology, mechanisms of formation and subglacial bed conditions for the Bornhöved tunnel valley. *Sedimentary Geology* 89: 107–41.
Rappol, M., Haldorsen, S., Jörgensen, P., van der Meer, J. J. M., and Stoltenberg, H. M. P. (1989), Composition and origin of petrographically-stratified thick till in the northern Netherlands and a Saalian glaciation model for the North Sea basin. *Mededelingen van de Werkgroep voor Tertiaire en Kwartaire Geologie* 26: 31–64.
Reille, M., and de Beaulieu, J.-L. (1995), Long Pleistocene Pollen Records from the Praclaux Crater, South-Central France. *Quaternary Research* 44: 205–15.
Schaefer, I. (1953), Die donaueiszeitlichen Ablagerungen an Lech und Wertach. *Geologica Bavarica* 19: 13–64.
—— (1973), Das Grönenbacher Feld. Ein Beispiel für Wandel und Fortschritt der Eiszeitforschung seit Albrecht Penck. *Eiszeitalter und Gegenwart* 23/4: 168–200.
Schreiber, U., and Müller, D. (1991), Mittel- und jungpleistozäne Ablagerungen zwischen Landsberg und Augsburg (Lech). *Sonderveröffentlichungen Geologisches Institut der Universität zu Köln* 82: 265–82.
Schreiner, A. (1992), *Einführung in die Quartärgeologie*. Schweizerbart, Stuttgart.
Schwarz, C. (1996), Neue Befunde zur Verbreitung und Dimension pleistozäner Rinnensysteme auf dem deutschen Nordseeschelf. *Geologisches Jahrbuch A* 146: 233–44.
Sejrup, H. P., Larsen, E., Landvik, J., King, E. L., Haflidason, H., and Nesje, A. (2000), Quaternary glaciations in southern Fennoscandia, evidence from southwestern Norway and the northern North Sea region. *Quaternary Science Reviews* 19: 667–85.
Shackleton, N. J. (1987), Oxygen isotopes, ice volume and sea level. *Quaternary Science Reviews* 6: 183–90.
—— Berger, A., and Peltier, W. A. (1990), An alternative astronomical calibration of the lower Pleistocene timescale based on ODP Site 677. *Transactions of the Royal Society of Edinburgh, Earth Sciences* 81: 251–61.
Sjørring, S. (1983), The glacial history of Denmark. In: J. Ehlers (ed.) *Glacial deposits in north-west Europe*. Balkema, Rotterdam, 163–79.
Skupin, K., Speetzen, E., and Zandstra, J. G. (eds.) (1993), *Die Eiszeit in Nordwestdeutschland*. Geologisches Landesamt Nordrhein-Westfalen, Krefeld.
Stoker, M. S., and Bent, A. (1985), Middle Pleistocene glacial and glaciomarine sedimentation in the west central North Sea. *Boreas* 14: 325–32.
Svendsen, J. I., and Mangerud, J. (1987), Late Weichselian and Holocene sea-level history for a cross-section of western Norway. *Journal of Quaternary Science* 2: 113–32.
ter Wee, M. W. (1962), The Saalian glaciation in The Netherlands. *Mededelingen Geologische Stichting* NS 15: 57–74.
Turner, C. (2000), The Eemian interglacial in the North European plain and adjacent areas. *Geologie en Mijnbouw/The Netherlands, Journal of Geosciences* 79: 217–31.
Urban, B. (1995), Palynological evidence of younger Middle Pleistocene Interglacials (Holsteinian, Reinsdorf and Schöningen) in the Schöningen open cast lignite mine (eastern Lower Saxony, Germany). *Mededelingen Rijks Geologische Dienst* 52, 175–86.
—— Lenhard, R., Mania, D., and Albrecht, B. (1991), Mittelpleistozän im Tagebau Schöningen. *Zeitschrift der Deutschen Geologischen Gesellschaft* 142: 351–72.
Vandenberghe, J. (2000), A global perspective of the European chronostratigraphy for the past 650 ka. *Quaternary Science Reviews* 19: 1701–7.
van der Wateren, F. M. (1992), *Structural Geology and Sedimentology of Push Moraines. Processes of soft sediment deformation in a glacial environment and the distribution of glaciotectonic styles*. Ph.D. thesis, Amsterdam.
van Gijssel, K. (1987), A lithostratigraphic and glaciotectonic reconstruction of the Lamstedt Moraine, Lower Saxony (FRG). In: J. J. M. van der Meer (ed.) *Tills and glaciotectonics*. Balkema, Rotterdam, 145–55.
van Kolfschoten, T. (1990), The evolution of the mammal fauna in the Netherlands and the Middle Rhine area (Western Germany) during the Late Middle Pleistocene. *Mededelingen Rijks Geologische Dienst* 43/3.
van Kolfschoten, T. (2000), The Eemian mammal fauna of central Europe. *Geologie en Mijnbouw/The Netherlands Journal of Geosciences* 79: 269–81.

von Koenigswald, W. (1988), Paläoklimatische Aussage letztinterglazialer Säugetiere aus der nördlichen Oberrheinebene. In: W. von Koenigswald (ed.) *Zur Paläoklimatologie des letzten Interglazials im Nordteil der Oberrheinebene*. Gustav Fischer, Stuttgart, 205–314.

Wansa, St., and Wimmer, R. (1990), *Geologie des Jungpleistozäns der Becken von Gröbern und Grabschütz*. Altenburger Naturwissenschaftliche Forschungen 5. Mauritianum, Altenburg, 49–91.

Winsemann, J., and Asprion, U. (2001), Glazilakustrine Deltas am Südhang des Wesergebirges, Aufbau, Entwicklung und Kontrollfaktoren. *Geologische Beiträge Hannover* 2: 139–57.

Woldstedt, P., and Duphorn, K. (1974), *Norddeutschland und angrenzende Gebiete im Eiszeitalter*, 3rd edn. Köhler, Stuttgart.

Zagwijn, W. H. (1989), The Netherlands during the Tertiary and the Quaternary, A case history of Coastal Lowland evolution. *Geologie en Mijnbouw* 68: 107–20.

Zandstra, J. G. (1983), Fine gravel, heavy mineral and grain-size analyses of Pleistocene, mainly glacigenic deposits in the Netherlands. In: J. Ehlers (ed.) *Glacial deposits in north-west Europe*. Balkema, Rotterdam, 361–77.

Ziegler, P. A. (1982), Geological Atlas of Western and Central Europe. Shell International, The Hague.

4 The Last Glaciation and Geomorphology

Margot Böse

Introduction

The Weichselian ice sheets were smaller than those of the preceding Middle Pleistocene and covered only the north-eastern part of the German lowland, as well as the eastern and northern parts of the Jutland peninsula (Figs. 4.1 and 4.2). The Late Weichselian (Late Devensian) ice sheet also covered the northern part of the North Sea Basin (Ehlers and Wingfield 1991). The young morainic landscape still has distinct morphological features distinguishing it from the old morainic areas with relief shaped by the long-lasting periglacial processes during the Weichselian. The area of the last glaciation is easily recognizable on topographical maps owing to the irregular relief and the numerous lakes that are still preserved today. This scenic landscape also features end moraines and kames, till plains, and huge meltwater valleys (*pradolinas* or *Urstromtäler*), as well as a complex drainage system.

The general climatic development of the last interglacial-glacial-Holocene cycle can be interpreted from curves showing the ratio of the oxygen isotopes ^{18}O and ^{16}O in deep-sea sediments, which is indirectly indicative of the growth of ice masses worldwide. The different OIS (Oxygen Isotope Stages) are numbered, with uneven numbers representing the relatively warmer phases (Fig. 4.3). OIS 5e corresponds to the Eemian Interglacial. The stages 5d–5a represent the Early Weichselian, which is colder than the preceding interglacial but is subdivided into stadials and interstadials. The Brørup interstadial s.l., including both the Amersfoort (5c) and the Odderade interstadial (5a), is characterized by boreal forests in this part of Europe. OIS 4 marks the onset of the Middle Weichselian or Pleniglacial, and is the first stage involving much lower temperatures and the growth of considerable ice masses. This is probably the time when a real inland ice sheet started to grow in Fennoscandia. OIS 3, the middle part of the Pleniglacial, is represented by several climatic changes. The interstadials—or intervals, as they are also called—such as Oerel, Glinde, Moershoofd, Hengelo, and Denekamp—are well documented by terrestrial palynological records of a herb and shrub bush vegetation (van der Hammen *et al.* 1967; Behre 1989; Caspers and Freund 2001). All these Weichselian interstadials are called after type localities in The Netherlands, Germany, and Denmark. OIS 2 includes the classical LGM (Last Glacial Maximum) in Germany and Denmark, which shaped the young morainic landscape.

The Middle Weichselian Ice Advance

Denmark

For a long time the LGM Weichselian ice advance was considered to be the first to have reached the southern Baltic Sea and the Jutland peninsula. However, thermoluminescence (TL) dates of meltwater deposits and related tills of previously unclarified stratigraphical position in Denmark have supplied evidence of an Early Middle Weichselian Ice Advance (OIS 4, 70,000 to 60,000 years BP) (cf. Petersen 1985). The petrographical composition of the till—which includes flint, Cretaceous limestone, and Palaeozoic limestone—indicates that this ice stream came from the south-east, thus following the depression of the Baltic. The till, called Rinstinge Klint Till in eastern Denmark, contains redeposited Eemian foraminifera and has therefore been ascribed

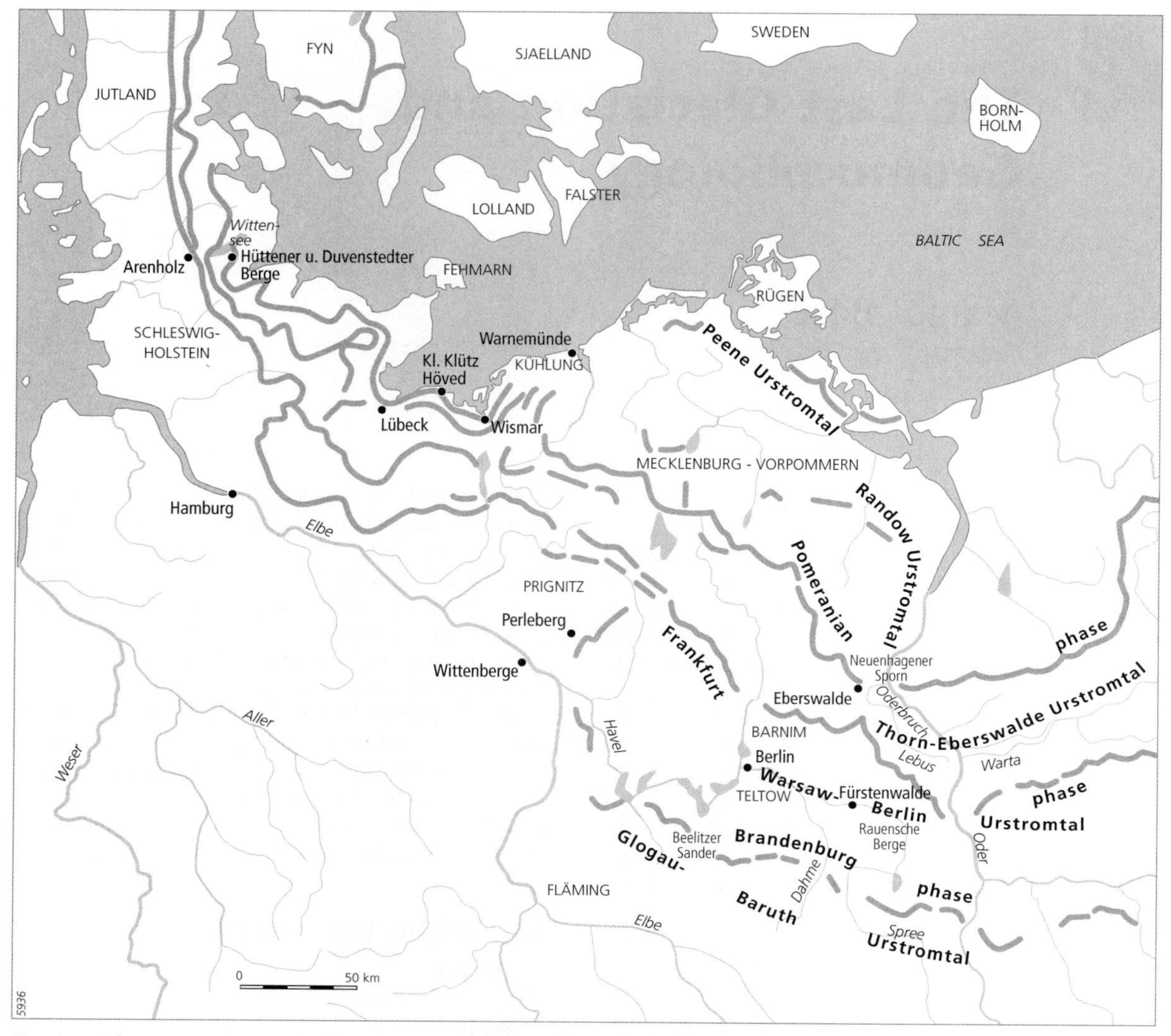

Fig. 4.1. The main ice marginal positions in northern Germany and adjacent part of Poland, and locations of sites mentioned in the text.

to the Weichselian (Houmark-Nielsen 1987, 2000). The outwash fan faced west and can be traced on the island of Fyn and in southern Jylland. The outermost limit is difficult to reconstruct, as there is only lithological and glacitectonic evidence but no morphological trace of this ice advance (Fig. 4.2).

The stratigraphical position of other tills is less clear. The deposition of a till (the Klinholm till) between the Hengelo and Denekamp interstadials cannot be excluded (Houmark-Nielsen 2000). Ice advances from the north first calved only into the Skærumhede Sea, at that time also covering the northernmost part of Jutland called Vendsyssel (Fig. 4.4). An advance from the north, represented by a Norwegian till, is still of unknown age, but it predates the LGM. For a while it blocked a northward palaeocurrent pattern, and the meltwaters drained towards the south and west (Houmark-Nielsen 1987, 2000).

Schleswig-Holstein

In Schleswig-Holstein, the oldest Weichselian ice cover —the Brügge advance—formed an extended lobe in

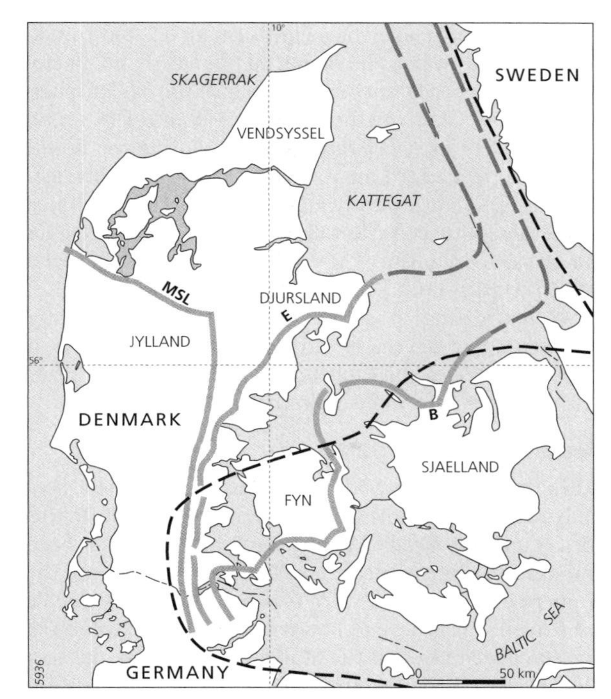

Fig. 4.2. The main Weichselian ice marginal positions in Denmark. MSL: Main Stationary Line; E: East Jylland ice border line; B: Bælthav ice margin; dotted line: possible extent of the inland ice sheet during OIS 4 (= Early Middle Weichselian) (after Houmark-Nielsen 1989, 2000).

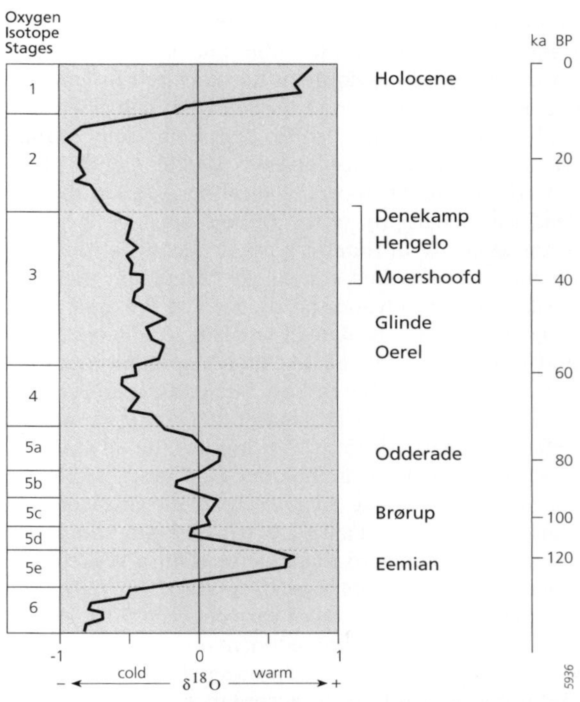

Fig. 4.3. Oxygen Isotope Stages, $^{18}O/^{16}O$ ratio of deep-sea records and the biostratigraphical units in north and central Europe (modified after Behre 1989 and literature cited herein).

Fig. 4.4. Ice-pushed morainic and fluvioglacial deposits exposed in a beach cliff section in northern Jutland, Denmark (photo: Eduard Koster).

the southwest, coming close to the Hamburg area; it probably also represented the maximum extent in northern Schleswig-Holstein. Its till is rich in flint and Cretaceous limestone from the southern Baltic Sea area (Kabel 1982; Stephan 1995). Morphologically distinct terminal moraines are absent. Menke (1991: 347) reports the notable morphological observation that no dead ice topography of the Brügge advance is visible in the area east of Hamburg. He deduces that the dead ice may have melted completely before the younger, Bordesholm ice advance partly covered the area of the Brügge advance. The date of the Brügge advance is still under discussion: in all likelihood, glaciation started about 25,000 BP at the earliest, but a time span between 70,000 and 50,000 BP cannot be excluded (Menke 1991; Stephan 1995: 8) owing to the absence of geochronological data in this area. Piotrowski (1996) still correlates the Brügge advance with the Brandenburg phase at 20,000 BP. First GLSL (Green Light Stimulated Luminescence) data from an outcrop of intra-Weichselian meltwater sands, published by Preusser (1999), give inconclusive results, as ages vary between 92 ± 19 and 31 ± 6 ka. A glaciolimnic sediment on top of the second Weichselian till in this area has been dated to 26 ± 4 ka and is considered to be in accordance with a supposed ice advance at about 25 ka (Fig. 4.5).

Further north in Schleswig-Holstein (cf. Piotrowski 1996), tills as well as fluvioglacial (Fig. 4.6) and glaciolacustrine deposits have been dated to the Weichselian by TL and GLSL methods (Marks *et al.* 1995 and Preusser 1999 respectively). The results suggest an ice cover and related sediments during OIS 4, probably lasting until the lower part of OIS 3. This can be correlated along the Baltic coast to Poland, where similar TL ages have been published (Drozdowski and Federowicz 1987; Marks 1988). Considering the methodological problems arising in dating glacigenic and meltwater sediments and the uncertain reliability of the results, it is nevertheless surprising that few samples give results which are in agreement with the classical LGM.

Mecklenburg-Vorpommern

Although the Early Middle Weichselian ice sheet possibly reached the southern Baltic coast (Böse 1989), this idea of an Early Middle Weichselian ice advance has been a matter of some controversy. The key localities are the Stoltera cliff coast close to Warnemünde and the profile on Rügen. According to Ludwig (1964), the position of the so-called m2-till at the Stoltera cliff corresponds to a Weichselian (Brandenburg) age, whereas Cepek (1973) even attributed it to an uppermost Saalian till (SIII); however, von Bülow *et al.* (1977) again classified this till as

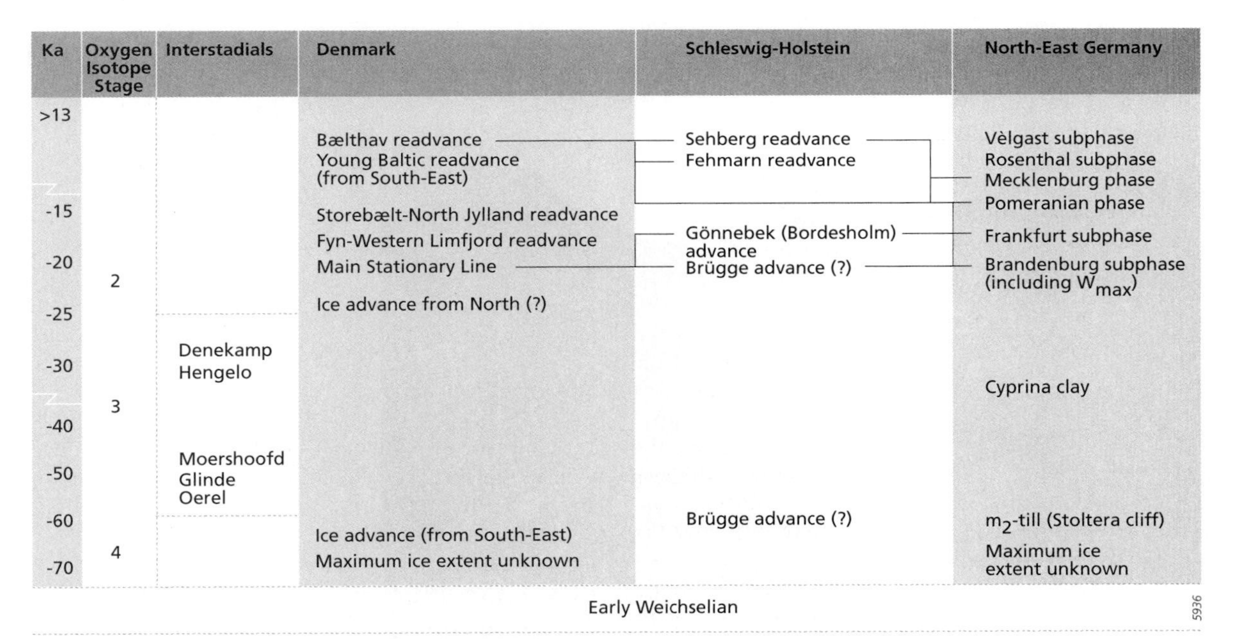

Fig. 4.5. The possible correlation of Middle Weichselian ice advances on the Jutland peninsula and in north-east Germany (compiled after: Houmark-Nielsen 1987, 1989, 2000; Böse 1989; Menke 1991; Petersen and Kronborg 1991; Steinich 1992; Rühberg *et al.* 1995; Stephan 1995; and Piotrowski 1996).

Fig. 4.6. Coarse-grained sandy fluvioglacial sediments with channel infillings and ice wedge casts, Schleswig-Holstein, Germany (photo: Eduard Koster).

Weichselian (Brandenburg). The problem was that the lithologic composition of the m2-till differed from a typical Weichselian till; according to Cepek (1973) it contained less Palaeozoic siltstone than the overlying tills.

So far, a long sedimentary gap has been assumed between the Eemian (for example, represented in the cliffs of Klein Klütz Höved and Rügen) and the LGM deposits of the Brandenburg till. It was only in 1989 that Böse presented a first interpretation of the m2-till as being of Early Middle Weichselian age on the basis of comparison with results from Denmark and Poland. In the meantime, reinvestigations at coastal cliff sections of Mecklenburg-Vorpommern, mainly on the island of Rügen (Steinich 1992), as well as interpretations of drillings, paved the way for the ongoing discussion on the existence of a Middle Weichselian m2-till, also called WO-till (cf. Rühberg *et al.* 1995 and literature cited herein). Nevertheless, Krienke (2002 and pers. comm.) did not find any till between Eemian deposits and the Upper Weichselian tills studying the south-eastern part of the island of Rügen.

No detailed studies have been published about the extent of this first Weichselian ice advance. Rühberg *et al.* (1995) briefly discuss the idea that this ice cover might have been even more extensive than that of the LGM (Brandenburg phase), reaching as far as Perleberg to the south in the Prignitz area. They even mention probable post-Eemian glacitectonic push features at the northern fringe of the Fläming, an area which Cepek (1967) had attributed to the youngest Saalian subphase (SIII). By contrast, Lippstreu *et al.* (1995) clearly stated that the first Weichselian ice cover in Brandenburg was that of the Brandenburg phase (LGM at about 20,000 BP). In addition, the almost complete lack of Middle Weichselian interstadial deposits complicates the stratigraphical interpretation. In coastal cliffs of Mecklenburg-Vorpommern, clay and fine-grained sands of interstadial character have been investigated (Rühberg *et al.* 1995). Besides other fossils, they contain *Azolla filiculoides*, which is not represented in typical Eemian deposits. Also, the much-discussed Cyprina clay at different outcrops on the island of Rügen, which Cepek (1975) interpreted as of Saalian interstadial age, has been reinvestigated (Steinich 1992). The marine fossil assemblage as well as radiocarbon and thermoluminescence dating gives results of Middle Weichselian time, and no hiatus was detected to the underlying till, which is now attributed to the Early Middle Weichselian ice advance. However, a correlation with one of the Middle Weichselian interstadials of OIS 3, such as Oerel, Glinde, Moershoofd, Hengelo, or Denekamp (Houmark-Nielsen 1989; Behre 1989), is not possible. In conclusion, the question of the existence and the extent of an Early Middle Weichselian glaciation in northern Germany and adjacent countries is therefore still unsolved and remains under discussion.

The Late Middle Weichselian Ice Advances

The Last Glacial Maximum ice advance has shaped the young morainic area in north-eastern and northern

Fig. 4.7. An example of a tunnel valley in central Jutland (photo: Eduard Koster).

Germany as well as the main part of Denmark. Two maps by Woldstedt (1935) and Liedtke (1980) present an overview of the topographical situation and the spatial distribution of landforms related to the glaciation. Glacigenic forms and deposits as well as meltwater valleys constitute the main structures of the landscape, which were modified in detail by aeolian, fluvial, and hill slope (periglacial) processes during Late Glacial and Holocene time (Ehlers 1983; Ehlers *et al.* 1995). Pre-Weichselian relief elements are nevertheless recognizable in some areas; for instance, pre-Weichselian push moraines of the bigger and more powerful Saalian ice cover have not been completely obliterated as morphological forms. The effects of the ice cover, as well as the resulting morphological pattern on the Jutland peninsula, are different from those south of the Baltic Sea. The difference is predominantly due to the proglacial drainage pattern, oriented westward to the basin of the present North Sea in Jutland, whereas south of the Baltic Sea, the unique system of east–west trending *Urstromtäler* subdivides the glacigenic landforms (Fig. 4.7). Steep-sided depressions ('solle') in the morainic plateaus testify to the presence of dead-ice blocks after retreat of the ice sheet (Fig. 4.8).

The Jutland Peninsula

At about 21,000 BP the marine area of the Skagerrak, Kattegat, and Vendsyssel turned into a fluvial to lacustrine environment while the inland ice sheet advanced from northerly directions. These deposits were overridden by a north-east advance as far as central Denmark (Houmark-Nielsen 1989), depositing the Hantklitt till (Houmark-Nielsen 2000). This ice advance underwent a short retreat before reaching its maximum extent, known as the Main Stationary Line. In Denmark, the classical Main Stationary Line was first described by Ussing (1907) as the 'junction between the sand and gravel plains in the west and the hilly country with boulder clay deposits towards the east' (Houmark-Nielsen 1987) (Figs. 4.2 and 4.4). The deposition of the Mid-Danish till as well as the distinct outwash plains from central Jutland to the west belong to this event. It was formed at about 20,000 or 18,000 BP (Houmark-Nielsen 1987, 2000; Peterson and Kronborg 1991), in the latter case corresponding to the Frankfurt subphase south of the Baltic Sea. The main direction of the ice advance was from the north-east as indicated by the erratics from Sweden and the Kattegat region. Between 18,000 and 17,000 BP, the ice had retreated already from northern Jutland, as shown by the marine Younger Yoldia clay that was deposited at that time (Petersen and Kronborg 1991). Critical comments on the identification of Main Weichselian tills in the Kattegat and Vendsyssel region, as well as on the deglaciation patterns presented by different authors, are given by Lykke-Anderson *et al.* (1993). They are not convinced of the existence of an ice cover during that time and attribute the absence of evidence to 'a lack of pertinent observation'. In Denmark, Houmark-Nielsen (1987, 1989, 2000) classifies several readvances such as the East Jylland advance from the

Fig. 4.8. A 20-m deep dead-ice depression (*soll*) in a morainic plateau in northern Jutland, Denmark (photo: Eduard Koster).

south-east, with the dominant ice stream along the Baltic Sea depression, and the Bælthav advance from the same direction apart from other minor oscillations.

Schleswig-Holstein

In Schleswig-Holstein, the landforms created by the Main Weichselian ice advance at about 20,000 BP are still a question of interpretation. As already described above, the age of the Brügge advance is still under discussion. Stephan (1995) correlates it with the Brandenburg phase. As there are very few morphological features related to the maximum ice extent, it has been reconstructed mainly from dead ice hollows in (younger) outwash plains, but also from till patches. But the possibility that these morphological features were due to icings is also discussed (cf. Walther 1990). An example of the uncertainty with respect to the exact extent of the ice advance can be observed to the west of the town of Schleswig. Here, the Arenholzer See and the esker of Arenholz are situated to the west of the morphologically evident end moraines of the last glaciation and as such probably indicate that the ice locally advanced some 2–7 km beyond the morphological maximum line in this central part of Schleswig-Holstein (Walther 1990: fig. 57).

Stephan (1995) correlates the following Weichselian ice advance (Gönnebek or Bordesholm advance) with the Pomeranian phase south of the Baltic Sea (Hölting 1958; Stephan *et al.* 1983). This advance formed terminal moraines in some areas, but its extent corresponds to the western ends of the subsequent *Tunneltäler* (tunnel valleys). Nevertheless it cannot be ruled out that this advance was parallel in time to the Brandenburg phase or even the Frankfurt oscillation (cf. Piotrowski 1996). Characteristic for this advance are proglacial sediments of the transgressing ice. During oscillations these deposits were reshaped to push moraines, of which the Hüttener Berge and the Duvenstedter Berge are conspicuous examples (Fig. 4.1); they have erosional basins at their eastern sides, one of which is occupied by the almost rectangular Wittensee.

Mecklenburg-Vorpommern, Brandenburg

Further to the south, the end moraines of the different advances diverge. The ice margin turned from NNE–SSW to W–E. But there is no well-established correlation of the end moraines between Schleswig-Holstein, including the region around Lübeck, and Mecklenburg-Vorpommern. In any case, the depression of the Lübeck bay provided good preconditions for the formation of a south-west directed lobate structure of the ice sheet.

The end moraines of the Gönnebek advance, attributed to the Pomeranian advance by Stephan (1995) and to the Frankfurt subphase by Rühberg *et al.* (1995; cf. Liedtke 1980), continue in ice marginal positions in Mecklenburg-Vorpommern. In south-western Mecklenburg the terminal moraines of the Brandenburg phase and the Frankfurt subphase are located 10–20 km apart. The ice marginal position of the Brandenburg phase is difficult to identify on the basis of morphological features, as the meltwater of the more or less parallel Frankfurt subphase formed extensive outwash plains covering landforms and deposits of the previous phase. A few small remnants of till plains have been preserved. The ice in this area was part of the Belt glacier, passing west of the island of Rügen, which formed an obstacle in the glacier's path, inducing a temporary bifurcation of the ice sheet.

The ice margin of the Brandenburg phase makes a distinct bend to the south-west, passing just east of the confluence of Havel and Elbe, and continues in a south-easterly direction to Poland. In the region south-east of Berlin the Brandenburg ice margin runs some 50–70 km south of the end moraines of the Frankfurt subphase, and belongs to the Oder ice stream extending far to the south and passing east of the island of Rügen. In some places the ice probably reached beyond the morphologically distinct ice marginal positions up to 10 km into the Glogau-Baruther *Urstromtal*. Meltwater deposits of this maximum advance form a sharp erosional edge at the southern border of the *Urstromtal* separating it from older sediments and periglacial valleys of the Fläming (Liedtke 1980; Juschus 1999, 2000; *Physische Geographie* 1991: 561).

The lobate character of the margin of the ice sheet is only occasionally recorded by terminal moraines, mostly consisting of melt-out till. More typical are the outwash plains, sometimes with a distinct step at their inner fringe. The 'Beelitzer Sander' is a classical meltwater fan, dipping from the northern edge at 70 m a.s.l. over a distance of 14 km to the southwest to about 45 m a.s.l. at the Baruth *Urstromtal*.

Between the Brandenburg phase and Frankfurt subphase the landscape is characterized by:

- till plains subdivided by (subglacial) meltwater channels,
- the western part of the Warsaw–Berlin *Urstromtal*,
- kame deposits,
- minor ice marginal features.

The *Urstromtäler* are more or less parallel to the original ice marginal positions. The broadness of the *Urstromtäler* in general results from the huge supply of meltwater from the decaying ice sheet from the north and major rivers coming from southerly directions between Warsaw and Berlin, including the Weichsel, the Warthe, and the Oder. As the way to the Baltic Sea was blocked, river discharge had to follow the *Urstromtäler* to the west. It is estimated that maximum summer discharge of the lower Elbe as the receiving stream may have been as much as 300,000 m^3/sec, which corresponds to the discharge of the river Niger (Lied 1953). The river patterns at that time were braided ones.

In general, the landscape in Brandenburg inclines gently from east to west. Consequently, the bigger and higher till plains such as Lebus, Barnim, and Teltow in the neighbourhood of Berlin are situated in the east, whereas to the west the till plains are smaller and are partly covered by meltwater deposits. In the eastern part Saalian push moraines, which were only partly reshaped by the Weichselian ice, occupy the highest positions in the landscape (Hannemann 1969). Among these landforms are the Rauensche Berge south of Fürstenwalde (154 m) (Fig. 4.9). This figure illustrates the highly complex inner structure of such ice-pushed morainic ridges, containing a glacitectonically deformed sequence of Miocene, Elsterian, Saalian, and Weichselian sediments. The Baruth *Urstromtal* has two different levels and its sediments consist mainly of sand and some gravel. After partial deglaciation of the region, meltwater streams left some sections of the *Urstromtal* and occupied deeper-lying meltwater channels more to the north (Franz 1962; *Geomorphologische Übersichtskarte* 1970). This resulted in the present-day uneven gradient of the *Urstromtal* floor with steps in the longitudinal profile. As a consequence some sections show no Holocene fluvial activity.

Isolated hills to the north of the Brandenburg ice margin are interpreted either as morainic remnants of subphase marginal positions or as kames. Especially in the Berlin–Potsdam area, these hills consist of layered sands sometimes pushed or squeezed by the glacier (Weisse 1977). Morphological features that represent such a retreat phase, crossing the till plain south of the Warsaw–Berlin *Urstromtal* (Liedtke 1980; Pachur and Schulz 1983), are isolated hills and meltwater channels with dead ice depressions. The till plains are generally rather flat areas. Kettle holes are sometimes visible as isolated forms, but more often they follow a linear pattern, probably reflecting the position of minor subglacial meltwater streams which were filled by dead ice during ice decay (Fig. 4.8). The number of dead ice hollows has been diminished by agricultural activity, and the majority of them presently show steep edges due to ploughing.

In general, the ice sheet of the Brandenburg phase was not very powerful. The resulting landforms have been shaped by minor oscillations and by meltwater activity; dead ice forms are widespread. A system of mostly SSW trending subglacial meltwater channels, including the

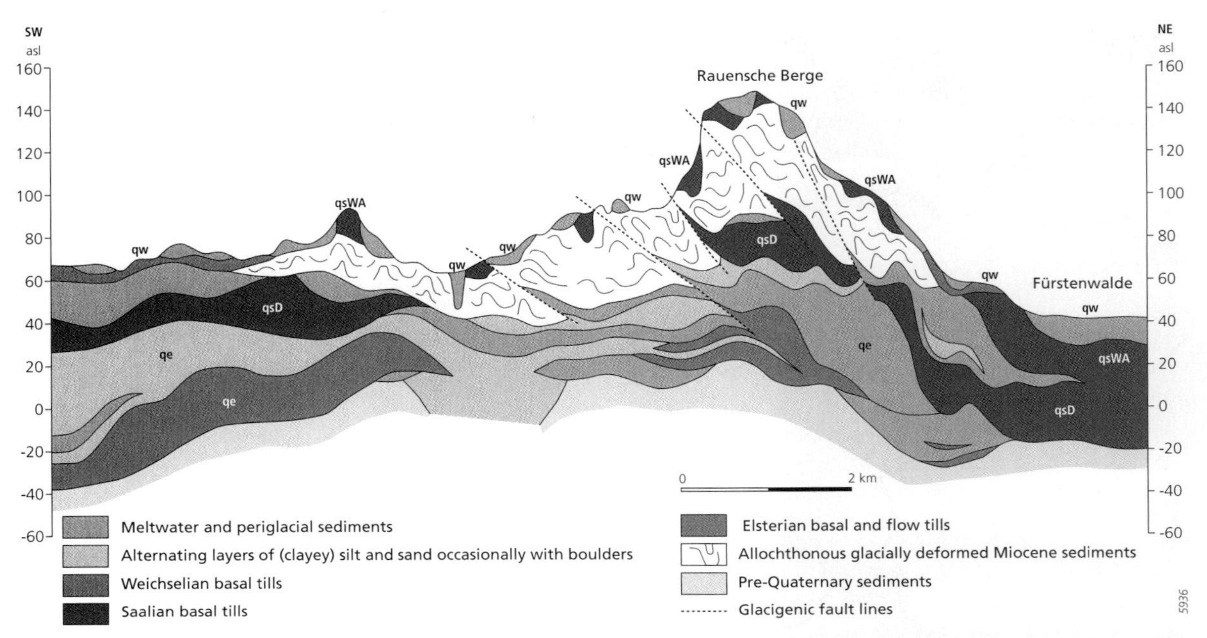

Fig. 4.9. Section through a Saalian ice-pushed moraine in the Weichselian young morainic area, location Rauensche Berge, south-east of Berlin (simplified after Lippstreu 1995). qw: Quaternary, Weichselian; qsWA/qsD: Quaternary, Saalian (Warthe and Drenthe stage); qe: Quaternary, Elsterian.

lake system of the river Havel and of the river Dahme, subdivides the landscape. Some meltwater channels can be traced from areas north of the Warsaw–Berlin *Urstromtal*, right across the *Urstromtal* itself, almost as far as the retreat positions of the Brandenburg phase. In the *Urstromtal* they are preserved as lake-rich sections probably due to the long duration of dead ice conservation.

The SE–NW trending Warsaw–Berlin *Urstromtal* ranges in width between 2 and 10 km and has at least two terraces. Like the Baruth *Urstromtal*, the Warsaw–Berlin *Urstromtal* is only partly occupied by a river, the Spree, which first crosses the Baruth *Urstromtal* from south to north and later follows the Warsaw–Berlin *Urstromtal* to the west. The Spree is then tributary to the river Havel, which crosses the *Urstromtal* from north to south. The western part of the *Urstromtal* as far as the river Elbe has sections without any rivers at all; in other parts some streamlets follow the broad meltwater valley. Figure 4.1 clearly shows that the major rivers in the north-east German Lowlands recovered their original south–north direction after deglaciation.

The Frankfurt subphase is a long-lasting marginal position during the downmelting of the glacier. Though it is important in the context of morphostratigraphical interpretation, it is here not classified as a readvance. Oscillations of some kilometres just resulted in a local separation of the Brandenburg and the Frankfurt lodgement tills. Extensive outwash plains in front of the ice marginal position and meltwater valleys incised into the till plains are tributary to the Berlin *Urstromtal*. At the north-western edge of the Barnim till plain, the Frankfurt ice margin crosses the Thorn–Eberswalde *Urstromtal* in a north-westerly direction. Some meltwater streams followed this north-westerly course, other streams were directed southward through a broad north–south oriented meltwater valley, today used by the river Havel, to reach the Berlin *Urstromtal*. One of the most remarkable landforms of this region is the Oderbruch, an almost rectangular, NW–SE oriented, flat-bottomed basin, 57 km long and up to 27 km wide (Fig. 4.10). It used to be part of the Thorn-Eberswalde *Urstromtal*; today the river Oder flows through it. It has repeatedly been suggested that this unique feature was created by neotectonic activity, but no substantiation for this theory has been provided yet (Hannemann 1970, 1995). In general, neotectonic influence on the glacial landscape cannot be excluded, but evidence is scarce. Liedtke (1996) attributed the origin of the Oderbruch to glacial exaration, induced by a proglacial lake in front of a glacier advancing over unfrozen subsoil in this area. The glaciotectonic disturbances are deep

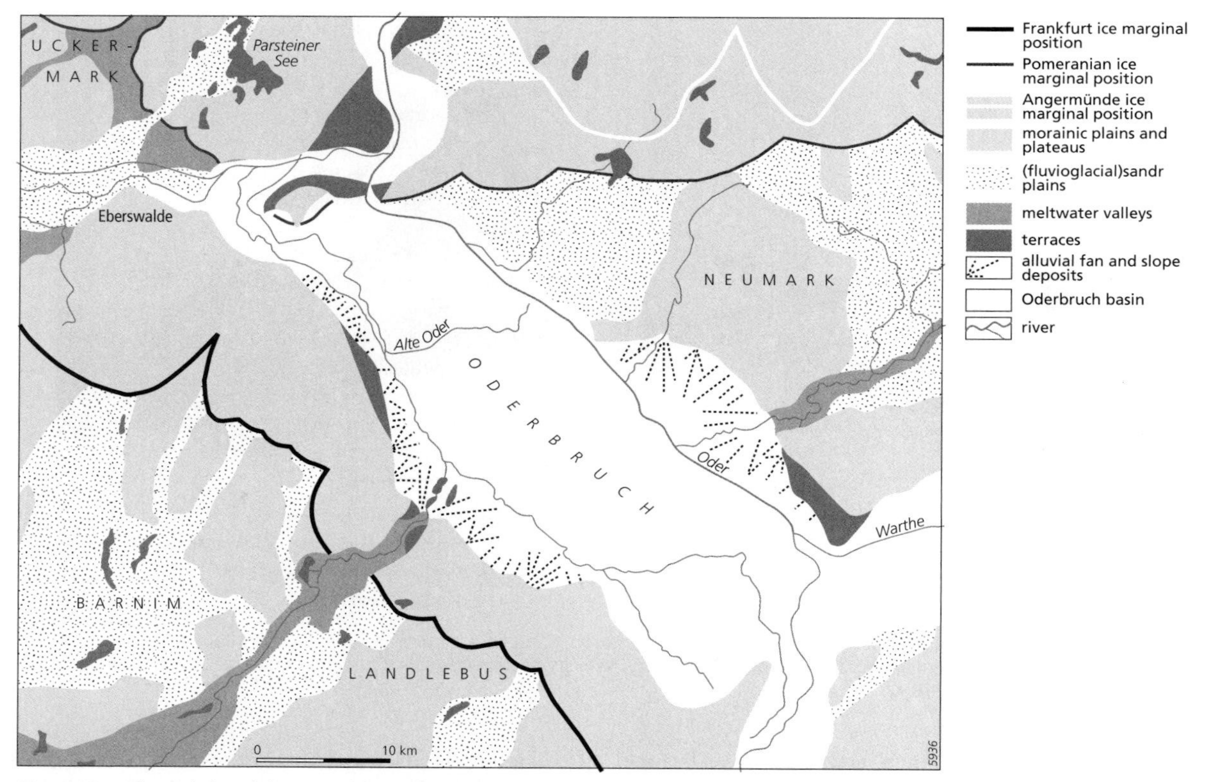

Fig. 4.10. The Oderbruch basin and its surroundings at the German–Polish border, east of Berlin (simplified after Liedtke 1996).

(Fig. 4.11) and the depression was shaped like a trough valley. After the onset of deglaciation, masses of dead ice about 50 m thick hindered the formation of fluvial cones by the meltwater running along the *Urstromtal* and filling the basin. The meltwater flowed as a braided river mainly at the western fringe and even covered the dead ice with fluvioglacial sediments.

The Pomeranian Phase and Younger Readvances

After the Frankfurt subphase a large-scale downmelting of the ice sheet took place during what has been termed the Blankenberg interphase, and the margin of the ice sheet was situated somewhere in the Baltic Sea basin. The readvance redeposited its own meltwater sediments as well as its own till. The general direction of the ice margin runs from the area of Lübeck in the north-west to the Neuhagener Sporn at the Oder in the south-east. Like the Brandenburg phase, the Pomeranian phase has a maximum position, whose end moraines are not well preserved, having been partly destroyed by meltwater activity parallel to the main ice marginal position. This main marginal position has the best-developed end moraines in north-eastern Germany. The ice margin consisted of many small lobes encircled by end moraines, and lateral moraines consisting of ablation moraines, push moraines, and ridges of boulder concentrations. Deep glacier-tongue basins, shaped by both glacial exaration processes and by concentrated subglacial meltwater erosion, are formed in the hinterland. These basins are often filled with lakes that are several tens of kilometres in length. This abundance of lakes is characteristic for the landscape of the Pomeranian phase and is easy to recognize on topographic maps. Probably the ground was not—or not deeply—frozen when it was overridden by the readvance ice, therefore the glacier and its meltwater could easily remodel the landscape. Because of the conspicuous relief, with hills often higher than 100–120 m a.s.l., these end moraines are called *nördlicher Landrücken*.

In the east, the Thorn–Eberswalde *Urstromtal* was blocked for a short time by the ice sheet in the northern part of the Oderbruch and close to Eberswalde (compare

Fig. 4.11. Weichselian, silty, and sandy fluvioglacial deposits plastically deformed due to glaciotectonic activity (photo: Eduard Koster).

Figs. 4.1 and 4.10). As a result meltwater streams coming from the east along the *Urstromtal*, as well as from the ice margin in the north had to follow the Oder river valley east of the Lebus plain in a southerly direction. At the same time relatively narrow meltwater channels formed between the till plains of Lebus and Barnim and across the Barnim (Fig. 4.10) and reached the Berlin *Urstromtal* (Liedtke 1956/7). West of Eberswalde, the north-west oriented *Urstromtal* led the meltwater of the Pomeranian ice margin to the river Elbe. West of Wittenberge on the Elbe river, the meltwater stream converged and flowed directly into the Elbe valley *Urstromtal* after crossing the elevated area of the Prignitz in narrow incised channels. As soon as downmelting of the Pomeranian ice sheet started, the Thorn–Eberswalde *Urstromtal* came into function.

In the surroundings of Wismar, near the present Baltic Sea coast, a huge interlobate push moraine was formed between two large ice lobes. Called the 'Kühlung' (130 m a.s.l.), the push moraine was up to 12 km broad and trended NW–SE over about 45 km. This large glacitectonic structural ridge was not overridden during a later readvance. Smaller push moraines of similar structure can be found at other sites to the west and the east of this area. Especially in western Mecklenburg, the hinterland of the Pomeranian phase is characterized by elevated till plains and glaciofluvial deposits forming distinct kame elevations.

Whereas south of the Baltic Sea the morphostratigraphy shows a clear pattern of oscillations and readvances, at the south-western side of the ice sheet the chronology and morphostratigraphy of the readvances are much more complicated because the ice marginal features of different phases and subphases overlap; the ice margin repeatedly reached similar positions. Moreover, the correlation is complicated because it depends on the previously mentioned problem concerning the age of the Brügge advance. This advance is either of Early Middle Weichselian age or it corresponds to the Brandenburg phase. If the Brügge advance is much older than OIS 2, then the Bordesholm advance in Schleswig-Holstein probably correlates with the Brandenburg phase to the south-east and the Main Stationary Line in Denmark. In Schleswig-Holstein, the density of ice marginal positions of several readvances inhibited the development of a proper drainage network and produced numerous meltwater lakes. These lakes are mainly filled by fine-grained glaciolacustrine sediments. Further north, the subglacial meltwater system was linked to more distinct subglacial tunnel systems, partly transformed by additional glacial exaration. Some of these former meltwater systems form the present-day *Förden* (German) or *Fjorde* (Danish) at the Baltic Sea coast. In Denmark, the East Jylland advance—according to Houmark-Nielsen (1987, 2000)—has a different direction of ice flow, coming from the south-east along the Baltic Sea depression. This is well documented by the quartz-poor till composition associated with Baltic indicator boulders. This advance is supposed to have taken place between 14,000 and 13,000 BP (Houmark-Nielsen 1987; Petersen and Kronborg 1991). But these dates do not allow the chronostratigraphic correlation with the Pomeranian ice; on the other hand, a morphostratigraphic correlation has been discussed repeatedly.

The following readvance in Schleswig-Hostein is the Fehmarn readvance (Fig. 4.5). Prior to this, the active ice margin had withdrawn into the interior of the Baltic Sea basin and the readvance of the ice sheet is documented by a separate till in the coastal cliffs of western Schleswig-Holstein. This readvance covered only the easternmost part of Schleswig-Holstein, and may be correlated with the Bælthav readvance of Houmark-Nielsen (1987) in Denmark (Stephan *et al.* 1983). This ice marginal phase is characterized by a strongly lobate structure following the valleys and depressions in the pre-existing landscape. The morphostratigraphic position of a still younger but morphologically very distinct ice marginal line of the Sehberg readvance (Fig. 4.5) is also still heavily debated (Stephan 1995; Piotrowski 1996). In the north-east of Denmark, the ice margins of the East Jylland and the Bælthav readvances were in contact with the transgressing Younger Yoldia Sea, which induced a more rapid ice discharge by calving.

The Mecklenburg readvance (Fig. 4.5) south of the Baltic Sea has its own till. As mentioned earlier, the end moraines are best preserved around Wismar. From the large push moraine of the Kühlung ridge, its extent can be traced in an easterly direction by means of tills with a characteristic petrographical composition, but few end moraines can be identified. Two additional subphases have been identified on the basis of ridges of end moraines, the Rosenthal and Velgast subphases (Rühberg *et al.* 1995; Fig. 4.5). As the ice margin retreated to deeper-lying areas in the north-east, the meltwater streams could not reach the *Urstromtäler* in the south any more and consequently flowed in a westerly direction to the Baltic Sea. At the same time, parts of the lower Oder valley north of the Oderbruch became free of ice, enabling river water from the south to join the meltwater streams of the Velgast subphase in the Randow *Urstromtal*. Finally, the Peene *Urstromtal* formed, draining the meltwater from the outwash plains of the eastern Velgast ice margin.

Liedtke (1980) reconstructs a younger distinct ice margin on the island of Rügen, which, however, cannot be followed in the area presently covered by the Baltic Sea. It has been suggested that this ice margin corresponds with end moraines west of Copenhagen. Lagerlund (1987) and Lagerlund and Houmark-Nielsen (1993) propose a local ice dome which developed on stagnant ice in the Baltic Sea depression south of the province of Skane in southern Sweden. This ice dome theory is also documented in the *National Atlas of Sweden* of 1994 (Geology sheet, p. 133). This ice dome, which may have covered parts of Schleswig-Holstein (Fehmarn readvance?) and parts of the islands of Sjælland and Fyn in Denmark, had vanished before 14,000 BP (Lagerlund and Houmark-Nielsen 1993). Krienke (2002), who studied indicator boulders and glacitectonic features, suggests that the youngest till on the island of Rügen is of a north-easterly origin.

It is remarkable that drumlins and eskers are much more frequent north of the Pomeranian ice border line than in the areas of the Brandenburg phase and the Frankfurt subphase (Liedtke 1980; *Geol. Karte von Mecklenburg-Vorpommern* 1997). According to studies by Piotrowski (1996) in Schleswig-Holstein this is due to a reduced permeability for groundwater, owing to either subsurface till layers or the presence of permafrost. The areal distribution of subglacial permafrost, its impact on erosional subglacial processes and on resulting geomorphological features requires further study.

Fig. 4.12. An example of a large erratic boulder (the scale on the boulder is 1 m long) found in very coarse-grained morainic sediments (photo: Eduard Koster).

Aspects of Further Geomorphological Development

After the downmelting of the ice sheet buried dead ice was probably preserved for about 7,000 years in the Brandenburg phase area to 2,000 years in the coastal areas. Though periglacial processes were partly reshaping the glacial landscape (Böse 1991, 1995), morphological features—mainly former glaciofluvial channels and kettle holes—came into existence only at the end of the Oldest Dryas, as was the case in the Oderbruch basin (Liedtke 1996). Former subglacial channels sometimes cross *Urstromtäler*, therefore they did not exist as morphological forms during subaerial meltwater runoff. Likewise, examples of Late Weichselian dune forms occurring on both sides of such channels show that the latter were not yet present as morphological features during the main period of aeolian activity. Limnic and/or organic sedimentation in these dead ice forms generally did not start before the Bölling interstadial (Firbas 1949). Further to the north, melting of dead ice can even be attributed to the later part of the Bölling/Alleröd interstadial complex (Nitz *et al.* 1995; Strahl and Keding 1996). In this respect it must be stressed that dead ice of glacial origin and ground ice related to permafrost formation should be carefully distinguished, as they had different effects on the evolution of the landscape during and after deglaciation.

Besides various human impacts on the glacial landforms, the decrease in the amount of boulders in the young morainic area should be mentioned. The surfaces of till plains and end moraines were rich in huge erratics (Fig. 4.12), that have been drastically reduced in number by man. The quantity of boulders found in subsurface glacial sediments in sand and gravel pits may give some indication of the transport of blocks from the Fennoscandian hardrock regions. Careful and very elaborate analyses of indicator boulders and other ice-contact phenomena have enabled the detailed reconstruction of ice stream directions and the determination of the regions of provenance of erratic boulders in Fennoscandia.

References

Behre, K.-E. (1989), Biostratigraphy of the Last Glacial Period in Europe. *Quaternary Science Reviews* 8: 25–44.

Böse, M. (1989), Methodisch-stratigraphische Studien und paläomorphologische Untersuchungen zum Pleistozän südlich der Ostsee. *Berliner Geographische Abhandlungen* 54.

—— (1991), A palaeoclimatic interpretation of frost wedge casts and aeolian sand deposits in the lowlands between Rhine and Vistula in the Upper Pleniglacial and Late Glacial. In: S. Kozarski (ed.) Late Vistulian (= Weichselian) and Holocene Aeolian Phenomena in Central and Northern Europe. *Zeitschrift für Geomorphologie* NS, Suppl. 90: 15–28.

—— (1995), Problems of dead ice and ground ice in the central part of the North European Plain. *Quaternary International* 28: 123–5.

Caspers, G., and Freund, H. (2001), Vegetation and climate in the Early- and Pleni-Weichselian in northern central Europe. *Journal of Quaternary Science* 16: 31–48.

Cepek, A. G. (1967), Stand und Probleme der Quartärstratigraphie im Nordteil der DDR. *Berichte der Deutschen Geologischen Gesellschaft A, Geol. Paläont.* 12/3–4: 375–404.

—— (1973), Zur stratigraphischen Interpretation des Quartärs der Stoltera bei Warnemünde mit neuen Geschiebeanalysen. *Zeitschrift für Geologische Wissenschaften* 1/9: 1155–71.

—— (1975), Zur Stratigraphie des Quartärs in den Kliffprofilen nördlich Saßnitz/Rügen. *Wissenschaftliche Zeitschrift der Ernst-Moritz-Arndt-Universität Greifswald, XXIV*, math.-nat. 3/4: 171–4.

Drozdowski, E., and Fedorowicz, St. (1987), Stratigraphy of Vistulian glaciogenic deposits and corresponding thermoluminescence dates in the lower Vistula region, northern Poland. *Boreas* 16/2: 139–53.

Ehlers, J. (ed.) (1983), Glacial deposits in north-west Europe. Rotterdam, Balkema, 470.

—— and Wingfield, R. (1991), The extension of the Late Weichselian/Late Devensian ice sheets in the North Sea Basin. *Journal of Quaternary Science* 6: 313–26.

—— Kozarski, S., Gibbard, P. L. (eds.) 1995, *Glacial deposits in north-east Europe*. Rotterdam, Balkema.

Firbas, F. (1949), *Spät- und nacheiszeitliche Waldgeschichte Mitteleuropas nördlich der Alpen*, i. *Allgemeine Waldgeschichte*. Jena, Gustav Fischer.

Franz, H.-J. (1962), Morphogenese der Jungmoränenlandschaft des westlichen Brandenburger Stadiums. *Wissenschaftliche Zeitschrift der pädagogischen Hochschule Potsdam*, math.-nat. 7/1–2, Potsdam, pt. 1. Die Eisrandlagen, 29–48; pt. 2. Die Schmelzwasserabflüsse und die durch sie geschaffenen Ablagerungen und Formen, 49–60.

Geologische Karte von Mecklenburg-Vorpommern, Übersichtskarte 1:500 000, Geologische Sehenswürdigkeiten. 1st edn., Schwerin 1997.

Geomorphologische Übersichtskarte 1:200 000, Berlin-Potsdam und Frankfurt-Eberswalde (ed. J. F. Gellert and E. Scholz), Erläuterungen für die Kartenblätter Berlin-Potsdam und Frankfurt-Eberswalde (H.-J. Franz, R. Schneider and E. Scholz, Gotha/Leipzig, 1970).

Hannemann, M. (1969), Saale- und weichselzeitliche glazigene Dynamik und Alter der Lagerungsstörungen im Jungmoränengebiet Brandenburgs. *Geologie* 18/2: 168–87.

—— (1970), Grundzüge der Reliefentwicklung und der Entstehung von Großformen im Jungmoränengebiet Brandenburgs. *Petermanns Geographische Mitteilungen* 114: 103–16.

—— (1995), Über Intensität und Verbreitung glazigener Lagerungsstörungen im tieferen Quartär und Tertiär Brandenburgs. *Brandenburger Geowissenschaftliche Beiträge* 2/1: 51–9.

Hölting, B. (1958), Die Entwässerung des würmzeitlichen Eisrandes in Mittelholstein. *Meyniana* 7: 61–98.

Houmark-Nielsen, M. (1987), Pleistocene stratigraphy and glacial history of the central part of Denmark. *Bulletin of the Geological Society of Denmark* 36/1–2.

—— (1989), The last Interglacial–Glacial Cycle in Denmark. *Quaternary International* 3/4: 31–9.

—— (2000), A lithostratigraphy of Weichselian glacial and interstadial deposits in Denmark. *Bulletin of the Geological Society of Denmark* 46: 101–14.

Juschus, O. (1999), Zur Entwicklung des Urstromtales zwischen Baruth und Luckenwalde. 66. *Tagung der Arbeitsgemeinschaft*

Nordwestdeutscher Geologen, Tagungsband und Exkursionsführer, 15 f., Halle.

—— (2000), *Das Jungmoränenland südlich von Berlin—Untersuchungen zur jung-quartären Landschaftsentwicklung zwischen Unterspreewald und Nuthe*. Dissertation, Humboldt-Universität zu Berlin.

Kabel, Ch. (1982), *Geschiebestratigraphische Untersuchungen im Pleistozän Schleswig-Holsteins und angrenzender Gebiete*. Dissertation mathematisch-naturwissenschaftliche Fakultät, Christian-Albrechts-Universität zu Kiel.

Krienke, K. (2002), *Südostrügen im Weichsel-Hochglazial. Lithostratigraphische, lithofazielle, strukturgeologische und landschaftsgenetische Studien im Küstenraum Vorpommerns*. Dissertation, Universität Greifswald.

Lagerlund, E. (1987), An alternative Weichselian glaciation model, with special reference to the glacial history of Skåne, South Sweden. *Boreas* 16: 433–59.

—— and Houmark-Nielsen, M. (1993), Timing and pattern of the last deglaciation in the Kattegat region, southwest Scandinavia. *Boreas* 22: 337–47.

Lied, H. (1953), Der Abfluß des Glogau-Baruth-Hamburger Urstromtales. *Petermanns Geographische Mitteilungen* 97/2: 89–96.

Liedtke, H. (1956/57), Beiträge zur geomorphologischen Entwicklung des Thorn-Eberswalder Urstromtales zwischen Oder und Havel. *Wissenschaftliche Zeitschrift Humboldt-Universität Berlin*, math-nat. 6: 3–49.

—— (1980), *Die nordischen Vereisungen in Mitteleuropa*, Karte im Maßstab 1:1.000.000. Frankfurt am Main.

—— (1996), Die eiszeitliche Gestaltung des Oderbruches. In: R. Mäusbacher and A. Schulte (eds.) *Heidelberger Geographische Arbeiten (Festschrift für Dietrich Barsch)*. Geographischen Instituts der Universität Heidelberg, 104: 327–51.

Lippstreu, L., with contributions by F. Brose and J. Marcinek (1995), Brandenburg. In: L. Benda (ed.) *Das Quartär Deutschlands*. Bornträger, Stuttgart, 116–47.

Ludwig, A. (1964), Stratigraphische Untersuchungen des Pleistozäns der Ostseeküste von der Lübecker Bucht bis Rügen. *Geologie* 13, Suppl. 42.

Lykke-Anderson, H., Knudsen, K. L., and Christiansen, C. (1993), The Quaternary of the Kattegat area, Scandinavia: a review. *Boreas* 22/4: 269–81.

Marks, L. (1988), Relation of substrate to the Quaternary paleorelief and sediments, western Mazury and Warmia (northern Poland). *Zeszyty Naukowe Akademii Gorniczo-Hutniczej* 1165, *Geologia* 14/1: 1–76.

—— Piotrowski, J. A., Stephan, H.-J., Fedorowicz, S., and Butrym, J. (1995), Thermoluminescence indications of the Middle Weichselian (Vistulian) Glaciation in northwest Germany. *Meyniana* 47: 69–82.

Menke, B. (1991), Zur stratigraphischen Stellung der ältesten Weichsel-Moränen in Schleswig-Holstein. In: B. Frenzel (ed.) *Klimageschichtliche Probleme der letzten 130,000 Jahre*. Stuttgart, Gustav Fischer, 343–51.

National Atlas of Sweden (1994), Geology, Spec. Edn.: Curt Fredén, Theme Manager: Geological Survey of Sweden.

Nitz, B., Schirrmeister, L., and Klessen, R. (1995), Spätglazial-altholozäne Landschafts-geschichte auf dem nördlichen Barnim—zur Beckenentwicklung im nordostdeutschen Tiefland. *Petermanns Geographische Mitteilungen* 139/3: 143–58.

Pachur, H.-J., and Schulz, G. (1983), *Erläuterungen zur Geomorphologischen Karte 1:25.000 der Bundesrepublik Deutschland*, GMK 25 Sheet 13, 3545 Berlin-Zehlendorf, GMK Schwerpunktprogramm Geomorphologische Detailkartierung in der Bundesrepublik Deutschland, Berlin.

Petersen, K.-S. (1985), The Late Quaternary History of Denmark. *Journal of Danish Archaeology* 4: 7–22.

—— and Kronborg, Ch. (1991), Late Pleistocene history of the inland glaciation in Denmark. In: B. Frenzel, *Klimageschichtliche Probleme der letzten 130,000 Jahre*. Stuttgart, Gustav Fischer, 331–42.

Physische Geographie—Mecklenburg-Vorpommern, Brandenburg, Sachsen-Anhalt, Sachsen, Thüringen (1991), Collective authors: H. Bramer, M. Hendl, J. Marcinek, B. Nitz, K. Ruchholz, S. Slobodda. Gotha, Herrmann Haack.

Piotrowski, J. A. (1996), Dynamik und subglaziale Paläohydrogeologie der weichselzeit-lichen Eiskappe in Zentral-Schleswig-Holstein. *Berichte-Reports*, Institute of Geology and Palaeontology, University of Kiel, 79.

Preusser, F. (1999), Lumineszenzdatierung fluviatiler Sedimente, Fallbeispiele aus der Schweiz und Norddeutschland. *Kölner Forum für Geologie und Paläontologie* 3.

Rühberg, N., Schulz, W., von Bülow, W., Müller, U., Krienke, H.-D., Bremer, F., and Dann, T. (1995), Mecklenburg-Vorpommern. In: L. Benda (ed.) *Das Quartär Deutschlands*. Stuttgart, Bornträger, 95–115.

Steinich, G. (1992), Die stratigraphische Einordnung der Rügen-Warmzeit. *Zeitschrift für Geologische Wissenschaften* 20/1–2: 125–54.

Stephan, H.-J. (1995), Schleswig-Holstein. In: L. Benda (ed.) *Das Quartär Deutschlands*. Stuttgart, Bornträger, 1–22.

—— Kabel, Ch., and Schlüter, G. (1983), Stratigraphical problems in the glacial deposits of Schleswig-Holstein. In: J. Ehlers (ed.) *Glacial deposits in north-west Europe*, 305–20.

Strahl, J., and Keding, E. (1996), Pollenanalytische und karpologische Untersuchung des Aufschlusses 'Hölle' unterhalb Park Dwasieden (Halbinsel Jasmund, Insel Rügen), Mecklenburg-Vorpommern. *Meyniana* 48: 165–84.

Ussing, N. V. (1907), Omfloddale og randmoræner i Jylland. *Overs. K. danske Videnks. Selsk. Forh.* 4: 161–213.

van der Hammen, T., Maarleveld, G. C., Vogel, J. C., and Zagwijn, W. H. (1967), Stratigraphy, climatic succession and radiocarbon dating of the Last Glacial in The Netherlands. *Geologie en Mijnbouw* 46: 79–95.

von Bülow, W., Harff, J., and Müller, U. (1977), Gedanken zur Auswertung von Geschiebeanalysen an Hand numerisch klassifizierter Zählergebnisse der Stoltera (Kreis Rostock). *Zeitschrift für Geologische Wissenschaften* 5/1: 39–49.

Walther, M. (1990), Untersuchungsergebnisse zur jungpleistozänen Landschaftsentwicklung Schwansens (Schleswig-Holstein). *Berliner Geographische Abhandlungen* 52.

Weisse, R. (1977), Struktur und Morphologie von Kames und Endmoränen in den mittleren Bezirken der Deutschen Demokratischen Republik. *Zeitschrift für Geomorphologie*, NS Suppl. 27: 29–45.

Woldstedt, P. (1935), *Geologisch-morphologische Übersichtskarte des norddeutschen Vereisungsgebietes im Maßstab* 1:1 500 000 mit Erläuterungen. Preußisch-geologische Landesanstalt, Berlin.

5 Periglacial Geomorphology

Else Kolstrup

Introduction

Many present landscape elements in western and central Europe (Fig. 5.1) are to a large extent the result of periglacial processes that prevailed during cold periods more than 10,000 years ago. As with the glacial chapter, this account of the periglacial geomorphology also needs to base itself upon processes that no longer or only to a limited extent take place in the areas today. Consequently, this chapter will include an overview of some of the most important periglacial processes and deposits and their effects upon landscape development as influenced by variations in periglacial environmental conditions, lithology, and vegetation cover. Most landscapes that were glaciated during the Weichselian have accentuated relief, especially where subsequent human modification has been relatively modest. It is probable that the glaciated parts of Europe were also accentuated after the Saalian glaciation, but today smooth surfaces and gentle slopes characterize the Saalian areas. During interglacial periods relatively little landscape modification has taken place, and the difference in morphology between the Weichselian glacial landscape and the areas beyond is mainly due to the activity of periglacial processes. As a consequence these European landscapes can be regarded as periglacial.

Even where a periglacial overprinting can be strongly demonstrated in the geomorphology of many western European landscapes the expression 'periglacial landscape' has not been widely used. There may be two main reasons for this. First, even if landforms resulting from periglacial processes may be geographically widespread, they are not normally as eye-catching and morphologically diverse on a local scale as are those resulting from glacial activity (Fig. 5.2). Secondly, it is difficult to geographically delimit a periglacial area: in relation to a glaciated area where the criterion is whether the ice was there or not, the delineation of a periglacial area is dependent on much more subtle features and arbitrary criteria. Further, landscapes that show general imprints of past periglacial conditions often contain areas that bear identifiable imprints of the dominant activity of a single agent, such as water, wind, or gravity. Even if some of these activities may be particularly efficient in cold climates, they are nevertheless of a wider occurrence. Thus areas with these characteristics that can be related to a certain agent may therefore often get a more specific genetic landscape classification. It follows that an association with cold climate conditions will mainly be revealed when a region is seen in overview. To further complicate the matter, periglacial processes may have taken place in formerly glaciated areas and thus locally overlap geographically with a glacial geomorphology. Likewise they can overprint or become overprinted by fluvial landforms that developed under non-periglacial conditions. Therefore, large-scale periglacial European landscapes often represent complex situations with a range of operating regimes and associated processes and they are not easily defined and delimited. However, in some cases very characteristic and conspicuous forms and deposits can result and become preserved. In areas that have experienced repeated cycles of periglacial activity the phases of built-up, continuation, and decay in these activities imposed a series of changes through time upon these landscapes. Formerly periglacial landscapes therefore reflect complex histories of alternating prevailing processes and associated landscape development in response to climate change. Also, the periglacial imprints often form discontinuous series that are poorly

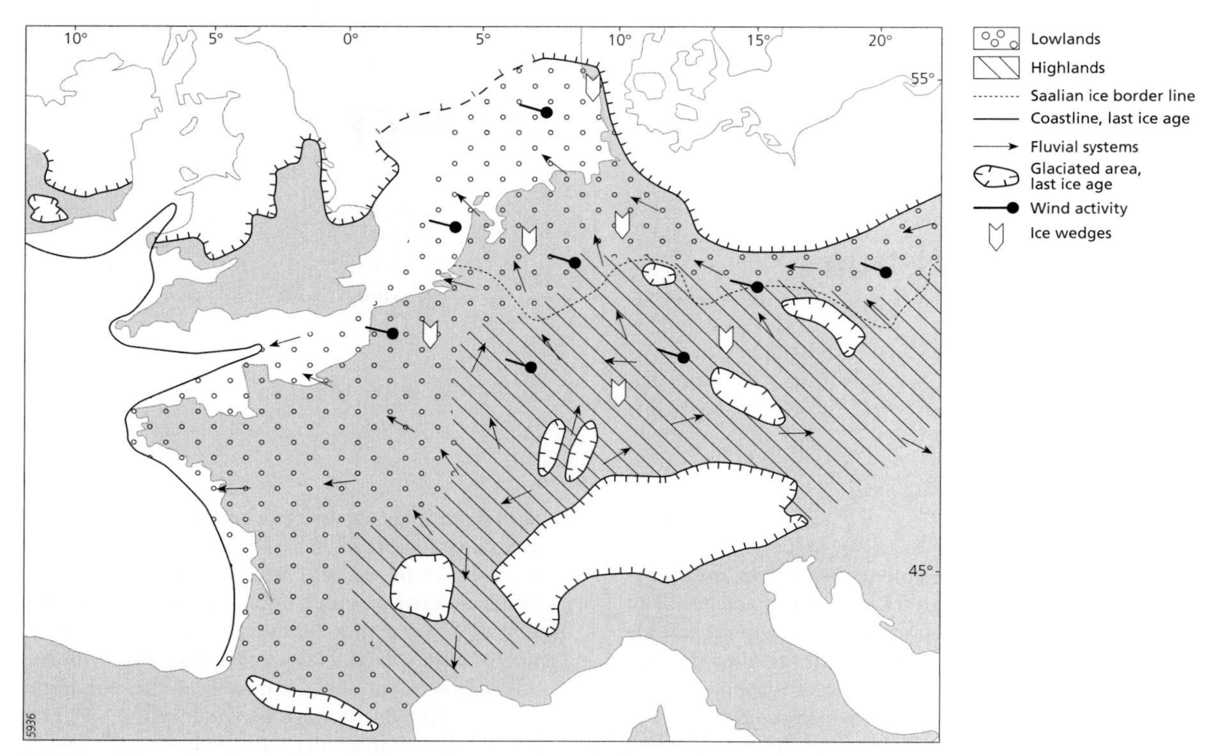

Fig. 5.1. Western and central Europe during the time of Weichselian sea level lowstand and Weichselian maximum ice extension when the whole area periodically experienced periglacial conditions. Traces of former permafrost have been found in the area between the Scandinavian ice cap and the Alps. Permafrost probably extended to the northern part of France. The location of the former shoreline is approximate.

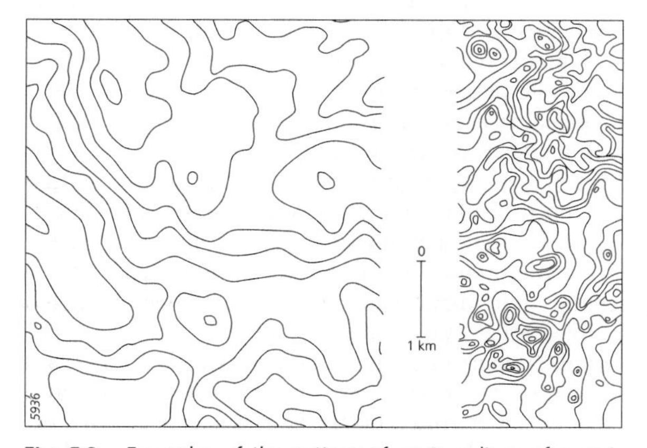

Fig. 5.2. Examples of the pattern of contour lines of a periglacially smoothed Saalian land surface to the left and a Weichselian glacial landscape to the right. The contour line interval is 5 m (based upon maps in *Danmark 1:100 000 Topografisk Atlas*. 1982: 43, 48. 1st edn. Copenhagen, Geodætisk Institut.

dated. As a consequence it would be a futile effort to attempt to map successive European periglacial palaeoreconstructions. Instead this chapter focuses on indicator forms and land-forming processes as well as on the underlying principles for landscape development during long periods of pronounced and rapid cold-climate changes during the (late) Quaternary.

Terminology

The term 'periglacial' originated in central Europe. Von Lozinski (1909) recognized that well-developed block fields in the eastern Carpathians were a result of mechanical (physical) weathering and found descriptions of similar forms from near the firn line in the Alps and along the Greenland inland ice. Von Lozinski therefore suggested that the forms he had found had developed in a similarly raw climate during the ice ages in areas south of the Pleistocene ice border. The climate in such areas he termed a periglacial climate and he referred to the stone shattering as the periglacial facies of the mechanical weathering (translated). Later von Lozinski (1912) stated that the intensive weathering must have taken place at a very early time when the glaciers started to move from the high North into central Europe and that thus the cooling of the climate was the steering factor for development of the frost shattering.

With time the word 'periglacial' has come to be used in a rather loose way. It covers areas with cold climate conditions and processes in all parts of the world where permafrost exists or has existed, with or without proximity to glaciers or ice caps, and/or areas where frost-related processes strongly dominate or dominated: 'Periglacial environments . . . those in which frost-action and permafrost-related processes dominate' (French 1996). A periglacial environment thus includes areas with permafrost, i.e. areas where the year-round ground temperature is below 0 °C for at least two succeeding years (Bates and Jackson 1987). Formerly, some scientists advocated that the term 'periglacial' should be reserved for areas with permafrost but today the looser definition that includes a wider range of situations, such as intensive frost action, is generally preferred.

The word 'periglacial' is attributive and may precede environment, climate, process, landform, zone, area, domain, or facies. It would have been convenient if a general term had been agreed upon for periglacial sciences. Periglaziologie (e.g. Weise 1983) ('periglaciology') has not become accepted although 'periglaciation' has been used in the title of a textbook to denote the results of the processes upon landscapes (Ballantyne and Harris 1994). 'Geocryology' (Washburn 1979) (cryo—icy cold, frost) is being increasingly used, e.g. Yershov (1998), but is still not generally accepted.

It is hard to define a periglacial climate in terms of specific temperatures because periglacial conditions may exist under a variety of temperature amplitudes (e.g. French 1996) and because a number of non-temperature parameters such as hydrology, vegetation, and lithology can exert great influence locally (Williams and Smith 1989). Since mean annual air temperatures are usually a little colder than the mean annual ground temperatures of the same area (Taber 1943; Brown and Péwé 1973; Black 1976), and since the periglacial regime normally extends into warmer areas than those underlain by permafrost, the upper mean annual air temperature limit for the definition of a periglacial climate would be a few degrees above 0 °C. A definition of periglacial processes becomes somewhat arbitrary because many processes that are particularly efficient in cold regions, such as for example frost shattering and ice segregation (see below), are not unique to the periglacial zone. However, generally speaking these processes have in common that they are dependent on or at least strongly favoured by frost, so that a periglacial landscape bears the imprint of high intensity and high frequency frost-related processes. Towards the colder part of the periglacial zone there are features that are unique to cold environments and indicative of the presence of permafrost, for example, deep thermal contraction cracking with associated wedge growth. In the following sections the use of 'periglacial processes, landscapes and landforms' will be used rather loosely in accordance with the above.

Periglacial Areas

Periglacial conditions are found today in Arctic, Antarctic, as well as in high altitudinal regions. Figure 5.3 gives a simplified overview of the subdivision and distribution of northern hemisphere periglacial areas mainly based upon the map of Brown *et al.* (1997). Areas with continuous permafrost are found in the coldest areas. Towards warmer conditions the percentage of the ground that is underlain by permafrost decreases through discontinuous to sporadic occurrences and at lower latitudinal or lower altitudinal limits only isolated patches of permafrost are found. Beyond these occurrences, the periglacial zone is expressed by the dominance of frost-related processes and landforms.

In Fig. 5.4 a schematic vertical transect between a cold and a warmer area is presented. In the coldest part there is thick and almost continuous permafrost while only isolated patches are present in the warmer areas (cf. also Brown 1970, referred to in Washburn 1979). Also indicated is the presence of still unthawed frozen ground that has remained from the last ice age. As a rule of thumb it can be stated that the colder the mean annual air temperature is and the longer the negative temperatures last during the year and have lasted over time, the colder will the ground be, and the deeper will the negative temperature conditions penetrate. However, local factors such as the presence/absence of vegetation, peat, winter snow cover, water bodies, salts, human activity as well as lithology, aspects of slope and local climatic conditions can be of great importance. Another phenomenon illustrated schematically in Fig. 5.4 is the spatial difference in the thickness of the active layer (i.e. that part of the ground above permafrost that thaws in the summer and freezes in winter) and of the seasonally frozen ground. Where the summer is cool and the period of positive temperatures is short, the active layer may be only 1–2 dm thick. It may increase in thickness to more than 2 m in areas with longer and

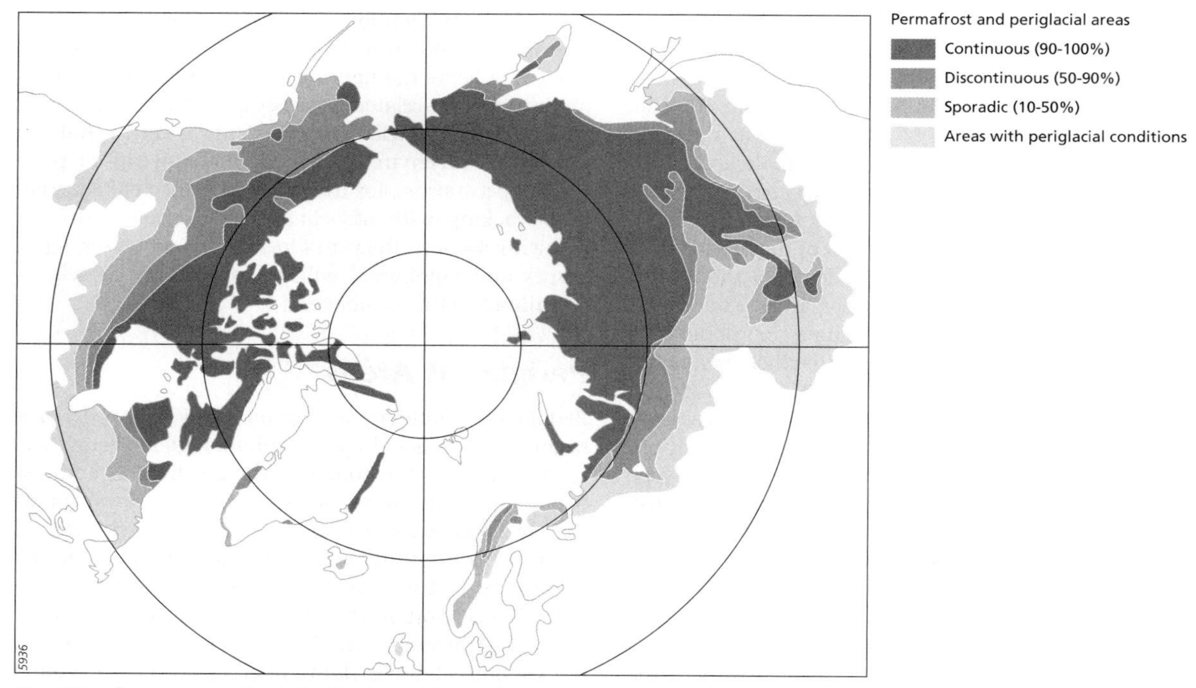

Fig. 5.3. Present permafrost distribution in the northern hemisphere. Simplified from Brown *et al.* (1997) and with a tentative extension of areas with periglacial conditions.

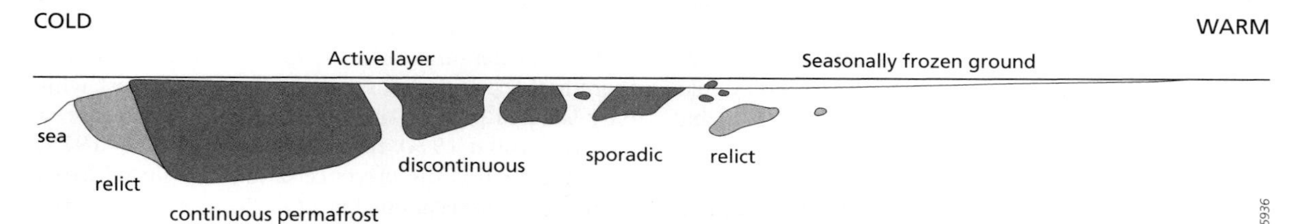

Fig. 5.4. Schematic transect through continuous, discontinuous, and sporadic permafrost in northern latitudes without regard to altitude. Note that the thickness of the active layer and the seasonally frozen ground are proportionally too thick (up to 2–3 m) as compared to the permafrost thickness (up to some hundred metres).

warmer summers. However, the active layer, the layer of freezing and thawing, is difficult to assess in relation to former periglacial conditions.

Basic Processes and their Prerequisites

Traditionally, phase changes between the liquid and solid state of H_2O have been regarded as being of great importance for mechanical rock disintegration processes (physical weathering). This is because the transition from thawed to frozen pure water is accompanied by a volumetric expansion in the order of 9% so that when water freezes in confined rock spaces the increase of volume helps with further disintegration of the rock. However, newer investigations show that in some cases other factors, including the effect of wetting and drying, may be more efficient (e.g. Hall 1993). This rock disintegration results in fragmented rocks and blockfields such as reported by von Lozinski (1909).

Most materials contract and expand when they are exposed to temperature changes. The linear expansion coefficient differs for different rocks, for granite for example it is around 8×10^{-6}/°C at 20 °C (Andersen *et al.* 1989). Pure ice has a higher value, between 30 and 60×10^{-6}/°C (Yershov 1998) and Mackay (1986) found values between 50 and 340×10^{-6}/°C in frozen ground consisting of a mixture of ice and soil particles in Arctic Canada. As a consequence of thermal contraction, cracks can develop in frozen ground during a temperature decrease. During the season of negative air temperatures cold will penetrate from the surface into unprotected ground and moisture will freeze at the freezing front. In areas without permafrost the freezing front is from the top only, but in areas with underlying permafrost there may also be freezing from below although at a slower rate, i.e. two-sided freezing (e.g. Mackay 1980). In unconsolidated sediments and permeable rocks unfrozen water can move through open spaces. Two main principles apply to the freezing of unconsolidated sediments. In sand and gravel pore water will freeze at or just below 0 °C and become pore ice that binds mineral particles together. If the soil is water-saturated from the start, the 9% volumetric expansion may cause the ground surface to heave and/or cause pore water expulsion at the freezing front (Mackay 1998). If the water cannot escape and the unfrozen zone at the same time becomes increasingly narrow, a high-pore water pressure can build up beneath frozen surface sediments. The ground becomes supersaturated as it contains more water than the pore space can contain. As a consequence the particles lose contact and the sediment becomes unstable and/or water moves upwards under pressure.

In silty and clayey soils there can still be unfrozen H_2O at −3 to −5 °C (Beskow 1935; Williams and Smith 1989) owing to an adsorbed water film around individual particles. In silty soils the pore spaces are sufficiently large for water movement but at the same time sufficiently small to make the capillary effect important so that unfrozen water and vapour can move in all directions by suction. In freezing fine-grained soils, unfrozen H_2O will move towards the freezing front and add volume to already existing ice so that, if sufficient moisture is available, lenses or layers of so-called segregation ice develop sometimes with water depleted areas in between (e.g. Mackay 1980). At the ground surface this kind of local subsurface ice accumulation can be reflected as higher areas. Upon melting of such ice concentrations, the excess water cannot always find its way back to where it came from, and the surface layers of such areas can become very unstable.

When positive air temperatures prevail, frozen ground thaws from the top downwards. If an area has high ground moisture content and low permeability a water-rich situation may develop, in particular if there is underlying impermeable frozen ground. In flat terrain the result will be a boggy situation. However, geomorphological consequences of the freeze–thaw processes and related hydrological changes in the ground are particularly effective and well seen in sloping terrain where unstable top layers result, especially where there is underlying permafrost that can act as a sliding plane.

Frost sorting is widespread in cold areas and can result in a large variety of patterned ground (Fig. 5.5), different from the ice wedges mentioned below. As pointed out by French (1996), frost sorting is the result of a complex of processes and van Vliet-Lanoë (1991: 133) presents a table that outlines various factors of importance for this. Frost sorting into patterned ground may include upfreezing of stones, as well as differential frost heave of some areas relative to others with associated lateral movement of gravel and stone-sized particles from higher to lower areas owing to frost creep and gravitational movement. In this way relatively fine-grained higher areas become surrounded by lower-lying concentrations of coarser particles that form nets.

In order that the periglacial processes can take place and be efficient it is necessary that there is sufficient energy exchange between the air and the ground, a situation that is optimal when the ground is free of snow and vegetation. Snow acts as an insulator that prevents exchange of heat, but the degree of insulation depends on air temperature, and rate and frequency of temperature oscillations and their duration. Vegetation plays a role as it helps to trap the snow, and also in other ways, as discussed below.

Vegetation

Vegetation binds the soil and dampens the activity of deflation as well as the effect of slope wash and mass movement processes and, as mentioned above, it helps to trap the snow in winter thus further protecting the ground from cold spells of short duration and/or small amplitude. In the coldest periglacial areas there is short, often discontinuous vegetation where those part of the plants that extend above the snow in winter die off. The species composition in the harshest areas where both cold and dry conditions prevail is made up of lichens and some annual field layer plants. In less harsh environments there are annuals and some perennials, including dwarfed forms of birch and willow, that can survive the winter under the snow and make use of the short summer for their continued existence. With warmer and longer summers the vegetations become denser, more species-rich, and taller, in particular where the

Fig. 5.5. Sorted circles in gravel and stones with some silt in an area of sporadic permafrost (northern Sweden). Presence of vegetation on the surface suggests low periglacial activity. The walking stick in the centre is *c.*1.7 m long.

hydrological conditions are favourable. There is a gradual transition from almost barren areas towards the tree line where tree birch, pine and other tall-growing species gradually take over. The more completely the vegetation covers the ground, the taller it usually is and the more its root net is developed. It is then more efficient in its physical preservation effect on the landscape. This means that where the vegetation is discontinuous or fragmented periglacial processes may become substantially more efficient than in surrounding more sheltered areas.

In present periglacial areas the vegetation is more or less in balance with the general climate regime because fairly constant conditions have existed during the Holocene. Fossil finds from periods with periglacial conditions in Europe show that at that time the vegetation was rich in short plants often of a pioneer type, that could inhabit newly exposed ground before soil development. Yet, some of the species that grew in Europe then do not belong to the high arctic or alpine flora today; rather the European periglacial vegetations seem to have been relatively species-rich with some taxa that require mean July temperatures above 10 °C. This means that there is no direct parallel between past vegetations and those of present periglacial areas (Kolstrup 1990). Instead it seems as if the conditions for the plants in the past have been unstable with repeated disruption of the vegetation cover and the transport of much sediment by wind and water. These conditions were due to rapid and strong climatic changes. As a consequence stable plant communities often might not have had time to become established compared with cold areas today under the same temperature conditions (Kolstrup 1990). It follows that the combination of discontinuous, disrupted vegetation covers and unstable ground surface conditions enhanced frost action and permafrost-related processes, as well as the particular efficiency of wind and running water and so favoured a high rate of landscape change.

Climate and Stratigraphic Background

Continuous records from deep-sea cores and long ice cores show numerous oscillations, interpreted as temperature changes, of different amplitude and duration over the last 2–3 Ma. For example, from records derived from the Greenland ice sheet (Johnsen *et al.* 1995; see also Fig. 5.6) there are more than fifty clearly recognizable, often sudden shifts from colder to warmer conditions and vice versa between the last and present interglacial, i.e. as a mean, a marked change in temperature would have taken place at intervals of less than *c.*1,000 years during the last Ice Age. Several of the temperature shifts were of an order of 5–10 °C. During the coldest periods of the Weichselian northern Poland, northern Germany, and The Netherlands, and possibly also the most northern part of France, are thought to have been sufficiently cold for the development of permafrost at least once (Kolstrup 1980; Ran and van Huissteden 1990; Vandenberghe and Pissart 1993; Lautridou and Coutard 1995) (Figs. 5.1, 5.6). Periodically, the mean annual air temperature may have been up to or even more than 10 °C lower than at present in these areas. There is also evidence that the climate was more continental: mean July temperatures seem to have been below or around 10 °C; winter temperatures decreased even more than mean annual temperatures; precipitation was reduced probably in part due to the lowered sea-level and the presence of the Scandinavian and Alpine ice caps. During earlier ice ages when the ice caps had a larger extension than during the Weichselian, western and central Europe would have experienced even harsher conditions.

One of the coldest parts of the Weichselian is a period of a few thousand years around 20,000 BP. It is not known whether this was a single, relatively long permafrost period or whether fluctuations around critical

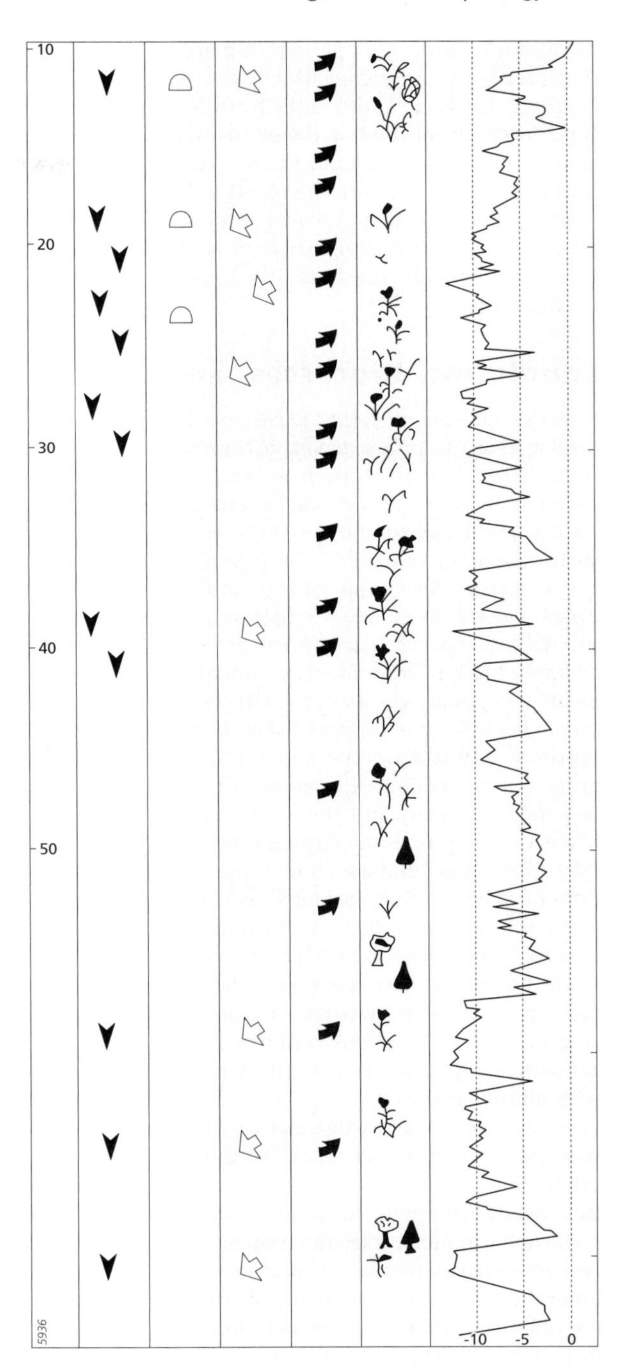

Fig. 5.6. (*right*) Weichselian chronologic table on a timescale in 1,000 years with decreasing certainty of age of forms back to 50,000 BP and extrapolated values beyond that except for the right-hand curve. The columns indicate from left to right: arrowheads: thermal contraction cracks (ice and sand wedges); domes: frost mounds; white arrows: mass movement; black arrows: wind activity; tentative vegetation outline with herbs and, in the older and youngest part, also single conifers and deciduous pioneer trees; to the right is a curve showing the temperature deviation in °C from the present temperature through time from the Greenland (GRIP) inland ice core. The decreasing density of symbols back in time indicates lack of information possibly in addition to less periglacial activity (based upon data from Lautridou 1985; Vandenberghe and Pissart 1993; Johnsen *et al.* 1995; Kolstrup 1995).

values took place. Other periods that are known to have had pronounced periglacial activities are parts of the Younger Dryas (11,000 and 10,000 BP), and some time intervals around and before 40,000 BP whose exact number and duration are not known. Obviously, European landscapes are the result of long time developments during cold and warm periods, but periglacial processes during the coldest parts of the Weichselian have contributed most to the present landscape configuration.

Landforms, Processes, and Climate

Some periglacial processes leave characteristic, recognizable imprints and/or deposits after the cold conditions have ceased to exist. If such imprints and deposits are found in an area without cold climate conditions today, former periglacial conditions can be reconstructed from them. There are two categories of periglacial form. The first is restricted to areas with permafrost. To this category belong large ice wedges and perennial frost mounds as well as thermokarst features. The second category has a wider cold-environmental distribution. It includes slope deposits, aeolian sand sheets and loess, some fluvial deposits, and fragmented rock and stones resulting from physical weathering. Features of the second group are not delimited to areas of permafrost or even to periglacial areas, but they are particularly frequent in cold-climate regions. Remains of the first category reflect former permafrost conditions. Once these conditions have been established for specific time intervals, coexistent forms from the second category can be related to a cold-climate setting with more confidence.

The transition between non-periglacial and periglacial areas is today somewhat arbitrary but the feeling of entering a periglacial environment can most easily be achieved in alpine areas where the temperature decrease with altitude causes relatively major landscape changes over comparably short distances. Across such a transect the frequency and areal distribution increases of solifluction and other active slope deposits on slopes that would otherwise have been stable, of weathered rock rubble in areas with outcropping bedrock, of sorted and unsorted patterned ground, and of the effects of intensified wind activity and fluvial peak events. It is conceivable that similar transitions from non-periglacial to periglacial areas existed in the European past when periglacial conditions prevailed in areas that are now temperate.

It is worth stressing that transitions between permafrost and non-permafrost conditions, as well as between periglacial and non-periglacial conditions, have occurred with variable duration and magnitude. As a consequence cold-dependent landforms could become inactive (fossilized) or decayed until the next cold period sets in. For example, ice wedges might have remained inactive without melting at depth then become reactivated during the next cold phase or, alternatively, they may have melted completely and still become recreated in the same place. Such changing conditions and associated process-landform responses will have occurred repeatedly during the Pleistocene (Fig. 5.7).

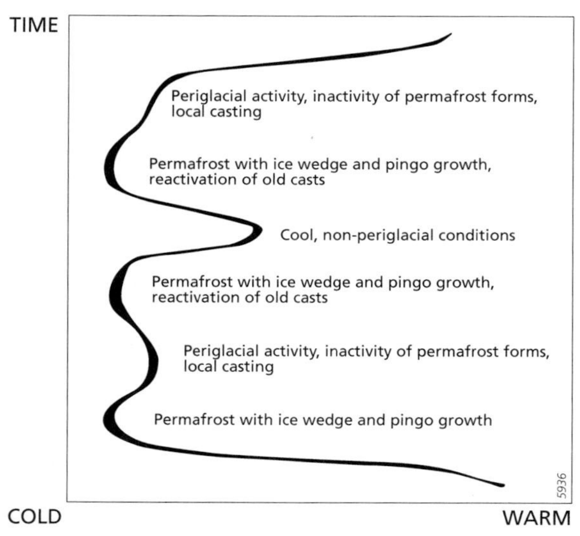

Fig. 5.7. Schematic outline of changes in periglacial activity in relation to temperature change. The timescale is arbitrary.

Landforms Dependent on Permafrost

Ice and Sand Wedges

Pseudomorphs of former ice wedges are more easily recognized than other periglacial forms. Besides, they are relatively widely distributed in formerly periglaciated areas and are regarded as having high climate indicator value because of their restriction to permafrost areas today. Owing to this combination they have attracted much attention and will therefore be treated in relative detail. Open cracks that cut more or less vertically into frozen ground can develop at low negative temperatures due to thermal contraction of the ground once the temperature gradient is sufficiently high. During the summer the temperature in the upper part of the ground increases and causes the sediments to increase in volume again so that the open cracks become narrower or close entirely. If snow or clastic material has entered the cracks before they closed such filling material will

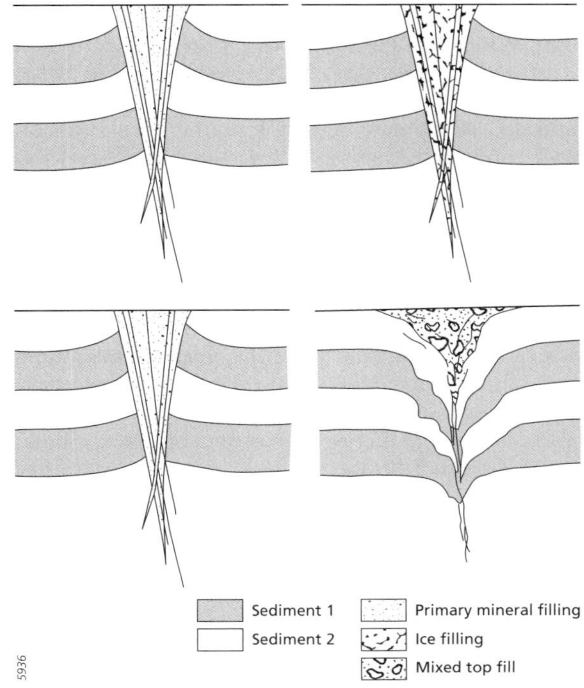

Fig. 5.8. Generalized outline of ice and sand wedges from fully developed wedges with several generations of infill and growth of cracks (upper half) to casts (lower part). The host sediments often have upturned layers towards wedges with original, primary infilling. If the filling was ice-rich the ice melts upon a climatic warming and the surrounding sediments will collapse or be washed into the void as secondary filling.

Fig. 5.9. Weichselian ice wedge cast in sand and gravel (central Jutland, Denmark). Note downturned layers of host sediment along the wedge.

remain. Over time new thermal cracking events can make use of the old zones of weakness and wedges that receive filling regularly can therefore grow in width with time (Fig. 5.8) becoming some decimetres or even more than a metre wide at the top (e.g. Eissmann 1981). The cracks can extend some metres into the ground and when seen from the air they form the outlines of polygonal patterns with net widths of metres to tens of metres.

If sedimentary structures are present in the surrounding host material distinct deformation structures can be recognized in vertical sections at the contact with the wedges (Figs. 5.8, 5.9; Romanovskij 1973). These are due to the pressure caused by wedge growth versus expansion of host sediment. The original (primary) filling material can be pure ice, pure non-ice material (often wind blown sand but also in-washed material from the surface), or any mixture of these (Gozdzik 1973; Romanovskij 1973) dependent on what material is available to enter the open wedge first (Kolstrup 1986, 1995). Following a temperature increase of a sufficient magnitude and duration to thaw the frozen ground, wedges that were filled by clastic material retain their primary infilling and shape (Fig. 5.8) In contrast, wedges with former icy fillings show signs of pressure release of the host sediments and have down-turned host layers and secondary fillings of clastic material where the ice used to be (Gozdzik 1973; Romanovskij 1973; Fig. 5.8). The appearances of the pseudomorphs depend not only on the former proportion of ice to non-ice fill in the wedge and the surrounding ground but also on the structure and texture of the host material. Sand and gravel will normally be relatively stable upon thawing and therefore produce the most regular and easily recognizable casts (Fig. 5.9). On the other hand, silty and clayey sediments may become highly liquefied and thus provide a low preservation potential for casts.

Ice wedge casts forming patterned ground can still be recognized at the ground surface, in particular when seen from the air and if the filling differs in texture from the host sediment. The pattern can also be reflected in crops as these show growth differences on different sedimentological (and nutritional) substrates (e.g. Christensen 1974). However, pseudomorphs of ice wedges are best known from vertical cuttings where the structural and textural characteristics can be studied and where wedges with primary non-ice filling can be observed, sometimes even better than in presently frozen environments. The wedges often exhibit (sub)vertical foliation of the filling that may reflect successive filling events.

The temperature requirements for development of ice and sand wedges are currently under debate. Originally Péwé (1966) proposed that the mean annual air temperature (MAAT) needed to be −6 to −8 °C or colder for the formation of ice wedges. According to Romanovskij (1976), the overall minimum temperature requirements for development of ice wedges in permafrost in Siberia depends on the composition of the host sediment. He suggests (Fig. 2, p. 292) (mean annual ground) temperature requirements of −3 to −5 °C for development of ice wedges in loam and peat, while lower temperatures are required with coarser grain size and it is about −10 °C or below in gravel. Newer investigations show that active perennial wedges can locally be found in areas of MAAT of −4 °C (Burn 1990). Where conditions are particularly favourable for cracking, active wedges can even be found in areas of local marginal permafrost in Lapland where the MAAT is as high as around −1 °C (Westin and Zuidhoff 2001).

During the coldest part of the Weichselian, i.e. around 20,000 BP, ice wedges developed in northern France and in Belgium, and large ice-wedge casts from that time are also found in central Germany (Kolstrup 1980; Karte 1981). Chronologically best confined are the relatively small periglacial phenomena from the Younger Dryas, whereas the age of growth and decay of many other wedge casts from other periods of the Weichselian (Fig. 5.6) are still under investigation.

Frost Mounds

Another group of permafrost indictors are remnants of frost mounds. Active mounds have a sediment cover over an ice(rich) core and can be subdivided into the main forms: hydrostatic (closed system) pingos, hydraulic (open system) pingos, and palsas (including lithalsas). Hydrostatic pingos owe their growth to pore water expulsion and ice aggradation at the freezing front in saturated sand where freezing from the top and underlying permafrost confine excess water in a two-sided freezing system. This causes heave of the overlying ground (Mackay 1978, 1998). Such forms occur in slight depressions, often in groups, in waterlogged localities with sandy and silty sediments (Mackay 1978). The presence of this type is regarded as indicative of continuous permafrost conditions. Hydraulic pingos receive their artesian water from upslope areas from where the water moves in unfrozen parts of the otherwise frozen rock or sediment (Müller 1959; Mackay 1978). As a consequence open system pingos are found in the cold part of the discontinuous permafrost zone; they are often single and located at the lower part of slopes.

Palsas grow in waterlogged areas with peat and owe their elevation to the growth of local segregation ice lenses formed by water suction (e.g. Wramner 1967; Åhman 1977) in areas of relatively little snow cover (Seppäla 1982). During the summer, the peat cover upon the elevated palsas dries out and thereby insulates the underlying icy core from melting (Wramner 1967). In the winter, on the other hand, the peat becomes moist and thereby conductive for the cold to enter the core. Palsas usually occur in groups and the peat cover in the bog and upon the palsas can be from a few centimetres to more than a metre in thickness. For a long time (Wramner 1972) it has been known that there are frost mounds with the same mechanism of growth (i.e. segregation ice) as the palsas but without peat cover. Harris (1993) introduced the term 'lithalsa' to distinguish such forms from those that are peat covered. Palsas and lithalsas can exist in relatively warm permafrost and both have been reported from Swedish Lapland in areas with a MAAT as high as around −1 °C (Westin and Zuidhoff 2001) but they are most frequently found in wet areas in somewhat colder parts of the discontinuous permafrost zone (e.g. Åhman 1977).

Pingos can attain heights of about 50 m and diameters of up to half a kilometre while palsas and lithalsas are smaller, usually less than 10 m in height (Lundqvist 1969; Mackay 1978). The reason that former frost mounds can become preserved as remnants is that when frost mounds grow their sides become steeper and mass movement take place down their sides so that sediments accumulate as a bulge around the periphery. After the ice core of a mound melts the accumulated sediments remain as a rampart around a depression, often water-filled where the ice used to be. Where the accumulated rampart consists of clastic material such forms can remain for millennia after the permafrost conditions have ceased to exist.

Owing to the different environmental requirements of the three frost mound types, it is important that the remnants are properly classified in relation to active forms, a task that has caused some discussion (e.g. Pissart *et al.* 1998). Yet, while the interpretation of what type of mound the fossil forms actually represent has been debated (and thus the more specific temperature conditions related to them) the former existence of permafrost deduced from them is not queried. Ramparts of various sizes and shapes have been reported from different parts of the European continent, for example The Netherlands (Maarleveld and van den Toorn 1955), Belgium (Pissart 1983), and Poland (Rutkowski *et al.* 1998). The present state of interpretation is that the Late Glacial remnants in the Belgian Ardennes most likely represent former palsas of the lithalsas type (Pissart *et al.* 1998), while large ramparts dating from between 25,000 and 19,000 BP in The Netherlands are thought to represent hydraulic system pingos (de Gans 1981).

Thermokarst

Thermokarst features are the third landform type confined to areas with permafrost but it is more difficult to attribute former cold conditions to these features than is the case with the previous two landforms. Permafrost with a thick top layer of almost pure ice is known from arctic areas today. Where the ice melts locally, thermokarst lakes, or alas, can develop. Traces of thermokarst depressions may not leave diagnostic characteristics and are therefore difficult to recognize as such, especially because such areas may be expected to end up as relatively high-lying areas in the terrain once the ice in the surrounding ground has melted. Also, thermokarst remnants can be confused with forms of another origin, in particular in dead ice terrain. The possibility of traces of former thermokarst lakes has been discussed for a few localities in Denmark (Funder 1983) and in The Netherlands (de Gans 1981).

Landforms Related to the Active Layer and Seasonally Frozen Ground

The forms mentioned above are dependent on the presence of permafrost. They are usually rather conspicuous but do not modify the topography to a large extent. This is in contrast to landforms described below, which are favoured in their development by the presence of an active layer (over permafrost) and/or influenced by pronounced seasonal freezing. Slope processes, expressed in a variety of ways, are one of the most important landforming agents in periglacial regions since they cause widespread denudation in high areas and accumulation on lower slope segments and in valleys. Another process that contributes to periglacial landscape modification is that which produces sorted and unsorted patterned ground. Such forms are rarely preserved as relict surface patterns in Europe but it is possible that some forms of involuted (cryoturbated) ground (Fig. 5.10) are related to such patterns. The precise periglacial environmental requirements and confinements for this type of patterned ground are still insufficiently known (e.g. van Vliet-Lanoë 1988).

Slopes, Mass-Related Processes

Mass-related slope processes are dependent on gravity, and their speed and efficiency depends on slope angle, water content of the soil, lithology, and vegetation cover. In periglacial areas rapid mass movement preferably occurs in areas where free drainage into deeper soil layers is prevented. This may be due, for example, to the presence of impermeable ground such as permafrost or areas with remnants of annual ice layers beneath

Fig. 5.10. Deformation of sand and gravel deposits due to cryoturbation (location in west Jutland, Denmark).

Fig. 5.11. Active layer earthflows over impermeable permafrost on a large valley slope (Ellesmere Island, Canada).

thawing ground and snow in the spring. In such situations a water-saturated horizon can develop on top of the frozen ground that acts as a sliding plane (Fig. 5.11). Due to the gravitational forces topsoil from large areas can slide or slump downwards or the top layer may come down as a liquefied mass in the case where it is less cohesive. If slopes are steep and have a high water content loose material may come down in the form of a slurry or debris flow that ploughs its way downslope at high speed and accumulates at the foot of the slope and further into the valley. Slope failure can occur repeatedly on the same slope, in particular in areas where permafrost or deep seasonal freezing exist. This is because each time the thawed top layer has been removed the process can be repeated in the next decimetres of ground.

In areas of outcropping bedrock, weathering processes may help to break up the rock and thus provide new material that will slide, fall, or roll downslope, particularly once periglacial conditions take over. In areas with such conditions as in the northern Alps and the central European mountains this method of large-scale mass wasting seems to have been widespread and abundant during Pleistocene cold intervals as demonstrated by, for example, Terhorst (2001).

Frost creep and solifluction cause slow slope movement. Frost creep is the result of frost heave perpendicular to the soil surface during freezing and (sub)vertical settling upon thawing (Washburn 1979). Solifluction is 'the slow viscous downslope flow of waterlogged soil and other unsorted and saturated surficial material' (Bates and Jackson 1987). Often the word gelifluction is used for the combined effects of the two in periglacial environments and it can occur upon slopes as gentle as 1–2°. In western Europe there are many gently sloping landforms and it is estimated that slope deposits from mass wasting are probably one of the most common groups of sediments. There are a number of reports on vertical exposures through mass-related sediments that drape the lower slope segments and fill adjacent valleys in central and western Europe (Klosterman 1995; Tricart and Raynal 1969; Woldstedt and Duphorn 1974). More recently, geotechnical methods have been applied to the lateral and vertical distribution of 'head' deposits. In combination with vertical exposures through the slope deposits, the importance of cryoturbation and solifluction could be demonstrated, in combination with aeolian input, to have been common in the highlands of central Germany (Völkel *et al.* 2001). Another example that demonstrates the existence of former mass movement is that of Michel (1975) who recognized overriding lobate forms of solifluction lobes with diverse lithology in the Paris area. Some of the deposits mentioned in the literature are several metres thick and may derive from various parts of the Pleistocene, but many of them are from the cold periods of the Weichselian.

Notwithstanding the widespread occurrence of slope deposits in Europe, there is still a relative scarcity of detailed structural and textural descriptions of mass-related slope deposits. This may partly be due to a lack of good exposures as such deposits are not usually industrially useful and therefore not generally exposed. To some extent there may also be difficulties in distinguishing mass-derived deposits from other sediments, especially when they are often mixed or interdigitated with water lain or aeolian deposits. The importance of slope processes under periglacial conditions is consequently primarily seen from the widespread occurrence of the smoothed landforms they helped to create.

Slopes and Surface Runoff

In areas with frequent slope failure, the vegetation will be disturbed and discontinuous and therefore running water can become an efficient agent. In periglacial areas with snow patches the spring melt period is usually sudden (e.g. French 1996), resulting in winter precipitation that runs off over short periods with high discharges. On slopes over (still) frozen soil the snow meltwater will often run off as unconcentrated sheet flow. This can transport mineral particles downslope even where slope angles are very gentle, thus further contributing to the levelling-out of the landscape. Where water comes down concentrated in gullies high discharges may cause efficient and deep erosion in areas with relief. It appears that slope wash has been a particularly efficient process in levelling the landscape. For example, at the transition of the Dutch high push moraines of Saalian age and the surrounding lower-lying plains large alluvial fans with exceptionally gentle slopes of less than $^1/_4$° have been build up (Kolstrup 1980).

In the context of slope deposits, the presence of *Grèzes litées* should also be mentioned. These deposits are not widespread but have drawn relatively much attention. They are bedded slope deposits that can be several metres thick and they are thought to be the result of a combination of frost weathering of calcareous rock and slope wash or solifluction or debris flow or other slope processes in a periglacial environment. The deposits are found in parts of France and Germany (Ozouf *et al.* 1995; Karte 1983), but modern parallels to ascertain their development are extremely scarce.

Valleys and Drainage

The effect of running water in periglacial areas is most clearly marked in low-lying areas where major drainage systems cause remodelling of valleys and riverbeds. Rivers are characterized by a peak discharge period of a duration that reflects the snow melt period of the drainage basin; and in present periglacial areas snow accumulated over several months can melt and be drained from the area in a few days (French 1996). Also during the Pleistocene in Europe it seems as if riverbeds could be filled to the brim in spring and that lateral, thermal erosion of (still) frozen riverbanks could occur (e.g. Dylik 1969; Kolstrup 1985).

In the North-west European Lowlands the most marked and extensive valley systems are kilometre-wide, flat-bottomed, so-called *pradolinas* or *Urstromtäler*. These forms can be seen as transitional between a glacial and periglacial origin, with water from the melting Scandinavian ice sheet as well as from the periglacial drainage area of snowbeds, local ice caps, mountain glaciers, and rainwater.

Marked changes over time in the fluvial regime and the sea level resulted in alternating incision and accumulation of sediment in many riverbeds and, as a result, there are sets of terraces along many present and former drainage systems (e.g. Lautridou 1985). Within former wide valley systems where large riverbeds could develop over a downstream stretch of only a few kilometres (e.g. Büdel 1977) the rivers of today seem merely symbolic.

During periods of discontinuous vegetation cover and the presence of permafrost, surface water would run off quickly. This resulted in strong sediment transport including transport of coarse gravelly material supplied in part from periglacial weathering in the area (Gibbard 1988). During most of the cold phases there does not seem to have been forest in Europe, except possibly in southern France (e.g. Büdel 1977: 44; fig. 17 for the last ice age). Further north the vegetation may have been insufficient to protect the ground for long periods at a time (Kolstrup 1990). As a consequence fluvial transport and sedimentation will have been high, not only during times of permafrost conditions, but also during intervening warmer intervals when the vegetation had not had sufficient time to establish. According to Gibbard (1988), 90% of the sediments in the north-west European fluvial sequences are of cold-climate origin.

Dependent on vegetation cover, frozen ground, snowbed conditions, and sediment load and relief, the fluvial regimes changed over time. The precise interrelationship between these parameters and fluvial activity is complex, but generally speaking braided river systems with wide valley tracks existed during the coldest periods with permafrost conditions, while ephemeral anastomosing rivers might have prevailed during moderately cold periods when denser, yet limited, vegetation was present in the floodplains (e.g. Mol *et al.* 2000).

From present data it is hard to estimate the importance of the geographic configuration of Europe, for example the effect of earlier thaw in the southern part of a northwards-directed drainage pattern. Flooding by (melt)water from the southern regions has most probably been frequent over extensive, still-frozen areas, and during times of permafrost the effects might have been accentuated, as a modest parallel to present-day Siberian situations, so that large parts of the North-west European Lowlands would have been swamped during spring. Owing to the level topography, the unstable vegetation situation and the periodically water-rich environment, the picture of the area during times of permafrost or deep seasonal freezing could well have been without present-day parallels.

Valleys of Mixed Origin

In north-west Europe and the Alpine foreland asymmetric, usually dry, valleys that no longer experience fluvial processes have attracted some attention. The overview in Weise (1983) gives a good impression of their variety as to aspect and the diversity of explanations for their development. Similar forms are also found in The Netherlands, Germany, and northern France (Lautridou 1985). In high latitudinal periglacial areas asymmetric valleys are also found (e.g. French 1996). The valleys are characterized by a steeper, 12–30° slope facing a more gentle one with an angle of 4–6°. In Europe most asymmetric valleys have the steeper slope facing towards the east, but this picture is far from consistent and it even seems as if a difference in altitude within the same area can be of importance (Weise 1983: 124). As a consequence their origin is being debated and the factors responsible for them include the combined effects of: distribution of snowbeds in relation to wind direction, melting of the snow and the effects of the meltwater, effects of slope aspect in relation to the sun and associated thawing of frozen ground, solifluction in relation to all these, runoff, erosive effect of former drainage at the valley bottom, and more. However, there is general agreement that they are the result of processes that prevailed in former periglacial environments.

Aeolian Landforms and Processes

Widespread deposits of loess and so-called cover sand, that are both of aeolian origin, seem strongly related to periods with periglacial environmental conditions. In cold areas today there is aeolian activity in, for example, Alaska (e.g. Trainer 1961) and northern Canada, and accumulation of sediment can be very rapid locally (Lewkowicz and Young 1991). Nevertheless, aeolian activity in cold regions today is far less widespread than seems to have been the case during parts of the Pleistocene in Europe.

As mentioned above, abundant slope processes exposed the ground, and repeated activity by running (melt)water caused erosion and deposition of sediment during the cold phases in Europe. In combination with quick changes in climate this made it difficult for vegetation to become established even if the climate was not always particularly harsh. Periglacial weathering would have been frequent in exposed areas causing *in situ* breakdown of stones into smaller particles (cf. Lautridou 1985). As large parts of Europe are built up of unconsolidated sediments, and as ground surfaces became repeatedly exposed, there would have been abundant availability of sediments for physical weathering and transport. The source areas for aeolian sand include local (e.g. Crommelin 1964) as well as more distant areas such as the exposed North Sea flats. It is therefore not surprising that Weichselian deposits of aeolian origin are geographically widespread.

The aeolian deposits that are characteristic of the periglacial phases can be subdivided into sand sheets, the so-called cover sands, and loess. These sediments together with dunes of various age and origin are treated more extensively in Ch. 8. The cover sands are (sub)horizontally laminated sands (Fig. 5.12) of which the more or less rounded particles of each successive, well-sorted lamina are normally of medium sand size (the so-called younger cover sand type) or there may be alternations between laminae of sand and silt (older cover sand type) (van der Hammen 1951). The sheet-like appearance of the aeolian deposits is more common in western Europe than in the eastern part where aeolian sands tend to form dunes (Nowaczyk 1986; Böse

Fig. 5.12. Aeolian cover sand in central Jutland (Denmark). The photo is by the author and is reproduced with permission from VARV. The trench is 1.8 m deep.

1991). Towards the north, within the areas that were glaciated during the Weichselian, aeolian deposits are most common in the areas that became ice-free first. Possibly owing to the fact that similar sand sheets do not seem to accumulate in temperate areas today, no satisfactory explanation has been found for the layering of the cover sands.

Loess normally forms massive deposits with particles in the silt fraction and they are found in a region extending from north-western France through Germany to central Europe (e.g. Rohdenburg and Meyer 1979; Lautridou 1985). The geographical transition between the cover sands and loesses is gradual and alternations between these deposits are found in a number of localities.

Aeolian deposits are most abundant from the middle and late part of the Weichselian (e.g. van Vliet-Lanoë 1989). They are between a few centimetres and several metres thick, and in closed depressions thicknesses up to several tens of metres have been ascertained. They often interdigitate with other periglacial deposits on slopes and along rivers and occasionally thermal contraction cracks and involutions are found within the sequences. The sands drape all types of landforms. On exposed areas deflation surfaces with wind-polished stones are found at the surface (Christensen and Svensson 1998). The aeolian activity has thus helped to further level out the topography. Beds with faceted and polished pebbles are found at several levels within Weichselian lithological sequences and there is thus indication for repeated phases of aeolian deflation and associated accumulation. Probably the best impression of the frequency and importance of aeolian deposition at large may be gained from low-lying loess accumulation areas (Rohdenburg and Meyer 1979; Lautridou 1985) where sediments might not have been as easily eroded and where the sequences are therefore relatively thick.

Present and Former Periglacial Processes and Landscapes: A Discussion

Processes that took place during the European periglaciation are also active today, but the conditions under which they were active may have differed substantially from present environments and, as a consequence, the results of their activity may have been expressed in different ways or happened at different rates. In present high-latitude areas of Alaska, Canada, and Siberia the climate has experienced only minor temperature changes in the order of 1–2 °C over the last *c.*10,000 years. This is in contrast to the quick shifts of temperature of substantial amplitude during the Late Pleistocene when continuous, discontinuous, and non-permafrost conditions have succeeded each other in the same area within relatively short time intervals.

In environmental reconstructions the stable 'normal' environmental conditions of today have traditionally been used as a standard for the reconstruction of environmental conditions for forms of similar origin in the past. It is, however, likely that stable conditions similar to those in present-day cold climates may have been virtually non-existent during the periglacial phases in Europe. This makes parallels between general environmental conditions for development of similar forms of the past and the present of dubious value. The 'normal' Pleistocene palaeo-situation with quick, dramatic climate changes may form a very rare or even non-existent situation in periglacial areas today. The implications are many. For example, thermal contraction cracking of the ground could happen in the past once the conditions of frozen ground, appropriate temperature gradient, and other factors were satisfied. But it is uncertain how far the distribution and abundance of relict forms can be seen in parallel with the distribution and abundance of similar forms in active situations. Cracking is dependent on specific conditions in relation to frost penetration into frozen ground, and from the sedimentological and palaeobotanical records it can be proposed that the soil surface conditions in Europe were much more exposed than in present cold areas. This is attested to by the vegetation that had difficulty in becoming established and, once it did, was destroyed again. The relatively modest vegetation cover would have influenced the capacity of the vegetation to catch a snow cover to protect the ground. Instead the snow could easily blow off, leaving barren soil surfaces open to cold spells as well as to deflation and surface wash, even under cool and temperate temperature conditions. This would not happen today without dramatic, often man-made disturbances. It follows that periglacial processes may have had an exceptionally strong impact upon the landscapes. It might also be suggested that, for some periods that were not particularly cold, scarcity of vegetation and drought could have been more important than low temperature.

Also physical weathering might have been particularly efficient under environmental conditions characterized by accelerated timescales and amplitudes. Set against this background, reliable reconstructions of past periglacial environmental and landscape developments are very difficult to make. Many questions cannot be answered by comparison with mid-latitude areas of alpine permafrost today because these areas, after all, also represent situations where the environments have been relatively stable for millennia. The only way to approach the problem seems to be to take into account all possible combinations within the boundary conditions for each landform or phenomenon and its requirements, from one extreme over the norm to the other, including the time factor and the implications of changes over time at a given locality. An example of an environmental situation that is very unusual and 'extreme' today, but which might have been the norm in the past, could be an area where there is lack of vegetation to catch the snow and where formation of new permafrost into unprotected ground with subsequent thermal contraction cracking can create exceptionally wide cracks within 2–3 years. However, such a situation might have been common in sparsely vegetated European landscapes during the sudden changes of warm to cold conditions during the Pleistocene. Today a similar situation is restricted to local, recently drained or eroded areas before they become colonized by surrounding vegetation (Mackay 1986). Conversely, the 'normal' situation of today in many areas, for example the presence of thick peat layers or massive ice accumulations at the top of permafrost build-up over long periods in extensive arctic areas, may have been an extreme or even non-existent situation in the past and so, as a consequence, would landform developments dependent on the presence of such deposits.

Concluding Remarks

The present overview of periglacial landforms and processes does not pretend to be complete. Forms of dubious origin and/or without clear periglacial significance, for example cryoplanation terraces (see e.g. Weise 1983), have been left out. Likewise chemical weathering and its effects on the landscapes is not included even though there is increasing awareness of the importance of this factor in cold environments. Instead the overview has chosen to concentrate on forms that are thought to be of high environmental indicator value and to outline how periglacial conditions remodelled the landscape configuration. At the same time the dynamics of temperature and precipitation changes have been stressed. These influenced the vegetation which in turn influenced the rate at which physical weathering, slope-related processes, and fluvial and wind activity could have acted. The dynamics also have importance for the debate on the presence or absence of permafrost and to set the background for alternative interpretations of the development of various frost-related phenomena. A lack of parallels with present day 'normal' periglacial situations necessitates the use of a deductive piecemeal approach that combines all thinkable prerequisites and processes with the time factor for an understanding of how European landscapes developed during periods of cold-climate change.

References

Åhman, R. (1977), Palsar i Nordnorge. *Meddelanden från Lunds Universitets Geografiska Institution*, Avhandlingar LXXVIII.

Andersen, E. S., Jespersgaard, P., and Östergaard, O. G. (1989), Databog fysik kemi. F&K Forlaget, Copenhagen.

Ballantyne, C. K., and Harris, C. (1994), *The periglaciation of Great Britain*. Cambridge University Press, Cambridge.

Bates, R. L., and Jackson, J. A. (1987), *Glossary of geology*. 3rd edn. American Geological Institute, Alexandra.

Beskow, G. (1935), Tjälbildningen och Tjällyftningen med särskild hänsyn till vägar och järnvägar. *Sveriges Geologiska Undersökning*, C.375, *Årsbok 26 (1932)*, 3. 1–242.

Black, R. F. (1976), Periglacial features indicative of permafrost: ice and soil wedges. *Quaternary Research* 6: 3–26.

Böse, M. (1991), A palaeoclimatic interpretation of frost-wedge casts and aeolian sand deposits in the lowlands between Rhine and Vistula in the Upper Pleniglacial and Late Glacial. *Zeitschrift für Geomorphologie*, Suppl. 90: 15–28.

Brown, J., Ferrians, O. J., Jr., Heginbottom, J. A., and Melnikov, E. S. (1997), *Circum-Arctic map of permafrost and ground-ice conditions*, 1:10,000,000, Map CP-45. Department of the Interior, United States Geological Survey, Washington, DC.

Brown, R. J. E., and Péwé, T. L. (1973), Distribution of permafrost in North America and its relationship to the environment: A review, 1963–1973. In: *Permafrost: North America contribution to the second international conference on permafrost, Yakutsk, Siberia*. National Academy of Sciences, Washington, 71–100.

Burn, C. R. (1990), Implications for palaeoenvironmental reconstruction of recent ice-wedge development at Mayo, Yukon Territory. *Permafrost and Periglacial Processes* 1: 3–14.

Büdel, J. (1977), *Klima Geomorphologie*. Gebrüder Borntraeger, Berlin.

Christensen, H. H., and Svensson, H. (1998), Windpolished boulders as indicators of a late Weichselian wind regime in Denmark in relation to neighbouring areas. *Permafrost and Periglacial Processes* 9: 1–21.

Christensen, L. (1974), Crop-marks revealing large-scale patterned ground structures in cultivated areas, southwestern Jutland, Denmark. *Boreas* 3: 153–80.

Crommelin, R. D. (1964), A contribution to the sedimentary petrology and provenance of young Pleistocene cover sand in The Netherlands. *Geologie en Mijnbouw* 43/9: 389–402.

de Gans, W. (1981), *The Drentsche Aa Valley System*. Ph.D. thesis, Vrije University, Amsterdam.

Dylik, J. (1969), Slow development under periglacial conditions in the Lódz region. *Biuletyn Peryglacjalny* 18: 381–410.

Eissmann, L. (1981), Periglaziäre Prozesse und Permafroststrukturen aus sechs kaltzeiten des Quartärs. *Altenburger naturwissenschaftliche Forschungen*, 1. Mauritianum, Naturkundliches Museum, Altenburg.

French, H. M. (1996), *The Periglacial Environment*. 2nd edn. Longman, London.

Funder, S. (1983), Sibiriens permafrost—og Danmarks. *VARV* 2: 50–63.

Gibbard, P. L. (1988), The history of the great northwest European rivers during the past three million years. *Phil. Trans. R. Society London B* 318: 559–602.

Gozdzik, J. (1973), Geneza i Pozycja stratygraficzna struktur peryglacjalnych w srodkowej polsche (with English summary). Lodzkie towarzystwo naukowe Soc. Sc. Lodz. *Acta Geographicae Lodziensia*, 31.

Hall, K. (1993), Rock moisture data from Livingston Island (Maritime Antarctic) and implications for weathering processes. *Permafrost and Periglacial Processes* 4: 245–53.

Harris, S. A. (1993), Palsa-like mounds developed in a mineral substrate, Fox Lake, Yukon Territory. Proceedings of the 6th International Conference on permafrost, Beijing, 238–43.

Johnsen, S. J., Dahl-Jensen, D., Dansgaard, W., and Gundestrup, N. (1995), Greenland palaeotemperatures derived from GRIP bore hole temperature and ice core isotope profiles. *Tellus* 47B: 624–9.

Karte, J. (1981), Zur Rekonstruktion des weichselhochglazialen Dauerfrostbodens im westlichen Mitteleuropa. *Bochumer Geographische Arbeiten* 40: 59–71.

—— (1983), Grèzes Litées as a special type of periglacial slope sediments in the German Highlands. *Polarforschung* 53: 67–74.

Klosterman, J. (1995), Nordrhein-Westfalen. In: L. Benda (ed.) *Das Quartär Deutschlands*. Bornträger, Stuttgart, 59–94.

Kolstrup, E. (1980), Climate and stratigraphy in northwestern Europe between 30,000 BP and 13,000 BP, with special reference to the Netherlands. *Mededelingen Rijks Geologische Dienst* 32/15: 181–238.

—— (1983), Cover sands in southern Jutland (Denmark). *Proceedings of the Fourth International Conference of Permafrost, Washington*. National Academy Press, 639–44.

—— (1985), Late Pleistocene periglacial conditions in Blaksmark near Varde (Denmark). *Geologie en Mijnbouw* 64: 263–9.

—— (1986), Reappraisal of the upper Weichselian periglacial environment from Danish frost wedge casts. *Palaeogeography, Palaeoclimatology, Palaeoecology* 56: 237–49.

—— (1990), The puzzle of Weichselian vegetation types poor in trees. *Geologie en Mijnbouw* 69: 253–62.

—— (1995), Palaeoenvironments in the north European lowlands between 50 and 10 ka BP. *Acta zool. Cracoviensia* (Poland), 38/1: 35–44.

Lautridou, J.-P. (1985), *Le Cycle périglaciaire Pleistocene en Europe du nord-ouest et plus particulierement en Normandie*. University of Caen, i and ii.

—— and Coutard, J.-P. (1995), Le problème de l'extension et de la profondeur du pergélisol Pléistocène en Normandie (France du nord-ouest) *Quaestiones Geographicae*, Special Issue 4: 201–3.

Lewkowicz, A. G., and Young, K. L. (1991), Observations of aeolian transport and niveo-aeolian deposition at three lowland sites, Canadian Arctic Archipelago. *Permafrost and Periglacial Processes* 2: 197–210.

Lundqvist, J. (1969), Earth and ice mounds: a terminological discussion. In: T. L. Péwé (ed.) *The Periglacial Environment, Past and Present*. McGill-Queens University Press, Montreal, 205–15.

Maarleveld, G. C., and van den Toorn, J. C. (1955), Pseudo-sölle in Noord-Nederland. *Tijdschrift van het Koninklijk Nederlandsch Aardrijkskundig Genootschap* 72: 344–60.

Mackay, J. R. (1978), Contemporary pingos: a discussion. *Biuletyn Periglacjalny* 27: 133–54.

—— (1980), The origin of hummocks, western Arctic coast, Canada. *Canadian Journal of Earth Sciences* 17: 996–1006.

—— (1986), The first 7 years (1978–1985) of ice wedge growth, Illisarvik experimental drained lake site, western Arctic coast. *Canadian Journal of Earth Sciences* 23: 1782–95.

—— (1998), Pingo growth and collapse, Tuktoyaktuk Peninsula area, western arctic coast, Canada: a long-term study. *Géographie physique et Quaternaire* 52: 271–323.

Michel, J. P. (1975), Périglaciaire des environs de Paris. *Biuletyn Peryglacjalny* 24: 259–352.

Mol, J., Vandenberghe, J., and Kasse, C. (2000), River response to variations of periglacial climate in mid-latitude Europe. *Geomorphology* 33: 131–48.

Müller, F. (1959), Beobachtungen über Pingos. *Meddelelser om Grönland* 153/3: 1–127.

Nowaczyk, B. (1986) Wiek Wydm w Polsce. Universyet im. Adama Mickiewicza w Poznaniu, *Seria Geografia 28. Poznan*, 230–45. English abstract: The age of dunes, their textural and structural properties against atmospheric circulation pattern of Poland during the Late Vistulian and Holocene.

Ozouf, J.-C., Texier, J.-P., Bertran, P., and Coutard, J.-P. (1995), Quelques coupes caractéristiques dans les dépôts de versant d'Aquitaine septentrionale: faciès et interprétation dynamique. *Permafrost and Periglacial Processes* 6: 89–101.

Péwé, T. L. (1966), Paleoclimatic significance of fossil ice wedges. *Biuletyn Peryglacjalny* 15: 65–73.

Pissart, A. (1983), Remnants of periglacial mounds in the Hautes Fagnes (Belgium): structure and age of the ramparts. *Geologie en Mijnbouw* 62: 551–5.

—— Harris, S., Prick, A., and van Vliet-Lanoë, B. (1998), La Signification paleoclimatique des lithalses (palses minerales). *Biuletyn Peryglacjalny* 37: 141–54.

Ran, E. T. H., and van Huissteden, J. (1990), The Dinkel valley in the middle Pleniglacial: dynamics of a tundra river system. *Mededelingen Rijks Geologische Dienst* 44/3: 209–20.

Rohdenburg, H., and Meyer, B. (1979), Zur Feinstratigraphie und Paläopedologie des Jungpleistozäns nach Untersuchungen an Südniedersächsischen und nordhessischen Lössprofilen. *Landschaftsgenese und Landschaftsökologie* 3: 1–89.

Romanovskij, N. N. (1973), Regularities in formation of frost-fissures and development of frost-fissure polygons. *Biuletyn Peryglacjalny* 23: 237–77.

—— (1976), The scheme of correlation of polygonal wedge structures. *Biuletyn Peryglacjalny* 26: 287–94.

Rutkowski, J., Krol, K., and Lemberger, K. (1998), The pingo remnants in the Suwalki lake region (NE Poland). *Quaternary Studies in Poland* 15: 55–60.

Seppälä, M. (1982), An experimental study of the formation of palsas. In: H. M. French (ed.) *The Roger J. E. Brown Memorial Volume. Proceedings Fourth Canadian Permafrost Conference*. Calgary, Alberta: National Research Council Canada, Ottawa, 36–42.

Taber, S. (1943), Perennially frozen ground in Alaska: its origin and history. *Bulletin of the Geological Society of America* 54: 1433–548.

Terhorst, B. (2001), Mass movements of various ages on the Swabian Jurassic escarpment: geomorphologic processes and their causes. *Zeitschrift für Geomorphologie*, NS Suppl. 125: 105–27.

Trainer, F. W. (1961), Eolian deposits of the Matanuska Valley agricultural area Alaska. *United States Geological Survey Bulletin*, 1121-C: 1–35.

Tricart, J., and Raynal, R. (1969), *Périglaciaire de l'Alsace à la Méditerranée*. Livret-guide de l'excursion C 15. VIIIe Congrès INQUA, Paris.

Vandenberghe, J., and Pissart, A. (1993), Permafrost changes in Europe during the Last Glacial. *Permafrost and Periglacial Processes* 4: 121–35.

van der Hammen, T. (1951), Late-Glacial flora and periglacial phenomena in the Netherlands. *Leidse Geologische Mededelingen* 17: 71–183.

van Vliet-Lanoë, B. (1988), The significance of cryoturbation phenomena in environmental reconstruction. *Journal of Quaternary Science* 3: 85–96.

—— (1989), Dynamics and extent of the Weichselian permafrost in western Europe (substage 5e to stage 1). *Quaternary International* 3/4: 109–13.

—— (1991), Differential frost heave, load casting and convection: Converging mechanisms; a discussion of the origin of cryoturbations. *Permafrost and Periglacial Processes* 2: 123–39.

Völkel, J., Leopold, M., and Roberts, M. C. (2001), The Radar signatures and age of periglacial slope deposits, central highlands of Germany. *Permafrost and Periglacial Processes* 12: 379–87.

von Lozinski, W. (1909), Über die mechanische Verwitterung der Sandsteine im gemässigten Klima. *Bulletin International de l'Académie des Sciences de Cracovie. Classe des Sciences Mathematique et Naturelles* 1: 1–25.

—— (1912), Die Periglaziale Fazies der mechanischen Verwitterung. *Compte rendu de la XIe session du Congrès Internationale Géologie, Stockholm 1910*, 1039–53.

Washburn, A. L. (1979), *Geocryology. A survey of periglacial processes and environments*. Edward Arnold, London.

Weise, O. R. (1983) *Das Periglazial. Geomorphologie und Klima in gletscherfreien, kalten Regionen*. Gebr. Borntraeger, Berlin.

Westin, B., and Zuidhoff, F. S. (2001), Ground thermal conditions in a frost-crack polygon, a palsa and a mineral palsa (lithalsa) in the discontinuous permafrost zone, Northern Sweden. *Permafrost and Periglacial Processes* 12: 325–35.

Williams, P. J., and Smith, M. W. (1989), *The frozen earth, fundamentals of geocryology*. Cambridge University Press, Cambridge.

Woldstedt, P., and Duphorn, K. (1974), *Norddeutschland und angrenzende Gebiete im Eiszeitalter*. Koehler, Stuttgart.

Wramner, P. (1967), Studier av palsmyrar i Laivadalen, Lappland. *Meddelanden från Göteborgs Universitets Geografiska Institution* 86: 435–49.

—— (1972), *Palslika bildningar i mineraljord*. GUNI Report 1, Göteborg, Göteborg University, Institute of Physical Geography.

Yershov, E. D. (1998), *General Geocryology. Studies in Polar Research*. Cambridge University Press, Cambridge.

6 River Environments, Climate Change, and Human Impact

Eduard Koster

Introduction

In this chapter a short overview of the evolution, geomorphological expression, sedimentary records, and discharge and sediment regimes of the major rivers in western Europe is presented. The rivers Elbe, Weser, Rhine, Meuse, Scheldt, Seine, Loire, Garonne, Rhône, and Danube (Fig. 6.1A and Table 6.1) will be separately reviewed but not necessarily in this order and not with equal attention. Emphasis is placed on the Quaternary record and most issues are exemplified by a discussion on phenomena and processes in the Rhine–Meuse delta (Fig. 6.1B). As almost all these rivers are strongly influenced by man's activities, attention is also focused on river management practices, both in a historic context and at present. Finally, modern concepts and plans concerning river conservation and rehabilitation are briefly examined.

TABLE 6.1. *Physical characteristics of major rivers in western Europe*

	L	D	Q_m	Q_{m^*}	Q_v		
Elbe	1,150	149,000	850	—	150	—	3,000
Weser	725	46,400	320	—	105	—	1,350
Danube	2,860	817,000	6,425	—	3,000	—	8,000
Rhine	1,320	185,000	—	2,200	575	—	12,600
Meuse	935	36,000	—	250	app. 0	—	3,000
Scheldt	430	22,000	130	—	10	—	575
Seine	760	76,000	270*	—	30	—	1,300
Loire	1,010	115,000	875	—	app. 0	—	1,350
Garonne	580	57,000	630	—	70	—	8,000
Rhône	812	99,000	1,700	—	420	—	5,000

Note: All figures are approximations (compiled from various sources).

L = length (km); D = drainage area (km^3); Q_m = mean discharge (m^3/s) near the mouth; * near Paris; Q_{m^*} = mean discharge (m^3/s) at the Dutch–German border; Q_v = discharge variability (min.-max.)(m^3/s).

Development of the Major Drainage Systems

The foundations of the modern drainage system in north-western Europe were laid in the Miocene when earth movements associated with Alpine orogenesis and the opening of the North Atlantic were at their height (Gibbard 1988). During the Late Tertiary–Early Quaternary the North Sea basin was dominated by an extensive fluvial system that drained the Fennoscandian and Baltic shield through the present Baltic Sea (Overeem *et al.* 2001; Fig. 6.2). The dimensions of this (former) drainage system were enormous; through empirical relationships based on recent fluvio-deltaic systems the drainage area is estimated to have been in the order of 1.1×10^6 km^2. Cenozoic marine and fluvial sediments reach a thickness of more than 3,500 m in the North Sea basin. Quaternary sediments with a thickness of over 1,000 m imply a tenfold increase in sedimentation during this period in comparison to the Tertiary infilling. The fluvial system of Miocene to Middle Pleistocene age has been referred to as the Baltic River system (Bijlsma 1981). It is also designated as the Eridanos delta system by Overeem *et al.* (2001) named after the legendary Eridanos river in northern Europe mentioned in Greek records (7th century BC). In a seismo-stratigraphic study Overeem *et al.* (2001) have documented the large-scale basin-fill architecture in terms of external forcing by tectonics, sea-level variations, and climate. The development of this drainage system

Fig. 6.1. A. The major rivers and river catchments in western Europe.

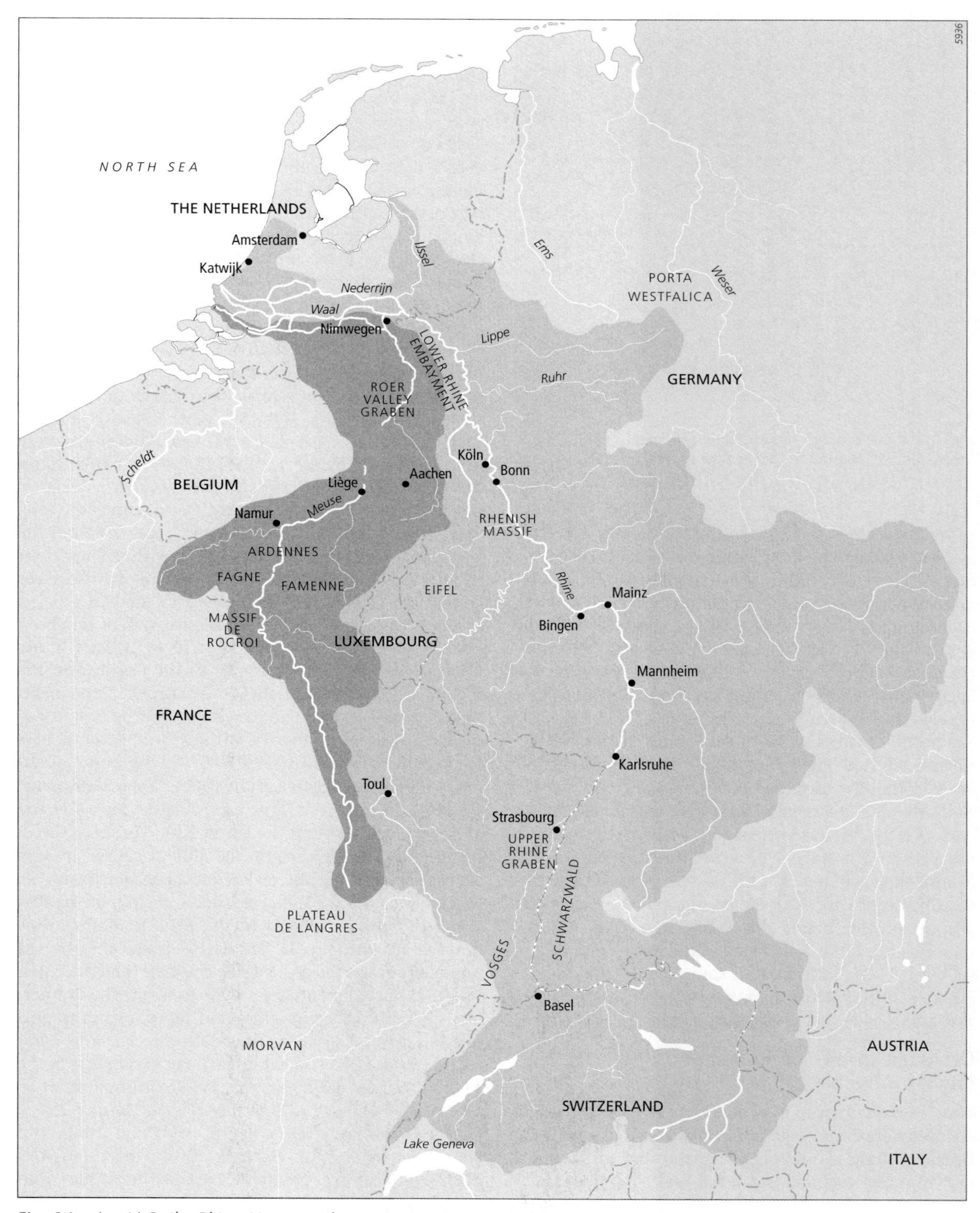

Fig. 6.1. (cont.) B. the Rhine–Meuse catchment.

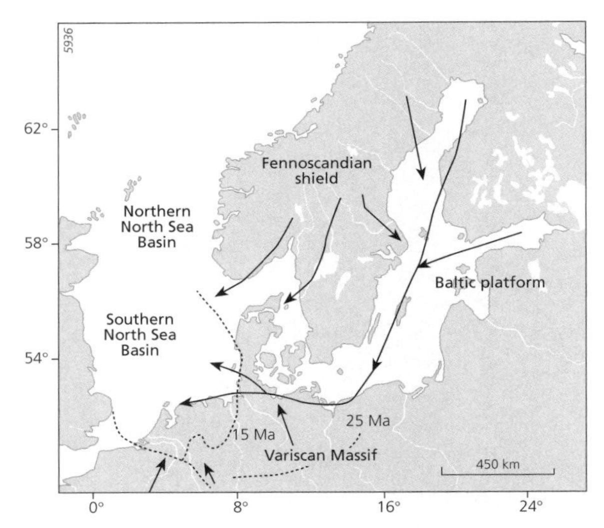

Fig. 6.2. The Eridanos fluvio-deltaic system and coast lines at 25 Ma and 15 Ma (after Overeem *et al.* 2001).

is attributed to the simultaneous Neogene uplift of the Fennoscandian Shield and the accelerated subsidence of the North Sea basin. The German proto-Elbe and proto-Weser rivers formed tributaries to the main system, according to Bijlsma (1981). Heavy mineral and gravel assemblages of these fluvial deposits clearly indicate a Fennoscandian provenance. During the Pliocene and Early Pleistocene sand and gravel were deposited by a braided river system in The Netherlands and the North Sea basin. During the Early Pleistocene (Waalian) stage this river system, including the German rivers as tributaries to the Eridanos, still covered a large part of The Netherlands. This huge river system was joined by precursors of the Rhine and Meuse at this time (Bijlsma 1981; Gibbard 1988). The North German rivers continued to deposit their sediments in the eastern Netherlands until the Middle Pleistocene. The changes in the palaeogeography of the major drainage lines from the late Pliocene until the late Pleistocene were compiled by Gibbard (1988) from various sources, and illustrated in Fig. 6.3. In this figure reconstructed coastlines and the extension of ice sheets are also shown.

Drainage to the North Sea and the English Channel

The drainage lines of most large rivers in western Europe were more or less already established in the Miocene. Drastic changes, however, occurred due to the interplay of major tectonic, sea-level, and climate changes, as explained above. In particular repeated ice sheet advances during the Middle and Late Pleistocene strongly changed the courses of the Elbe and Weser rivers. This was also the case for the proto-Rhine and proto-Meuse river systems.

The first traces of the valley systems of the Elbe and Weser date back to the Miocene. During most of the Pleistocene the proto-Elbe river system drained a large part of the Central European Uplands. This was either in a north-east direction towards the region of the present Oder valley, or to the north into the western Baltic Sea, or to the west joining the Eridanos system. It is only since the Late Saalian–Early Weichselian glacial periods that a connection was formed between the upper Elbe region (the catchment of the Saale and Elster rivers) and the north-west-oriented, lower reach near Hamburg with its present outlet in the North Sea (Lüttig and Meyer, 1974). The evolution of the Weser drainage system and that of the Ems river since the Late Miocene is very similar to that of the Elbe (Lüttig 1974; Thiermann 1974). Before reaching the Porta Westfalica (location indicated in Fig. 6.1B), where the Weser enters the low-lying North German Lowlands, the upper reach of the Weser is an antecedent upland river with unimportant tributaries. Similarly to the Elbe, the present Weser river course was established during the Late Saalian.

The precursor of the Rhine–Meuse system has been equally active since the Late Miocene (≈ 10 Ma). In the Miocene the coastline retreated to the north from the region around Köln in the Lower Rhine Embayment (LRE). During this time extensive lignite (brown coal) layers formed here (up to 135 m thick) in a fluvio-deltaic environment (Schirmer 1990). In the Pliocene a connection developed between the Rhenish shield drainage system and the rivers flowing from the Alps and the Molasse basin through the Upper Rhine graben (URG). During the Late Pliocene the first fluvial, coarse-grained deposits covered the LRE region and adjacent parts of The Netherlands. Heavy mineral studies document the first influx of Alpine material in the URG from that time onwards. Eventually, in the Early Pleistocene the sea retreated beyond the present-day coastline, and since that time fluvial sedimentation dominated in The Netherlands. The lithostratigraphic and morphostratigraphic division of the Pleistocene fluvial deposits is highly complex and has been studied intensively in all parts of the Rhine drainage area (Quitzow 1974; Brunnacker *et al.* 1982; Gibbard 1988; Schirmer 1990; Hantke 1993; Ruegg 1994; Fetzer *et al.* 1995; Bridgland 2000; van Huissteden *et al.* 2001; Boenigk 2002; Meyer and Stets 2002). The upstream Alpine part of the Rhine is an

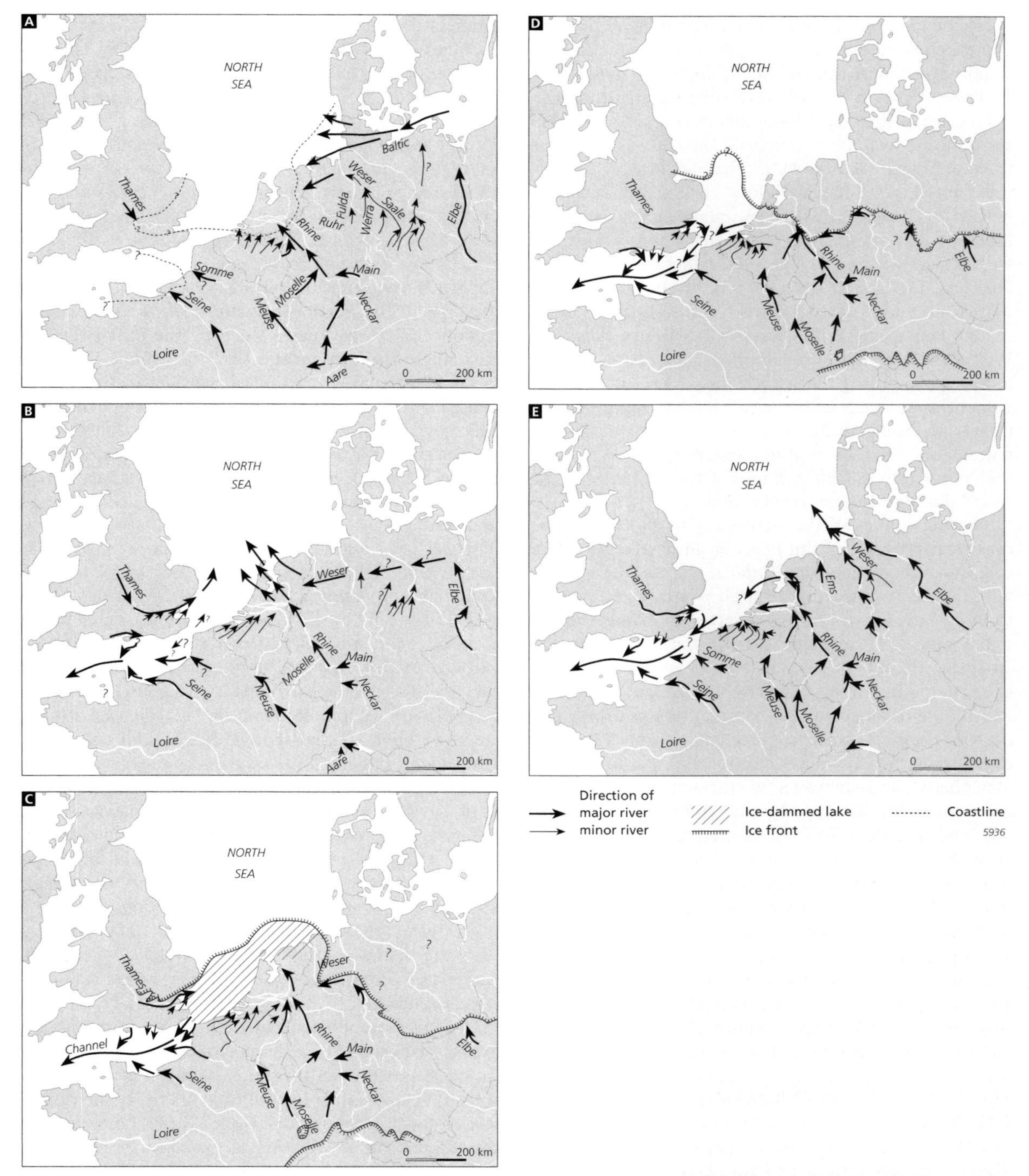

Fig. 6.3. Schematic palaeogeographical reconstruction of the major drainage lines; A. during the late Pliocene to early Pleistocene stages at high sea-level stand; B. Middle Pleistocene 'Cromerian Complex' at low sea-level stand; C. Elsterian-Anglian Stage at the glacial maximum; D. Saalian-Wolstonian Stage at the glacial maximum; E. early Weichselian-Devensian Stage at low sea-level stand; (after Gibbard 1988).

energetic mountain river, which has experienced a very complex evolution. A proto-Alpine Rhine connected to a proto-Danube during the Miocene and most of the Pliocene. Only at the Pliocene–Pleistocene transition did the proto-Rhine near Basle shift its course to the north through the tectonic structure of the URG as far as the town of Bingen (Hantke 1993; Boenigk 2002). The section from Basle to Bingen follows the tectonic structure of the URG between the Vosges uplands to the west and the Schwarzwald uplands to the east (for locations, see Fig. 6.1B). Downstream of Bingen the character of the river changes. The Middle Rhine, as this section is called, is incised in a deep narrow canyon through the Rhenish massif, consisting of the Hunsrück, Taunus, Eifel, and Westerwald uplands. The Rhenish massif was uplifted more rapidly during the last 800 ka, as a result of which volcanism revived and a river terrace flight formed (Meyer and Stets 2002). Near Bonn, where the landscape flattens out, the Lower Rhine enters the Lower Rhine Embayment. In the flat lowlands of The Netherlands the river bifurcates into several branches.

The main lines of the hydrographic pattern in the north-eastern part of the Paris basin developed on an Oligocene-Miocene erosion surface. A proto-Meuse probably flowed to the west through northern France. In the Pliocene, captures have diverted the waters of the upper reach of the Meuse to the north through the Ardennes (Pissart 1974). The Meuse, which had been a tributary of the Rhine until the Cromerian in the Middle Pleistocene (Zonneveld 1974), presently has a distinctly narrow drainage basin of a peculiar form. The river now rises in the Plateau of Langres in France and follows a northerly course between two Jurassic questa ridges of the Paris basin. Between Namur and Liège, the Meuse has eroded a deep, SW–NE oriented valley perpendicularly through the Hercynian Ardennes which consist of a series of resistant rocks, such as the Palaeozoic (Cambrian, Upper Devonian, and Carboniferous) quartzites, sandstones, and limestones of the Massif of Rocroi and of the Fagne-Famenne region. It is generally believed that river incision into these very resistant rocks could keep pace with the epirogenetic upheaval of the Ardennes, thereby forming a clear example of an antecedent valley. Over time, the Meuse has lost considerable parts of its original drainage basin through river captures: in the west by tributaries of the Seine and the Scheldt and in the east by tributaries of the Rhine; e.g. the Mosel, which is now a major tributary of the Rhine (with a length of 545 km and a catchment area of 28,300 km^2). The Mosel originally was part of the Meuse drainage basin; it was captured by the Rhine near the town of Toul in north-eastern France. This occurred probably during the Holsteinian interglacial in the Middle Pleistocene. As a result, the Meuse has lost a large part of the headwaters of the former Upper Meuse that drained the crystalline Vosges area.

The drainage basin of the Scheldt covers a major part of the low-lying regions of central and western Belgium. Its origin is very different from that of the Meuse river. The drainage pattern with many, more or less parallel, south–north flowing tributaries dates from the late Miocene, when the sea finally retreated from this part of Belgium. These rivers, which flowed over a gently sloping (marine erosion) surface, followed the retreating coastline to the north. Thus, the course of these consequent valleys was determined by the initial slope of the land. In the northernmost part of Belgium the rivers show an east–west orientation parallel to the former coastline, which maintained its position during the Pliocene and early Pleistocene. During the Middle Pleistocene incision was accompanied by the replacement of the consequent river system in the eastern Scheldt basin by a system more adapted to the geological structure (Vandenberghe and de smedt 1979). Deep river incisions occurred and terrace flights developed (Tavernier and de Moor 1974). Most of these deep incisions were again filled up during the Last Glacial period (Kiden 1991). It is only in the lower reach of the Scheldt, the so-called Flemish Valley, that terrace levels formed partly related to sea-level rises during the Middle and Late Pleistocene (Holsteinian and Eemian) transgressions (Tavernier and de Moor 1974). Likewise, during the second part of the Holocene the Scheldt catchment again came under the influence of sea-level rise. This caused an accelerated rise of the groundwater table and an extensive peat layer developed filling palaeochannels and spreading over the Late-glacial flood plain (Kiden 1991). From AD 1000–1100 onwards tidal influence rapidly penetrated the lower part of the catchment, resulting in erosion of the peat and deposition of clay. The lower reach of the Scheldt is strongly influenced by tides; the tidal amplitude varies from 4 to 5 m. In The Netherlands the river now widens to a broad estuary called Westerschelde.

The courses of the Seine and Somme rivers are strongly determined by the geological structure of the region. The valley of the Upper Seine crosses an area of tectonically stable, low-altitude plains in the north-eastern part of the Paris Basin. The course of the Lower Seine and the Somme occupy the axes of NW–SE trending synclinal structures of supposedly Miocene age. During the Pleistocene cold stages both rivers were tributaries of the River Manche or 'Channel River', which exported sediments into the central deeps of

Fig. 6.4. Incised meanders of the Seine in the neighbourhood of Rouen (photo: Yvette Dewolf).

the Channel (Antoine *et al.* 2000). The rivers exhibit well-developed, primarily climate-controlled, terrace systems recording incision that began around 1 Ma. The uplift rate deduced from the chronology of these terraces is *c.*55–60 m/Ma since the end of the Early Pleistocene (Antoine *et al.* 2000); these values are in good accordance with those determined for the Meuse valley near Maastricht for the same period (van den Berg and van Hoof 2001). It is also possible that, in north-western France, tectonic uplift has been enhanced during glacial stages by the development of a forebulge (i.e. upward bulging of the crust at the margin of an ice sheet to compensate for glacio-isostatic depression of the crust under the ice sheet). Similar to the Somme, the present Lower Seine, downstream of Paris, is characterized by a spectacular series of meanders (Fig. 6.4). These are deeply incised into Tertiary sands and clays and Jurassic limestones forming the gentle rolling chalk of the Upper Normandy Plateau (Gibbard 1988). The Somme is strictly not one of the 'great rivers' in western Europe, but it is renowned for the abundance of mammalian remains and Palaeolithic artefacts recovered from its deposits. On the alluvial plain of the Upper Seine numerous human settlements have also been found from the younger Magdalénien (13,000 BP) onwards. Several periods of erosion and deposition have been identified since that time by archaeological excavations that yielded Late Palaeolithic, Mesolithic, Neolithic, and Bronze Age artefacts in association with palaeobotanical and lithostratigraphical analyses (Bourdier 1974; Roblin-Jouve 1995).

Drainage to the Atlantic Ocean, the Mediterranean, and the Black Sea

The large catchment of the Loire is geologically very heterogeneous, consisting of deformed Proterozoic and Palaeozoic crystalline and sedimentary rocks, around which are wrapped a succession of Mesozoic and late Cenozoic marine and continental strata (Straffin *et al.* 1999). Volcanic rocks cover parts of the catchment in the Massif Central and Morvan regions. Tertiary fluvial and lacustrine sediments locally fill rift valleys, which form the principle valley axes of the upstream reach of the Loire and its tributaries, the Arroux, and the Allier. During the Pliocene–Early Pleistocene the Loire was still a tributary of the Seine.

The drainage area of the Garonne covers about 57,000 km^2. The river receives water from the major part of the northern Pyrenees, from the southern part of the Massif Central by large tributaries such as the Tarn, the Lot, and the Dordogne, and from the Aquitaine basin. The river course follows a wide valley with a complex sequence of Quaternary sediments. A major part of the catchment consists of Tertiary deposits: predominantly Eocene sediments in the upstream part, Oligocene sediments in the northern half of the catchment, and mainly Miocene and Pliocene sediments south of the river, including the foothills of the Pyrenees.

In the upstream part of the Rhône, especially the 200-km-long section from Lake Geneva to Lyons, the river crosses the southernmost extension of the Cretaceous and Jurassic rocks of the Jura mountains. This part of the river is partly incised in morainic and fluvioglacial deposits forming deep gorges. At Lyons the river is joined by the Saône, which drains the south-eastern part of the Parisian basin. The rivers follow the Saône–Rhône graben (or 'Sillon Rhodanien') extending from the relatively flat region of Bresse in the north to the Rhone delta in the Camargue. South of Lyons the river is joined by large tributaries, such as the Isère, the Drôme, and the Durance, draining the French Alps and alpine forelands. The river also receives water from the eastern part of the Massif Central and the Cévennes.

In the Late Tertiary, during deposition of the upper freshwater molasse in what is now the region of the South German Alpine foothills, drainage of the Danube to the west still prevailed. In the Pliocene the drainage direction reversed, and the river flowed eastwards towards the Black Sea. In the Pliocene and Early Pleistocene the catchment of the Danube extended much further north than today; the river Main was a tributary of the Danube and not of the Rhine as today. During the successive Middle and Late Pleistocene glaciations the Danube became a northern Alpine ice-marginal stream, collecting meltwater from a large part of the Alpine glaciated area (Ehlers 1996). The geological succession of the Last Glacial Danube terrace deposits is relatively well known (Buch 1989). In contrast to other investigations, Buch (1989) concludes that Late Glacial and Holocene terrace formations were not only determined by the climate-controlled developments of the Alpine glaciation, but were mostly due to self-controlled mechanisms of the river system itself.

Terrace Chronology and Climate

The interplay of climate and tectonics was the main forcing factor in the upper and middle reaches of most rivers in western Europe, whereas climate and sea level played a dominant role in the river dynamics of the lower reaches (Tebbens *et al.* 2000). Bridgland (2000) stipulated that cyclic fluctuations of climate during the Quaternary have driven the generation of terraces, through the direct and indirect influence of both temperature and precipitation on fluvial activity. In theory it is assumed that the base of each terrace marks the early part of the cooling phase of a climate cycle, shortly after each interglacial. The top of each terrace is assumed to represent the early part of the next warming phase, shortly after the subsequent glacial maximum (Westaway 2002). In particular, Middle and Late Pleistocene terraces in north-western Europe beyond the reach of the erosive activities of Pleistocene ice sheets can commonly be seen to have formed in synchrony with glacial–interglacial cycles or with longer-periodicity megacycles (Bridgland 2000; van den Berg and van Hoof 2001; Meyer and Stets 2002). However, climate forcing alone is not enough, uplift is also necessary. Many regions show evidence of an uplift in the order of several hundred metres since the Late Pliocene. Bridgland (2000) concludes his overview of terrace formation in north-western Europe by suggesting that the redistribution of eroded material from fluvial catchments to depositional basins may also contribute to the uplift by causing an isostatic response to crustal unloading. Finally, a comparison of many terrace records from different regions in Europe has revealed that terrace sequences appear best developed when regions are uplifting at rates of ≈ 0.07 mm/a^{-1}, a rate that is frequently observed for the Late and Middle Pleistocene (van den Berg and van Hoof 2001; Westaway 2002).

The Rhine drained the Lower Rhine Embayment more or less parallel to the main tectonic faults. The extension of the main terrace deposits in this region, according to Boenigk (2002), is illustrated in Fig. 6.5. As there are very extensive exposures in the Lower Rhine basin, owing to the presence of huge lignite opencast mines, the drainage patterns and sediment characteristics of the fluvial deposits are known in detail.

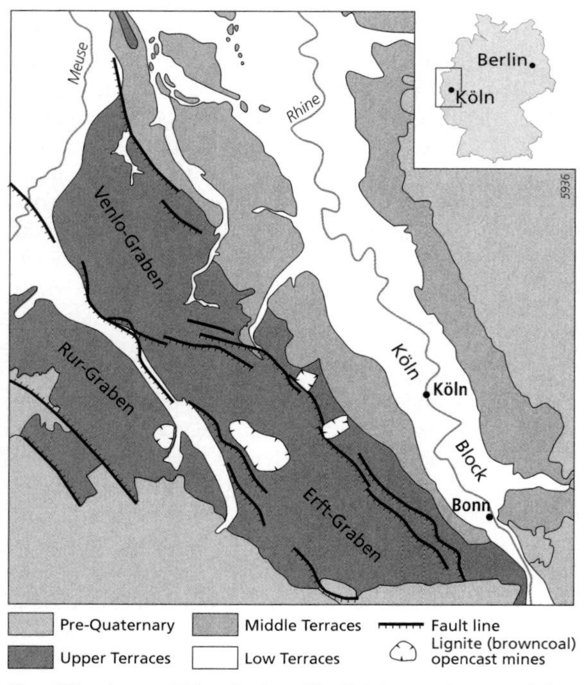

Fig. 6.5. Lower Rhine basin with Pleistocene terraces (after Boenigk 2002).

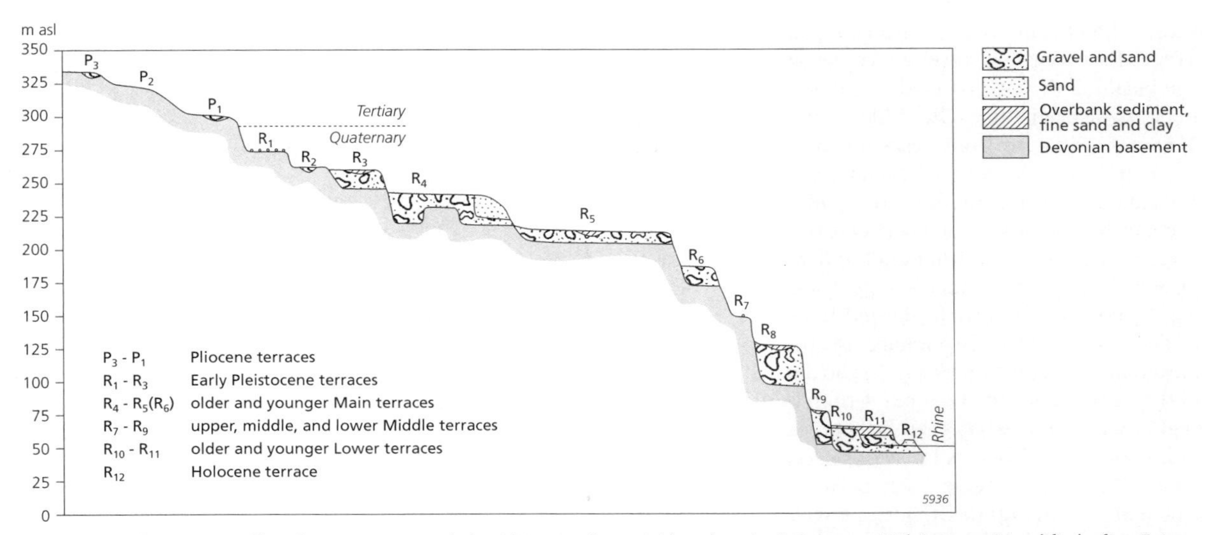

Fig. 6.6. Schematic profile of river terraces of the Rhine in the neighbourhood of Cologne (Köln)-Bonn (simplified after Fetzer *et al.* 1995)

Fig. 6.7. Coarse-grained, Late Weichselian (late Allerød/early Younger Dryas) Rhine deposits exposed near Krefeld, Germany; pumice from the Laacher See eruption is found at the base (photo: Leo Tebbens).

Successive glacial–interglacial cycles resulted in an intricate system of river terraces, the so-called Older Lower Pleistocene terraces, the Older and Younger Main Terraces (*Ältere* and *Jüngere Hauptterrassen*) of Early to Middle Pleistocene age, the Upper, Middle, and Lower Middle Terraces (*Obere, Mittlere,* and *Untere Mittelterrassen*) of the second part of the Middle Pleistocene, and finally the Older and Younger Lower Terraces (*Ältere* and *Jüngere Niederterrassen*) of Late Pleistocene (Weichselian) age (Hantke 1993; Fetzer *et al.* 1995; Hoselmann 1996). A schematic profile of these terraces is presented in Fig. 6.6. Geomorphology, sedimentological features, and petrographic (heavy mineral and gravel) composition of the sediments enabled this division. Moreover, the volcanic activity in the Eifel region, which was at its maximum in the Oligocene, was reactivated after the Brunhes-Matuyama transition (≈ 780 ka), resulting in a series of tephra layers intercalated in these Middle and Late Pleistocene fluvial sequences. The Older Lower Pleistocene terraces appear to predate the Jaramillo Event (≈ 1,000 ka), based on palaeomagnetic measurements. The Older and Younger Main Terraces originate from respectively before and after the Brunhes-Matuyama boundary (Hoselmann 1996). Volcanic eruptions continued during the Late Pleistocene; the latest eruptions in the Eifel occurred around 11,000 BP during the Allerød Interstadial, depositing the well-known Laacher See pumice (Fig. 6.7). These volcanic ash layers have greatly facilitated the establishment of a detailed terrace chronology in the Lower Rhine Embayment (Schirmer 1990). In the downstream reach of the Rhine, near the German–Dutch border, there is a region where there has been little preservation of fluviatile sediments either as terraces or beneath the valley floor. This is the hinge area between the subsiding North Sea basin and the uplifting area in Germany. The progressive tilting of the Rhine catchment towards the North Sea basin has given rise to a clear convergence of the Rhine terraces downstream towards the hinge area. Figure 6.8 illustrates the reconstructed gradient lines of the Rhine for the last *c*.800,000 years to the present.

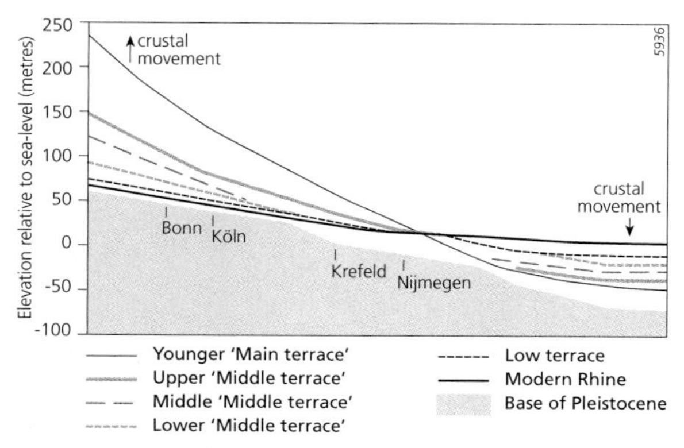

Fig. 6.8. Longitudinal profiles of the Rhine terraces, showing their convergence downstream and eventual disappearance beneath younger sediments in the subsiding area of The Netherlands (after Bridgland 2000).

Continuous Late Cenozoic uplift of the macro-scale alluvial fan of the Meuse downstream of the Ardennes, in connection with aggradation during glacial periods and dissection during interglacial periods, resulted in a flight of no less than thirty-one terraces since the Late Miocene. This contrasts with a maximum of sixteen terraces recognized along any reach of the Rhine (Veldkamp and van den Berg 1993; Tebbens *et al.* 2000; van den Berg and van Hoof 2001). A cross-section through the Meuse valley from Maastricht to Aachen provides a good example of the intricate terrace morphology of the Meuse from Early to Late Pleistocene age (Zonneveld 1974; Pissart 1974; Fig. 6.9 after Ruegg 1994). The main tectonic factor controlling the distribution of the terraces in this region is the interplay of the compression dynamics of the Ardennes massif and the extensional dynamics of the Roer valley rift system (van den Berg and van Hoof 2001). The amount of incision indicates *c*.200 m of uplift, the rate of which fluctuated between 2 and 11 cm/ka. The exceptionally long and well-dated series of terraces, representing the upper Pliocene and the whole Quaternary period (≈ 4 Ma), enables a good correlation with the north-west European pollen stages and marine isotope records. The number of steps in this staircase is probably one of the highest for any river in the world.

Major aggradational terraces in various regions have revealed important artefactual evidence for the presence

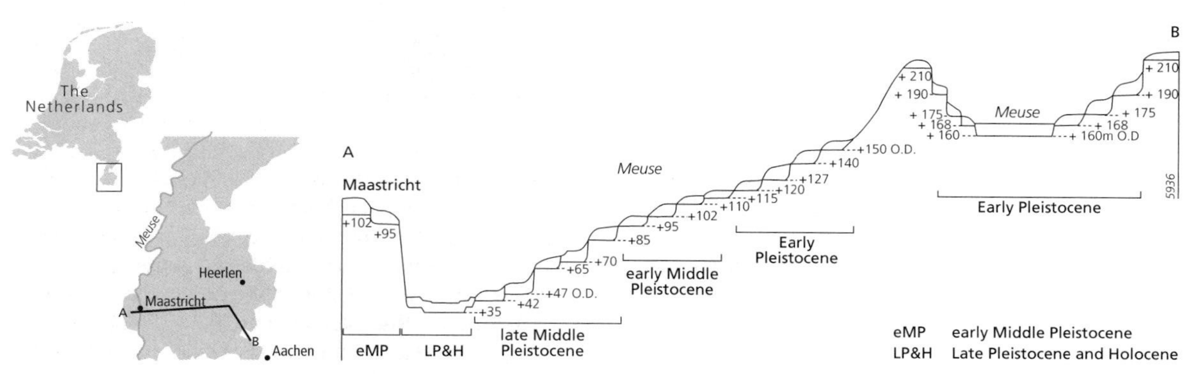

Fig. 6.9. Quaternary terrace sequence of the river Meuse near Maastricht and Aachen (after Ruegg 1994); the terrace staircase is composed of many terrace fragments that have been reconstructed into terrace levels (van den Berg and van Hoof 2001).

of Lower Palaeolithic hunter-gatherers. Mammalian fossils can even offer a powerful tool for correlation of fluvial sequences with the global marine record of glacials and interglacials, according to Bridgland (2000) and Schreve and Bridgland (2002). The first appearance of artefacts in terrace sediments, e.g. in the Rhine and Somme rivers, marking the arrival of tool-making hominids, probably dates back to the transition of the Early to the Middle Pleistocene (Cromerian Complex). Sites in the Somme river, excavated by the famous archeologist l'abbé Henri Breuil in the 1920s and 1930s, have drawn international attention. The reason for this was the recovery of artefacts produced by a special flint-processing technique known as the Levallois technique, which is also dated to the Cromerian Complex (Bourdier 1974).

Another remarkable achievement of Quaternary fluvial research is the establishment of an 11,000-year German oak and pine dendrochronology, used for radiocarbon calibration (Becker 1993). In gravel pits, especially along the Upper Rhine, Main, and Danube rivers and their tributaries, huge numbers of oak trunks—remnants of riparian Holocene oak forests (*Quercus robur, Q. petraea*)—were found. The trees were washed into the rivers by the undercutting of meander banks and by erosion of river channels during floods, and subsequently preserved in alluvial sediments below the water table for thousands of years. Most trees lived less than 300–400 years, which explains why many thousands of subfossil tree trunks were needed to establish a continuous Holocene dendrochronology, which is now one of the best in the world and comparable to the famous US bristlecone pine series. The chronology has also been extended into the Late Glacial by analysing well-preserved pine (*Pinus sylvestris*) trunks. These subfossil pines are the remnants of Late Glacial forests, where both pine and birch (*Betula verrucosa, B. pubescens*) covered the west and central European valley plains before the arrival of oaks (Becker 1993).

Factors Influencing Palaeogeographic Evolution

Throughout most of the Quaternary almost the entire region of The Netherlands was part of a delta formed by the rivers Rhine, Meuse, Scheldt, Weser, and Elbe. Due to continued subsidence of the North Sea basin a Quaternary (onshore) sequence of fluvial and intercalated marine, (fluvio-)glacial and aeolian deposits with a thickness of up to 500 m has accumulated. This huge sediment wedge forms one of the most complete, primarily terrestrial, records of the glacial–interglacial cycles of the Quaternary. The chrono- and lithostratigraphic subdivision of this sediment wedge, largely based upon extensive sedimentpetrological, mineralogical, and palaeontological (pollen) analyses, represents one of the most detailed and extensive terrestrial sequences in Europe (Zagwijn 1985, 1989).

The evolution of the (French/Belgian/Dutch) river Meuse during the Last Glacial–Interglacial Transition (LGIT) provides one of the best cases for studying the impact of climate on river dynamics (Kasse *et al.* 1995; Tebbens *et al.* 1999; Bogaart *et al.* 2003). The chronology of events during this time interval, based on many detailed sedimentological, litho- and morphostratigraphical, and palaeobotanical investigations, including large numbers of conventional and AMS-radiocarbon dates, has been summarized as follows by Bogaart *et al.* 2003 (absolute dates according to the GRIP ice core chronology):

- during the cold and dry Pleniglacial (≈ 16,9–14,7 ka) the Meuse was an aggrading braided river;
- during the warm and wet Bølling (≈ 14,7–14 ka) the river changed towards a single-channel meandering one;
- during the Allerød (≈ 14–12,6 ka) the river had a highly sinuous meandering state;
- during the Younger Dryas (≈ 12,6–11,5 ka) a braided river pattern returned;
- finally, during the relatively warm and wet early Holocene the river incised and changed again into a low sinuosity, single-channel type.

More or less similar sequences of events have been recorded from a variety of rivers in north-western Europe, e.g. for tributaries of the Seine by Pastre *et al.* (2003), for tributaries of the Somme by Antoine *et al.* (2003), by Mol (1995) and Mäusbacher *et al.* (2001) for tributaries of the Elbe and Weser, and by Andres *et al.* (2001) and Schirmer (1988) for the Rhine and Main in central Germany. This similarity in evolution reflects a broad regional pattern of fluvial response, which illustrates its sensitivity to the rapid climatic changes which characterize the LGIT. According to Tebbens *et al.* (1999) and Antoine *et al.* (2003), incision events, that occurred during the Bølling and in the early Holocene, represent short periods of rapid climate change from cold/arid conditions towards a much more temperate/humid climate. The downcutting can be related to sudden increases in water discharge contemporaneous with sharp reductions in sediment supply. This is probably related to a combination of factors such as (*a*) the disappearance of permafrost, (*b*) a reduction of masswasting (slope) processes, (*c*) a cessation of aeolian sedimentation, and (*d*) the development of soils. Permafrost, which was generally present in the Meuse catchment during the Last Glacial (Weichselian) maximum, probably completely disappeared during the short Bølling interstadial; however, discontinuous permafrost returned during the Younger Dryas cold phase. Contrary to earlier statements, Antoine *et al.* (2003) conclude that downcutting clearly predates the main modifications of the vegetation cover.

These findings agree with the general concept that during glacial periods excessive sediment supply over river discharge causes floodplains to aggrade, provided long-term tectonic uplift cannot be compensated for by river downcutting. As soon as the climate ameliorates and sediment supply fades, the rivers will respond with degradation, ultimately resulting in floodplain lowering (Tebbens *et al.* 1999). It should be noted, however, that river response to climate change is also strongly controlled by other independent and dependent drainage system variables and that consequently exceptions from this model occur. Based on case studies of tributaries of the Elbe (Elster) and Oder (Spree) rivers in eastern Germany, Mol *et al.* (2000) discuss the river evolution during the Last Glacial in response to climate change. This study clearly illustrates that exceptions to the model, presented above, can be caused by specific catchment characteristics. These characteristics include low-energy conditions in the headwaters of a catchment, valley geometry, or accommodation space, and especially grain size of the sediment supplied to the river. Another example of a fluvial sequence that is thought to be primarily the result of self-control mechanisms and not of climate fluctuations is described by Buch (1989) for the Danube river. This river changed during the Last Glacial–Holocene time from a braided type to an anastomosing one, followed by a renewed braided pattern and eventually by a single-channel meandering system. This was explained by Buch (1989) as a complex reponse to the interaction between discharge, sediment load, and slope. In the case of the Danube it is assumed that these parameters are largely determined by the existing segmentation of the river course into alternating narrow valley stretches and basins, and by the various tributaries joining the main river. Consequently, Buch (1989) questions the commonly accepted causal relationship between climatic changes and fluvial processes. On the other hand, Straffin and Blum (2002) conclude that the response of large alluvial systems, such as the Loire, should be less affected by localized and internal complex-response mechanisms. The Loire and its tributaries, the Allier and the Arroux, show a remarkable synchroneity in alluvial deposition during the Last Glacial and Holocene period (Veldkamp and Kroonenberg 1993; Straffin *et al.* 1999). Thus, in this case the regional similarity in response of these streams, and the lack of tectonic activity, provides a strong argument for a climatic control on changes in fluvial facies and depositional style (Straffin and Blum 2002).

The fluvial history of the northern, subsiding, Upper Rhine-Rift Valley in south-western Germany (Dambeck and Thiemeyer 2002) typifies a characteristic evolution for many rivers in western Europe. This part of the Rhine graben is filled by up to 400 m of Tertiary and Quaternary sediments, mainly as fluvial gravels and sands. The uppermost sediments of this infilling are preserved in two Last Glacial (Weichselian/Würm) river terraces. The upper level (*Obere Niederterrasse*) is locally covered by aeolian dune fields of Younger Dryas age, whilst the lower level (*Untere Niederterrasse*) lacks these. Both levels contain overbank fines, which were deposited during the early Holocene (Preboreal to early Boreal period). The present floodplain of the Rhine is incised in this lower level. The overbank deposits can be related to an early meandering channel of the Rhine, which developed after the Allerød from the former braided pattern as was the case in many western European rivers. River dunes formed especially during the second half of the Younger Dryas period, which was characterized by a somewhat drier and warmer climate. The same is observed along the eastern banks of the river Meuse in The Netherlands as well as in many other German rivers. Human impact, e.g. by the clearing of forests during the late Neolithic, may have caused the revival of aeolian activities in some areas. During the Holocene, the Rhine retained its meandering pattern; three meander generations have been distinguished that differ in morphological, lithological, and pedogenetical criteria. The oldest generation dates back to the Boreal–Atlantic period, the next generation formed during the late-Atlantic to late-Subboreal, and the younger generation dates from the Subboreal–Subatlantic transition to the present (Dambeck and Thiemeyer 2002).

The Late Glacial–Holocene Rhine–Meuse delta is certainly the most closely studied deltaic region in the world. About 200,000 lithological borehole descriptions, 1,200 ^{14}C dates, 36,000 dated archeological remains, and gradient lines of a multitude of palaeochannels were used to reconstruct the Holocene evolution of this delta (Berendsen and Stouthamer 2000, 2002). The spatial distribution and ages of more than 200 different Holocene channel belts were distinguished and stored in a GIS-database. This has enabled the reconstruction of palaeogeographic (or morphogenetic) maps for the entire Holocene period with a time resolution of 200 years.

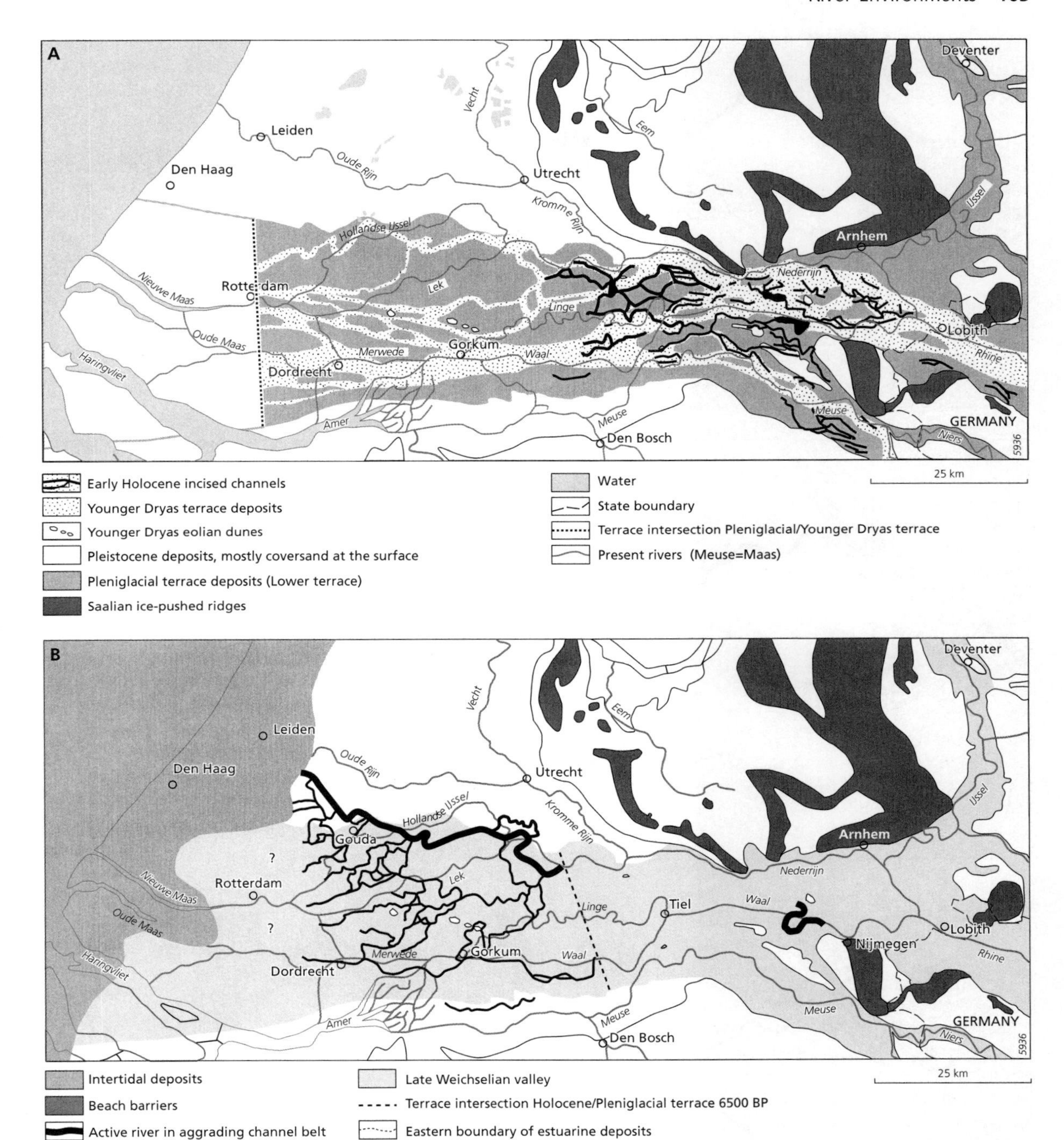

Fig. 6.10. Palaeogeographical reconstructions of the Rhine–Meuse delta (after Berendsen and Stouthamer 2000): A. palaeogeography around 10,000 BP, showing Pleniglacial and Younger Dryas braidplains with incised channels that became filled with peat in the early Holocene; B. palaeogeography around 6500 BP, backfilling of the Late Weichselian valley had progressed eastward to the dashed line that indicates the terrace intersection of Holocene deposits and the Pleniglacial terrace; C. palaeogeography around 4000 BP, the river Meuse followed the southern margin of the Late Weichselian valley; D. palaeogeography around 2500 BP, a new river course (the river Vecht) flowed to the NW, the exact location of river courses near the Rotterdam estuary is unknown, due to later erosion; E. palaeogeography around 1000 BP, the present rivers Lek, Waal, Meuse, and IJssel had formed, older distributaries silted up and were dammed between AD 1100 and 1300, a major marine flood in AD 1421 caused large-scale erosion in the south-western part of the delta resulting in extensive estuaries.

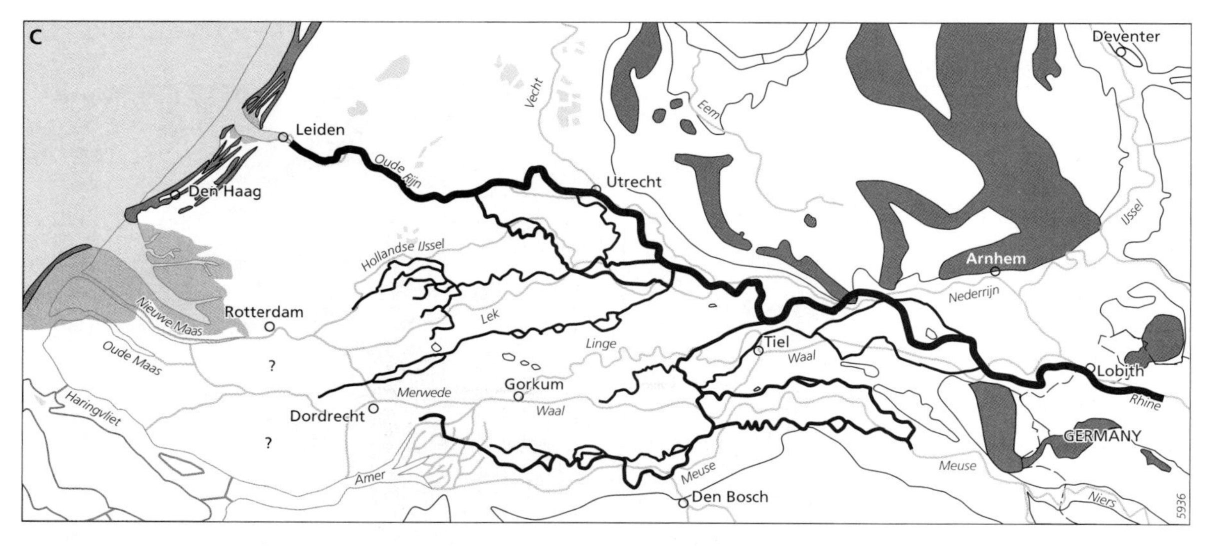
C
Deventer
Leiden
Vecht
Eem
Oude Rijn
Utrecht
Den Haag
IJssel
Hollandse IJssel
Arnhem
Nederrijn
Nieuwe Maas
Rotterdam
Lek
Linge
Tiel
Oude Maas
?
Waal
Lobith
Gorkum
Merwede
Waal
Rhine
Haringvliet
Dordrecht
?
GERMANY
Meuse
Amer
Meuse
Den Bosch
Niers
5936

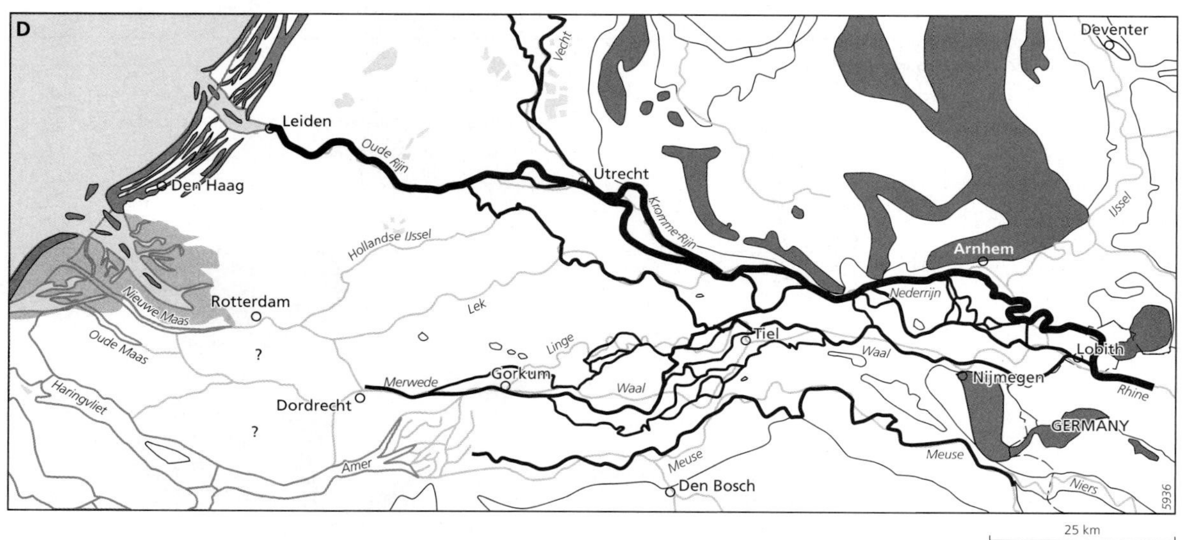
D
Deventer
Vecht
Leiden
Oude Rijn
Utrecht
Den Haag
Kromme-Rijn
IJssel
Hollandse IJssel
Arnhem
Nederrijn
Rotterdam
Nieuwe Maas
Lek
Tiel
Oude Maas
?
Linge
Lobith
Waal
Nijmegen
Merwede
Gorkum
Waal
Rhine
Haringvliet
Dordrecht
?
GERMANY
Amer
Meuse
Meuse
Den Bosch
Niers
5936
25 km

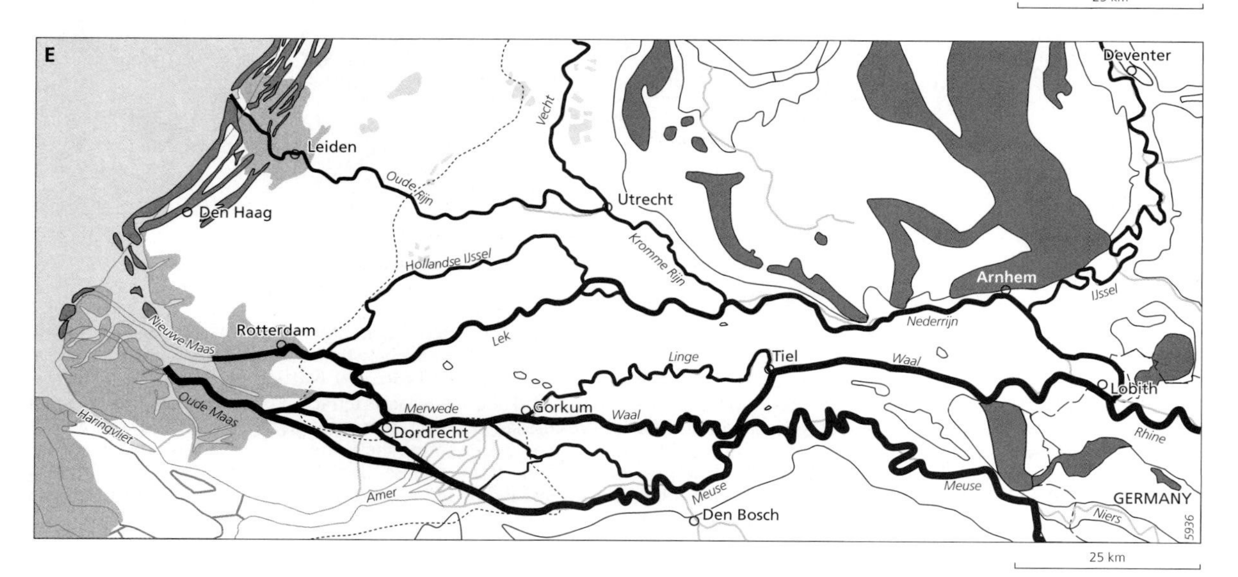
E
Deventer
Vecht
Leiden
Oude Rijn
Den Haag
Utrecht
Kromme Rijn
Hollandse IJssel
Arnhem
IJssel
Nieuwe Maas
Nederrijn
Rotterdam
Lek
Linge
Tiel
Waal
Lobith
Oude Maas
Merwede
Gorkum
Waal
Haringvliet
Dordrecht
Rhine
Meuse
Amer
Meuse
GERMANY
Den Bosch
Niers
5936
25 km

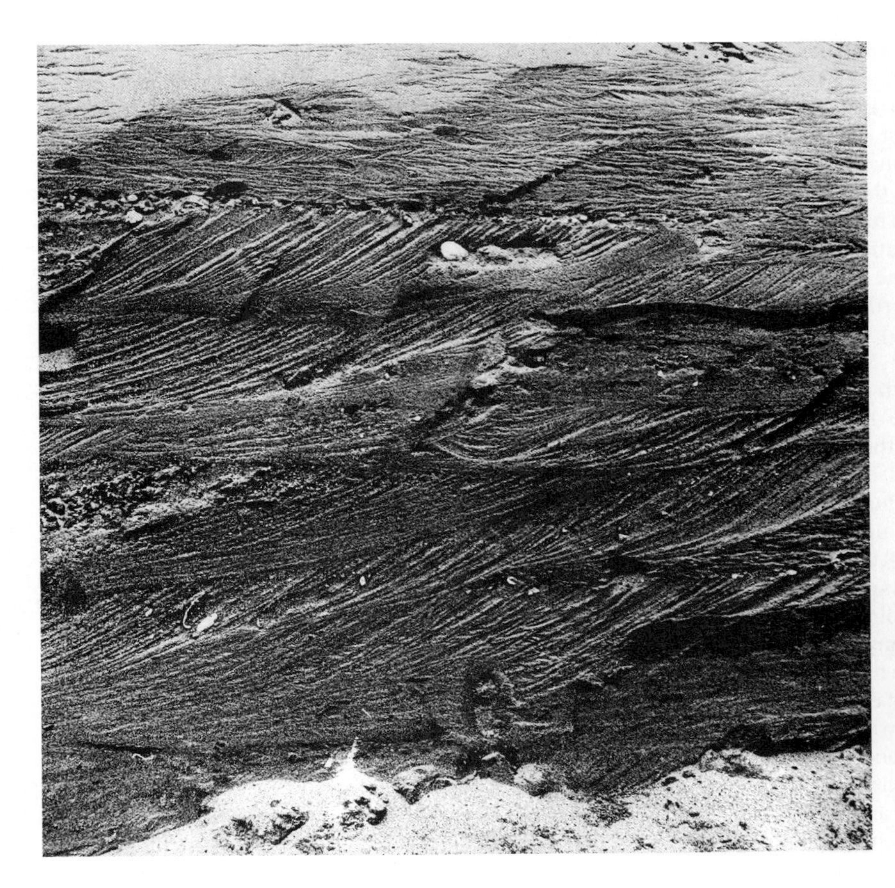

Fig. 6.11. Sequence of a large-scale, trough cross-stratified unit of fluvial origin, overlain by a bimodal small-scale, trough cross-stratified estuarine unit. The units are separated by an erosional surface with lag deposit, composed of gravel and shell fragments. Location along the river Meuse near Rotterdam.

A simplified overview of several of these palaeogeographic channel belt reconstructions is presented in Fig. 6.10. Near the present coast, in the vicinity of Rotterdam (see Fig. 6.10E), fluvial sediments intercalate with estuarine deposits, and are distinctly reflected in the sediments by sedimentary structures resulting from tidal movements (Fig. 6.11).

The Holocene palaeogeographic evolution of the Rhine–Meuse delta is governed by complex interactions between the following factors, according to Berendsen and Stouthamer (2000, 2002) and Stouthamer and Berendsen (2000):

- location and shape of the previous (Late Weichselian) palaeovalley, controlling river migration;
- sea-level rise, resulting in backfilling of the palaeovalley;
- spatial and temporal changes in river channel patterns;
- differential tectonic movements, probably also influencing the location of avulsion nodes;
- increases in discharge, sediment load and/or within-channel sedimentation, especially after *c*.2000 BP;
- coastal evolution, i.e. reduction of the number of tidal inlets and closure of the coastal barrier system;
- textural composition of the substrate and river banks, determining channel migration, formation of crevasses, and the initiation of avulsions;
- marine ingressions, causing large-scale erosion;
- and finally, human activities, instrumental in fixation of the river channels.

The interaction of these factors resulted in a series of avulsions, which were responsible for the frequent shifts of areas of clastic sedimentation. Avulsions are defined as the abandonment of a part or an entire channel belt by a stream in favour of a new course. It has often been suggested that high aggradation rates, related to base-level (sea-level) rise, are reflected by high avulsion frequencies. On the other hand, Stouthamer and Berendsen (2000) demonstrated that an increase of avulsion frequency can coincide with a decrease in the rate of sea-level rise, which contradicts the previous statement. In this case a more likely explanation might be increased discharge and/or within-channel sedimentation. Avulsions strongly influence the geomorphology and alluvial architecture, because they determine the channel recurrence interval on the floodplain, and, as a result, channel density and interconnectedness (ibid).

A feature of particular interest is the fact that neotectonic movements appear to have influenced the Late Quaternary fluvial dynamics in the Rhine–Meuse delta. The youngest and north-westernmost extension of the continental-scale west and central European rift system consists of the Lower Rhine Embayment (LRE) in Germany extending to the north-west into the Roer Valley graben (RVG) and the North Sea basin. In other words, the rivers in the Rhine–Meuse delta presently cross a complex fault system with a SE–NW orientation. Seismic data suggest that the fault-structure of the RVG originally formed during the Late Carboniferous. Tectonic movements continued through the Mesozoic and Cenozoic. Since the early Quaternary the LRE and the RVG have been areas of mainly fluvial deposition; the Quaternary sequence in the RVG reaches a thickness of 200 m. Cohen *et al.* (2002) have shown evidence for neotectonic influence in the Holocene fluvial deposits, exemplified by deformations in gradient lines of the top of sandy channel deposits together with subtle differences in floodbasin aggradation rates of the RVG and the adjacent horst regions. Moreover, it was shown that several middle Holocene avulsion sites concentrate at or near the Peel Boundary Fault, which forms the north-eastern edge of the RVG (Stouthamer and Berendsen 2000). This is explained by rather sudden changes in gradients in the area where the fault system and the river channels cross, making it a favourable location for avulsion (Cohen *et al.* 2002).

Man-Made Changes

It is a well-known fact that almost all large European rivers are strongly influenced by man. Table 6.2 summarizes the main man-made changes that occurred in the Rhine–Meuse delta. An extensive overview of river engineering works, shipping, and water management has been published by the International Commission for the Hydrology of the Rhine Basin (Buck *et al.* 1993). During the Neolithic period the influence of man was mainly restricted to local deforestation and expansion of arable lands. However, it is generally assumed that, from 3000–2000 BP onwards, human activities in the upstream areas of the rivers, especially deforestation, contributed significantly to an increase in river discharges. There is also ample evidence for a coeval increase in deposition of sediment in the downstream reaches of various rivers.

It is highly probable that the Romans in the first century AD carried out the first river engineering works in the Rhine–Meuse delta to strengthen the northern limit of their empire (van Urk and Smit 1989; Havinga and Smits 2000). The Roman emperor Claudius declared the Rhine as the northern border (Latin: *limes*) of the Empire in AD 47. To defend this *limes* the Romans constructed many tens of *castella* along the river banks all the way from southern Germany to the former mouth of the river Rhine near Katwijk in The Netherlands. Another example of river engineering work by the Romans about 2,000 years ago is found along the river Lippe, a small tributary of the Rhine. During the Roman campaign against the German tribes, the Romans built a towpath along the river and may have changed the channel planform from its natural, anabranching pattern to a meandering form by building small dams on local distributary forks, according to morphological studies by Herget (2000). During their struggles against the German tribes, the Roman conquerors often used the rivers for supply and transport. To improve navigation secondary channels were probably dammed to concentrate flow in a single channel. Moreover, towpaths facilitated the movement of vessels.

TABLE 6.2. *Main events induced by man in the Rhine–Meuse delta*

Period	Measure/activity	Effect
< 10,000 BP	hunting large grazers	extinction of large grazers, extension of forests in the delta
7300–4000 BP	beginning of agriculture	grazing by domesticated animals leading to a more open river delta
3000–2000 BP	upstream deforestation	increased discharge of water and sediments
1st century	start of Roman occupation	first alterations of river discharge over different river branches
5th–10th century	rise of trade cities	intensified shipping on the rivers
10th–15th century	construction of winter dykes	disconnecting the flood basin, silting up of embanked flood plains
17th–19th century	construction of summer dykes	less frequent but more abrupt flooding
18th–20th century	cutting of meanders	increased flow velocities, incision of main channels
19th–20th century	construction of groynes	less bank erosion and channel migration, narrowing and incision of main channels
20th century	construction of sluice-dams, barriers, sand, and gravel excavations	control of water level fluctuations, formation of unnaturally large and deep stagnant waters in flood plains

Source: After Lenders 2003.

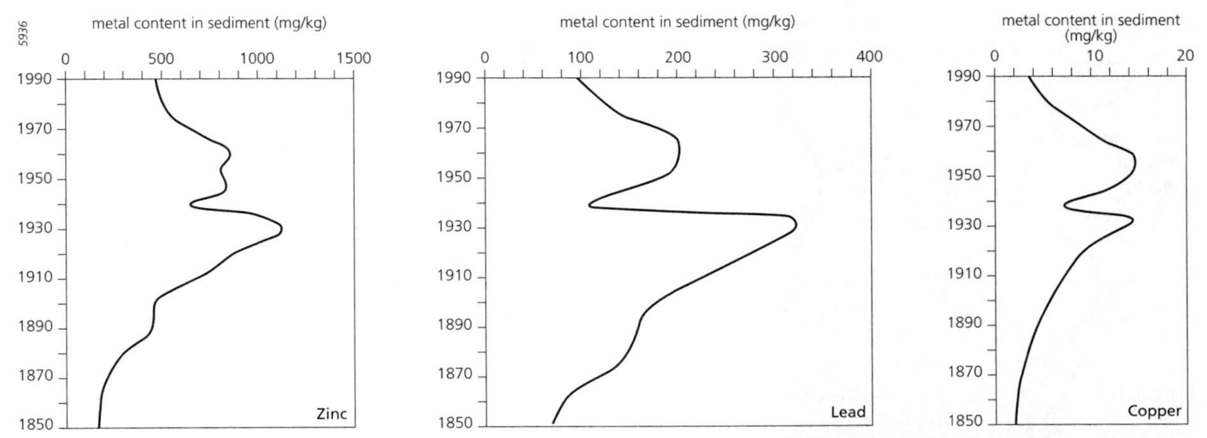

Fig. 6.12. The history of metal pollution of the Rhine as reflected in sediment deposited in a dyke-breach pond within the embanked Waal floodplain (after Middelkoop 1999).

Much later, the building of flood defence dykes from about AD 1000 onwards resulted in drastic changes in river flow dynamics, sediment transport and sedimentation, and floodplain ecosystem development. All river sections were completely embanked by the end of the fourteenth century. In the last two centuries extensive canalization works took place in many river reaches, followed by an increasing number of flood control dams, river locks, barrages, and hydroelectric power stations. Nowadays large parts of rivers such as the Elbe, Weser, Rhine, Mosel, Main, Danube, and Meuse are completely man-manipulated. Unlike other large European rivers, such as parts of the river Loire and its tributary the Allier, not a single section of the larger German rivers has remained untouched. Today river regulation still continues, e.g. on the Danube the missing link of the Rhine–Main–Danube canal is completed, which now connects the North Sea with the Black Sea.

Until the Middle Ages the composition of the water in the rivers was hardly influenced by humans. Later deforestation and land reclamation introduced accelerated soil erosion and changed the river sediment load. Originally, river valleys in western and central Europe were covered with multi-species deciduous riverine forests. Most of these have gradually been replaced by man-made crop and grassland ecosystems, e.g. 74% of the originally forested area of the river Elbe catchment has been replaced by cropland (Nienhuis *et al.* 2000). Finally, industrialization and the rapid increase of human population in the western European river catchments during the nineteenth and twentieth centuries caused very severe pollution of river waters. The maximum pollution level in most rivers was reached by the 1960s and early 1970s after which water quality improved (Fig. 6.12).

Major interventions have been implemented not only in the downstream sections of the rivers. For example, along the upper Rhine major canalization works were executed especially in the nineteenth century. Originally, the upper Rhine comprised two different morphological river patterns. Between Basle and Strasburg the Rhine was of a braided type, whereas a meandering pattern prevailed north of Strasburg up to the city of Mainz. In both river sections floods and a shifting riverbed, together with poor hygienic conditions, formed a continuous threat, and even led to waterborne diseases such as malaria and typhoid (Kern 1992; Havinga and Smits 2000). To eliminate these plagues and to increase flood protection, the German hydraulic engineer Johann Tulla developed a 'normalization plan'. This plan consisted of the straightening of the meandering river reaches and the closing of many braided river channels. It also involved the construction of some 440 km of artificial levees and dams. The execution of this enormous plan was completed in 1878 (Havinga and Smits 2000; Nijland and Cals 2000). Figure 6.13, for example, illustrates a nineteenth-century regulation work in the meandering section of the Rhine north of Karlsruhe. Supported by diversion structures, the river naturally cut its meanders. This resulted in erosion of the new riverbeds and only the river banks had to be stabilized in order to maintain the projected channel (Kern 1992). Tulla was an ingenious engineer with foresight as he planned no levees to be constructed close to the new channel, but instead he calculated the dimensions of the new channel to accommodate only about 1,000 m^3s^{-1}. At this threshold of 1,000 m^3s^{-1} the pre-existing meanders were thought to be flooded, whereas at a threshold of about 2,000 m^3s^{-1} the former floodplain could act as an overflow. Tulla started his planning in 1815 but he died

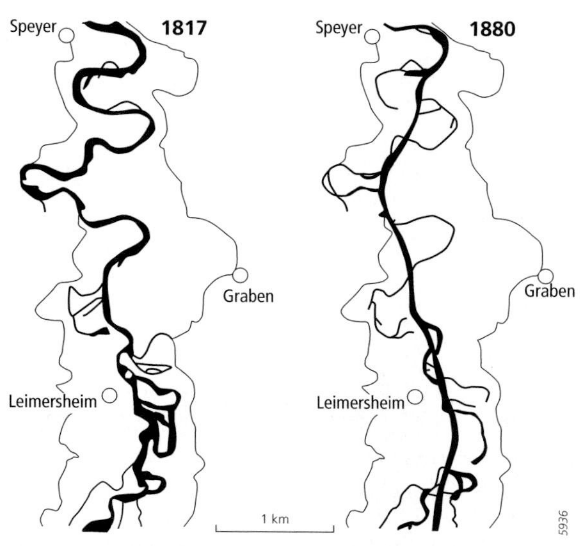

Fig 6.13. River regulation at the upper Rhine meander zone between Karlsruhe and Mannheim (after Kern 1992).

in 1828, long before completion of his plans. Unfortunately his successors deviated from his original plans and constructed levees quite close to the new channel as a means to obtain new land for agricultural purposes. Inevitably this resulted in unwanted side-effects. These channelization works caused a strong increase of stream velocity that led to increasing erosion of the riverbed, thereby also causing a severe lowering of groundwater levels in the surrounding arable lands.

The first major hydraulic engineering works since the Roman era were undertaken at the lower Rhine bifurcations in the early eighteenth century. Before that time interventions were restricted to bend corrections, which took place in the seventeenth century (van Urk and Smit 1989). Around 1775 restructuring works at the bifurcations in The Netherlands were completed and the water distribution along the side branches of the lower Rhine became strictly controlled. Additional construction of groynes (Fig. 6.14), weirs (Fig. 6.15) and sluices ensured that sufficient water quantities for shipping, counteraction of salt-water intrusion, industrial and agricultural use, cooling water for hydropower installations, and for drinking water supply became available in the different river branches. Two-thirds of the Rhine discharge went to the Waal, because this was and is the major shipping route. The remaining third was divided between the side branches of the Nederrijn and the IJssel, in the proportion 2 to 1 respectively. Although the works at the bifurcation point regulated the water distribution, floods continued to be hazardous. In the case of the Rhine numerous dyke breaches occurred, such as those during the 1809 and 1820 floods when large river polders became inundated (van Urk and Smit 1989). In the twentieth century navigation became an increasingly important function of the river. Although protection from severe floods was provided by the regulation works, the channel bed of the lower Rhine still contained irregular gravel bars and scour holes, making this river section unsuitable for navigation. In addition, this configuration of the channel promoted the formation of ice dams in winter, which were repeatedly instrumental in the formation of dyke breaches. The last extreme high flow event in the Rhine and Meuse, leading to extensive flooding and huge loss of life and

Fig. 6.14. Oblique photograph of the Waal river with groynes (photo: Henk Berendsen).

Fig. 6.15. Oblique photograph of the Nederrijn river showing the weir at Amerongen (photo: Henk Berendsen).

properties, took place in 1926, when river dykes failed at various places.

Flooding due to ice dams was a common phenomenon in north-western European rivers. For example, historical sources in southern Germany show that, particularly during the 'Little Ice Age' from 1560 to 1860, winter floods combined with ice jams were very frequent in the Danube (Buch 1989). Between 1907 and 1939 groynes were constructed in the riverbed of the upper Rhine, thereby reducing the free flow channel width to a mere 75–150 m (Kern 1992). After World War I a series of hydroelectric dams was built. This necessitated digging lateral canals for shipping together with many navigation locks. All these interventions along the upper Rhine have led to a very considerable loss of potential flooding area, in the order of 1,000 km^2 since 1800 (Havinga and Smits 2000). Such a development has obviously resulted in a significant increase of flooding risks in the more downstream reaches of the river. Nowadays flood waves, representing an identical volume of water, travel faster and reach much higher peaks than before the normalization and regulation works. Kern (1992), for example, mentioned that the flood protection level for the cities of Karlsruhe and Mannheim changed from a 200- to a 50-year event.

Around 1800 the Meuse was brought into the European network of river communications. Since that time canalization of the channel, the construction of mobile weirs with lateral locks, straightening of river sections, and the building of hydroelectric power stations have greatly altered the morphology and ecology of the river (Micha and Borlee 1989). Also around 1800 organized regulation started along the Weser. The channel was narrowed and deepened, longitudinal dykes and groynes as well as dams and locks were constructed. At present the lower reach of the Weser is one of the most regulated rivers used as an ocean-going shipway (Busch *et al.* 1989).

Man's impact on the Garonne has been progressive through the ages and has greatly changed the river over the last 300 years (Decamps *et al.* 1989). The Garonne, which receives water from the Pyrenees, the southern Massif Central, and the Aquitaine basin, is a river of ill repute because of the frequent occurrence of flash floods, sometimes with catastrophic effects (see Table 6.2). A variety of measures have been taken for flood protection, for improvement of navigation, and for agricultural purposes. Because of the rather limited extent of embankments the isolation of the Garonne from its floodplain has not been complete, potentially enabling reversible trends in, for example, floodplain forest succession.

The Rhône and its tributaries form another example of a river system that has been drastically changed, especially in the twentieth century under the direction of the Compagnie Nationale du Rhône (CNR) (Vivian 1989). In the downstream half of the river an impressive number of twenty power plants, fourteen sluices, and twenty-four dams on a regulated river section 480 km long have been constructed and are now operated by the CNR. Additional flood protection, irrigation, and agricultural land drainage schemes accompany these developments. The same is true for most tributaries of the Rhône, for example along the middle section of the

Durance where dam construction started in 1959 and now ten power plants are in operation (ibid.). In spite of all these regulation measures and consequent changes in annual and seasonal discharge characteristics, this region often suffers from flash floods particularly in the smaller tributaries. The upstream part of the Rhône, especially the 200-km-long section from Lake Geneva to Lyons, has undergone similar changes (Roux *et al.* 1989). This part of the river has been partly incised in morainic and fluvioglacial deposits, forming deep gorges. The river has also infilled (formerly) overdeepened lake areas, resulting in braided river reaches. Presently this river section has also seen the development of retention dams, bypassed or lateral channels, and artifical reservoirs, as well as a series of conventional and nuclear power plants. It is generally accepted that hillslope reafforestation in mountainous areas may significantly decrease the supply of sediment to the rivers as well as reduce the height of peak flood events. Liébault and Piégay (2001) present observations made in the watershed of a small tributary of the Rhône where an increase in forest cover following rural depopulation of the area during the twentieth century could be linked to a strong decrease in bedload supply and a correlated decrease in peak flows.

Climate Change and Flood Control

Many examinations of the possible hydrological response to climate changes have emphasized changes in average conditions, rather than individual flooding events. Historical evidence indicates, however, that such events may have had a considerable regional impact even in the face of any relatively modest climate change (Knox 1993). Based on detailed sedimentological analyses of the alluvial records of palaeofloods in the Mississippi and Colorado river basins Knox (1985, 1993) has clearly shown that recurrence frequencies of large floods have been subject to abrupt changes over time. Flood chronologies from various regions suggest that times of rapid climate change have a tendency to be associated with more frequent occurrences of extreme floods (Knox 2000).

The unusual high frequency of large flood events observed in some regions since the early 1950s are often attributed to land-use changes, but might also be a result of the rapid climate forcing from the increase of atmospheric greenhouse gases. In the last two decades numerous studies on the sensitivity of the seasonal and annual discharge distribution of rivers to global warming, including possible changes in frequency of extreme high and low flow events, have been executed. These studies relate to catchments varying in size from a few km^2 to 10^5–10^6 km^2. Overviews of these scenario analyses, including studies on several rivers in western Europe, are presented by Arnell (1996) and can be found in the latest IPCC Reports (McCarthy *et al.* 2001). Major efforts have been devoted to the assessment of the impact of climate change on the discharge regime and water resources in the Rhine basin, under the co-ordination of the International Commission for the Hydrology of the Rhine (CHR). Spatially distributed water balance models with monthly, daily, and even hourly time steps have been developed for the entire Rhine catchment, as well as for several representative subcatchments (varying in size from 150 to 4,000 km^2). The potential effects of climate change on the discharge regime were calculated using the results of GCM-experiments. To enable the assessment of changes in peak flow, a high-temporal model resolution of one day was necessary. All models indicate the same trends in the changes: higher winter discharge as a result of intensified snowmelt and increased winter precipitation, and lower summer discharge due to the reduced snow storage in winter and an increase of evapotranspiration (Kwadijk 1991; Grabs 1997; Middelkoop *et al.* 2001). In other words, the results indicate that the river regime of the Rhine might shift from mixed snowmelt/rainfall to a rainfall-dominated regime. Looking at the results in more detail it becomes perfectly clear that regional differences occur. These depend on the physical characteristics of the subcatchments. Equally, uncertainties in the modelling and the representation of water balance parameters, such as evapotranspiration and snowmelt, will influence the assessment of changes in river regimes. The largest increases in winter flow are found for the Alpine part of the catchment; in the German part of the drainage basin the winter increase is damped by the smaller increase in winter flow from the tributaries Neckar, Main, and Mosel. Likewise, the reduction in summer flow is largest in the Alpine region. In general, the simulation results show a maximum discharge reduction in September, and a considerable increase in high and maximum daily discharges in the January–March period (Grabs 1997). The results also confirm the conclusion by Knox (2000) that more frequent and more extreme (low and high) daily flows will occur as a direct response to climate change. In conclusion, the hydrological changes will increase flood risks during winter, whilst more frequent low flows of probably longer duration during summer will negatively affect inland navigation, and reduce water availability for agriculture, industry, and drinking water companies (Grabs 1997; Middelkoop *et al.* 2001).

River Conservation and Rehabilitation

European floodplain river ecosystems originally were characterized by a high biodiversity and productivity reflecting dynamic and complex habitat structures. The fluvial dynamics of flooding led to high flux rates of nutrients and organic matter, rapid shifts in environmental conditions and the evolution of adaptive strategies to exploit the spatio-heterogeneity (Nienhuis and Leuven 1998). However, as has been shown, rivers and floodplains in western Europe have been modified drastically during the past centuries. It is estimated that almost 80% of the total water discharge of the main European rivers is more or less strongly affected by water regulation measures. Especially in those regions where riverine reaches have long been regulated and constrained by engineering works there is, since the late 1970s, an increased interest in river protection, conservation, and rehabilitation. Integrated water management programmes at first concentrated on wastewater treatment. These have significantly reduced water pollution by point sources. For example, between 1972 and 1985 the mean mercury concentration in the sediments of the Rhine had decreased from 8.9 to 0.9 mg kg^{-1}, whereas in the Elbe an increase from 13.9 to 17.0 mg kg^{-1} was still being registered (Nienhuis *et al.* 2000). Only after the German reunification in 1989 did the former German Democratic Republic close down most heavy metal point sources. Consequently, soon afterwards, heavy metal concentrations started to decrease rapidly. The same is true for inorganic nutrients and chlorinated hydrocarbons. From around 1930 right up to the 1960s, heavy metal concentrations in the Rhine rose almost constantly, apart from a clear dip during World War II (Middelkoop 1999). During the last few decades heavy metal concentrations have dropped sharply (Fig. 6.12). However, heavy metal pollution is still a problem as metals attach themselves to suspended clay and silt particles, which settle to the bottom in slow-moving stretches of water. Cadmium, mercury, copper, and zinc concentrations in dredge spoils are often so high that the spoil cannot be freely disposed of (Middelkoop 1999). Attention must now also be focused on the more difficult and persistent problems of the effects of diffuse sources, containing dispersed pesticides, fertilizers, and toxicants on the receiving water bodies (Nienhuis and Leuven 1998). Finally, the accumulation of pollutants in the underwater sediments 'archive' still form a particular unsolved problem.

Parallel in time with measures to improve the water quality and to attain a sustainable ecological development of river ecosystems it has been realized that a variety of flood control measures must be taken to ensure safety for the population and its property. Notwithstanding all the intrinsic uncertainties in models of climatic change and subsequent water balance models, as has been discussed above, it is generally accepted that more extreme high discharge events are likely to occur in the future. This will lead to inadmissibly high water levels if no further action is taken. Consequently, in The Netherlands the present-day governmental policy is to allow the rivers to take more physical space, in order to avoid emergency scenarios during extreme high flood events. This means that every development or action of man that could raise the water level in the winter bed of the river (building of houses, reclamation of wetlands, forest growth, etc.) must be compensated for. Compensating measures might take the form of heightening and strengthening the winter dykes, which can be called 'business as usual'. This has been executed on a large scale, for example immediately following the 1993 and 1995 high flow events in the Rhine and Meuse rivers. These high flow events were near disasters as more than 200,000 people had to be evacuated from the threatened polders. Fortunately, no flooding occurred. However, this process cannot be continued forever. Higher and higher river dykes separating the low-lying, ever compacting, heavily populated polders from the higher situated silted-up riverbeds and floodplains constitute too great a risk. In some parts of the Rhine delta land subsidence, both due to natural and anthropogenic causes, has reached values of 3 to 4 m. If a dyke breach were to occur the enormous resulting damage and possible loss of life is now considered unacceptable. A wide range of river landscape planning measures is presently available for the improvement of the various, sometimes conflicting, river functions (Fig. 6.16).

Recently, the 'design discharge' for the river Rhine in The Netherlands had to be raised from 15,000 m^3s^{-1} to 16,000 m^3s^{-1} because the two very high flood events of 1993 and 1995 had to be taken into account. This design discharge is a theoretical calculated discharge, based on historical data. It acts as a standard for the height of the main river dykes, guaranteeing an accepted degree of protection against flooding. It is a discharge for which the risk of occurrence in a given year is one in 1,250 (Middelkoop 1999). Whatever the flood control measures that are taken, the river must be able to accommodate this modelled maximum discharge. Lowering the level of the winter bed over large stretches seems to be a logical solution. However, apart from obvious negative ecological consequences, the lowering of the embanked floodplain surface will inevitably lead to

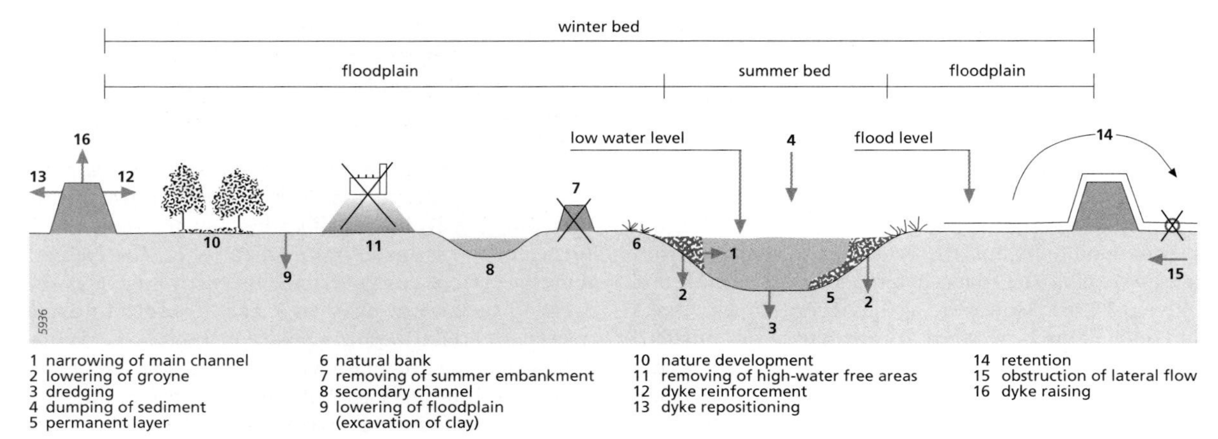

Fig. 6.16. Possible measures to reduce flooding risks (after Nijland and Cals 2000).

increased sedimentation of fine sand and silt on this surface due to increased flooding frequency and duration. This means that this process of lowering the floodplain surface will have to be repeated every few decades or so. In this respect the term 'cyclic rejuvenation' has been introduced by analogy with the natural rejuvenation of floodplain morphology and vegetation succession, which occurs due to channel migration under natural conditions (Nijland and Cals 2000). Other options, such as creating spillways and overflows and/or the construction of retention polders may be good solutions from a purely hydrological point of view (Havinga and Smits 2000), but will undoubtedly encounter serious opposition in heavily populated areas along the river. Another issue that conflicts with the ecological rehabilitation of the river is shipping. The Rhine is by far the busiest inland waterway in western Europe. Likewise, the river Meuse is a waterway of major importance. To protect the viability of the shipping industry, huge amounts of money are spent each year on navigational improvements. Climate change scenarios not only 'predict' higher flows during winter but also extreme low flow events of longer duration during the summer season. Consequently, the river channel must regularly be deepened and widened to sustain the economically very important freight traffic by ships during summer. Clearly, there are many more dilemmas that will face the river managers in the near future in their efforts to combine the many functions of the river.

The restoration of more natural and dynamic erosion and sedimentation processes in rivers and (formerly) embanked floodplains or polders has become an additional objective in modern river management in all western European countries. Along most of the large rivers and their tributaries a multitude of 'river rehabilitation' projects are being executed. These involve major engineering activities, such as lowering parts of the floodplain by removal of clay deposits, (re)construction of lateral channels and floodplain lakes, and removal of artificial levees (summer dykes) in the floodplains (see Fig. 6.16; Kern 1992; Nijland and Cals 2000). In response to an initiative of the Council of Europe to promote scientific and technical co-operation in environmental and resource management of rivers, the European Large Alluvial Rivers Network was formed in 1986 (Petts 1989).

References

Andres, W., Bos, J. A. A., Houben, P., Kalis, A. J., Nolte, S., Rittweger, H., and Wunderlich, J. (2001), Environmental change and fluvial activity during the Younger Dryas in central Germany. *Quaternary International* 79: 89–100.

Antoine, P., Lautridou, J. P., and Laurant, M. (2000), Long-term fluvial archives in NW France: response of the Seine and Somme rivers to tectonic movements, climatic variations and sea-level change. *Geomorphology* 33: 183–207.

—— Munaut, A.-V., Limondin-Lozouet, N., Ponel, P., Dupéron, J., and Dupéron, M. (2003), Response of the Selle river to climatic modifications during the Lateglacial and Early Holocene (Somme Basin-Northern France). *Quaternary Science Reviews* 22: 2061–76.

Arnell, N. W. (1996), *Global warming, river flows and water resources.* Wiley, Chichester.

Becker, B. (1993), An 11,000-year German oak and pine dendrochronology for radiocarbon calibration. *Radiocarbon* 35: 201–13.

Berendsen, H. J. A., and Stouthamer, E. (2000), Late Weichselian and Holocene palaeogeography of the Rhine–Meuse delta, The Netherlands. *Palaeogeography, Palaeoclimatology, Palaeoecology* 161: 311–35.

—— —— (2002), Paleogeographic evolution and avulsion history of the Holocene Rhine–Meuse delta, The Netherlands. *Netherlands Journal of Geosciences / Geologie en Mijnbouw* 81: 97–112.

Bijlsma, S. (1981), Fluvial sedimentation from the Fennoscandian area into the North-West European Basin during the late Cenozoic. *Geologie en Mijnbouw* 60: 337–45.

Boenigk, W. (2002), The Pleistocene drainage pattern in the Lower Rhine Basin. *Netherlands Journal of Geosciences / Geologie en Mijnbouw* 81: 201–9.

Bogaart, P. W., van Balen, R. T., Kasse, C., and Vandenberghe, J. (2003), Process-based modelling of fluvial system reponse to rapid climate change II. Application to the river Maas (The Netherlands) during the Last Glacial–Interglacial transition. *Quaternary Science Reviews* 22: 2097–110.

Bourdier, F. (1974), Essai sur le creusement de la Vallée de la Somme au Quaternaire. In: P. Macar (ed.) *L'Évolution Quaternaire des bassins fluviaux de la Mer du Nord Méridionale.* Société Géologique de Belgique (Liège), 233–40.

Bridgland, D. R. (2000), River terrace systems in north-west Europe: an archive of environmental change, uplift and early human occupation. *Quaternary Science Reviews* 19: 1293–303.

Brunnacker, K., Löscher, M., Tillmans, W., and Urban, B. (1982), Correlation of the Quaternary terrace sequence in the lower Rhine valley and northern Alpine foothills of central Europe. *Quaternary Research* 18: 152–73.

Buch, M. W. (1989), Late Pleistocene and Holocene development of the Danube valley east of Regensburg. *Catena Supplement* 15: 279–87.

Buck, W., Felkel, K., Gerhard, H., Kalweit, H., van Malde, J., Nippes, K.-R., Ploeger, B., and Schmitz, W. (1993), *Der Rhein unter der Einwirkung des Menschen—Ausbau, Schiffart, Wasserwirtschaft.* Internationale Kommission für die Hydrologie des Rheingebietes (KHR), Report I-11, Lelystad.

Busch, D., Schirmer, M., Schuchardt, B., and Ullrich, P. (1989), Historical changes of the river Weser. In: G. E. Petts, H. Möller, and A. L. Roux (eds.) *Historical changes of large alluvial rivers: Western Europe.* Wiley, Chichester, 297–321.

Cohen, K. M., Stouthamer, E., and Berendsen, H. J. A. (2002), Fluvial deposits as a record for Late Quaternary neotectonic activity in the Rhine–Meuse delta, The Netherlands. *Netherlands Journal of Geosciences / Geologie en Mijnbouw* 81: 389–405.

Dambeck, R., and Thiemeyer, H. (2002), Fluvial history of the northern Upper Rhine River (southwestern Germany) during the Lateglacial and Holocene times. *Quaternary International* 93/4: 53–63.

Decamps, H., Fortune, M., and Gazelle, F. (1989), Historical changes of the Garonne river, southern France. In: G. E. Petts, H. Möller, and A. L. Roux (eds.) *Historical changes of large alluvial rivers: Western Europe.* Wiley, Chichester, 249–67.

Ehlers, J. (1996), *Quaternary and glacial geology.* Wiley, Chichester.

Fetzer, K. D., Larres, K., Sabel, K.-J., Spies, E.-D., and Weidenfeller, M. (1995), Hessen, Rheinland-Pfalz, Saarland. Fluviatile Sedimente des Rheins und seiner Nebenflüsse. In: L. Benda (ed.) *Das Quartär Deutschlands.* Bornträger, Berlin, 220–40.

Gibbard, P. L. (1988), The history of the great northwest European rivers during the past three million years. *Phil. Trans. R. Soc. London* B318: 559–602.

Grabs, W. (ed.) (1997), *Impact of climate change on hydrological regimes and water resources management in the Rhine Basin.* Intern. Comm. for the Hydrology of the Rhine Basin (CHR), Report I-16.

Hantke, R. (1993), *Flussgeschichte Mitteleuropas. Skizzen zu einer Erd-, Vegetations- und Klimageschichte der letzten 40 Millionen Jahre.* F. Enke, Stuttgart.

Havinga, H., and Smits, A. J. M. (2000), River management along the Rhine: a retrospective view. In: A. J. M. Smits, P. H. Nienhuis, and R. S. E. W. Leuven (eds.) *New approaches to river management.* Backhuys, Leiden, 15–32.

Herget, J. (2000), Holocene development of the River Lippe Valley, Germany: A case study of anthropogenic influence. *Earth Surface Processes and Landforms* 25: 293–305.

Hoselmann, C. (1996), Der Hauptterrassen-Komplex am unteren Mittelrhein. *Zeitschrift deutschen geologischen Gesellschaft 147*: 481–97.

Kasse, K., Vandenberghe, J., and Bohncke, S. (1995), Climatic change and fluvial dynamics of the Maas during the late Weichselian and early Holocene. In: B. Frenzel (ed.) *European river activity and climatic change during the Lateglacial and early Holocene.* Palaeoclimatic Research 14, European Science Foundation, Strasburg, 123–50.

Kern, K. (1992), Restoration of lowland rivers: the German experience. In: P. A. Carling and G. E. Petts (eds.) *Lowland floodplain rivers: Geomorphological perspectives.* Wiley, Chichester, 279–97.

Kiden, P. (1991), The late glacial and Holocene evolution of the Middle and Lower River Scheldt, Belgium. In: L. Starkel, K. J. Gregory, and J. B. Thornes (eds.) *Temperate palaeohydrology. Fluvial processes in the temperate zone during the last 15,000 years.* Wiley, Chichester, 283–99.

Knox, J. C. (1985), Responses of floods to Holocene climatic change in the Upper Mississippi Valley. *Quaternary Research* 23: 287–300.

—— (1993), Large increases in flood magnitude in response to modest changes in climate. *Nature* 361: 430–2.

—— (2000), Sensitivity of modern and Holocene floods to climate change. *Quaternary Science Reviews* 19: 439–57.

Kwadijk, J. C. J. (1991), Sensitivity of the River Rhine discharge to environmental change, a first tentative assessment. *Earth Surface Processes and Landforms* 16: 627–37.

Lenders, H. J. R. (2003), *Environmental rehabilitation of the river landscape in the Netherlands.* Thesis, University of Nijmegen.

Liébault, F., and Piégay, H. (2001), Assessment of channel changes due to long-term bedload supply decrease, Roubion River, France. *Geomorphology* 36: 167–86.

Lüttig, G. (1974), Geological history of the River Weser (Northwest Germany). In: P. Macar (ed.) *L'Évolution Quaternaire des bassins fluviaux de la Mer du Nord Méridionale.* Société Géologique de Belgique, Liège, 21–34.

Lüttig, G., and Meyer, K. D. (1974), Geological history of the River Elbe, mainly of its lower course. In: P. Macar (ed.) *L'Évolution Quaternaire des bassins fluviaux de la Mer du Nord Méridionale.* Société Géologique de Belgique, Liège, 1–19.

McCarthy, J. J., Canziani, O. F., Leary, N. A., Dokken, D. J., and White, K. S. (eds.) (2001), *Climate Change 2001: Impacts, Adaptation, and Vulnerability.* IPCC, Cambridge University Press, Cambridge.

Mäusbacher, R., Schneider, H., and Igl, M. (2001), Influence of late glacial climate changes on sediment transport in the River Werra (Thuringia, Germany). *Quaternary International* 79: 101–9.

Meyer, W., and Stets, J. (2002), Pleistocene to Recent tectonics in the Rhenish Massif (Germany). *Netherlands Journal of Geosciences / Geologie en Mijnbouw* 81: 217–22.

Micha, J.-C., and Borlee, M.-C. (1989), Recent historical changes on the Belgian Meuse. In: G. E Petts, H. Möller, and A. L. Roux (eds.) *Historical changes of large alluvial rivers: Western Europe.* Wiley, Chichester, 269–95.

Middelkoop, H. (ed.) (1999), *Twice a river, Rhine and Meuse in the Netherlands.* RIZA (National Inst. for Inland Water Management and Waste Water Treatment) Report 99.003, Arnhem-Lelystad.

—— Daamen, K., Gellens, D., Grabs, W., Kwadijk, J. C. J., Lang, H., Parmet, B. W. A. H., Schädler, B., Schulla, J., and Wilke, K. (2001), Impact of climate change on hydrological regimes and water resources management in the Rhine Basin. *Climatic Change* 49: 105–28.

Mol, J. (1995), Weichselian and Holocene river dynamics in relation to climate change in the Halle-Leipziger Tieflandsbucht (Germany). *Eiszeitalter und Gegenwart* 45: 32–41.

—— Vandenberghe, J., and Kasse, C. (2000), River response to variations of periglacial climate in mid-latitude Europe. *Geomorphology* 33: 131–48.

Nienhuis, P. H., and Leuven, R. S. E. W. (1998), Ecological concepts for the sustainable management of lowland river basins: a review. In: P. H. Nienhuis, R. S. E. W. Leuven, and A. M. J. Ragas (eds.) *New concepts for sustainable management of river basins*. Backhuys, Leiden, 7–33.

—— Chojnacki, J. C., Harms, O., Majewski, W., Parzonka, W., and Prus, T. (2000), Elbe, Odra, and Vistula: reference rivers for the restoration of biodiversity and habitat quality. In: A. J. M. Smits, P. H. Nienhuis, and R. S. E. W. Leuven (eds.) *New approaches to river management*. Backhuys, Leiden, 65–84.

Nijland, H. J., and Cals, M. J. R. (eds.) (2000), *River restoration in Europe. Practical approaches*. RIZA (National Inst. for Inland Water Management and Waste Water Treatment) Report 2001.023, Lelystad.

Overeem, I., Bishop, C., Weltje, G. J., and Kroonenberg, S. B. (2001), The Late Cenozoic Eridanos delta system in the Southern North Sea Basin: a climate signal in sediment supply? *Basin Research* 13: 293–312.

Pastre, J.-F., Limondin-Lozouet, N., Leroyer, C., Ponel, P., and Fontugne, M. (2003), River system evolution and environmental changes during the Lateglacial in the Paris Basin (France). *Quaternary Science Reviews* 22: 2177–88.

Petts, G. (1989), Historical analysis of fluvial hydrosystems. In: G. E. Petts, H. Möller, and A. L. Roux (eds.) *Historical changes of large alluvial rivers: Western Europe*. Wiley, Chichester, 1–18.

Pissart, A. (1974), La Meuse en France et en Belgique. Formation du bassin hydrographique. Les terrasses et leurs enseignements. In: P. Macar (ed.) *L'Évolution Quaternaire des bassins fluviaux de la Mer du Nord Méridionale*. Société Géologique de Belgique, Liège, 105–31.

Quitzow, H. W. (1974), Das Rheintal und seine Entstehung, Bestandsaufname und Versuch einer Synthese. In: P. Macar (ed.) *L'Évolution Quaternaire des bassins fluviaux de la Mer du Nord Méridionale*. Société Géologique de Belgique, Liège, 53–104.

Roblin-Jouve, A. (1995), Late glacial and early Holocene geomorphology of the upper Seine river valley. In: B. Frenzel (ed.) *European river activity and climatic change during the Lateglacial and early Holocene*. Palaeoclimatic Research 14, European Science Foundation, Strasburg, 191–203.

Roux, A. L., Bravard, J. P., Amoros, C., and Pautou, G. (1989), Ecological changes of the French Upper Rhône river since 1750. In: G. E. Petts, H. Möller, and A. L. Roux (eds.) *Historical changes of large alluvial rivers: Western Europe*. Wiley, Chichester, 323–50.

Ruegg, G. H. J. (1994), Alluvial architecture of the Quaternary Rhine–Meuse river system in the Netherlands. *Geologie en Mijnbouw* 72: 321–30.

Schirmer, W. (1988), Holocene valley development on the Upper Rhine and Main. In: G. Lang and C. Schlüchter (eds.) *Lake, mire and river environments*. Balkema, Rotterdam, 153–60.

—— (1990), *Rheingeschichte zwischen Mosel und Maas. Der känozoische Werdegang des Exkursionsgebietes*. DEUQUA-Exkursionsführer, Hannover, 9–16 Sept. 1990, 1–33.

Schreve, D. C., and Bridgland, D. R. (2002), Correlation of English and German Middle Pleistocene fluvial sequences based on mammalian biostratigraphy. *Netherlands Journal of Geosciences / Geologie en Mijnbouw* 81: 357–73.

Stouthamer, E., and Berendsen, H. J. A. (2000), Factors controlling the Holocene avulsion history of the Rhine–Meuse delta (The Netherlands). *Journal of Sedimentary Research* 70: 1051–64.

Straffin, E. C., Blum, M. D., Colls, A., and Stokes, S. (1999), Alluvial stratigraphy of the Loire and Arroux Rivers, Burgundy, France. *Quaternaire* 10: 271–82.

—— and Blum, M. D. (2002), Holocene fluvial response to climate change and human activities; Burgundy, France. *Netherlands Journal of Geosciences / Geologie en Mijnbouw* 81: 417–30.

Tavernier, R., and de Moor, G. (1974), L'Évolution du bassin de L'Escaut. In: P. Macar (ed.) *L'Évolution Quaternaire des bassins fluviaux de la Mer du Nord Méridionale*. Société Géologique de Belgique, Liège, 159–231.

Tebbens, L. A., Veldkamp, A., Westerhof, W., and Kroonenberg, S. B. (1999), Fluvial incision and channel downcutting as a response to Late-glacial and Early Holocene climate change: the lower reach of the River Meuse (Maas), The Netherlands. *Journal of Quaternary Science* 14: 59–75.

—— Veldkamp, A., Van Dijke, J. J., and Schoorl, J. M. (2000), Modeling longitudinal-profile development in response to Late Quaternary tectonics, climate and sea-level changes: the River Meuse. *Global and Planetary Change* 27: 165–86.

Thiermann, A. (1974), Zur Flussgeschichte der Ems Nordwestdeutschland. In: P. Macar (ed.) *L'Évolution Quaternaire des bassins fluviaux de la Mer du Nord Méridionale*. Société Géologique de Belgique, Liège, 35–51.

van den Berg, M. W. and van Hoof, T. (2001), The Maas terrace sequence at Maastricht, SE Netherlands: evidence for 200 m of late Neogene and Quaternary surface uplift. In: D. Maddy, M. G. Macklin, and J. C. Woodward (eds.) *River basin sediment systems: Archives of environmental change*. Balkema, Lisse, 45–86.

Vandenberghe, J., and de Smet, P. (1979), Paleogeomorphology in the Eastern Scheldt Basin (Central Belgium). The Dijle-Demer-Grote Nete confluence area. *Catena* 6: 73–105.

van Huissteden, K. J., Gibbard, P. L., and Briant, R. M. (2001), Periglacial fluvial systems in northwest Europe during marine isotope stages 4 and 3. *Quaternary International* 79: 75–88.

van Urk, G., and Smit, H. (1989), The lower Rhine geomorphological changes. In: G. E. Petts, H. Möller, and A. L. Roux (eds.) *Historical changes of large alluvial rivers: Western Europe*. Wiley, Chichester, 167–82.

Veldkamp, A., and Kroonenberg, S. B. (1993), Late Quaternary chronology of the Allier terrace sediments (Massif Central, France). *Geologie en Mijnbouw* 72: 179–92.

—— and van den Berg, M. W. (1993), Three-dimensional modelling of Quaternary fluvial dynamics in a climo-tectonic dependent system. A case study of the Maas record (Maastricht, The Netherlands). *Global and Plenetary Change* 8: 203–18.

Vivian, H. (1989), Hydrological changes of the Rhône river. In: G. E. Petts, H. Möller, and A. L. Roux (eds.) *Historical changes of large alluvial rivers: Western Europe*. Wiley, Chichester, 57–77.

Westaway, R. (2002), Long-term river terrace sequences: Evidence for global increases in surface uplift rates in the Late Pliocene and early Middle Pleistocene caused by flow in the lower continental crust induced by surface processes. *Netherlands Journal of Geosciences / Geologie en Mijnbouw* 81: 305–28.

Zagwijn, W. H. (1985), An outline of the Quaternary stratigraphy of the Netherlands. *Geologie en Mijnbouw* 64: 17–24.

—— (1989), The Netherlands during the Tertiary and the Quaternary: A case history of Coastal Lowland evolution. *Geologie en Mijnbouw* 68: 107–20.

Zonneveld, J. I. S. (1974), The terraces of the Maas (and the Rhine) downstream of Maastricht. In: P. Macar (ed.) *L'Évolution Quaternaire des bassins fluviaux de la Mer du Nord Méridionale*. Société Géologique de Belgique, Liège, 133–57.

7 Marine and Coastal Environments

Aart Kroon

Introduction

The present coastline of western Europe is shaped by physical processes such as wind, waves, and tidal currents, which cause the marine and coastal sediment transport. Spatial gradients in sediment transport rates induce the morphological adaptation, reflected by either an accumulation or erosion of material. All mutual interactions between these physical processes, and the resultant gradients in sediment transport together with the morphological adaptations, constitute the coastal morphodynamics. The specific initial stage of the morphology and the availability of sediment influence the direction of the morphological adaptation, whereas the rate of the morphological adaptation mostly depends on the energy input into the system (Fig. 7.1).

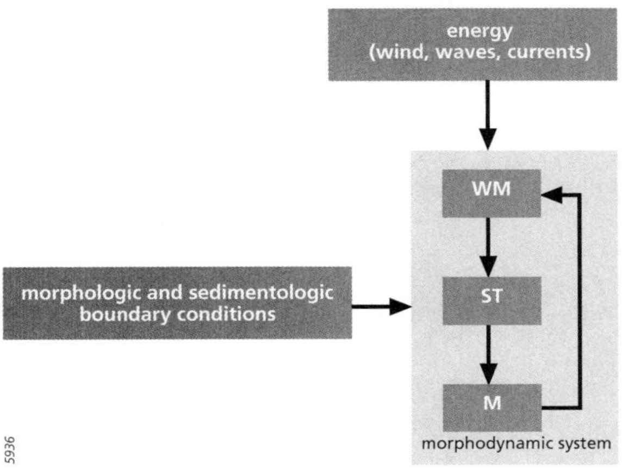

Fig. 7.1. Morphodynamics of coastal systems with its energy input and boundary conditions. WM is the water motion, ST is the sediment transport, and M is the morphology.

Chemical processes are of less importance in coastal environments of the high to mid-latitudes. Here most geochemical reactions are far too slow to influence the coastal morphology. However, biological processes sometimes play an important role. For example, flocculation of fine sediments by algae in estuaries (Ten Brinke 1993) or filtering by salt-marsh vegetation (Houwing 2000*a*, *b*) both positively influence the sediment accumulation rates.

The long-term boundary conditions upon which the physical processes act are often determined by geology. The nature of the drainage basin that delivers fresh water and sediments into coastal waters and the nature of the shoreline can be considered as static boundary conditions for short-term morphodynamics. Tectonic forces and global sea-level rise are typical long-term geophysical forces. They will slowly change these boundary conditions, but they hardly influence the short-term adaptations of the morphology.

The western European shelf fringes a series of integrated coastal environments that vary from coastal dunes and sandy beaches to estuaries and tidal basins and to sea cliffs and shore platforms. In this chapter a general description of the location and dimensions of the shores of western Europe is presented, followed by a brief summary of its geological history. The geology is focused on present-day deposits, the local lithology with sinks and sources of sediments, and with reference to some geophysical processes such as the relative sea-level rise. Thereafter, the actual coastal processes are discussed. This section covers the climatic and oceanic factors relevant to western Europe, with a summary of the physical processes and the morphodynamics of each characteristic coastal environment. Finally, a regional characterization of the coastal environments from southern France to northernmost Denmark is presented.

Geographical Setting of Western European Shores

The shoreline of western Europe has a total length of 2,500 km and extends from France to Belgium, The Netherlands, Germany, and Denmark (Fig. 7.2A). The border between Spain and France marks the most southern end and Skagen Odde in Denmark forms the most northern end. The west European shelf extends seaward of these shores. The shelf edge, at about 100 m below mean sea level (MSL), is close to the present coastline along the southern part of the French Atlantic coast near Les Landes and at the northern tip of the Danish North Sea coast being about 10 km offshore in these localities. All other western European shores are much further from the shelf edge, facing relatively shallow semi-enclosed seas like the English Channel and the North Sea. Physical processes on the shelf and near the shelf edge do not directly influence the major sediment budgets and transports near the shorelines. The shores of France face the Atlantic Ocean in the west and the English Channel in the north. All other shores are along the semi-enclosed North Sea. The shores of western Europe are located in the temperate latitudes between 43° 20′ and 57° 80′ and are strongly influenced by the Mid Atlantic Gulf Stream that brings a continuous flux of relatively warm water from the Caribbean into the area.

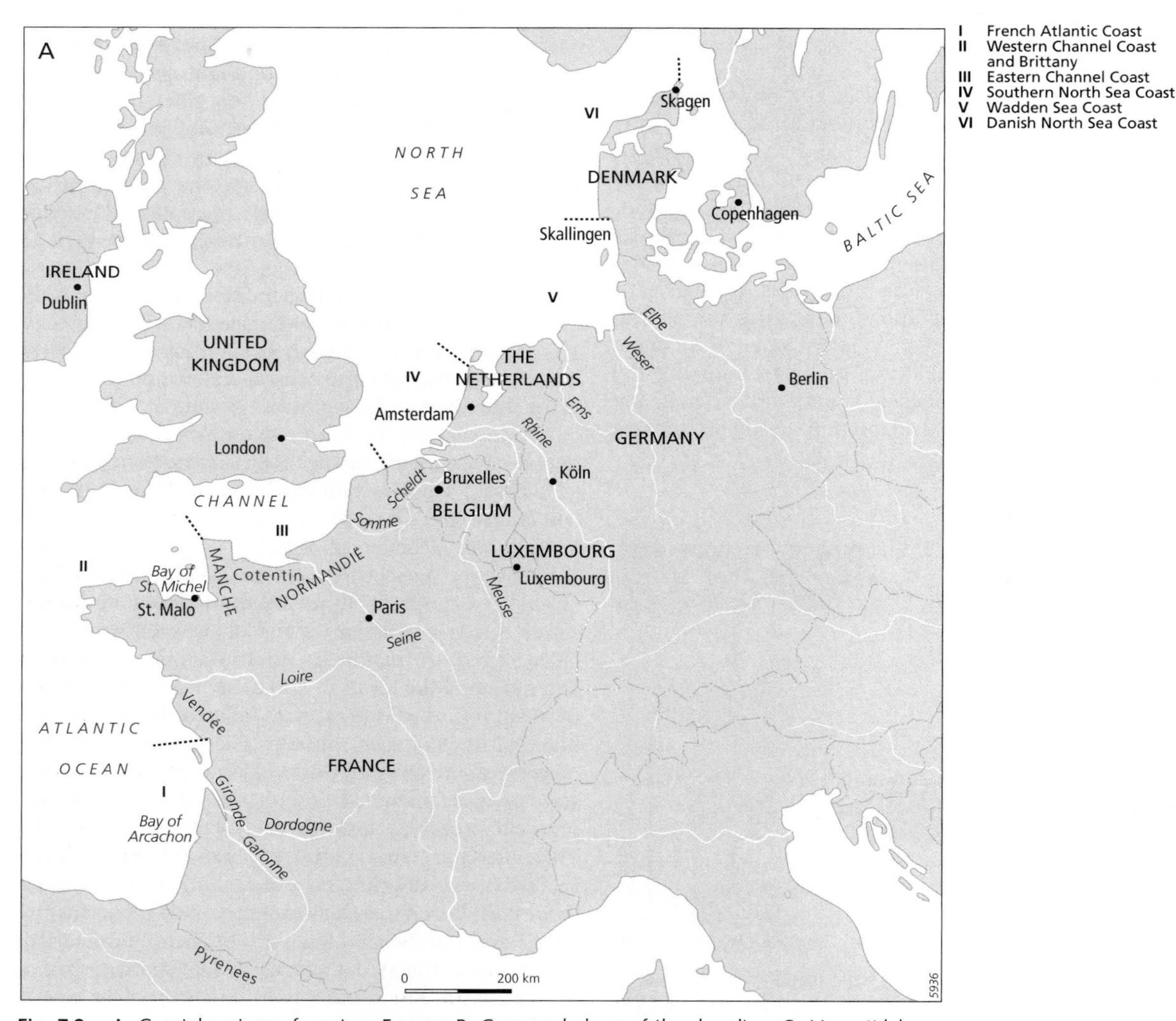

Fig. 7.2. A. Coastal regions of western Europe; B. Geomorphology of the shoreline; C. Mean tidal range.

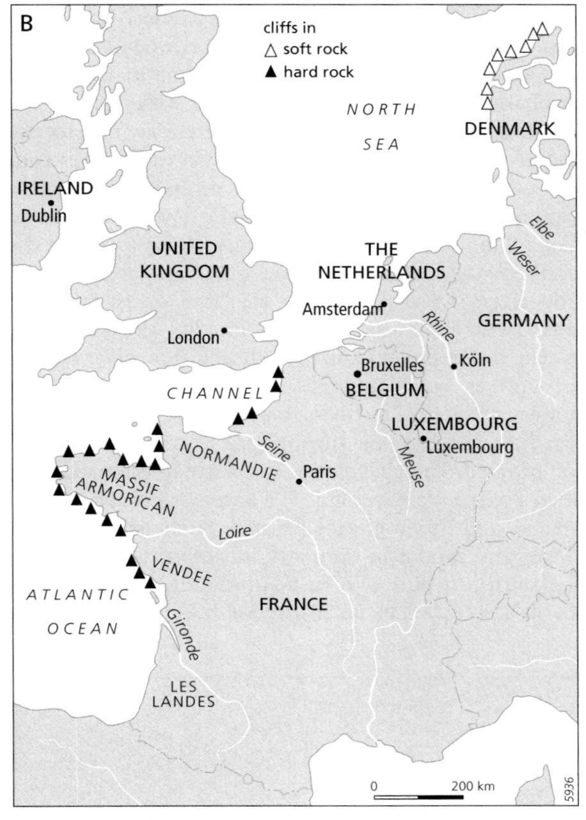

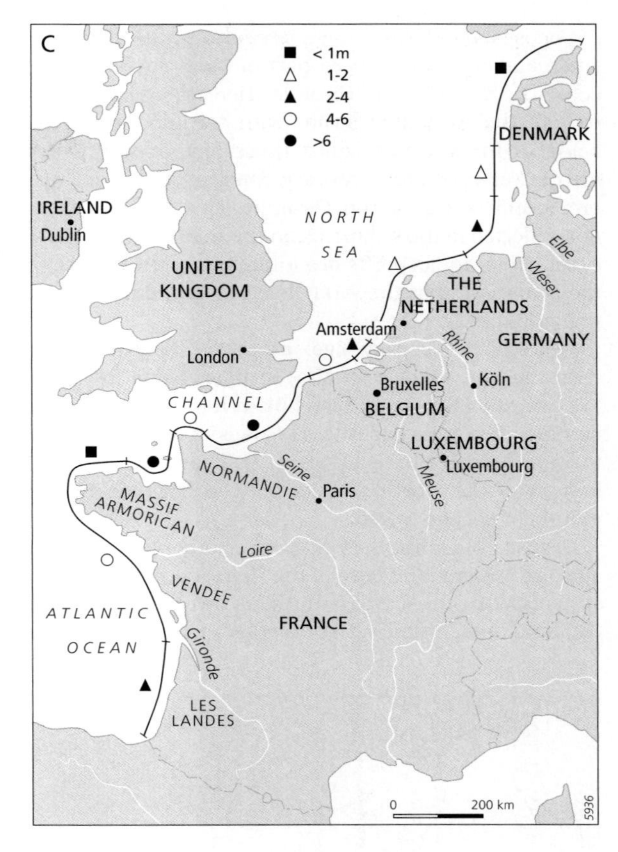

Fig. 7.2. (cont.)

These shorelines show a wide variety of coastal environments: from sandy beaches with dunes or glacial cliffs; to barrier islands and tidal inlets with tidal basins; to estuaries of major river mouths (sometimes blocked); and to shore platforms and small pocket beaches. The western European shore can be divided into six regions, each one reflecting a specific environment including a certain uniformity in coastal morphodynamics (Fig. 7.2A).

Geology

Deposits

The coastal areas predominantly consist of a series of low elevation terrains. Most of these areas experienced Tertiary subsidence, as is well demonstrated for the western parts of the Aquitaine basin, the Paris basin, the Tertiary Belgium basin and the North German basin (Fig. 7.2B). The slightly uplifted Armorican massif is much older and has the oldest deposits of any section of the western European shores. This Armorican massif is located in the Western Channel and Brittany coast (II) region. Here, Proterozoic and Palaeozoic deposits are at the present surface, together with local intrusive and metamorphic rocks such as granite and gneiss. This whole area was reshaped during the Hercynian orogeny. More recently, Holocene sediments were deposited along its coastal fringes in the Bay of Mont St Michel and near the estuary of the Loire. The deposits on the sea bottom in this area are mostly of Tertiary age and consist of Eocene sediments.

Along the Eastern Channel Coast (III) the deposits near the surface are of Mesozoic age. They form part of the Paris basin. Jurassic deposits are found along its fringes and Cretaceous deposits occur in its centre. This is also reflected in the sediments on the sea bottom of the Channel. The deposits further seaward in the centre of the Channel are of Tertiary, in particular, Eocene age. Holocene deposits are found only at the mouths of the estuaries in this area, especially those of the Somme and the Seine rivers.

Deposits of Mesozoic age are also situated at the surface along the northern part of the French Atlantic coast (I), north of the Gironde. Here the deposits are part of the Aquitaine basin, with the most northern being of Jurassic age whilst those further southward are of Cretaceous age. Tertiary deposits of Pliocene age are found south of the Gironde. This alternation is also reflected in the sediments on the shelf seaward of the coast. Holocene deposits are found along the fringes of the estuary and in the sandy beaches and dunes in the region of Les Landes.

Deposits of Tertiary age, especially Eocene sands, occur at the surface in the Tertiary Belgium basin. Westward a Holocene coastal plain has developed with fluvial and marine deposits. This Belgian coast is part of the Southern North Sea Coast (IV). The deposits on the bottom of the North Sea seaward of Vlaanderen are mainly of Eocene origin.

Deposits of Quaternary age, especially of Pleistocene age, are present landward of the Holocene barrier island systems along the western and south-western provinces (Holland and Zeeland) of The Netherlands as part of the Southern North Sea Coast (IV) region, the Wadden Sea Coast (V) region and the Danish North Sea Coast (VI) region. The palaeogeographic situation in the North Sea basin and English Channel at four stages during the Pleistocene is shown in Fig. 7.3. In the early Pleistocene the ice caps of Britain and Scandinavia were not in contact with each other and the rivers drained to the northwest into the Atlantic Ocean. The Elsterian and Saalian glaciations covered western Europe as far south as the northern and central part of The Netherlands, respectively. The rivers were flowing to the south-west in these periods through the Channel. Finally, the ice sheets during the last glaciation (Weichselian) reached only as far south as central Jutland on the Danish west coast. The major part of the North Sea basin was not covered with ice, but river drainage through the Straits of Dover was maintained. The glacial deposits relate to these different stages of the ice coverage. The Saale glaciation created glacial and fluvio-glacial features such as ice-pushed ridges and sandar in Denmark, northern Germany and the northern part of The Netherlands. The Weichsel ice sheet reached as far as the Main Stationary Line in

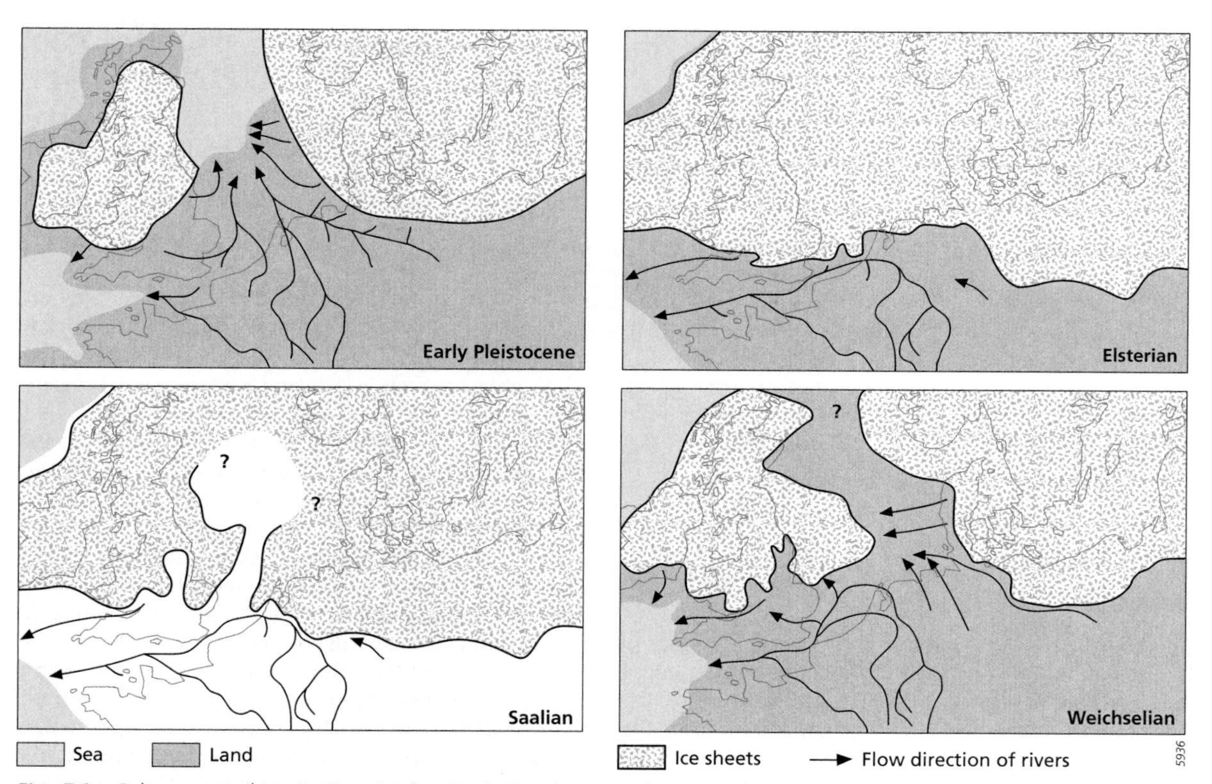

Fig. 7.3. Palaeogeographic situation in the North Sea basin at four periods during the Pleistocene (after Berendsen and Stouthamer 2001).

central Jutland in Denmark. Glacial features of this glaciation are observed north of this line, whereas periglacial phenomena are present all along the southern Danish North Sea Coast (VI). These glacial deposits of both the Saale and Weichsel ice ages are exposed in the cliffs of Denmark. An extensive Holocene coastal plain is well developed along the Holland Coast, where it is interrupted by tidal inlets and estuaries, especially along the Zeeland and Wadden Sea coastal sections. Large tidal basins are located in these areas. However, the deposits on the bottom of the North Sea in front of these areas are predominantly of Pleistocene age.

Lithology: Sinks and Sources

The shores of western Europe consist of consolidated rocks and of unconsolidated sediments, the latter mainly of Quaternary age. The abrasion of the consolidated rocks is a relatively slow process that does not deliver much sediment to the present shores. The unconsolidated sediments were transported during the Pleistocene and Holocene by a range of mechanisms related to marine, fluvial, (peri-)glacial, and aeolian processes. These deposits are sensitive to erosion and reworking by marine processes, which may operate on a large scale.

Deposits of glacial origin are mainly found in the northern part of the area from Denmark to The Netherlands. During different phases of the Pleistocene (Fig. 7.3) glacial and fluvioglacial deposits accumulated in front of the ice sheets in the form of moraines, ice-pushed ridges, and sandar. The valleys between these higher parts in the landscape were filled by periglacial deposits, and by fluvial and aeolian deposits during the ice-free periods. The final shape of the present coastline is often the result of the reworking of this glacial and periglacial morphology by marine processes. Some of the coastal cliffs along the Danish North Sea Coast consist, for instance, of eroded ice-pushed ridges.

Deposits of fluvial origin are widely found, especially along the shores of Germany, The Netherlands, and France. The main rivers in the Holocene that contributed their fluvial sediments to the coastal waters were the Elbe, Weser, and Ems in Germany; the Rhine, Meuse, and Scheldt in The Netherlands; and the Somme, Seine, Loire, and Gironde in France. Estuaries are often found at the mouth of these rivers. These estuaries partly deliver sandy sediments to the marine environments, but also trap fine cohesive sediments along the fringes of their basins.

The marine deposits mainly originate from drowned morphologies of Eocene and Pleistocene origin, such as glacial or braided river deposits, or come from Holocene deposits of the present rivers. In additon, the reworked sediments also originate from former outer deltas in front of (former) tidal inlets, as is the case along the Holland coast of The Netherlands. Late Holocene to Recent aeolian deposits are found along the fringes of several western European shores and form extensive dune areas. These deposits were often originally of marine or fluvial origin and were later reworked by the wind.

Holocene Evolution and Relative Sea-Level Rise

The Holocene evolution of the fluvial and coastal plains area is determined by a number of factors (Jacobsen 1997; Berendsen and Stouthamer 2001). These factors are: (1) the morphology of the (Pleistocene) subsurface and the nature of the sediments; (2) the size and speed of relative sea level fluctuations; (3) the availability of sand; (4) the tidal range; (5) the climatic fluctuations, including the frequency of storm events; (6) and the inundations of the rivers. The balance between these factors determines the evolution of a coastal stretch. In general, cold climatic phases are characterized by marine regressions and low sea-level stands, whereas warmer phases are characterized by marine transgressions and a high mean sea level. The Holocene had two major warmer phases in western Europe, the Atlantic (8000–5000 ^{14}C years BP) and the Subatlantic (3000 ^{14}C years BP–present).

The relative sea-level rise on the geological timescale is affected by components related to the dynamics of the earth and by components related to the dynamics of the water masses on the earth. The former include tectonic subsidence of the North Sea basin, isostatic subsidence of the land and the glacial forebulge, and compaction of underlying sediments. The latter include eustatic sea-level rise, the heating of the atmosphere, and the amount of precipitation in relation to the volume of ice masses in glaciers and polar regions.

The Holocene (relative) sea-level change along the French Atlantic coast, the southern North Sea coast, and the coast of northernmost Denmark is spatially different because of the response of the crust to changing ice and water load in Late Pleistocene and Holocene time (Fig. 7.4). Sea-level changes decrease from north to south due to the glacio-isostatic response to the melting of the Fennoscandian ice sheet, and from east to west due to the hydro-isostatic response to the addition of meltwater in the Atlantic Ocean (Lambeck 1997). Sea level was −100 to −120 m during the Last Glacial Maximum (LGM) at 18,000–20,000 ^{14}C years BP, about −15 m at 8,000 ^{14}C years BP and about −10 m at 6000 ^{14}C years BP. The eustatic sea level increased by about 3 m over the last 6,000 years and in-situ marine deposits are mostly not older than 6000 ^{14}C years BP (Lambeck 1997). The

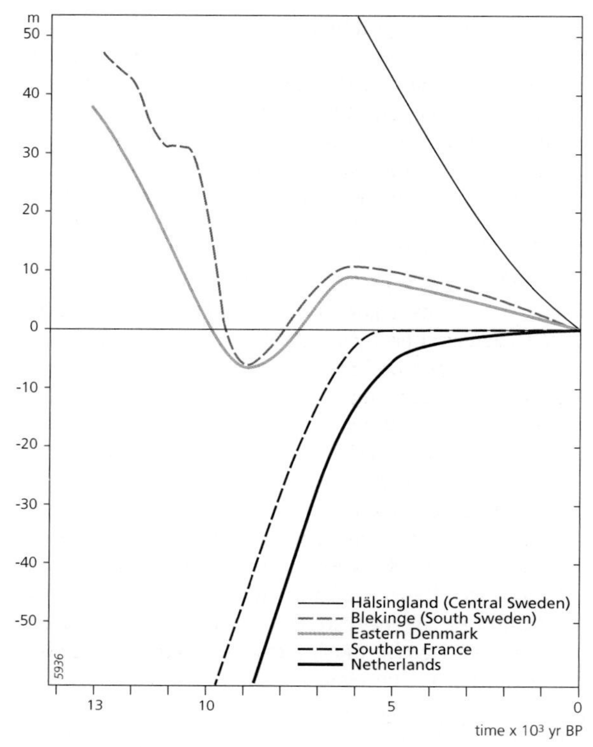

Fig. 7.4. Schematic relative sea level curves for Sweden, Denmark, The Netherlands and southern France, illustrating the differences in isostatic movements between these areas.

relative sea-level curves in Belgium, The Netherlands, and Germany also show a deceleration of the rise with rates of 0.6–1.0 m/century in the early phases of the Holocene (Atlantic) declining to 0.10–0.25 m/century at present (Beets *et al.* 2003; Eitner 1996). Especially in the early phases of the Holocene up to 4000 ^{14}C years BP, the Pleistocene surface of Belgium was higher and indicates smaller isostatic adjustments due to the glacial forebulge (Denys and Baeteman 1995). Sedimentation in the present coastal plain started in the Atlantic period (Berendsen and Stouthamer 2001). However, the barriers along the fringes of the North Sea were not permanently transgressing. Most of the Holland coast beach ridges in the Subatlantic were regressive, due to the large influx of sediments by the rivers and the availability of large quantities of sediment in the former ebb-tidal deltas in front of inlets that were just closed (Beets *et al.* 1992). The ice cover of the last Weichsel glaciation reached as far as the central part of Jutland in Denmark. Consequently, the northern part of the Danish North Sea Coast is still in a phase of unloading of the ice masses, and the relative sea level has decreased over the Holocene. An example of the morphological result of this sea-level decrease is a series of beach ridges in the area near Skagen. Here, the oldest deposits in the early Holocene are at about +10 m above present MSL. The southern part of the Danish North Sea Coast and the shores of Germany and The Netherlands are in the area of the former glacial forebulge. Consequently, the periglacial subsurface is sinking and a relative sea-level rise is observed (Jacobsen 1997).

Coastal Processes

Climatic Factors

Currently, the coastal regions of western Europe experience a maritime climate with precipitation in all seasons. The mean annual precipitation ranges between 500 and 1,000 mm/year. Summer and winter temperatures are moderate with a mean annual temperature ranging from 12.5 °C in the south to 8.5 °C northwards. Depressions coming from the Atlantic Ocean determine the usual weather pattern. Wind directions during the eastward passage of these depressions are mainly westerly (between SSW and NNW). Low-pressure periods are alternated by more stable weather patterns with high-pressure cells over Scandinavia. In these cases winds are predominantly from easterly directions and have significantly lower velocities. Sea-surface temperatures and salinity values reflect the spatial variability in solar radiation, air temperature, freshwater inflow, and surface winds. In general, the solar radiation and the air temperature decrease from south to north. The freshwater inflow is locally centred at the mouths of the major rivers (Fig. 7.2A). The surface winds are strongest over open water surfaces, especially over the Atlantic Ocean and, to a lesser extent, the North Sea. All these aspects cause a decrease in sea-water temperature from the south to the north and a decreased salinity from open sea (about 30 ppm) to the mouths of the rivers.

Oceanic Factors

The major water circulation of the North Atlantic brings a North Atlantic Drift with relatively warm water masses to the western European shelf. This North Atlantic Drift has its origin in the Caribbean. From here it travels to the north, across the North Atlantic Ocean and then follows the eastern shores to the north. This circulation system is the main reason for the maritime climate of the temperate latitudes in western Europe.

Regular variations in the water levels of the ocean and along the shores are caused by the tides. Common tidal

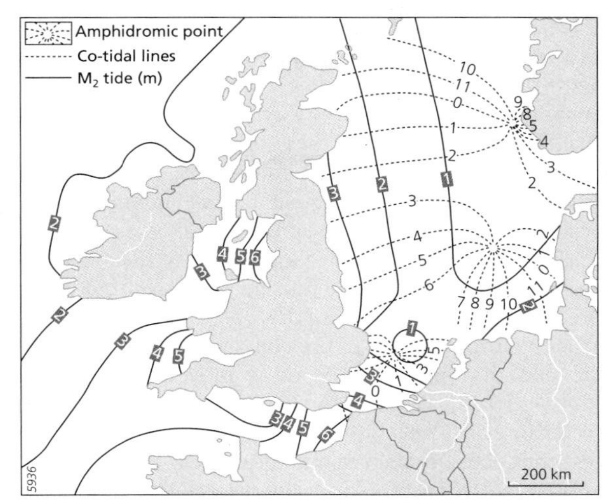

Fig. 7.5. The amphidromic points and tidal propagation of the M2 tide.

water-level fluctuations range over periods of hours to weeks. The general pattern of the tides in the North Sea, the English Channel, and the western Atlantic Ocean is semi-diurnal. The main component of the tide is the M2 and there are three amphidromic points in the North Sea basin (Fig. 7.5). The tide comes in from the south and travels to the north along the shores of France and the eastern shores of the North Sea. The tidal range, defined as the difference between mean high water level and mean low water level, is not constant along the shores. Maximum values of the tidal range are observed along the southern shores of the English Channel and are about 11 m at St. Malo in France. Values along the eastern shores of the North Sea range from macro-tidal (over 4 m) at the Belgium shore near the entrance of the English Channel and in the German Bight to micro-tidal (less than 2 m) along the central coast of The Netherlands and the Danish North Sea Coast (Fig. 7.2c). Fortnightly cycles of spring tide and neap tide are also commonly present along the shores. The differences in tidal ranges within these cycles are of the order of decimetres in the micro-tidal areas of The Netherlands and Denmark and metres in the macro-tidal areas.

Sea waves and swell are usually observed along the Atlantic Ocean shores and the northern part of the North Sea. The sea waves are locally generated and strongly increase with increasing wind velocity, wind duration, and fetch. Swell waves do not have a local origin. The almost permanent background of swell from the Atlantic Ocean or northern North Sea places these areas in a moderate- to high-energy wave environment. High-energy wave conditions occur during the passage of storms associated with passing depressions. The shores along the semi-enclosed seas of the English Channel and the southern part of the North Sea are mainly approached by sea waves. This means that there is a large variety in wave energy climate through time, with low-energy wave conditions during easterly winds and moderate- to high-energy wave conditions during the westerly storm passages. Storm surges, caused by water masses that are pushed into the funnel shape of the enclosed seas, can further increase the mean water levels of the astronomical tide by 1 to 4 metres along the shores. This increase can even be higher near the seaward entrance of estuaries.

The relative sea-level rise is a background process on a geological timescale. Currently, the relative sea-level rise in western Europe is estimated at about 0.10–0.25 m/century (Eitner 1997; Berendsen and Stouthamer 2001). However, human interference within the global atmosphere may create an increase of sea-level rise with values of up to 0.5 m/century. This factor could make this aspect more relevant for long-term coastal morphodynamics and sediment budgets (over periods of decades).

Hydrologic Factors

The maritime climate of the temperate latitudes has precipitation in all seasons. The precipitation exceeds the evaporation and fresh water and sediment are transported from inland to the coastal areas by major rivers such as the Elbe, Weser, Ems, Rhine, Meuse, Scheldt, Somme, Seine, Loire, and Gironde (Fig. 7.2A). The average discharge of some of these rivers is presented in Table 6.1 in Ch. 6. The discharge has a seasonal variation with major discharges in the spring and minor discharges in the fall. The peak flow in the spring is caused by a combination of snowmelt, especially in the higher areas of the Alps and Pyrenees, and additional rainfall in the lower parts of the drainage basins. The shorelines of Brittany and the Danish North Sea Coast are the only ones in western Europe that lack the influx of fresh water and sediment from the major drainage basins of European rivers.

Sediment Transport

Tidal processes cause sediment transport in the marine environment near the bed of the shallow seas. The current velocities near the bed induce a flux of sediment that is variable in rate and direction over the tide. The influence of the orbital motion of the waves near the seabed of the shallow seas is only present during a storm. Such waves will stir the sediment so that it can be transported by tidal and other oceanographic currents.

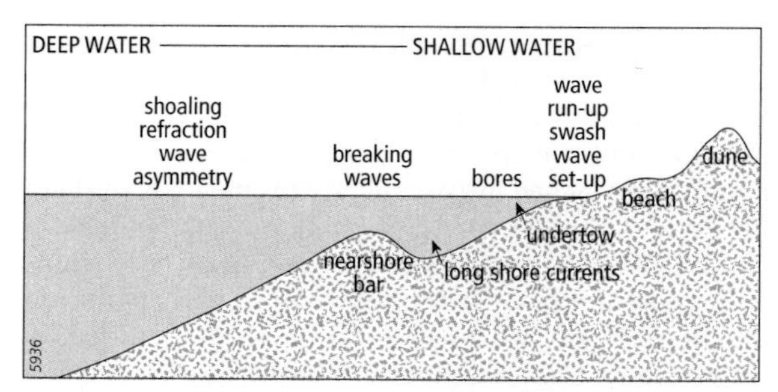

Fig. 7.6. A definition sketch of wave-related processes in a coastal environment.

The impact of tidal currents near the shore in coastal environments is less pronounced and decreases for environments with low tidal ranges. The main influence of the tide near the shore is the variation in water levels. The intensity and duration of wave-related processes on the coastal profile are a function of the water-level fluctuations due to the semi-diurnal tide and the spring–neap tidal cycle. The duration of high-intensity wave-related processes, such as bore propagation or swash run-up, is often limited on macro-tidal beaches.

Waves are the main forcing mechanism in the coastal environment of sandy beaches. The sinusoidal orbital motion of the waves in deep water is transformed to an asymmetric orbital motion in intermediate water depths. The onshore-directed velocity under the wave crest exceeds the offshore-directed velocity under the wave trough and the net effect is an onshore-directed transport, mainly as bed load. Shoaling, wave refraction, and wave set-down occur in the same intermediate water depths (Fig. 7.6). Wave breaking, as spillers or plungers, occurs further landward at shallower water depths. The increased velocities of the wave motion and the extra vortex of the plunging waves will stir a lot of extra sediment in the water column. Most of this sediment will be suspended in the water column and can be transported by a mean longshore current or a mean cross-shore current, the offshore-directed undertow. Finally, the broken waves can further advance shorewards as bores and run up the beach. Sediments are mainly transported onshore as bed load and suspended load, and can be transported back to deeper water by the undertow or in concentrated rip currents. Infragravity waves and wave set-up may also play an additional role in the sediment fluxes in the area between the breaking waves and the waterline. The actual water level at the beach is a summation of the mean sea level with its tidal fluctuations and the wave-related water-level adaptations through wave set-up, wave run-up and storm surges. The wave-related adaptations of the water level can be as large as 4 to 5 m during storms (e.g. in Denmark). The waves over these increased water levels progress further onshore and may cause a wave attack on the dunes or cliffs. The net effect of these processes is an offshore movement of sediments to the nearshore zone. Aeolian sediment transport can move the sediments on the beaches and may create the dune areas.

The rate of cliff erosion is a function of the sediment properties of the cliffs, the marine processes that erode the toe of the cliff (including the water level surges), the beach volume in front of the cliff, and the hydrology of the cliff that may induce mass movements as well (Komar 1998). Waves will also cause abrasion of hard rock shores. However, abrasion is a relatively slow process that does not deliver much sediment to the shores. Consequently, morphologic changes of cliffs and shore platforms are small.

Coastal Geomorphology

The seabed of the shallow semi-enclosed seas of the North Sea and the Channel shows a hierarchy of sandy bed forms (Fig. 7.7). These range from small-scale ripples with heights in centimetres and lengths in decimetres, to sand waves with heights in metres and lengths in tens of metres, to sand banks with heights in metres and lengths of hundreds of metres, and to shoreface-connected ridges with heights in metres, widths of 100 metres and lengths of kilometres. The net effect of these bed-form adaptations is the result of sediment fluxes over the tidal ellipse. Sand waves migrate several m/year in the direction of the major tidal current velocity. Shoreface-connected ridges show hardly any migration over years and are supposed to be relict forms more or less in equilibrium with the present-day conditions.

Morphological features such as sandy coastal stretches, barrier islands, spits, salients (cuspate forelands), and

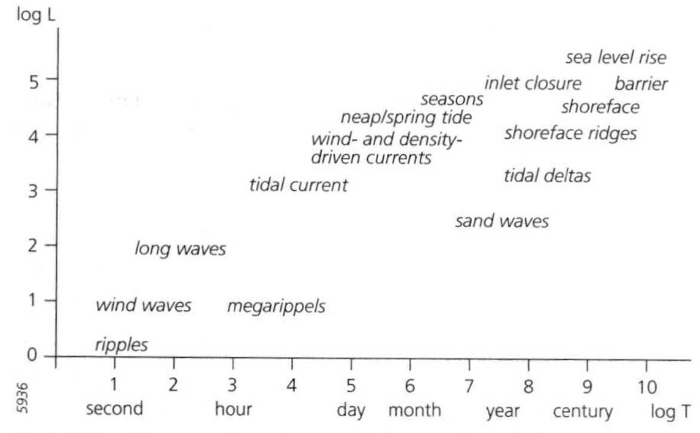

Fig. 7.7. Temporal and spatial hierarchy of bed forms on sandy seabeds, North Sea (after van de Meene 1994).

tombolos occur along the sandy shorelines of western Europe. Where sandy coastal stretches are almost uninterrupted by tidal inlets or estuaries wave processes dominate over tidal influences. In these cases, the longshore wave-generated currents produce a longshore sediment transport that straightens the coastline. Small rivers entering the coastal area will find a major obstruction across their mouth formed by a spit. Spits develop by a unidirectional longshore current. However, the straight coastline is divided into smaller sections, as barrier islands, when the influence of the tides is larger (e.g. in meso- to macro-tidal environments). Salients are formed by converging longshore currents or specific refraction patterns over subtidal morphologies.

Sandy beaches and sandy nearshore zones are often found on Holocene sandy coastlines and in front of glacial cliffs. Gravel beaches, or a mixture of gravel and sand on the beaches, are often found in small embayments between two headlands (pocket beaches) or in front of cliffs. A well-known classification scheme of sandy beaches with six classes is based on the wave energy input from offshore (Wright and Short 1984). Wide and flat beaches with a number of longshore bars in the cross-shore direction are formed under high-energetic wave conditions and swells (dissipative beaches). Narrow and steep beaches with no bars are formed under low-energetic wave conditions (reflective). These two end-stages are often uniform in the longshore direction. Intermediate stages show more longshore irregularities like rips. These rips are formed in periods with low mean water levels when the onshore mass transport of water cannot be compensated by only the undertow. At this time, a horizontal circulation generates the feeder and rip channels. Figure 7.8 presents a schematic view of the beach and nearshore area of a typical sandy beach. Two types of intertidal bar occur on the beaches: low-amplitude ridges and slip-face bars (Wijnberg and Kroon 2002). The low-amplitude ridges are typical features in meso- to macro-tidal environments, whereas the slip-face bars are common on micro- to meso-tidal beaches. The beaches with low-amplitude ridges are often multiple-barred and the dynamics of the bars are strongly coupled to the standstill phases in the tidal curve. The duration of certain wave-related processes on a specific place on the beach determines the rate of morphological change. The beaches with slip-face bars are often single-barred and the dynamics are strongly coupled to the intensity of bores and swash run-up processes near the waterline under low to moderate wave-energy conditions. Three phases are distinguished: (1) generation near the low-water line; (2) onshore migration or stabilization on the intertidal beach; (3) and

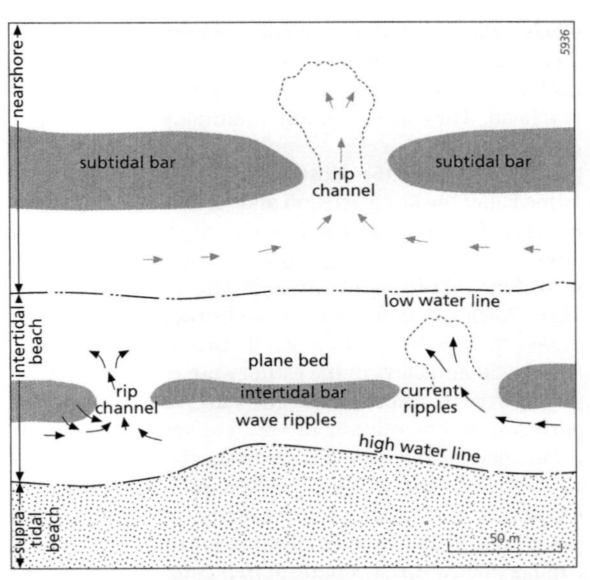

Fig. 7.8. Schematic representation of major morphological features on a sandy (barred) beach, at the North Sea shores, Holland coast near Egmond.

merging with the upper beach at the high-water line. These phases can be interrupted by a storm, when high water levels and a strong undertow will erode the bar and flatten the beach. Nearshore bars are subtidal (Fig. 7.8) and can be straight or crescentic in a longshore direction (Wijnberg 1995). Low to moderate wave energy conditions may move these bars onshore and enhance the crescentic features. Moderate to high wave energy conditions will move the bars offshore and straighten their crest. However, the net response of these bars over years is almost always seaward, with three phases: (1) generation near the low-water line; (2) offshore migration to the seaward boundary of the nearshore; (3) and flattening of the bar structure at this seaward end (Ruessink and Kroon 1994; Wijnberg 1995, 2002).

The morphology of wind-blown features is classified into two main types as either free dunes or vegetated dunes. The type of dune is a function of the sediment supply and the rate and direction of the drift potentials. Free dunes are often observed on sandy beaches. However, the main part of the sandy dunes in such areas is often (artificially) vegetated by marram grasses and shrubs. The shape of most of the dunes is parabolic.

Processes and Geomorphology of Estuaries and Tidal Basins

Most western Europe rivers have an estuary. Estuaries occur at the river mouth where there are low sediment yields relative to the dissipative forces. They have a characteristic funnel-shaped basin that exponentially decreases in width from the estuary mouth to the estuary head. They are generally maintained under meso- and macro-tidal range conditions with large tidal prisms (Perillo 1995). Most of these estuaries have been formed after the last glaciation and serve as sediment traps between the fluvial and marine environments. Estuarine environments have large spatial gradients in current velocities, salinity, and sediment characteristics. Fresh water flowing in from the rivers is obstructed by salt water driven inshore by the tidal flood current. The fine sediments (silt and clay) in the estuary are cohesive and are transported in suspension at the same rate as the water particles. Sedimentation of the fine sediments mostly occurs during standstill phases of the tides (slack water) when the current velocities are at a minimum. The sedimentation rate of the fine sediment particles is also influenced by flocculation. Two important aspects play a major role in this mechanism that causes the sediment particles to stick together: the presence of algae and the presence of filter feeders. Algae in the water column create the sticking polysaccharides. This stuff enlarges the sediment particle size and increases the settling rate (ten Brinke 1993). Filter feeders like mussels incorporate the fine sediments into faecal pellets (Augustinus 2001). These faecal pellets are also larger than the original fine sediment particles and settle more easily. The rate of flocculation depends on the collision of suspended particles, and the sticking capacity of the particles (Dyer 1995). The net sedimentation of the fine sediments is thus determined by the tidal asymmetry in the estuary, reflected in the larger flood velocities and longer slack waters at high tide, and by threshold values for sediment entrainment, settling lags and scour lags, (Dyer 1995). Maximum suspended sediment concentration occurs at the position of the turbidity maximum where the fresh water mixes with the salt water. This position often changes with the spring-tide neap-tide cycle. The amount of mixing between the fluvial discharge and the tidal volumes of marine water depends on the relative speeds and volumes of the two water masses. Three types of estuaries are distinguished with respect to the mixing and spatial distribution of current velocities and salinity (Dyer 1995): (1) highly-stratified or salt wedge estuaries; (2) partially-mixed estuaries; (3) and well-mixed or fully mixed estuaries. Highly stratified estuaries occur with a large inflow of river water and small tidal ranges (micro-tidal with tidal ranges < 2 m). They are fluvially dominated and the mixing is almost absent. This results in seaward-flowing fresh water (low density) over the stationary salt wedge (high density) at the bottom. The sediments are mainly of fluvial origin and the sediment size decreases seaward. This type of estuary quite often has a delta and barrier islands at the mouth. Partially mixed estuaries occur under moderate tidal ranges (meso-tidal with tidal ranges of 2–4 m) and have a two-way mixing. Seaward-flowing fresh water at the surface slowly mixes with the landward-flowing salt water near the bottom. The sediments are mainly of marine origin and the sediment size decreases landward. The tidal channel is often meandering near the estuary head and the estuary has ebb-tidal deltas and flood-tidal deltas near the estuary mouth. Well-mixed estuaries occur under large tidal ranges (macro-tidal with tidal ranges > 4 m) and with strong tidal currents. The water column is completely mixed and there is no vertical stratification. The Coriolis effect is responsible for lateral stratification of landward-flowing salt water during flood tides and the seaward-flowing brackish water during ebb tides. Fine sediments accumulate at the head of the estuary near the river mouth in the form of mudflats and salt marshes, and coarser sediments accumulate along the channels and at the estuary mouth in the form of linear shoals and sand banks. The estuary has separate ebb- and flood-dominated channels with linear sand bars parallel to the tidal flow (Perillo 1995; Fig. 7.9).

A tidal basin or coastal lagoon with a significant tidal range is protected from wave activity on the open sea by spits or barrier islands. The inflow and outflow of the tidal wave runs through the tidal inlet, a channel with a length of a few kilometres and a width of hundreds of metres (Fig. 7.10). Extensive salt marshes or other supra-tidal deposits usually form the fringes of the tidal basin. Single tidal basins have only one tidal inlet and all other boundaries of the embayment are fully closed for sediment exchange. Multiple tidal basins have multiple inlets between the barrier islands and also have inter-tidal watersheds between the distinct tidal basins (e.g. the multiple tidal basins of the Wadden Sea). The horizontal tidal velocities on these watersheds are almost zero and do not induce a sediment flux between the tidal basins. A relatively shallow sandy outer delta with depths between 1 and 5 m is present on the seaward site of the tidal inlet. The water motion in a tidal inlet is not merely driven by tides and waves, but wind and density differences also play a role. Typical current velocities through an inlet are in the order of 1 m/s. Observations indicate that these currents transport large amounts of fine and coarse sediments. As a direct result the morphological features of channels and shoals occur with length scales of 500 m–5 km and timescales of several decades. The outer deltas of tidal inlets are important elements of the barrier coastal system, because of their

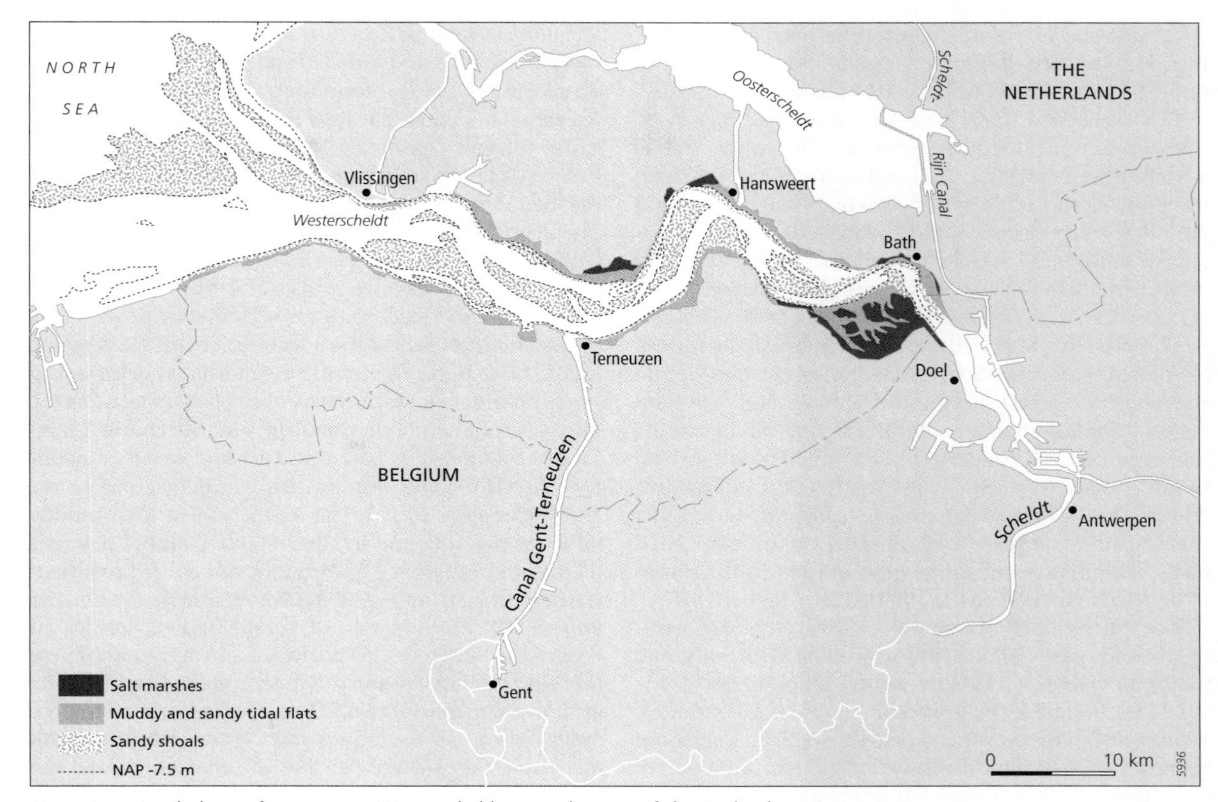

Fig. 7.9. Morphology of an estuary, Westerscheldt example, part of the Zeeland coast.

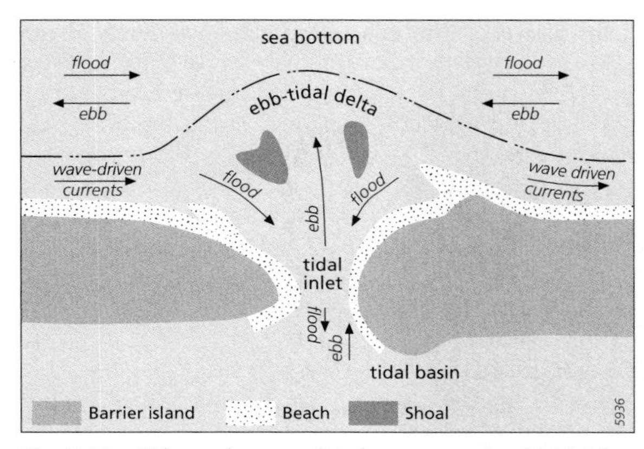

Fig. 7.10. Tide- and wave-related processes at a tidal inlet.

capacity to store or provide large amounts of sand to or from the coastal system. Wave activity is important along the seaward fringes of the outer delta where waves reach shallow water and create an onshore flux of sediment to the upper part of the outer delta. The width of the tidal inlets, the length of the barrier islands, and the volume of the outer deltas is often related to the tidal range and the dimension of the tidal basin, or to the combined parameter, i.e. the tidal prism. A macro-tidal range will result in wide inlets with large basins and small shoals between the successive inlets (German Bight). Meso- to micro-tidal ranges will result in smaller tidal basins and inlets and longer barrier islands (Dutch and Danish Wadden Sea).

Coastal Environments

The French Atlantic Coast

The French Atlantic Coast is located between the Pyrenees in the south and the Vendée in the north. The Gironde estuary is the seaward end of the Garonne and Dordogne rivers that drain the Massif Central mountains in central France and the Pyrenees in the south. The coastal plain north of the Gironde is filled with sediments from Jurassic and further southward of Cretaceous age. The coastal plain south of the Gironde is filled with Pliocene and Early Pleistocene deposits. Most of these deposits are Quaternary aeolian sands (Tastet and Pontee 1998). Holocene deposits form the coastline.

The Gironde estuary is a well-mixed estuary with a mean annual discharge of about 900 m^3/s. It has a well-defined turbidity maximum that causes mud-trapping in the main channel which shows seasonal variations in position coupled to the seasonal river discharge variations (Sottolichio and Castaing 1999). The tidal currents in the estuary cause a residual circulation of non-cohesive sediments (Mallet *et al.* 2000). This can be observed in the periodic migration and behaviour of elongated estuarine bars (Fenies and Tastet 1998). The sediments at the estuary mouth with a small ebb-tidal delta are highly dynamic and the strong tidal currents and oceanic swell transport the sediments offshore and to the adjacent coastal stretches by longshore drift (Howa 1997). Fine sediments are transported even further offshore and are deposited in continental shelf fields, with maximum sedimentation rates in the centre of 0.5 cm/year (Lesueur *et al.* 2001).

The north–south oriented coastal stretch just south of the Gironde is almost uninterrupted with only one major tidal basin, the Bay of Arcachon (Figs. 7.11, 7.12), in the centre (Bellessort and Migniot 1987: pt. C). Waves and wave-related processes such as longshore currents are the main transport agents. Incident waves are often coming from western directions and are most energetic during storms. In addition, an almost permanent background of energetic swell waves from the Atlantic Ocean exists. The net result of the waves and longshore currents is a straight coastline with a slight net southward-directed transport over years. This is reflected in the spit formation near Cap Ferret at the Arcachon Inlet and spits near the mouth of the Loire. The whole Gironde coastline just south of the estuary suffers from erosion by longshore currents to the north near the inlet and to the south in the direction of Cap Ferret. The average coastal retreat is between 1 and 2 m/year (Aubié and Tastet 2000). The inlet morphology of the Arcachon basin has two units: (1) the outer inlet south of the spit Cap Ferret with an ebb-tidal delta with one or two channels, where the littoral drift bypasses the inlet, and (2) the inner inlet with a small flood delta and a single channel along the bank of a spit (Cayocca 2001). The progradation of the sand spit and the channel migrations follow cyclic patterns with two cycles of about 80 years over the last centuries (Michel and Howa 1997; Cayocca 2001). The basin is filled with muddy tidal flat deposits and exhibits stable channel patterns (Fenies and Faugères 1998). Sandy dunes are prominent features and are among the highest in Europe (Tastet and Pontee 1998). A series of small lagoons (*étangs* in French) is situated just landward of the dunes. These are filled mainly by fresh water from the small rivers draining the coastal plain. The beaches and nearshore area in front of the dunes have intertidal ridges and nearshore bars. These beaches are meso-tidal and the ridges show hardly any migration during low-energy wave conditions (Michel and Howa 1999). Rhythmic patterns with longshore interruptions by rip channels are often observed. The coastal stretch just north of the

Fig. 7.11. The inlet of the estuary of the Bay of Arcachon, France (photo: Philippe Larroudé).

Fig. 7.12. Sandy shoals and salt marshes in the Arcachon basin, France (photo: Philippe Larroudé).

Gironde consists of several islands (e.g. Île d'Oléron with Cretaceous deposits in its centre) with tidal flats along the shore (Bellesort and Migniot 1987: pt. B).

Human interferences are evident in the Gironde estuary. Many tidal flats along the fringes of the estuary have been reclaimed.

The Western Channel Coast and Brittany, Including the Vendée

The deposits of the Armorican massif form the centre of the Western Channel Coast and Brittany. This massif consists of Proterozoic and Palaeozoic deposits, together with local intrusive and metamorphic rocks like granite and gneiss. The waves from the Atlantic Ocean slowly erode these hard rock sediments by the process of abrasion. The rate of coastal erosion is small (< 0.1 m/year). The whole area was reshaped during the Hercynian orogeny. Holocene sediments are scarce but are found in the estuary of the Loire, on pocket beaches along the coast of Brittany, in the Bay of St. Michel and in the north along the western shores of the Cotentin (Bellesort and Migniot 1987: pt. A).

The tidal basins have extensive tidal flats sometimes behind a sandy spit. Fine sediments accumulate in these macro-tidal embayments (Gouleau *et al.* 2000). The sediment transport patterns over the intertidal mudflats also show a permanent presence of fine sediments in the water column, either as high suspension rates during spring tide or as fluid mud in the runnels during neap tide (Bassoullet *et al.* 2000). The embayment north-east of St. Nazaire is filled in by Holocene sediments from the Loire. Rocky shorelines with cliffs and shore platforms are most common in Brittany. Sediment supply from small rivers that drain the massif is small and will only fill in some pocket beaches. However, most of the pocket beaches are closed systems and experience some sedimentation. Regnauld and Louboutin (2002) studied the variability of sediment transport in the beach and coastal dune environments in some pocket beaches in Brittany and found a net accretion in most of the bays. Easterly, offshore-directed winds tend to feed the tidal sediment settling on the sea floor. This sediment is attacked by onshore-directed storms that finally feed the upper beach and dune area. Aeolian features are rare, mainly due to the scarcity of sediments.

The tide is of major importance in the Bay of Mont St. Michel and produces a net deposition of fine sediments along its fringes. The Bay of Mont St. Michel has a macro-tidal range of up to 15 m and can be divided into three morphological units (Marchand *et al.* 1998): (1) western mudflats with swell waves and relatively low tidal ranges; (2) eastern mudflats with tidal channels and three small rivers entering the embayment; accumulation rates are highest in this area where both the open sea waves and the macrotidal environment enhance sedimentation; (3) the dune and spit area in the north, formed by longshore drift from the Manche coast. The sediments in the bay come from the Channel and to a lesser extent from small rivers draining the north-eastern part of Brittany. Longer stretches of sandy beaches with nearshore and intertidal bars are found along the western shores of Cotentin, Normandy (Levoy *et al.* 2000, 2001) in front of Devonian deposits. The waves and wave-related longshore currents play a major role in cases where there is abundant sediment supply. Here, they smooth the shoreline and create elongated structures such as spits. The tide also influences the barred morphology along the western shores of Cotentin. The spring tidal range in this region is macro-tidal (> 4 m) and creates smaller inlets in the sandy beaches and increases the amount of barred features. The bar behaviour on the beaches is dominated by wave processes and the tidal levels more or less determine the position and migration rates of the bar crests.

Human interferences with the natural system are seen at the mouth of the Loire and in the Bay of Mont St. Michel. This latter bay area is still silting up and the original character is changing. Recently, engineering works were planned to increase the tidal processes in this area and hinder the natural sedimentation.

The Eastern Channel Coast: Normandy and Northern France

The Eastern Channel Coast extends from the Bay of the Seine up to the Belgian border. The deposits of the coastal area form part of the Paris basin. Jurassic deposits are found along the fringes of this area of the basin and Cretaceous deposits are encountered in its centre. Holocene deposits form the coastline (Clique and Lepetit 1986). Two major rivers drain the fresh water from the Paris basin: the Seine and the Somme. The Seine estuary is partially mixed and open to the Channel. The estuary is about 160 km long. The tidal range is about 7 m at the mouth and the mean river discharge is 480 m^3/s. The intertidal mudflats near the river mouth have been separated from the main channel by dykes over the last decades. The turbidity maximum on the river results from accumulation of suspended sediment due to tidal pumping and river flow, and to salinity gradients (Brenon and LeHir 1999). The turbidity maximum also changes with the spring-neap tidal cycle and with the river discharge (Dyer 1995). The plume of the Seine river delivers fine sediments that partially accumulate in the eastern bay of the Seine or is transported to the Channel (Cugier and LeHir 2000). Both the Seine and the Somme have distinct outer deltas with characteristic channels and shoals. Spits, generated by longshore drift, are also observed near the estuary mouth of the Somme, especially on the southern shore. A small coastal plain with sandy dunes at its seaward side is situated just south and north of the Somme estuary up to the Boulognais.

Most of the other shorelines consist of cliffs with sandy beaches. Many of these cliffs erode at the toe due to marine processes (wave activity during storm surges) and/or by mass movements caused by saturation of the soils (slides and slumps). Steep cliffs with toe erosion are found along the north-western French coastline (Duperret *et al.* 2002). Rotational and planar slides and slumps are well developed along the coast of Calvados. Retreating cliffs with fringing gravel and sandy beaches line much of the shorelines. Aeolian dunes are found only near the embayments near the Somme estuary. These embayments experienced a continuous deposition of relatively homogeneous sand during the Holocene, with dunes overlying beach/washover and estuarine sediments (Anthony 2002). Three factors determine the position of these gravelly and sandy shorelines: (1) the inherited coastal lithology and morphology of soft-rock cliffs; (2) the coastal orientation to the large-scale tide- and wind-driven marine bedload, and (3) the coastal orientation to the westerly wind fields that enhanced dune formation. If one of these factors is not favourable, there is a lack of large-scale sand trapping. As a result, the shorelines are characterized by retreating cliffs and fringing beaches, with a few major gravel barriers formed in local downdrift sinks. The sediments originating from offshore gravel fields and from cliff retreat are reworked during storms. The morphodynamics of the ridges and runnels on the macro-tidal beaches in this area show that the ridges have a fairly stable position during low-energy wave conditions and migrate only during storms (e.g. Corbeau *et al.* 1999; Masselink and Anthony 2001). The shorelines of the coasts of north-western France show almost no erosion over the last few decades (Battiau-Queney *et al.* 2003).

The sea bottom of the Channel coast of Belgium has shoreface banks that are asymmetric and almost emergent during low-tide spring. Their maximum length, width, and height reach 8 km, 1.5 km, and 12 m respectively. They are supposed to be initiated and constructed by residual circulation resulting from an interaction between tidal currents, bottom friction, and the Coriolis force (Pattiaratchi and Collins 1987). However, it is assumed that the dynamics of these banks in the macro-tidal environment of the Belgian coast are influenced by onshore movements due to north-western storm wave action, and tidally controlled longshore and seaward movements (Tessier *et al.* 1999).

The Southern North Sea Coast: Belgium, Zeeland, and Holland

The Southern North Sea Coast of Belgium and the provinces of Zeeland and Holland were built up in the Holocene and form extensive coastal plains. These Holocene coastal plains are in front of the Tertiary Belgium. This basin consists of Eocene sands in Belgium, and fluvial, aeolian, or glacial deposits from the Pleistocene in The Netherlands (Denys and Baeteman 1995; Berendsen and Stouthamer 2001). Relative sea-level changes played a prominent role in the history of this coastal plain. The sea level was about −10 m in the early Holocene and a rapid sea-level rise of 0.6–1.0 m/century occurred in the Atlantic period. Barriers were built seaward of the present coastline and peat, fluvial, and marine sediments were deposited behind these barriers. These barriers moved onshore during the whole Atlantic period and the beginning of the Subatlantic, when they reached a position of about 5 to 10 km inshore of the present shoreline. At that stage, the sediment supply was increased by the inflow of rivers and reworked marine deposits of former ebb-tidal deltas of inlets that had recently closed (Baeteman *et al.* 1999; Berendsen and Stouthamer 2001). Relative sea-level rise then became much smaller (0.2–0.5 m/century) and the coastal plain moved in a seaward direction, forming a set of beach ridges in the coastal plain (Eitner 1996; Beets *et al.* 2003). Finally, the coastline of the Zeeland area was breached in the late Holocene and a series of large tidal basins and estuaries was formed along the tributaries of the river Scheldt.

The coastline of Belgium and Holland is characterized by uninterrupted sandy stretches with sandy beaches, extensive dune areas on its landward side, and a flat sloping shoreface with nearshore bars on its upper part (Fig. 7.13). The dynamics of all these features is widely studied. Most of the dunes are vegetated but due to human management now hardly show any natural dune development. The beaches on the Belgian coast are in macro-tidal environments and have multiple ridges along the shore (de Moor 1979). The beaches on the Holland coast are in a micro- to macro-tidal environment and have developed slip-face ridges on the intertidal beach. These have their own morphodynamic sequence (Wijnberg and Kroon 2002). Nearshore bars are present along the whole Holland coast and show distinct regional differences (Short 1992; Wijnberg 2002). However, the net response of these bars over years is almost always seaward, with three phases: (1) generation near the low-water line; (2) offshore migration to the seaward boundary of the nearshore; and (3) flattening of the bar structure at this seaward end (Ruessink and Kroon 1994; Wijnberg and Terwindt 1995; Wijnberg 2002). Spatial and temporal oscillations in the shoreline position have magnitudes of 2–3 km length and a periodicity of 4–15 years. These oscillations are controlled by the influence of the nearshore bar systems and the

Fig. 7.13. A sandy beach with near-shore bars and dunes at the Holland coast near Egmond, The Netherlands.

changes in storm conditions reaching the coast (Guillén *et al.* 1999). The sediment budget of the central part of the Holland coast is almost neutral. However, net losses of sediment with shoreline retreat in the order of 0.1–1 m/year are observed in the northern and southern part (van Rijn 1997). The outflow of the Rhine river mainly contains fine sediments that will disperse in a northward direction towards the Wadden Sea (de Ruiter *et al.* 1997).

An estuary (Westerschelde), a tidal basin (Oosterschelde), and a series of former tidal basins (Haringvliet, Grevelingen) interrupt the shoreline of the Zeeland coast (Fig. 7.14). The Westerschelde is a well-mixed estuary where tidal pumping as a result of tidal-wave asymmetry plays a major role (Cancino and Neves 1999; Wang *et al.* 2002). This wave asymmetry has changed over the Holocene due to variations in the shape of the estuary (van der Spek 1997). The estuary has two major channels and is funnel-shaped towards the Antwerp harbour. Salt marshes with silty sediments are found along its fringes and sandy shoals are in between the major channels. The discharge of the estuary during ebb transports a flux of sediments into the southern North Sea basin that will partly accumulate on the bottom of the North Sea (Fettweiss and van den Eynde 2003).

Since the 1950s the Delta Project has been executed in the south-western part of The Netherlands (Fig. 7.14). The implementation of this huge project was hastened by the occurrence of the storm surge that flooded the low-lying polders of Holland and Zeeland on 1 February 1953; over 1,800 people drowned and 200,000 ha of land was flooded. This disaster was due to the regular fortnightly spring tide which coincided with an extremely severe north-westerly storm; moreover, the maintenance of the dykes had been neglected during World War II and the post-war years. The Delta Project consisted of the construction of four primary dams directly closing estuaries, and five secondary barriers further inland. By their construction the danger of floods was significantly reduced, the water management of the area was improved, and better connections between the former islands were provided. Work on the Delta Project started soon after the 1953 disaster and was completed in 1997 by the construction of a mobile storm surge barrier in the important navigation gateway to the harbour of Rotterdam, called the Nieuwe Waterweg. The characteristic tidal basin morphologies with large tidal channels and ebb-tidal deltas changed dramatically by all these constructions. Fine sediments and sandy sediments from the shoals filled in the tidal channels. The ebb-tidal delta eroded at the seaward side by wave processes (asymmetry of shoaling waves). The related sediments were transported landward and settled on top of the former delta, creating islands that emerge during low-tide (Louters *et al.* 1991). The storm surge barrier that was built in the Oosterschelde reduced the tidal prism of the basin and caused adaptations in the basin by the infill of sediments in the channels and erosion of the intertidal flats and salt marshes (Louters *et al.* 1991). Shoreline oscillations, reflected in the dunefoot position, occur along the former islands in the Zeeland area. These oscillations or sandwaves only occur where the shore faces the tidal shoals. Their length is in the order of 5 km, their spacing in the order of 200–1,200 m and their migration rate about 65 m/year (Ruessink and Jeuken 2002).

Sandwaves, sandbanks, and shoreface-connected ridges are common features on the southern North Sea bed (Laban and Schüttenhelm 1981; Hulscher *et al.* 1993; van de Meene *et al.* 1995; Vincent *et al.* 1998; Trentesaux *et al.* 1999). Sandbanks are formed by residual circulation resulting from an interaction between tidal currents, bottom friction, and the Coriolis force. Storm waves may stir the sediments and reshape the banks, but are supposed to be of minor importance for shoreface-connected ridges (Vincent *et al.* 1998).

The Wadden Sea Coast

The Wadden Sea extends from the north coast of The Netherlands, round the German coast, to the southwest coast of Denmark. The tidal ranges on the North Sea, just north of the Wadden Sea basin, increase from meso-tidal in the west near the Holland coast to macrotidal in the German Bight, and then decrease to microtidal along the Danish coast (Ehlers 1988). Three major rivers have their estuaries in the Wadden Sea area: the Ems, Weser, and Elbe. The estuaries are of the partially mixed type and show a hypersynchronous gradient of water levels and velocities, due to the dominance of the estuary convergence over the friction of the estuary fringes (e.g. Dyer 1995). All these estuaries have a characteristic turbidity maximum (Grabemann *et al.* 1997; Ridderinkhof *et al.* 2000) and drain a suspended sediment plume towards the North Sea.

The Wadden Sea is characterized by a series of tidal basins with sandy channels and shoals. The tidal basins are sheltered from the wave action of the North Sea by barrier islands (meso-tidal) or shoals (macro-tidal). Most of the sediments were deposited in the Holocene. The tidal inlets through the barriers were formed during relative sea-level rise during the middle Holocene (e.g. Ehlers 1988; Eitner 1996; Gerdes *et al.* 2003). The successive tidal basins are separated by sandy intertidal watersheds, where tidal current velocities are very

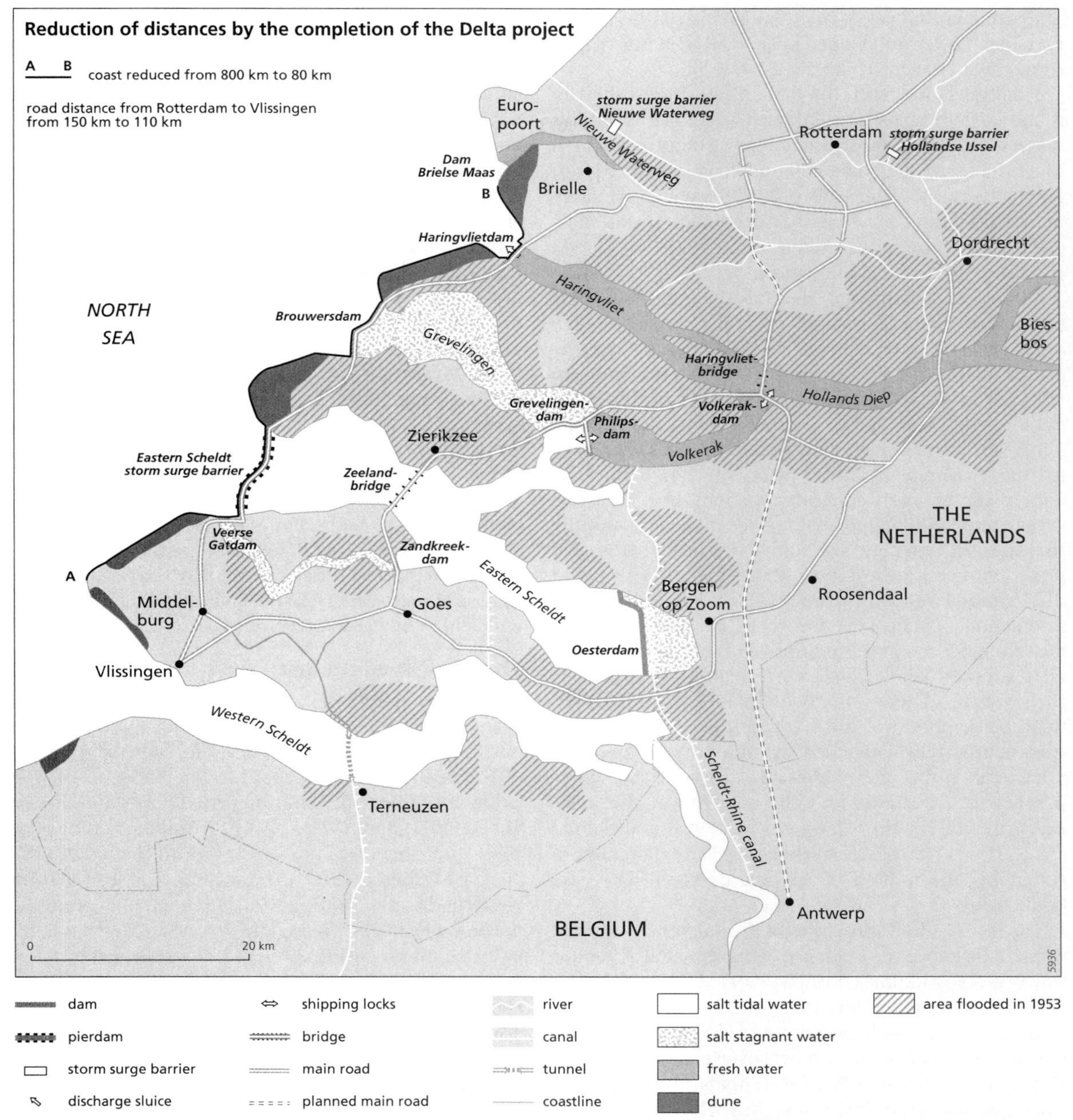

Fig. 7.14. The estuaries in the south-western part of The Netherlands and the coastal defence system of the Delta-Works (after KNAG 2001).

small. Unvegetated tidal flats and salt marshes form the landward fringe of the tidal basins. The dynamics of these muddy deposits and the sediment deposition (by flocculation) and erosion processes have been widely studied in The Netherlands and Denmark (e.g. Dyer *et al.* 2000; Ridderinkhof *et al.* 2000; Pejrup and Andersen 2000). The salt marshes are still accreting and only locally does wave erosion of the cliff edge take place.

The sediment budget of the tidal flats is spatially more variable. Lower tidal flats near tidal channels often show a net-erosion (Houwing 2000*a*, *b*).

Tidal inlets interrupt the Frisian and Danish barrier islands. These inlets have large ebb-tidal deltas at their seaward end. The most prominent structures on such outer deltas are the main ebb and flood channels (with tidally averaged water transport in the ebb/flood direction) which have typical length scales of 2–5 km and show cyclic behaviour on timescales of decades (Sha 1990; Jacobsen 1998; Israel and Dunsbergen 1999; Hofstede 1999*a*). A conceptual model for the dynamic behaviour (growth, migration, decline) of tidal channels on the outer delta of the Texel Inlet (The Netherlands) was presented by Sha (1989). His model assumes the presence of a well-developed ebb-channel. The flow in this channel interrupts the tidal longshore current, and also the wind- and wave-driven longshore current. This results in the development of shoals on the downdrift side, which subsequently migrate onshore due to wave action and aeolian processes. Recent studies indicate that the cyclic process is strongly controlled by the tidal prism (van de Kreeke 1996). Ridderinkhof (1990) and Ridderinkhof and Zimmerman (1992) showed that the interaction of tidal currents with the sloping bottom near the coast causes the generation of tidal residual eddies on the outer delta, which provide an explanation for the presence of ebb- and flood channels. The observed updrift orientation of the ebb-channel on outer deltas was heuristically explained by Sha (1990) and was attributed to the phase difference between the shore-parallel tidal currents and the currents in the inlet. Ridderinkhof (1990), Wang *et al.* (1995), and Stanev *et al.* (2003) have extensively studied the dynamics of the inflow and outflow of the tidal currents into these tidal basins.

The beaches and shorefaces of the barrier islands and shoals in between the influence of the ebb-tidal deltas are slowly eroding in time (Hofstede 1997, 1999*b*; Ruessink *et al.* 1998; Raudkivi and Dette 2002). Nearshore bars in front of the beach and dune areas of the Wadden barrier islands often show a cyclic development in time (Ruessink and Kroon 1994). The shoals on the ebb-tidal delta are also periodically merging with the adjacent coastline and give a periodic influx of extra sediment to the coastal zone (Ehlers 1988). The net-longshore drift is flood-directed to the east and forces some East Frisian islands to migrate in an eastward direction.

The earliest human settlements that have been recovered on natural levees of the Ems and Weser rivers and its tributaries in the coastal lowlands of Germany date from the late Bronze Age to the beginning of the Iron Age, i.e. *c*.tenth–ninth century BC (Gerdes *et al.* 2003). However, all these settlements had to be abandoned in due time as shown by covers of clastic sediments of either fluvial or marine origin. More widespread settlement of the coastal marshlands started at about 100 BC and soon afterwards people started constructing the first dwelling mounds enabling more or less permanent occupation. A second, more extensive phase of construction of dwelling mounds is dated from the seventh century until about AD 1100 (Gerdes *et al.* 2003). At that time people began to erect ring dykes around isolated areas, protecting their lands against inundation, and by the thirteenth century most of the coastal lowlands along the Dutch–German–Danish coastline were protected by a continuous system of sea dykes.

More recent human interference with the natural system is clearly seen in the reduction of tidal basins due to works such as the Afsluitdijk, which closed off the former Zuiderzee in the centre and the Lauwersdijk that closed off the Lauwerszee (Biegel and Hoekstra 1995) in the north of The Netherlands, the storm surge barrier in the Eider (Ehlers 1988) in Germany, or the dyke over the watersheds to the barrier islands of Sylt in Germany and Rømø in Denmark. In addition, many salt marshes in the whole area have been reclaimed since the Middle Ages.

The Danish North Sea Coast

The Danish North Sea coast is located between the Wadden Sea in the south and the Skagerrak in the north. The whole area is micro-tidal with tidal ranges of 1 m in the south and about 0.5 m in the north. The isostatic rise since the Litorina period (late-Atlantic and Subboreal) is almost zero at the central part of the west coast and rises to values over 10 m to the north near Skagen (Jacobsen 1993). The sea waves and swell from westerly directions off the North Sea mainly determine the present processes along the shore. Skallingen is a spit in the southern part that migrates to the south by longshore wave-generated drift and erodes shorewards with about 0.5–1.0 m/year over the last decades (Aagaard 2002). The beaches on Skallingen have intertidal slipface ridges and rips that closely resemble the morphodynamics of those along the Holland coast (Aagaard *et al.* 2002; Aagaard 2002). The sandy sediments with gravels originate from a glacial headland. This headland can be followed some kilometres offshore as a shallow shoal. North of this headland near Skallingen is an alternation of glacial cliffs (Fig. 7.15) and glaciofluvial deposits in the form of drowned sandar from the Saalian period. The waves are still eroding the beaches and longshore currents create spits with lagoons over the drowned sandar. Tidal inlets are absent because of the

Fig. 7.15. Cliff erosion along the Danish North Sea coast, northwestern Jutland.

strong longshore drifts and the small tidal ranges (micro-tidal). The influence of the Weichselian glaciation starts halfway along the Danish west coast. The relief becomes more pronounced. Marine and coastal deposits, sometimes overlain by aeolian deposits, now fill in the areas between the glacial headlands and fluvioglacial deposits. The relative sea level in this area has fallen during the last 4,000–5,000 years due to the unloading of the Fennoscandian ice masses and former cliffs and spits are found at about 8–9 metres above present sea level (Bowman *et al.* 1989). The shoreline in the northern part has some concave bights with headlands. Locally, severe erosion of the glacial cliffs by heavy rainfall and storm waves occurs (Christiansen and Møller, 1979). The northern part of the Danish North Sea Coast is still in a phase of unloading the ice masses and the relative sea level has decreased over the Holocene. A series of beach ridges occurs on the peninsula in northernmost Jutland with the oldest and most landward deposited ridges dating from the early Holocene and at about +10 m above the present sea level. Large parabolic dunes are found at several places along the south-west coast of Jutland (Clemensen *et al.* 1996). An exceptionally large and still active dune ridge is formed on top of a glacial cliff near Lønstrup in the north (Fig. 8.14), and an extensive migrating dune field is located on Skagen Odde (Anthonsen *et al.* 1996).

Human interference with the natural system is closely related to the eroding beaches and shorefaces. Recently, some large shoreface nourishments have been carried out. Moreover, the inlets to the large lagoons and Danish *fjorden* are a major problem for coastal management. Redistribution of sediments threatens to close these inlets, and construction works need to be applied to keep the entrances open to flushing water and to enable shipping activities.

Concluding Remarks

The shorelines of western Europe show a large variety of coastal and marine environments. The Atlantic shores in the south are wave-dominated with a sandy origin and extensive dune fields and have some major estuaries like the Gironde and the Loire. Rocky shorelines are mainly found along the Armorican massif in Brittany. Recent sedimentation of fine sediments is found only in embayments. Sandy beaches are rare and enclosed by headlands. The shoreline of the channel coast shows an alternation of beaches in front of cliffs and the major Holocene plains of the Seine and Somme rivers. The tide is also of major importance in these areas and even determines the dynamics of typical wave-related features such as bars on the beach. The coastal plain of the southern North Sea is an almost uninterrupted sand body with sandy intertidal and nearshore bars, which landward are fringed by large dune areas. However, major rivers such as the Rhine and Scheldt enter the coastal waters in this area. Due to human interference the former tidal basins of the southwest of The Netherlands are rapidly changing their typical inlet-basin-related features and already show typical wave-related bars on top of the former ebb-tidal deltas. Sand banks and shoreface connected ridges frequently occur in the southern part of the North Sea. Tidal inlets and basins with barrier islands (meso-tidal) or shoals (macro-tidal) are found in the Wadden Sea area. Finally, the west coast of Denmark was heavily glaciated in the Saale and Weichsel periods and the structure of the glacial and fluvioglacial deposits still determines the major configuration of the shorelines. In addition, specific features such as raised spits and beach ridges and cliffs occur in the north of Denmark due to the unloading of the ice masses and the relative sea-level fall in the Holocene.

References

Aagaard, T. (2002), Modulation of surf zone processes on a barred beach due to changing water levels; Skallingen, Denmark. *Journal of Coastal Research* 18: 25–38.

—— Black, K. P., and Greenwood, B. (2002), Cross-shore suspended sediment transport in the surf zone: a field-based parameterization. *Marine Geology* 185: 283–302.

Anthonsen, K. L., Clemmensen, L. B., and Jensen, J. H. (1996), Evolution of a dune from crescentic to parabolic form in response to short-term climatic changes: Råbjerg Mile, Sagen Odde, Denmark. *Geomorphology* 17: 63–77.

Anthony, E. J. (2002), Long-term marine bedload segregation, and sandy versus gravelly Holocene shorelines in the eastern English Channel. *Marine Geology* 187: 221–34.

Aubié, S., and Tastet, J.-P. (2000), Coastal erosion, processes and rates: an historical study of the Gironde coastline, Southwestern France. *Journal of Coastal Research* 16: 756–67.

Augustinus, P. G. E. F. (2001), Biochemical factors influencing deposition and erosion of fine grained sediment. In: T. Healy, Y. Wang, and J-A. Healy (eds.) *Muddy coasts of the world: processes, deposits and function.* Elsevier Science, Amsterdam, 203–28.

Baeteman, C., Beets, D. J., and van Strydonck, M. (1999), Tidal crevasse splays as the cause of rapid changes in the rate of aggradation in the Holocene tidal deposits of the Belgian Coastal Plain. *Quaternary International* 56: 3–13.

Bassoullet, Ph., Le Hir, P., Gouleau, D., and Robert, S. (2000), Sediment transport over an intertidal mudflat: field investigations and estimation of fluxes within the 'Baie de Marennes-Oleron' (France). *Continental Shelf Research* 20: 1635–53.

Battiau-Queney, Y., Billet, J. F., Chaverot, S., and Lanoy-Ratel, P. (2003), Recent shoreline mobility and geomorphologic evolution of macro-tidal sandy beaches in the north of France. *Marine Geology* 194: 31–45.

Beets, D. J., van der Valk, L., and Stive, M. J. F. (1992), Holocene evolution of the coast of Holland. *Marine Geology* 103: 423–43.

—— de Groot, T. A. M., and Davies, H. A. (2003), Holocene tidal back-barrier development at decelerating sea-level rise: a 5 millenia record, exposed in the western Netherlands. *Sedimentary Geology* 158: 117–44.

Bellessort, M. B., and Migniot, M. C. (1987), Catalogue sédimentologique des côtes françaises, Côtes de la Manche et de l'Atlantique. Éditions Eyrolles, Paris (in French).

Berendsen, H. J. A., and Stouthamer, E. (2001). *Palaeogeographic development of the Rhine–Meuse delta, The Netherlands.* Koninklijke Van Gorcum, Assen.

Biegel, E., and Hoekstra, P. (1995), Morphological response of the Zoutkamperlaag, Frisian Inlet, The Netherlands. In: B. W. Flemming and A. Bartholomä (eds.) *Tidal signatures in modern and ancient sediments.* Blackwell, Oxford. Special Publications of the International Association of Sedimentologists 24: 84–99.

Bowman, D., Christiansen, C., and Magaritz, M. (1989), Late-Holocene coastal evolution in the Hanstholm-Hjardemaal region, NW Denmark. Morphology, sediments and dating. *Geografisk Tidsskrift* 89: 49–57.

Brenon, I., and Le Hir, P. (1999), Modelling the turbidity maximum in the Seine Estuary (France): identification of formation processes. *Estuarine, Coastal and Shelf Science* 49: 525–44.

Cancino, L., and Neves, R. (1999), Hydrodynamic and sediment suspension modelling in estuarine systems, Part II: Application to the Western Scheldt and Gironde estuaries. *Journal of Marine Systems* 22: 117–31.

Cayocca, F. (2001), Long-term morphological modelling of a tidal inlet: the Arcachon Basin, France. *Coastal Engineering* 42: 115–42.

Christiansen, C., and Møller, J. T. (1979), Beach erosion at Klim, Denmark. A ten-year record. *Coastal Engineering* 3: 283–96.

Clemmensen, L. B., Andreasen, F., Nielsen, S. T., and Sten, E. (1996), The late Holocene coastal dunefield at Vejers, Denmark: characteristics, sand budget and depositional dynamics. *Geomorphology* 17: 79–98.

Clique, M. P.-M., and Lepetit, M. J.-P. (1986), Catalogue sédimentologique des côtes françaises, Côtes de la Mer du Nor et de la Manche. Éditions Eyrolles, Paris (in French).

Corbeau, C., Tessier, B., and Chamley, H. (1999), Seasonal evolution of shoreface and beach system morphology in a macrotidal environment, Dunkerque Area, Northern France. *Journal of Coastal Research* 15: 97–110.

Cugier, P., and Le Hir, P. (2000), Three dimensional modelling of suspended matters in the eastern 'baie de Seine' (English Channel, France). *Earth and Planetary Sciences* 331: 287–94.

de Moor, G. (1979), Recent beach evolution along the Belgian North Sea Coast. *Bulletin. Belgique Société Geologique* 88: 143–57.

Denys, L., and Baeteman, C. (1995), Holocene evolution of relative sea level and local mean high water spring tides in Belgium—a first assessment. *Marine Geology* 124: 1–19.

de Ruijter, W. P. M., Visser, A. W., and Bos, W. G. (1997), The Rhine outflow: a prototypical pulsed discharge plume in a high energy shallow sea. *Journal of Marine Systems* 12: 263–76.

Duperret, A., Genter, A., Mortimore, R. N., Delacourt, B., and de Pomerai, M. R. (2002), Coastal rock cliff erosion by collapse at Puys, France: the role of impervious marl seams within chalk of NW Europe. *Journal of Coastal Research* 18: 52–62.

Dyer, K. R. (1995), Sediment transport processes in estuaries. In: G. M. E. Perillo (ed.) Geomorphology and Sedimentology of Estuaries. *Developments in Sedimentology* 53: 423–49.

—— Christie, M. C., Feates, N., Fennessy, M. J., Pejrup, M., and van der Lee, W. (2000), An investigation into processes influencing the morphodynamics of an intertidal mudflat, the Dollard Estuary, The Netherlands: I. Hydrodynamics and suspended sediment. *Estuarine, Coastal and Shelf Science* 50: 607–25.

Ehlers, J. (1988), *The morphodynamics of the Wadden Sea.* Balkema, Rotterdam.

Eitner, V. (1996), Geomorphological response of the East Frisian barrier islands to sea-level rise: an investigation of past and future evolution. *Geomorphology* 15: 57–65.

Fenies, H., and Faugères, J. C. (1998), Facies and geometry of tidal channel-fill deposits (Arcachon Lagoon, SW France). *Marine Geology* 150: 131–48.

—— and Tastet, J. P. (1998), Facies and architecture of an estuarine tidal bar (the Trompeloup bar, Gironde Estuary, SW France). *Marine Geology* 150: 149–69.

Fettweiss, M., and van den Eynde, D. (2003), The mud deposits and the high turbidity in the Belgian–Dutch coastal zone, southern bight of the North Sea. *Continental Shelf Research* 23: 669–91.

Gerdes, G., Petzelberger, B. E. M., Scholz-Böttcher, B. M., and Streif, H. (2003), The record of climatic change in the geological archives of shallow marine, coastal, and adjacent lowland areas of Northern Germany. *Quaternary Science Reviews* 22: 101–24.

Gouleau, D., Jouanneau, J. M., Weber, O., and Sauriau, P. G. (2000), Short- and long-term sedimentation on Montportail-Brouage intertidal mudflat, Marennes-Oléron Bay (France). *Continental Shelf Research* 20: 1513–30.

Grabemann, I., Uncles, R. J., Krause, G., and Stephens, J. A. (1997), Behaviour of turbidity maxima in the Tamar (U.K.) and Weser (F.R.G.) estuaries. *Estuarine, Coastal and Shelf Science* 45: 235–46.

Guillén, J., Stive, M. J. F., and Capobianco, M. (1999), Shoreline evolution of the Holland Coast on a decadal scale. *Earth Surface Processes and Landforms* 24: 517–36.

Hofstede, J. L. A. (1997), Process-response analysis for the North Frisian supratidal sands (Germany). *Journal of Coastal Research* 13: 1–7.

—— (1999*a*), Regional differences in the morphologic behaviour of four German Wadden Sea barriers. *Quaternary International* 56: 99–106.

—— (1999*b*), Process-response analysis for Hörnum tidal inlet in the German sector of the Wadden Sea. *Quaternary International* 60: 107–17.

Houwing, E. J. (2000*a*), Morphodynamics of intertidal mudflats: consequences for the extension of the pioneer zone. *Continental Shelf Research* 20: 1735–48.

—— (2000*b*), *Sediment dynamics in the pioneer zone in the land reclamation area of the Wadden Sea, Groningen, The Netherlands.* Ph.D. thesis, Utrecht University.

Howa, H. L. (1997), Sediment budget in the southern inlet of the Gironde Estuary (SW France). *Phys. Chem. Earth* 22: 373–5.

Hulscher, S. J. M. H., de Swart, H. E., and de Vriend, H. J. (1993), The generation of offshore tidal sand banks and sand waves. *Continental Shelf Research* 13: 1183–204.

Israel, C. G., and Dunsbergen, D. W. (1999), Cyclic morphological development of the Ameland Inlet, The Netherlands. In: G. Seminara (ed.) *Proc. IAHR Symposium on river, coastal and estuarine morphodynamics*, University of Genova, Genova.

Jacobsen, N. K. (1993), Shoreline development and sea-level rise in the Danish Wadden Sea. *Journal of Coastal Research* 9: 721–9.

—— (1997), The periglacial subsurface topography of the West Coast of Jutland, Denmark. *Journal of Coastal Research* 13: 1238–44.

—— (1998), The high sands of the Danish Wadden Sea—especially the ebb-tide delta, Søren Jessens Sande, and its incorporation with the island of Fanø. *Journal of Coastal Research* 14: 175–84.

KNAG (Royal Dutch Geographical Society) (2001), *Compact Geography of The Netherlands.* Utrecht University, Utrecht.

Komar, P. D. (1998), Beach processes and sedimentation. 2nd edn. Prentice-Hall, Upper Saddle River, NJ.

Laban, C., and Schüttenhelm, R. T. E. (1981), *Some new evidence on the origin of the Zeeland ridges.* Special Publications of the International Association of Sedimentologists 5: 239–45.

Lambeck, K. (1997), Sea-level change along the French Atlantic and Channel coasts since the time of the Last Glacial Maximum. *Palaeogeography, Palaeoclimatology, Palaeoecology* 129: 1–22.

Lesueur, P., Jouanneau, J.-M., Boust, D., Tastet, J.-T., and Weber, O. (2001), Sedimentation rates and fluxes in the continental shelf mud fields in the Bay of Biscay (France). *Continental Shelf Research* 21: 1383–401.

Levoy, F., Anthony, E. J., Monfort, O., and Larsonneur, C. (2000), The morphodynamics of megatidal beaches in Normandy, France. *Marine Geology* 171: 39–59.

—— Monfort, O., and Larsonneur, C. (2001), Hydrodynamic variability on megatidal beaches, Normandy, France. *Continental Shelf Research* 21: 563–86.

Louters, T., Mulder, J. P. M., Postma, R., and Hallie, F. P. (1991), Changes in coastal morphological processes due to the closure of tidal inlets in the SW Netherlands. *Journal of Coastal Research* 7: 635–52.

—— van den Berg, J. H., and Mulder, J. P. M. (1998), Geomorphological changes of the Oosterschelde tidal system during and after the implementation of the Delta project. *Journal of Coastal Research* 14: 1134–51.

Mallet, C., Howa, H., Garlan, T., Sottolichio, A., Le Hir, P., and Michel, D. (2000), Utilisation of numerical and statistical techniques to describe sedimentary circulation patterns in the mouth of the Gironde estuary. *Earth and Planetary Sciences* 331: 491–7.

Marchand, Y., Auffret, J. P., and Deroin, J. P. (1998), Morphodynamics of the Mont-Saint-Michel bay (West France) since 1986 by remote sensing data. *Earth and Planetary Sciences* 327: 155–9.

Masselink, G., and Anthony, E. J. (2001), Location and height of intertidal bars on macrotidal ridge and runnel beaches. *Earth Surface Processes and Landforms* 26: 759–74.

Michel, D. and Howa, H. L. (1997), Morphodynamic behaviour of a tidal inlet system in a mixed-energy environment. *Phys. Chem. Earth* 22, 339–43.

—— (1999), Short-term morphodynamic response of a ridge and runnel system on a mesotidal sandy beach. *Journal of Coastal Research* 15: 428–37.

Pattiaratchi, C. B., and Collins, M. B. (1987), Mechanisms for linear sandbank formation and maintenance, in relation to dynamical oceanographic observations. *Progress in Oceanography* 19: 117–76.

Pejrup, M., and Andersen, T. J. (2000), The influence of ice on sediment transport, deposition and reworking in a temperate mudflat area, the Danish Wadden Sea. *Continental Shelf Research* 20: 1621–34.

Perillo, G. M. E. (1995), Definitions and geomorphologic classifications of estuaries. In: G. M. E. Perillo (ed.) Geomorphology and Sedimentology of Estuaries. *Developments in Sedimentology* 53: 17–47.

Raudkivi, A. J., and Dette, H.-H. (2002), Reduction of sand demand for shore protection. *Coastal Engineering* 45: 239–59.

Regnauld, H., and Louboutin, R. (2002), Variability of sediment transport in beach and coastal dune environments, Brittany, France. *Sedimentary Geology* 150: 17–29.

Ridderinkhof, H. (1990), *Residual currents and mixing in the Wadden Sea.* Ph.D. thesis, Utrecht University.

—— and Zimmerman, J. T. F. (1992), Chaotic stirring in a tidal system. *Science* 258: 1107–11.

—— van der Ham, R., and van der Lee, W. (2000), Temporal variations in concentration and transport of suspended sediments in a channel-flat system in the Ems-Dollard estuary. *Continental Shelf Research* 20: 1479–93.

Ruessink, B. G., and Jeuken, M. C. J. L. (2002), Dunefoot dynamics along the Dutch Coast. *Earth Surface Processes and Landforms* 27: 1043–56.

Ruessink, B. G., and Kroon, A. (1994), The behaviour of a multiple bar system in the nearshore zone of Terschelling, The Netherlands: 1965–1993. *Marine Geology* 121: 187–97.
—— Houwman, K. T., and Hoekstra, P. (1998). The systematic contribution of transporting mechanisms to the cross-shore sediment transport in water depths of 3 to 9 m. *Marine Geology* 152: 295–324.
Sha, L. P. (1989), Cyclic morphological changes of the ebb-tidal delta, Texel Inlet, The Netherlands. *Geologie en Mijnbouw* 68: 35–49.
—— (1990), *Sedimentological studies of the ebb-tidal deltas along the West Frisian Islands, The Netherlands*. Ph.D. thesis, Utrecht University.
Short, A. D. (1992), Beach systems of the central Netherlands coast: Processes, morphology and structural impacts in a storm-driven multi-bar system. *Marine Geology* 107: 103–37.
Sottolichio, A., and Castaing, P. (1999), A synthesis on seasonal dynamics of highly-concentrated structures in the Gironde estuary. *Earth and Planetary Sciences* 329: 795–800.
Stanev, E. V., Wolff, J.-O., Burchard, H., Bolding, K., and Flöser, G. (2003), On the circulation in the East Frisian Wadden Sea: numerical modeling and data analysis. *Ocean Dynamics* 53: 27–51.
Tastet, J. P., and Pontee, N. I. (1998), Morpho-chronology of coastal dunes in Médoc. A new interpretation of Holocene dunes in Southwestern France. *Geomorphology* 25: 93–109.
ten Brinke, W. B. M. (1993), *The impact of biological factors on the deposition of fine-grained sediment in the Eastern Scheldt (The Netherlands)*. Ph.D. thesis, Utrecht University.
Tessier, B., Corbau, C., Chamley, H., and Auffret, J.-P. (1999), Internal structure of shoreface banks revealed by high-resolution seismic reflection in a macrotidal environment (Dunkerque Area, Northern France). *Journal of Coastal Research* 15: 593–606.
Trentesaux, A., Stolk, A., and Berné, S. (1999), Sedimentology and stratigraphy of a tidal sand bank in the southern North Sea. *Marine Geology* 159: 253–72.
van de Kreeke, J. (1996), Morphological changes on a decadal time scale in tidal inlets: modelling approaches. *Journal of Coastal Research* SI23: 73–81.
van de Meene, J. W. H. (1994), *The shoreface-connected ridges along the central Dutch coast*. Ph.D. thesis, Utrecht University.
—— Boersma, J. R., and Terwindt, J. H. J. (1995), Sedimentary structures of combined flow deposits from the shoreface-connected ridges along the central Dutch coast. *Marine Geology* 131: 151–75.
van der Spek, A. J. F. (1997), Tidal asymmetry and long-term evolution of Holocene tidal basins in The Netherlands: simulation of palaeo-tides in the Schelde estuary. *Marine Geology* 141: 71–90.
van Rijn, L. C. (1997). Sediment transport and budget of the central coastal zone of Holland. *Coastal Engineering* 32: 61–90.
Vincent, C. E., Stolk, A., and Porter, C. F. C. (1998), Sand suspension and transport on the Middelkerke Bank (southern North Sea) by storms and tidal currents. *Marine Geology* 150: 113–29.
Wang, Z. B., Louters, T., and de Vriend, H. J. (1995), Morphodynamic modelling for a tidal inlet in the Wadden Sea. *Marine Geology* 126: 289–300.
—— Jeuken, M. C. J. L., Gerritsen, H., de Vriend, H. J., Kornman, B. A. (2002), Morphology and asymmetry of the vertical tide in the Westerschelde estuary. *Continental Shelf Research* 22: 2599–609.
Wijnberg, K. M. (1995), Morphologic behaviour of a barred coast over a period of decades. Ph.D. thesis, Utrecht University, *Netherlands Geographical Studies* 195, KNAG.
—— (2002), Environmental controls on decadal morphologic behaviour of the Holland coast. *Marine Geology* 189: 227–47.
—— and Kroon, A. (2002), Barred beaches. *Geomorphology* 48: 103–20.
—— and Terwindt, J. H. J. (1995), Extracting decadal morphological behaviour from high-resolution, long-term bathymetric surveys along the Holland coast using eigen-function analysis. *Marine Geology* 126: 301–30.
Wright, L. D., and Short, A. D. (1984), Morphodynamic variability of surf zones and beaches: a synthesis. *Marine Geology* 56: 93–118.

8 Aeolian Environments

Eduard Koster

Introduction

The literature on aeolian processes and on aeolian morphological and sedimentological features has shown a dramatic increase during the last decade. A variety of textbooks, extensive reviews, and special issues of journal volumes devoted to aeolian research have been published (Nordstrom *et al.* 1990; Pye and Tsoar 1990; Kozarski 1991; Pye 1993; Pye and Lancaster 1993; Cooke *et al.* 1993; Lancaster 1995; Tchakerian 1995; Livingstone and Warren 1996; Goudie *et al.* 1999). However, not surprisingly the majority of these studies discuss aeolian processes and phenomena in the extensive warm arid regions of the world. The results of aeolian research in the less extensive, but still impressive, cold arid environments of the world are only available in a diversity of articles. At best they are only briefly mentioned in textbooks on aeolian geomorphology (Koster 1988, 1995; McKenna-Neuman 1993). Likewise, the literature with respect to wind-driven deposits in western Europe is scattered and not easily accessible.

The European 'Sand Belt', Dune Fields, and Sand Sheets

The aeolian geological record for Europe, as reflected in the 'European sand belt' (Fig. 8.1) in the north-western and central European Lowlands, which extends from Britain to the Polish–Russian border, is known in great detail (Koster 1988; van Geel *et al.* 1989; Böse 1991). Zeeberg (1998) showed that extensive aeolian deposits progress with two separate arms into the Baltic Region, and into Belorussia and northernmost Ukraine. Recently, Mangerud *et al.* (1999) concluded that the sand belt extends even to the Pechora lowlands close to the north-western border of the Ural mountain range in Russia. Sand dunes and cover sands are widespread and well developed in this easternmost extension of the European sand belt. The northerly edges of this sand belt more or less coincide with the maximal position of the Late Weichselian (Devensian, Vistulian) ice sheet (Fig. 8.1), while the southern edges grade into coverloams or sandy loess and loess (Mücher 1986; Siebertz 1988; Antoine *et al.* 1999). However, along these southern edges the dune fields and sand sheets regionally are derived from different sources, such as the sands of the Keuper Formation or the floodplains of the Rhine and Main rivers. The westernmost occurrences of dune and cover sands are encountered in the UK in Breckland, North Lincolnshire, the Vale of York, and South Lancashire (Fig. 8.1). In all of these UK regions wind-driven deposits of comparable origin, composition, and age to their contiguous European counterparts have been studied (e.g. Bateman 1998; Bateman *et al.* 1999, 2000). Although the majority of the 'sand belt' deposits is of last glacial (Weichselian) age, Ruegg (1983) concluded that sandy aeolian deposits in The Netherlands—a region of continued regional subsidence throughout the Quaternary combined with extensive delta-building—originated in at least six glacial stages, forming stacked sequences locally up to forty metres thick.

The total area covered by surficial aeolian sands, including extensive sheets of late Weichselian cover sand and late Holocene drift sands not showing a dune relief, is of the order of several tens of thousands of square kilometres (Koster 1978, 1988; Niessen *et al.* 1984). Going from west to east in the north-western European Lowlands the proportion of cover sands to dune sands changes. Cover sands are found mainly in the western part, whereas large dune fields predominate

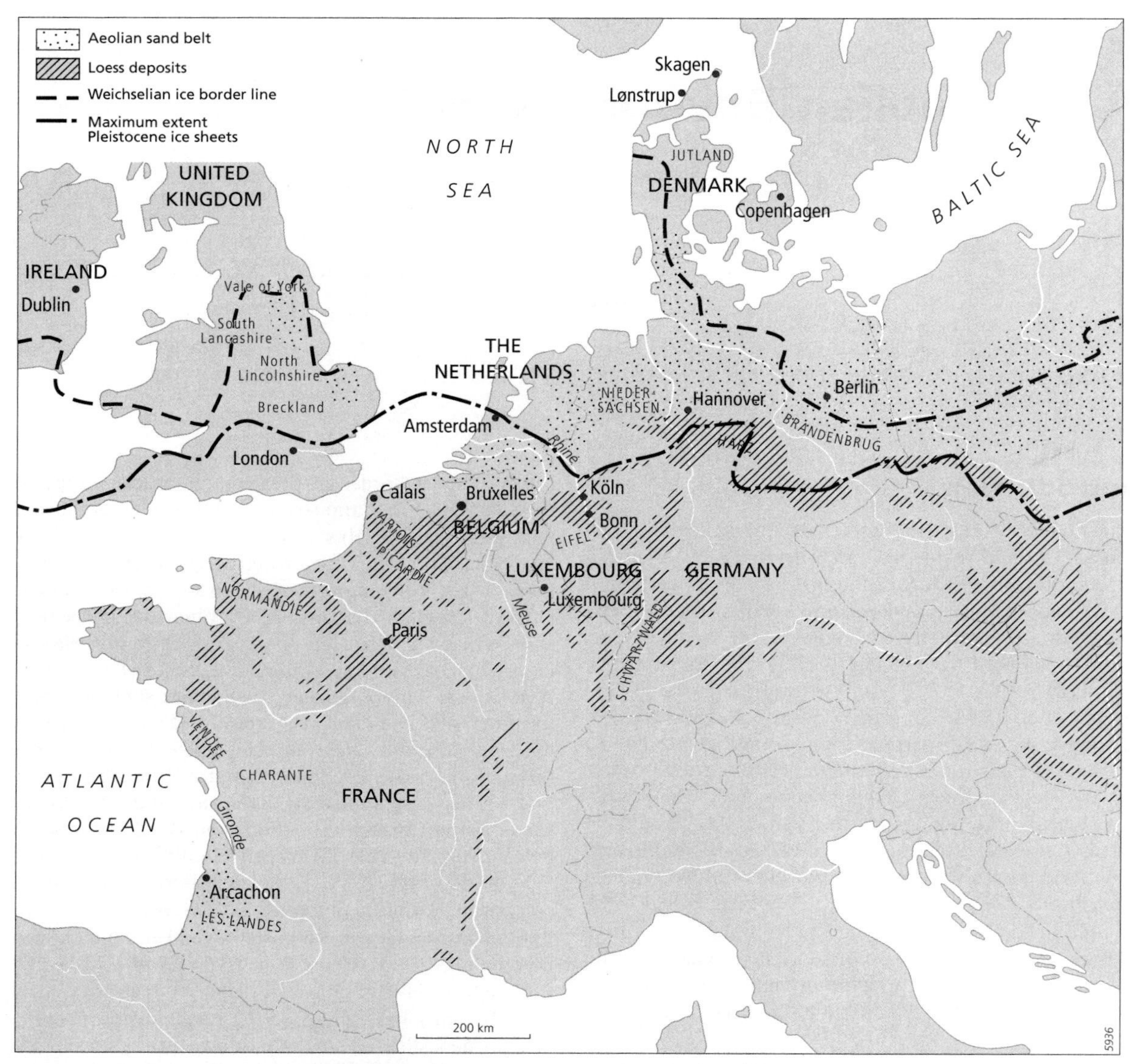

Fig. 8.1. Areal distribution of the European 'sand belt', comprising inland and river dunes, cover and drift sands, and sandy loess and loess deposits. Coastal dunes are not shown with the exception of the extensive coastal dune fields of Les Landes in south-western France. The maximum extent of the continental Pleistocene and of the Weichselian (Devensian, Vistulian) ice sheet are indicated (compiled from various sources).

in the eastern regions. According to some authors (Schwan 1988; Böse 1991) this may be due to eastward-increasing aridity during the time of formation. The extensive dune fields in western and central Europe attracted the attention of German earth scientists, such as J. Wessely, F. Solger, and H. Poser, in the late nineteenth and early twentieth century, but the studies of Pyritz (1972) can be regarded as the first modern analysis of the origin, form, and sediment properties of inland dunes in Germany. The largest areal extent of aeolian sediments within the European sand belt is found within Poland. Not surprisingly therefore, the number of publications on 'aeolian studies' in that country runs to many hundreds. As this is outside the geographical scope of this chapter, reference is made to only one, fairly recent review on the origin, sediment characteristics, and chronology of aeolian deposits in Poland by Manikowska (1995).

Coastal dunes extend in an almost continuous zone, including a series of Wadden Islands, from northern France to the northernmost point of Denmark, a distance of over 1,100 km. Moreover, large coastal dune fields occur along the Normandy coasts, the coast of the Vendée and Charante, in the region of Les Landes (van der Meulen *et al.* 1989; Bakker *et al.* 1990) and along the southern edges of the Baltic Sea.

Without going into any detailed discussions about controversies regarding terminology, the following simplified division of (surficial) aeolian forms and deposits, based on a combination of geomorphology, material, and depositional environment, is used throughout this chapter (see Koster 1982 and Pye and Tsoar 1990):

- coastal dunes, sometimes divided into Younger (late Middle Ages and younger) and Older (early Middle Ages and older) coastal dunes;
- inland dunes, often divided into Younger (Holocene) and Older (pre-Holocene) inland dunes;
- river dunes, occasionally divided into Younger (Holocene) and Older (pre-Holocene) river dunes;
- aeolian sand sheets or cover sands, usually divided into Younger (Late Glacial to early Holocene) and Older (pre-Late Glacial) cover sands;
- aeolian deposits, irrespective of their form as sheets or dunes, resulting from reactivation of sandy deposits by human impact during the late Holocene, are often called (aeolian) drift sands;
- additionally, sandy loess (or sand loam) and loess deposits are distinguished.

Note that the distinction between coastal, inland, and river dunes cannot be made in many instances because of the many transitional forms. Likewise, the distinction between cover and drift sands, when they show a similar geomorphology and soil formation, can be difficult to make. Moreover, loess, sandy loess, loamy cover sand, and cover sand also form a continuum in grain-size distribution. A comparable terminological overview of aeolian forms and deposits in the German language is published by Alisch (1995). As most countries use their own lithostratigraphic divisions, there is no need to concern ourselves with that in this chapter.

Before discussing the various aeolian forms the main, and to a large extent interdependent, factors that govern the form and scale of aeolian accumulations, according to Pye and Tsoar (1990), are listed as: (*a*) sand availability; (*b*) grain size distribution; (*c*) wind energy, velocity, distribution, and directional variability; (*d*) vegetation cover; (*e*) topographic obstacles; and (*f*) sequential climatic changes. In the case of coastal dunes additional factors such as beach and nearshore conditions and sea-level change must be taken into account.

Inland and River Dunes

Maps showing the extent of aeolian sands in the northwest European Lowlands clearly indicate that the majority of dune fields are closely linked to the many large and small river courses, including periglacial floodplains (German: *Talsandflächen*) with their presorted fluvial sediments. Whether a particular dune complex should be designated as an inland dune or as a river dune field often remains a matter of conjecture. The most complete map of dune occurrence has been compiled by Pyritz (1972), and was some years later supplemented by Koster (1978; Fig. 8.1) The areal extent of the aeolian deposits in Belgium and The Netherlands is illustrated in more detail in Fig. 8.2. The classic study by Pyritz (1972) on inland dunes and sand sheets in Niedersachsen in Germany has been followed by a number of Ph.D. studies on various aspects of dune morphology and sedimentology, but these are too many to discuss in this chapter. An exception is made, however, for the comprehensive study of the morphogenesis of Older and Younger inland dunes in the region east of Hannover by Alisch (1995). His observations on the anthropogenic factors leading to late-Holocene reactivation of aeolian sands, the destruction of dune forms by excavations of sand, and nature conservation value of those remaining, deserve special attention.

Extensive river dune fields consisting of an array of large, SW–NE orientated, parabolic dunes on the east bank of the Meuse valley have accumulated in a surprisingly short time, according to Bohncke *et al.* (1993), mainly between 10,500 and 10,150 BP, i.e. in the second, relatively dry part of the Younger Dryas interval. Similarities in granulometric and mineralogic properties of fluvial and aeolian sediments confirm that the nearby floodplain sediments formed the source material for these dunes. Modal grain-size values of 250–300 μm make these river dune sands somewhat coarser than the majority of cover sands. Similar river dune occurrences are encountered in many places in The Netherlands, Belgium, and northern parts of Germany. But, the most extensive inland and river dune fields, covering many thousands of square kilometres, are found in the vicinity of the large river valleys, including former glacial meltwater valleys (*pradolinas* or *Urstromtäler*), in Germany and Poland. Major dune forms in these dune fields are parabolic, transverse, and longitudinal dunes.

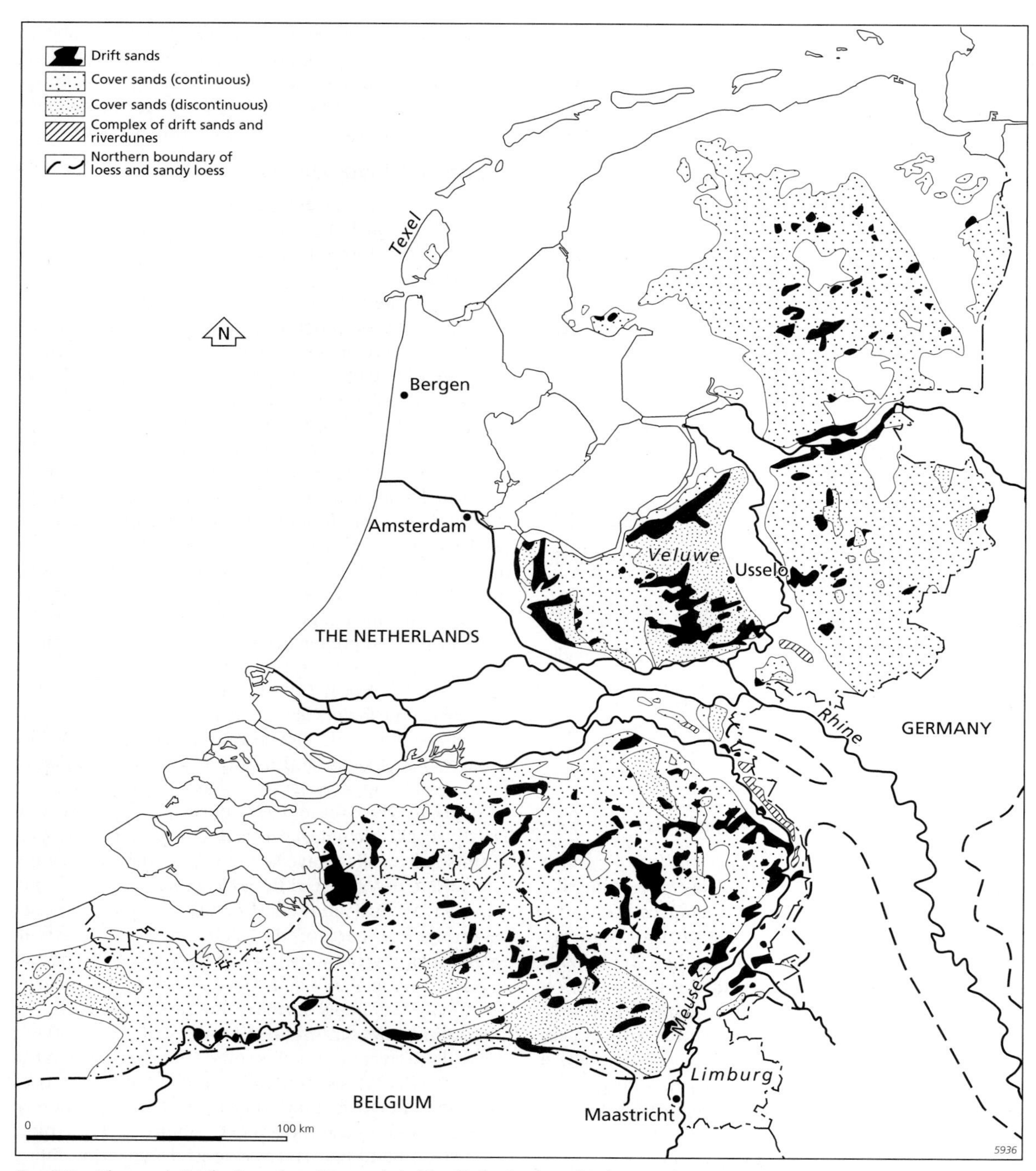

Fig. 8.2. The areal distribution of aeolian sands in The Netherlands and Belgium (after Koster 1978).

Cover Sands

The litho- and chronostratigraphy of cover sands in the north-west European Lowlands is well established since early studies by, amongst others, Dücker and Maarleveld (1957). Summaries are found in Nowaczyk (1976), Koster (1988), van Geel *et al.* (1989), Vandenberghe (1991), and Kasse (1999). Hundreds of publications presenting details of the cover sand stratigraphy, in many cases only of local/regional significance, have been published in the last few decades. The number of radiocarbon dates in some way or another related to the fine-tuning of the aeolian stratigraphy is enormous. Nevertheless, the chronology is mainly based upon dates reflecting phases of non-deposition, as will be explained in the section on geochronology. A firm chronology of phases of aeolian activity of supra-regional significance, even within this region of exceptionally manifold and detailed studies, is still lacking.

Vandenberghe (1991) distinguishes two particular models for the aeolian activity at the end of the Last Glacial. Firstly, the final Pleniglacial was characterized by widespread areas of loose surface sediments. Under cold desert-like conditions aeolian transport took place over long distances and probably all year round. Deflation was important on the interfluves, and accumulation dominated in depressions and wet areas. This resulted in typical, evenly horizontal laminated cover sands or sand sheets. Secondly, Late Glacial activity was restricted to particular areas (river flats and local dry regions), while transport distances were relatively short. An irregular dune morphology resulted.

The type section of the well-known stratigraphic marker of the Usselo bed (Veluwe, central Netherlands) —or Usselo layer or Usselo soil, as it is sometimes called—of Allerød interstadial age, usually dated at 11,900–10,950 ^{14}C yrs BP, was re-examined in an elaborate study by van Geel *et al.* (1989). In addition to pollen, other microfossils, insect assemblages, and seeds, have been described and their palaeoecological significance is discussed. The Usselo bed, which has been found and dated in the European sand belt from The Netherlands to Poland, separates the Younger cover sand I and II deposits. Important contributions to the litho- and chronostratigraphy of the extensive cover sands in Niedersachsen (Germany) were made by Meyer (1981).

Factors that favoured sand sheet formation were: (*a*) ample availability of thick, unconsolidated, sandy, fluvial, and fluvioglacial deposits; (*b*) a sparse vegetation cover; (*c*) generally low relief and absence of topographic barriers; (*d*) periodically low sand availability due to wet, frozen, or cemented condition of the surface, and (*e*) the occurrence of permafrost degradation coinciding with increased aridity during the transition from the Weichselian Pleniglacial to Lateglacial time (Kasse 1997).

Cover sand deposition dominated from *c.*14–12.4 ka, but locally deposition in the form of low dunes on the sand sheet surface continued during the late Bølling and Older Dryas period until 11.9 ka (Kasse 1997). Concurrent with cover sand deposition deflation occurred at other places. This resulted in a characteristic gravel-sized lag deposits with frequent occurrence of ventifacts, called the Beuningen Gravel Bed in The Netherlands (van Huissteden *et al.* 2001). A typical case study showing the idealized cover sand sequence, including periglacial phenomena, is presented in Fig. 8.3. OSL-based chronology by Bateman and van Huissteden (1999) permitted a tentative age of 22–17 ka to be assigned for the first

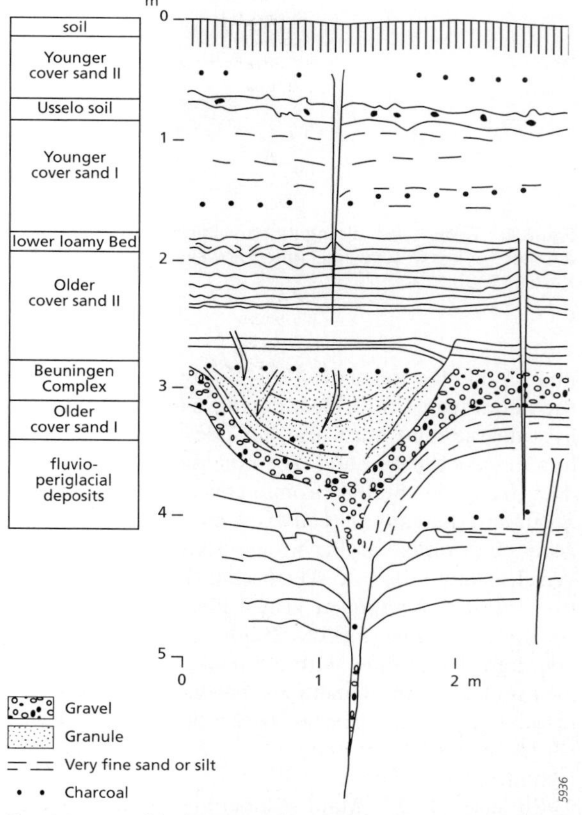

Fig. 8.3. Simplified section of a sandpit in the central Netherlands, showing a typical sequence of periglacial and aeolian sediments, including characteristic frost phenomena, exemplifying the usual cover sand succession (after van Geel *et al.* 1989, compiled by G. C. Maarleveld).

FACIES	DEPOSITIONAL ENVIRONMENT	STRUCTURES
6 dune sand facies (Older inland dunes, and river dunes)	dry aeolian	dune-foreset cross-bedding, (sub)horizontal lamination
5 sand sheet facies a (Younger cover sands)	deflation surface, desert pavement	evenly laminated even horizontal or slightly inclined parallel lamination, rarely cross-bedded/low-angle cross-lamination, granule and adhesion ripples/occasional strings of small pebbles, deflation levels, small frost cracks and cryogenic deformations
	dry aeolian (seasonal frost?)	
4 sand sheet facies b (Older cover sands)	moist aeolian (permafrost?)	evenly laminated ('layer-cake')/horizontal alternating bedding, silty laminae or silt layers, adhesion ripples, frost wedges and major cryogenic deformations, 'vertical-platy' cracks
	wet aeolian (permafrost?)	
3 (aeolian-)lacustrine facies	shallow pools, aeolian supplied material	evenly or wave-ripple laminated, silt and gyttja layers, adhesion lamination
2 (local) flowing-water 1 or fluviatile facies a and b	water current velocity low	fining-upward sequence of
	low energy (ripple phase) fast running water	climbing-ripple cross-lamination, scour troughs, horizontal lamination, adhesion lamination
	low energy (dune phase) fast running water	on large-scale trough cross-bedding, sand with scattered granules, no cryogenic deformations
	high energy very fast running water	

Fig. 8.4. Facies types distinguished within Weichselian dune and cover sand deposits in north-western Europe. The facies types are more or less in sequential order; however, types 6 and 5 can be synchronous and types 4 and 3 can be in reverse order (after Koster 1988).

time to the development of the stratigraphic marker of the Beuningen Gravel Bed, during which time permafrost degradation must also have occurred. Additional luminescence (IRSL and TL) and radiocarbon (AMS) dates, recently obtained from a cover sand sequence in north-east Belgium (Frechen *et al.* 2001) reveal five cover sand units varying in age from Saalian or Early Weichselian to the Late Weichselian/Early Holocene. In this case the Beuningen gravel layer was assigned a somewhat younger age of *c.*16 ka. The cover sands are generally fine grained with a modal grain size between 105 and 210 μm. Sometimes, particularly in the case of Older cover sands, a second peak is present in the coarse silt to very fine sand fraction (16–75 μm), forming alternating loamy laminae within the fine sand (Schwan 1986; kasse 1997). Main sedimentary structures have been described by Ruegg (1983), Schwan (1986), and Koster (1988). The dominant types are: (*a*) low-angle cross-bedding to horizontal bedding, which is attributed to deposition by small wind ripples or in plane beds on a dry surface; (*b*) horizontal alternating bedding of fine sand and loamy very fine sand, generally attributed to deposition and adhesion of sediment on an alternating dry and wet surface. Investigations of sedimentary structures at several cover sand locations by Ruegg (1983) led to an idealized representation of a sequential trend, starting with flowing-water subfacies, via lacustrine subfacies to wet, moist, and eventually dry aeolian subfacies. Schwan (1986, 1988) in a series of publications, refined the detailed description of these cover sand structures, and included frequently occurring periglacial deformations. A summary of these facies types is given in Fig. 8.4.

Tentative hypotheses were developed by Ruegg (1983) and Schwan (1986, 1988) with respect to the environmental conditions responsible for transport and sedimentation during Weichselian cold-climate periods. Factors like the density of the vegetation cover, the importance of snow as a transporting agent, meltwater activity leading to secondary structures, the presence or absence of permafrost and/or seasonal frost leading to deformational phenomena, were interpreted on the basis of textural and structural analysis. However, conformation or rejection of these genetic interpretations was only possible when more information became available from (sub-)arctic regions where present-day wind transport and its environmental boundary conditions can be observed. As this topic is outside the scope of this chapter, the reader is referred to reviews on these issues by McKenna-Neuman (1993) and Koster (1988, 1995). Suffice it to say, that in general the dominance of sand sheet deposits over dune deposits mainly reflects the limited availability of loose, dry sand for transportation. Immobilization of sand after deposition as low-relief sheets is thought to result from a seasonally variable combination of ice cementation, sparse vegetation, high water tables, and snow cover (Koster 1995).

Drift Sands

Local resedimentation by wind of terrestrial (mainly aeolian) deposits, resulting in so-called drift sands, occurred on a large scale from the beginning of the Neolithic up to the present (Castel *et al.* 1989) mainly in the western part of the sand belt. The morphogenesis, sediment characteristics, provenance, and age of these late-Holocene drift sands have been studied by, amongst others, Pyritz (1972) in Germany; Koster (1978); van Mourik (1988) and Castel (1991) in The Netherlands; de Ploey (1977) in Belgium; and Kolstrup (1991) in Denmark. Drift sand regions usually display a rather chaotic pattern of erosional and accumulational parts that strongly vary in outline and size. Individual, often isolated, dunes greatly vary in height (up to *c.*20 m) and slope values (max. *c.*15–33°, in extreme cases up to 47°; Fig. 8.5). In contrast to most coastal or river dunes, drift sand forms often lack a system of distinct morphogenetic units in a downwind direction. The typical drift sand relief, consisting of an irregular alternation of deflational and accumulational parts is shown in Fig. 8.6.

In order to determine a *post quem* age of drift sand accumulation, radiocarbon ages of the topmost part of buried peat sections and humic podzol soils at locations in The Netherlands, Germany, and Denmark have been

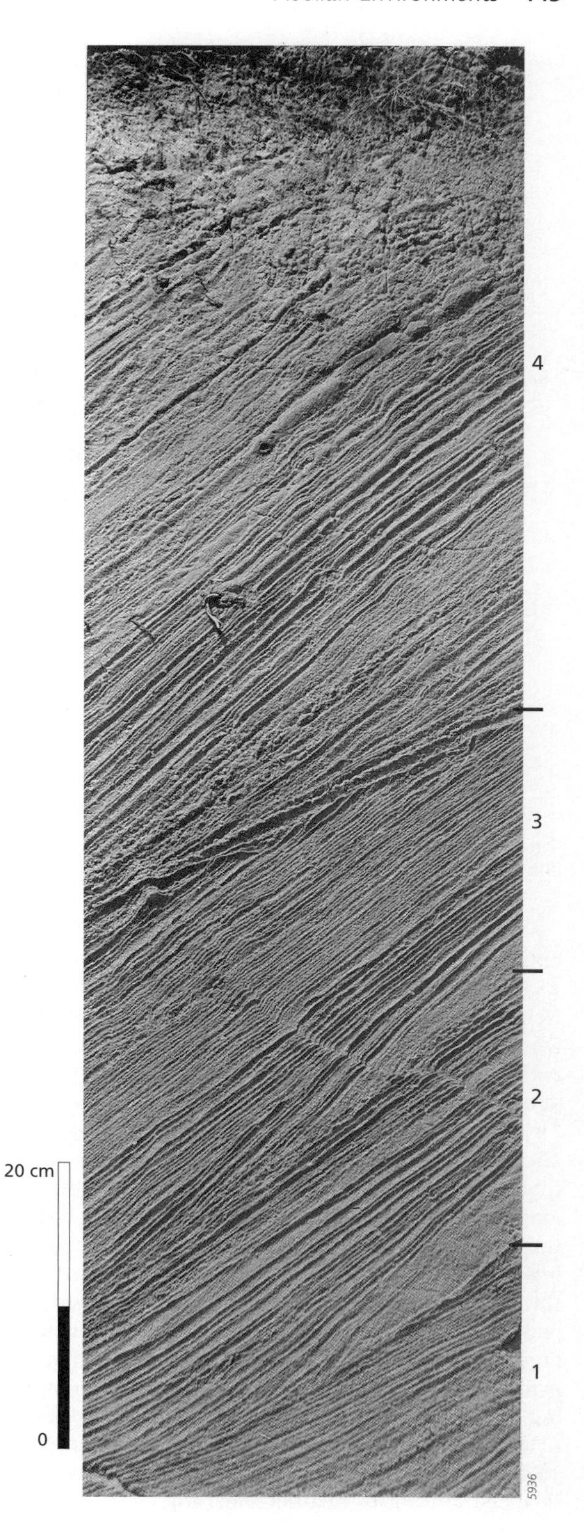

Fig. 8.5. (*right*) Lacquer peel showing four thick sets of high-angle, large-scale cross-stratification on the distal side of an inland dune. The presence of crinkly laminae (e.g. in the lower part of set 4) as well as the abundance of very steeply dipping laminae probably indicate dune migration under relatively moist environmental conditions. The wedge-shaped structureless features at the base of sets 2 and 3 are interpreted as the result of grainfall deposition.

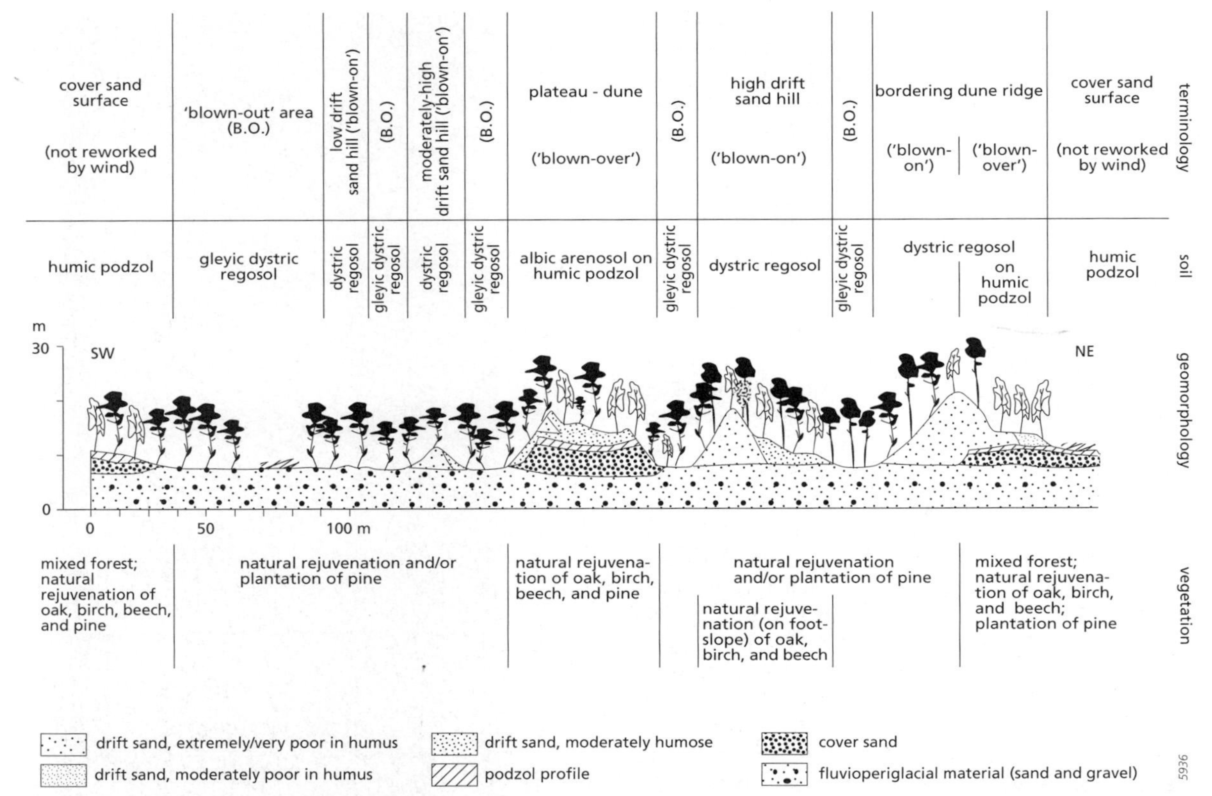

Fig. 8.6. Schematic cross-section of a drift sand landscape (central Netherlands), emphasizing the interrelations between landform, soil, and vegetation development (after Castel *et al.* 1989).

obtained. These are summarized in Koster *et al.* (1993: 251). Although there is evidence for local sand drifts from Neolithic to Roman age, the beginning of really extensive sand drifting in north-western Europe is related, both in time and space, to the rapid expansion of agriculture, together with the use of plaggen fertilizer (Pape 1970) and the (over)exploitation of heathlands, mainly after AD 950. Conry (1974) defined a plaggen soil as a thick man-made epipedon which has been produced by long-continued manuring with substances (heather-sods, grass-sods, forest litter, humic sand, etc.) containing clastic material, and which is more than 50 cm thick. The objective of replenishing nutrient losses in the poor sandy soils of north-western Europe, including England and Ireland, is largely responsible for the large extent of plaggen soils. The practice of plaggen manuring essentially started on a large scale around the sixth to the eighth century AD, although an earlier origin has been documented at some places. It is believed that repeated cutting of heather-sods, together with other forms of overexploitation, will have initiated sand drifting at many places. Some authors (e.g. Heidinga 1984) have suggested a relationship between sand drifting and a drier climate in the 'climatic optimum' in the Middle Ages (*c.*AD 950 to 1250). But no evidence for this climatic oscillation has been found in many peat sections, all of which encompass this period. Thus far all available evidence seems to confirm the conclusion of Castel *et al.* (1989) that, the majority of drift sand accumulation originated during the early part of the Late Middle Ages, but important sedimentation also occurred as late as the eighteenth and nineteenth century as indicated by very young soil profiles (micropodzols). However, regionally minor sedimentation occurred throughout most of the Holocene. Specific drift sand phases of more than local significance cannot yet be distinguished. Even in a small country like The Netherlands Koster (1978), van Mourik (1988), and Castel (1991) did not find a synchronicity in sand drifting phases in central, southern, and northern parts of the country. Similarly,

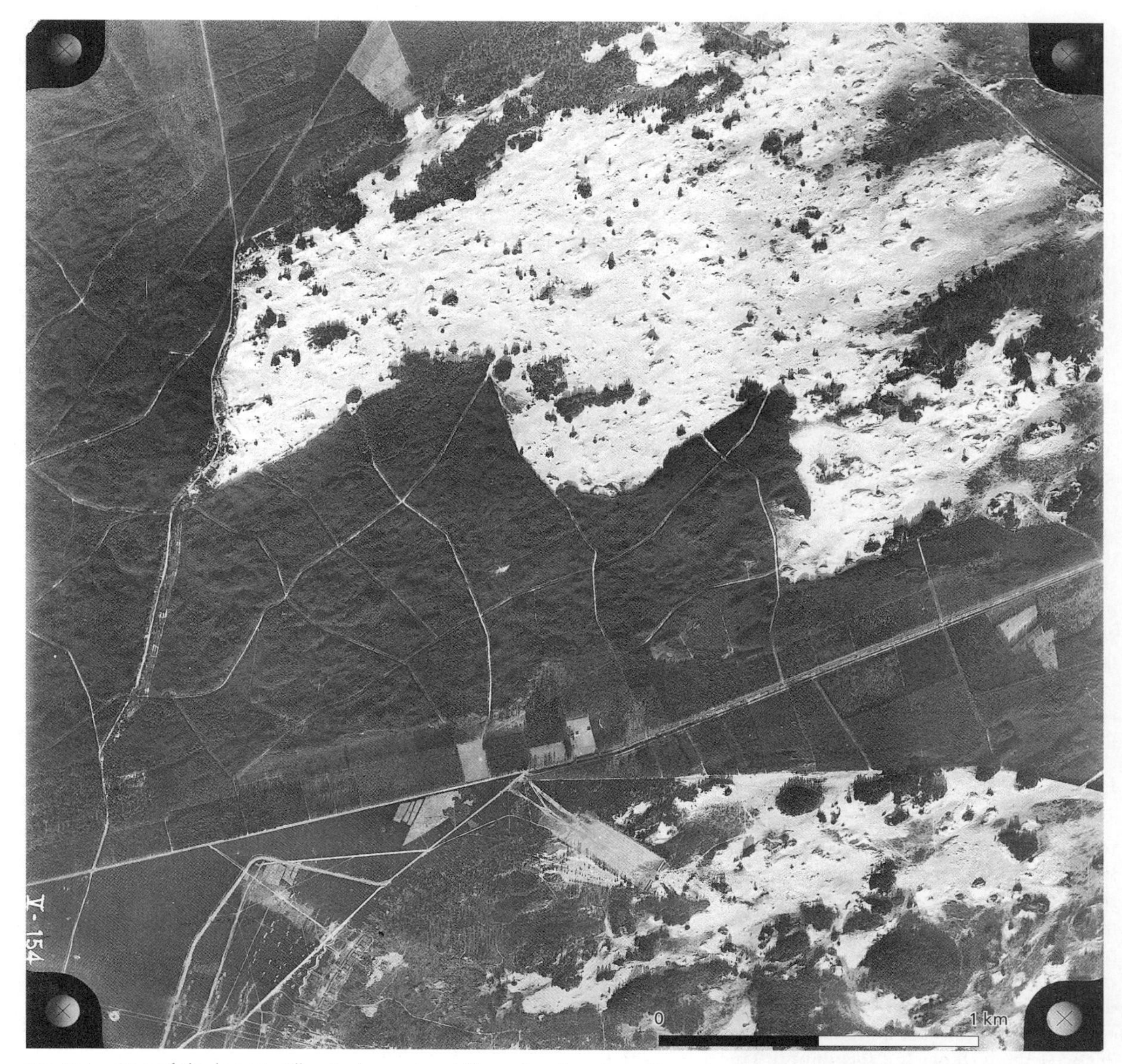

Fig. 8.7. One of the largest still actively moving drift sand regions in north-western Europe (central Veluwe, The Netherlands); the dune fields are surrounded by pine plantations dating from the beginning of the twentieth century (photo: Topographic Survey of The Netherlands).

chronologies determined for dune fields all over Germany point in the same direction. In all cases, a causal relationship seems to exist between human cultural activities and changing local environmental conditions. The majority of inland dune, including drift sand, regions in Europe have now been stabilized by a natural or artificial vegetation cover. For example, many dune fields were transformed into pine (*Pinus sylvestris*) plantations during the late nineteenth and early twentieth centuries, the wood being used for the formerly important coal mining industry in north-western Europe at that time. Only a few actively moving drift sand regions remained today (Fig. 8.7).

Although the sediment attributes of drift sands usually reflect their local provenance quite well, they mainly consist of very fine to moderately fine, non-loamy, well-sorted, and well-rounded unimodal sands. Moreover, they are characterized by an absence of carbonate, a dominance of quartz and to a lesser extent feldspar particles, a variable but very low content of heavy minerals and a usually low organic matter content. Koster *et al.* (1993) discussed characteristic primary sedimentary structures of drift sands indicative of tractional or grainfall deposition, and secondary (postgenetic) structures. Specific combinations of these features are interpreted in terms of dry, moist, or wet conditions during transport and accumulation. De Ploey (1977), in an experimental study on present-day wind transport in an unvegetated drift sand area near the Belgian–Dutch border, observed that deflation occurred not only during dry spells but also on wetted dune sands during or just after heavy rainstorms at wind speeds of more than 30 km/h or *c.*8 m/s. He also reported that gravel (2–5 mm) movement occurred by saltation up to a height of 50 cm during heavy winter storms. The observed silt and very fine sand fractions concentrated in a basal, dense flow skimming over the dune surface suggests that sedimentation of this flow, by dumping, could explain the intermingling of a silt or loess fraction in some Weichselian cover sands instead of supposing its fall-out from a high suspension related to long distance transport. Obviously, these kinds of observations are relevant to the interpretation of palaeoenvironments, as discussed elsewhere in this chapter.

Loess

The term 'loess' was originally used to denote the widespread fine soil in the Upper Rhine graben area. It found its way into the scientific literature in the early nineteenth century, but its origin remained a matter of debate for a very long time. Provenance, modes of transport, sediment attributes, and practical issues such as (accelerated) soil erosion in loess regions are extensively discussed by Mücher (1986) and Pye (1995). A large extension of loess sediments is found in north-western France (Artois, Picardy, Normandy). This belongs to the loess zone, including sandy loess deposits or sand-loam, which extends from France through Belgium, through the southernmost tip of The Netherlands, to Germany (see Fig. 1), and further east and south-eastwards through central Europe with its maximum extension close to the Ural mountains (Antoine *et al.* 1999; Haesaerts and Mestdagh 2000). In central Europe, the northern limit of loess distribution lies approximately along the edge of the uplands. Thus a nearly loess-free zone of about 150 km wide remains in northern Germany and Poland between the margin of the Weichselian glaciation and the loess belt. The loess distribution also has an upper altitudinal boundary. Upland regions such as the Harz, the surroundings of Cologne, or the Schwarzwald are free of loess. An altitudinal limit somewhere around 400–500 m can be interpreted as a vegetation limit, as might also be the case with the northern limit. The westernmost, more isolated occurrences of loess are found in southern England. The distribution and age structure of loess deposits strongly suggest that the material was mainly derived from ice-marginal areas and fluvio-glacial outwash channels during successive Pleistocene glaciations (e.g. Mücher 1986). Braided river plains and local sources also provided fine sediments for aeolian transport.

Loess deposits can be viewed as the source-distal equivalent of the cover sands (Singhvi *et al.* 2001). These sandy loess or loess deposits often lie adjacent to, grade into, or intercalate with cover sand deposits (Koster 1988; Siebertz 1988; Hagedorn *et al.* 1991; also see Pye and Tsoar 1990: 68). The schematic model showing four conditions under which loess deposits may accumulate (Fig. 8.8, after Pye 1995) can be

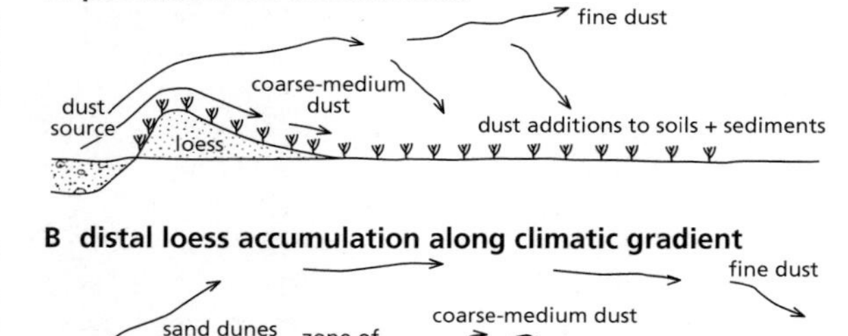

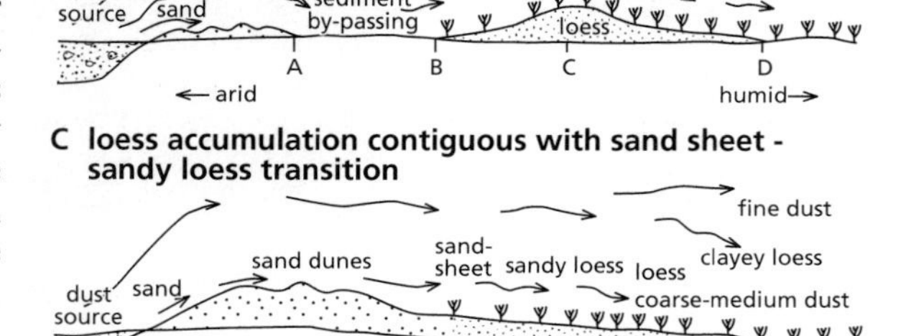

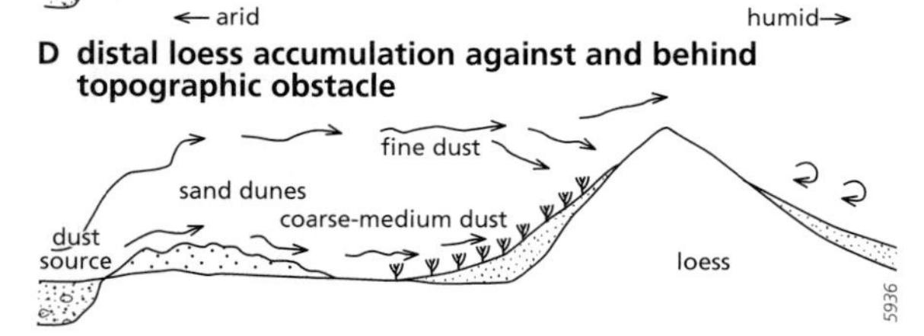

Fig. 8.8. Schematic model showing four conditions under which loess deposits may accumulate (after Pye 1995).

applied in many instances in the European aeolian sand and loess region.

Tephra layers in loess sections are often investigated because of their value as stratigraphic markers. The presence of tephra layers, like the Laacher See tephra (Late Weichselian) and the Eltville tuff layer (Middle Weichselian), in German and Belgian loess facilitated the fixing of the stratigraphic position of several Weichselian interstadial and Eemian (the Rocourt soil is usually attributed to this period) interglacial palaeosols. However, in a series of detailed pedostratigraphic and palaeoecological studies the discussion on the interglacial or interstadial character of the Rocourt soil continues (van den Haute *et al.* 1998; Vandenberghe *et al.* 1998). Thick sequences of loess palaeosols in western Europe occur in the Eifel-Rhine region. In this classical loess region, for example in the famous Kärlich clay pits in the Middle Rhine region, long time sequences of loess deposits alternating with palaeosols and tephra layers, encompassing probably four glacial–interglacial cycles, have been documented. Although luminescence and K/Ar dates are still in dispute, it is believed that the oldest parts of the loess sequence date back to the Cromerian Complex. The significance of alternating phases of sedimentation and soil formation in the loess region of southern Limburg (The Netherlands) and adjacent parts of Belgium is exemplified in the sequences at the Belvédère-Maastricht loess and gravel quarry, and the brickyards of Kesselt and Rocourt south of Maastricht in Belgium (Vandenberghe *et al.* 1985; van den Haute *et al.* 1998; Vandenberghe *et al.* 1998; Antoine *et al.* 2001). The sometimes cryoturbated Nagelbeek or Kesselt Horizon is an important stratigraphic marker with an age of *c.*20–22 ka, reflecting conditions during the Last Glacial maximum. Comparison of loess sequences at sites located in the loess belt from Belgium through Germany and Poland to Russia reveal a remarkable consistent picture of palaeosol successions according to Haesaerts and Mestdagh (2000). An absolute timescale for these successions is being constructed, based on TL and OSL dates. Palaeosols of last interglacial age are correlated with oxygen isotope stages (OIS) 5e and 5c, whereas several palaeosols of Weichselian age are positioned within stages 5a and 4. Moreover, there is evidence that relatively high loess sedimentation rates prevailed during the Upper Pleniglacial, starting several ka before the OIS 3 to OIS 2 transition (Antoine *et al.* 2001).

Reviews of loess studies have shown that the loess record is seldom continuous and, even during the last interglacial–glacial cycle, large amplitude changes in loess sedimentation rates occurred (Singhvi *et al.* 2001). This was already established by Wintle *et al.* (1984), who presented luminescence ages for loess-like sediments, or *limons à doublet*, in Normandy (France). Their data suggest that loess accumulation occurred rapidly during relatively short time intervals, in this case during Early and Late Weichselian time. Distinctive characteristics of the typical loess are a dominant silt fraction (*c.*16–60 μm), a fine porous structure lacking an apparent bedding, a brownish-yellow colour, and the frequent presence of carbonate in a diffuse state (Lautridou *et al.* 1984). Notable within the group of bedded loesses is the non-carbonate type known as *limons à doublet*, which has been decalcified and contains thin bands formed by clay and iron dislocation. Absence of carbonate may be either original or due to post-depositional decalcification.

The excellent properties of loess for agriculture together with their extensive areal extent within western and central Europe make these deposits important. Consequently, the number of publications on the genesis, the sediment properties, and especially on the problems of (accelerated) soil erosion of loess regions is enormous. Mücher (1986) compiled an excellent review of loess research. Experimental and micromorphological analyses enabled him to determine the relative impact of splash erosion, overland flow, through-flow, and meltwater flow on erosion and redeposition of loess. Microstructures indicative for (*a*) desiccation under dry and warm conditions on the one hand, (*b*) the effects of repeated freezing and thawing cycles, (*c*) the process of denivation on the other hand, and (*d*) redeposition by aeolian, flowing water or mass movements, were investigated and reported in a series of publications by H. J. Mücher and colleages (see Mücher 1986). Loess soils were among the first to be cultivated in north-western Europe because of their relatively high nutrient availability and favourable soil structure. Since man began to disturb the natural landscape by deforestation for agricultural purposes in the Neolithic, erosion became extensive and accelerated, resulting in widespread anthropogenic slope deposits or colluvium and in silty alluvial plain sediments (German: *Auelehm*). This process had already started locally around 6000 BP, and was intensified especially during the Roman period and the Late Middle Ages. Accelerated soil erosion on intensively used arable land in loess regions continues to be an actual problem.

Coastal Dune Systems

In north-western Europe an almost continuous coastal dune system extends from the neighbourhood of Calais in north-western France to the northernmost point of Jutland at the village of Skagen in Denmark. Moreover, coastal dunes along the shores of the Baltic Sea extend all the way from northern Germany to the Baltic States. The reader is referred to Nordstrom *et al.* (1990) for general information on coastal dune form classifications, the complex interactions of geomorphic processes and dune forms, the human influences on dune development, and methods of dune management. The formation of (Younger) coastal dunes in The Netherlands began in the period between the tenth and the twelfth century. Three phases of dune building have been recognized, based upon a large series of radiocarbon dates of dune palaeosols and elaborate palynological and archeological evidence (Jelgersma *et al.* 1970; Klijn 1990): (*a*) an initial phase, which ended in the thirteenth century, and mainly consisted of filling up the pre-existing undulating relief of the coastal barriers; (*b*) the main phase of formation of large parabolic dune fields between *c.* AD 1450 and 1750; and (*c*) a minor phase starting in the nineteenth century, including the formation of the largely artificial narrow strip of foredunes. Almost the complete dune surface of both coastal and inland dunes has been protected against further wind action by afforestation with pine (mainly *Pinus sylvestris*, but also *P. nigra* and *P. pinaster*) and by planting marram grass (*Ammophila arenaria*) and sea lime grass (*Elymus arenarius*). It is remarkable that the formation of Older coastal dunes, which developed on top of the coastal barrier system of Boreal and Atlantic age, lasted only until Roman times. The material for the build-up of the Younger dunes mainly came from the submarine part of this barrier system.

Recent AMS radiocarbon ages obtained from the different humic fractions in buried podzols in Danish coastal dunes (Dalsgaard and Odgaard 2001) are in agreement with the dune-forming phases distinguished in The Netherlands. Major sand accumulations occurred at around 700 BC and from about AD 1200–1400 until quite recently. Luminescence dating of dunes at the west coast of Denmark (Murray and Clemmensen 2001) shows comparable results. Three main periods of aeolian sand movement have been identified, starting at about 4,200, 2,700, and 900 years ago. The youngest phase of dune building continued until the Little Ice Age. Clemmensen *et al.* (1996) obtained ages of around AD 300 for a lower coastal dune unit in western Jutland, and a more prominant upper unit formed between AD 1550 and 1850. Obviously, local differences in dune-forming episodes in Denmark, as in other western European countries, can be derived from the various sources (see e.g. Bakker *et al.* 1990).

By far the largest coastal dune and aeolian sand system in France occurs in the Aquitaine basin in the departments of Gironde and Les Landes. The *c.* 10-km-wide Holocene coastal dunes are developed upon a very extensive region of 'Sables des Landes', which stretches up to 100 km inland. This development is believed to be of late Weichselian age, when large parts of the adjacent continental shelf were exposed during low sea-level stands (Bressolier *et al.* 1990; Clarke *et al.* 1999). Palaeosol and radiocarbon analyses initially defined a series of four dune generations of Subboreal–Subatlantic age, differing in geomorphology and with the youngest one near the present littoral foredunes and the oldest located furthest inland. IRSL dating by Clarke *et al.* (1999 and 2002) has shown that episodic aeolian activity in the Aquitaine region has occurred mainly during the following periods: around 4,000–3,000 years ago, around 1,300–900 years ago, and 550–250 years ago, the latter at least partly associated with the Little Ice Age. As is the case in many instances, Clarke *et al.* (1999) demonstrate that geomorphology cannot be used as a surrogate for chronology. The chronology observed in the spectacular asymmetrical transverse 'dune du Pilat', 2.5 km long, 500 m wide, and 100 m high, situated near the entrance of the Bay of Arcachon, serves as a case study for a major part of the French Atlantic dune formations (Bressolier *et al.* 1990).

The links between marine sediment supply and coastal dune construction, especially on century to millennium evolutionary timescales, are well recognized. However, at the level of processes studies linkages between the marine-dominated beach system and the aeolian-dominated dune system are far from understood (Nordstrom *et al.* 1990). Pye (1990) has presented an illustrative schematic model showing the relationships between sediment budget, wind regime, vegetation, and resultant dune morphology over time (Fig. 8.9). To be more specific, this model takes into account (*a*) the rate of littoral sand supply to the shoreline, (*b*) the wind energy available to transport sand inland, and (*c*) the effectiveness of sand-trapping vegetation, which in turn reflects the nature of the natural flora and the extent of its destruction due to human pressures (Pye 1990). In most instances sand is supplied to the beach zone by waves and coastal currents. Sand transport rates from the beach to the foredunes on long timescales are a reflection of the availability of sand in the shore zone as influenced by differences in coastal erosion or accumulation, wind climate, and/or sea level. Sand transport rates for dunes of Subatlantic age in the order of 20–25 m^3 (m width)$^{-1}$ yr^{-1} have been calculated by Clemmensen *et al.* (1996) for a dunefield at the west coast of Jutland. For dunes formed in the period from AD 1550–1850, that is, during the Little Ice Age, Clemmensen *et al.*

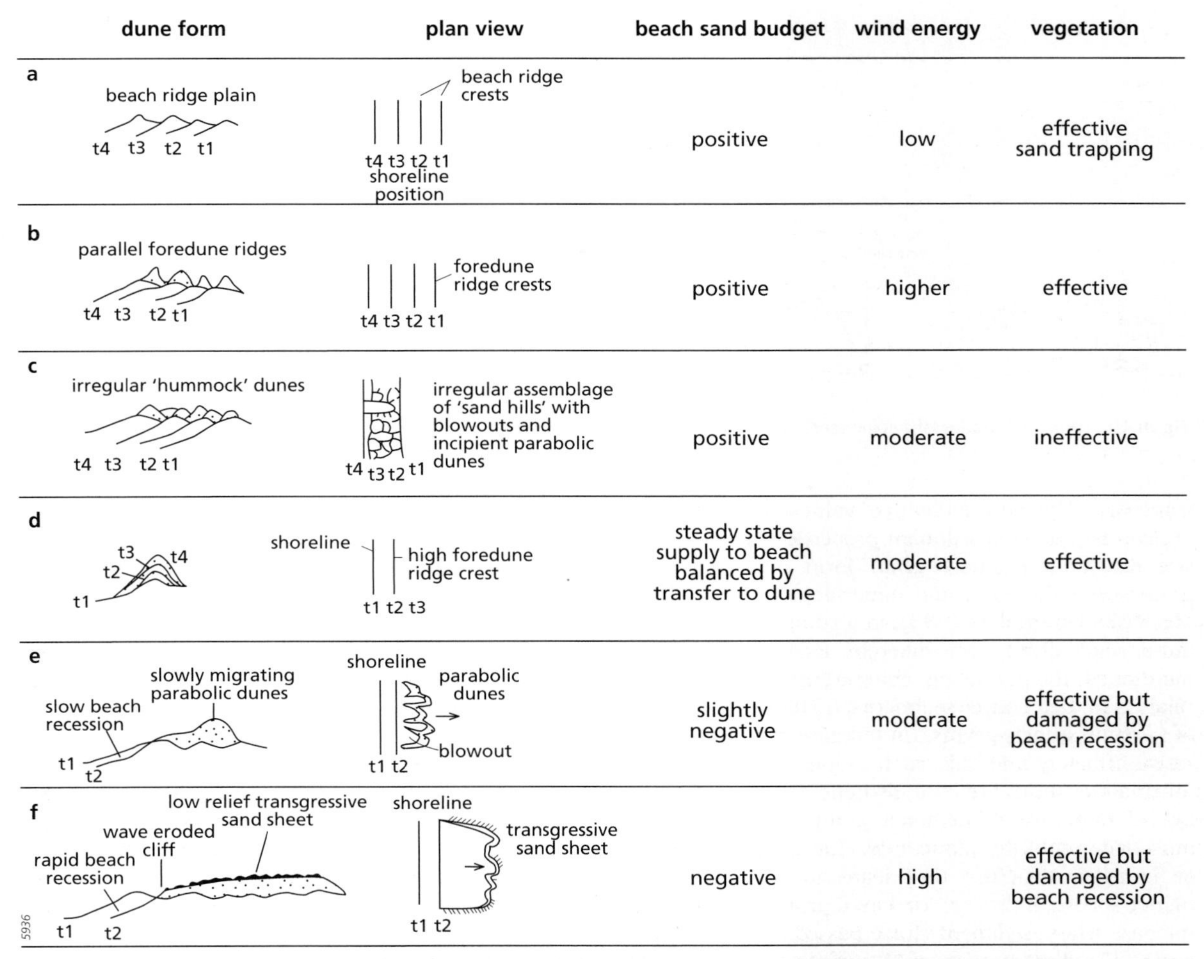

Fig. 8.9. Schematic diagram showing relationships between sediment budget, wind energy, vegetation, and resultant dune morphology over time t1–t4 (after Goudie *et al.* 1999, adapted from Pye 1990).

(1996) even found sand transport rates up to 83 m^3 (m width)$^{-1}$ yr^{-1}. Such a short-term rate of sand transport is very high, but not impossible during a period of increased storminess and a high supply of beach sand. For severely eroded parts of the Wadden Island of Texel in The Netherlands Arens (1994) calculated a comparable maximum aeolian transport rate in the order of 50 m^3 (m width)$^{-1}$ yr^{-1}. In this case large amounts of sand are removed from the dune system during storm surges (negative sediment budget), part of which is re-entered into the foredunes afterwards. On a longer timescale this process is comparable to the situation of a retreating coastline during the main period of formation of the Younger Dunes in the fifteenth to eighteenth century. Arens (1994) has shown that even with onshore winds actual transport rates are much lower than potential rates. The main cause of this large deviation lies in the absence of transport during very wet conditions of prolonged rainfall regardless of wind speed, and the variation of threshold velocity with time. As a consequence a general conclusion can be drawn that the bulk of aeolian transport in temperate humid climates is related to moderate conditions and not to extreme events, in conformity to the principle of magnitude and frequency.

Sediment Attributes

The degree of transformation of source sediments into aeolian ones may help to estimate the duration of aeolian

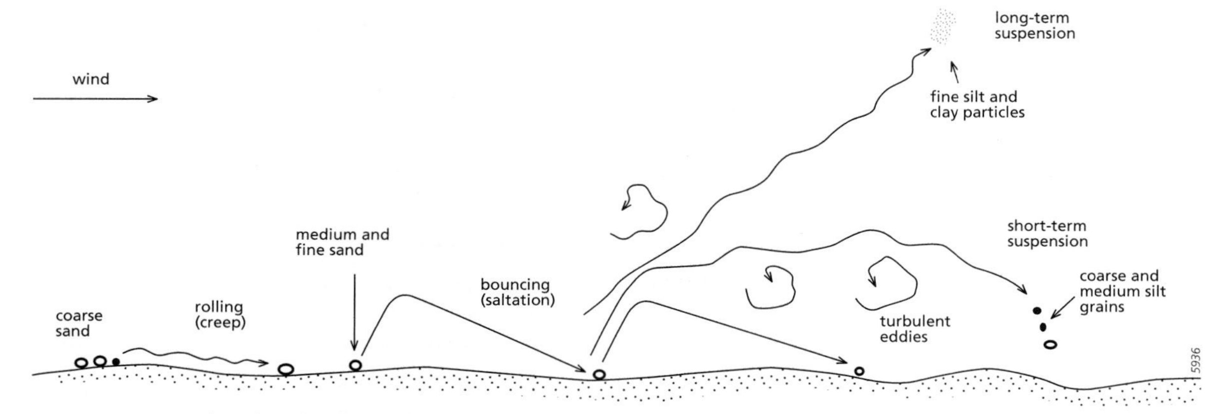

Fig. 8.10. Principal modes of aeolian sediment transport (after Pye 1995).

transport. The best indicative values in this respect are combinations of sediment properties such as grain-size distribution, quartz-grain form and roundness, grain surface texture, and mineralogical composition. Mycielska-Dowgiallo (1993), in a comparative study of dune sands from very different European localities, mentioned the following characteristic properties in relation to their source sediments: (*a*) higher percentage of resistant quartz grains, (*b*) increase in percentage of resistant heavy minerals, such as garnet, zircon, rutile, tourmaline, staurolite, and disthene, (*c*) lower percentage of minerals of the mica group, such as biotite, muscovite, chlorite, glauconite, due to less resistance or lamellar structure that leads to winnowing, (*d*) higher sorting index, (*e*) increased grain roundness. In the case where sediments have passed through several cycles of sedimentation grain surface features, as observed by scanning electron microscopy (SEM), are most likely to reflect different transport modes. In a particular region it might well be possible to distinguish aeolian facies types only on the basis of grain-size distribution (see e.g. Siebertz 1988), but this certainly does not mean that grain size can be used as a diagnostic criterium in general. Nevertheless there is an obvious relationship between grain size and the different modes of particle transport by wind (Fig. 8.10).

Palaeoenvironmental Significance

Geomorphological and sedimentological characteristics from aeolian environments can provide information on palaeowind direction, in particular during high wind-energy events, on wind velocity, and on the seasonality of wind regimes. These characteristics include dune form and orientation, dune and ripple migration rates and spacing, outline of dune fields and of sand or loess sheets, orientation of deflational forms and ventifacts, trends in sediment attributes and relationships to source areas, and periodic discontinuities in sediment texture and structure (Koster 1988; Isarin *et al.* 1997). Aeolian phenomena not only reflect former wind patterns but provide qualitative information as well about the moisture content of the soil, effective precipitation, and snow cover.

As stated before, reconstruction of the circulation by aeolian dune forms is not a simple resultant of daily prevailing surface winds, but of the less frequent high-magnitude winds that are often linked with the passage of fronts. It has been repeatedly shown that dune orientations closely represent sand-transporting wind directions. However, determination of wind velocity from dune forms is not possible in any detail, although the degree of unidirectionality of the airflow can be judged. Aeolian sediment textures will only reflect variations in sand-transporting wind velocities. No indications of average or extreme wind velocities can be obtained this way.

Early attempts to reconstruct the general (summer) atmospheric circulation in Europe during the Last Glacial Maximum (*c.*20–18 ka) and during the Late Weichselian (*c.*13–10 ka) based on aeolian landforms have been published by Poser (1950), Kottmeier and Meyer (1988), and Meyer and Kottmeier (1989). However, it appears that Poser's interpretation was not correct as his reconstruction was based on dunes of varying age, even including dunes of Holocene origin (Seraphim 1986). Moreover, Poser's reconstruction comprised stadial and interstadial conditions in the time interval from *c.*13–10 ka, during which time circulation patterns may have varied strongly. Meyer and

Kottmeier's wind regime reconstructions are based upon a combination of a simple analytical flow model and interpretations of palaeowind directions in hundreds of publications. However, it is well known that absolute dating of aeolian formations and other palaeo-indicators during these time intervals is inaccurate, to say the least. Consequently, although valuable as first approximations, their reconstructions must be viewed with some scepticism. A recent attempt to reconstruct the surface wind climate during the Younger Dryas (10.95–10.15 ^{14}C ka BP) in Europe, based upon a comparison of the detailed aeolian record and an Atmospheric General Circulation Model (AGCM-ECHAM3), was published by Isarin *et al.* (1997). During this marked cold interval depositional winds in north-western and central Europe came from westerly quadrants. Regional contrasts between aeolian accumulations in The Netherlands and in Poland can be explained by differences in surface wind climate, in sediment supply by dissimilar river types (braiding in the west but more meandering in the east) and differences in the vegetation cover (as influenced by the distance to the Weichselian ice sheet margin). The model simulations show that wind speeds were higher in the western part of the sand belt due to the proximity of the ocean. Aeolian landforms indicate that during the Younger Dryas in The Netherlands and Belgium depositional winds came from the south-west to west, whereas in central Poland westerly winds ranged from south-west to north-west. Model experiments support these observations. Moreover, Isarin *et al.* (1997) found evidence that aeolian transport and deposition primarily took place during the winter season. In contrast to suggestions by various authors, no indication of a shift towards an atmospheric circulation with more frequent easterly winds was found in the proxy data, nor in the simulation experiments for the Younger Dryas. The formation of ventifacts indicating easterly palaeowind directions, as have been found in southern Sweden, was probably restricted to the immediate vicinity of the retreating ice sheets. Zeeberg (1998) concludes that dune formation in the European sand belt appears to be primarily controlled by sand supply in ice-marginal environments by strong more or less unidirectional westerlies over the periglacial regions, in particular during the winter season. Van Huissteden *et al.* (2001) also attributed cover sand deposition mainly to westerly sand-transporting winds. However, they argue that during winter aeolian deposition also occurred from frequent and strong easterly winds as well. If this is correct, then Late Glacial wind regime conditions in the north-west European Lowlands closely resembled those of today, where sand transport by dominating strong westerly winds alternates with occasional sand drifting from bare sand surfaces and arable fields that takes place during dry spells with easterly winds in spring. Bateman (1998) presented evidence for comparable westerly palaeowind directions inferred from sand sheet geometry and sediment provenance in the UK.

Geochronology

In palaeoenvironmental reconstructions one of the main problems that often remains is the accurate dating of aeolian sediments and landforms, or more specifically the establishment of the duration of the aeolian 'effective' periods. In most cases age determinations of aeolian phases are still based on dating of intercalating organic (peat) layers and palaeosols that represent the non-aeolian phases (Koster 1988). Luminescence dating techniques might shed new light on this matter in the near future. However, early attempts to date Late Glacial cover sands by TL (Dijkmans 1990; Dijkmans and Wintle 1991) resulted in an apparent underestimation of the ages of the samples by 20–50%, when compared with the existing chronology (Bateman and van Huissteden 1999). Lately, a strongly increasing number of publications is devoted to thermoluminescence (TL), and optically stimulated luminescence (OSL) studies of aeolian (and other) sediments (e.g. see *Special Issue Quaternary Science Reviews* 20 (2001) edited by Grün and Wintle). Radtke *et al.* (2001) applied luminescence dating in a study of dune sands in central Germany, which could be tested with an independent age control as it contained a Laacher See tephra layer (12.88 ka, based on varve chronology). The authors conclude that the results of the luminescence measurements are not yet satisfactory. Although measurements of the quartz fraction provide promising results with respect to the presumed Late Weichselian age, the results for potassium feldspars in contrast show an underestimation of ages of up to 25%. Hilgers *et al.* (2001) are more satisfied with their luminescence ages obtained from Late Glacial and Holocene dune sands in the vicinity of Berlin (Germany). These authors used OSL measurements of the quartz fraction. In this case the same Laacher See tephra layer as well as the marker horizon of the Allerød 'Finow' palaeosol, equivalent to the Usselo soil of the Dutch stratigraphy, enabled an independent age control. Other marker horizons which have aided in establishing an absolute chronology of German and Belgian loess sequences are the Eltville and Rambach tephra layers, dated by Juvigné and Wintle (1988) using a self-consistent set of TL and ^{14}C ages, at *c.* 16.2 and 19.4 ka BP respectively. A multitude of other marker horizons, such as the Kesselt soil of Weichselian-Pleniglacial age and the Rocourt soil of Eemian age, is documented in Belgian, Dutch, and German loess studies. Hilgers *et al.* (2001) even report on very young luminescence dates of only a few hundred years, which have been obtained from recently remobilized sands near the dune surface. It is considered that these young ages provide strong evidence that the resetting process in these aeolian sediments was complete, and that the last mobilization of these sands probably took place towards the end of the Little Ice Age. Similar Little Ice Age results have been reported using the single-aliquot regenerative-dose (SAR) protocol from the top of an aeolian dune sequence on the west coast of Denmark by Murray and Clemmensen (2001). Janotta *et al.* (1997) applied TL and IRSL methods to Late Glacial and Holocene dune sands and sandy loess near Bonn (Germany) aiming to establish a subdivision of different phases of aeolian activity. As expected they found a clustering of ages around the Younger Dryas, but surprisingly they also tentatively concluded that there are indications for continued accumulation of dune sands in this region during the Boreal and Atlantic (from about 8–6.5 ka BP). As the evidence they present is not very convincing this statement needs further confirmation. Some support for this somewhat exceptional standpoint comes from the combined radiocarbon and TL dates of dune and cover sands in Brandenburg in north-east Germany (Bussemer *et al.* 1998) which suggest that Holocene aeolian accumulation continued from the late Preboreal to the Boreal. All younger aeolian reactivation phases can be ascribed to human interference in the landscape.

A wealth of informative luminescence data on dune sand and loess ages from localities all over Germany is made available in the elaborate study of Radtke (1998) and his collaborators. Early to Late Weichselian loess and sandy loess deposits, Late Weichselian to early Holocene dunes, and phases of aeolian reactivation during the Medieval period and the last few centuries, have been dated and compared to radiocarbon ages and other proxies, like Palaeolithic artefacts. Radtke (1998) also presents a concise overview of luminescence studies of aeolian sediments following the review by Wintle (1993). The division of Late Glacial–Holocene time into a series of phases of strong and weak aeolian morphodynamics in German dune fields (Radtke 1998) agrees very well with existing chronologies summarized for other regions in Europe by Koster (1988), Kolstrup *et al.* (1990), Vandenberghe (1991), and Kozarski and Nowaczyk (1991). Wintle *et al.* (1984) presented preliminary TL-ages of loess sediments and soils, sampled at a reference section for the earlier-mentioned *limons à doublet* in the Pays de Caux (Normandy, France). A logical series of Saalian to Late Weichselian ages has been determined and at the same time it could be established that there are apparently large time gaps in the loess accumulation. TL age determinations of cover sand deposition in the UK (Bateman 1998; Bateman *et al.* 2000) clearly show a coincidence with the Late Weichselian (Devensian) Older Coversand II / Younger Coversand I depositional phase found on the continent and even more so with the predominant Younger Coversand II depositional phase during the Younger Dryas (or Loch Lomond) Stadial.

Absolute dating of phases of loess sedimentation is notoriously difficult. Interpretation of loess ages often requires inputs of micromorphological analysis, especially with reference to the extent of pedoturbation. Luminescence age of A-horizons reflects the age of the soil formation episode, whereas the ages of B- and C-horizons, that experience clay and carbonate deposition, reflect the event of loess sedimentation more closely. These two events might well be separated by several thousand years.

Singhvi *et al.* (2001) compiled a dataset containing more than 300 previously published luminescence dates on aeolian (cover)sand deposits (Fig. 8.11). Subsets refer to ages obtained by: thermoluminescence (TL) on quartz, TL on feldspars, green light stimulated luminescence (GLSL) on quartz, and infrared stimulated luminescence (IRSL) on feldspars. Figure 8.11 clearly illustrates that (*a*) there still is a large, partly unexplained, variation with respect to number of clusters and their timing, and (*b*) cover sand deposition was minimal during the last glacial maximum, but was initiated at *c.*15 ka and subsequently peaked during the Younger Dryas climate oscillation. Minor aeolian reactivation, i.e. sand drifting due to human influences, is shown by small peaks in historic time. Note that, whilst the clustering of ages at particular times is significant, the height of a peak reflects the number of analyses and has no climatic connotation. A similar probability curve has been plotted, based on an equally large number of luminescence ages from (mainly European) loess/palaeosol sequences.

The reliability of conventional radiocarbon dates of bulk samples from organic remains in palaeosols is inherently problematic, e.g. due to large age differences obtained by dating the various organic fractions, the possible inclusion of old carbon derived from deflation of older surfaces, or the rejuvenation by contamination from younger roots (Castel 1991; Koster *et al.* 1993). To overcome the difficulties of diverging age determinations resulting from the analyses of the different humic fractions contained in peat or soil samples, nowadays

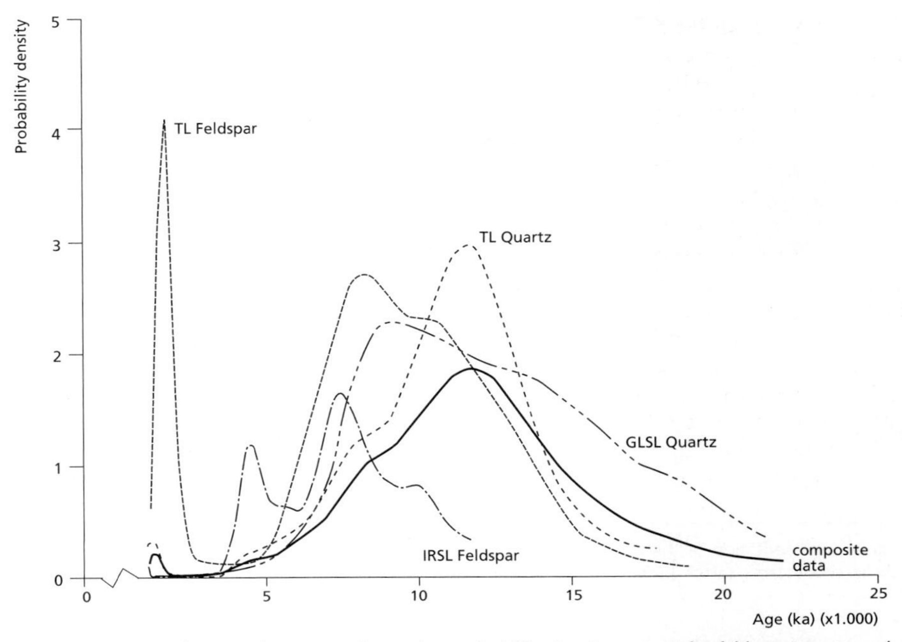

Fig. 8.11. Compilations of cover sand ages (as probability density curves) for feldspar/quartz and TL/IRSL/GLSL analyses (after Singhvi *et al.* 2001); the composite data curve combines quartz-based TL and GLSL ages.

AMS-dating techniques on individual seeds or other organic particles are favoured. Van Mourik (1988) and Castel (1991) have reviewed the difficulties in interpreting the apparent differences in radiocarbon ages obtained from the fulvic acid, humic acid, and residual or humine fractions from A1- and B2-horizon of buried podzolic soils in dune sands. Castel (1991) also compared conventional radiocarbon ages with AMS-ages, which led to the suggestion that the most reliable ages were those obtained from the residue fraction.

Dune Conservation and Management

To attain a sustainable development of coastal dune areas, also called dynamic coastal management, a subtle balance has to be found with respect to the many functions these areas fulfil. The main functions for society are: coastal defence, nature conservation, infiltration and extraction of public drinking water in reservoirs, and recreation. In van der Meulen *et al.* (1989) simple systems of categories of dune forms and dune systems, following both ecological and morphological criteria, are presented as a basis for dune inventories. It will be obvious, that a feasible and integrated nature management of coastal dunes cannot consist of striving towards a so-called natural situation, whatever that may be. It should rather aim at the safeguarding and, if possible, promotion of an optimal diversity in life forms, species, and ecosystems. New perceptions in nature management have also led to the view that stabilization of moving sand surfaces in coastal or inland dune fields is not always necessary. On the contrary, nature management practices are now aimed in particular instances at facilitating processes of wind transport in dune areas because this creates very specific environmental conditions for flora and fauna.

The wide variety of landforms and vegetation cover of dune landscapes results from its dynamics. To express these dynamics, Jungerius and van der Meulen (1988) have distinguished three compartments in dune ecosystems (Fig. 8.12). In the first compartment geomorphological processes dominate. These include blow-out formation, dune migration, and erosion by flowing water (in some instances aided by the water-repellency of sand surfaces). Biological processes are of minor importance and no soil profile is formed in this compartment. In the third compartment geomorphic processes have been brought to a stop by the vegetation and now soil development and biological processes dominate. Both compartments are linked by a transitionary phase

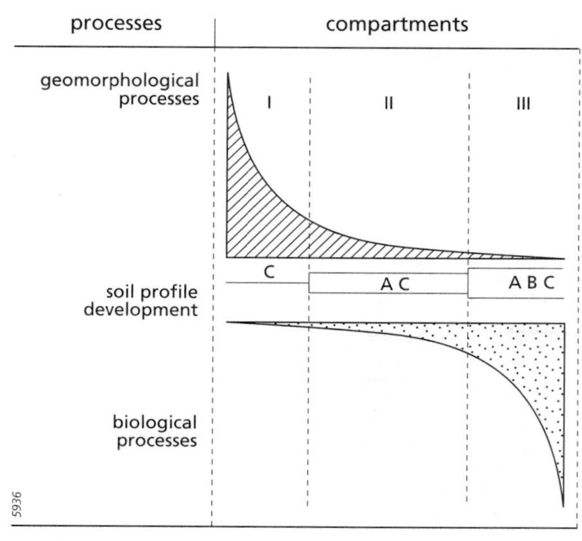

Fig. 8.12. Schematic representation of the relative importance of processes in dune ecosystems (after Jungerius and van der Meulen 1988).

(compartment two). The equilibrium between the geomorphological and biological processes is not static. Any change in the environmental factors will cause a shift either towards more stability or towards more instability (van der Meulen *et al.* 1989). This in turn affects the distribution of the compartments in the dune landscape. The pronounced gradient structure of the dune landscape, which includes many small-scale gradients and high concentration of species, is in part dependent on the continuous action of dynamic processes, both abiotic and biotic (van der Meulen and van der Maarel 1993). Consequently, management policies presently try to restore natural values and processes by focusing on: (*a*) (re)introduction of aeolian dynamics and even marine processes by artificial breaches in the foredunes (Fig. 8.13); (*b*) development of blow-outs and limited dune migration; and (*c*) regeneration of wet dune valleys. In each of the dune compartments detailed studies on the efficiency of foredunes in trapping sand (Arens 1994), on the growing of blow-outs (Jungerius and Verheggen 1981), and on the monitoring of dune dynamics and coastline retreat or advance, have been or are being carried out to assist dune management procedures. These modern attempts to reintroduce morphodynamics in dune ecosystems, albeit on a local scale, are obviously in strong contrast to the practices of stabilizing dune surfaces by planting vegetation on a large regional scale, which proved to be necessary during the past centuries to prevent coastal villages becoming completely covered by migrating dune sands. Nevertheless, due to coastal retreat as well as migrating coastal dune fields several villages, e.g. in the Netherlands and Denmark, have been either abandoned or rebuilt further inland. Figure 8.14 shows an example of a lighthouse on the north Jutland coast that has become useless as it has become shielded from the sea by the accumulation of cliff-top dunes.

Fig. 8.13. An artificial breach in the foredune near Bergen (Holland coast, The Netherlands), excavated in 1997 to promote the development of ecosystem diversity. The location of the site has been determined by the fact that previously (in 1928, 1953, and 1974) natural breaches of the foredunes occurred at this place (photo: Simon Smit Fotografie).

Fig. 8.14. High cliff-top dunes encroaching upon a lighthouse near Lønstrup, north-western Jutland, Denmark.

Unvegetated drift sand areas constitute a very specific habitat for plants and insects due to the continuously changing environmental conditions and the exceptional microclimate in which daily temperature ranges of more than 50 °C and extreme changes in soil moisture content can occur. Alisch (1995) is one of the few authors who has explicitly paid attention to the mapping and evaluation of inland aeolian forms in Germany from the point of view of nature conservation. Because of the scarcity and ecological importance of drift sands that are still actively moving, projects to reactivate sand drifting have been promoted and executed by the Nature Conservation Board in The Netherlands. To enable the prediction of sand transport rates and directions under present-day climatic conditions, drift potentials have been calculated and a simulation model for sand transport has been developed (Castel 1988). As little sand transport can be expected under actual conditions, only special management measures, like locally removing all vegetation, can guarantee mobility of the dunes. Likewise, other projects were aimed at promoting actual river dune accumulation during high flood events as a means of nature rehabilitation. Landscapes with active aeolian processes are uncommon in Europe and need careful management, which requires knowledge of the interaction of the many geomorphological, ecological, and land-use processes.

References

Alisch, M. (1995). Das äolische Relief der mittleren Oberen Allerniederung (Ostniedersachsen)—spät- und postglaziale Morphogenese, Ausdehnung und Festlegung historischer Wehsande, Sandabgrabungen und Schutzaspekte. *Kölner Geographische Arbeiten* 62.

Antoine, P., Rousseau, D-D., Lautridou, J.-P., and Hatté, C. (1999). Last interglacial–glacial climatic cycle in loess-palaeosol successions of north-western France. *Boreas* 28: 551–63.

—— Rousseau, D-D., Zöller, L., Lang, A., Munaut, A-V., Hatté, C., and Fontugne, M. (2001). High-resolution record of the last Interglacial–glacial cycle in the Nussloch loess-palaeosol sequences, Upper Rhine Area, Germany. *Quaternary International* 76/7: 211–29.

Arens, S. M. (1994). *Aeolian processes in the Dutch foredunes*. Thesis, Department of Physical Geography and Soil Science, University of Amsterdam.

Bakker, Th. W., Jungerius, P. D., and Klijn, J. A. (eds.) (1990). Dunes of the European coasts. Geomorphology—hydrology—soils. *Catena* Suppl. 18.

Bateman, M. D. (1998). The origin and age of coversand in North Lincolnshire, United Kingdom. *Permafrost and Periglacial Processes* 9: 313–25.

—— and van Huissteden, J. (1999). The timing of last-glacial periglacial and aeolian events, Twente, eastern Netherlands. *Journal of Quaternary Science* 14: 277–83.

—— Hannam, J., and Livingstone, I. (1999). Late Quaternary dunes at Twigmoor Woods, Lincolnshire, United Kingdom: a preliminary investigation. *Zeitschrift für Geomorphologie*, Suppl. 116: 131–46.

—— Murton, J. B., and Crowe, W. (2000). Late Devensian and Holocene depositional environments associated with the coversand around Caistor, north Lincolnshire, United Kingdom. *Boreas* 29: 1–15.

Bohncke, S., Vandenberghe, J., and Huijzer, A. S. (1993). Periglacial environments during the Weichselian Late Glacial in the Maas valley, The Netherlands. *Geologie en Mijnbouw* 72: 193–210.

Böse, M. (1991). A palaeoclimatic interpretation of frost-wedge casts and aeolian sand deposits in the lowlands between Rhine and Vistula in the Upper Pleniglacial and Late Glacial. In: S. Kozarski (ed.) Late Vistulian (=Weichselian) and Holocene aeolian phenomena in Central and Northern Europe. *Zeitschrift für Geomorphologie*, Suppl. 90: 15–28.

Bressolier, C., Froidefond, J. M., and Thomas, Y. F. (1990), Chronology of coastal dunes in the south-west of France. In: Th. W. Bakker, P. D. Jungerius, and J. A. Klijn (eds.) Dunes of the European coasts, Geomorphology—hydrology—soils. *Catena*, Suppl. 18: 101–7.

Bussemer, S., Gärtner, P., and Schlaak, N. (1998), Stratigraphie, Stoffbestand und Reliefwirksamkeit der Flugsande im brandenburgischen Jungmoränenland. *Petermanns Geographische Mitteilungen* 142: 115–25.

Castel, I. I. Y. (1988), A simulation model of wind erosion and sedimentation as a basis for management of a drift sand area in The Netherlands. *Earth Surface Processes and Landforms* 13: 501–9.

—— (1991), *Late Holocene aeolian drift sands in Drenthe (The Netherlands)*. Thesis, Department of Physical Geography, University of Utrecht.

—— Koster, E. A., and Slotboom, R. T. (1989), Morphogenetic aspects and age of Late Holocene eolian drift sands in North-west Europe. *Zeitschrift für Geomorphologie* 33: 1–26.

Clarke, M. L., Rendell, H. M., Pye, K., Tastet, J.-P., Pontee, N. I., and Massé, L. (1999), Evidence for the timing of dune development on the Aquitaine coast, south-west France. *Zeitschrift für Geomorphologie*, Suppl. 116: 147–63.

—— Rendell, H., Tastet, J.-P., Clavé, B., and Massé, L. (2002), Late-Holocene sand invasion and North Atlantic storminess along the Aquitaine Coast, southwest France. *The Holocene* 12: 231–8.

Clemmensen, L. B., Andreasen, F., Nielsen, S. T., and Sten, E. (1996), The late Holocene coastal dunefield at Vejers, Denmark: characteristics, sand budget and depositional dynamics. *Geomorphology* 17: 79–98.

Conry, M. J. (1974), Plaggen soils. A review of man-made raised soils. *Soils and Fertilizers* 37: 319–26.

Cooke, R. U., Warren, A., and Goudie, A. S. (1993), *Desert geomorphology*. UCL Press, London.

Dalsgaard, K., and Odgaard, B. V. (2001), Dating sequences of buried horizons of podzols developed in wind-blown sand at Ulfborg, Western Jutland. *Quaternary International* 78: 53–60.

de Ploey, J. (1977), Some experimental data on slopewash and wind action with reference to Quaternary morphogenesis in Belgium. *Earth Surface Processes* 2: 101–15.

Dücker, A., and Maarleveld, G. C. (1957), Hoch- und spätglaziale äolische Sande in Nordwestdeutschland und in den Niederlanden. *Geologisches Jahrbuch (Hannover)* 83: 215–34.

Dijkmans, J. W. A. (1990), *Aspects of geomorphology and thermoluminescence dating of cold-climate eolian sands*. Thesis, Department of Physical Geography, University of Utrecht.

—— and Wintle, A. G. (1991), Methodological problems in thermoluminescence dating of Weichselian cover sand and late Holocene drift sand from the Lutterzand area, eastern Netherlands. *Geologie en Mijnbouw* 70: 21–33.

Frechen, M., Vanneste, K., Verbeeck, K., Paulissen, E., and Camelbeeck, T. (2001), The deposition history of the coversands along the Bree Fault Escarpment, north-eastern Belgium. *Netherlands Journal of Geosciences / Geologie en Mijnbouw* 80: 171–85.

Goudie, A. S., Livingstone, I., and Stokes, S. (1999), *Aeolian environments, sediments and landforms*. Wiley and Sons, Chichester.

Grün, R., and Wintle, A. G. (eds.) (2001), Proceedings of the Ninth International Conference on Luminescence and Electron Spin Resonance Dating LED 99. *Quaternary Science Reviews* 20.

Haesaerts, P., and Mestdagh, H. (2000), Pedosedimentary evolution of the last interglacial and early glacial sequence in the European loess belt from Belgium to central Russia. *Geologie en Mijnbouw / Netherlands Journal of Geosciences* 79: 313–24.

Hagedorn, H., Rösner, R., Kurz, J., and Busche, D. (1991), Loesses and aeolian sands in Franconia, F.R.G. In: S. Kozarski (ed.) Late Vistulian (=Weichselian) and Holocene aeolian phenomena in Central and Northern Europe. *Zeitschrift für Geomorphologie*, Suppl. 90: 61–76.

Heidinga, H. A. (1984), Indications of severe drought during the 10th century AD from an inland dune area in the Central Netherlands, *Geologie en Mijnbouw* 63: 241–8.

Hilgers, A., Murray, A. S., Schlaak, N., and Radtke, U. (2001), Comparison of quartz OSL protocols using Lateglacial and Holocene dune sands from Brandenburg, Germany. *Quaternary Science Reviews* 20: 731–6.

Isarin, R. F. B., Renssen, H., and Koster, E. A. (1997), Surface wind climate during the Younger Dryas in Europe as inferred from aeolian records and model simulations. *Palaeogeography, Palaeoclimatology, Palaeoecology* 134: 127–48.

Janotta, A., Radtke, U., Czwielung, K., and Heidger, M. (1997), Luminescence dating (IRSL/TL) of Lateglacial and Holocene dune sands and sandy loesses near Bonn, Gifhorn and Diepholz (Germany). *Quaternary Science Reviews (Quaternary Geochronology)* 16: 349–55.

Jelgersma, S., de Jong, J., Zagwijn, W. H., and van Regteren Altena, J. F. (1970), The coastal dunes of the western Netherlands; geology, vegetational history and archeology. *Mededelingen Rijks Geologische Dienst* 21: 93–167.

Jungerius, P. D., and Verheggen, A. J. T. (1981), The development of blowouts in 'De Blink', a coastal dune area near Noordwijkerhout, The Netherlands. *Earth Surface Processes and Landforms* 6: 375–96.

—— and van der Meulen, F. (1988), Erosion processes in a dune landscape along the Dutch coast. *Catena* 15: 217–28.

Juvigné, E. H., and Wintle, A. G. (1988), A new chronostratigraphy of the late Weichselian loess units in middle Europe based on thermoluminescence dating. *Eiszeitalter und Gegenwart* 38: 94–105.

Kasse, C. (1997), Cold-climate aeolian sand-sheet formation in North-western Europe (*c.*14–12.4 ka): a response to permafrost degradation and increased aridity. *Permafrost and Periglacial Processes* 8: 295–311.

—— (1999), Late Pleniglacial and Late Glacial aeolian phases in The Netherlands. In: W. Schirmer (ed.) *Dunes and fossil soils. GeoArchaeoRhein* 3: 61–82.

Klijn, J. A. (1990), The Younger Dunes in The Netherlands; chronology and causation. In: Th. W. Bakker, P. D. Jungerius, and J. A. Klijn (eds.) Dunes of the European coasts. Geomorphology—hydrology—soils. *Catena*, Suppl. 18: 89–100.

Kolstrup, E. (1991), Danish Weichselian and Holocene aeolian deposits and their environment: a preliminary account. In: S. Kozarski (ed.) Late Vistulian (=Weichselian) and Holocene aeolian phenomena in Central and Northern Europe. *Zeitschrift für Geomorphologie*, Suppl. 90: 89–97.

—— Grün, R., Mejdahl, V., Packman, S. C., and Wintle, A. G. (1990), Stratigraphy and thermoluminescence dating of Late Glacial coversand in Denmark. *Journal of Quaternary Science* 5: 207–24.

Koster, E. A. (1978), *De stuifzanden van de Veluwe: een fysisch-geografische studie. (The eolian drift sands of the Veluwe (Central Netherlands): a physical geographical study.)* Thesis, University of Amsterdam, Publicatie Fysisch Geografisch en Bodemkundig Laboratorium 27.

—— (1982), Terminology and lithostratigraphic division of (surficial) sandy eolian deposits in The Netherlands: an evaluation. *Geologie en Mijnbouw* 61: 121–9.

—— (1988), Ancient and modern cold-climate aeolian sand deposition: a review. *Journal of Quaternary Science* 3: 69–83.

—— (1995), Progress in cold-climate aeolian research. *Quaestiones Geographicae (Poznan)*, Special Issue 4: 155–63.

—— Castel, I. I. Y., and Nap, R. L. (1993), Genesis and sedimentary structures of late Holocene aeolian drift sands in north-west Europe. In: K. Pye (ed.) *The dynamics and environmental context of aeolian sedimentary systems*. Geological Society Special Publication (London) 72: 247–67.

Kottmeier, C., and Meyer, H. H. (1988), Ein einfaches analytisches Modell zur Darstellung der atmosphärischen Zirkulation in Europa im Weichsel-Hochglazial. *Erdkunde* 42: 261–73.

Kozarski, S. (ed.) (1991), Late Vistulian (=Weichselian) and Holocene aeolian phenomena in Central and Northern Europe. *Zeitschrift für Geomorphologie*, Suppl. 90.

—— and Nowaczyk, B. (1991), Lithofacies variation and chronostratigraphy of Late Vistulian and Holocene aeolian phenomena in north-western Poland. In: S. Kozarski (ed.) Late Vistulian (=Weichselian) and Holocene aeolian phenomena in Central and Northern Europe. *Zeitschrift für Geomorphologie*, Suppl. 90: 107–22.

Lancaster, N. (1995), *Geomorphology of desert dunes*. Routledge, London.

Lautridou, J. P., Sommé, J., and Jamagne, M. (1984), Sedimentological, mineralogical and geochemical characteristics of the loesses of North-West France. In: M. Pécsi (ed.) *Lithology and stratigraphy of loess and paleosols*. INQUA Commission on Loess and Paleosols, Geographical Research Institute Hungarian Academy of Sciences, 121–32.

Livingstone, I. and Warren, A. (1996), *Aeolian geomorphology. An introduction*. Addison Wesley Longman, Harlow.

McKenna-Neuman, C. M. (1993), A review of aeolian transport processes in cold environments. *Progress in Physical Geography* 17: 137–55.

Mangerud, J., Svendsen, J. I., and Astakhov, V. I. (1999), Age and extent of the Barents and Kara ice sheets in Northern Russia. *Boreas* 28: 46–80.

Manikowska, B. (1995), Aeolian activity differentiation in the area of Poland during the period 20–8 ka BP. *Biuletyn Peryglacjalny* 34: 125–65.

Meyer, H. H. (1981), Zur klimastratigraphischen und morphogenetischen Auswertbarkeit von Flugdecksandprofilen im Norddeutschen Altmoränengebiet—erläutert an Beispielen aus der Kellenberg-Endmoräne (Landkreis Diepholz). *Bochumer Geographische Arbeiten (Paderborn)* 40: 21–30.

—— and Kottmeier, C. (1989), Die atmosphärische Zirkulation in Europa im Hochglazial der Weichsel-Eiszeit—abgeleitet von Paläowind-Indikatoren und Modellsimulationen. *Eiszeitalter und Gegenwart* 39: 10–18.

Mücher, H. J. (1986), Aspects of loess and loess-derived slope deposits: an experimental and micromorphological approach. *Netherlands Geographical Studies* 23.

Murray, A. S. and Clemmensen, L. B. (2001), Luminescence dating of Holocene aeolian sand movement, Thy, Denmark. *Quaternary Science Reviews* 20: 751–4.

Mycielska-Dowgiallo, E. (1993), Estimates of late glacial and Holocene activity in Belgium, Poland and Sweden. *Boreas* 22: 165–70.

Niessen, A. C. H. M., Koster, E. A., and Galloway, J. P. (1984), Periglacial sand dunes and eolian sand sheets. An annotated bibliography. *U.S. Geological Survey Open-File Report*. US Geological Survey, Menlo Park, Calif., 84–167.

Nordstrom, K., Psuty, N., and Carter, B. (eds.) (1990), *Coastal dunes. Form and process*. Chapman and Hall, London.

Nowaczyk, B. (1976), Eolian cover sands in central-west Poland. *Quaestiones Geographicae (Poznan)* 3: 57–77.

Pape, J. C. (1970), Plaggen soils in The Netherlands. *Geoderma* 4: 229–55.

Poser, H. (1950), Zur Rekonstruktion der spätglazialen Luftdruckverhältnisse in Mittel- und Westeuropa auf Grund der vorzeitlichen Binnendünen. *Erdkunde* 4: 81–8.

Pye, K. (1990), Physical and human influences on coastal dune development between the Ribble and Mersey estuaries, north-west England. In: K. F. Nordstrom, N. Psuty, and R. W. G. Carter (eds.) *Coastal dunes. Form and process*. Chapman & Hall, London, 339–59.

—— (ed.) (1993), *The dynamics and environmental context of aeolian sedimentary systems*. Geological Society Special Publication (London) 72.

—— (1995), The nature, origin and accumulation of loess. *Quaternary Science Reviews* 14: 653–67.

—— and Lancaster, N. (eds.) (1993), *Aeolian sediments. Ancient and modern*. International Association of Sedimentologists, Special Publication 16.

—— and Tsoar, H. (1990), Aeolian sand and sand dunes. Unwin Hyman, London.

Pyritz, E. (1972), Binnendünen und Flugsandebenen im Niedersächsischen Tiefland. *Göttinger Geographische Abhandlungen* 61.

Radtke, U. (ed.) (1998), Lumineszenzdatierung äolischer Sedimente. Beiträge zur Genese und Altersstellung jungquartärer Dünen und Lösse in Deutschland. *Kölner Geographische Arbeiten* 70.

—— Janotta, A., Hilgers, A., and Murray, A. S. (2001), The potential of OSL and TL for dating Lateglacial and Holocene dune sands tested with independent age control of the Laacher See tephra (12 880 a) at the Section 'Mainz-Gonsenheim'. *Quaternary Science Reviews* 20: 719–24.

Ruegg, G. H. J. (1983), Periglacial eolian evenly laminated sandy deposits in the Late Pleistocene of NW Europe, a facies unrecorded in modern sedimentological handbooks. In: M. E. Brookfield and T. S. Ahlbrandt (eds.) *Eolian sediments and processes*. Developments in Sedimentology 38: 455–82.

Schwan, J. (1986), The origin of horizontal alternating bedding in Weichselian aeolian sands in north-western Europe. *Sedimentary Geology* 49: 73–108.

—— (1988), The structure and genesis of Weichselian to Early Holocene aeolian sand sheets in western Europe. *Sedimentary Geology* 55: 197–232.

Seraphim, E. T. (1986), Spätglazial und Dünenforschung. Eine kritische Erörterung des spätglazialen Luftdruck-Wind-Systems H. Posers. *Westfälische Geographische Studien (Münster)* 42: 119–36.

Siebertz, H. (1988), Die Beziehung der äolischen Decksedimente in Nordwestdeutschland zur nördlichen Lössgrenze. *Eiszeitalter und Gegenwart* 3: 106–14.

Singhvi, A. K., Bluszcz, A., Bateman, M. D., and Someshwar Rao, M. (2001), Luminescence dating of loess-palaeosol sequences and coversands: methodological aspects and palaeoclimatic implications. *Earth-Science Reviews* 54: 193–211.

Tchakerian, V. (ed.) (1995), *Desert aeolian processes*. Chapman & Hall, London.

Vandenberghe, J. (1991), Changing conditions of aeolian sand deposition during the last deglaciation period. In: S. Kozarski (ed.) *Late Vistulian (=Weichselian) and Holocene aeolian phenomena in Central and Northern Europe*. Zeitschrift für Geomorphologie Suppl. 90: 193–207.

—— Mücher, H. J., Roebroeks, W., and Gemke, D. (1985), Lithostratigraphy and palaeoenvironment of the Pleistocene deposits at Maastricht-Belvédère, southern Limburg, The Netherlands. In: T. van Kolfschoten and W. Roebroeks (eds.) Maastricht- Belvédère:

stratigraphy, palaeoenvironment and archaeology of the Middle and Late Pleistocene deposits. *Mededelingen Rijks Geologische Dienst* 39/1.

Vandenberghe, J., Huijzer, B. S., Mücher, H., and Laan, W. (1998), Short climatic oscillations in a western European loess sequence (Kesselt, Belgium). *Journal of Quaternary Science* 13: 471–85.

van den Haute, P., Vancraeynest, L., and de Corte, F. (1998), The Late Pleistocene loess deposits and palaeosols of eastern Belgium: new TL age determinations. *Journal of Quaternary Science* 13: 487–97.

van der Meulen, F., and van der Maarel, E. (1993), Dry coastal ecosystems of the central and south-western Netherlands. In: E. van der Maarel (ed.) *Dry coastal ecosystems, Polar regions and Europe*. Ecosystems of the world 2A: 271–306.

—— Jungerius, P. D., and Visser, J. (eds.) (1989), *Perspectives in coastal dune management*. SPB Academic Publishing, The Hague.

van Geel, B., Coope, G. R., and van der Hammen, T. (1989), Palaeoecology and stratigraphy of the Lateglacial type section at Usselo (The Netherlands). *Review of Palaeobotany and Palynology* 60: 25–130.

van Huissteden, K. J., Schwan, J. C. G., and Bateman, M. D. (2001), Environmental conditions and paleowind directions at the end of the Weichselian Late Pleniglacial recorded in aeolian sediments and geomorphology (Twente, Eastern Netherlands). *Geologie en Mijnbouw / Netherlands Journal of Geosciences* 80: 1–18.

van Mourik, J. M. (ed.) (1988), Landschap in beweging: Ontwikkeling en bewoning van een stuifzandlandschap in de Kempen. *Netherlands Geographical Studies* 74.

Wintle, A. G. (1993), Luminescence dating of aeolian sands: an overview. In: K. Pye (ed.) *The dynamics and environmental context of aeolian sedimentary systems*. Geological Society Special Publication 72: 49–58.

—— Shackleton, N. J., and Lautridou, J. P. (1984), Thermoluminescence dating of periods of loess deposition and soil formation in Normandy. *Nature* 310: 491–3.

Zeeberg, J. J. (1998), The European sand belt in eastern Europe and comparison of Late Glacial dune orientation with GCM simulation results. *Boreas* 27: 127–39.

9 Peatlands, Past and Present

Eduard Koster and Tim Favier

Introduction

Peatlands are fascinating wetland ecosystems. They provide a habitat for a wide range of highly adapted plant and animal species. In addition to the floristic and ornithological richness, peatlands have been recognized for many other values. For instance, drained peatland soils often have good agricultural properties, and peat has been and still is in some places extensively used as fuel. In coastal wetlands peat has even been used for salt extraction. Furthermore, peat is an interesting material for science, as it contains information on the palaeoecological environment, climate change, carbon history, and archaeology.

In north-western Europe, peatlands were once quite extensive, covering tens of thousands of square kilometres. However, most of them have been strongly exploited by humans during past centuries. Many peatlands have been cultivated for agriculture and forestry, or have been exploited by commercial or domestic peat extraction for fuel. As a result, only a very small part of north-western Europe's peatlands remains today in a more or less natural state.

This chapter focuses on the peat deposits and peatlands in north-western Europe that have formed since the Late Glacial (*c.*13 ka BP). First, the most common concepts in peatland terminology are explained, and the distribution of peatlands is described. Next, processes of peat formation and the relationship between peat-forming processes and climate, hydrology, vegetation, and other factors are discussed. In the following section, frequently used classification methods are presented. A historical overview of the cultivation and exploitation of peatlands is given and the present land use and characteristics of peatland soils are discussed. The following section deals with methods of conservation and rehabilitation of the remaining mires. The importance of peatlands as palaeoecological archives is examplified. Finally, the role of peatlands as a source and/or sink of CO_2 and the relations with climate change are briefly explained.

Peatland Terminology

Peat is the unconsolidated material that predominantly consists of slightly decomposed or undecomposed organic material in which the original cellular and tissue structures can often be identified. Peat forms in lakes and mires under waterlogged, anaerobic conditions. There are three general categories of accumulating peat and organic deposits: *limnic peat*, *telmatic peat*, and *terrestrial peat*. Limnic peat is in fact an organic deposit that forms by accumulation of plant material on the bottom of lakes. The plant material may originate from floating vegetation or algae in the lake, but limnic peat may also form by accumulation of organic material transported into the lake from elsewhere. Telmatic peat and terrestrial peat on the other hand are sedentary deposits. In other words, they form by accumulation of organic remains from plants that grew at that site, without being transported. Both these two types originate in mires. Telmatic peat forms in the swamp zone, with vegetation partly submerged, partly above the water; terrestrial peat forms at or above the high water level.

Peatland is a general term referring to all kinds of drained or undrained areas with a minimal thickness of peat of at least several decimetres. In Europe the English term *mire* is used for undrained virgin peatlands with living peat-forming vegetation. The word 'mire' originates from the Swedish *myr* (Moore 1984), and is used synonymously with the English-German word *moor*. The term

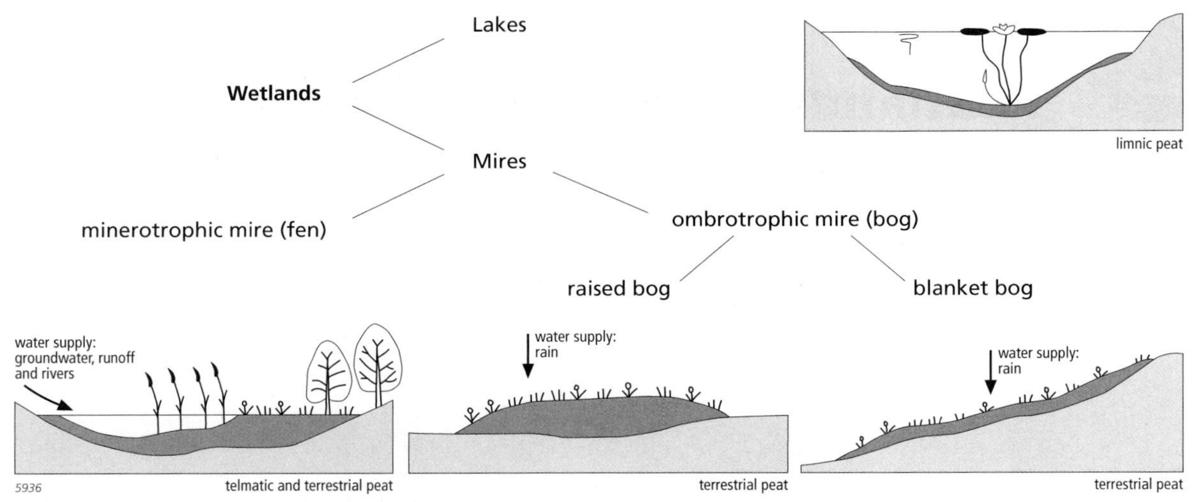

Fig. 9.1. Subdivision of wetlands and important features of natural fens and bogs.

TABLE 9.1. *Mire terminology, with equivalents in German, French, Dutch, and Danish*

English	mire	fen	bog
German	*Moor*	*Niedermoor*	*Hochmoor*
French	*tourbière*	*bas-marais*	*tourbière haute*
Dutch	*veen*	*laagveen*	*hoogveen*
Danish	*mose*	*lav mose*	*højmose*

'wetland' is used in a broader sense than the term peatland. It refers to all kinds of wet soils or shallow waters, from freshwater lacustrine to salt marine environments. Wetlands include peatlands, but also areas where peat is not formed or preserved, as for instance salt marshes, rice paddies, and mangrove swamps. A detailed review and discussion of wetland and peatland terminology is presented in Moore and Bellamy (1974) and Gore (1978). French equivalents for the terminology used in this chapter are found in Manneville (1999: 301). The terminology is complex, and terms used in Europe are often different from those used in North America (Mitsch and Gosselink 1986; Charman 2002).

Mires can be subdivided on the basis of origin and chemistry of their water supply into minerotrophic mires, called *fens*, and ombrotrophic mires, called *bogs* (Fig. 9.1; Table 9.1). Here the terms 'minerotrophic' and 'ombrotrophic' refer to the source of the water of the mire. Minerotrophic mires (fens) are supplied by groundwater, runoff, or rivers. The water varies widely in composition, and water conditions in the fen may be *oligotrophic* (nutrient poor) or, more often, *eutrophic* (nutrient rich) or *mesotrophic* (intermediate). Conversely, ombrotrophic mires (bogs) are fed exclusively by precipitation, which is always mineral poor. Most bogs in north-western Europe are *raised bogs*, characterized by a convex dome of bog-peat that rises up to several metres above the original relatively flat surface and, more important, above the original water table. Another type is the *blanket bog*, which is not confined to depressions or flat surfaces. Blanket bogs follow the relief of the landscape, covering minor slope and hill irregularities. The thickness of the peat layer in blanket bogs is usually less than 1 m. Two other types of wetland ecosystems are *swamps* and *marshes* (Charman 2002). Both types are highly productive freshwater ecosystems, standing under shallow water at least part of the year. In contrast to the term 'mire', 'swamp' and 'marsh' are used for wetland ecosystems developing on mineral soils, with no net peat accumulation. Swamps are wooded while marshes have an open, grassy vegetation.

Distribution of Peatlands

Peatlands occur in all but the drier regions of the world, from the tropics to the tundra (Gore 1978). Estimates of global peatland resources vary considerably and none is very accurate. Values between 4 million and 5 million km^2 are suggested (e.g. Kivinen *et al.* 1979; Sjörs 1980; Matthews and Fung 1987; Bather and Miller 1991; Charman 2002), which is 2.8–3.5% of the Earth's land surface. The larger part is found in the boreal zones of Russia, Scandinavia, and North America, but peatlands are also quite widespread in the temperate zone. For north-western Europe, published estimates of the total surface of peatlands highly deviate from each other. According to Succow and Joosten (2001), peatlands cover approximately 41,200 km^2, more than 4% of the surface of north-western Europe. They are especially widespread in Denmark, Northern Germany, and The Netherlands. However, since medieval times, cultivation and exploitation of peatlands has strongly reduced the surface area of living mires. Today, a mere 1% of the former extent of the peatlands can be classified as undisturbed or (near-)natural, that still have living mire vegetation and active peat accumulation. Peatlands are especially extensive in Great Britain and Ireland, but these are not dealt with in this chapter as the focus is on the mainland of north-western Europe.

In general, the distribution of peatlands is mainly determined by climate, with additional local influences of topography and geology. The relation between climate and the geographical distribution of different mire types in Europe has been described by Moore and Bellamy (1974) and Succow and Joosten (2001). As raised and blanket bogs are fed by rainwater, their distribution is more strongly constrained by climate than is the case with fens (Wheeler and Shaw 1995). Bogs are restricted to landscapes having a positive water balance, where total precipitation exceeds evaporation and runoff. Raised bogs predominantly occur in the high rainfall coastal areas of The Netherlands, north-west Germany, and Denmark and in some humid low mountain ranges (Barkman 1992; Succow and Joosten 2001). As one moves south, evaporation increases and bogs become rare. For example, in central and southern France, raised bogs are found only at isolated sites in the low mountain ranges and in the Pyrenees. The distribution of blanket bogs depends even more on rainfall. They are restricted to very humid, cool-oceanic climates, and are exclusively found in areas with an annual precipitation exceeding 1,250 mm (Schouten 1984; Schouten and Nooren 1990). Blanket bogs are very common in the British Isles and Norway, but on the continent in north-western Europe they are quite rare. According to Barkman (1992), they are to be found in the Vosges (France) and Harz mountains (Germany).

Distribution of Peatlands per Country

Peatlands in Denmark are small and widely scattered, with an areal extent of *c.*1,200 km^2 (Charman 2002). Their distribution is strongly related to the geology. For instance, in Fyn, Sjaelland, and the north and east of Jylland (Fig. 9.2) most of the peatlands are of the eutrophic fen type. They are confined to valley systems and depressions in the glacial landscape—especially on tills—and in the Holocene marine sand and clay areas (Andersen *et al.* 1996). In the south-west of Jylland though, raised bogs and oligotrophic fens are dominant (Andersen *et al.* 1996). In this area, mineral-poor Weichselian fluvioglacial and aeolian sands are at the surface. Denmark's largest, remaining peatlands are the raised bogs of Store and Lille Vildmose in northern Jylland, which cover several hundreds of square kilometres.

Germany's peatlands cover 14,000–16,000 km^2, approximately 4% of the country, according to Succow and Joosten (2001) and Charman (2002). Of this area, 3% are fens and 1% raised bogs. Fens occur predominantly in Mecklenburg-Vorpommern, where they make up over 12% of the land surface, and in Schleswig-Holstein, Niedersachsen, and Brandenburg. Raised bogs occur mainly in the high rainfall coastal areas of Niedersachsen, covering more than 5% of the land surface. The surficial extent of the originally more than 6,500 km^2 extensive peat bogs in Niedersachsen has been significantly reduced in size by human activities since medieval times. These bogs in Niedersachsen have been systematically mapped by Schneekloth *et al.* (1970/83). Here Germany's largest raised bogs can be found: the Teufelsmoor (near Bremen, 800 km^2) and the Bourtanger Moor (at the Dutch–German border, 1,000 km^2) (Fig. 9.3). Raised bogs are also found in some upland areas, such as the Harz, Schwarzwald, and Oberbayern. As in other north-west European countries, the present mire area in Germany is much smaller than it used to be. Today about 7% of the original mire area is covered by mire vegetation, but only a few square kilometres have active peat accumulation (Succow and Joosten 2001).

Relative to its surficial extent peatlands are most abundant in The Netherlands, where they cover about 4,500 km^2, which is approximately 11% of the surface. This area was much larger in Roman times. It has been estimated that mires covered about 35% of The Netherlands around 2000 BP. However, much of the Dutch peatlands became eroded by sea incursions or covered by younger marine and fluvial deposits. Widespread peat deposits are still present in the coastal plain (in the west of The Netherlands) and the Rhine–Meuse fluvial plain (in the centre). The peat deposits in the coastal plain are either eutrophic fens or drowned raised bogs. Raised bogs also occur in the stream valleys in the north of The Netherlands (Fig. 9.3). Nowadays, almost all peatlands in The Netherlands have been cultivated or otherwise exploited, and today only about 80 km^2 of raised bogs with ombotrophic vegetation remain (Westhoff 1990). However, even this area mostly

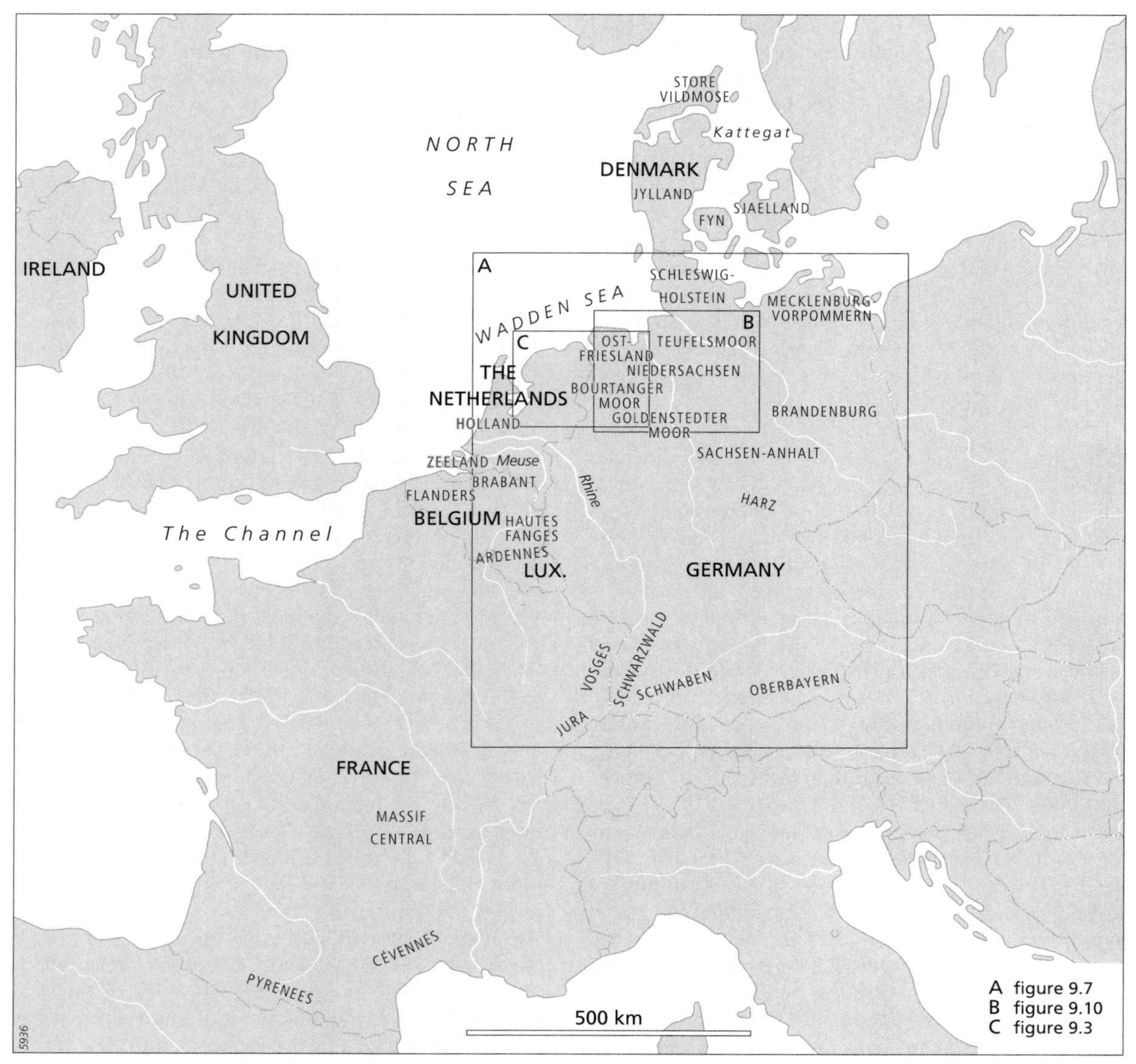

Fig. 9.2. Location map.

concerns degraded, partially cut, or dead bog with no net peat accumulation, except for a few very small natural remnants (Barkman 1992).

Belgium's peatlands cover about 1,000 km^2, which is about 3% of the country (Succow and Joosten 2001). They are found in the coastal plain, in the stream valleys, and in the Ardennes. The coastal plain peatlands can be seen as the southern extension of the large mires that once covered almost the entire Coastal Plain of the Low Countries, which stretches from West Flanders, via the west and north of The Netherlands to north-eastern Germany. Initially eutrophic fens developed, later raised bogs. Peat accumulation occurred between approximately 7500 and 2000 BP, after which the sea invaded the area and the mires became covered by marine clays (Verbruggen *et al.* 1996). Most of the peatlands in the Ardennes are raised bogs. They are confined to small, widely scattered patches, especially near the German border. Most of the raised bogs in the Ardennes are cultivated, though parts of the Hautes Fagnes, Belgium's

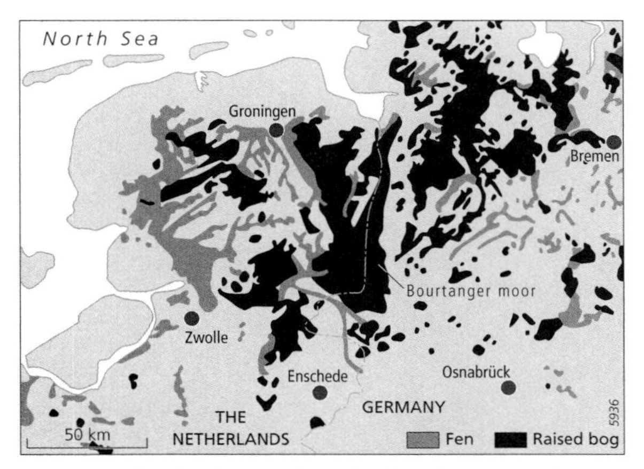

Fig. 9.3. The distribution of peatlands and mire complexes in the north of The Netherlands, East Friesland, and part of Niedersachsen (Germany). The Bourtanger Moor, at the Dutch–German border, is western Europe's largest raised bog. In its natural state (until the seventeenth century), it had an estimated extent of 1,000 km^2 (Barkman 1992). The location of the area is indicated in Fig. 9.2.

largest peatland, remain in a more or less natural state.

Peatlands in France cover 1,000 to 1,200 km^2, which is about 0.2% of the country (Succow and Joosten 2001; Charman 2002). Although the peatlands are numerous, most of them are small, isolated and strongly exploited. Raised bogs are especially found in the higher regions of the Massif Central, Vosges, Cévennes, Jura, and in the Pyrenees mountains. In the south of France, mires are very rare. Here they are restricted to river valleys (fens) and the mountain ranges (bogs). Manneville (1999) gives detailed descriptions, including areal extent and floral and faunal characteristics, for a very large number of peatland sites in France, Belgium, and Luxembourg.

Processes of Peat Formation

In mires, dead leaves, branches, and stems fall into the water and accumulate on the bottom and roots may grow down into the substrate and die in situ. The water at the water table is oxygenated, resulting in aerobic decay. This reduces the oxygen content. As oxygen can be replenished only by diffusion from above, and as this is a very slow process, continuing decay therefore causes conditions below the water table to become anaerobic. Further decay is only by anaerobic processes, which are much slower than aerobic ones. Consequently, the organic matter does not (or will only partially) decay but instead accumulates, forming peat.

Mire Flora and Related Peat Types

The type of peat that is formed basically depends on the type of vegetation in the lake or mire, which in turn is closely related to the chemical composition of the mire water and water depth. In lakes deeper than 2 metres, *hydrophytes* (floating and submerged aquatic plants) are the only higher plants, as the water is too deep for *helophytes* (plants that root in the soil and grow above the water). In this type of environment organic deposits are formed. Typical organic deposits are *diatom earth*, *marl*, *gyttja*, and *dy*. Diatom earth is a white siliceous mud formed from diatom frustules. Marl is a calcium carbonate precipitation from algae and Charophytes. Gyttja is a calcareous organic deposit formed in eutrophic lakes with abundant floating vegetation, like water lilies (*Nymphaea alba*). Dy is a muddy deposit formed from colloidal suspensions. It forms in oligotrophic lakes with low biological production. The accumulation of organic deposits is a very slow process, especially in oligotrophic lakes. The formation of one metre of organic deposits may take thousands of years. In cases where there is a supply of clastic material to the lake, layers of organic sediment may alternate with layers of clay or silicate-mud.

In some lakes, especially in oligotrophic and mesotrophic lakes, a *quagmire* may develop along the sheltered shores. In this case floating vegetation overgrows the lake and forms a firm mat. The mat is usually several decimetres thick, and overlies a water body of a few metres. Below the mat, deposition of organic matter continues. In oligotrophic quagmires organic deposits called *detritus mud* and *dy* form below the mat (Fig. 9.4). In eutrophic lakes a calcareous and nutrient-rich organic mud, called *organosapropel*, forms. The sedimentation in eutrophic quagmires can be very fast. Accumulation rates of more than 5 cm per year, prior to compaction, have been mentioned for lakes in Germany (Succow and Joosten 2001).

If the lake is less than 2 m deep, helophytes are able to colonize the mire. Reed (*Phragmites australis*) is a typical species in eutrophic and mesotrophic lakes with a water depth ranging between 2 and 0.5 m. It grows in fresh waters as well as in slightly brackish waters. Reed often grows on clayey substrate and, therefore, reed peat is often found at the transition from clayey deposits to other types of peat. This peat type is especially abundant in coastal wetlands and floodplains of lowland rivers. Other typical mire plants are sedges (*Carex* species), which are found in eutrophic as well as oligotrophic water environments (Crum 1988). Most species of sedge grow at a water depth of 0.5 to 0 m. As soon as

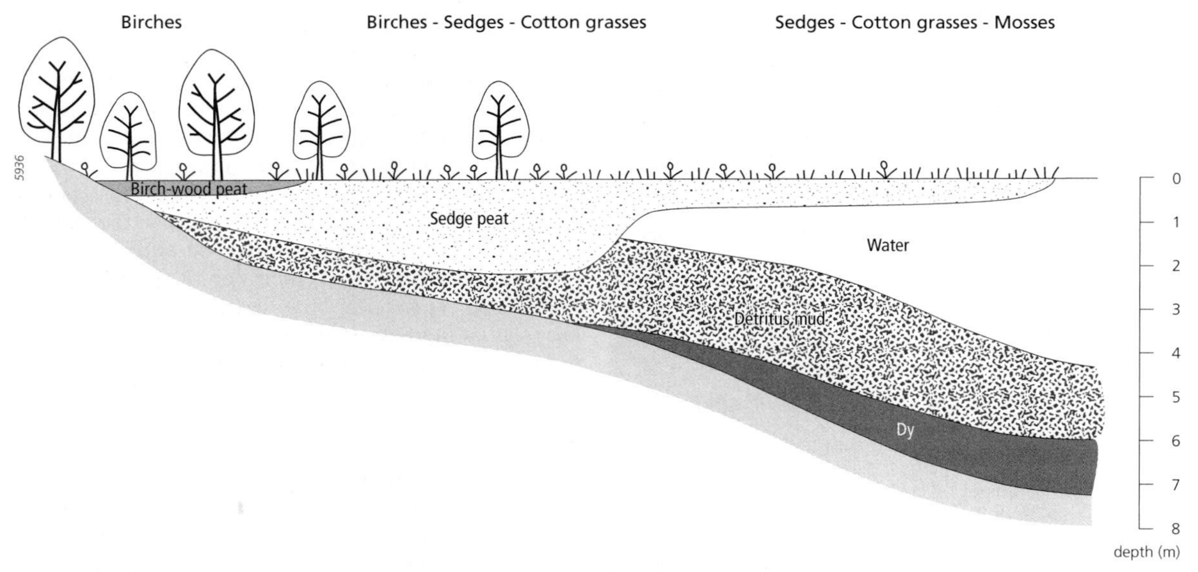

Fig. 9.4. Schematic cross-section over a quagmire in an oligotrophic lake (modified after Succow and Jeschke 1986).

the peat rises above the mean water level and the surface of the mire dries out during summer, trees will be able to colonize the mire. Initially willows (*Salix* spec.) will start to grow, and finally alders (*Alnus glutinosa*) (Heathwaite and Göttlich 1993). In oligotrophic mires, birches (*Betula* spec.) are more common. Trees grow preferably on the higher parts, the *hummocks*. The hollows between the tree hummocks are often flooded in winter and spring and therefore maintain a sedge and reed vegetation. Oligothropic fens generally have a low biodiversity. They are characterized by fewer than thirty species, while mesotrophic and eutrophic mires may contain around one hundred species. For a detailed description of the mire flora in western Europe, see Manneville (1999) and Succow and Joosten (2001).

In raised bogs, *Sphagnum* mosses are the ubiquitous species. They are referred to as the 'key species' since their dominance is essential for bog formation. One feature of *Sphagnum* mosses is of particular importance: they are able to acidify the water around them by a process of cation exchange. This makes the environment unsuitable for most other species (Clymo 1991). Therefore, colonization of a mire system by *Sphagnum* often marks a profound ecological change. As a result of limited differences in habitat, low pH and low nutrient concentration of the water, the vegetation cover in raised bogs usually consists of fewer plant species than that of fens (Crum 1988). Besides *Sphagnum*, shrubs of the heath family (Ericaceae) and cottongrasses (*Eriophorum* species) may be present (Clymo 1991) but in general, biodiversity is low.

Ingram (1978) distinguished two functional horizons in raised bogs: the *acrotelm* and the *catotelm*. The acrotelm is the active, oxygenated surface layer of approximately 30 cm, which is relatively permeable. The catotelm refers to the subsurface, fossilized layer of low hydraulic conductivity. Water table fluctuations are restricted to the acrotelm; the catotelm is permanently saturated. Furthermore, the acrotelm has a capacity for vertical movement, swelling in wet conditions, shrinkage in dry conditions. Rain quickly sinks through the acrotelm, because hydraulic conductivity is high, but stagnates on the catotelm, which has a much lower conductivity (Ingram 1982). The immediate agents of aerobic decay—fungal mycelia and aerobic bacteria—are present in the acrotelm, but living fungi and bacteria become rare as one moves down into the anoxic catotelm (Clymo 1991). Peat accumulation takes place in the top of the catotelm. The stems of *Sphagnum* grow up to a length of one or two metres beneath the water table, and when the lower parts die off, they largely retain their structure and position. Once bog formation has started the influence of groundwater will gradually decline. Jensen (1961) showed that the influence of groundwater in a raised bog in the Harz (Germany) terminated when 100 to 150 cm of peat had accumulated. The *Sphagnum* vegetation then isolates itself from the underlying substrate, and the mire becomes a self-organizing system with a hydrology of its own.

Paludification and Terrestrialization

Two classical concepts in peatland formation are *paludification* and *terrestrialization*. The term 'paludification' literally means swamping (German: *Versumpfung*). It is used for the formation of a mire over previously forested land, grassland, or even bare rock, due to a natural rise in the groundwater table and subsequent drowning of the surface. The paludification of surfaces can often be attributed to a change to wetter or colder (less evaporation) climatic conditions (Crum 1988).

According to Casparie and Streefkerk (1992), a period of widespread mire initiation occurred in northwest Europe in the Early Atlantic. Around 7000 BP, many areas became paludificated and raised bogs started to form as a result of increased precipitation. Before the Atlantic (in the Preboreal and Boreal), a relatively dry continental climate suppressed the formation of raised bogs. Paludification can also take place as a result of changes in the vegetation cover. For instance, deforestation causes a decrease in evaporation, and therefore leads to a rise in the groundwater table. Damming of a river (local base level rise) may cause paludification of floodplains in hilly areas. A rise in sea level (widespread base level rise) has often been cited as a key factor initiating paludification in coastal plains (Heathwaite and Göttlich 1993). Paludification took place on a large scale in the Coastal Plain of the Low Countries during the Early Holocene. As a result of the post-glacial sea level rise, the groundwater table rose and the Pleistocene sandy deposits became paludificated. Consequently, a thin layer of peat formed, usually directly on top of a podzolic soil. This peat layer, called *basal peat*, has been identified at many locations. Basal peats are widespread in the coastal regions from Belgium, through The Netherlands and Germany, to Denmark. They have also been identified in submarine position in the Wadden Sea, North Sea, The English Channel, and even in the Kattegat (Denmark). At most locations the peat layer is located between 20 and 0 m below MSL, and is covered by younger marine or fluvial deposits. As the progressively higher position with decreasing age reflects the Holocene sea-level rise, basal peat layers have often been used for constructing curves of sea-level rise in great detail (van de Plassche 1982).

'Terrestrialization' (German: *Verlandung*) literally means the transition from water into (peat)land. Sooner or later, all lakes become filled in with allochthonous sediments or autochthonous material and change from open water to wetland. When a lake slowly becomes filled-in, it often shows a typical sequence of peat types, related to the succession of vegetation communities. This sequence is called *hydroseral succession*. A hydroseral succession for eutrophic lakes usually starts with diatom earth, followed by gyttja, reed peat, sedge peat, and alder wood peat (Fig. 9.5). Lakes filled in by eutrophic hydroseral successions are particularly abundant in the Weichselian moraine landscapes of northern Germany and Denmark (Succow and Joosten 2001). They also occur in residual channel lakes in lowland floodplains, which formed after meander cut-offs or avulsions. In oligotrophic lakes, the hydroseral succession

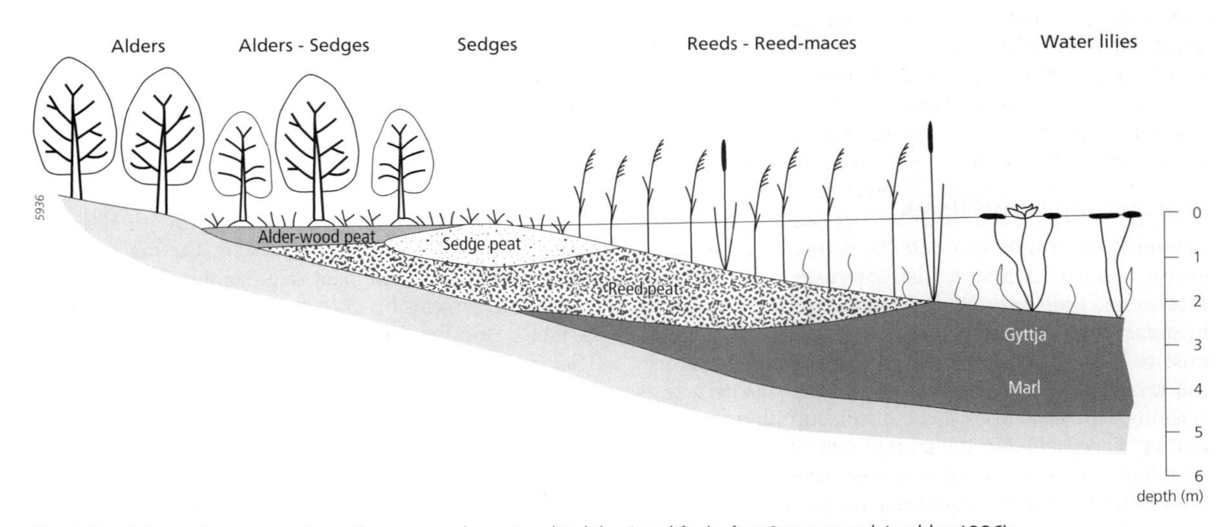

Fig. 9.5. Schematic cross-section of a mire and a eutrophic lake (modified after Succow and Jeschke 1986).

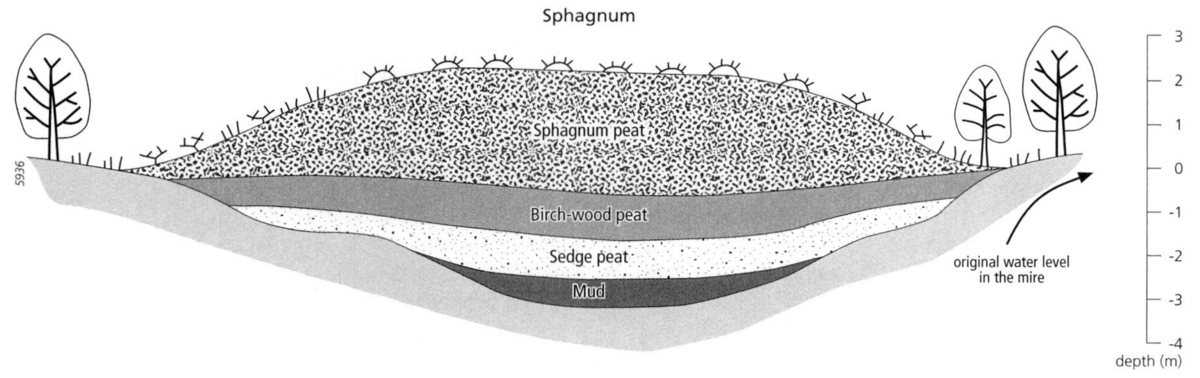

Fig. 9.6. Schematic cross-section of a raised bog developed on top of a terrestrialized oligotrophic lake infilling (modified after Succow and Jeschke 1986).

starts with dy, followed by sedge peat and birch wood peat. Lakes filled in by oligotrophic hydroseral succession are far less common than the eutrophic variant, and deposits like dy are rarely found in this part of the world. Furthermore, the terrestrialization in nutrient-rich lakes is much more rapid than in oligotrophic lakes. This is mainly because the primary production is much higher in nutrient-rich conditions. Lakes filled in by oligotrophic hydroseral succession can be found throughout the mineral-poor Weichselian outwash plains in northern Germany and Denmark (Succow and Joosten 2001).

In completely filled-in lakes, a raised bog may develop on top of fen peat. In this case, *Sphagnum* peat is the last step of the hydroseral succession. In time, the peat body may even grow over the margins of the former lake (Fig. 9.6). Raised bogs may also develop on top of eutrophic filled-in lakes. This happens because the mire environment often becomes nutrient-poor due to increased influence of rainwater at the end of the terrestrialization process. Then the vertical sequence shows eutrophic fen peat, followed by mesotrophic fen peat, and finally oligotrophic raised bog peat.

Peat Accumulation Rates

The peat accumulation rate is usually expressed in grams of dry matter per square metre per year. It can be determined indirectly by dating peat profiles or directly by measuring the accumulation rate of organic material in mires. With the latter method, one actually measures the gross accumulation rate or apparent growth rate. Although the decomposition of the accumulated organic matter is very small in the anoxic zone, it still affects the long-term (net) accumulation rate because peat accumulation often proceeds for hundreds or thousands of years. The long-term (net) peat accumulation is therefore smaller than the gross or apparent accumulation rate. Expressed in cm yr^{-1} peat accumulation rates strongly vary. Charman (2002) mentions values between 4 and 500 cm per 1,000 years, although the majority of the rates for raised mire sites in Europe fall between 20–100 cm per 1,000 years. Numerous studies have demonstrated that the long-term peat accumulation rate varies during the development of the mire (Malmer 1992). Rates generally decrease with age. The accumulation rate may be over 100 $g/m^2/yr$ for young mires of less than 1,000 years old, but is often less than 50 $g/m^2/yr$ for mires older than 5,000 years (Succow and Joosten 2001). Furthermore, the accumulation rate also depends on the mire type. Although primary productivity is higher in fens than in bogs, the decomposition processes are more effective due to the higher pH and the more easily decomposed plant litter present (Malmer 1992). As a result, the long-term (net) peat accumulation in fens is on average slightly smaller (Table 9.2).

Classification of Mires and Peat Deposits

Peat deposits may be classified on the basis of the degree of decomposition (Stanek and Worley 1983), the chemical properties (Clymo 1978), or the macrobotanical composition of the peat (e.g. Damman and French 1987). Mires may be classified on the basis of many variables, for instance mire morphology (Moore and Bellamy 1974), vegetation composition, or topography (see also Manneville 1999). Two of the most important and widely used mire classification systems are the *ecological classification* and the *geogenetic classification*, whilst the more recently developed *hydrogeomorphic classification* and the *hydrogenetic classification* are now gaining more importance.

TABLE 9.2. *Peat accumulation rates*

mire type	age (yr)	location	peat accumulation rate ($gm^{-2}a^{-1}$)
raised bog	500	Finland	165
raised bog	1,000	Finland	56–86
raised bog	4,000–10,000	Finland and Estland	26–82
raised bog	10,000–11,000	Finland	10–40
fen	100–200	Finland	83–99
fen	4,000–10,000	Finland and Estland	16–50
fen	10,000–10,500	Finland	20–40

Sources: After Tolonen and Turunen (1996); Succow and Joosten (2001).

Ecological Classification

The ecological classification divides mires on the basis of the nutrient content into eutrophic mires (mineral-rich), mesotrophic mires (intermediate), and oligotrophic mires (mineral-poor). The classification is not unambiguous, and different variables are used to distinguish the ecological mire types. For instance, classification can be done on the basis of the total nitrogen content in the water, with values ranging between 0.01 and 0.25 mg/l for oligotrophic mires, 0.25 and 0.60 mg/l for mesotrophic mires and 0.6 and 1.00 mg/l for eutrophic mires. Classification can also be done on the basis of the C/N rate of the peat. In oligotroph mires peat has a C/N value above 33/1, in mesotrophic mires between 33/1 and 20/1, and in eutrophic mire below 20/1 (Succow and Joosten 2001). The pH is strongly related to the trophic state. Oligotroph mires have a low pH and eutrophic mires a high pH. A useful approximation is that oligotrophic fens commonly have a pH of about 4–5, bogs about 3–4, and eutrophic fens of 6–7.5 (Crum 1988). Mesotrophic mires are often subdivided into mesotrophic-acid mires, mesotrophic-neutral mires, and mesotrophic-alkaline mires, with boundaries at pH 4.8 and pH 6.0.

Geogenetic Classification

The geogenetic classification distinguishes mires on the basis of the landforms in which they develop (Damman and French 1987). Four classes of mires are distinguished: *ombrogenous mires*, *topogenous mires*, *limnogenous mires*, and *soligenous mires*. Ombrogenous mires are mires that grow over surface depressions and are fed by rainwater only; topogenous mires form in topographic depressions that are fed by groundwater; limnogenous mires develop along lakes and slow-moving streams; soligenous mires form on slopes and depend on moisture seeping from the layers that form the slope.

Hydrogenetic Classification

Succow and Lange (1984) proposed a hydrogenetic classification, which was later adjusted by Succow and Joosten (2001), in order to classify mires in Germany and other central European countries (Fig. 9.7). This classification is based on the position in the landscape, the hydrological condition (water input), and the

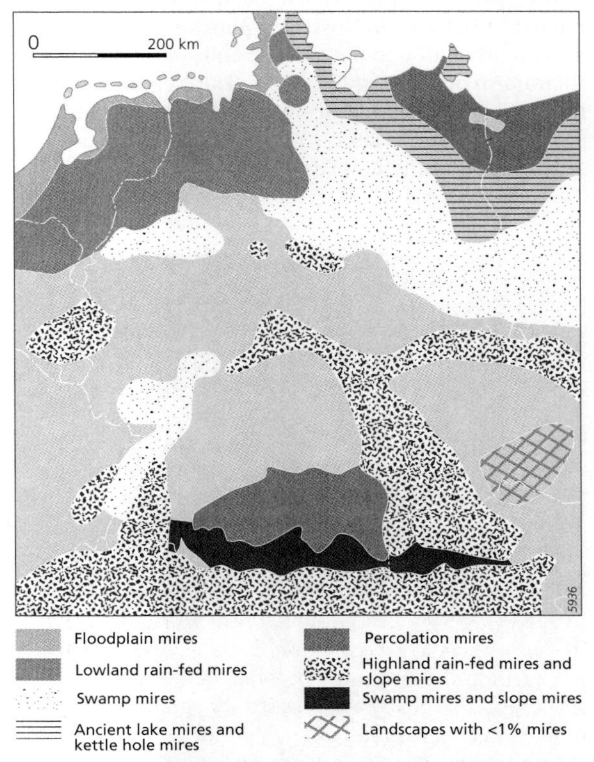

Fig. 9.7. The distribution of the hydrogenetic mire types in Germany and parts of adjacent countries. The location of the area is indicated in Fig. 9.2. (After Succow 1988.)

peat-forming processes related to it. This results in a subdivision of eight mire types: former lake mires (German: *Verlandungsmoore*), swamp mires (*Versumpfungsmoore*), floodplain mires (*Überflutungsmoore*), percolation mires (*Durchströmungsmoore*), slope mires (*Hangmoore*), spring mires (*Quellmoore*), kettle hole mires (*Kesselmoore*), and rain-fed mires (*Regenmoore*). Former lake mires are formed by terrestrialization; swamp mires by paludification. Floodplain mires are related to low-lying surfaces that are flooded by rivers, lakes, or sea. Percolation mires develop in the case of a strong and even water supply from the mineral soil. Rain-fed mires are in fact raised bogs while slope mires are the same as blanket bogs.

Hydrogeomorphic Classification

The hydrogeomorphic classification method distinguishes mires on the basis of their form and topography (Charman 2002). Examples are domed bogs, valley bogs, plateau raised bogs, palsa bogs, blanket bogs, flooded fens, aapa fens, etc. Moore and Bellamy (1974) present an overview of the distribution of the different mire types in Europe. Their occurrence is primarily related to climate. For instance, plateau raised bogs are particularly found in the oceanic and sub-oceanic zones (Denmark, north-west Germany, The Netherlands, and Belgium) and continental raised bogs are dominant in the subcontinental zone (the low mountain ranges of southern Germany and France).

Peatland Cultivation and Exploitation

The exploitation of north-western Europe's peatlands started in the Late Neolithic (since approximately 4000 BP) (Succow and Joosten 2001), but was most intense during the past millennium. Four categories of cultivation and human exploitation of peatlands are recognized: draining, reclaiming, and colonizing of mires; digging and dredging of peat for fuel; digging of peat for salt making; peat extraction for soil fertilizer.

Cultivation of Natural Mires: Draining, Reclaiming, and Colonizing

In order to use peatlands as arable land or grazing meadow, a dry and aerated surface layer is absolutely required. Therefore, drainage is usually the first step in the cultivation of mires. By constructing ditches, the water table in the mire is lowered and the top layer dries out. When mires are drained, the physical and chemical characteristics of the peat change. The water-filled pores are emptied and subsequently compressed and, as a result, the permeability decreases. Furthermore, the top layer becomes better aerated, which accelerates the decomposition processes. Humification processes transform the organic fragments into dark-coloured humic substances, whereby the original cellular structures and tissue structures are lost (Fig. 9.8). Mineralization processes finally convert the organic matter to

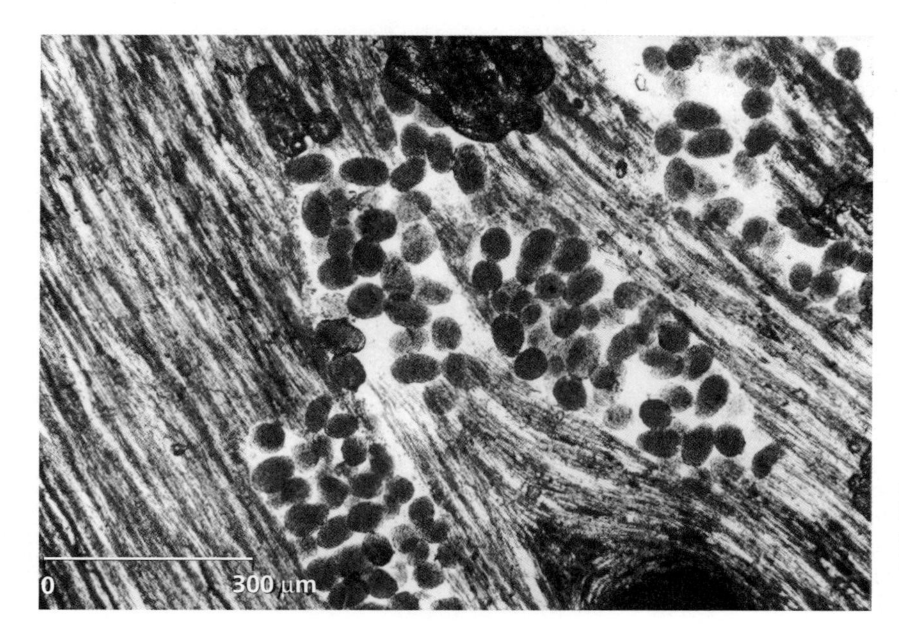

Fig. 9.8. Example of humification: detail of a large plant fragment with organic fecal pellets *c.*50 µm in diameter.

TABLE 9.3. *Chemical composition of* Sphagnum *peat types*

peat	Carbon (%)	Oxygen (%)	Hydrogen (%)	Nitrogen (%)	Sulphur (%)
undecomposed *Sphagnum* peat (H1–H3)	48–53	40–46	5.0–6.0	0.5–1.0	0.1–0.2
decomposed *Sphagnum* peat (H8–H10)	59–63	31–34	5.0–6.5	0.9–1.9	0.2–0.5

Sources: After Heathwaite and Göttlich (1993). The H-values are from the decomposition index by von Post and Granlund (1926).

TABLE 9.4. *C/N ratios for different peat types*

	C/N value
young *Sphagnum* peat	80–90
old *Sphagnum* peat	35–55
reed peat	15–25
sedge peat	25–30
birch-woodpeat	20–25
alder-woodpeat	15–20
decomposed *Sphagnum* peat	30–40
decomposed fen peat	10–20

TABLE 9.5. *Average nutrient contents (kg/ha) for the upper 20 cm of a fen and a raised bog*

	N	P_2O_5	K_2O	CaO
fen	15,000	1,250	500	20,000
raised bog	7,000	250	125	1,000

Source: After Heathwaite and Göttlich (1993).

simple inorganic compounds. As a result carbon dioxide, methane, water, and nutrients such as potassium, phosphorus, and nitrogen are released and become available for plants. The carbon content increases while the oxygen content decreases (Table 9.3). Furthermore, the C/N ratio becomes smaller (Table 9.4). Fen soils are particularly suitable for agricultural conversion, as they are relatively rich in nitrogen, phosphorous, and other nutrients (Table 9.5). Bog soils on the other hand are typically acidic and nutrient-poor (Wheeler and Shaw 1995). For this reason, fens were put into cultivation earlier than raised bogs (Heathwaite and Göttlich 1993). Cultivation of raised bogs often involved the application of lime and fertilizer, and sometimes also sand mixing.

During the drainage process, the solid volume increases from less than 3% (for peat in natural mires) to more than 12% (long and intensely drained peatlands) (Heathwaite and Göttlich 1993). As a result, peat shrinks. This is partly through compaction and partly through oxidation, as a result of better aeration of the soil (Borger 1992). Peat shrinkage is an irreversible process, and as a consequence of shrinkage, subsidence of the surface takes place. The rate of subsidence depends on factors as drainage depth, peat type and moisture content, but also on climate. Heathwaite and Göttlich (1993) suggest that peat loss in a humid-temperate climate can be as much as 2 to 3 cm per year. For example, in certain parts of the peat district in the western Netherlands, the rate of subsidence is calculated at *c*.1 m/century. In this area, some peatlands have been subjected to drainage since reclamation of the mires started in the eleventh and twelfth centuries. Originally, the top of the mires was a few metres above mean sea level, but prolonged draining caused a subsidence of the surface to a level of several metres below mean sea level. The embankment of rivers, creeks, and canals is now often a few metres above the surface of these peatlands. This required, and continues to require, careful maintenance of an artificial drainage system to keep the peatlands dry. Furthermore, the land subsidence also made the area prone to sea incursions. The severe floods and losses of land in The Netherlands since the Late Middle Ages are therefore at least partly due to the reclamation of mires by man (Borger 1992).

Historically, a number of different methods developed to drain, reclaim, and colonize peatlands. These depended on the mire type, the thickness of the peat and the state of the technology and organization at that time. Most of these cultivation methods developed in Germany and The Netherlands and were later introduced in other parts of Europe (Heathwaite and Göttlich 1993).

One of the oldest reclamation methods is *mire burning*. The method dates back to the Late Neolithic and continued to be used up to the twentieth century. In the eighteenth century, mire burning became widely used in many parts of Europe (Succow and Joosten 2001). With the mire-burning cultivation method, vegetation and top soil of raised bogs were burned, ploughed, and seeded. The use of fire not only eliminated the unwanted vegetation, it also worked as a fertilizing agent (Borger 1992). Crop raising (mainly rye and buckwheat, but also oats and potatoes) could take place for seven to ten years, followed by a thirty-year rest period to allow mire regeneration (Fig. 9.9). However, repeated cycles of

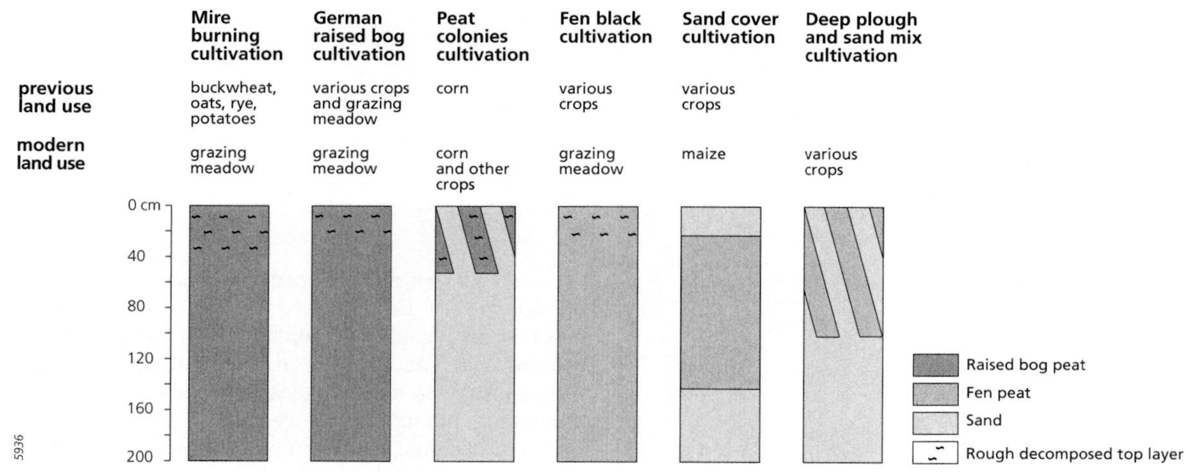

Fig. 9.9. Peatland soil profiles for six cultivation methods. With the mire-burning cultivation, fen black cultivation, and German raised bog cultivation the soil profile is largely kept intact. The peat colonies cultivation, sand cover cultivation, sand mix cultivation and the deep plough cultivation produce completely new soil profiles. (Modified after Heathwaite and Göttlich 1993.)

burning and crop-raising led to the destruction of the surface layer of many bogs and accelerated the shrinkage of peatlands. According to Heathwaite and Göttlich (1993), each burning results in a loss of 2–4 cm of peat. Furthermore, mire burning caused severe air pollution, which could darken the sun and aggravate respiration for days. For this reason, mire burning became forbidden in the early twentieth century. In Germany, mire burning was rapidly replaced by the *German raised bog cultivation method*. This method involved the systematic drainage, levelling of the surface, and liming of the top layer (and thus increasing the pH) of raised bogs. Often nutrients were added to fertilize the soil. Originally, the soils were used for agriculture and grazing, but after tens of years the physical conditions deteriorated. Nowadays, such soils are often only suitable for use as permanent grassland.

In the seventeenth century, the Dutch started to set up *peat colonies*, which implied the planned and systematic exploitation of bog peat accompanied by colonization and cultivation of the areas. At the initial stage, large straight canals were dug into the bog peat and the underlying sandy deposits, which served to drain the peatlands. The peat was then excavated and transported via the canals to the cities, except for the rough and humified topmost layer, which was put aside. Colonists subsequently transformed the excavated areas into arable land by mixing the humified peat with the underlying sand, which served as a soil improver and fertilizer (Borger 1992). The first Dutch peat colonies were set up in order to fulfill the rising demand for grain and peat in the economically booming cities of Holland (Borger 1992). Later in the seventeenth century, the peat colonies method was introduced in East Friesland, north-west Germany (Heathwaite and Göttlich 1993). Activity was especially high during the eighteenth and nineteenth centuries, when several thousands of square kilometres of raised bogs were excavated and changed into arable land (Fig. 9.10).

Fens were traditionally cultivated by the *fen black cultivation method*. This method was first used in the reclamation schemes of late-Medieval German monasteries. In Schwaben (south-west Germany), cultivation of fens was recorded as early as the eleventh and twelfth centuries. With the fen black cultivation method, fens were drained to a depth of at least 150 cm. After being drained, the fen was fertilized with phosphorous and potassium. Because of the low C/N ratio (10–20) and the often high calcium content in drained fens, the mineralization and humification could proceed rapidly. As a result of the high degree of decomposition, soils formed after reclamation of fens therefore have a typical dark colour.

Fens with a thin layer of peat on top of sandy deposits were locally cultivated with the *sand cover cultivation method*. This method was developed in Sachsen-Anhalt (Germany) in the nineteenth century. After draining the fens with ditches, a 10-cm-thick layer of sand was manually spread over the mire, which killed the vegetation. The peat and sand layer was subsequently ploughed and cultivated. In the twentieth century, the *sand mix cultivation* or *deep plough cultivation method* was

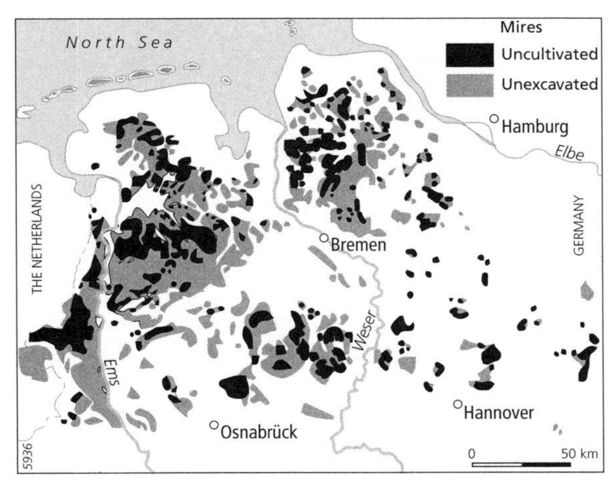

Fig. 9.10. The distribution of uncultivated and unexcavated mires, predominantly raised bogs, in Niedersachsen (Germany) at the end of the eighteenth century (hatched area) and in the mid-twentieth century (black area). The location of the area is indicated in Fig. 9.2. Mires in The Netherlands are not indicated in this figure. (After Succow and Jeschke 1986; modified after Overbeck 1975.)

developed. The fen peat and underlying sand was mixed by deep ploughing, resulting in a new soil profile with excellent properties for agriculture.

Digging and Dredging of Peat for Fuel

Peat is a precursor to coal. When dried and burned, it can be a significant energy source. The combustion value is slightly greater than wood, but much lower than that of coal (Table 9.6). This is mainly due to the high water content. *Sphagnum* peat is most suitable for use as a fuel, as it has a lower ash content than other types of peat. Furthermore, thick roots and branches of trees are absent, which makes it easy to cut. Therefore peat exploitation for fuel was focused on lowland raised bogs.

TABLE 9.6. *Water content and combustion values of peat and coal*

	water (%)	combustion value (kcal/kg)
peat	90–99	—
dried peat	~ 70	< 2,500
brown coal	~ 10	5,000–7,000
coal	< 8	7,000–8,000

The traditional method of peat cutting in raised bogs was the *block trench cutting method.* Prior to the peat extraction the bogs were drained. Usually the humified top layer was stripped from the surface and deposited on the cut-over areas (Heathwaite and Göttlich 1993). Traditionally, the raw peat was excavated with spade-like tools and the sods were stacked on baulks on a 'spread field' to dry for 8–12 weeks (Fig. 9.11). Peat cutting was done by hand for centuries, but from around 1830 attempts were made to mechanize the peat extraction. Small-scale block cutting usually produced a regular system of trenches, flats, and baulks, while

Fig. 9.11. Peat cutting in the Goldenstedter Moor near Vechta, northern Germany (May 1983). Blocks of peat (turves) are cut and subsequently dried on a spreading field.

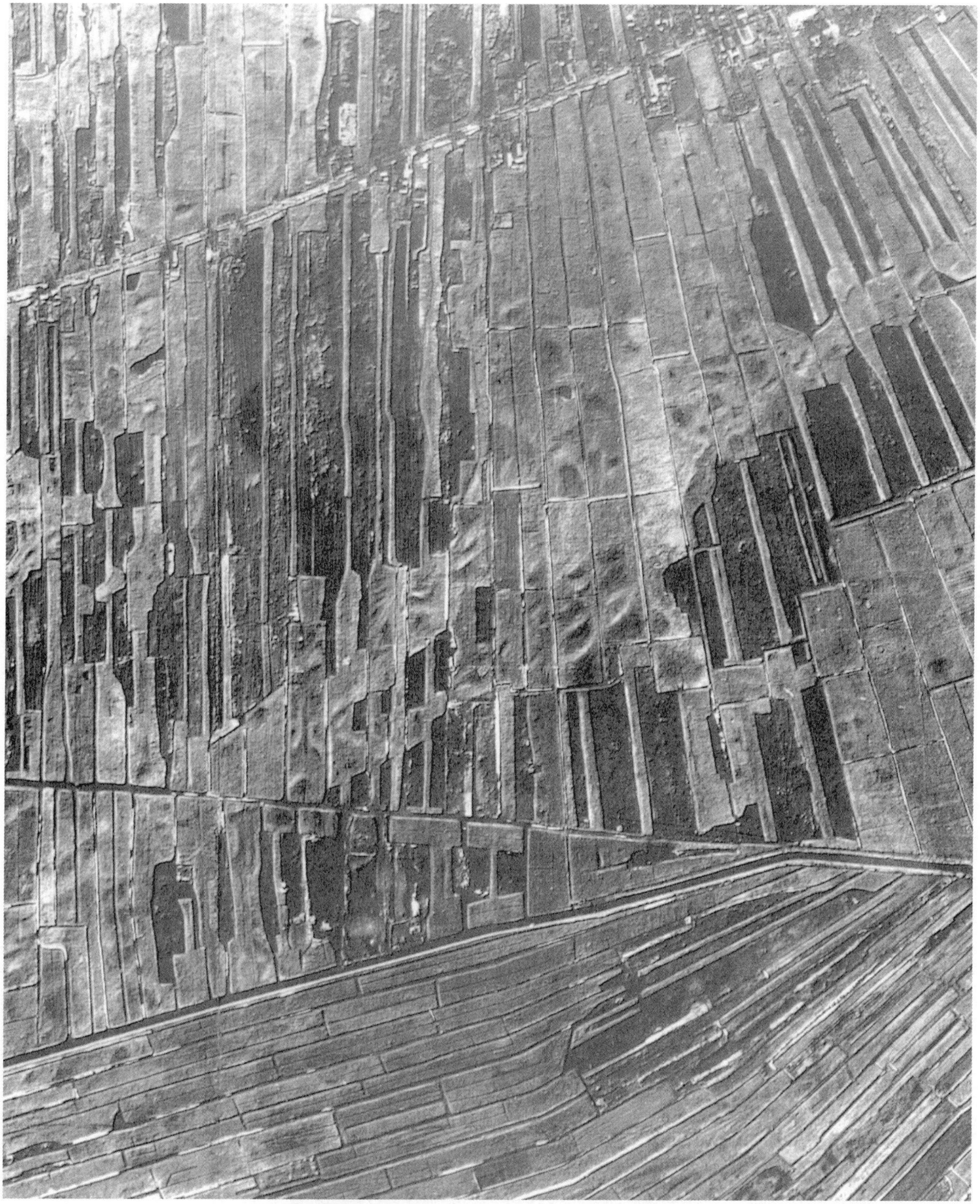

Fig. 9.12. Aerial photograph of excavated peatlands in the western part of The Netherlands. From south-west to north-east an (aeolian) sand ridge occurs on top of which the peat layer was thin or absent. (*Source*: J. N. B. Poelman 1966.)

extensive mechanized block cutting often produced a largely flat surface.

In western Europe, peat cutting for fuel started earlier than 2000 BP. Peat was used by peasants for domestic warming, despite the availability of extensive forests at that time. However, until the late Middle Ages, the use of peat was limited and only local peat extraction took place. Many of the peatlands were left largely untouched. In the late Middle Ages, urban and commercial expansion created great demands for fuel. This demand, together with shrinking woodlands, stimulated the development of commercial exploitation of the lowland bogs (Borger 1990). For example, the growth of towns in Flanders and Brabant (Belgium) was the driving power for systematic digging of bog peat, starting in the thirteenth century.

In The Netherlands peat cutting produced a typical landscape (Fig. 9.12). When peat extraction started in the fifteenth century, it was first concentrated on the excavation of the upper layers of the extensive coastal plain peatlands, which were already under cultivation at that time. Later, during the sixteenth century, peat cutting was replaced by peat dredging (Borger 1992). Systematic peat dredging resulted in the formation of rectangular ponds, called *petgaten*. As a result of wave erosion, many of the *petgaten* and ditches increased in size, finally forming large lakes, which threatened surrounding agricultural land and villages. In the course of the seventeenth to nineteenth centuries most of the lakes were reclaimed and the polders were put to agricultural use.

In other parts of western Europe, peat cutting took place in haphazard fashion until the seventeenth century. At this time the peat colonies cultivation method (see above) was introduced and peat cutting became regulated by law. Peat extraction has considerably declined since the middle of the twentieth century due to increasing availability of other, superior fuels. In France the annual production of peat for fuel decreased from *c.*600,000 tonnes in 1845 to 45,000 tonnes in 1913 and 20,000 tonnes just before 1940 with short, steep increases during the two world wars (Manneville 1999). Nowadays, peat is not used as a fuel anymore in the countries covered by this chapter (Succow and Joosten 2001), but in Ireland, Fennoscandia, and central and eastern European countries excavation continues. In these countries energy generation from peat still is a significant proportion of their total energy supply (Charman 2002).

Digging of Peat for Salt-Making

In coastal areas salt extraction has been an important motive for peatland exploitation from the early Middle Ages to the fifteenth century. In the process, salt-impregnated peat was cut and burned, and the ashes were dissolved in water. The salt was then obtained from this brine by evaporation (Borger 1992). Peat suitable for salt-making was present in peatland areas recently flooded by sea water. Especially in the province of Zeeland (The Netherlands) along the estuaries an important salt industry existed in the Middle Ages. Although the salt production was important for the local economy, it also contributed to further destruction of the peatlands in Zeeland. Eventually, large areas of Zeeland became flooded repeatedly and covered by marine clay deposits. Salt production by peat burning also took place in the northern provinces of The Netherlands and in the western coastal area of Schleswig-Holstein (Germany).

Excavation of Peat for Fertilizer

On a limited scale peat extraction still continues in some places. It is mainly used for soil fertilization in agriculture, horticulture, and gardening. Peat, and especially *Sphagnum* peat, is an excellent material for soil improvement. It is known for its favourable hydrological characteristics, high cation exchange capacity, good aeration, and high content of humic complexes (Boron *et al.* 1986). The pH and nutrient content can be adjusted as required (Heathwaite and Göttlich 1993). The use of peat in commercial agriculture has gained importance over the past twenty to thirty years. Today most of the peat harvesting is done with the *milled peat extraction method* (Bather and Miller 1991). After the bog is drained and the vegetation is stripped off, the bog surface is loosened to a depth of 15–50 mm, allowed to dry and then collected with vacuum or scraping harvesters (Wheeler and Shaw 1995). For example, in 1995 approx. 400,000 tonnes of peat were excavated at the few remaining large sites in France for use in horticulture, etc. (Manneville 1999).

Conservation of the Remaining Mires

The overwhelming majority of natural mires in northwestern Europe have been drained, cultivated, and/or excavated during the past centuries. Of the remaining mires, most are degraded: biodiversity has declined and peat growth has come to an end. This deterioration is mainly caused by hydrological degradation (water table lowering), fragmentation of the landscape, and by acidification, eutrophication, and pollution of the mire water with chemicals (herbicides, pesticides, and other toxic substances) (Bootsma 2000). Eutrophication (nutrient enrichment) is the most serious problem for many of the remaining mires. Since the 1950s the mires have been subject to increased atmospheric deposition of NO_x and SO_2, inflow of nutrient-rich water from agricultural fields and infiltration of nutrient-rich river water (Bootsma 2000). As a result, primary productivity has increased and the typical oligotrophic species have been replaced by plants preferring eutrophic conditions. For example, in north-east Germany many oligotrophic mires have evolved to eutrophic and polytrophic (nutrient-overburdened) mires. They are now dominated by reed (Succow and Joosten 2001).

Recovery and Restoration

Recovery is the process by which a degraded or disturbed mire ecosystem returns more or less to its original state (Schouten and Nooren 1990). It is important to distinguish transient disturbances (e.g. flooding, fire) from persistent environmental changes (e.g. increase in atmospheric N deposition, lowering of the groundwater table). With respect to persistent environmental changes, mires cannot return to their original state. Through adaptation to new environmental conditions, which will take an unknown time, these mires will reach a new state. If the perturbation is temporal, mire ecosystems can return to the original state. Mire ecosystems where only the flora and fauna are affected will recover faster than mires where the a-biotic components are also affected (Bootsma 2000). For instance, the recovery of the natural mire vegetation from acid rain will take several decades, while the recovery of the natural hydrology after drainage will take at least several decades up to several centuries. In many cases removal of the 'stressor' only is not sufficient for mire recovery. Simply stopping the drainage or supply of nutrients will not bring back the original mire vegetation, and additional measures have to be taken. Activities that enable, accelerate, and guide the recovery processes are called *restoration*. Important aspects are rewetting, that is the re-establishment and maintenance of wet conditions at the surface, and renaturation or redevelopment of the mire vegetation.

Verhoeven (1992), Wheeler and Shaw (1995), Manneville (1999), and Charman (2002) present guidelines for restoration strategies for degraded and damaged mires. These include improvement of the water balance and in many cases also a reduction of the nutrient supply and reduction of the nutrient load in the mire. Reducing eutrofication by atmospheric N deposition is difficult, as atmospheric deposition itself cannot be stopped. Industrial pollution since the mid-nineteenth century has not only caused acid rain, resulting in decreased pH levels, and increased supply of sulphur, phosphorus, and nitrates, but also the deposition of significant quantities of heavy metals (Charman 2002). As long as emission continues curative restoration measures remain necessary for the survival of threatened species. During the past decades, various methods have been tried to reduce the nutrient load in mires. For example, Bootsma (2000) showed that an acidified quagmire in The Netherlands was successfully restored by digging trenches and removing the moss layer (upper 10–25 cm). These measures led to an increase in pH and calcium content, and in turn led to increase of species-richness. Obviously, restoration of cultivated peatlands is a nearly impossible task. As most of them have been drained and fertilized for centuries, it could take hundreds of years to change them back into peat-accumulating mire ecosystems.

Peatlands as Palaeoecological Archives

Peat deposits provide a repository of environmental changes during the late Quaternary (Smol *et al.* 2001). This is mainly due to the stratified character, often continuous accumulation, and high degree of preservation of organic remains. Sources of information to reconstruct the palaeoecological environment include botanical data, such as pollen and spores, fungal remains (van Geel 1986), macrofossils, diatoms (Battarbee 1986), algae (Cronberg 1986; van Geel 1986), and other (micro)fossils such as rhizopods (Tolonen 1986) and faunal remains such as those of beetles (Coope 1986). Charcoal, humified horizons, and archeological artefacts may provide additional information on past environments (Birks and Birks 1980).

Pollen and Spores

Pollen grains and spores are produced and dispersed in very large numbers and often transported over great distances. Mires are excellent repositories for 'pollen rain'. Statistical examination of the pollen in stratified peat profiles will reveal information about the history of the local vegetation in the lake or mire as well as the regional vegetation (Birks and Birks 1980). The calculation of absolute accumulation rates of pollen grains per taxon can contribute to the precise determination of the timing of arrival, expansion, and sometimes contraction of different plant species, which in their turn usually reflect environmental changes of various origin.

Regional syntheses of palaeoecological studies of lake and mire deposits, including elaborate description of reference sites, are presented by Berglund *et al.* (1996). These syntheses provide well-documented data on palaeoecological changes over the last 15,000 years, related to vegetation, mire hydrology, soil erosion, human impact, etc. The reference sites preferably contain continuous sedimentary or peat growth records covering the entire Late Glacial and Holocene time. Radiocarbon-dated pollen diagrams illustrate the natural climate-driven environmental changes and the increasingly intensive anthropogenic impact on the European landscape since the early Neolithic, which started at around 5000 BP coinciding with the Atlantic-Subboreal transition. Evidence for forest clearance soon followed by heathland expansion, and the first appearance of cultural indicators such as Cerealia pollen is documented for sites in Denmark (Andersen *et al.* 1996), Germany (Behre *et al.* 1996), Belgium (Verbruggen *et al.* 1996), and France (Visset *et al.* 1996). Moreover, these pollen diagrams also present clear evidence of the times of expansion and/or immigration of tree species during the course of the Holocene. Although ages of appearance or expansion obviously differ from region to region, e.g. pollen diagrams from sites in the Pyrenees (southern France) very clearly show the following order: *Quercus*, *Corylus*, and *Ulmus* during the early Holocene, followed by *Abies*, *Tilia*, *Alnus*, *Fraxinus*, and *Fagus* during the Atlantic, and *Castanea*, *Vitis*, *Olea*, and *Juglans* as anthropogenic indicators since the Subboreal–Subatlantic (Visset *et al.* 1996; see also Manneville 1999 for the Jura and Massif Central).

Numerous studies of Late Glacial-Holocene peat deposits in north-western Europe have been executed. A typical example of a detailed palaeoenvironmental reconstruction discussing not only pollen and spores assemblages, but also the presence of other macro- and microfossil remains of the extensive peat bogs at the German–Dutch border is published by van Geel (1978). Palaeogeographical maps of former vegetation distribution, based upon a large number of these kind of studies, are presented for The Netherlands by Hoek (1997). Huntley and Birks (1983) constructed maps of the spatial movements of plant species since the Late Glacial based on a large number of pollen diagrams. Following the deglaciation, land ice, polar desert, and tundra, vegetation gave way to boreal birch and pine woodland and finally deciduous forest. During this period, trees spread northward with an annual rate between 300 and 500 m/year. Birches (*Betula*) and pines (*Pinus*) spread over western Europe during the Allerød and Early Holocene. Later hazel (*Corylus*) arrived, followed by alder (*Alnus*), oak (*Quercus*), and beech (*Fagus*). In general, the migration rate of tree species was lower than in North America during the Late Glacial and Early Holocene. The difference has been explained by the orientation of the mountain ranges. In Europe, the east–west-orientated Alps and Pyrenees formed natural barriers for the northward migration of the trees, which slowed down their dispersal. In many cases, migration rates of trees varied with time. For instance, the spreading rate of lime (*Tilia*) was much higher between 10,000 and 8000 BP (up to 500 m/year) than between 7000 and 5000 BP (50–130 m/year). This is probably related to changing climatic factors or to increased competitive interactions with other species.

In addition to radiometric dating techniques there is a wide range of time markers that can provide a detailed chronology of peat development. The use of tephra layers from volcanic activity has been known for a long time and has been applied in many areas. A recent development is the study of microscopic tephras that are becoming especially important in peatland palaeoecology in north-western Europe, where there appears to be a number of tephras of Icelandic origin throughout the Holocene (Charman 2002). Other useful time markers are pollen markers, that is: major changes in pollen composition over very short periods known to be consistent across large areas. Sudden declines of certain taxa in the pollen assemblages may be associated with climate change or the arrival of new species. It may also be the result of outbreaks of pathogens or human activities. For example, the decline of the elm in Europe at around 5000 BP was probably caused by the spread of a fungal pathogen, which was possibly connected to human movements and activities in the forests (Aaby 1986). This early Neolithic elm decline is also attributed to the interplay of human activities and climate change by Parker *et al.* (2002). It is proposed that humans played an important role in dispersing disease via clearances that increased the habitat of the elm bark beetle.

Human impact on the vegetation cover becomes increasingly noticeable in the pollen diagrams since the Neolithic. Aaby (1986) concluded that the alterations of the Subboreal and Subatlantic tree pollen assemblages are induced mainly by man, whereas soil conditions and other natural environmental parameters, including climate, have been of minor importance in Denmark. Agricultural activities and other human practices (such as fires and forest clearings) are reflected by a decline of the proportion of trees (destruction of the natural vegetation), introduction of crop species and increase of weed species that are associated with crop-raising. The recovery of the natural vegetation after abandonment of the site can often be determined in the pollen diagrams as well. Behre (1981) developed a detailed tabulation of the association between key indicator pollen types and different agricultural activities and other disturbances practised by humans.

As the vegetation cover is largely determined by climate, apart from the influence of man, the record of changing vegetation as determined by pollen analysis can be a useful source of evidence of past climates. Webb (1980) showed that it is possible to set up a mathematical relation between pollen frequencies (for example the percentage of beech pollen) and mean July temperature. The relation can be expressed as a numerical function, called *transfer function* (Howe and Webb 1983). By the use of the transfer functions, Huntley and Prentice (1988) have mapped the July temperatures across Europe at 6000 BP. Another approach to climate reconstruction on the base of pollen analysis is the use of individual taxa with very precise environmental requirements. Such taxa are called *indicator species*. Their presence in the pollen record may provide reliable information on past conditions. However, complications arise because of the slow response of vegetation to climate change. The arrival of beech in north-western Europe lags behind climate change for several thousands of years. The speed of migration has to be considered. Furthermore, local soil and hydrological conditions must also be taken into account. Based upon a considerable number of palynological investigations Hoek (1997) constructed high-resolution iso-pollen and abundance maps for selected taxa of Late Glacial and Early Holocene time intervals in The Netherlands. In this way the vegetational development of this region could be visualized in great detail.

Humification Horizons

In stratigraphic sequences of raised bogs, layers of dark, humified peat, sometimes with a layer of tree stumps of birch or pine, are considered to be indicators of dry and warm periods. Layers of *Sphagnum* peat with a low degree of decomposition on the other hand are considered to reflect periods of fast peat growth and, therefore, wet conditions (Birks and Birks 1980). A boundary where dark humified peat is directly overlain by fresh *Sphagnum* peat is called a *recurrence surface*. It is assumed to indicate a restart of peat growth after a period of slow accumulation. A well-known study by Aaby (1976), showed a large number of recurrence surfaces in a section through Draved Mose, a raised bog in the south-west of Denmark. Whether or not a synchroneity exists in the occurrence of these recurrence surfaces over large regions and whether or not these features reflect a (*c*.260-year) periodicity in climate shifts is still subject to debate. Nowadays, studies of the continuous variations in indicators of surface wetness or dryness, like humification, plant macrofossils, and hydrogen, oxygen, and carbon isotopes, provide a more realistic portrayal of past climate changes.

The subdivision of the Holocene into climatic periods (Preboreal, Boreal, Atlantic, Subboreal, and Subatlantic) is based on peat stratigraphy. The different periods were distinguished on the basis of climatic shifts identified in the peat stratigraphy of raised bogs in northern Europe. The most important recurrence surface is the *Grenzhorizont*, which was assumed to reflect the transition from Subboreal to Subatlantic (Birks and Birks 1980). It marks a sudden deterioration of climate, starting around 800 BC. The *Grenzhorizont* has been identified in many raised bogs in north-western Europe, and has been substantiated with evidence from pollen analysis and datings. Other horizons on which the subdivision of the Holocene is based are not supported by independent climatic evidence and dates, but their names are still used.

Wiggle-Match Dating

Modern investigations using accelerator mass spectrometry (AMS) wiggle-match dating of selected macrofossils from European raised bog sections have revealed a sharp decline in decomposition of the peat and also a changing species composition at *c*.2700 BP. This indicates a more or less sudden change from relatively dry and warm to cool, moist climatic conditions. This change, for which evidence has also been found on other continents, coincides with an equally sudden and sharp rise in the atmospheric ^{14}C content, which is ascribed to reduced solar activity (van Geel *et al.* 1996, 1998). The inverse relationship between solar activity and radiocarbon production raises a highly relevant discussion on the possible solar influence on phases of strong climate change, including the present-day global warming. Apart from that discussion, ^{14}C wiggle-matching provides opportunities for much more precise dating of organic deposits. Recently, analyses of peat samples from west European ombrotrophic bogs have shown that climatic deteriorations during the 'Little Ice Age' can also be associated with solar-driven changes in atmospheric ^{14}C content (Mauquoy *et al.* 2002*a*, *b*).

Mire Corpses

In the past centuries, remains of hundreds of human bodies have been found during peat-cutting activities, especially in The Netherlands, northern Germany, and Denmark. These remains are known as 'bog bodies' or 'mire corpses'. The state of preservation shows great variation, from skeletons to well-preserved more or less complete bodies. The anoxic and acidic conditions found

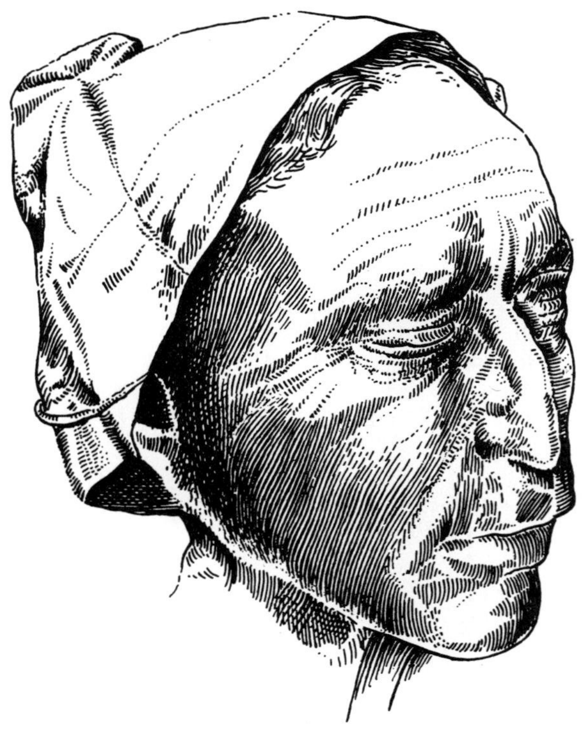

Fig. 9.13. The face of the Tollund man. The man was killed by strangling and buried in a mire in Denmark, approximately 2000 BP. (*Source*: Succow and Jeschke 1986.)

in bogs serve to preserve organic materials. Denmark's Tollund man (Fig. 9.13) is one of the most famous and well-preserved mire corpses, complete with clothing and even stomach contents. His bones were dissolved by the acids in the peat and his skin was tanned like leather (Succow and Jeschke 1986). Ages of the mire corpses range from 8000 BP to the early Medieval period, but most date from the first centuries AD. Many show marks of violent death. Mire corpses are interesting objects for archaeologists, as they provide a wealth of information about past cultures.

Of comparable interest are the remains of extensive systems of so-called 'mire or bog roads' or wooden trackways. As the large peatlands in north-western Europe formed impassable regions, except during cold winters when frozen, traffic was severely hindered. Already around 3000 BP, but especially during the Iron and Roman age, people started systematic construction of wooden roads on poles driven into the peat to connect villages. Eventually some of these roads became part of the network of trade routes in north-western Europe.

Peatlands and the Global Carbon Cycle

The role of peatlands on global climate change has gradually gained more attention during the past decades. Peatlands influence the climate through their role as carbon sources or sinks. Climate in its turn affects the carbon balance through the effect of temperature on the intensity of the decay processes. Substantial amounts of carbon are stored in peat deposits. Most of it has accumulated since the last glaciation. Malmer (1992) suggested that the net flow of carbon in peatlands reached a maximum during the Early and Middle Holocene when peatlands were still rapidly expanding. Globally, accumulation in peat bogs over thousands of years has resulted in a vast store of carbon of *c*.450.10^{15} gC (Gorham 1991), representing about one-fifth of the global carbon storage in terrestrial ecosystems.

Virgin mires are assumed to be net sinks for atmospheric carbon. The vegetation takes up carbon dioxide (CO_2) from the atmosphere and, when the vegetation dies off, the carbon is stored in the accumulating peat deposits. However, peatlands also release carbon in the form of carbon dioxide (CO_2) and methane (CH_4). Decay of plant litter in the aerobic surface zone releases CO_2. Decay in the anoxic zone below releases CH_4, in proportion to the thickness of the peat layer (Malmer 1992). On average, about 25% of the carbon released to the atmosphere consists of CH_4. Altogether, the carbon dioxide uptake usually exceeds the CO_2 and CH_4 release, which makes natural mires a carbon sink. Under stable hydrological and climatological conditions, carbon can be stored for many thousands of years in peat deposits.

Estimates of the average long-term carbon accumulation in mire ecosystems may show great variations. The accumulation rate depends on the mire type (bogs generally have a higher carbon average accumulation rate than fens), the age of the mire, climate, and local hydrological conditions. The occurrence of fluctuations in the carbon accumulation rate is clearly shown in a profile (Fig. 9.14) through Store Mosse, a raised bog in southern Sweden (Svensson 1988). This profile shows a raised bog that developed in three stages on top of a fen deposit. The apparent peat growth rate varied from 0.2 to 1.8 mmy^{-1}. High growth rates proved to be connected to periods of wet climate, while low growth rates and humification horizons reflect drier conditions. The net carbon accumulation rate varied in a same way, but in bog stage I it was slightly higher than in bog stage III. The yearly input of carbon into the catotelm during the last stage has probably not been more than 50% of what it was during the first stage (Malmer 1992).

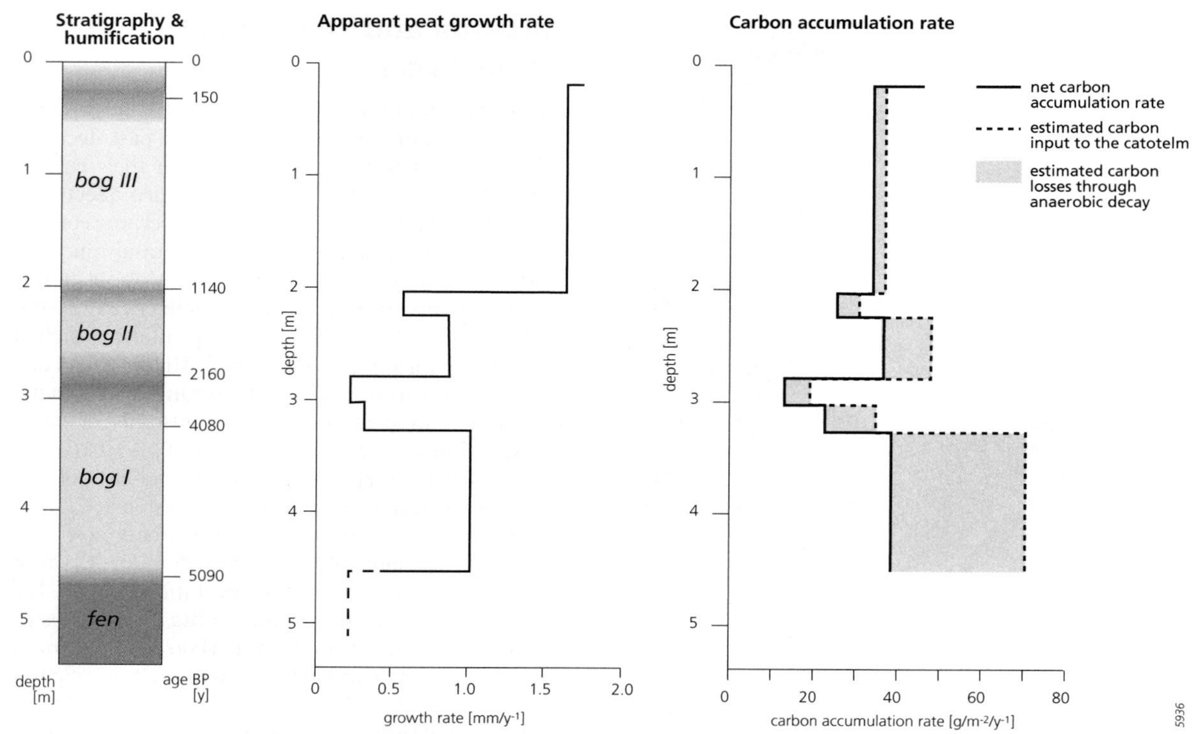

Fig. 9.14. The stratigraphy, and the peat growth rate and carbon accumulation rate in the Store Mosse raised bog (southern Sweden) since 5000 BP. The profile shows four stages of development: a fen stage and three stages of bog development, separated from each other by highly humified peat layers (dark grey). The figure at the right shows the net carbon accumulation rate (solid line) and the estimated annual input of carbon to the catotelm (dashed line). The hatched areas indicate the estimated carbon losses due to anaerobic decay of the peat in the catotelm. (After Malmer 1992; modified after Svensson 1988.)

However, a large part of the carbon that accumulated in the catotelm during the first stage has gradually been released through anaerobic decay as CH_4. Recently, Mauquoy *et al.* (2002*b*) showed, on the basis of detailed peat stratigraphic analyses and by using ^{14}C wiggle-match dating, that there are distinct relations between long-term climate change (changes in effective precipitation) and changes in species composition on ombrotrophic raised peat bogs. For example, peat-bog primary productivity may have been reduced during the Little Ice Age (LIA) among others, due to lower summer temperatures and a consequently shorter growing season for peat-forming plants.

According to Clymo *et al.* (1998), optimal conditions for carbon sequestration appear to be in areas with mean annual temperatures between 4 and 10 °C, which are the southern boreal and the temperate climate zones. During the past decade, extensive research has been carried out on carbon accumulation rates in natural mires in the boreal zones (e.g. in West Siberia (Karpenko 1996; Glebov *et al.* 1997; Bleuten and Lapshina 2001), North America (Gorham 1991), and Finland (Tolonen and Turunen 1996; Turunen 1999)). However, little reliable data is available on carbon accumulation rates in peatlands in the temperate zone, and estimates of the carbon accumulation rate in natural mires in north-western Europe are scarce. However, most peatlands here are drained anyway. As a result, carbon flow to the peatland has stopped, and the stored carbon gradually disappears through emission of CO_2 and CH_4. Thus, by large-scale and prolonged drainage of peatlands, digging of peat for fuel, and previously also by mire-burning, man has turned over the carbon balance of north-west European peatlands and made them a net carbon source. The drained peatlands now contribute to global warming. Furthermore, a large part of Europe's drained peatlands is now in use for intensive cattle-raising, especially in Denmark and The Netherlands, by which activity large quantities of methane are released.

In summary, there is still considerable debate about what the net effects of future climate change might be on peatlands, including negative or positive feedbacks. An overview of these effects together with an assessment of the confidence in each of these is presented by Charman (2002). On the other hand, peatland restoration measures have attracted more attention as peatlands are one of the few effective terrestrial carbon sinks in the longer term (> 100 years).

References

Aaby, B. (1976), Cyclic climatic variations in climate over the past 5,500 yr reflected in raised bogs. *Nature* 263: 281–4.

—— (1986), Trees as anthropogenic indicators in regional pollen diagrams from eastern Denmark. In: K-E. Behre (ed.) *Anthropogenic indicators in pollen diagrams*. Balkema, Rotterdam, 73–93.

Andersen, S. Th., Aaby, B., and Odgaard, B. (1996), Denmark. In: B. E. Berglund, H. J. B. Birks, M. Ralska-Jasiewiczowa, and H. E. Wright (eds.) *Palaeoecological events during the last 15 000 years*. Wiley, Chichester, 215–31.

Barkman, J. J. (1992), Plant communities and synecology of bogs and heath pools in The Netherlands. In: J. T. A. Verhoeven (ed.) *Fens and bogs in the Netherlands: vegetation, nutrient dynamics and conservation*. Kluwer Academic, Dordrecht, 173–235.

Bather, D. M., and Miller, F. A. (1991), *Peatland utilization in the British Isles*. Centre for Agricultural Strategy, Reading.

Battarbee, R. W. (1986), Diatom analysis. In: B. E. Berglund (ed.) *Handbook of Holocene Palaeoecology and Palaeohydrology*. Wiley, Chichester, 527–70.

Behre, K-E. (1981), The interpretation of anthropogenic indicators in pollen diagrams. *Pollen and Spores* 23: 225–45.

—— Brande, A., Küster, H., and Rösch, M. (1996), Germany. In: B. E. Berglund, H. J. B. Birks, M. Ralska-Jasiewiczowa, and H. E. Wright (eds.) *Palaeoecological events during the last 15 000 years*. Balkema, Rotterdam, 507–51.

Berglund, B. E., Birks, H. J. B., Ralska-Jasiewiczowa, M., and Wright, H. E. (eds.) (1996), *Palaeoecological events during the last 15 000 years*. Balkema, Rotterdam.

Birks, H. J. B., and Birks, H. H. (1980), *Quaternary Palaeoecology*. Edward Arnold, London.

Bleuten, W., and Lapshina, E. D. (2001), Carbon storage and atmospheric exchange by West Siberian peatlands. *Scientific Reports*, Department of Physical Geography, Utrecht University, Utrecht.

Bootsma, M. G. (2000), *Stress and recovery in wetland ecosystems*. Thesis, Faculty of Geographical Sciences, Utrecht University.

Borger, G. J. (1990), Peatland exploitation in The Low Countries. In: M. G. C. Schouten and M. J. Nooren (eds.) *Peatlands, economy and conservation*. SPB Academic, The Hague, 15–21.

—— (1992), Draining—digging—dredging; the creation of a new landscape in the peat areas of the Low Countries. In: J. T. A. Verhoeven (ed.) *Fens and bogs in the Netherlands: vegetation, nutrient dynamics and conservation*. Kluwer Academic, Dordrecht, 131–71.

Boron, D. J., Eveans, W. E., and Peterson, J. M. (1986), An overview of peat research, utilization, and environmental considerations. *International Journal of Coal Geology* 8: 1–31.

Casparie, W. A., and Streefkerk, J. G. (1992), Climatological, stratigraphic and palaeo-ecological aspects of mire development. In: J. T. A. Verhoeven (ed.) *Fens and bogs in the Netherlands: vegetation, nutrient dynamics and conservation*. Kluwer Academic, Dordrecht, 81–130.

Charman, D. (2002), *Peatlands and environmental change*. Wiley, Chichester.

Clymo, R. S. (1978), Peat. In: D. W. Goodall (ed.) *Ecosystems of the World* 4a. Elsevier, Amsterdam, 159–224.

—— (1991), Peat Growth. In: L. C. K. Shane and E. J. Cushing (eds.) *Quaternary landscapes*. University of Minnesota Press, Minneapolis, 76–112.

—— Turunen, J., and Tolonen, K. (1998), Carbon accumulation in peatland. *Oikos* 81: 368–88.

Coope, G. R. (1986), Coleoptera analysis. In: B. E. Berglund (ed.) *Handbook of Holocene Palaeoecology and Palaeohydrology*. Wiley, Chichester, 527–70.

Cronberg, G. (1986), Blue-green algae, green algae and Chrysophyceae in sediments. In: B. E. Berglund, (ed.) *Handbook of Holocene Palaeoecology and Palaeohydrology*. Wiley, Chichester, 507–26.

Crum, H. A. (1988), *Peatlands and peat mosses*. University of Michigan Press, Minneapolis.

Damman, A. W. H., and French, T. W. (1987), The ecology of peat bogs of the glaciated northeastern United States. *United States Fish and Wildlife Service Biological Report* 85.

Glebov, F. Z., Karpenko, L. V., Klimanov V. A., and Mindeeva, T. N. (1997), Palaeoecological characteristics of the Holocene between Ob and Vasyugan on the data of peat section 'Vodorazdel'. *Ecology* 6: 412–18.

Gore, A. J. P. (1978), Introduction. In: D. W. Goodall (ed.) *Ecosystems of the World* 4a. Elsevier, Amsterdam, 1–34.

Gorham, E. (1991), Northern peatlands: role in the carbon cycle and probable responses to climatic warming. *Ecological Applications* 1: 182–95.

Heathwaite, A. L., and Göttlich, K. (1993), *Mires, process, exploitation and conservation*. Wiley, Chichester.

Hoek, W. (1997), *Atlas to Palaeogeography of Lateglacial vegetations; Maps of Lateglacial and Early Holocene vegetation, abiotic landscape, and climate in The Netherlands*. Thesis, Faculty of Earth Sciences, Vrije Universiteit Amsterdam.

Howe, S., and Webb, T. (1983), Calibrating pollen data in climatic terms: improving the methods. *Quaternary Science Reviews* 2: 17–51.

Huntley, B., and Birks, H. J. B. (1983), *An atlas of past and present pollen maps for Europe: 0–13,000 years ago*. Cambridge University Press, Cambridge.

—— and Prentice, I. C. (1988), July temperatures in Europe from pollen data, 6000 years before present. *Science* 241: 687–90.

Ingram, H. A. P. (1978), Soil layers in mires: Function and terminology. *Journal of Soil Sciences* 29: 224–7.

—— (1982), Hydrology. In: D. W. Goodall (ed.) *Ecosystems of the World* 4a. Elsevier, Amsterdam, 67–158.

Jensen, U. (1961), Die Vegetation des Sonnenberger Moores im Oberharz und ihre ökologischen Bedingungen. *Naturschutz und Landschaftspflege* 1: 1–85.

Karpenko, L. V. (1996), Dynamics of vegetation, peat and carbon accumulation in Tugulanskaya basin (middle taiga, left bank of Yenissey). *Geography and Nature Resources* 3: 74–80.

Kivinen, E., Hairkurainen, L., and Pakarinen, P. (1979), Geographic distribution of peat resources and major peatland complexes in the world. *Annales Academiae Scientiarum Fennicae* 132: 1–28.

Malmer, N. (1992), Peat accumulation and the global carbon cycle. In: M. Boer and E. Koster (eds.) Greenhouse-impact on cold-climate ecosystems and landscapes. *Catena* Suppl. 22: 97–110.

Manneville, O. (ed.) (1999), Le Monde des tourbières et des marais, France, Suisse, Belgique et Luxembourg. Delachaux & Niestlé, Lausanne.

Matthews, E., and Fung, I. (1987), Methane emission from natural wetlands: global distribution area, and environmental characteristics of sources. *Global Biochemical Cycles* 1: 61–86.

Mauquoy, D., van Geel, B., Blaauw, M., and van der Plicht, J. (2002*a*), Evidence from northwest European bogs shows 'Little Ice Age' climatic changes driven by variations in solar activity. *The Holocene* 12: 1–6.

—— Engelkes, T., Groot, M. H. M., Markesteijn, F., Oudejans, M. G., van der Plicht, J., and van Geel, B. (2002*b*), High-resolution records of late-Holocene climate change and carbon accumulation in two north-west European ombrotrophic peat bogs. *Palaeogeography, Palaeoclimatology, Palaeoecology* 186: 275–310.

Mitsch, W. J., and Gosselink, J. G. (1986), *Wetlands*. Wiley, New York.

Moore, P. D. (1984), *European Mires*. Academic Press, London.

—— and Bellamy, D. J. (1974), *Peatlands*. Elek Science, London.

Overbeck, F. (1975), *Botanisch-geologische Moorkunde unter besonderer Berücksichtigung der Moore Nordwest-Deutschlands als Quellen zur Vegetations-, Klima- und Siedlungsgeschichte*. Karl Wachholtz, Neumünster.

Parker, A. G., Goudie, A. S., Anderson, D. E., Robinson, M. A., and Bonsall, C. (2002), A review of the mid-Holocene elm decline in the British Isles. *Progress in Physical Geography* 26: 1–45.

Poelman, J. N. B. (1966), *De bodem van Utrecht*. Stichting voor Bodem kartering, Wageningen.

Post, L. von, and Granlund, E. (1926), *Södra Sveriges torvtillgångar*. I. *Sveriges geol. unders.*, ser. C 335. Stockholm.

Schouten, M. G. C. (1984), Some aspects of the ecogeographical gradient in the Irish ombrotrophic bogs. *1st Peat Congress, Dublin*, 414–32.

—— and Nooren, M. J. (eds.) (1990), *Peatlands, economy and conservation*. SPB Academic, The Hague.

Schneekloth, H., *et al.* (1970/83), *Die Moore in Niedersachsen*. Schriften der wirtschaftswissenschaftlichen Gesellschaft zum Studium Niedersachsens e.V. AI.96, Göttingen, Hannover.

Sjörs, H. (1980), Peat on earth: Multiple use or conservation. *Ambio* 6: 303–8.

Smol, J. P., Birks, H. J. B., and Last, W. L. (eds.) (2001), *Tracking environmental change using lake sediments*. 4 vols. Kluwer Academic, Dordrecht.

Stanek, W., and Worley, I. A. (1983), A terminology of virgin peat and peatlands. In: C. H. Fuchsman and S. A. Spigarelli (eds.) *International symposium on peat utilization*. Bemidji State University, Minnesota, 10–13.

Succow, M. (1988), *Landschaftsökologische Moorkunde*. Gustav Fischer, Jena.

—— and Jeschke, L. (1986), *Moore in der Landschaft*. Urania, Leipzig.

—— and Joosten, H. (2001), *Moorkunde*. E. Schweizerbart, Stuttgart.

—— and Lange, E. (1984), The mire types of the German Democratic Republic. In: P. D. Moore (ed.) *European Mires*, Gustav Fischer, Jena, 149–75.

Svensson, G. (1988), Bog development and environmental conditions as shown by the stratigraphy of Store Mosse in southern Sweden. *Boreas* 17: 89–111.

Tolonen, K. (1986), Rhizopod analysis. In: B. E. Berglund (ed.) *Handbook of Holocene Palaeoecology and Palaeohydrology*. Wiley, Chichester, 645–66.

—— and Turunen, K. (1996), Carbon accumulation in mires in Finland. In: R. Laiho, J. Laine, and H. Vasander (eds.) *Northern peatlands in global climatic change*. Publications of the University of Finland 1/96, Edita, Helsinki, 250–5.

Turunen, J. (1999), *Carbon accumulation of natural mire ecosystems in Finland—application to boreal and subarctic mires*. University of Joensuu Publications 55.

van de Plassche, O. (1982), Sea-level change and water level movements in the Netherlands during the Holocene. *Mededelingen Rijks Geologische Dienst* 36/1.

van Geel, B. (1978), A palaeoecological study of Holocene peat bog sections in Germany and The Netherlands, based on the analysis of pollen, spores and macro- and microscopic remains of fungi, algae, cormophytes and animals. *Review of Palaeobotany and Palynology* 25: 1–120.

—— (1986), Application of fungal and algal remains and other microfossils in palynological analyses. In: B. E. Berglund (ed.) *Handbook of Holocene Palaeoecology and Palaeohydrology*. Wiley, Chichester, 497–505.

—— Buurman, J., and Waterbolk, H. T. (1996), Archaeological and palaeoecological indications of an abrupt climate change in The Netherlands, and evidence for climatological teleconnections around 2650 BP. *Journal of Quaternary Science* 11: 451–60.

—— van der Plicht, J., Kilian, M. R., Klaver, E. R., Kouwenberg, J. H. M., Renssen, H., Reynaud-Farrera, I., and Waterbolk, H. T. (1998), The sharp rise of Δ^{14}C ca. 800 cal BC: possible causes, related climatic teleconnections and the impact on human environments. In: W. G. Mook and J. van der Plicht (eds.) *Proceedings of the 16th International ^{14}C Conference. Radiocarbon* 40: 535–50.

Verbruggen, C., Denys, L., and Kiden, P. (1996), Belgium. In: B. E. Berglund, H. J. B. Birks, M. Ralska-Jasiewiczowa, and H. E. Wright (eds.) *Palaeoecological events during the last 15 000 years*. Balkema, Rotterdam, 553–74.

Verhoeven, J. T. A. (ed.) (1992), *Fens and bogs in The Netherlands: Vegetation, history, nutrient dynamics and conservation*. Kluwer, Dordrecht.

Visset, L., Aubert, S., Belet, J. M., David, F., Fontugne, M., Galop, D., Jalut, G., Janssen, C. R., Voeltzel, D., and Huault, M. F. (1996), France. In: B. E. Berglund, H. J. B. Birks, M. Ralska-Jasiewiczowa, and H. E. Wright (eds.) *Palaeoecological events during the last 15 000 years*. Balkema, Rotterdam, 575–645.

Webb, T. (1980), The reconstruction of climatic sequences from botanical data. *Journal of Interdisciplinary History* 10: 749–72.

Westhoff, V. (1990), Bogs in world perspective. In: M. G. J. Schouten and M. J. Nooren (ed.) *Peatlands, economy and conservation*. SPB Academic, The Hague, 5–13.

Wheeler, B. D., and Shaw, S. C. (1995), *Restoration of damaged peatlands*. Department of the Environment, HMSO, London.

Regional Environments

10 Danish–German–Dutch Wadden Environments

Jacobus Hofstede

Introduction

The Wadden Sea environment is a coastal tidal environment situated between the North Sea and the northwestern European Lowlands. It stretches over a distance of about 450 km from Den Helder in The Netherlands to the peninsula of Skallingen in Denmark (Fig. 10.1). The approximately 10,000 km^2 large Wadden Sea is a coastal sediment sink that developed in the course of the Holocene transgression. It resulted from a specific combination of sediment availability (mainly from the North Sea) and a hydrodynamic regime of tides and waves. In its present state, the Wadden Sea environment consists of extensive tidal flats (the wadden), tidal gullies and inlets, salt marshes, and about twenty-four sandy barrier islands. Further, four estuaries exist that discharge into the Wadden Sea.

The Wadden Sea may best be characterized by the words 'dynamic' and 'extreme'; dynamic from a geomorphological point of view, extreme in its biology. According to Spiegel (1997), with each flood phase a tidal energy input in the order of 2.2 thousand MW occurs in the Wadden Sea of Schleswig-Holstein (Germany). This energy input, combined with the energy impact of wind, waves, and storm surges, results in strong morphological processes. Flora and fauna in the Wadden Sea have to adapt to these intense morphodynamics. Further, they have to endure the permanent change of flood and ebb and fluctuations in salinity, as well as high water temperatures during summer and occasional ice cover during winter. As a result of these extreme environmental conditions, a highly specialized biosystem with about 4,800 species has developed (Heydemann 1998).

In its present state the Wadden Sea is one of the last remaining near-natural large-scale ecosystems in central Europe. Its ecological significance is underlined by the fact that 250 animal species live exclusively here (Heydemann 1998). Furthermore, nowhere else in Europe is an ecosystem of this size visited by more birds per surface area for the purpose of feeding. However, the Wadden Sea is subjected to considerable human influences, e.g. the input of nutrients and pollutants, fisheries, dredging, boat traffic, and tourism (de Jong *et al.* 1999). In order to manage these human stresses in an ecologically sustainable way large efforts are being made, e.g. the declaration of national parks and the establishment of the Trilateral Wadden Sea Plan (CWSS 1997). Another major challenge for the Wadden Sea is the encounter of possible negative effects of climate change. Changes in precipitation and in wind climate as well as an increase of global mean temperatures and sea level will occur (IPCC 2001). Investigations on how the predicted changes may affect structure and functions of the Wadden Sea ecosystem are in progress.

After an introductory regional description, the Holocene evolution as well as the present-day structure and functions of the Wadden Sea are described. Based on this systems analysis, possible future developments will be discussed. The last part of this chapter concentrates on the human factor in the Wadden Sea, its historical development, the stresses it exerts on the ecosystem, and its prospects within the framework of an integrated coastal management.

Regional Description

The Wadden Sea fringes the Danish, German, and Dutch mainland coastlines between Skallingen in Denmark and Den Helder in The Netherlands over a distance of about 450 km (Fig. 10.1). The width of this approximately 10,000 km^2 large coastal ecosystem varies from 10 km

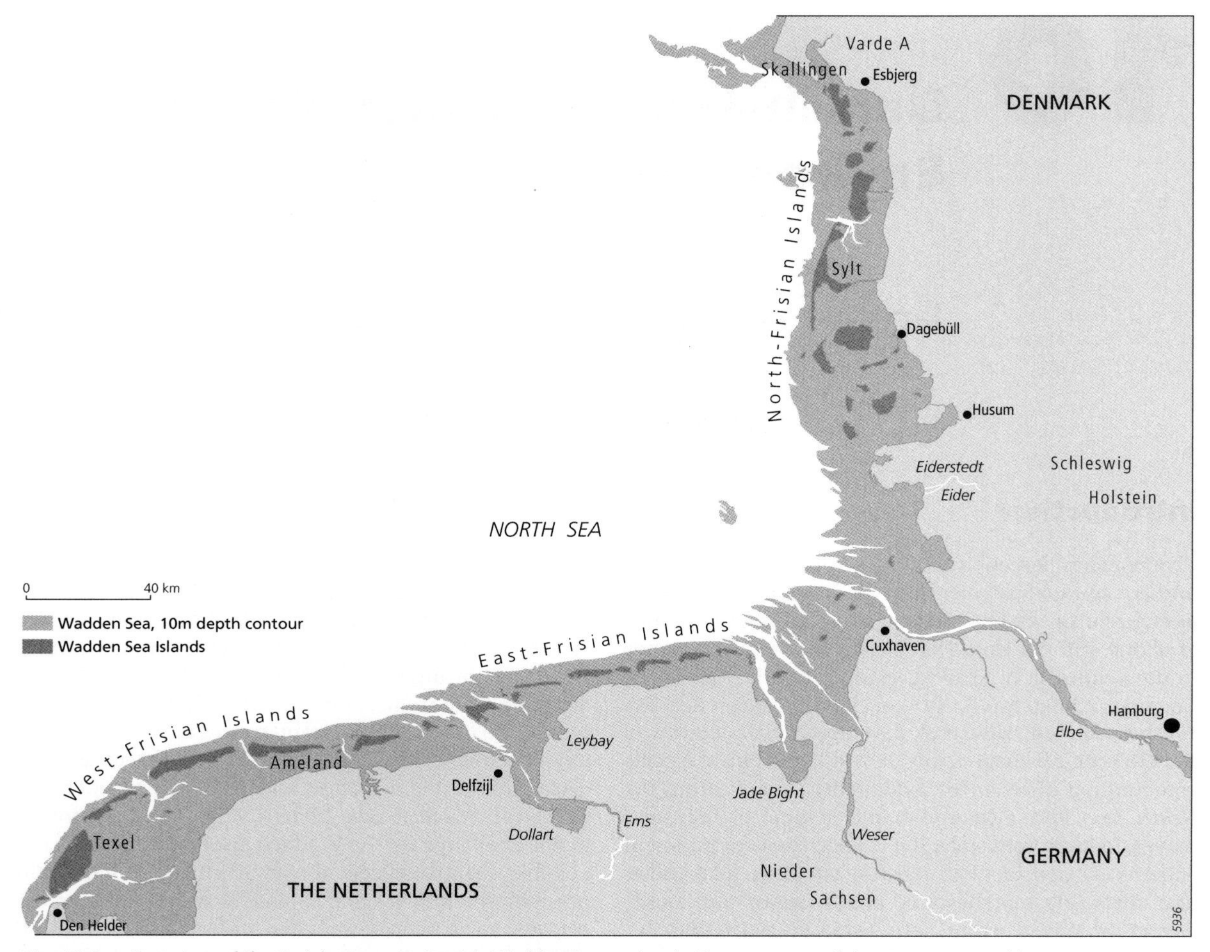

Fig. 10.1. Overview of the Dutch–German–Danish Wadden Sea region (with permission of the Common Wadden Sea Secretariat). The Wadden Sea is defined by the −10 m depth contour.

near Ameland to up to 50 km in the Dollart and the Jade Bight. More than half the area is intertidal, i.e. is flooded regularly by the tide. The geomorphology is characterized by extensive tidal flats, tidal channels and inlets, sandy barrier islands, salt marshes, and the estuary mouths of the rivers Ems, Weser, Elbe, and Eider. In its prehistoric natural state, the terrestrial border was situated at the transition from brackish to fresh water ecotopes. Over the last 1,000 years, about one-third (the marsh lands) of the Wadden Sea has been reclaimed (Lozan *et al.* 1994). Only in a few remaining areas, such as the Varde Å in the Danish Wadden Sea, does there still exist a natural gradient from salt to brackish to freshwater ecotopes. Consequently, the landward border of the coastal ecosystem is presently situated along the outer slope of the dykes. The seaward border of the Wadden Sea is defined as the water depth at which the forces of tides and wind waves may induce sediment transport, i.e. may influence the morphology. According to Oost (1995), during storms, wind waves may stir up material from about 13 m below the water surface. Correspondingly, the seaward border is defined as the 13-m depth contour.

In the Wadden Sea, no hard structures, such as bedrock, are present. The largest part of the surface sediments are sandy, the rest is fine-grained material (Oost 1995). The sands have a mean diameter of 170–190 μm, and are largely composed of quartz (more than 80%) with some feldspar, $CaCO_3$, and mica. The fine-grained sediments in the Wadden Sea consist mostly of clay minerals, organic matter, and $CaCO_3$ (van Straaten 1964). Most of the sediments originate from Pleistocene deposits on the bottom of the North Sea. Part of the fine-grained material originates from the rivers, and another part is biogenic (Postma 1981; Dieckmann 1985; Hoselmann and Streif 1998).

The salinity of the water shows typical values of 30 psu (= practical salinity unit in ‰) (in the North Sea about 34 psu) with a more or less pronounced seasonal fluctuation. For example, salinity varies between about 25 psu in winter and 32 psu in summer at List on the island of Sylt (Becker 1998*a*). Like salinity, water temperatures show a strong seasonal fluctuation. In the long term, mean temperatures vary from about 2 °C in February to 17 °C in August (Becker 1998*b*). However, maximum water temperatures of 30 °C as well as up to three months of ice cover have been recorded.

According to Mengelkamp *et al.* (1998) the yearly mean wind velocities in the German sector of the Wadden Sea varies from about 8 m/s at the barrier islands to 6.5 m/s along the mainland coast. During winter storms, maximum wind speeds of well over 30 m/s have been recorded. Prevailing wind directions in the region are from south-west to north-west, and the mean significant wave height in the North Sea in front of the Wadden Sea is in the order of 1 m. However, during severe storms, maximum wave heights of more than 6 m were recorded in front of the Isle of Sylt. Within the Wadden Sea, locally generated wave heights strongly depend upon water depths and fetch. Only in the major tidal streams might the waves exceed 2.5 m during storms. The tidal wave in the North Sea moves from the south-west (Den Helder) to the north-east (Skallingen). The smallest tidal ranges of about 1.3 m are found near Skallingen and Den Helder. From these two positions, the tidal range increases towards the inner German Bight (the estuary of the Elbe) where yearly mean values of up to 3.5 m are recorded.

A large number of long-term gauge stations register water levels in the Wadden Sea. The oldest gauge world-wide is situated in Amsterdam. Here, water levels were recorded from 1700 until 1925 (Fig. 10.2), when construction activities to cut off the Zuiderzee from the Wadden Sea started. Other long-term gauge stations that still record water levels are, amongst others, Den Helder (since 1832), Cuxhaven (1843), and Esbjerg (1890). With these data, it is possible to establish the history of relative sea-level change in the region since about 1850 in great detail (Töppe 1993; Jensen *et al.* 1993). During the twentieth century, the MSL rise varied spatially from 10 to 20 cm/100 years with an overall mean of about 15 cm/100 years. A part of MSL rise might have been caused by global factors such as

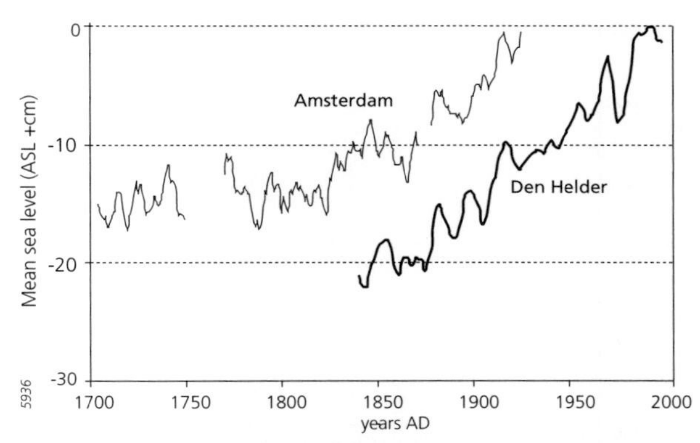

Fig. 10.2. Development of mean sea level (5-year running means) at the gauge stations Amsterdam (1700–1925) and Den Helder (1840–1999), The Netherlands.

melting of glaciers and thermal expansion of the upper layer of the oceans (Oerlemans 1993), the rest by regional and local factors such as ocean and atmospheric circulation, tectonics, compaction, gas winning, land reclamation, etc. As with MSL, the relative trend in mean high water (MHW) was not constant, neither spatially nor temporally. The causes for these variations might be natural (e.g. salt tectonics, changes in storm tracks, morphological changes) or artificial (e.g. gas exploitation, land reclamation, dredging activities). However, a mean rise of about 20–25 cm/100 years has been observed at most stations with an acceleration during the 1960s, 1970s, and 1980s. The mean low water (MLW) did not show significant changes between 1900 and 2000, and, finally, the mean tidal range (MTR) increased by about 20–25 cm/100 years. Figure 10.3 shows the development of MHW, MLW, and MTR at the gauge stations Delfzijl, Cuxhaven, and Husum. Apart from a common signal, large local variations in the development of tidal water levels become obvious. The strong increase in the values of MTR between about 1960 and 1985 is, at least partly, caused by human activities, i.e. dredging of the main shipping channels.

Geomorphology of the Wadden Sea

Already during former interglacial periods, wadden environments existed in the area of the present Wadden Sea (Streif 1993). A compilation of geological data about the spatial extent, amplitudes, and dynamics of marine transgressions in the southern North Sea basin and adjacent lowlands during warm stages of the Pleistocene is presented by Streif (2004). With falling sea

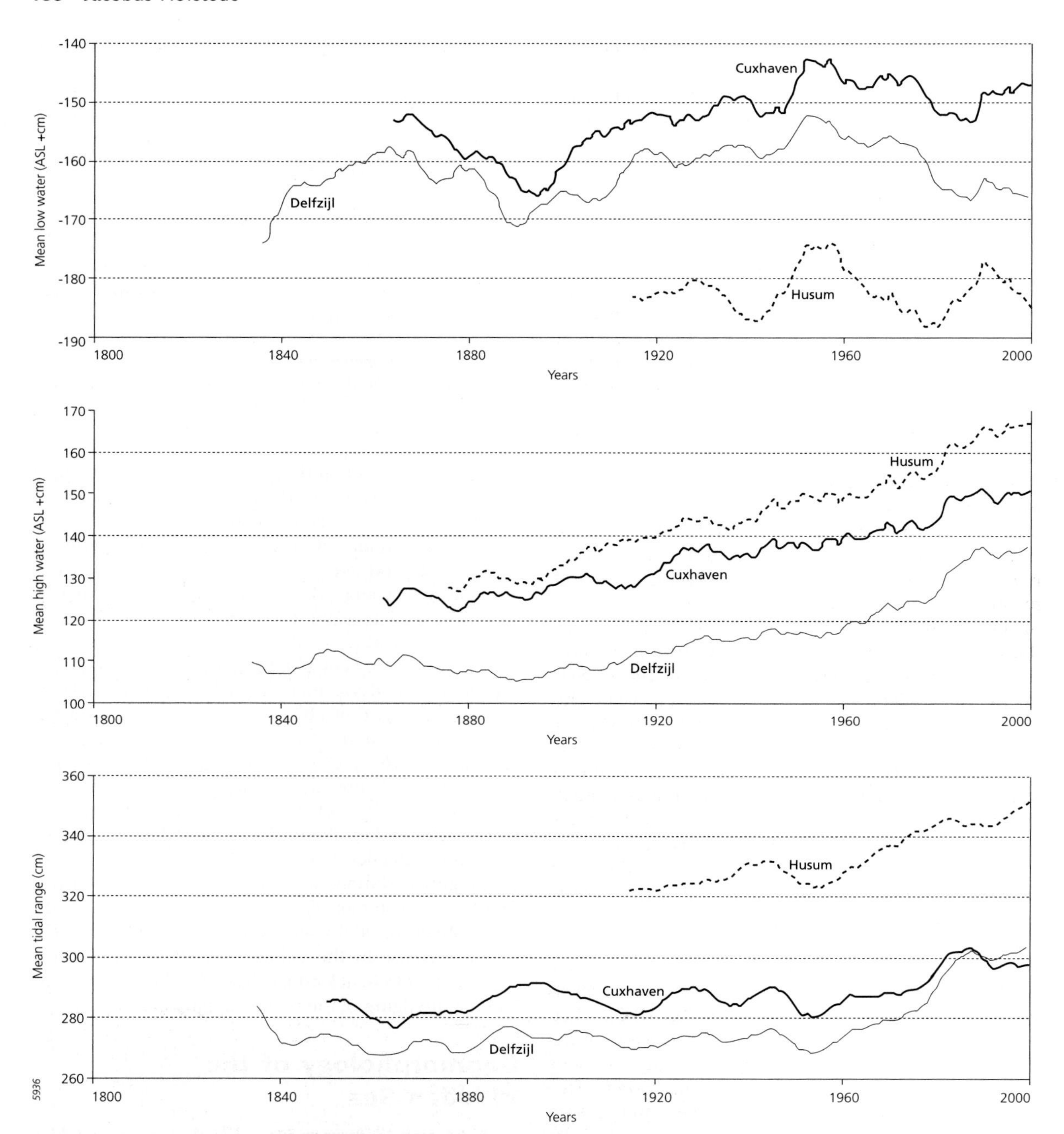

Fig. 10.3. Long-term development (10-year running mean) of mean low water (MLW), mean high water (MHW), and mean tidal range (MTR) at gauge stations Delfzijl (The Netherlands), and Cuxhaven and Husum (Germany). Note that German data is given in GOL/NN ± cm (German Ordnance Level), Dutch data is given in NAP ± cm (Dutch Ordnance Level), GOL ≈ NN ≈ NAP ≈ MSL ≈ a.s.l.

level during the succeeding ice ages, these coastal ecosystems were replaced by terrestrial periglacial environments. The genesis of the modern Wadden Sea ecosystem started about 8000–7500 BP when the Holocene marine transgression reached the area. From this time, marine processes started to reorganize the former terrestrial environment.

Origin and Holocene Evolution

The geomorphological development of the Wadden Sea is strongly governed by the Holocene sea-level rise. According to Streif (1989, 1993), sea level rose from about −46 to −15 m in the time interval between 8600 and 7100 BP (i.e. 2.1 m century). This ingression of the North Sea occurred in a single transgressive overlap. First indications for a reduction in the rate of sea-level rise date from about 6500 BP. After about 5000 BP, sea-level rise seems to have occurred in transgressive and regressive phases. However, sea-level curves from The Netherlands, Niedersachsen, and Schleswig-Holstein indicate that these phases did not occur synchronously throughout the region. Data from the north-west German islands suggest that the North Sea might have reached its present level about 600 years ago (Hoselmann and Streif 1998). After a small sea-level fall (in the order of 0.3 m) during the Little Ice Age (Hofstede 1991), sea level is again rising, as described in the introduction to this chapter.

During the rapid early Holocene sea-level rise, the Pleistocene land surface was literally drowned. As a result, the coastline receded over several hundreds of kilometres. The present-day coastal environment with its barrier islands, tidal gullies, flats, and marshes started to develop after about 7500 BP. Although the geomorphological development did not occur synchronously everywhere (e.g. van der Spek and Beets 1992; Bartholdy and Pejrup 1994; Oost and de Boer 1994), a general accumulative history becomes evident (Streif 1989). The resulting sedimentary sequence is shown schematically in Fig. 10.4. On the seaward side, continuous sequences of clastic, shallow marine and intertidal deposits overlie pre-Holocene terrestrial sediments and in some places basal peat. In the zone bordering the Pleistocene hinterland, thick sequences of peat occur that indicate continuous bog development. In between these two zones, an alternation of lacustrine deposits and semi-terrestrial peat layers indicate a cyclic alternation of transgressive and regressive overlaps. The development of this locally more than 35-m thick layer of sediments was more or less finished 2,000 years ago (Gerdes *et al.* 2003). Moreover, Gerdes *et al.* (2003) conclude that only a minor proportion of this huge sediment wedge is supplied to the coastal region by the rivers Ems, Weser, and Elbe. Afterwards, on the barrier islands, dunes, locally up to 35-m high, evolved by aeolian processes. Hoselmann and Streif (1998) calculated a Holocene sediment volume for the approximately 3,000 km^2 large Jade–Weser area (including the reclaimed marshes) of about 25 thousand million m^3, i.e. a yearly net accumulation of about 3.3 million m^3. However, large spatial and temporal differences in the accumulation of sediments exist in the Wadden Sea. Using an index E describing the trapping efficiency of fine-grained sediments in tidal basins, Pejrup *et al.* (1997) calculated a difference in the values of E of the order of 12 among two tidal basins in the Danish sector of the Wadden Sea.

In general, the following mechanisms caused the accumulation of sediments in the Wadden Sea during the

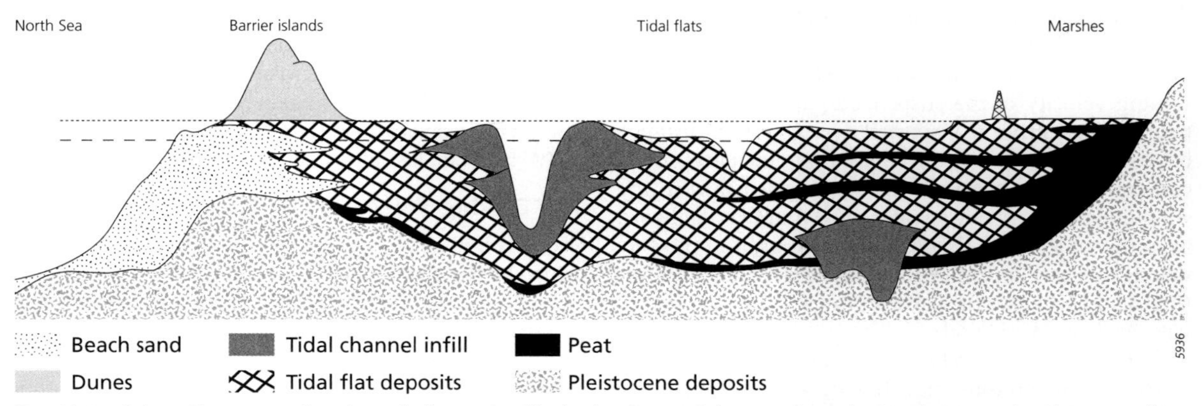

Fig. 10.4. Schematic cross-section through the wedge-like body of coastal deposits from the North Sea to the Pleistocene hinterland (after Streif 1989).

Holocene. Davis (1994) proposed three modes of barrier islands formation: (1) the emergence and upward-shoaling of shallow sand bars, (2) the development and subsequent breaching of spits, and (3) the drowning of coastal ridges. In general, the barrier islands in the Wadden Sea originate from mode (1). However, a number of barrier islands (e.g. Texel, Sylt) have a Pleistocene core that partly reaches up to the surface. At least, the actual position of these islands is codetermined by these resistant cores. The development of shallow sandbars and their subsequent transformation into barrier islands is made possible by the landward transport of sand through wave action, as described by, amongst others, Dean (1987) and Carter (1988). After the shoals emerged above sea level, aeolian processes started and dunes developed. The long-term net result of these wave and aeolian processes is large sand accumulations along the seaward side of the Wadden Sea (Fig. 10.4).

The major part of the sediments, however, was imported from the North Sea by tidal currents. Various tidal mechanisms contribute to the accretion of sediments into the Wadden Sea (e.g. van Straaten and Kuenen 1958; Postma 1961; Oost 1995):

1. The mean water depth (in the areas covered with water) is much less at high tide, when the shallow tidal flats are covered with water, than at low tide when the water fills only the deep channels. The probability of suspended matter settling to the bottom over the tidal flats at high tide is, therefore, much greater than over the channel beds during low tide.

2. This probability is further increased, as the tidal wave becomes more asymmetrical in the inner parts of the back-barrier bays. Here the tide turns quickly much more around low water than around high tide. In consequence, the period with low current velocities at the turn of the tides (slack water), during which sediments may settle, is much longer around high tide than around low water.

3. After the tidal current velocity drops below the settling velocity of the suspended particles, these particle will be further transported over a certain distance before reaching the bottom (the settling lag effect). As a consequence, the particles can be eroded during the subsequent ebb only by a water mass that was landward from the water mass that originally brought in the particles. This mechanism, in combination with mechanisms 1 and 4, causes the particles to settle more landward with each successive tide.

4. Higher current velocities are needed to erode particles from the bottom once they have settled, than the velocity at which these particles settle from suspension (the scour lag effect). This effect is enhanced if the deposited particles have some time to consolidate, a process significantly increasing the shear strength after as little as one hour.

Apart from these tidal mechanisms, organisms in the Wadden Sea play an important role in the deposition of suspended matter as well (Eisma 1993; ten Brinke 1993). Organic matter released by algae and bacteria acts as 'biological glue'. It stimulates the flocculation of fine particles and causes the formation of aggregates. These are larger than the original particles and therefore sink faster. Filter feeders such as cockles and mussels filter fine particles from the water and produce comparatively coherent faecal pellets and, occasionally, less coherent pseudo-faeces. Bed flora and fauna such as diatoms and bacteria produce slime, which binds sediment particles, resulting in a strong increase in shear strength. According to Führböter (1984), the current velocities needed to entrain particles may be up to five times larger in biologically stabilized material as in sterile sand. Finally, in the salt marshes deposited particles are retained by the vegetation. As described before, peat may, at the landward side of the ecosystem, contribute a significant part to the Holocene accumulative body.

After the barrier islands in the Wadden Sea came into existence (at the latest before 5000 BP), they retreated landward in response to the Holocene sea-level rise by erosion on their seaward sides and deposition on their landward sides (Sha 1992). The following process is responsible for this migration. During storms, sand is stirred up from the nearshore and the beach by breaking waves. Part of this material will be transported to the back-barrier by overwash through breaches in the dunes where it settles in washovers (Carter 1988). According to Davis (1994), this process (rollover) represents the primary agent for landward migration of barrier islands in response to sea-level rise. In a steady state between sea-level rise (cause) and geomorphology (response), the barrier island does not change its size, but migrates landward. In the Wadden Sea, this mechanism was shown for some German barrier islands that have migrated landward in response to the observed rise in MHW over the last fifty years (Hofstede 1999*a*).

The Driving Forces

Geomorphological processes in the Wadden Sea are driven by the energy of tidal and wind-induced currents (including waves). A small proportion of this energy performs work on the bed (e.g. Carter 1988). As a result, particles may become stirred up from the bed and transported (either as bedload or suspended) to other places where they settle again under diminishing current velocities. A comprehensive description of the erosion, transport, and sedimentation processes of particles in the water column is given in Zanke (1982) and van Rijn (1993). The totality of all (interacting) currents in the Wadden Sea may be seen as the driving forces. The most important are described below.

Meteorological Forces

A number of driving forces in the Wadden Sea result from meteorological processes, i.e. the wind. First, the wind itself is an important driving force, as it may cause significant aeolian sand transport on the beaches and in the dunes. The significance of this force is underlined by the fact that almost all barrier islands in the Wadden Sea possess extensive, up to 35-m high, dune fields. Secondly, the transfer of wind energy from the air to the water surface induces waves that are connected to several morphologically significant currents. Along the outer part of the Wadden Sea a number of currents connected to waves, e.g. orbital, breaker, and longshore currents as well as undertow, cause significant sediment redistribution (e.g. Komar 1998). In the tidal basins of the Wadden Sea orbital motions of waves will stir up particles from the bed and prevent them from settling even after the tidal currents drop below the settling velocity of the particles. Finally, strong onshore winds in combination with wave effects can create storm surges that are negatively correlated with water depth. Storm surges induced by low-pressure centres moving over the North Sea basin in a south-easterly direction may cause a piling up of the water level by up to 4 m for several hours in the inner parts of the Wadden Sea. On the tidal flats, the transport of these huge water masses may result in (morphologically significant) surge currents of up to 1.5 m/s (Göhren 1968).

Astronomical Forces

Tides, or rather, tidal currents, are the second primary agent for morphologic processes in sandy coastal environments such as the Wadden Sea (e.g. Pethick 1984; Carter 1988; Komar 1998). The tidal wave affecting the Wadden Sea is a semi-diurnal resonant tide of the Atlantic, running counter-clockwise through the North Sea around an amphidromic point. One tidal cycle (one flood plus one ebb period) has a duration of about 12 h 25 min. Within the Wadden Sea, the flood and ebb periods are not equal in duration. This tidal asymmetry is due to the tidal phase shift (about 1 hour per day), bottom morphology, and the influence of the wind. Normally, the ebb period is longer than the flood period, e.g. in Husum the flood lasts 5 h 53 min, and the ebb 6 h 32 min. Further, the tidal range varies through time, depending on the relative positions of the sun, moon, and Earth. For example, as the main solar tidal constituent (S 2) lags about 50 min behind the main lunar tidal constituent (M 2), they are in-phase only once every 14.8 days, creating spring tides. When they are out of phase then neap tides occur. The difference between spring and neap tides may attain a maximum of 0.9 m at Husum. The movement of the tidal wave around the amphidromic point causes currents, the direction of which alternates with each tidal period (ebb or flood). In the North Sea, these currents are normally less than about 0.05 m/s. However, in the Wadden Sea tidal current velocities are up to 1.5 m/s in the tidal inlets and around 0.5 and 1.0 m/s in the tidal gullies. On the flats, tidal current velocities normally do not exceed 0.35 m/s. As a result of tidal asymmetry, flood currents in the Wadden Sea are normally stronger than ebb currents and this has significant effects on net water and sediment fluxes.

Artificial Forces

Artificial forces increasingly affect structure and functioning of the Wadden Sea. Like natural forces, they result from the transfer of (fossil) energy into transporting agents causing sediment redistribution. The results (responses) may be either accumulative, e.g. sea walls, dams, and beach nourishment, or erosive, e.g. deeper (shipping) channels or sediment extraction holes. At the 38-km-long beach at the Isle of Sylt, about 30 million m^3 of sand has been added since 1972 (Fig. 10.5). Further, as a consequence of gas exploitation in the eastern part of the Dutch Wadden Sea, it is expected that the surface elevation of the tidal flats may locally reduce by up to 0.3 m (Oost and Dijkema 1993). Apart from these direct influences, human actions may work indirectly, in that they influence natural forces, e.g. salt-marsh works (Dijkema *et al.* 1990; Houwink *et al.* 1999; Fig. 10.6).

Structure and Functions

The Wadden Sea as we envisage it today is the result of the driving forces acting under a moderately rising sea level. In its present state, the system may be divided into thirty-three tidal systems and four estuaries. Similar to terrestrial (fluvial) drainage basins, each of the tidal systems is drained by one unitary stream system and possesses a more or less similar morphometry. In Fig. 10.7, the major elements (beach and nearshore zone, barrier island, ebb-tidal delta, tidal inlet, tidal flats, tidal gullies, and salt marsh) contributing to a tidal system are shown schematically. Tidal flats, tidal gullies (the back-barrier part of) tidal inlets and salt marshes are often combined into one tidal subsystem: a back-barrier bay or tidal basin. On a lower hierarchical level, each of the morphologic elements may be defined as a system with its

Fig. 10.5. Sand supplementation on Sylt, Germany (view from the south towards the north-spit of the island). In the middle of the nourishment the pipe through which the sand–water mixture is flushed on the beach is visible. Note the prominent coastline resulting from the nourishment (photo: Bernd Probst).

Fig. 10.6. Salt-marsh works, brushwood groins, and drainage furrows, in the Wadden Sea of Schleswig-Holstein (view from the sea wall, one of the small Halligen islands, Hallig Habel, in the background).

own elements, etc. In this chapter, only the higher-order elements are described. A detailed description of each of the elements is given in Ehlers (1988) and Oost (1995).

At the seaward border in front of the barrier islands lies the *beach and nearshore zone*. This element reaches from the wave base at about 13 m water depth up to the MHW-line. In response to wave-impact a concave profile develops (Bruun 1988). Superimposed on this profile, nearshore bars may develop that respond sensitively to the wave regime (Kroon 1994; Wolf 1997). The function of this element is a buffer to wave energy. The next element in the direction of the mainland is the *barrier island*. Its seaward border is the already mentioned MHW-line, whilst the back-barrier boundary to the tidal flats or salt marsh normally is, at least in a natural state, transient. The barrier island is an accumulative

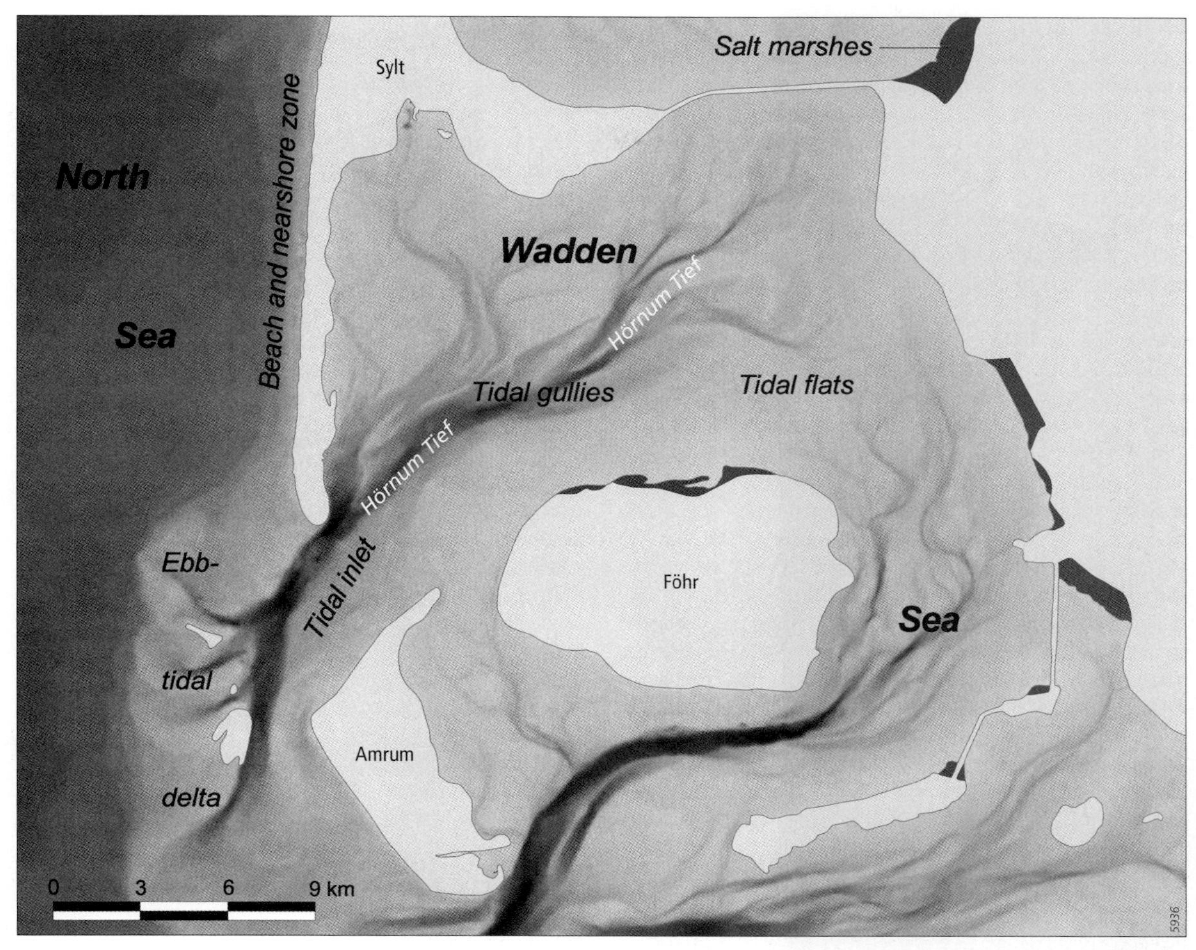

Fig. 10.7. Digital terrain model of a part of the North-German sector of the Wadden Sea displaying the main morphologic elements of a tidal system.

structure where material, stirred up during storms by waves in the nearshore and/or by wind on the beach, may settle. Some prominent morphologic features on barrier islands are dunes, washovers, and spits. In a longshore direction the beach and nearshore zones are interrupted by *ebb-tidal deltas*. An ebb-tidal delta represents an accumulative sand body situated above the wave base (Fig. 10.8). Its highest elevations may exceed the MHW-line. An ebb-tidal delta consists of swash bars situated at a swash platform and a terminal lobe. In this element, material transported by the ebb-tidal currents through the tidal inlet may settle as tidal current velocities decrease (Oertel 1988). The tidal sand body is shaped into its typical structure by tidal currents as well as by wave-impact. Hence, both tidal and wave forces play an important role in the formation and development of ebb-tidal deltas (de Vriend *et al.* 1994; Dombrowsky and Mehta 1996). Landward of the ebb-tidal delta and intercalated between the barrier islands lie the *tidal inlets*. As the name indicates, they act as passages through which the tidal water masses enter and leave the tidal basins during flood and ebb respectively. Usually, an inlet consists of one main ebb channel that may reach maximum depths of about 35 m, and some shallower lateral flood channels (Fig. 10.8). In a seaward direction, the inlet channels cut into the ebb-tidal delta. Because they are strongly interrelated, the ebb-tidal delta and the channels of the tidal inlet are often investigated as one element. Towards the mainland, the tidal channels gradually pass over into tidal gullies. Pending on the

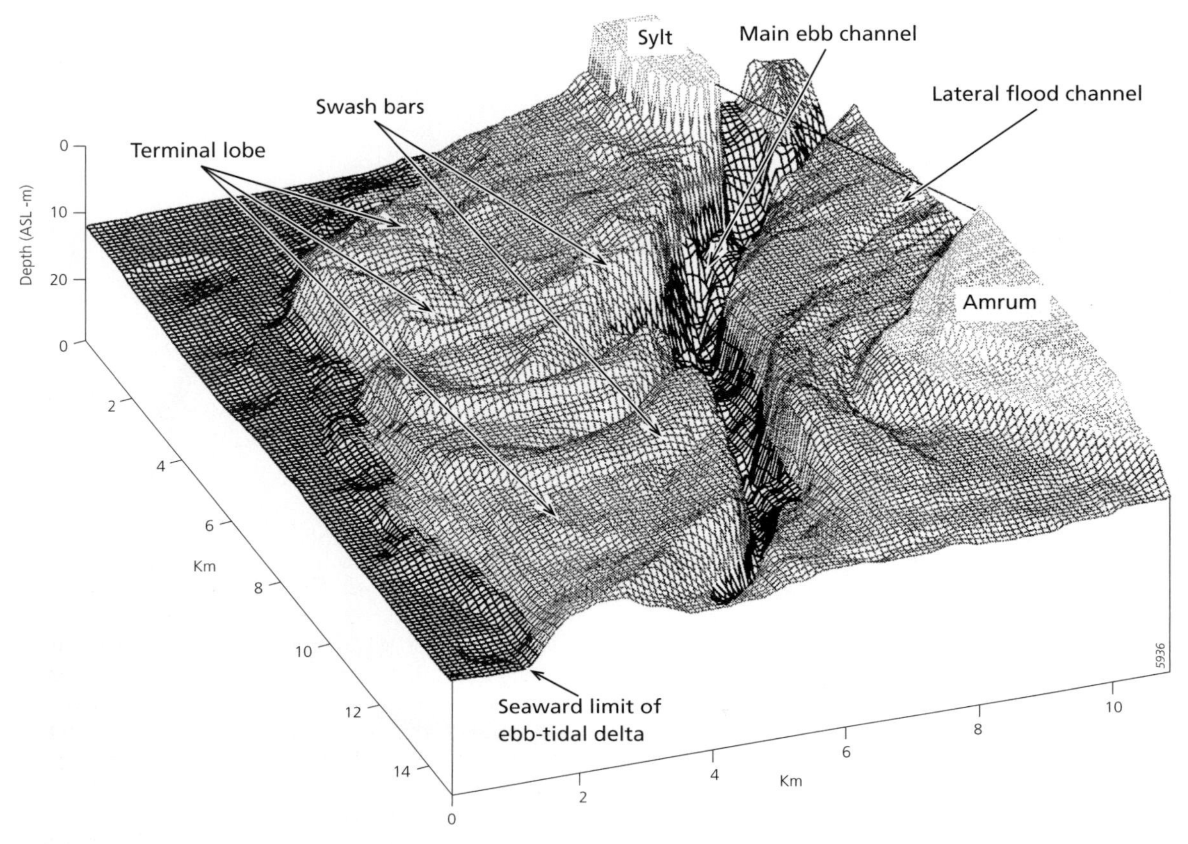

Fig. 10.8. Three-dimensional image of the ebb-tidal delta and tidal inlet Hörnum Tief (see Fig. 10.7) situated between the barriers Sylt and Amrum in the North-German sector of the Wadden Sea (modified after Hofstede 1999*b*).

relative importance of tidal versus wave forces, the seaward orientation of the main ebb channels along the West and East Frisian islands may be asymmetrically in a downdrift (wave dominance) or an updrift (tidal dominance) direction (Sha 1989). Behind the barrier islands the *tidal flats* are situated. This (flat) element normally occupies the largest part of a tidal system. The upward boundary is the MHW-line, the lower boundary is transitional towards the tidal gullies. Pending on local wave regime, the tidal flats may be divided into sandy, mixed, and muddy flats. In the long-term, they represent tidal accumulative structures where particles, brought into the area by flood currents, settle. Significant short-term sediment redistribution occurs during storm surges. The maximum elevation of the tidal flats is defined by the ratio between sedimentation during fair weather and erosion during storms (Göhren 1968). In sheltered areas, sedimentation may prevail, and the tidal flats can evolve into salt marshes.

Tidal currents scour *tidal gullies* into the tidal flats. They may be divided into subtidal channels and (strongly meandering) intertidal creeks. As the tidal inlet, they function as a transport route for the tidal water masses to and from the tidal flats and salt marshes. Tidal gullies are characterized by high meandering or migration rates. During one year they may shift their position over several hundred metres. If sedimentation on the tidal flats exceeds erosion, eventually a *salt marsh* may form. Salt marshes are inter- and supratidal areas of fine sediments stabilized by a halophytic vegetation cover. Boundary conditions for establishing and sustaining salt marshes are an adequate supply of fine sediments, a low-energy environment that allows for sedimentation, regular saltwater inundation, and, finally, a moderate sea-level rise to balance the accumulation and prohibit vegetation succession towards a fresh-water biotope. Their lower boundary is near the MHW-line, their upper boundary is, in a natural state, transitional to brackish and fresh-water biotopes. They represent a tidal accumulative structure whose maximal elevation is limited by the time of tidal inundation. Nowadays, most of the mainland salt marshes in the Wadden Sea are artificial, i.e. developed by salt marsh accretion enhancement techniques (Fig. 10.6). Comprehensive descriptions of the development and dynamics of (artificial) salt marshes along the mainland coast of the Dutch Wadden sea may be found in Houwink *et al.* (1999) and Jansen-Stelder (2000).

All elements are situated in a specific height interval relative to sea level. Hence, they are affected by changes in tidal water levels. Further, each element is driven by a specific combination of natural forces. The beach and nearshore zones and the barrier islands are wave-dominated, whereas the tidal inlet, tidal gullies, tidal flats, and salt marshes are governed by the tide. The ebb-tidal delta presents a mixed element. Tidal inlets and tidal gullies are erosive forms, whereas barrier islands, ebb-tidal deltas, tidal flats, and salt marshes are accumulative structures. Apart from the barrier islands, tidal currents are a positive force in the Wadden Sea, i.e. they conserve the present morphology. Storm wave currents, on the other hand, work negatively on all elements but the barrier islands, i.e. they may cause erosion in accumulative structures and sedimentation in erosive structures.

Each tidal system can be considered to form a more or less separate (closed) sand-sharing system (Dean 1988; Oost 1995). All elements of a sand-sharing system are interrelated and can be in, or strive towards, a dynamic equilibrium (steady state) with the hydrodynamic conditions. Changes in any part of the system will primarily be compensated by sediment transport to or from the other parts of the same system. When changes are temporary and limited, the old steady state will eventually be restored. For example, a moderate increase in sea-level rise induces a stronger accumulation on tidal flats and salt marshes as a result of longer tidal inundation (i.e. the sediment having more time to settle). As a result, the elevation of the flats and salt marshes increases and the time of tidal inundation decreases again until the old dynamic equilibrium is restored. If changes are more permanent or intense (e.g. by loss of area through land reclamation), a new equilibrium will be established after a certain response time. Especially in the last situation, sediment may be imported from or exported to areas outside the sand-sharing system.

Large research efforts have been undertaken to analyse and quantify the complex interactions between the hydrological processes and the morphological responses as a basis for morphodynamic modelling (e.g. de Vriend

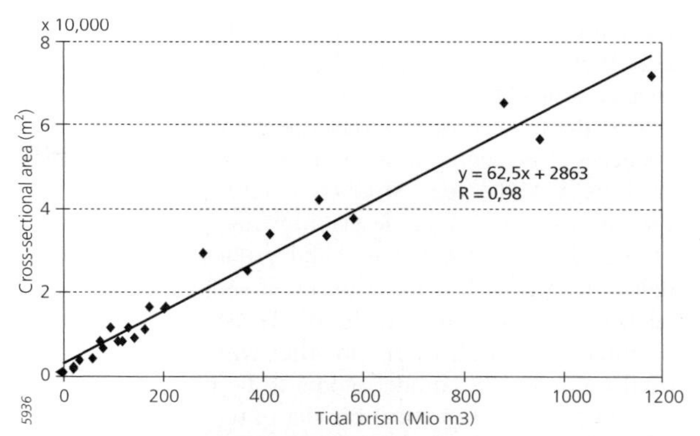

Fig. 10.9. Empirical relation between tidal prism and cross-sectional area of a tidal inlet for twenty-four tidal systems in the Wadden Sea (after Eysink and Biegel 1992).

1992; 1997; de Vriend *et al.* 1993; Ribberink *et al.* 1995). Despite these efforts, still not enough is known about the basic underlying processes to produce robust numerical tools for practical applications (Wang and van der Weck 2002). Hence, most investigations on the geomorphology of the Wadden Sea analyse empirical data. As a result, numerous empirical relations between hydrological and morphological state variables have been established for the Wadden Sea (e.g. Eysink and Biegel 1992; Niemeyer *et al.* 1995). The most significant relations are elaborated by Oost (1995).

Functional relations with strong correlation coefficients were established between the tidal prism (the volume of water that enters a tidal basin during flood and respectively leaves the basin during ebb) and a number of morphological state variables, e.g. cross-sectional area of tidal inlets (Fig. 10.9), mean water depth of tidal gullies, water volume of tidal gullies, intertidal area, and intertidal sediment volume (e.g. Eysink and Biegel 1992). Further, an empirical relation between tidal prism and volume of sediment stored in ebb-tidal deltas for different wave regimes was established by Dean and Walton (1975), and further developed by, amongst others, Walton and Adams (1976), Dean (1988), and Dombrowsky and Mehta (1996). For wave-dominated sandy beach and nearshore zones, Bruun (1962, 1988) formulated an equilibrium profile independent of wave regime and mean particle diameter, which has been much discussed. Finally, Hayes (1979) developed a general model for barrier island morphology as a function of tidal and wave regime, which was further developed by Davis and Hayes (1984).

With the established relations, it is possible to estimate possible morphologic trends resulting from changes in sea level and the driving forces, as well as from human activities. However, these relations can describe only a specific state of the system as the actual development from one state to another cannot be simulated. Recently, progress has been made with semi-empirical models such as ASMITA (e.g. Stive *et al.* 1998). In addition to the empirical relations describing the equilibrium state of the elements tidal flats, tidal gullies, and ebb-tidal delta, in ASMITA a description of sediment transport processes is integrated in the model formulation, using sediment concentrations in the water column and tidal volume. This model seems to be more suitable to predict the morphological impact of sea-level rise to the Wadden Sea than the process-based numerical models (Oost *et al.* 1998; Wang and van der Weck 2002).

Future Developments in the Wadden Sea

The Intergovernmental Panel on Climate Change (IPCC 2001) predicts a global mean temperature rise from 1.4 to 5.8 °C from 1990 to 2100. The resulting heating of the ocean waters and melting of land-based glaciers might cause a sea-level rise of between 0.09 and 0.88 m, with an average between 0.30 and 0.40 m, over the same time period. This large range is due to the broad band of assumptions for the various socio-economic scenarios on which the modelling (with several climate models) was based. Local effects, influencing the tidal wave characteristics, may obscure global sea-level changes. Although regional changes in the atmospheric circulation and, therewith, in storminess may also be expected, climate models currently lack the spatial detail required to make confident predictions.

Using a nested hydro-numerical model for the Inner German Bight, Stengel and Zielke (1994) predict that a moderate rise in MSL will, in the German sector of the Wadden Sea, be accompanied by an increase in the values of mean tidal range of about 30% of the actual MSL rise. The expected rise in MHW will be about 15% larger, the rise in MLW about 15% lower than the rise in MSL. Further, Hoyme and Zielke (2001) concluded on the basis of climate simulations for double and triple CO_2-concentrations in the atmosphere no significant increase in the severity of storms in the German Bight. Von Storch and Reichardt (1997) and the WASA group (1998) calculated slight increases in storm surge heights (0.1 m at Cuxhaven) and storm wave heights (0.0 to 0.5 m) in the German Bight for double CO_2-concentrations. These small increases, however, are well within the range of natural variability observed in the past. Bijl (1997) investigated the possible effects on the storm surge heights of two wind climate scenarios: (1) northward shift of the climate system, and (2) increase in the intensity of storms. Scenario 1 only has a small impact on the storm surge heights in the area. Scenario 2, on the other hand, suggests a high sensitivity of storm surge heights on changes in the intensity of storms.

Morphological Consequences of Changes in Water Levels and Tidal Range

Research on the consequences of a rise in sea level on structure and functions of the Wadden Sea ecosystem is in progress (e.g. Misdorp *et al.* 1990; Eysink 1993; Flemming and Bartholomä 1997; Witez *et al.* 1998). On the basis of a detailed analysis of sediment patterns in the Wadden Sea, Flemming and Bartholomä (1997) predict possible changes in the sediment composition that might result from a rise in sea level. They postulate that a rise in sea level may cause stronger turbulence in the water column, thereby prohibiting finer particles to settle. As a result, the area of sandy tidal flats in the Wadden Sea will increase at the expense of mud flats. Eysink (1993) and Misdorp *et al.* (1990) established empirical models for different sectors of the Wadden Sea by which the morphologic responses of tidal basins to a rise in the values of MSL were simulated. Eysink (1993) expects that a moderate rise in sea level will result in a (temporal) reduction in the area of intertidal flats. According to Misdorp *et al.* (1990), the response of tidal basins depends on the extension of intertidal flats in the basin. If this proportion is high, sea-level rise will be accompanied by increasing tidal prisms. This may induce erosion in the channels and possible sedimentation on the flats. If no intertidal flats are present, no increase in tidal prism will occur. The increase in cross-sectional area, as a result of higher water levels, may cause sedimentation in the channels and possible erosion on the intertidal flats.

These models do not consider the possibility of differential developments in the levels of MHW and MLW or, rather, changes in MTR as predicted by Stengel and Zielke (1994). Witez *et al.* (1998) and Hofstede (2002) investigated the possible morphologic consequences of changes in tidal water levels and tidal range in tidal basins in the German sector of the Wadden Sea. On the basis of nine empirical relationships between hydrological and morphological state variables established for up to twenty-one tidal basins, Hofstede (2002) calculated the resulting morphological trend for the two tidal scenarios (Table 10.1).

Hofstede assumed that, under a moderate sea level rise, the system will strive towards its old steady state. However, with these empirical relationships it is not

TABLE 10.1. *Hydrological scenarios for tidal basins in the German sector of the Wadden Sea*

	MSL	MLW	MHW	MTR
Scenario I	+ 0.30 m	+ 0.25 m	+ 0.35 m	+ 0.10 m
Scenario II	+ 0.50 m	+ 0.40 m	+ 0.60 m	+ 0.20 m

possible to quantify the actual morphological development towards the old steady state. This is mainly the result of the complex morphologic feedback mechanism, which may vary as a function of time and place. Further, the hydrologic forcing is not constant but variable and declining when the old steady state is approached again.

Figure 10.10 clearly shows relationships between hydrological and morphological state variables. Using digital terrain models for twenty-one tidal basins in the German sector of the Wadden sea (Ferk 1995; Spiegel 1997), morphologic state variables were calculated and plotted in diagrams against MHW or tidal prism P. Through the single values regressions curves were fitted for the present (1975) situation as well as for the two tidal scenarios (assuming a constant topography). The correlation coefficients for the relationships are between 0.65 and 0.98. Figure 10.10A displays a weak positive relationship between mean high water (MHW) and the mean elevation of intertidal flats above MLW (Hi). With increasing values of MHW the values of Hi increase linearly. The regressions for both scenarios are situated below the regression curve for the present situation. In order to return to the old steady state, for both scenarios the values of the external hydrographic state variable MHW should decrease and/or the values of Hi should increase (i.e. sedimentation on the tidal flats). Figure 10.10B shows a third-order polynomial relationship between subtidal water volume (PV) and tidal prism (P). Both scenario regressions are situated below the present regression curve. Hence, an increase in the values of PV (i.e. erosion in the tidal gullies) and/or a decrease in the values of P (e.g. by sedimentation on the tidal flats) is needed to re-establish the present state. Figure 10.10C depicts an inverse semi-logarithmic relationship between the tidal prism (P) and the percentage intertidal sediment volume (ISV) to total intertidal volume (IV = ISV + P). With decreasing values of P, the ratio ISV/IV increases logarithmically. In very small tidal basins, the amount of intertidal sediment may reach up to 70% of the total intertidal volume. For tidal prisms larger than about 50 to 100 million m^3, the scenario regressions are situated below the 1975 regression curve. Hence, in tidal basins with values of P > 100 million m^3, an increase in the values of ISV (i.e. sedimentation on the tidal flats) is needed to re-establish the 1975 situation. In small tidal basins, the values of ISV should not change or diminish only slightly to arrive at the 1975 regression. The calculations were done on the basis of a constant topography, not considering the complex mechanism of hydrologic forcing and morphologic responses. Nevertheless, the model calculations allow the following general conclusions regarding the morphologic responses that may be expected from a moderate rise in tidal water levels and tidal range as predicted by the IPCC (2001) and Stengel and Zielke (1994): (1) sedimentation on the tidal flats, (2) vertical erosion in the tidal gullies (Fig. 10.11).

These morphologic trends may be explained by the following process–response mechanisms. Tidal gullies and inlets are governed by the tidal prism, i.e. tidal currents will scour channels that are proportional to the size of the respective tidal prism (Oertel 1988). An increase in tidal prism will induce erosion in the channels as a consequence of increased stress by stronger currents. Vice versa, a reduction in tidal prism will induce sedimentation in the channels. Göhren (1968) describes a positive relationship between the accumulation rate on intertidal flats and water depths or, rather, the time of tidal inundation. With increasing water depths, the time of inundation and, therewith, the period during which material may become settled, increases. By this mechanism the accumulation rate and, in consequence, the height of the tidal flats increase. Hence, larger water depths caused by an increase in the values of MHW results in higher tidal flats. However, with increasing elevation the water depths and, therewith, the time of inundation will decrease again, followed by a reduction in the accumulation rate. As a result of this negative feedback the steady state between the values of MHW and the mean elevation of the tidal flats is maintained. A similar relationship between the accumulation rate and the time of tidal inundation exists in salt marshes (Dijkema *et al.* 1990). This negative feedback is an example of self-regulation on a timescale of decades (Cowell and Thom 1994). However, there is a threshold in the rate of sea-level rise, after which the tidal flats and the salt marshes will not be able to balance sea-level rise by accretion, and a development in the direction of a new steady state begins (Wang and van der Weck 2002). This threshold may vary regionally, depending upon, amongst others, the amount of material available for accumulation.

The tidal process–response mechanisms for tidal gullies (the landward part of) tidal inlets and for tidal flats are combined in a systems approach in Fig. 10.12.

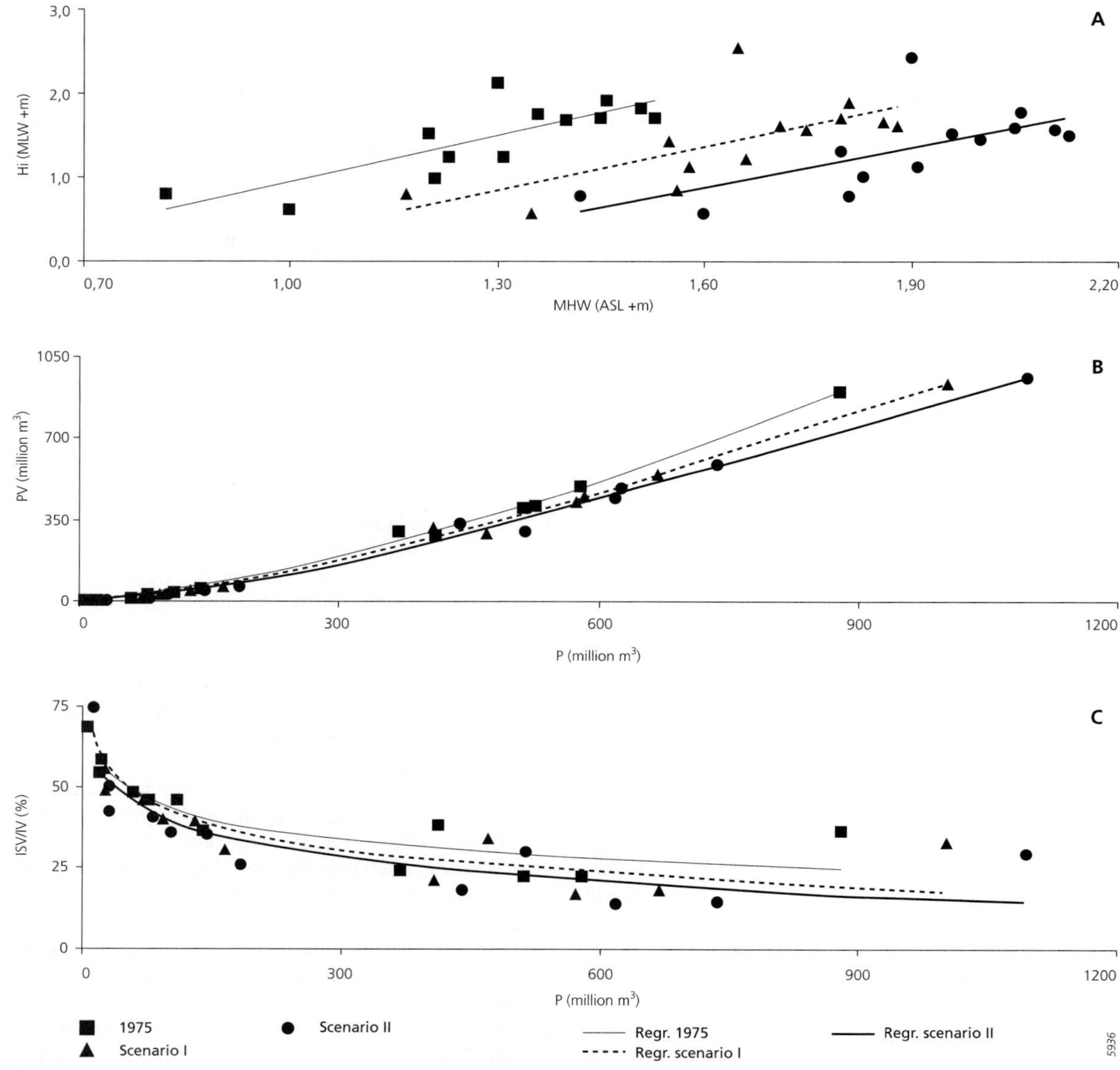

Fig. 10.10. Empirical relations between tidal and morphologic state variables for thirteen tidal systems in the German sector of the Wadden Sea for the situation 1975, for scenario I and for scenario II (for scenarios see text). A. Relationship between mean high water (MHW) and mean elevation of tidal flats above MLW (Hi). B. Relationship between tidal prism (P) and water volume below MLW (PV). C. Relationship between tidal prism (P) and the percentage of intertidal sediment volume (ISV) to total intertidal volume (ISV + P).

Under conditions of an increasing tidal prism, the material that is scoured from the channels by the stronger tidal currents may be transported to the tidal flats during flood. Here, the longer periods of tidal inundation cause an increased accumulation of this material. In consequence, the morphologic responses that result from an increase in tidal water levels and tidal range may, at least partly, be realized by internal sediment redistribution from the channels to the flats. The systems approach shown in Fig. 10.12, however, does not allow for quantitative estimates about the actual sediment redistribution.

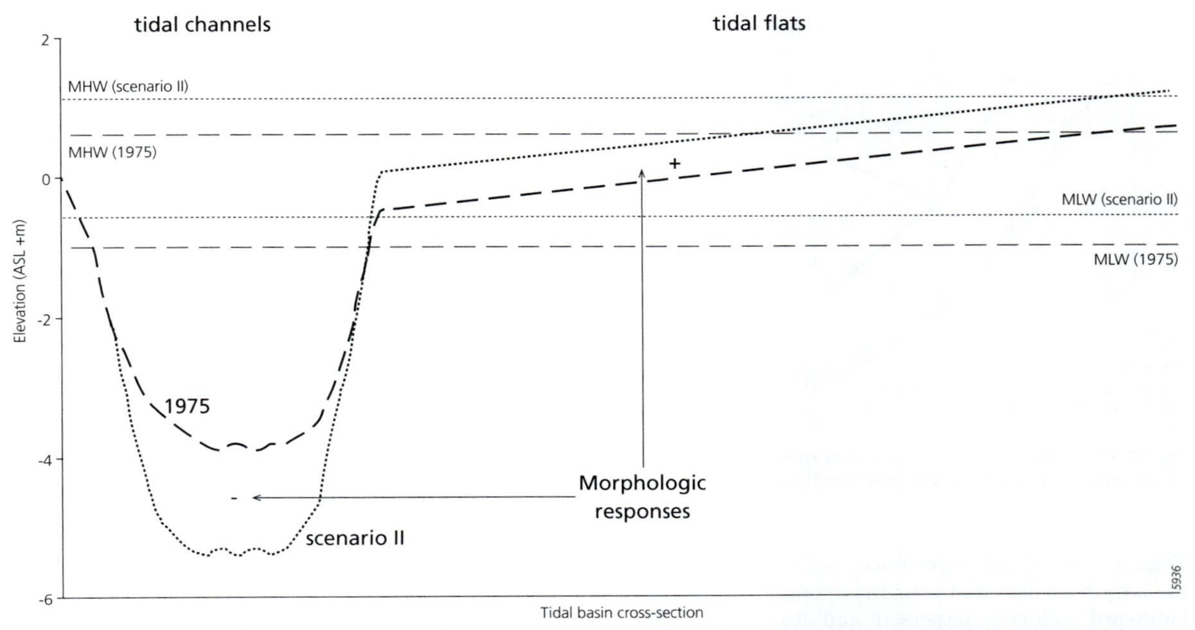

Fig. 10.11. Schematic illustration of the expected morphologic responses in tidal basins resulting from an increase in tidal water levels and tidal range (after Hofstede 2002).

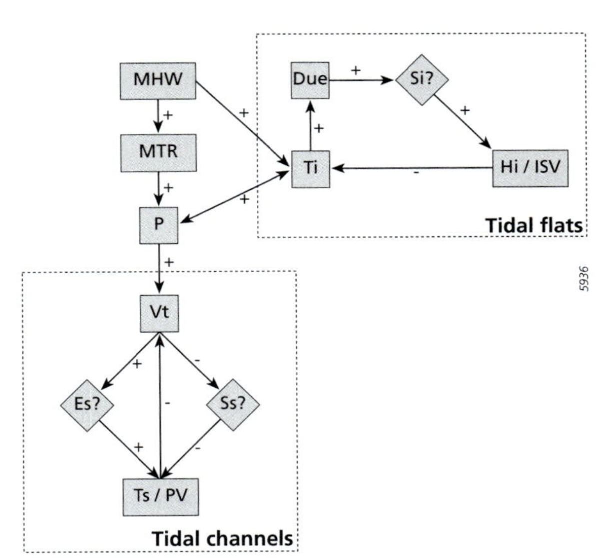

Fig. 10.12. Systems representation of a Wadden Sea tidal basin showing feedback loops between tidal forces and morphologic responses. Explanation: MHW = mean high water; MTR = mean tidal range; P = tidal prism; Ti = water depth tidal flats; Hi = surface elevation tidal flats; ISV = intertidal sediment volume; Due = duration of tidal inundation; Si? = sedimentation tidal flats?; Vt = tidal currents; Es? = erosion tidal channels?; Ss? = sedimentation tidal channels; Ts = water depth subtidal channels; PV = subtidal volume; + = positive correlation; – = negative correlation; → = one-directional correlation; ↔ = bi-directional correlation (after Hofstede 2002).

As described before, salt marshes need a moderate sea-level rise to balance accumulation and prohibit vegetation succession towards a fresh-water biotope. If sea level rise becomes too strong, however, accumulation cannot keep pace and the salt marshes will drown. According to Oost and Dijkema (1993), mainland salt marshes in the Wadden Sea may be able to balance a sea-level rise of up to 1.0 m per century, whereas the barrier salt marshes may balance a sea-level rise of about 0.5 m per century.

According to Bruun (1962, 1988), the morphological reaction of the beach and nearshore zones to sea-level rise, in a closed system, is: (1) erosion during storms in the upper parts of the profile, (2) seaward transport of this material by the undertow, and (3) sedimentation in the lower parts of the profile. By this mechanism, the same profile shape is restored/maintained in a more landward (upward) position. However, as hardly any closed systems exist in nature, import and export of material, either shore-parallel with the littoral drift or landward by washover, complicates the rule (Fig. 10.13). This washover process represents the primary agent for landward migration (rollover) of barrier islands in response to sea-level rise. In a steady state the barrier island does not change its size or shape, but migrates landward. In the Wadden Sea, a number of barrier islands do not migrate landward or even accrete, despite the observed rise in sea level.

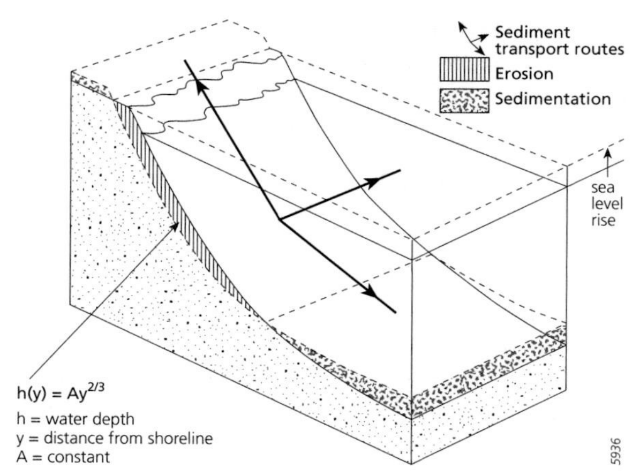

Fig. 10.13. Schematic presentation of morphological adaptations for a sandy coast to a rise in sea level (modified after Bruun 1962).

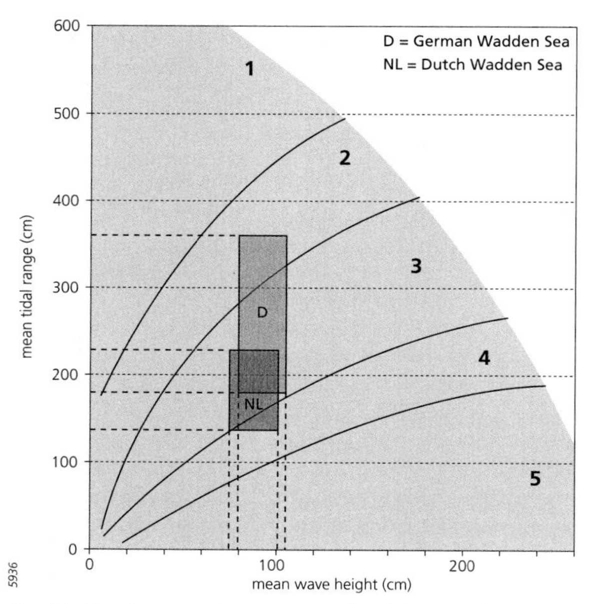

Fig. 10.14. Diagram showing wave height versus tidal range with various fields illustrating general morphologic conditions at sandy coasts: 1. tide-dominated (no barriers, extensive tidal flats). 2. tidal dominance (short 'embryo-like' barriers). 3. mixed energy (drumstick barriers, numerous tidal inlets). 4. wave dominance (long barriers, few tidal inlets). 5. wave-dominated (closed dune coast with occasional inlet/breach). (After Davis and Hayes 1984; included is the Wadden Sea situation.)

These barrier islands experience a sediment supply from the longshore drift that is strong enough to balance the landward rollover (Jespersen and Rasmussen 1994; Hofstede 1999*a*). If the longshore drift does not increase correspondingly, an accelerating sea-level rise might result in a landward retreat on these islands as well.

The effects of an increase in tidal range on barrier islands may be described by the model of Davis and Hayes (1984). Accordingly, the relative influence between wave processes and tidal processes determines the overall morphology of sandy coasts. A total dominance of waves results in a closed dune coast, not interrupted by tidal inlets. With increasing tidal influence, barrier islands intersected by tidal inlets and separated from the mainland by back-barrier bays develop. A further increase in tidal influence results in shorter barrier islands and more inlets until, under tidal dominance, no barrier islands are sustained anymore (Fig. 10.14). O'Brien (1969) introduced an 'inlet stability parameter' quantifying the ratio between wave and tidal energy. Hence, an increase in tidal range (under stable wave conditions) might result in the Wadden Sea in a shortening of the barrier islands. Already, in the Inner German Bight between the Weser and Eiderstedt, no barrier islands exist as a result of tidal ranges in excess of 3.0 m.

The morphological response of ebb-tidal deltas to an increase in tidal range may be evaluated using the model of Dombrowsky and Mehta (1996). They introduced a wave energy to tidal energy ratio for ebb-tidal deltas, similar to the one introduced by O'Brien (1969) for inlets. With increasing ratio, i.e. with increasing influence of tides over waves, the volume of the ebb-tidal delta increases. Hence, an increase in tidal range in the Wadden Sea, under stable wave conditions, may result in larger ebb-tidal deltas.

Morphological Consequences of Changes in Storminess and Storm Surges

Until now, the possible morphological effects of a rise in tidal water levels and tidal range in the different elements of a tidal system are elaborated. Although with the available models significant changes in storminess and storm surges are not expected (von Storch and Reichardt 1997), they cannot be excluded. With the presented empirical relationships of Davis and Hayes (1984), Bruun (1988), and Dombrowsky and Mehta (1996), it is possible to indicate possible morphologic responses of the beach and nearshore zones, ebb-tidal deltas, barrier islands, and tidal inlets resulting from an increase or a reduction in storminess and storm surges. An increase in storm waves and surges may induce erosion in the beach and nearshore zones and the ebb-tidal deltas as well as, at least in a natural situation without human influences, accretion on the barrier islands (washover) and at their spits (longshore drift). By this spit development, the tidal inlets will narrow. A reduction of storm waves and surges and, consequently, an increase in relative importance of the tidal currents, will cause barrier spits to erode and ebb-tidal deltas to grow. Finally, for the beach and nearshore zones, an increase in storm waves may result in flatter profiles, whereas a reduction in storm waves will induce steeper profiles (Wolf 1997).

The importance of storm waves and surges for the stability of tidal flats was formulated by Göhren (1968). He stresses that his tidal process–response mechanism for tidal flats is complicated by the erosive forces of storm waves. An increase in storminess will induce erosion on the tidal flats caused by stronger storm waves and surge currents or, rather, increased shear stress. As a consequence, the mean elevation of the tidal flats will diminish until a new steady state becomes established. With diminishing elevation of the tidal flats, i.e. increasing mean water depths during the tidal cycle, a larger proportion of the tidal prism may be transported over the tidal flats. Hence, less tidal channels would be needed.

Consequently, an increase in storm waves and surges may cause the tidal gullies to silt up. However, larger water depths and, therewith, larger water volumes, may cause higher current velocities during ebb tide, at least in the more seaward parts of the tidal channels. Hence, erosion rather than sedimentation might prevail here. According to Jansen-Stelder (2000), it depends on the severity of the storm whether the morphologic effect on the mud flats and pioneer zone in front of salt marshes is erosion or sedimentation. The vegetation cover of salt marshes presents a highly effective protection against storm waves and surge currents (Erchinger *et al.* 1996). However, an increase in storminess might result in higher storm wave impact by breaking waves along its outer (steeper) margins. Consequently, erosion might occur and cliffs develop. If these cliffs are not stabilized (by salt-marsh works, Fig. 10.6), they may migrate landward until they reach the sea walls, and then the salt marshes will have disappeared (Dijkema *et al.* 1990).

Effects of Hydrological Changes on the Biology

A similar estimation accounts for possible future development of the biology. As stated in the introduction, flora and fauna in the Wadden Sea have adapted to extreme environmental conditions including rapid and intense morphological changes. Hence, as long as the overall structure and functions of the ecosystem remain similar, the present biosystem (composition and mass) may persist as well. If a change from intertidal flats to a more open lagoonal environment should take place, biology would have to adapt to this new situation. Due to the extremely complex nature of interrelationships between the geo- and biosphere, it is not possible to give quantitative estimates of possible changes in flora and fauna, neither in composition nor in mass (Mulder 1993). In general, a strong increase in sea-level rise combined with an increase in storminess and, therewith, a shift from intertidal to lagoonal conditions may have the following effects on biology (CWSS 2001):

- a reduction of benthic biomass (due to an increase in currents and turbulence);
- a reduction in population size of bird species (breeding birds due to a reduction of breeding area, migratory birds due to a reduction of feeding area and time);
- a reduction of the fish nursery function (due to a disappearance of habitats);
- a reduction of the seal population (due to unfavourable haul-out conditions with increasing storminess);
- a decrease in sea grass coverage (due to a reduction in intertidal flats and an increase in turbulence); and
- an increase in the diversity of typical salt-marsh flora (due to a renewal of vegetation stages with increasing time of tidal inundation).

It should be stressed that these results are only general and qualitative. Certain species may even benefit from the reduction in other (competitive) species. Further, biology may be much more directly affected by other human stresses, e.g. pollution, over-exploiting, tourism. However, a growing awareness of the need to manage the Wadden Sea in a sustainable manner (CWSS 1997) will probably lead to a reduction in these negative impacts.

Synthesis

As a general conclusion, it may be stated that the Wadden Sea is a rather robust system. A moderate sea-level rise of less than about 0.5 m per century will probably not cause significant changes in the overall structure and functions of the Wadden Sea, i.e. the morphological and biological responses will remain within the natural variability observed during the last century. A stronger sea-level rise may induce the drowning of tidal flats and salt marshes, thereby changing the system from tidal basins with large intertidal areas into more open, subtidal lagoons (CWSS 2001). The same change in the system may occur if storminess increases significantly. Under this worst-case scenario, with a strong sea-level rise and/or increase in storminess, the biology of the Wadden Sea would be significantly influenced as well.

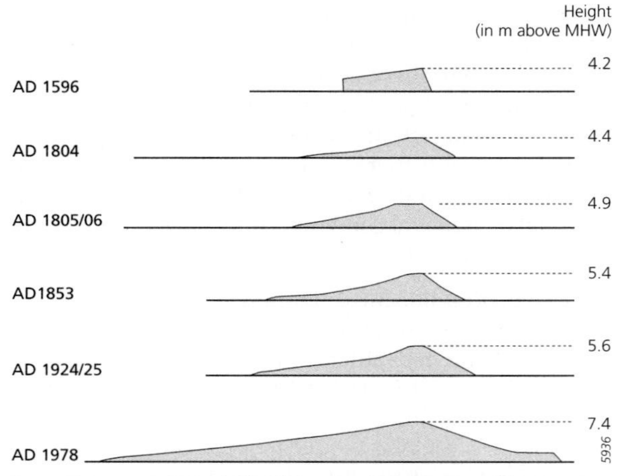

Fig. 10.15. Historical development of dyke profiles in Germany. Note the strong increase in height and width in the twentieth century.

Human Activity in the Wadden Sea

The history of more or less permanent human occupation in the coastal marshes of the Wadden Sea started around the tenth to ninth centuries BC, apart from earlier Neolithic settlements. Until about the tenth century AD, interference with the natural (wet and swampy) environment was relatively low. At first, people settled directly at the surface, e.g. on channel levees. During phases of rising sea level and/or storminess, people constructed dwelling mounds situated above the highest flood level to protect themselves and their cattle. Gerdes *et al.* (2003) distinguish two main phases of dwelling mound construction in northern Germany: from about AD 50–450 and 700–1000. A major breakpoint in human interference occurred at the turn of the first to the second millennium AD. At this time, farmers started to protect their cultivated land against flooding during (lower) summer storms by summer dykes. In the beginning, ring-dykes were constructed around the cultivated land (Meier 2000). Soon afterwards, these isolated ring-dyke systems were connected, and by the end of the fourteenth century a more or less complete sea wall system fronted the coastal marshes of the Wadden Sea. Due to the constricted possibilities these early dykes were rather low and steep (Fig. 10.15), and, as a result, breaches frequently occurred.

The following centuries were characterized by a continuous struggle of the local population against the sea. A number of catastrophic storm floods with considerable loss of land and human life occurred in the Wadden Sea area (e.g. Gottschalk 1975; Ehlers 1988; Oost, 1995). The greatest land loss resulted from the Second Marcellus Flood (*Erste große Mandränke*) in 1362. Jade Bight, Dollart, Harle Bay, and Ley Bay were enlarged, and in north Germany vast areas of land were lost. The All-Saints-Day flood of 1570 was one of the worst storm surges ever to hit the Dutch coast. A further catastrophic storm flood (*Zweite große Mandränke*) occurred in 1634. During these events, thousands of people lost their lives and huge areas of marshland were permanently lost to the sea. The Halligen islands are remainders of these marshes. During low water, indications of former human occupation (ploughing furrows, stone wells) may still be seen on the tidal flats. The Dollart and the Jade Bight reached their maximum extensions in the early sixteenth century. Afterwards, with increasing technical possibilities, more and more coastal marshes were reclaimed from the Wadden Sea. Nevertheless, extreme storm surges occasionally caused the breaching of dykes, e.g. in 1717, 1825, and 1962. During the catastrophic 1962 storm flood, more than 300 people in Hamburg and Niedersachsen lost their lives.

A second major phase in human interference with nature occurred with the industrial revolution in the nineteenth century. A number of significant socio-economic changes with major impacts for the Wadden Sea were initiated. Firstly, the technical capability to build and maintain strong sea walls increased tremendously (Fig. 10.15), allowing people to reclaim and use more land. Secondly, the population, especially of the (favourable) coastal zones strongly increased. As a consequence, the pressure on the environment, both as a resource for food, minerals, etc., and as a sink for contaminants and nutrients, increased as well. Further, the strong growth of the economy resulted in more goods to be transported, e.g. by shipping. Finally, the growing welfare and free time resulted in a (relatively recent) pressure on the coasts from tourism. Especially the high input of contaminants and nutrients, the intense exploitation of the resources, and the pressure of tourists present threats to a natural development of the Wadden Sea (Lozan *et al.* 1994).

Some of these problems may be overcome by integrated planning and management that aims at: (1) a sustainable socio-economic development, (2) the protection of the population against natural hazards (storm floods, land loss), and (3) the preservation of natural dynamics and biodiversity. To achieve these goals, new planning methods have been developed, such as the Integrated Coastal Zone Management—ICZM (European Commission 1999). ICZM is a dynamic, continuous, and iterative process by which decisions are made for the sustainable use, development, and protection of the coastal zones and their resources. Protection of the people against natural hazards in the Wadden Sea may be achieved by an integrated coastal defence management (Hofstede and Hamann 2002). The preservation of natural dynamics and biodiversity is realized in the Wadden Sea by an active conservation policy. Since 1978, the Ministers responsible for environmental protection in the Wadden Sea meet on a regular basis. These meetings resulted in the Trilateral Wadden Sea Plan (CWSS 1997) which is a framework for the overall environmental Wadden Sea management (de Jong *et al.* 1999). Apart from this environmental management plan, major parts of the Wadden Sea have, in the framework of the European Habitat Directive, been included in the ecological network NATURA 2000. Further, the German and Dutch parts of the Wadden Sea have been designated as Biosphere Reserves and, finally, major parts of the Wadden Sea are Ramsar Sites. Biosphere Reserves are protected areas of representative terrestrial and coastal environments that have been internationally recognized under the Man and Biosphere (MAB) program of UNESCO. The Ramsar Convention is a world-wide treaty for the conservation of wetlands.

In conclusion, the present Wadden Sea is one of the world's best-protected coastal environments. The aesthetic and ecological values of the Wadden Sea are increasingly considered as more important than its direct use as an economic resource (Reise *et al.* 1994). Land reclamation for socio-economic uses had already ended during the 1950s, input of contaminants and nutrients has levelled off or diminished, and the populations of a number of endangered species are recovering. However, these environmental successes should not result in a tendency to reduce effort. A number of challenges, e.g. increasing tourism, windmill power plants, gas exploitation, etc., remain that need to be managed in an ecologically sustainable way. The implementation of ICZM strategies that integrate socio-economic, cultural, and ecological principles in a practically oriented, acceptable way is a promising approach to achieve a sustainable development of the Wadden Sea.

References

Bartholdy, J., and Pejrup, M. (1994), Holocene evolution of the Danish Wadden Sea. *Senckenbergiana Maritima* 24: 187–209.

Becker, G. (1998*a*), Salinity and water temperatures in the Wadden Sea. In: J. Kohlus and H. Küpper (eds.) *Environmental Atlas of the Wadden Sea*, i. *Wadden Sea of North Frisia and Dithmarschen*. Stuttgart, Eugen Ulmer, 35–6.

—— (1998*b*), Water temperatures in the Wadden Sea. In: J. Kohlus and H. Küpper (eds.) *Environmental Atlas of the Wadden Sea*, i. *Wadden Sea of North Frisia and Dithmarschen*. Stuttgart, Eugen Ulmer, 36.

Bijl, W. (1997), Impact of a wind climate change on the surge in the southern North Sea. *Climate Research* 8: 45–59.

Brundtland-Commission/World Commission on Environment and Development (ed.) (1987), *Our Common Future*. WCED, Oxford.

Bruun, P. (1962), Sea level rise as a cause of shore erosion. *Journal of the Waterways and Harbours Division* 88: 107–30.

—— (1988), The Bruun rule of erosion by sea level rise: a discussion of large-scale two- and three-dimensional usages. *Journal of Coastal Research* 4: 627–48.

Carter, R. W. G. (1988), *Coastal Environments*. Academic Press, London.

Cowell, P. J., and Thom B. J. (1994), Morphodynamics of coastal evolution. In: R. W. G. Carter and C. D. Woodroffe (eds.) *Coastal evolution: Late Quaternary shoreline morphodynamics*. Cambridge University Press, Cambridge, 33–86.

CWSS (eds.) (1997), Ministerial Declaration of the Eighth Trilateral Governmental Conference on the Protection of the Wadden Sea, Stade, 1997, Annex 1: Trilateral Wadden Sea Plan, Wilhelmshaven, Common Wadden Sea Secretariat—CWSS, 13–83.

—— (eds.) (2001), Preliminary conclusions and recommendations of the trilateral working group on coastal protection and sea level rise. Internal report prepared for the TWG. Wilhelmshaven, Common Wadden Sea Secretariat—CWSS.

Davis, R. A. (1994), Barrier island systems—a geologic overview. In: R. A. Davis (ed.) *Geology of Holocene Barrier Island Systems*. Springer, Berlin, 1–46.

—— and Hayes M. O. (1984), What is a wave dominated coast? *Marine Geology* 60: 313–29.

Dean, R. G. (1987), Additional sediment input into the nearshore zone. *Shore Beach* 55: 76–81.

—— (1988), Sediment interaction at modified coastal inlets. *Lecture Notes on Coastal and Estuarine Studies* 29: 412–39.

—— and Walton, T. L. (1975), Sediment transport processes in the vicinity of inlets with special reference to sand trapping. In: J. L. Cronin (ed.) *Estuarine Research*, ii. *Geology and Engineering*. Academic Press, New York, 129–49.

de Jong, F., Bakker, F. J., van Berkel, C. J. M., Dankers, N. M. J. A., Dahl, K., Gätje, C., Marencic, H., and Potel, P. (1999), Wadden Sea Quality Status Report. *Wadden Sea Ecosystem* 9.

de Vriend, H. J. (1992), Mathematical modelling and large-scale coastal behaviour. Part 1. Physical processes; Part 2. Predictive models. *Journal of Hydraulic Research, Special Issue, Maritime Hydraulics*, 727–53.

—— (1997), Evolution of marine morphodynamic modelling: time for 3-D? *German Journal of Hydrography* 49: 331–41.

—— Zyserman, J., Nicholsen, J., Pechon, Ph., Roelvink, J. A., and Southgate, H. N. (1993), Medium-term 2-DH coastal area modelling. *Coastal Engineering* 21: 193–224.

—— Bakker, W. T., and Bilse, D. P. (1994), A morphological behaviour model for the outer delta of mixed-energy tidal inlets. *Coastal Engineering* 23: 305–27.

Dieckmann, R. (1985), Geomorphologie, Stabilitäts- und Langzeitverhalten von Watteinzugsgebieten der Deutschen Bucht. *Mitteilungen des Franzius-Instituts der Universität Hannover* 60: 133–361.

Dijkema, K. S., Bossinade, J. H., Bouwsema, P., and de Glopper, R. J. (1990), Salt marshes in the Netherlands Wadden Sea: rising high-tide levels and accretion enhancement. In: W. J. Wolff and J. J. W. M. Brouns (eds.) *Expected effects of climatic change on marine coastal ecosystems*. Dordrecht, Kluwer, 173–88.

Dombrowsky, M. R., and Mehta, A. J. (1996), Ebb tidal delta evolution of coastal inlets. *Proceedings of the 25th International Conference on Coastal Engineering*. American Society for Civil Engineers, New York, 3270–83.

Ehlers, J. (1988), *The morphodynamics of the Wadden Sea*. Balkema, Rotterdam.

Eisma, D. (1993), *Suspended matter in the aquatic environments*. Berlin, Springer-Verlag.

Erchinger, H. F., Coldewey, H.-G., and Meier, C. (1996), Interdisziplinäre Erforschung des Deichvorlandes im Forschungsprojekt 'Erosionsfestigkeit von Hellern'. *Die Küste* 58: 1–45.

European Commission (eds.) (1999), A European strategy for Integrated Coastal Zone Management (ICZM): general principles and policy options. http://europa.eu.int/comm/environment/iczm/home.htm, accessed 8 June 2004.

Eysink, W. D. (1993), Impact of sea-level rise on the morphology of the Wadden Sea within the scope of its ecological functioning. General considerations on hydraulic conditions, sediment transports, sand balance, bed composition and impact of sea-level rise on tidal flats. The Hague, Rijkswaterstaat Dienst Getijdewateren, ISOS*2 project, phase 4.

—— and Biegel, E. J. (1992), Impact of sea-level rise on the morphology of the Wadden Sea within the scope of its ecological functioning. Investigations on empirical morphological relations. The Hague, Rijkswaterstaat Dienst Getijdewateren, ISOS*2 project, phase 2.

Ferk, U. (1995), Folgen eines beschleunigten Meeresspiegelanstiegs für die Watteinzugsgebiete der niedersächsichen Nordseeküste. *Die Küste* 57: 135–56.

Flemming, B. W., and Bartholomä, A. (1997), Response of the Wadden Sea to a rising sea level: a predictive empirical model. *German Journal of Hydrography* 49: 343–53.

Führböter, A. (1984), Über mikrobiologische Einflüsse auf den Erosionsbeginn bei Sandwatten. *Wasser und Boden* 3: 106–16.

Gerdes, G., Petzelberger, B. E. M., Scholz-Böttcher, B. M., and Streif, H. (2003), The record of climatic change in the geological archives of shallow marine, coastal, and adjacent lowland areas of Northern Germany. *Quaternary Science Reviews* 22: 101–24.

Göhren, H. (1968), Triftströmungen im Wattenmeer. *Mitteilungen des Franzius-Instituts der TU Hannover* 30: 142–270.

Gottschalk, M. K. E. (1975), Stormvloeden en rivieroverstromingen in Nederland, deel 2, de periode 1400–1600. *Sociaal Geografische Studies* 13.

Hagemann, B. P. (1969), Development of the Western part of the Netherlands during the Holocene. *Geologie en Mijnbouw* 48: 373–88.

Hayes, M. O. (1979), Barrier island morphology as a function of tidal and wave regime. In: S. P. Leatherman (ed.) *Barrier Islands*. Academic Press, New York, 1–29.

Heydemann, B. (1998), Biology of the Wadden Sea. In: J. Kohlus and H. Küpper (eds.) *Environmental Atlas of the Wadden Sea, i. Wadden Sea of North Frisia and Dithmarschen*. Eugen Ulmer, Stuttgart, 44–7.

Hofstede, J. L. A. (1991), Sea level rise in the Inner German Bight since AD 600 and its implications on tidal flat geomorphology. In: H. Brückner and U. Radtke (eds.) *From the North Sea to the Indian Ocean*. Franz Steiner, Stuttgart, 10–27.

—— (1997), Process–response analyses for the North-Frisian supratidal sands (Germany). *Journal of Coastal Research* 13: 1–7.

—— (1999*a*), Regional differences in the morphologic behaviour of four German Wadden Sea barriers. *Quaternary International* 56: 99–106.

—— (1999*b*), Process–response analysis for Hörnum tidal inlet in the German sector of the Wadden Sea. *Quaternary International* 60: 107–17.

—— (1999*c*), Morphologische Auswirkungen eines Klimawandels im Wattenmeer. *Petermanns Geographische Mitteilungen* 142: 305–14.

—— (2002), Morphologic responses of Wadden Sea tidal basins to a rise in tidal water levels and tidal range. *Zeitschrift für Geomorphologie* 46: 93–108.

—— and Hamann, M. (2002), Integrated management of coastal defence in Schleswig-Holstein: experiences and challenges. In: G. Schernewski and U. Schiewer (eds.) *Baltic Coastal Ecosystems. Central and Eastern European Development Studies*. Springer, Berlin, 377–88.

Hoselmann, C., and Streif, H. (1998), Methods used in a mass-balance study of Holocene sediments accumulation on the southern North Sea coast. In: J. Harff, W. Lemke, and K. Stattegger (eds.) *Computerized modelling of sedimentary systems*. Springer, Berlin, 361–74.

Houwink, E. J., van Duin, W. E., Smit-van der Waaij, Y., Dijkema, K. S., and Terwindt, J. H. J. (1999), Biological and abiotic factors influencing the settlement and survival of *Salicornia dolichostachia* in the intertidal pioneer zone. *Mangroves and Salt Marshes* 43: 197–206.

Hoyme, H., and Zielke, W. (2001), Impact of climate changes on wind behaviour and water levels at the German North Sea coast. *Estuarine, Coastal and Shelf Science* 53: 451–8.

IPCC—Intergovernmental Panel on Climate Change (eds.) (2001), *Climate Change 2001, the scientific basis: summary for policy makers*. http://www.ipcc.ch, accessed 8 June 2004.

Jansen-Stelder, B. (2000), *A system analysis of salt marsh development along the mainland coast of the Dutch Wadden Sea*. Thesis, Department of Physical Geography, Utrecht University.

Jensen, J., Hofstede, J. L. A., Kunz, H., de Ronde, J., Heinen, P. F., and Siefert, W. (1993), Long term water level observations and variations. In: R. Hillen, and H. J. Verhagen, (eds.) *Coastlines of the southern North Sea*. New York, American Society for Civil Engineers, 100–30.

Jespersen, M., and Rasmussen, E. (1994), Koresand—die Entwicklung eines Außensandes vor dem dänischen Wattenmeer. *Die Küste* 56: 79–91.

Komar, P. D. (1998), *Beach processes and sedimentation*. 2nd edn. Prentice Hall, Englewood Cliffs, NJ.

Kroon, A. (1994) *Sediment transport and morphodynamics of the beach and nearshore zone near Egmond, The Netherlands*. Thesis, Department of Physical Geography, Utrecht University, Utrecht.

Lozan, J. L., Rachor, E., Reise, K., von Westernhagen, H., and Lenz, W. (eds.) (1994), *Warnsignale aus dem Wattenmeer*. Blackwell, Berlin.

Meier, D. (2000), Landschaftsgeschichte, Siedlungs- und Wirtschaftsweise der Marsch. In: Verein Ditmarschens Landeskunde (ed.) *Geschichte Dithmarschens*. 71–92.

Mengelkamp, H.-T., Bauer, M., and Schmidt, H. (1998), Wind and its uses. In: J. Kohlus and H. Küpper (eds.) *Environmental Atlas of the Wadden Sea, i. Wadden Sea of North Frisia and Dithmarschen*. Stuttgart, Eugen Ulmer, 39.

Misdorp, R., Steyaert, F., Hallie, F., and de Ronde, J. (1990), Climate change, sea level rise and morphological developments in the Dutch Wadden Sea, a marine wetland. In: J. J. Beukema, W. J. Wolff, and J. W. M. Brouns (eds.) *Expected effects of climate change on marine coastal ecosystems*. Kluwer Academic, Dordrecht, 123–31.

Mulder, J. P. M. (1993), Effects of an increased sea level rise on the geomorphology and ecological functioning of the Wadden Sea. *Wadden Sea News Letter* 1993/1: 3–5.

Niemeyer, H. D., Goldenbogen, R., Schroeder, R., and Kunz, H. (1995), Untersuchungen zur Morphodynamik des Wattenmeeres im Forschungsvorhaben WADE. *Die Küste* 57: 65–94.

O'Brien, M. P. (1969), Equilibrium flow areas of inlets on sandy coasts. American Society for Civil Engineers. *Journal of the Waterways and Harbours Division* 95: 43–52.

Oerlemans, J. (1993), Factors contributing to sea level rise on a decade to century time scale. In: S. Jelgersma *et al.* (eds.) *Sea level changes and their consequences for hydrology and water management*. The Hague, RIKZ, 43–8.

Oertel, G. F. (1988), Processes of sediment exchange between tidal inlets, ebb deltas and barrier islands. *Lecture Notes on Coastal and Estuarine Studies* 29: 297–318.

Oost, A. P. (1995), Dynamics and sedimentary development of the Dutch Wadden Sea with emphasis on the Frisian Inlet. *Geologica Ultraiectina* 126, Utrecht University.

—— and de Boer, P. L. (1994), Sedimentology and development of barrier islands, ebb-tidal deltas and backbarrier areas of the Dutch Wadden Sea. *Senckenbergiana Maritima* 24: 65–105.

—— and Dijkema, K. S. (1993), *Effecten van bodemdaling door gaswinning in de Waddenzee*. Texel, DLO-Instituut voor Bos- en Natuuronderzoek Texel, IBN-Rapport 025.

—— Ens, B. J., Brinkman, A. G., Dijkema, K. S., Eysink, W. D., Beukema, J. J., Gussinklo, H. J., Verboom, B. M. J., and Verburgh, J. J. (1998), *Integrale bodemdalingsstudie Waddenzee*. ISMN 90-804791-4-4.

Pejrup, M., Larsen, M., and Edelvang, K. (1997), A fine-grained sediment budget for the Sylt-Rømø tidal basin. *Helgoländer Meeresuntersuchungen* 51: 253–68.

Pethick, J. (1984), *An introduction to coastal geomorphology*. Arnold, London.

Postma, H. (1961), Transport and accumulation of suspended matter in the Dutch Wadden Sea. *Netherlands Journal of Sea Research* 1: 148–90.

—— (1981), Exchange of materials between the North Sea and the Wadden Sea. *Marine Geology* 40: 199–213.

Reise, K., Lozan, J. L., Rachor, E., und von Westernhagen, H. (1994), Ausblick: Wohin entwickelt sich das Wattenmeer? In: J. L. Lozan *et al.*, *Warnsignale aus dem Wattenmeer*. Blackwell, Berlin, 343–8.

Ribberink J. S., Negen, E. H., and Hartsuiker, G. (1995), Mathematical modelling of coastal morphodynamics near a tidal inlet system. In: W. R. Dally and R. B. Zeidler (eds.) *Coastal dynamics* '95, New York, American Society for Civil Engineers, 915–26.

Sha, L. P. (1989), Variation in ebb-delta morphologies along the West and East Frisian islands, The Netherlands and Germany. *Marine Geology* 89: 10–28.

—— (1992), *Geological research in the ebb-tidal delta of 'het Friese Zeegat', Wadden Sea, the Netherlands*. State Geological Survey-project 140010.

Spiegel, F. (1997), Zur Morphologie der Tidebecken im schleswig-holsteinischen Wattenmeer. *Die Küste* 59: 105–42.

Stengel, T., and Zielke, W. (1994), Der Einfluß eines Meeresspiegelanstieges auf Gezeiten und Sturmfluten in der deutschen Bucht. *Die Küste* 56: 93–108.

Stive, M. J. F., Capobianco, M., Wang, Z. B., Ruol, P., and Buijsman, M. C. (1998), Morphodynamics of a tidal lagoon and the adjacent coast. In: J. Dronkers and M. Scheffers (eds.) *Physics of estuaries and coastal seas*. Balkema, Rotterdam, 397–407.

Streif, H. (1989), Barrier islands, tidal flats, and coastal marshes resulting from a relative rise of sea level in East Frisia on the German North Sea coast, *Proceedings KNGMG Symposium Coastal Lowlands, Geology and Geotechnology*. Kluwer Academic, Dordrecht, 213–33.

—— (1993), Geologische Aspekte der Klimawirkungsforschung. In: H.-J. Schellnhuber and H. Sterr (eds.) *Klimaänderung und Küste, Einblick ins Treibhaus*. Kluwer Academic, Dordrecht, 77–96.

—— (2004), Sedimentary record of Pleistocene and Holocene marine inundations along the North Sea coast of Lower Saxony, Germany. *Quaternary International* 112: 3–28.

ten Brinke, W. B. M. (1993), *The impact of biological factors on the deposition of fine-grained sediment in the oosterschelde*. Thesis, Department of Physical Geography, Utrecht University.

Töppe, A. (1993), Longtime cycles in mean tidal levels, In: IHP/OHP National Committee of Germany (eds.) *Proceedings of the international workshop SEACHANGE '93—a contribution to the UNESCO IHP-IV project H-2-2*, Bundesanstalt für Gewässerkunde, Koblenz, I133–I143.

van der Spek, A. J. F., and Beets, D. J. (1992), Mid-Holocene evolution of a tidal basin in the western Netherlands: a model for future changes in the northern Netherlands under conditions of accelarated sea-level rise? *Sedimentary Geology* 80: 185–97.

van Rijn, L. C. (1993), *Principles of sediment transport in rivers, estuaries and coastal seas*. Aqua, Amsterdam.

van Straaten, L. M. J. U. (1964), De bodem der Waddenzee. In: W. F. Anderson, J. Abrahamse, J. D. Buwalda, and Ph. H. Kuenen (eds.) *Het Waddenboek*. Thieme & Co, Zutphen, 75–151.

—— and Ph. H. Kuenen (1958), Tidal action as a cause of clay accumulation. *Journal of Sedimentary Petrology* 28: 406–13.

von Storch, H., and Reichardt, H. (1997), A scenario of storm surge statistics for the German Bight at the expected time of doubled atmospheric carbon dioxide concentration. *Journal of Climate* 10: 2653–62.

Walton, T. L., and Adams, W. D. (1976), Capacity of inlet outer bars to store sand. *Proceedings of the 15th International Conference on Coastal Engineering*, New York, American Society For Civil Engineers, 1919–37.

Wang, Z. B., and van der Weck, A. (2002), *Sea-level rise and morphological development in the Wadden Sea, a desk study*. Report Z3441, Wl/Delft Hydraulics, The Netherlands.

WASA (1998), Changing waves and storms in the Northeast Atlantic? *Bulletin of the American Meteorological Society* 79/5: 741–60.

Witez, P., Bock, S., and Hofstede, J. L. A. (1998), *Modelluntersuchungen zur morphologischen Stabilität des Wattenmeeres bei einem beschleunigten Meeresspiegelanstieg*. Endbericht zum BMBF-Forschungsvorhaben MTK 0569, Kiel.

Wolf, F. C. J. (1997), *Hydrodynamics, sediment transport, and daily morphological development of a bar-beach system*. Thesis, Department of Physical Geography, Utrecht University, Utrecht.

Zanke, U. (1982), *Grundlagen der Sedimentbewegung*. Springer, Berlin.

11 German Uplands and Alpine Foreland

Eduard Koster

The Major Landform Regions of Germany

From north to south in Germany there is a rough symmetry in the distribution of the major geological and landform units (Fig. 11.1). Quaternary glacial and fluvioglacial deposits and landforms characterize the Northern Lowlands and the Alpine Foreland in the south. Relief in both these areas is relatively flat, mostly of the order of a few tens of metres to 200 metres. The central part of the country, roughly between a line from Bonn–Dortmund–Hannover–Leipzig–Dresden (Fig. 11.2) in the north and the river Danube in the south, is dominated by uplands and basins, mainly consisting of Palaeozoic and Mesozoic rocks, exhibiting a relief of several hundred metres. This central region is bordered in the western and eastern part by fault block mountains and massifs consisting of Palaeozoic, partly crystalline rocks. These massifs attain heights of *c.*500–1,500 m. Based on a combination of morphotectonic evolution and landform associations, most authors distinguish five major landform regions (Fig. 11.1):

- The North German Lowlands as a part of the North European Lowlands, extending from the north-western tip of France, through Belgium and The Netherlands to the Polish–Russian border and beyond. The southern border of this region more or less coincides with the 100–200 m contour lines as well as with the maximum extension of the Fennoscandian ice sheets. The usual thickness of the glacial/fluvioglacial sediment sequence is between 100 and 300 m; the maximum thickness is almost 500 m. In contrast to Ahnert (1989*b*), the Lower Rhine graben and the Münster Embayment are included in this region by Semmel (1996) and Liedtke and Marcinek (2002).
- The Central German Uplands. This region is characterized by a relief between 200 and 1,000 m, locally to 1,500 m, old Palaeozoic (Variscan) massifs, denudational landforms with planation surfaces, cuestas, hogbacks, basins, and deeply incised river valleys. Concerning the southern border of this region there also appears to be some difference of opinion. Semmel (1996) obviously includes the Saar-Nahe Upland and the Thüringer Wald, the Erzgebirge, the Bayerischer Wald, and Böhmer Wald. This is also the case with the geomorphic map in the *Nationalatlas* by Liedtke *et al.* (2003). Liedtke and Marcinek (2002), however, do not include the Saar-Nahe Upland nor the Bayerischer Wald and Böhmer Wald.
- The South German Scarplands (and the Upper Rhine graben). From a morphographical point of view this region resembles that of the Central German Uplands. However, in the central part of this region Mesozoic rocks predominate, and the landforms show an irregular pattern of uplifted plateaus and basins with characteristic cuesta and hogback forms. This central region is surrounded by partly crystalline mountain ranges in the west (Schwarzwald) and east (Bayerischer Wald and Böhmer Wald). The Upper Rhine graben is usually included in this fourth landform region. The southern boundary of this region almost perfectly coincides with the course of the Danube river.

The uplands in central and southern Germany are usually referred to as *Mittelgebirge*, meaning literally *middle mountains*, because they attain only moderate heights

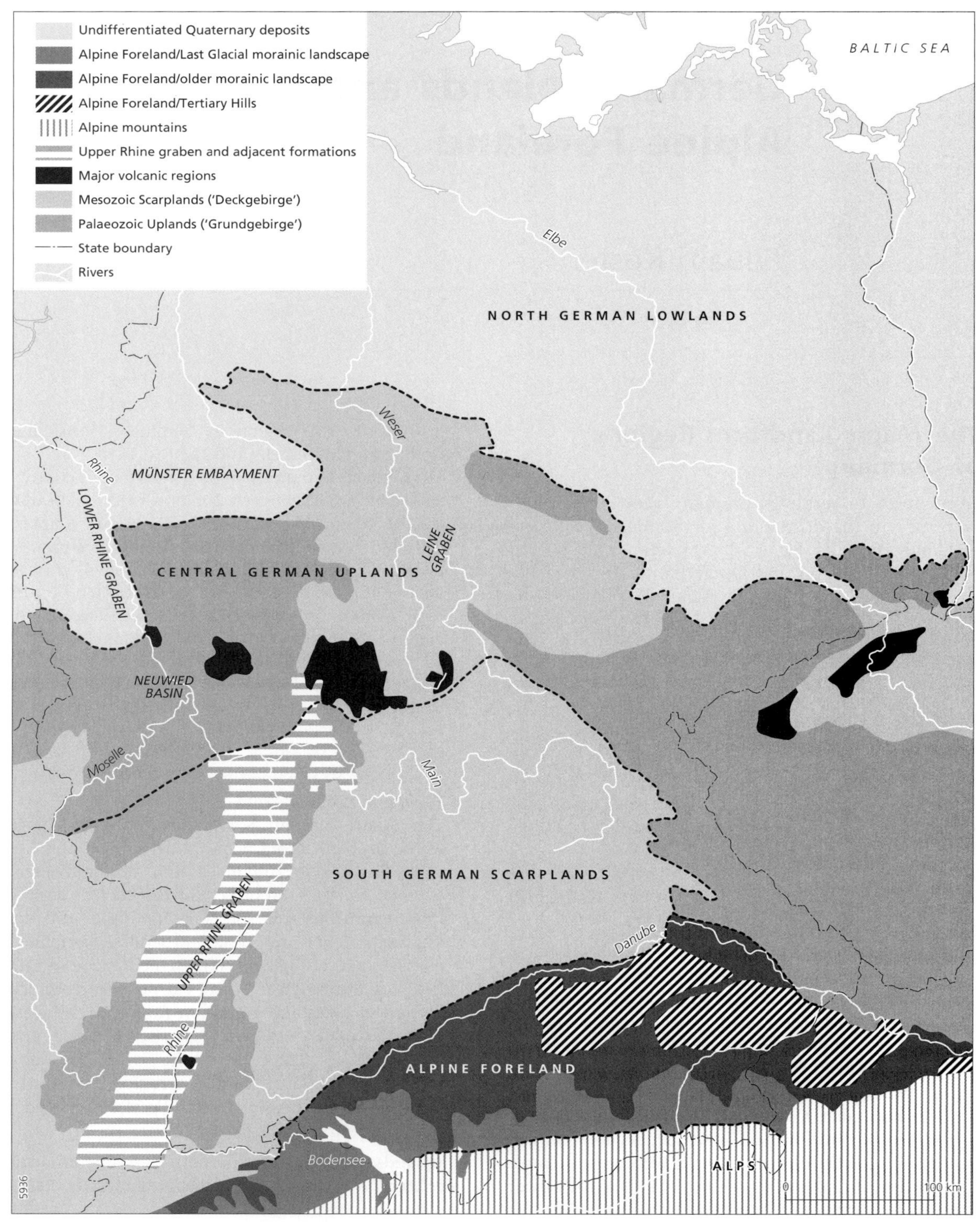

Fig. 11.1. Distribution of the main landform regions and geomorphological units in Germany (simplified after Semmel 1996 and Liedtke 2002).

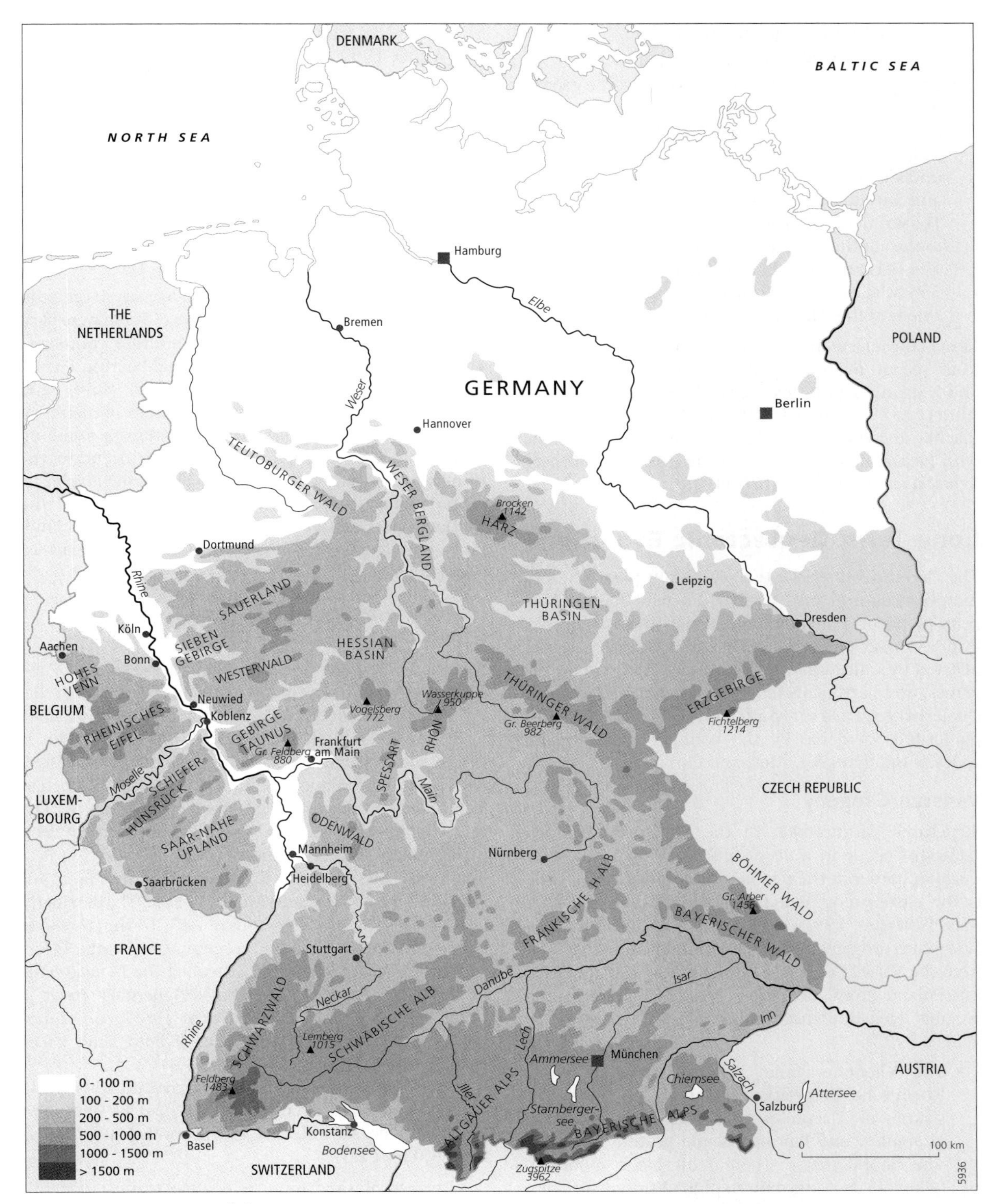

Fig. 11.2. Location map of Germany.

compared with the Alpine mountain chain to the south (Embleton and Demek 1984).

- The Alpine Foreland. Strictly speaking, the geomorphological boundary and the tectonic boundary between the Foreland and the Alps are not identical. The tectonic boundary is drawn between the undisturbed Molasse sediments (Foreland Molasse) and the faulted Molasse sediments (Alpine Molasse). The geomorphological edge of the Alps lies somewhat to the south of the tectonic edge (Fischer 2002).
- The German Alps. This is a narrow zone less than 35 km wide, which forms only *c*.2.2% of the areal extent of the Alpine mountain system.

Most of the information for this chapter has been obtained from recent overviews on the geology (Henningsen and Katzung 2002), geomorphology (Embleton 1984; Ahnert 1989*a*; Semmel 1996), and physical geography (Liedtke and Marcinek 2002) of Germany, supplemented with factual data from the *Nationalatlas Bundesrepublik Deutschland* (Liedtke *et al.* 2003).

Long-Term Geotectonic Evolution

The German *Mittelgebirge* represents a part of a large morphostructural unit termed the west-central European platform (Embleton and Demek 1984). This platform consists of pre-Cambrian and Palaeozoic rocks affected by Caledonian and Variscan (Hercynian) orogenies, and partly covered by sediments of Mesozoic to Quaternary age. On the south and south-eastern sides this platform is bordered by the younger epeirosynclinal zones of the Pyrenees, Alps, and Carpathians.

Variscan Orogeny

Structures dating back to the Caledonian orogeny (Silurian) occur in a very small area at the Belgian–German border in the region of the Hohes Venn, which is the easternmost extension of the London–Brabant massif (Walter 1992). The late Palaeozoic Hercynian (Variscan) orogeny formed a mountain chain extending from the Massif Central in France across central Europe into Poland. From north to south this Hercynian fold belt is subdivided into three major zones (Ahnert 1989*b*; Walter 1992):

- The Rhenohercynian zone, represented by the Rheinisches Schiefergebirge and Harz, mainly consisting of Devonian and Carboniferous slates, quartzites, and limestones and granite intrusions. The boundaries of these fault block mountains follow the tectonic pattern of the Variscan orogeny, orientated either SW–NE (Variscan orientation) or SE–NW (Hercynian orientation) (Andres 1989; Walter 1992).
- The Saxothuringian zone, including the Saar-Nahe Upland, the Odenwald, the Spessart, the Thüringer Wald, and the Erzgebirge. In addition to crystalline rocks and local volcanics, Carboniferous and Permian sedimentary rocks predominate.
- The Moldanubian zone, with extensive metamorphic, Precambrian, and Palaeozoic rocks, including granite intrusions, to which the Schwarzwald, the Bayerischer Wald, and the Böhmer Wald belong.

During and following the Variscan orogeny large sedimentary basins formed in central Europe. These were filled up by debris from the Variscan mountain complexes as well as by upper Carboniferous coal-bearing formations. These formations mainly contain clayey and sandy strata with a thickness of hundreds of metres to several kilometres, of which the coal layers make up only about 1.5%. Large coal resources occur in the Ruhr area just north of the Rheinisches Schiefergebirge, in the region around Aachen on both sides of the Netherlands–German border, in the Saar-Nahe Upland, and in various regions in eastern and south-eastern Germany.

Late Palaeozoic and Mesozoic Sedimentary Sequence

Denudation caused the mountain ranges of the Variscan orogeny to be worn down to landscapes of low relief by the early Permian, whereby low regions were filled in by Lower Permian sediments (Rotliegendes). This is called the Permian peneplain or *Rumpffläche*. Afterwards, extensive Upper Permian salt deposits (Zechstein formations) accumulated in northern Germany. The pre-Permian land surface (the unconformity between the denudated Variscan massifs and the overlying Permian and Mesozoic strata) represents an important structural boundary plane (Ahnert 1989*b*). The basement, i.e. all the *Mittelgebirge* mentioned in the previous subsection on the Variscan orogeny, consisting of Hercynian and older structures, is called the *Grundgebirge* in German. The post-Hercynian sedimentary cover is called the *Deckgebirge*. This distinction is of geomorphological significance as landforms of both units differ fundamentally (Fig. 11.1).

During the Mesozoic a long sequence of mainly marine sedimentary rocks accumulated in central Germany. These can be summarized in chronological order as follows (Fig. 11.3):

- Terrestrial sandstone and shale (Buntsandstein), marine limestone (Muschelkalk), and shallow

Era	Period	Epoch		Age (Ma)	Orogeny	Volcanic activity
Cenozoic	Quaternary	Holocene		0.01		
		Pleistocene		2.4		
	Tertiary	Pliocene	Neogene	5	Alpine	●
		Miocene	Neogene	24	Alpine	● ● ●
		Oligocene	Palaeogene	34	Alpine	
		Eocene	Palaeogene	53	Alpine	●
		Palaeocene	Palaeogene	65	Alpine	
Mesozoic	Cretaceous	Upper Cretaceous		96		
		Lower Cretaceous		135		
	Jurassic	Malm		154		
		Dogger		175		
		Lias		203		
	Triassic	Keuper		230		
		Muschelkalk		240		
		Buntsandstein		250		
Palaeozoic	Permian	Zechstein		272	Variscan (Hercynian)	●
		Rotliegendes		295	Variscan (Hercynian)	●
	Carboniferous			355	Variscan (Hercynian)	●
	Devonian			410		
	Silurian			435	Caledonian	
	Ordovician			500	Caledonian	
	Cambrian			540	Caledonian	
	Precambrian	Proterozoic		2500		
		Archaean				

Fig. 11.3. Simplified stratigraphical table (after Liedtke 2002).

marine sandstone, shale, marl and gypsum deposits (Keuper), all of Triassic age.

- Shale (Lias), brown limestone (Dogger) followed by light grey to white limestone (Malm), all of Jurassic age.
- Limestone, marl, sandstone, shale, siltstone, unconsolidated rocks, lignite, etc. of Cretaceous age.

Alpine Orogeny

During the Tertiary the northward-moving African plate collided with the Eurasian plate resulting in the Alpine orogeny. The complex, young fold mountain system of the Alps covers only a small strip of the German territory. It is a narrow zone along the outer margin of the Northern Limestone Alps, which are composed of the forward parts of nappes that have been thrust northwards. The nappes mainly consist of Mesozoic limestones and dolomites as well as early Tertiary sediments. In front of the Alps a large marginal basin developed, extending from the foot of the mountain chain to the (present course of the) Danube river. While the Alps were being uplifted the trough was filled by Molasse sediments derived from their denudation. These sediments reach a thickness of up to 5,000 m (Walter 1992).

Tectonic Uplifting and Graben Formation

During the uplift and northward extension of the Alpine chain crustal stresses were transmitted to the denuded mountain regions of the *Grundgebirge*. This resulted in a complex pattern of uplifted blocks and subsiding basins in central Germany. Palaeozoic faults were reactivated and new faults were formed. These uplifted block mountains are called *Rumpfschollengebirge* in German. Most of the fault lines were orientated either NNE–SSW (Rhenisch orientation) or NNW–SSE (Egge orientation). This pattern of tectonic lines had a major influence on the development of landforms (Andres 1989). The Mesozoic *Deckgebirge* was largely stripped from several of these uplifted blocks, exposing the old Palaeozoic rocks and structures again.

In the early Tertiary the subsidence of the Upper Rhine graben (URG) began. Subsidence continued during the Quaternary and evidently up to the present time. North of Frankfurt the tectonic rift zone can be followed in the direction of to the Hessian basin and the Leine graben. In the formation of the latter graben salt tectonics have also played a role. South of Hannover the subsidence zone disappears under the Cenozoic sediments of the North German Lowlands. Tertiary and Quaternary sediments are 2,000–3,000 m thick in the URG. Other subsidence areas that contain thick accumulations of Tertiary sediments are the Lower Rhine graben, the Molasse basin in the Alpine Foreland, smaller depressions in the Hessian and Thüringer basins, and a large region in the south-easternmost part of the German Lowlands (Henningsen and Katzung 2002).

Volcanic Activity and Landforms

The Alpine orogeny was accompanied by other endogenic phenomena such as a large number of volcanic eruptions in southern and central Germany during the Tertiary and Quaternary. Parts of the uplands of the Eifel, Siebengebirge, Westerwald, Vogelsberg, Rhön, and other regions in eastern and southern Germany are covered by volcanic deposits (Fig. 11.4). Tertiary volcanism in the Erzgebirge took place largely during the middle Oligocene to early Miocene with the accumulation of large basalt covers. Subsequent erosion has resulted in morphologically spectacular dome-shaped, cone-shaped, and table-shaped (mesa-like) mountains (Eissmann 1997). In several cases these landforms clearly reflect an inversion of the relief; moreover, they provide an indication of the measure of post-Miocene denudation. Denudation values range from 150–200 m

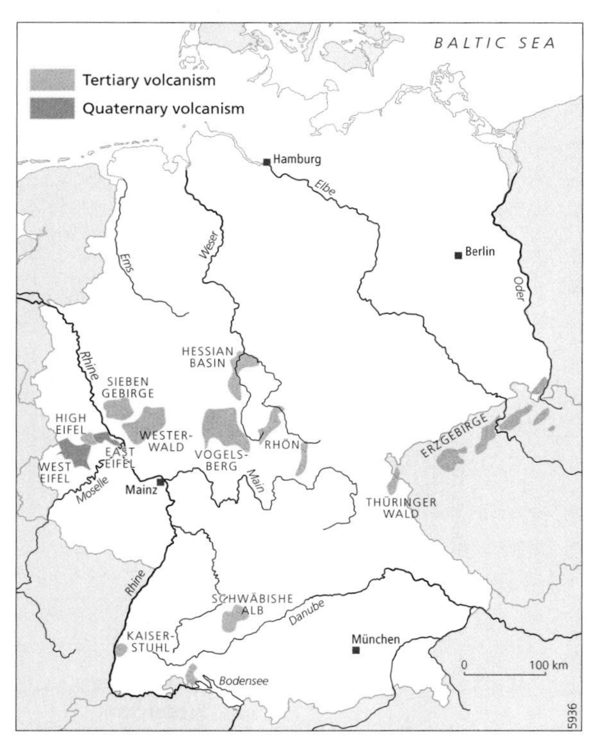

Fig. 11.4. Distribution of Tertiary and Quaternary volcanic regions in Germany (after Liedtke *et al.* 2003: 61).

on the higher planation surfaces to 250–300 m in the valleys. In the Central German Uplands volcanic landforms are frequent phenomena. Eruptions of basaltic volcanics mainly took place during the period 19–10 Ma ago. In the Hessian uplands they mostly occur in the form of small volcanic necks. Because of their greater resistance these necks rise above their surroundings, which were lowered by late Tertiary and Quaternary denudation. In the Rhön, where extensive sheets of basalt have been laid down, they appear as a group of dome-shaped mountains. The Vogelsberg is made up of a series of basaltic lava flows covering a region of approximately 2,500 km^2 and reaching a thickness of more than 300 m on top of Miocene sediments. Due to later uplift the Vogelsberg now stands up as a more or less circular upland with a radial drainage pattern and with a central summit nearly 800 m a.s.l. (Andres 1989). The Rhön and Vogelsberg volcanic areas lie on the margin of the south German crustal block. Eruptions started in the middle Miocene (Helvetian) and continued in the early Pliocene. In south Germany remnants of volcanic activity are found in the Schwäbische Alb. The Kaiserstuhl eruption in the Upper Rhine graben has been dated at 18–13 Ma BP. Older remnants of mainly basaltic volcanism occur in the Hunsrück, the Taunus, and the Spessart; K/Ar analyses here indicate an age range from late Cretaceous to early Tertiary (Walter 1992).

In the Eifel volcanic activity was at its maximum during the late Eocene/early Oligocene, whereas volcanism in the Siebengebirge and Westerwald continued from the late Oligocene to the early Pliocene (Walter 1992). The youngest volcanic features are found in the West and East Eifel regions, which contain hundreds of Quaternary eruption centres, dating from about the last 600,000 years (Fig. 11.5). The pattern of eruption centres, showing two parallel NW–SE oriented lines, suggests a tectonic origin. Large numbers of nearly circular explosion craters or *Maare* are scattered over the landscape. The depressions are often filled in with slope deposits, colluvium and/or peat, but sometimes remain in the form of open lakes (Fig. 11.6). Many of these Quaternary volcanoes are characterized by thick accumulations of volcanic bombs, lapilli, and ash (Fig. 11.7). The characteristics of the eruption and deposition of volcanic material (tephra) of the Laacher See volcano, about 40 km south of Bonn and just west of the river Rhine, have been subject to numerous studies (Schmincke *et al.* 1999; Litt *et al.* 2003). Most authors now agree that the eruption took place 12.9 ka BP. During the explosive, Plinian-type event *c.*6.3 km^3 of mafic phonolite magma, equivalent to *c.*20 km^3 of tephra, erupted (Schmincke *et al.* 1999). An oval-shaped depression resulted, presently occupied by a lake of a few km^2. Most of the Laacher See tephra (LST) was deposited east of the volcano in the Neuwied basin. This event resulted in a temporary damming of the Rhine, forming a large lake. The lake probably drained after a period of several weeks. The sudden dam collapse produced surge-like flood waves charged with volcanic ash, depositing this material more than 50 km downstream from the dam as far north as Bonn. Fallout deposits reach a maximum thickness of 50 m, quickly reducing in size with increasing distance from the eruption centre. At a distance of about 50 km east of the volcano tephra thickness is already reduced to about 1 m. As minute particles of the tephra are found over much of Europe, from southern Sweden to Poland, France, and northern Italy, the LST forms the most important stratigraphic marker in Late Glacial deposits in this region. The only volcanic eruption in the Eifel younger than this is that of the mafic Ulmener Maar, located about 30 km south-west of the Laacher See (Fig. 11.5). Its tephra layers, which have been found in several maar lake deposits, have been dated at 11 ka BP, based on detailed varve studies (Schmincke *et al.* 1999; Litt *et al.* 2003).

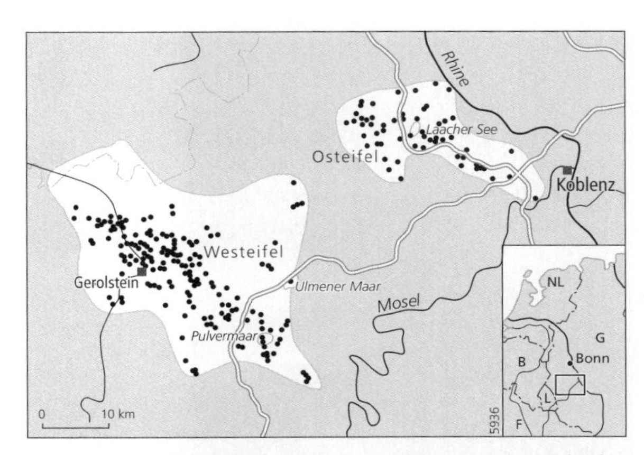

Fig. 11.5. Distribution of Quaternary volcanoes in the Eifel (after Mertes and Schmincke 1983, in Marcinek *et al.* 2002).

Geomorphological Evolution

Karst Phenomena

In the German Uplands karst phenomena are widespread as calcareous rock formations of varying ages and origin occupy large areas. The map of the *karst landscapes* in Liedtke *et al.* (2003: 95) shows karst features in Devonian limestones in the Rheinisches Schiefergebirge and Harz, in Zechstein deposits at the southern margin

Fig. 11.6. The volcanic explosion crater Pulvermaar in the Eifel. The nearly cicular crater (about 700 m in diameter and 70 m deep) lies in Palaeozoic bedrocks and is surrounded by a rim of volcanic ash of only a few metres thick (photo: Pim Beukenkamp, see also Andres 1989: 29).

Fig. 11.7. Quarry in the eruption cone of the Eppelsberg volcano, near the Laacher See in the Eifel, showing an alternation of layers with volcanic bombs, lapilli, and ash. The last eruption event of this volcano has been K/Ar-dated at around 223 ± 6 ka (photo: Pim Beukenkamp).

of the Harz mountains, in Muschelkalk outcrops all over central and southern Germany, in limestones and dolomites in the Schwäbische and Fränkische Albs, and finally in the northern Limestone Alps. The massive Zechstein evaporites are built up by sequences of rock salt, potash salt, and gypsum with intercalated limestone and claystone layers. In those places where Zechstein deposits come to the surface, or are covered by only thin overlying Buntsandstein, karstification took place (Fig. 11.8). The age and development of a variety of deep karst hollows, pits, and dolines in the limestone outcrops in the north-eastern part of the Rheinisches Schiefergebirge is discussed by Schmidt (1989). Another region with spectacular karst phenomena is the Fränkische Alb (the word 'Alb' has the same root as 'Alp', referring to a high summer pasture). After deposition of marine Jurassic limestones in this region terrestrial conditions returned and a karst relief, with steep karst domes, karst depressions, dolines and poljes, canyon-like valleys, and underground drainage, developed in the lower Cretaceous (Pfeffer 1989). In the late Cretaceous fluvial, and later marine, conditions returned and the landscape was covered by sandy and clayey deposits. Shortly afterwards a second phase of karstification

Fig. 11.8. Karst landscape with dolines just south of the Harz (photo: Pim Beukenkamp).

began, which continued throughout the Tertiary and Quaternary until the present. It appears that the dome-shaped hills that were partly or completely covered by sediments became exposed again by subsequent denudation. According to Pfeffer (1989), denudation was further favoured by three separate periods of tectonic uplift of the region leading to the development of three separate planation surface levels. The region is now characterized by dome-shaped hills, alternating with planation surfaces, dry valleys, dolines, ponors, and other features of underground drainage. Similar karst landscapes with karst domes, also called *Kuppenalb*, have developed in Malm limestones in the Schwäbische Alb (Liedtke *et al.* 2003: 95). Due to this strong karstification stream density in the Schwäbische and Fränkische Albs is exceptionally low in comparison to all other parts of the South German Scarplands, as clearly illustrated by Bremer (1989) and Liedtke *et al.* (2003: 25).

Planation Surfaces

In German geomorphological literature there has been and partly still is an intensive discussion on the possible influences of climate-related weathering and denudation on the formation of different kinds of planation surfaces (Bremer and Nitz 1993; Semmel 1996). Tertiary climates in central Europe have frequently been compared to present-day subhumid and semi-arid climates characteristic for the marginal tropics and subtropics, in which planation processes are thought to be particularly intensive. However, Semmel (1996), in his short overview on the origin of various types of planation surfaces in central Europe, holds the opinion that in spite of the manifold observations and theories still no conclusive and univocal interpretation is possible regarding the environmental conditions and prevailing processes. Moreover, Semmel (1991) expresses the opinion that the determination of the extent of tectonic deformation based upon differences in elevation of planation surfaces, must also be used with caution. Many of the Central German Uplands exhibit denudational surfaces that truncate underlying structures. These are called peneplains (*Rumpfflächen* or *Einebnungsflächen*) or primary erosion surfaces. Peneplains, sometimes also called etchplains, by definition approach base level and show no relation to geological structure. In cases where denudational planation has formed a series of steplike surfaces these are called piedmont benchlands (*Rumpftreppen* or *Piedmonttreppen*) or staircase-like planation surfaces. These landforms are assumed to be the result of periodic falls in base level induced by epirogenic uplift. All these planation surfaces represent old landform generations that are distinctly different from the Pleistocene valleys that dissect these surfaces. Planation processes can also produce erosion residuals, such as monadnocks or *Inselberge*. According to Ahnert (1989*b*), many of the planation surfaces in the German Uplands are relatively small in extent compared to the peneplains in the subhumid tropics. Consequently, it might be bettter to interpret most of these surfaces as locally developed pediments or coalescing pediments, called pediplains, rather than remnants of extensive peneplains. Normally, pediment formation is related to (semi-)arid climatic

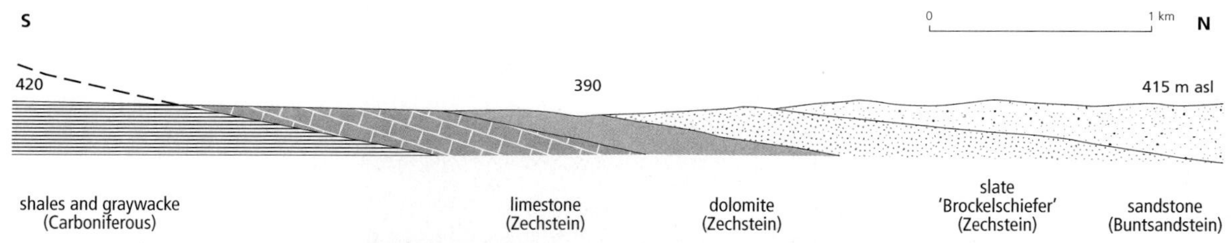

Fig. 11.9. Schematic profile of a peneplain (*Rumpffläche*) in the north-eastern part of the Rheinisches Schiefergebirge. The Permian planation surface between Carboniferous and Zechstein rocks crosses the peneplain with a slight angle (after Marcinek *et al.* 2002).

conditions and not to subhumid conditions. It seems that in many cases it appears virtually impossible to distinguish peneplanation processes, primarily characterized by areal down-wearing, from pedimentation processes, which are typified by slope retreat or back-wearing.

Typical examples of planation surfaces occur in the Rheinisches Schiefergebirge (Fig. 11.9) and in the Harz. Inselbergs in the Hunsrück and Taunus Uplands, consisting of resistant quartzitic rocks, stand up hundreds of metres above the planation surfaces. Andres (1989), in his discussion on the formation of various planation surfaces in the Rheinisches Schiefergebirge, identifies the highest and oldest surfaces, containing relics of deep weathering, as remnants of uplifted and partly deformed peneplains. Synchronously with planation, fluvial dissection of uplands during the Tertiary took place as well, particularly in those regions where uplift was relatively strong. Likewise, Brunotte and Garleff (1989) describe truncation surfaces over wide areas of the plateaus of Lower Saxony as peneplains. In the basins in Lower Saxony several levels of denudation surfaces have developed as stepped pediments, footslopes, or terraces on relatively weak rocks. Usually the pediment slopes are several kilometres long, they have inclinations of only about 2° and show weakly concave longitudinal profiles. Frequently there are several pediment levels above each other, separated by elevation differences of a few tens of metres. There appears to be a gradual transition from predominantly denudational pediment surfaces with only thin covers of debris of the order of decimetres to a few metres (at relatively high positions) to accumulational pediments with thicker slope deposits, often including loess (at lower positions). Finally, Spönemann (1989) in his discussion on planation surfaces and cuesta versus hogback formation in the Lower Saxonian and Hessian Uplands comes to the conclusion that the interrelation of pediment-like surfaces and ridges indicates that denudation by planation and the formation of cuestas and hogbacks must have occurred simultaneously. Moreover, Brunotte and Garleff (1980, 1989) have shown that the development of these structural landforms was primarily the result of intense denudation influenced by regional tectonic uplift, and that it was largely independent of the prevailing climatic conditions during the Tertiary.

Quaternary Surficial Deposits and Landforms

Alpine Glaciations

Multiple Alpine glaciations, as reflected in a stacking of a well-developed series of glacial and fluvioglacial sediments, have formed the basis for the classic subdivision of the Pleistocene in Europe. The early study by A. Penck and E. Brückner, *Die Alpen im Eiszeitalter* (1901–9), distinguishing four main Alpine glacial cycles, has become world-famous. Named the Günz, Mindel, Riss, and Würm, after small rivers in the Alpine Foreland, their subdivision essentially is still valid, notwithstanding many later modifications and additions (Habbe 1989; Ehlers 1996). However, correlation of these south German glacial periods (Brunnacker 1986; Habbe 2002) with the the multitude of glacial/interglacial and stadial/interstadial cycles recognized in the North European Lowlands (Ehlers 1996; Eissmann 1994, 2002), with the exception of the latest glaciation, remains full of uncertainties. Fan-like foreland glaciers spread out from the exits of Alpine valleys, depositing hummocky morainic ridges and fluvioglacial outwash fans and terraces. From west to east five individual glacier basins are distinguished (Fischer 1989; Jerz 1995) (Fig. 11.10):

- The Rhine glacier, which was the largest foreland glacier. The glacier probably reached its greatest extent during the Mindel glaciation. The northernmost extension just crossed the Danube river. Presently, a part of the former ice lobe basin is occupied by the Bodensee (Lake Constance).
- The much smaller Iller–Wertach–Lech glacier probably also reached its maximum size during the

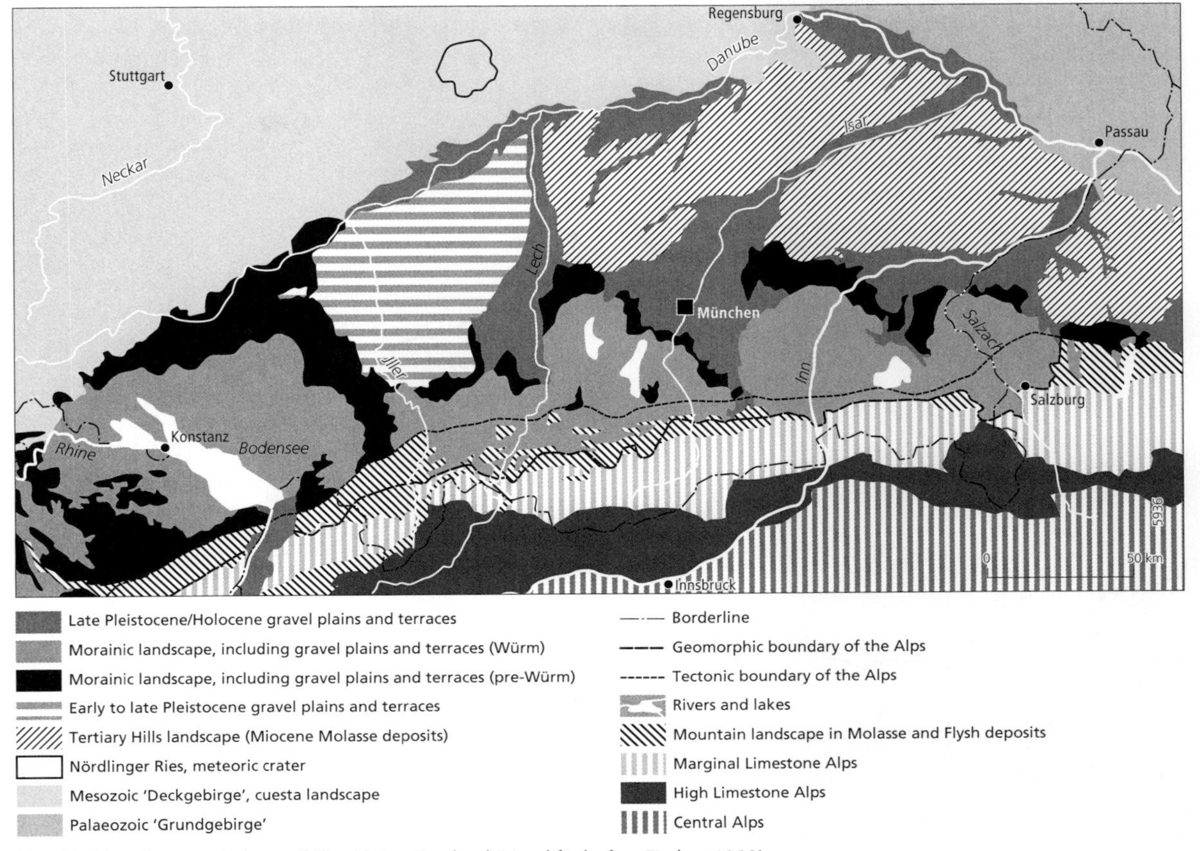

Fig. 11.10. Geomorphology of the Alpine Foreland (simplified after Fischer 1989).

Mindel glaciation. It is bordered by the Iller and Lech rivers.

- The Loisach–Isar glacier. Remnants of formerly much larger lakes that occupied this glacier basin, are the Ammersee and Starnberger See.
- The Inn–Chiemsee glacier. It is also assumed that this ice lobe reached its maximum extent during the Mindel glaciation. The Chiemsee is the remnant of an originally much larger lake that extended into the Alps. Thick varve clay deposits indicate the existence of another former lake, the Rosenheimer See, which has been drained by the Inn river.
- The Salzach glacier, which may have reached its greatest extension during the Günz glaciation, is the easternmost foreland glacier on German territory. As in all other foreland glacier basins (former) lakes clearly indicate impeded drainage caused by a more or less concentric series of morainic ridges and accompanying fluvioglacial landforms.

It is estimated that *c.*12,650 km^2 (= 3.5% of the German territory) was covered during the Last Glacial (Würm) by Alpine ice sheets, whereas *c.*47,450 km^2 (= 13.3%) was covered by the Weichselian ice cap in northern Germany (Liedtke *et al.* 2003). The maximal Fennoscandian ice border line more or less coincides with the northern boundary of the Central German Uplands. Together with the maximum extent of the Alpine ice sheet this means that *c.*50% of Germany experienced glacial conditions during the Pleistocene (Fig. 11.11). Moreover, all of Germany repeatedly experienced severe periglacial conditions. Morainic ridges of Last Glacial (Würm) age extend to *c.*50–60 km north of the Alps, whereas the maximum distance of glacial forms from the Alps related to earlier glaciations is approximately 70 km (Jerz 1995).

It is clear that the extent of the foreland glacier basins is related to the size of the alpine source areas. The Iller–Wertach–Lech had the smallest source area,

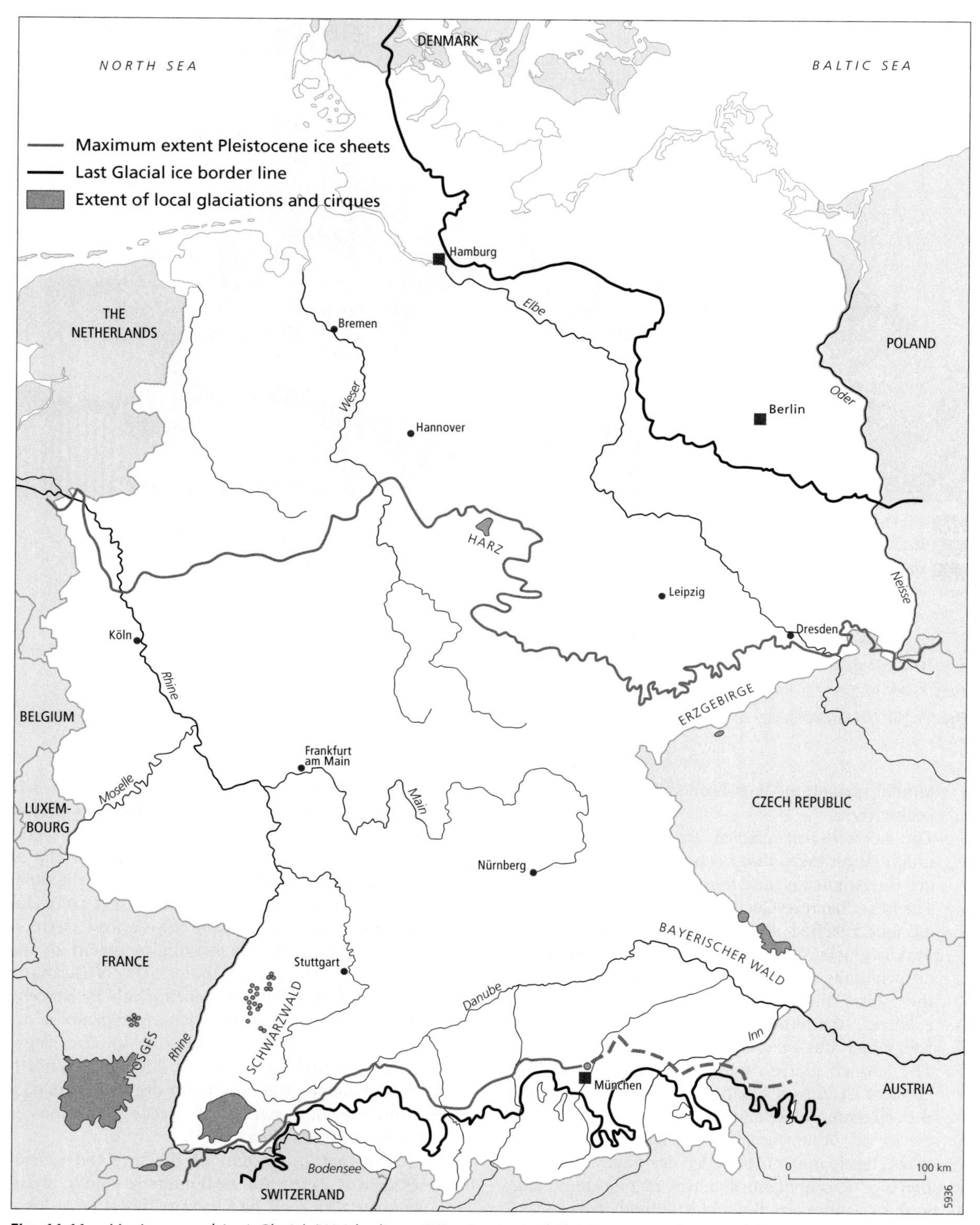

Fig. 11.11. Maximum and Last Glacial (Weichselian ≈ Würm) extent of Pleistocene ice sheets and locations of local glaciations (after Liedtke *et al.* 2003: 67 map).

restricted to the Allgäuer Alps, whereas the Rhine glacier drained a large part of the central Alps. Naturally, landforms and sediments of the Würm glaciation cover the largest parts of the glacier basins (van Husen 1987). Irregular, hummocky morainic ridges and relatively unaltered (fluvio)glacial landforms, such as drumlins, eskers, kames, kame terraces, and dead-ice features characterize these areas. Successive morainic ridges are interpreted as retreat stages, or short readvance stages of the ice, that remained within the maximum Würm ice limits (Habbe 1989). The main advance of the last glaciation occurred only after 27–25 ka BP during a single major ice advance. In other words, during its retreat several minor readvance phases left a series of recessional end moraines (Ehlers 1996). Each readvance phase was accompanied by outwash deposits. In some valleys glacial meltwater terraces were formed that can be linked to the moraines of individual ice retreat stages. Most of the lakes in the Würm morainic landscape represent depressions eroded by glacier lobes or depressions remaining after the melting of large dead ice blocks (Ehlers 1996). Landscapes resulting from earlier glaciations show a much more subdued relief, deeply weathered surficial materials, and widespread surficial deposits of periglacial or aeolian origin. Beyond the morainic systems of the Würm and Riss glaciation extensive gravel plains cover the Alpine foreland, especially the plateau-like interfluve regions. These deposits represent the outwash plains of older Pleistocene glaciations. As uplift of the Alpine Foreland continued during the Quaternary the oldest gravel deposits are found at the highest positions. Whether or not there is also a component of this uplift related to glacio-isostatic rebound remains to be seen, since the estimated volume of the Alpine ice sheets is relatively small, i.e. only 1/40 of that of the Fennoscandian ice sheet (Ehlers 1996). The highest and oldest of these gravel covers could be the result of cold periods prior to the Günz glaciation; pre-Günz cold periods have been identified as Danube and Biber glaciations. With respect to these older Pleistocene cold-climate periods, Habbe (1989) stresses the point that no chronostratigraphic significance should be attached to these terms. As the means of absolute dating are limited, even the morphostratigraphic significance of these assumed glacial episodes will remain uncertain. According to Jerz (1995), there is evidence of at least seven (and possibly ten) glaciations, during which glaciers advanced over the Alpine Foreland.

Local Glaciations

Because of their elevation both periglacial and glacial landforms developed during the Pleistocene in several *Mittelgebirge* ranges (Fig. 11.11); the most extensive examples occur in the Schwarzwald and in the Vosges (France). The local glaciers in the Schwarzwald, Harz, and Bayerischer Wald totalled an area of some 1,300–1,400 km^2. The highest point of the Vosges reaches 1,426 m a.s.l., almost as high as the Feldberg (1,493 m) in the Schwarzwald. Based on an inventory of glacial landforms in the Vosges and Schwarzwald by Zienert (1967), Blume and Remmele (1989) have mapped a large number of cirques, the formation of which seems to have depended not only on altitude and exposure, but also on structure and lithology. Evidence for the Würm glaciation of the high parts of the Schwarzwald is undisputed, and is represented by morainic ridges, glacier tongue basins, cirques, and glacier striae. Hantke (1978/92) presented an overview of glacial forms and deposits in the Schwarzwald as well as the Vosges. Locally, ice thicknesses of cirque- and valley-glaciers of up to 250 m have been interpreted. From the immediate surroundings of the Feldberg no less than five glacier ice front positions have been reported. All these positions are represented by morainic ridges dating from the Last Glacial maximum (Ellwanger *et al.* 1995). However, the extent of a possible earlier Riss glaciation, as presented by Hantke (1978/92) and others, is still debated (Semmel 1996).

The geomorphological influence of local glaciations on the high plateaus of the Harz is controversial. There is no agreement on a possible ice cover of the Harz during the Elster glaciation, but there is evidence of plateau glaciers during the Saalian and Weichselian glaciations (Hövermann 1987; Semmel 1996).

Naturally, much larger areas in Germany were repeatedly influenced by Pleistocene periglacial conditions, as witnessed by block fields, solifluction or gelifluction deposits (Scholten 2003), and cryoplanation features. Surficial deposits of cold-climate origin such as loess and sand sheet deposits, and sedimentary (deformational) structures such as ice wedge casts and cryoturbation features abound.

Holocene Landscape Development and Human Impact

The present-day landscape cannot be understood without taking into account the enormous impact agriculture has had during the course of the Holocene. The first farmers who reached German territory, and who settled on the loess regions in particular, were the people of the Early Neolithic Linear Bandkeramik culture about 5.5 ka BC (Kalis *et al.* 2003). However, the impact of these people on landscape and vegetation was local and rather limited. During the Atlantic period almost all of Germany was still covered with closed forest, and agriculture mostly involved very small forest clearings and woodland grazing. Early Neolithic life seems to have remained strongly adapted to the forest ecosystem. Following the introduction of the plough from the beginning of the fourth millennium BC human influence gradually increased. Indicators of early farming activities that are distinctly reflected in pollen diagrams are pollen from crop plants and from weed species. It seems that the commonly used AP/NAP ratio may not be the most appropriate for visualizing human impact during Neolithic times. It is only since the Late Bronze Age/Iron Age, when open grasslands expanded, that human impact becomes readily visible in the regional pollen rain. With growing intensity of farming and other forms of land use soil erosion also increased, leading to increases in sediment load in most river systems and a surficial cover of colluvium on most slopes. Processes that led to intense colluviation during the Iron Age as well as during the Middle Ages are not only a direct response to the extension of settlements and arable fields, but are also related to social and technological progress (Zolitschka *et al.* 2003). The complex history of human occupation in central Germany as reflected in peat records, fluvial, aeolian, and lacustrine sediment sequences and soils is known in great detail, but it is outside the scope of this chapter (see e.g. Bork *et al.* 1998, 2003). Obviously, human activities during the past centuries have completely remodified many landscape areas.

Regional Patterns

Central German Uplands

Rheinisches Schiefergebirge

This *Grundgebirge* is dissected by the Rhine valley. Westwards of the river it consists of the Hunsrück and Eifel regions separated by the Moselle river. To the east of the Rhine, the Taunus, Westerwald, and Sauerland regions occur from south to north. In the central part a small subsidence basin, the Neuwied basin, originated synchronously with the development of the rift valley of the Rhine. The southern edge of this upland is sharply delineated by a Variscan-trending fault-line scarp separating the Rheinisches Schiefergebirge from the Saar-Nahe Upland ('Nordpfälzer Bergland'). Devonian and Carboniferous slates, greywackes, and limestones are the predominant rocks of this *Grundgebirge*. More resistant beds of quartzite have led to the development of prominent ridges, particularly in the Hunsrück and Taunus (Andres 1989). Locally, the stratigraphic sequence includes Devonian volcanic rocks. In all parts of the Rheinisches Schiefergebirge there are extensive denudation surfaces at elevations of more than 400 m a.s.l. Apart from remnants of old (late Mesozoic to early Tertiary) planation surfaces, younger, Tertiary planation surfaces, also called trough surfaces or *Trogflächen* are present. These usually follow the lines of present-day drainage systems and are covered by fluvial quartz gravels, the Vallendar gravel deposits. These deposits are derived from the older planation surfaces by erosion, which was initiated by a phase of increased tectonic uplift. Finer material was largely transported into adjacent basins, forming widespread clayey deposits. Deposition of gravels and clay sediments probably took place during the late Eocene to early Oligocene. Shortly afterwards, during the middle and late Oligocene, parts of the Rheinisches Schiefergebirge experienced a marine transgression that advanced from the Mainz basin in the south and the Lower Rhine graben in the north. It is assumed that the development of the present drainage system, including the course of the Rhine through the Rheinisches Schiefergebirge, was initiated during the regression of the sea in the Miocene (Andres 1989; Schirmer 1995; Semmel 1996). This agrees with the fact that Boenigk (1982) found middle Miocene fluvial sediments in the Lower Rhine graben. The river course more or less linearly connected the tectonic depressions to the south and north with the small tectonic depression, the Neuwied basin or Middle Rhine basin, which had developed in the central part of the Schiefergebirge (Schirmer 1995). Pediment-like surfaces extend from the old planation levels to the nowadays deeply incised river valleys. Continued uplift of the uplands in combination with subsidence of marginal areas has resulted in a total amount of valley deepening of the Rhine of the order of 200 m since the end of the Tertiary (Fig. 11.12). During this period a complex series of fluvial terraces formed on both sides of the Rhine as well as in the valleys of many of its tributaries. The causes for the deep post-Tertiary incision of the Rhine are not completely clear. Semmel (1996) is of the opinion that subsidence of the Lower Rhine graben is of more importance for the deep incision than uplift of the Rheinisches Schiefergebirge. However, the suggestion that eustatic sea level changes during the Quaternary have influenced river valley and terrace morphology here seems implausible. Measurements show that vertical crustal movements continue to the present time. Fuchs *et al.* (1983) presented evidence for relatively high uplift rates in the Ardennes and the Eifel.

Fig. 11.12. Meander of the Rhine strongly incised in the Rheinisches Schiefergebirge. High Terraces (*Hochterrasse*) of the Rhine occur on both sides of the river at a height of 120–150 m above the water level (photo: Hans Middelkoop).

Saar-Nahe Upland

This region is mainly composed of upper Carboniferous and lower Permian (Rotliegendes) sedimentary rocks (each 3,000 to 4,000 m thick) that were derived from the Variscan fold mountains by denudation, and deposited in a late Palaeozoic intramontane basin (Walter 1992; Henningsen and Katzung 2002). It is separated from the Rheinisches Schiefergebirge by a major, southern marginal fault of the Hunsrück, delineating the boundary between the Rhenohercynian and Saxothuringian zones. Sedimentation was followed by post-orogenic (Permian) volcanism which left large masses of porphyric and melaphyric rocks (Andres 1989). The tectonic structure of the Saar-Nahe Upland is determined by a SW–NE oriented anticline. In the southwestern part of this region massif, upper Caboniferous coal layers come to the surface. Landform development, with older and younger Tertiary planation surfaces and Pliocene-Pleistocene valley incision, was quite similar to that in the Rheinisches Schiefergebirge.

Harz

The Harz is an elliptical-shaped block, some 90 km from east to west and 30 km wide, representing a horst in the Variscan basement. It consists of highly folded and faulted Palaeozoic rocks. The geological structure of the Harz resembles that of the previously mentioned *Grundgebirge* areas to a large extent, with the exception of the fact that two granite batholiths have intruded into the Variscan fold mountain (Walter 1992; Henningsen and Katzung 2002). A remnant of one of these granite complexes now forms the highest elevation in the Harz, the Brocken at 1,142 m a.s.l. (Fig. 11.13). Near the southern edge of the Harz outcrops of porphyric volcanics of Permian age occur. Karst landforms have developed in limestones and gypsum deposits of upper Permian (Zechstein) age. Especially along the southern margin of the Harz this has led to intensive karstification of the landscape (Fig. 11.8). During the late Tertiary and Quaternary the Harz was strongly uplifted asymmetrically along well-developed fault lines,

Fig. 11.13. Granite weathering, forming more or less spherical blocks of granite or *corestones*, in the northern part of the Harz (photo: Pim Beukenkamp).

leading among other things to deeply incised river valleys at the margin of the Harz. The Harz is sharply bounded by high fault scarps on all sides, except in the south-east. The old peneplain of the Harz, which truncates the Variscan geologic structures, has developed at several levels and is identified by Andres (1989) and others as a piedmont benchland or *Rumpffläche*. A younger planation surface (*Harzhochflache*) is thought to be equivalent to the trough surfaces of the Rheinisches Schiefergebirge. However, there appear to be no agreement yet on the origin of these planation surfaces (Marcinek *et al.* 2002).

Thüringer Wald, Erzgebirge, Bayerischer Wald, and Böhmer Wald

The distribution of crystalline and highly metamorphic rocks in the Thüringer Wald, Erzgebirge, Bayerischer Wald, and Böhmer Wald (the eastern German *Grundgebirge*) is greater than in the other Variscan mountain complexes. Metamorphic rock types alternate with igneous rocks containing a variety of ore deposits, extensive basaltic layers, as well as very coarse Oligocene gravels and other Tertiary sediments. Prominent mountain tops and ridges are related to the presence of solid, Palaeozoic, metamorphic rocks, such as gneiss, hornfels, phyllites, and mica schist, and to the presence of volcanic rocks. The Scheibenberg in the Erzgebirge close to the border of the Czech Republic is a famous example of a *basalt-table-mountain*, in which vertical prismatic basalt columns rest on top of Oligocene/Miocene quartz gravels and sands (Fig. 11.14). The regions in the eastern part of Germany have in common the presence of Tertiary planation surfaces, in some cases occurring in steplike series, which are usually indicated either as piedmont benchlands (*Rumpfflächen*) or as piedmont steps (*Rumpftreppen*). Strong middle Tertiary tectonic movements were responsible for denudation and deep valley incisions. Similar to the situation in the Rheinisches Schiefergebirge, marine transgressions reached and partly covered these *Grundgebirge* areas depositing mainly coarse quartz gravels.

Lower Saxonian and Hessian Uplands

The uplands in this region project for some distance into the North German Lowlands. These uplands were deformed mainly during the late Jurassic and Tertiary, resulting in synclinal, anticlinal, and monoclinal structures. In some areas, e.g. the Solling, uplift after the middle Tertiary amounted to 500–700 m relative to adjacent basins. In those cases where displacements were accompanied by large-scale deformation and diapiric uplift of Zechstein salt deposits (*halokinesis*) the tectonic phase is indicated as the Saxonic orogeny (Brunotte and Garleff 1980; Henningsen and Katzung 2002). The question of whether the deformation of salt formations is the cause or the result of tectonic displacements remains unanswered in many cases. The landscape in this region of strongly deformed Mesozoic *Deckgebirge* is characterized by an intricate pattern of cuestas, hogbacks, and monoclinal ridges. For example, the tectonic structure of the upland of the Teutoburger Wald is highly complex due to folding and faulting com-

Fig. 11.14. Vertical basalt columns of up to 60 m thick, representing remnants of extensive Upper Tertiary lava flow in the Erzgebirge, location Scheibenberg, 807 m a.s.l. (photo by courtesy of Lothar Eissmann 2000; see also Marcinek *et al.* 2002: 525).

Fig. 11.15. The *Externsteine*: remnants of limestone weathering in the Teutoburger Wald (photo: Pim Beukenkamp).

bined with halokinetic (salt) tectonics (Marcinek *et al.* 2002). Cuestas and hogbacks have formed in very resistant rock formations of the middle Trias (Muschelkalk), upper Trias (Keuper, quartzitic Röt Sandstone), upper Jura (Malm oolitic limestones and dolomites) and middle to upper Cretaceous deposits (Osning Sandstone, Cenoman Limestone) (Fig. 11.15). Remnants of Tertiary marine sediments indicate that intensified denudation and the formation of cuestas and hogbacks predominantly occurred during and after the late Tertiary (Brunotte and Garleff 1989). Buntsandstein deposits characterize the upland positions, and Muschelkalk is more often found in graben-like depressions. The Hessian basin extends through this area in a SSW–NNE direction. Mesozoic sedimentary rocks in this basin are covered by marine and limnic Tertiary deposits (Andres 1989).

Spönemann (1989) in his description of the structural landforms in the Lower Saxonian region makes a distinction between cuestas and homoclinal ridges. Although gradual transitions occur from cuestas to homoclinal ridges, a critical angle of 10–12° is suggested to distinguish these two landforms. In the northern parts of the Lower Saxonian and Hessian Uplands, such as the Teutoburger Wald and the Weserbergland, steeply dipping strata and thus homoclinal ridges are found. In contrast, in the southern parts of this region dip angles of Mesozoic rocks are generally low, which favours cuesta formation. According to Spönemann (1989) denudation will lead to the most resistant rock types becoming increasingly exposed with time. The initial form is a truncation surface, possibly part of a peneplain, which then develops into a swell, a round-crested ridge, and ultimately a hogback. All these stages of development are present in the Central German Uplands (Brunotte and Garleff 1980). The post-middle Tertiary landform development resulted in a Pliocene peneplain, cutting across Eocene and Miocene sediments as well as (non-resistant) Mesozoic rocks overtopped by well-developed cuestas, according to Brunotte and Garleff (1980).

Due to differences in weathering conditions and geochemical properties of the various Mesozoic rock outcrops and surficial sediment covers a clear relationship between rock types and soil profiles exists in these regions. This is illustrated by Schmidt (2002) with an example from the cuesta landscape in the Hessian Upland (Fig. 11.16). Parabraunerde (or Luvisols) development is characteristic for relatively dry parts of the Buntsandstein region, whereas Pseudogley soils (Gleysols) result from wetter conditions such as occur on impermeable rock types and massive loess covers. Soil types belonging to the Rendzinas (Leptosols) are linked to calcareous rocks.

South German Scarplands and Upper Rhine Graben

Odenwald and Spessart

The Odenwald and the western Spessart form part of the crystalline Hercynian basement or *Grundgebirge*, consisting of Precambrian and lower Palaeozoic rocks (Walter 1992; Henningsen and Katzung 2002). The

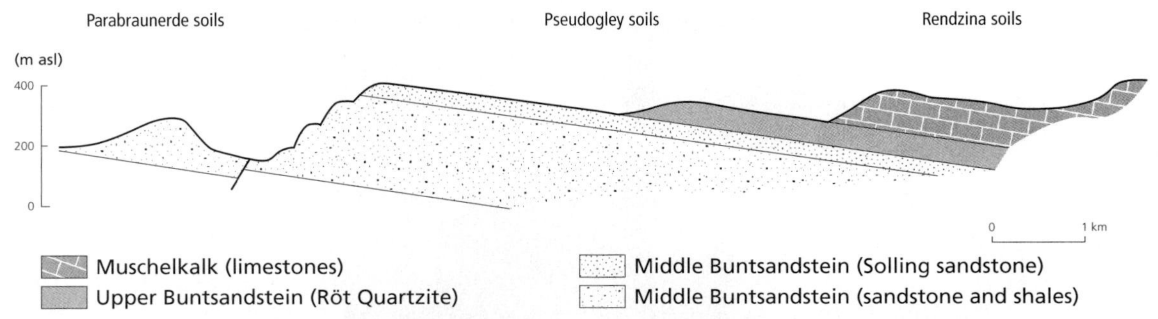

Fig. 11.16. Schematic profile illustrating the relationship between soil profile development and lithology in the south German cuesta landscape (after Schmidt 2002).

western edge of the Odenwald complex is separated from the Upper Rhine graben by a fault scarp with a maximum relative height of 450 m. The maximum vertical (tectonic) displacement, however, is *c*.4,000 m, two-thirds of which results from subsidence in the graben (Zienert 1989). He concluded, on the basis of a thickness of almost 400 m of upper Pliocene-Pleistocene deposits, found in a boring drilled in the Upper Rhine graben near Heidelberg, and an estimated 300-m uplift of the adjacent Odenwald, that a vertical displacement of about 700 m must have occurred since the late Pliocene. The crystalline Odenwald in its turn is divided into several structural blocks as a result of Cenozoic tectonics. Remnants of a Permian denudational surface and at least two younger planation surfaces could provide information about regional uplift, because they occur on each of the tectonic blocks at different altitudes. The eastern half of the Odenwald is mainly covered by Triassic Buntsandstein. Strata and planation surfaces generally descend in a south-easterly direction. As most strata dip more steeply than the planation surfaces, the latter cut across progressively younger formations in the downslope direction. The entrenched valley of the Main, a major tributary of the Rhine, forms the boundary between the uplands of the Odenwald and the Spessart. The latter consists of a series of blocks uplifted to different altitudes. The morphotectonic structure of the Spessart resembles that of the Odenwald. The crystalline part of the Spessart, the *Vorspessart*, remained covered by Buntsandstein probably till the early Miocene, but in the eastern part of the Spessart Buntsandstein deposits still reach a thickness of up to 500 m. Since that time at least two late Tertiary planation surfaces have been formed (Semmel 1996).

Schwarzwald (and Vosges)

The crystalline Schwarzwald *Grundgebirge* consists of highly metamorphic and granitic rock complexes, which are covered by mainly Permian and Triassic sedimentary covers of the Deckgebirge increasing in thickness to the east and south-east (Walter 1992; Henningsen and Katzung 2002). Prior to the Tertiary the Vosges and Schwarzwald formed the western and eastern flank of a mega-anticline of which the central section later subsided and formed the Upper Rhine graben (Fig. 11.17) (Hantke 1978/92; Embleton and Demek 1984). This explains why the Vosges now stands as a mountain block with a steep eastern face and a gradual western slope towards the Paris basin, whereas the western face of the Schwarzwald is represented by a striking fault scarp, and the upland slopes more gradually eastwards toward the south-west German scarplands. The subsidence of the Upper Rhine graben during the Tertiary and the simultaneous uplift of its flanks was responsible for the strong denudation of both regions (Blume and Remmele 1989). Subsidence and uplift continued in the Quaternary. Because of this strong tectonic uplift, especially in the southern part of the Schwarzwald, as well as of the Vosges (in France), the sedimentary cover, mainly composed of lower Triassic Buntsandstein, has been largely removed and the crystalline basement has become exposed. Only a few remnants of planation surfaces can be recognized because of the intensity of later dissection. Differences in the lithology of the gently inclined Buntsandstein strata, reaching a thickness of 250 to 300 m, have resulted in the development of cuesta scarps in the northern parts of the Schwarzwald and Vosges (Blume and Remmele 1989).

Upper Rhine Graben

One of the most striking morphotectonic features in central Europe is the Upper Rhine graben (URG). This is a subsidence zone 300 km in length and 35–45 km in width, with a Rhenisch (NNE–SSW) orientation, and bordered by uplands at both sides (Fig. 11.17). The URG is a part of the West European Rift system (WER), a

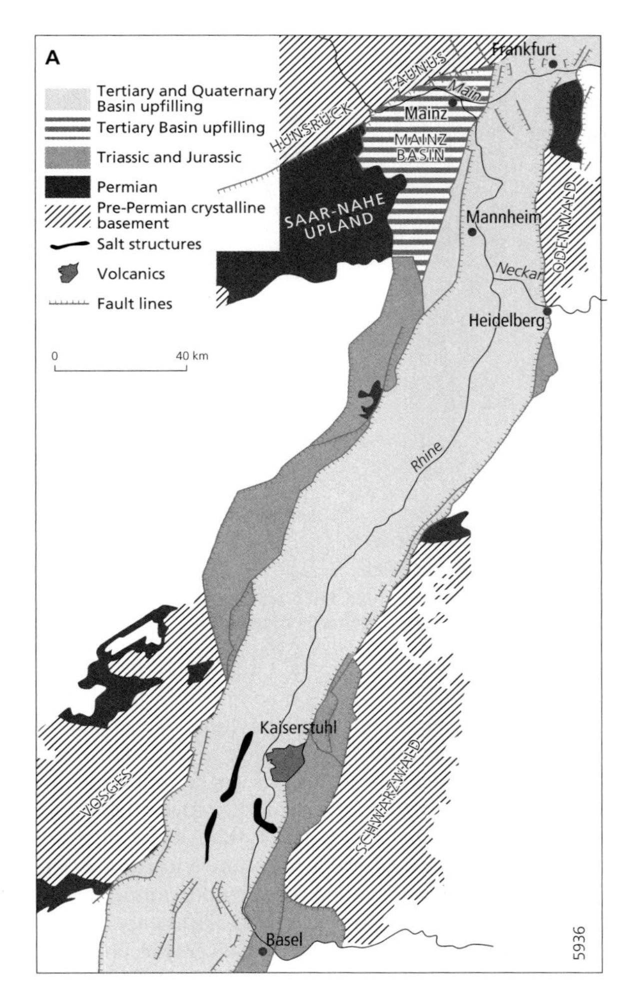

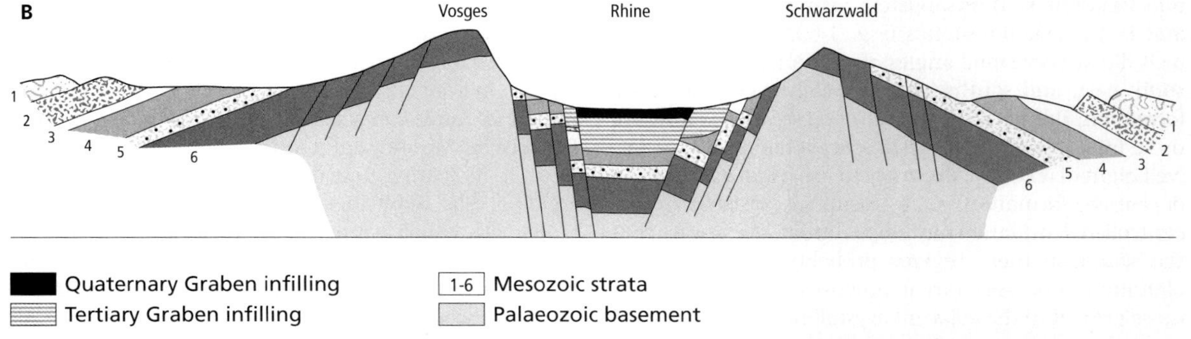

Fig. 11.17. A. Tectonic overview of the Upper Rhine graben and adjacent areas (after Walter 1992) B. Schematic cross profile (after Albrecht Penck).

tectonic rift zone crossing Europe. It can be traced from the Mediterranean to Denmark and the Oslo region in Norway over a length of no less than 2,000 km. It forms the Rhône–Saône corridor (Rhône graben and Bresse graben), continues in the URG, and then branches off in two directions. One branch follows the Hessian basin and the Leine graben to the north-east. Another branch extends to the north-west to the small Middle Rhine graben or Neuwied basin in the middle of the Rheinisches Schiefergebirge. It then follows the Lower Rhine graben, and the Roer Valley graben in The Netherlands towards the North Sea basin, and finally, the Viking graben west of Norway. The crystalline basement in the URG subsided along deep-seated faults and is today covered by 1,000–3,400 m of sediments. The base of the Tertiary sediments varies from a depth of *c.*2,000 m in the southern part of the graben just north of Basle to *c.*3,000 m in the northern part of the graben near Mannheim (Walter 1992; Liedtke *et al.* 2003). Quaternary sediments in the URG reach a thickness of 200 m (Ellwanger *et al.* 1995). Tectonic displacements since the beginning of the rift formation in the Eocene amount to about 4,400 m. The structure of the southern part of the rift zone is especially complicated as it consists of a system of horsts and grabens. At the flanks of the Upper Rhine graben, tectonic fault blocks (*Bruchschollen* or *Randstaffeln*) containing Mesozoic (Trias and Jura), Tertiary and early Quaternary rocks form the transition from the graben itself to the *Grundgebirge* at the sides. Subsidence of the graben still continues today; high-precision levelling has revealed local movements up to 0.7 mm/yr (Henningsen and Katzung 2002). Volcanic activity is related to the rift formation and occurred throughout the whole period of rift development, in some places continuing into the Quaternary. The Kaiserstuhl in the southern URG is the remnant of a huge Tertiary (late Eocene to early Miocene) strato-volcano. It has now been largely destroyed by erosion, but even the present remnants rise to 557 m and extend over 90 km² (see Liedtke *et al.* 2003: 64–5).

The geomorphology of the URG is characterized by large alluvial fans on both flanks of the basin, extensive late Pleistocene fluvial terraces, partly covered by loess and dune fields, and the present flood plain of the Rhine. The originally many kilometres-wide flood plain of the Rhine with its strongly anastomosing channel pattern has largely become inactive due to normalization and canalization works in the nineteenth and twentieth centuries (Hantke 1993; Liedtke *et al.* 2003: 93).

South-West German Cuesta Landscape

The cuesta landscape in between the rivers Main and Danube is surrounded by uplands and is made up of dissected old denudation surfaces, escarpments, basins, and valleys (Bremer 1989). The relief ranges from less than 200 m to just over 1,000 m in a few places. The region is crossed by the major continental divide between the Rhine and Danube catchments. These lead, respectively, to the North and Black Seas. Due to the proximity of the North Sea and the subsidence of the Lower Rhine basin and adjacent delta region in The Netherlands, the Rhine has been steadily extending its catchment at the expense of the Danube (Hantke 1993). Large areas of the Main and Neckar used to drain south-east to the Danube, but today the Danube receives only small leftbank tributaries draining the calcareous regions of the Schwäbische and Fränkische Albs. The thickness of the Mesozoic (Trias and Jura) sedimentary cover (*Deckgebirge*) ranges from *c.*1,200 m in regions of synsedimentary uplift to 2,000 m in basins of subsidence. The Mesozoic sequence is built up of resistant limestones and sandstones interbedded with less resistant shales and marls, and locally also salt and gypsum. Major scarps are formed in Triassic sandstones and limestones (Buntsandstein, Muschelkalk, Keuper) and in Jurassic limestones (Fig. 11.18). The Mesozoic beds dip at very small angles of only a few degrees east, south-east, and southwards, thereby favouring cuesta formation. Bremer (1989) mentions dip slopes that are up to 60 km wide, and cuesta scarps that attain a relative height of *c.*100 m. Contrary to the usual explanation of cuesta formation as a result of resistant rock-controlled denudation and slope retreat, she argues that the scarps in these regions probably developed from planation processes (areal down-wearing) similar to those present in the adjacent crystalline upland areas. In general, scarp retreat appears to have been limited. Uplift and denudation of the region began in the early Cretaceous. Widespread plateaus have developed in the Buntsandstein and Muschelkalk of the south-west German scarplands. These more or less flat or gently undulating regions, which are frequently covered by loess, are characterized by a high soil fertility. Consequently, they have been used as arable fields for centuries. In German these regions are called *Gäuflächen* (Semmel 1996, 2002). Further east, in the cuesta landscape of the Keuper, denudation of resistant rock layers has produced a major Keuper cuesta, consisting of at least five steps.

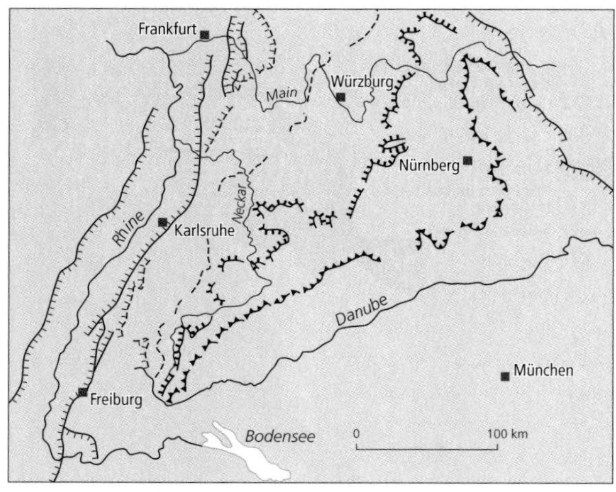

Fig. 11.18. Distribution of fault scarps and cuestas in the South German Scarplands (after Semmel 2002).

Schwäbische and Fränkische Alb

The eastern- and southernmost part of the German scarplands is characterized by cuesta development in Jurassic rocks (Lias limestones and sandstones, Dogger sandstone, and Malm limestones), as exemplified in Fig. 11.19. Erosional remnants, Inselbergs, or mesa-like outliers, in front of cuesta scarps provide evidence for scarp retreat. Denudation surfaces in the Schwäbische and Fränkische Albs presently lie at an altitude of 500–600 m a.s.l., indicating that this old relief has been uplifted en bloc. The Schwäbische Alb plateau is divided by faults with which young volcanic forms are associated. These include crater lakes, basalt, tuff, or phonolite hills, and volcanic necks (Embleton and Demek 1984). The main west-facing escarpment of the Fränkische Alb is a 150-m high edge of massive limestone. This cuesta is sharply incised by erosional features. Karst forms,

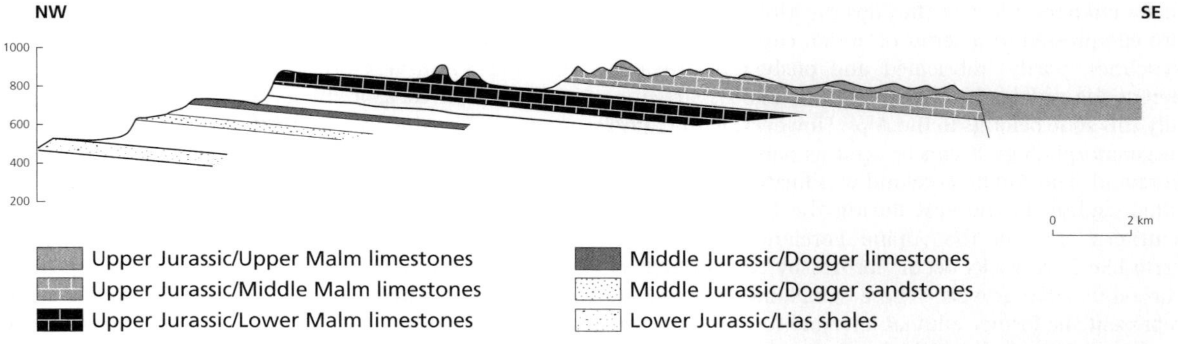

Fig. 11.19. Schematic profile of the cuesta landscape in the south-western part of the Schwäbische Alb (after Semmel 1996 and 2002).

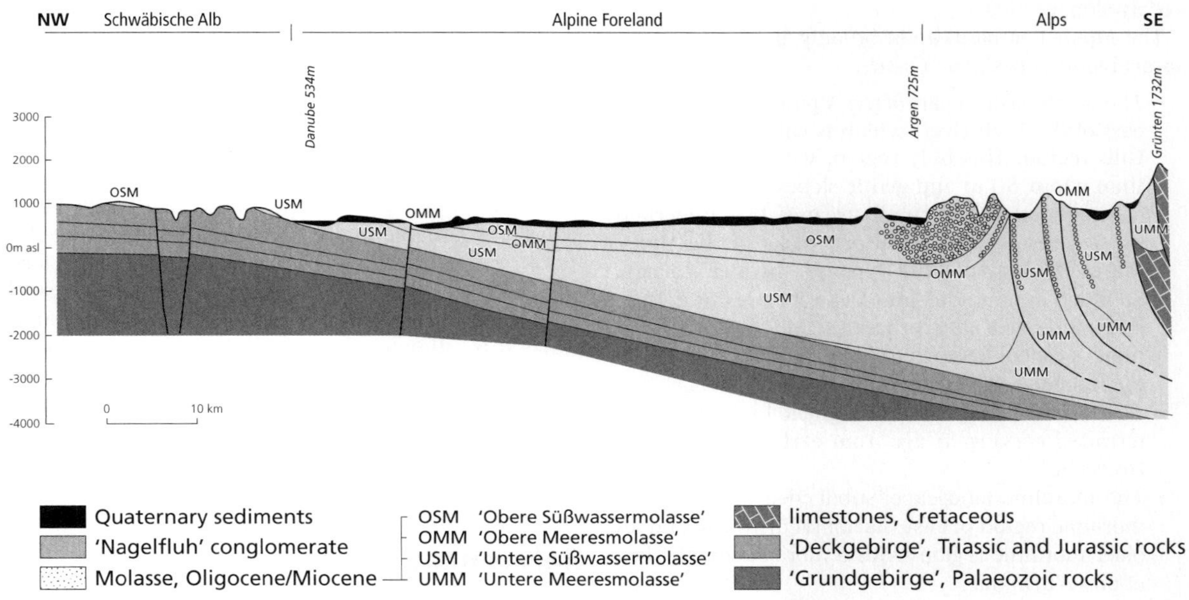

Fig. 11.20. Geological cross-section of the Alpine Foreland (simplified after Habbe 2002).

including dry valleys, sink holes, cave systems, and karst springs, are frequent phenomena.

A spectacular feature of extraterrestrial origin is the nearly circular depression (*c.*23 km in diameter) of the Nördlinger Ries between the Schwäbische and Fränkische Albs. It was formed in the middle Miocene (*c.*14.7 Ma) by the impact of a very large meteorite.

Alpine Foreland

Simultaneous with the uplift of the Alps in the Oligocene/Miocene, a pre-Alpine sedimentary basin formed. This depression was filled with debris coming from the rising mountain region. A sequence of Molasse sediments up to 5,000 m thick has been deposited. It appears to correlate with phases of uplift and dissection of the mountains (Fig. 11.20). The Molasse sediments were deposited below or just above sea level and therefore marine and fluvial/limnic facies alternate. They are traditionally indicated from bottom to top as Lower Marine Molasse (*Untere Meeresmolasse*), Lower Freshwater Molasse (*Untere Süßwassermolasse*), Upper Marine Molasse (*Obere Meeresmolasse*), and Upper Freshwater Molasse (*Obere Süßwassermolasse*). The sequence is Oligocene to Miocene in age. The extremely thick unfolded Molasse sediments contain clay, marl, sand, and gravel strata, reflecting frequent vertical and horizontal facies changes. Close to the Alps the Molasse sediments are compressed in a series of broad, east–west striking synclines, partly imbricated and pushed northwards across the unfolded Molasse (Fischer 1989). Structurally this zone belongs to the Alps. However, with regard to geomorphology it can be seen as part of the Alpine Foreland. The Alpine Foreland was further uplifted and tilted slightly to the east during the Pliocene. In the southern zone of the Alpine Foreland typical concrete-like hard rocks occur, formed by cementation of carbonate-rich gravels. These conglomerates, which represent the former alluvial fans and terraces of rivers draining the northern flank of the Alps, are called *Nagelfluh*. Because of their resistant character these rocks frequently stand out in the form of ridges or steeply sloping hills.

The Alpine Foreland can be broadly divided into three natural landscapes (Fig. 11.10):

- The north-eastern part of the Alpine foreland, to the east of the Lech river, which is called the Tertiary Hills region. This hilly region, with a relief of less than about 50 m and gentle slopes, is dissected by a dense network of small streams. It is underlain by non-resistant sedimentary rocks of the Upper Marine Molasse and the Upper Freshwater Molasse. Solifluction material and deposits resulting from other mass-wasting processes and fluvial dissection, as well as loess covers, make up most of the late Pleistocene surficial deposits (Fischer 1989).
- The landscape of fluvial and glaciofluvial fans and terraces, varying in age from early Pleistocene to Holocene.
- The morainic landscape, subdivided into a young morainic region of Last Glacial (Würm) age and an older morainic region, with a more subdued relief, of pre-Würm age.

Quaternary sediments reach a thickness of over 150 m. The sequence of Pleistocene sediments ranges from coarse gravel deposits of the Biber- and Donau-complex, possibly representing early Pleistocene glacial periods, to alternating glacial and fluvioglacial sediments of the middle and late Pleistocene glacial cycles (see Ehlers 1996 and Habbe 2002). The morphological expression of the foreland glaciations has been treated in the previous subsection on Alpine glaciations.

German Alps

During the last part of the Upper Cretaceous and the early Tertiary the main deformation and faulting of the Alps took place. Cretaceous rocks were moved northwards several hundreds of kilometres in huge nappes. Orogenesis was still incomplete at the beginning of the Quaternary, and uplift continues up to the present day of the order of 0.5–1.5 mm/yr. Horizontal movements continue as well; geodetic measurements show a northward movement of the Alps by 2 mm/yr. The German Alps are divided into the High Limestone Alps and the Marginal Alps. The latter are further subdivided into the Marginal Limestone Alps, the Flysh zone, and the Helveticum zone (Fischer 1989, 2002). The Flysh zone overlies the Helveticum zone and is in its turn overlain by three nappes of the Limestone Alps. Obviously, landforms in this high mountainous region are adapted to structure and lithology. Because of frequent changes of facies and the complex morphotectonic structure, a great variety of landforms over short distances has developed. The Limestone Alps reach heights above 2,000 m and show a sharp and strong relief including various forms of karstification. The Zugspitze, which is the highest summit in the German Alps (2,962 m), is made up of very resistant *Wetterstein*-limestone. The Flysh zone consists of a sequence of less resistant rocks, mainly deposited by turbidity currents in a narrow and deep sea trough of the east Alpine geosyncline. Consequently, the Flysh is easily weathered and eroded, and landforms exhibit more rounded ridges with moderate relief and slope angles (Fischer 1989). Glacial erosion and physical weathering have further sculptured this Alpine landscape. It is estimated that ice thicknesses in the Northern Limestone Alps were more than 1,000 m, locally resulting in intensive glacial erosion and overdeepening of valleys by hundreds of metres. In the valleys of the Bayerische Alps glacial sediments with a thickness of up to 500 m have been observed (Semmel 1996).

References

Ahnert, F. (1989*a*). Landforms and landform evolution in West Germany. *Catena*, Suppl. 15.

—— (1989*b*). The major landform regions. In: F. Ahnert (ed.) Landforms and landform evolution in West Germany. *Catena*, Suppl. 15: 1–9.

Andres, W. (1989). The Central German Uplands. In: F. Ahnert (ed.) Landforms and landform evolution in West Germany. *Catena*, Suppl. 15: 25–44.

Blume, H., and Remmele, G. (1989). A comparison of Bunter Sandstone scarps in the Black Forest and the Vosges. In: F. Ahnert (ed.) Landforms and landform evolution in West Germany. *Catena*, Suppl. 15: 229–42.

Boenigk, W. (1982). Der Einfluß des Rheingrabensystems auf die Flußgeschichte des Rheins. *Zeitschrift fur Geomorphologie*, Suppl. 42: 167–75.

Bork, H. R., Schmidtchen, G., and Dotterweich, M. (eds.) (2003). Bodenbildung, Bodenerosion und Reliefentwicklung im Mittel- und Jungholozän Deutschlands. *Forschungen zur deutschen Landeskunde* 253.

—— Bork, H., Dalchow, K., Faust, B., Piorr, H. P., and Schatz, T. (1998), *Landschaftsentwicklung in Mitteleuropa*. Klett-Perthes, Gotha and Stuttgart.

Bremer, H. (1989), On the geomorphology of the South German Scarplands. In: F. Ahnert (ed.) Landforms and landform evolution in West Germany. *Catena*, Suppl. 15: 45–67.

—— and Nitz, B. (1993), Geomorphology in Germany. In: H. J. Walker, and W. E. Grabau (eds.) *The evolution of Geomorphology*, 171–80. J. Wiley & Sons, Chichester.

Brunnacker, K. (1986), Quaternary stratigraphy in the Lower Rhine area and Northern Alpine Foothills. In: V. Šibrava, D. Q. Bowen, and G. M. Richmond (eds.) Quaternary glaciations in the Northern Hemisphere. *Quaternary Science Reviews* 5: 373–9.

Brunotte, E., and Garleff, K. (1980), Tectonic and climatic factors of landform development on the northern fringe of the German Hill Country (Deutsche Mittelgebirge) since the early Tertiary. *Zeitschrift fur Geomorphologie*, Suppl. 36: 104–12.

—— (1989), Structural landforms and planation surfaces in southern Lower Saxony. In: F. Ahnert (ed.) Landforms and landform evolution in West Germany. *Catena*, Suppl. 15: 151–64.

Ehlers, J. (1996), *Quaternary and glacial geology*. J. Wiley & Sons, Chichester.

Eissmann, L. (1994), Das Quartär Mitteldeutschlands. Ein Leitfaden und Exkursionsführer mit einer Übersicht über das Präquartär des Saale-Elbe-Gebietes. *Altenburger naturwissenschaftliche Forschungen* 7.

—— (1997), Die ältesten Berge Sachsens oder Die morphologische Beharrlichkeit geologischer Strukturen. *Altenburger naturwissenschaftliche Forschungen* 10.

—— (2000), *Die Erde hat Gedächtnis. 50 Millionen Jahre im Spiegel mitteldeutscher Tagebau*. Sax-Verlag, Beucha.

—— (2002), Quaternary geology of eastern Germany (Saxony, Saxon-Anhalt, South Brandenburg, Thüringia), type area of the Elsterian and Saalian Stages in Europe. *Quaternary Science Reviews* 21: 1275–346.

Ellwanger, D., Bibus, E., Bludau, W., Kösel, M., and Merkt, J. (1995), XI. Baden-Württemberg. In: L. Benda (ed.) *Das Quartär Deutschlands*. Bornträger, Berlin, 255–95.

Embleton, C. (ed.) (1984), *Geomorphology of Europe*. Macmillan Reference Books, London.

—— and Demek, J. (1984), Hercynian Europe. In: C. Embleton (ed.) *Geomorphology of Europe*. Macmillan Reference Books, London, 165–230.

Fischer, K. (1989), The landforms of the German Alps and the Alpine Foreland. In: F. Ahnert (ed.) Landforms and landform evolution in West Germany. *Catena*, Suppl. 15: 69–83.

—— (2002), Deutschlands Alpenanteil. In: H. Liedtke, and J. Marcinek (eds.) Physische Geographie Deutschlands. Gotha (Perthes Geographie Kolleg), 639–77.

Fuchs, K., von Gehlen, K., Mälzer, H., Murawski, H., and Semmel, A. (eds.) (1983), *Plateau uplift, the Rhenish Shield—a case history*. Springer Verlag, Berlin.

Habbe, K. A. (1989), Die pleistozänen Vergletscherungen des süddeutschen Alpenvorlandes. Ein Resümee. Mitteilungen der Geographischen Gesellschaft. *München* 74: 27–51.

—— (2002), Das deutsche Alpenvorland. In: H. Liedtke, and J. Marcinek (eds.) *Physische Geographie Deutschlands*. Gotha (Perthes Geographie Kolleg), 591–638.

Hantke, R. (1978/92), *Die jüngste Erdgeschichte der Schweiz und ihrer Nachbargebiete*, i. Thun: Ott., Landsberg am Lech: Ecomed.

—— (1993), *Flußgeschichte Mitteleuropas. Skizzen zu einer Erd-, Vegetations- und Klimageschichte der letzten 40 Millionen Jahre*. Enke, Stuttgart.

Henningsen, D., and Katzung, G. (2002), *Einführung in die Geologie Deutschlands*. 6th edn. Spektrum Akademischer, Heidelberg.

Hövermann, J. (1987), Neues zur pleistozänen Harzvergletscherung. *Eiszeitalter und Gegenwart* 37: 99–107.

Jerz, H. (1995), XII Bayern. In: L. Benda (ed.) *Das Quartär Deutschlands*. Bornträger, Berlin, 296–326.

Kalis, A. J., Merkt, J., and Wunderlich, J. (2003), Environmental changes during the Holocene climatic optimum in central Europe —human impact and natural causes. *Quaternary Science Reviews* 22: 33–79.

Liedtke, H. (2002), Oberflächenformen. In: H. Liedtke, and J. Marcinek (eds.) *Physische Geographie Deutschlands*. 3rd edn. Gotha (Perthes Geographie Kolleg), 127–56.

—— and Marcinek, J. (eds.) (2002), *Physische Geographie Deutschlands*. 3rd edn. Gotha (Perthes Geographie Kolleg).

—— Mäusbacher, R., and Schmidt, K-H. (eds.) (2003), *Nationalatlas Bundesrepublik Deutschland*, ii. Relief, Boden und Wasser. Institut für Landerkunde, Leipzig. Spektrum Akademischer, Heidelberg.

Litt, T., Schmincke, H. U., and Kromer, B. (2003), Environmental response to climatic and volcanic events in central Europe during the Weichselian Lateglacial. *Quaternary Science Reviews* 22: 7–32.

Marcinek, J., Richter, H., and Semmel, A. (2002), Die deutsche Mittelgebirgsschwelle. In: H. Liedtke, and J. Marcinek (eds.) *Physische Geographie Deutschlands*. 3rd edn. Gotha (Perthes Geographie Kolleg), 463–538.

Mertes, H., and Schmincke, H. U. (1983), Age distribution of Quaternary volcanoes in the West Eifel Volcanic Field. Jahrbuch für Geologie und Paläontologie. *Abhandlungen*, 260–93.

Pfeffer, K. H. (1989), The karst landforms of the northern Franconian Jura between the rivers Pegnitz and Vils. In: F. Ahnert (ed.) Landforms and landform evolution in West Germany. *Catena*, Suppl. 15: 253–60.

Schirmer, W. (1995), Rhein Traverse. In: W. Schirmer (ed.) INQUA 1995: *Quaternary field trips in Central Europe*. Pfeil, Munich, i. 475–558.

Schmidt, K. H. (1989), Geomorphology of limestone areas in the north-eastern Rhenish Slate Mountains. In: F. Ahnert (ed.) Landforms and landform evolution in West Germany. *Catena*, Suppl. 15: 165–77.

Schmidt, R. (2002), Böden. In: H. Liedtke, and J. Marcinek (eds.) *Physische Geographie Deutschlands*. 3rd edn. Gotha (Perthes Geographie Kolleg), 255–88.

Schmincke, H. U., Park, C., and Harms, E. (1999), Evolution and environmental impacts of the eruption of Laacher See Volcano (Germany) 12,900 a BP. *Quaternary International* 61: 61–72.

Scholten, T. (2003), Beitrag zur flächendeckenden Ableitung der Verbreitungssystematik und Eigenschaften periglaziärer Lagen in deutschen Mittelgebirgen. *Relief, Boden, Paläoklima*, 19.

Semmel, A. (1991), Neotectonics and geomorphology in the Rhenisch Massif and the Hessian Basin. *Tectonophysics* 195: 291–7.

—— (1996), *Geomorphologie der Bundesrepublik Deutschland. Grundzüge, Forschungsstand, aktuelle Fragen, erörtert an ausgewählten Landschaften*. 5th edn. Erdkundliches Wissen H. 30, Franz Steiner, Stuttgart.

—— (2002), Das Süddeutsche Stufenland mit seinen Grundgebirgsrändern. In: H. Liedtke, and J. Marcinek (ed.) *Physische Geographie Deutschlands*. 3rd edn. Gotha (Perthes Geographie Kolleg), 540–90.

Spönemann, J. (1989), Homoclinal ridges in Lower Saxony. In: F. Ahnert (ed.) Landforms and landform evolution in West Germany. *Catena*, Suppl. 15: 133–49.

van Husen, D. (1987), *Die Ostalpen in den Eiszeiten*. Geologische Bundesanstalt, Vienna.

Walter, R. (1992), *Geologie von Mitteleuropa*. 5th edn. Schweizerbart, Stuttgart.
Zienert, R. (1967), Vogesen- und Schwarzwald-Kare. *Eiszeitalter und Gegenwart* 18: 51–75.
—— (1989), Geomorphological aspects of the Odenwald. In: F. Ahnert (ed.) Landforms and landform evolution in West Germany. *Catena*, Suppl. 15: 199–210.
Zolitschka, B., Behre, K-E., and Schneider, J. (2003), Human and climatic impact on the environment as derived from colluvial, fluvial and lacustrine archives—examples from the Bronze Age to the Migration period, Germany. *Quaternary Science Reviews* 22: 81–100.

12 French and Belgian Uplands

Bernard Etlicher

Long-Term Evolution: Tertiary and Quaternary

The French Uplands were built by the Hercynian orogenesis. The French Massif Central occupies one-sixth of the area of France and shows various landscapes. It is the highest upland, 1,886 m at the Sancy, and the most complex. The Vosges massif is a small massif, quite similar to the Schwarzwald in Germany, from which it is separated by the Rhine Rift Valley. Near the border of France, Belgium, and Germany, the Ardennes upland has a very moderate elevation. The largest part of this massif lies in Belgium (Fig. 12.1). Though Brittany is partly made up of igneous and metamorphic rocks, it cannot be truly considered as an upland; in the main parts of Brittany, altitudes are lower than in the Parisian basin. Similarities of the landscape in the French and Belgian Uplands derive from two major events: the Oligocene rifting event and the Alpine tectonic phase.

Hercynian Basement and Post-Hercynian Peneplain Evolution

The Vosges and the Massif Central are located on the collision zone of the Variscan orogen. In contrast, the Ardennes is in a marginal position where primary sediments cover the igneous basement. Four main periods are defined during the Hercynian orogenesis (Bard *et al.* 1980; Autran 1984; Ledru *et al.* 1989; Faure *et al.* 1997). The early Variscan period corresponds to a subduction of oceanic and continental crust and a high-pressure metamorphism (450–400 Ma) The medio-Variscan period corresponds to a continent–continent collision of the chain (400–340 Ma). Metamorphism under middle pressure conditions took place and controlled the formation of many granite plutons: e.g. red granites (*granites rouges*), porphyroid granite, and granodiorite incorporated in a metamorphic complex basement of various rocks (Fig. 12.2). The neo-Variscan period (340–320 Ma) is characterized by a strong folding event: transcurrent shear zones affected the units of the previous periods and the first sedimentary basins appeared. At the end of this period, late-Variscan (330–280 Ma), autochthonous granites crystallized under low-pressure conditions related to a post-collision thinning of the crust. Velay and Montagne Noire granites are the main massifs generated by this event.

Sediment deposition in tectonic basins during Carboniferous and Permian times occurred in the Massif Central and the Vosges: facies are sandstone (Vosges), shale, coal, and sandstone in several Stephanian basins of the Massif Central, with red shale and clay 'Rougier' in the south-western part of the Massif Central. Consequently, the basement is very complex with a varied lithology, that was favourable for the development of differential weathering.

The formation of the Hercynian basement was concluded by a major event, the post-Hercynian peneplain, which saw erosion of igneous rocks and autunosaxonian sediments. This peneplain can be identified at the periphery of the Massif Central, the western Vosges, and the Ardennes (Baulig 1928; Perpillou 1940; Beaujeu Garnier 1951; Pissart 1962; Mandier 1989) and is preserved under Trias continental deposits or by Lias (Infra-Jurassic) marine deposits. The surface was worked under an arid continental climate, and is regarded by recent authors (Klein 1990) as a long (50 Ma) phase of pedimentation. From a recent interpretation of saprolites (Simon-Coinçon *et al.* 1983), this surface can be identified by dolomitization, albitization, and the presence of ferruginous deposits. The

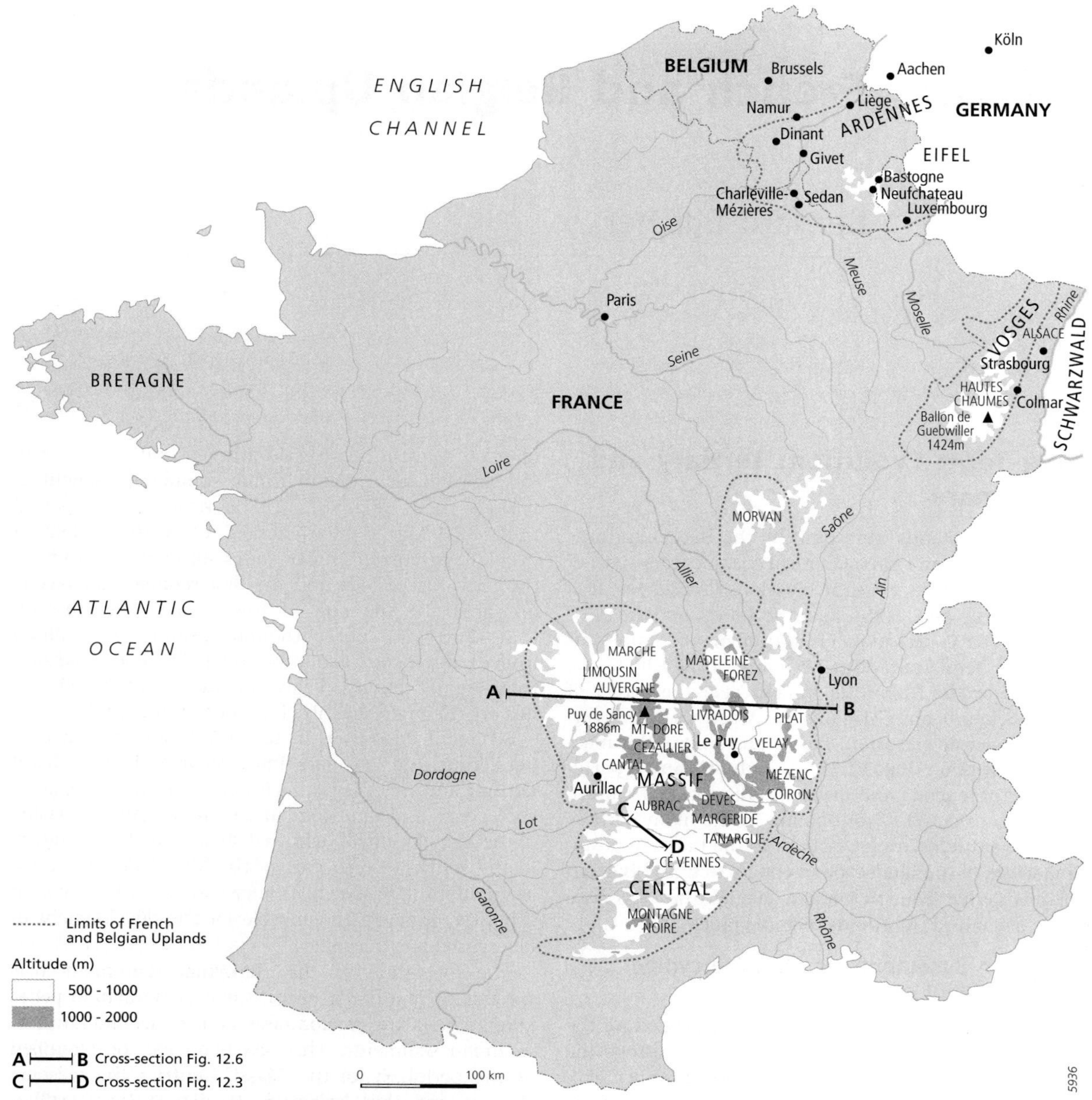

Fig. 12.1. Map showing locations in the French and Belgian uplands.

palaeoenvironment must be considered as a 'hill and basin system' with albitization and pedimentation in the upper parts, and evaporation basins in the lower parts (Fig. 12.3). A relative tectonic stability allowed a long duration of this system. Marine transgressions and regressions locally interrupted the erosion cycle (Simon-Coinçon 1987). A Triassic transgression is obvious in the northern Vosges and Cévennes where several residual sandstone outliers were preserved from erosion cycles (Fig. 12.4). These uplands gradually emerge during the Mesozoic Era, in relation to a regional tectonic uplift which occurred earlier in Auvergne, and later in the Causses basins and the north-eastern part of the Massif Central.

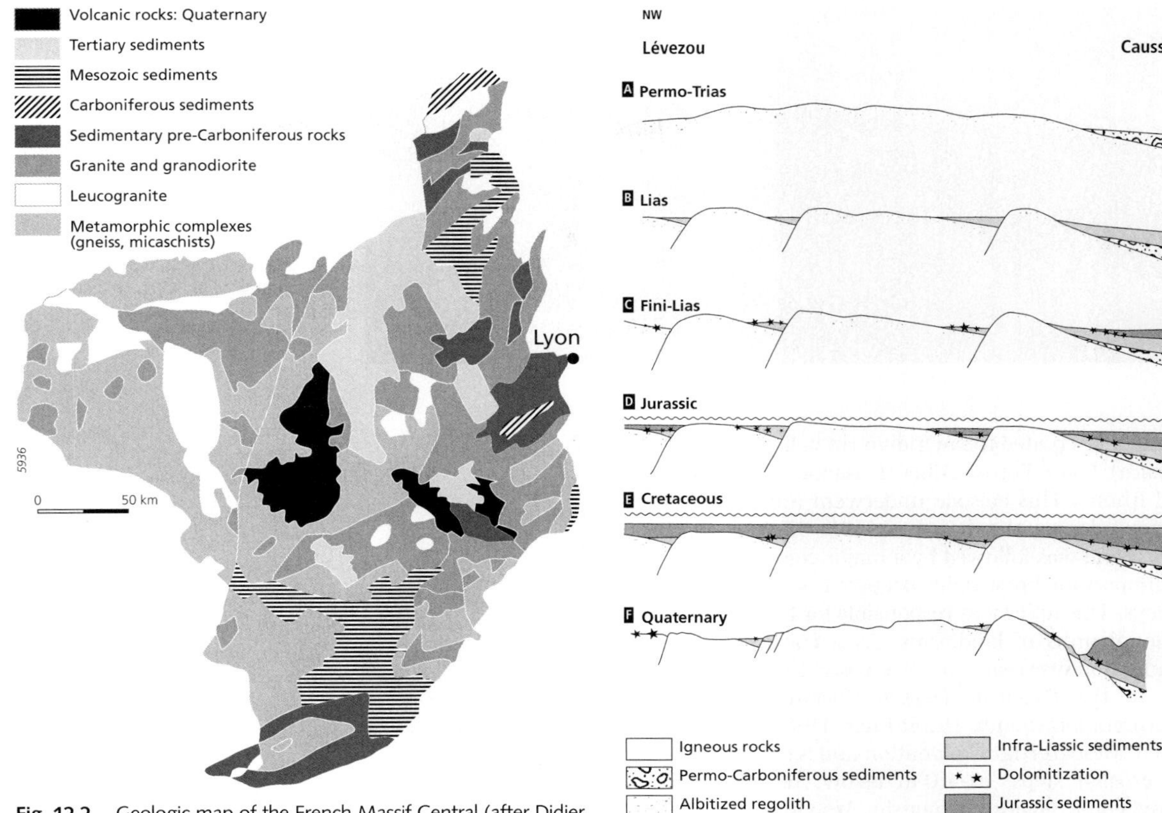

Fig. 12.2. Geologic map of the French Massif Central (after Didier and Lameyre 1971).

Fig. 12.3. Cross-section of the post-Hercynian peneplain in the southern Massif Central (after Simon Coinçon *et al.* 1983). For the location see Fig. 12.1.

The Main Early Tertiary Erosion Event

This episode is well known and called the Eogene planation, *aplanissement éogène* (Baulig 1928). According to Baulig, several authors consider summit landforms in Forez, Morvan, Limousin (Massif Central), the plateau de Rocroi (Ardennes), and the Hautes Chaumes (Vosges) as derived from a general planation level at the beginning of the Tertiary. Nevertheless, this planation is conserved only where the basement is formed by homogenous and resistant material. This planation occurred during 50 Ma at the end of the Mesozoic and the beginning of the Cenozoic. It seems to have been completed by the end of the Eocene. Klein (1990) called it 'Eo-Tertiaire' planation and interpreted it as an acyclic palaeo-landform. In such a model, a tectonic uplift affecting a pre-existent peneplain does not involve a new erosion cycle with significant entrenchment of the valleys. However, there are local rearrangements of the peneplain: a new surface was worked including large parts of the previous one and both parts cannot be distinctly identified. This erosion cycle resulted from a kaolinite and gibbsite alteration of the basement, under continental and tropical climatic conditions. A correlative sedimentation is identified around the Massif Central: siderolithic sand and clay deposits with lateritic induration are exposed on the borders of the Massif Central, and also on the borders of Marche, Limousin (Klein 1990), Margeride (Coque-Delhuille 1979), Velay (Larqué and Weber 1978), Morvan (Nieuwenhuis 1971), and Ardennes (Voisin 1977; Pierre 1999).

Graben, Tectonic Rifting (Eocene to Early Miocene)

The peneplain was completed when a major event, the Oligocene rifting, disrupted the Variscan basement of western Europe. Related to the opening of the Atlantic Ocean, a rifting began during the Eocene as a result of a Pyrenean compression phase. Tectonic faulting and

Fig. 12.4. The post-Hercynian peneplain in the Cévennes. Flat summits are elements of the uplifted (1,100 m), post-Hercynian peneplain (dotted line). Some outliers of Triassic sandstones (arrowed) are forested. In the background, the Rhone Rift Valley (R) and the Lozère horst, defined by a major fault scarp (F), are shown.

subsidence created the meridian rift valleys of the Rhine (Alsace), Loire (Forez), Allier (Limagne), Saône (Bresse), and Rhône. This episode underwent a maximal development during the Oligocene (35–30 Ma) (Bergerat 1987) and was followed by a major compressive phase. An important horst uplift occurred, separating the rift valleys. This uplift was responsible for the present main regional units of landforms. As a consequence, local denudation was reactivated: a new cycle began, according to the *Piedmont Treppen* (Piedmont benchland) interpretation (Penck 1925; Klein 1990). This event is called 'Meso-Tertiaire' planation and is recognized by local erosion steps 30–40 m below the peneplain, in Velay, Forez, Vivarais, Limousin, Ardennes (Klein 1990). For the first time, a tendency towards valley deepening is noticeable. The end of the acyclic evolution and the beginning of a new denudation cycle represented an essential change in the evolution of the upland palaeoenvironment. The Eocene surface is faulted and denudation causes sediment accumulation in the graben during the Oligocene. Sediment deposition is significant (300–400 m) in the main basins of Limagne, Forez, Alsace, and Saône valleys. Miocene sedimentation is characterized by the presence of strongly weathered residual deposits, as well as pebbles and gravel benches, which were due to fluvial activity (alluvial fans) at the periphery of the basins. Fine sediments, mainly sands and clay, fill the centre of the basins. Evaporites and pedologic concretions (calcrete, dolocrete) are interpreted as the result of a lacustrine sedimentation in a semi-arid warm temperate or subtropical environment. This episode does not offer a possibility for the development of a general denudation level or peneplain. An insufficient duration and tectonic instability are pointed out by Mandier (1989), Klein (1990), and le Griel (1991) to explain the absence of a general peneplain cycle. Valley deepening was more important than denudation.

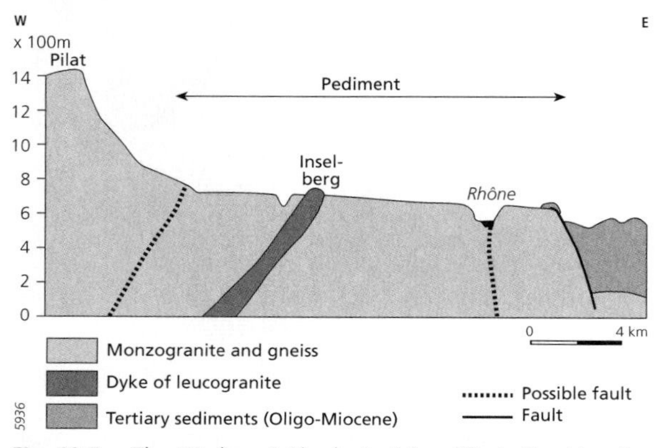

Fig. 12.5. The 'Piedmont Rhodanien' (modified after Mandier 1989).

At the boundary between horsts and grabens, a system of large valley-basins connected with local piedmont steps are interpreted as pediments (Mandier 1989). A pediment 100 km × 15 km wide, 320 m a.s.l. (Piedmont Rhodanien) is recognized along the Rhône valley, south of Lyons, with embayments and inselbergs. Locally, it is covered by scattered molassic deposits of Miocene age (Burdigalian) (Fig. 12.5). The possibility of the presence of a major fault system at the foot of the pediment is discussed by le Griel (1991) and Klein (1990). Klein considers the Piedmont Rhodanien as an equivalent of the Danubian Piedmont, at the foot of the Bavarian upland (Bayerischer Wald). Similar forms are described at the foot of other uplands, especially Forez (Etlicher 1986), Margeride (Coque-Delhuille 1979), and Vosges (Vogt 1992). Few of them are dated except in Forez, where an alluvial deposit was preserved by a volcanic lava dated 17.8 Ma (Etlicher 1986).

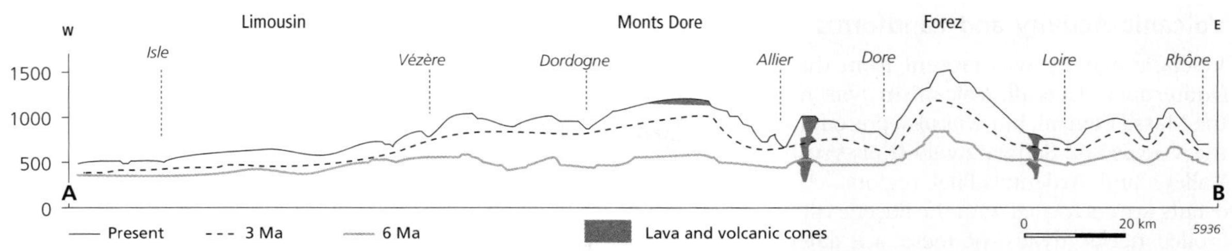

Fig. 12.6. Cross-section showing Mio-Pliocene tectonic uplift in the Massif Central (after le Griel 1991). For the location see Fig. 12.1.

Tectonic Uplifting (Mio-Pliocene)

A complete change of tectonic style and environmental conditions occurred during the Miocene. A strong compressive tectonic phase replaced the Eocene–Oligocene rifting episode. This phase was contemporaneous with uplifting of the Alps and took place from the Miocene (8 Ma) to the Quaternary (Fig. 12.6). Horst uplift was more active in the vicinity of the Alps, but affected all the massifs (Vosges, Massif Central, and Ardennes). Chemical weathering changed after 8 Ma (Pierre 1990) in relation with the transition from a subtropical to a temperate climate. Sedimentation in the main grabens was mainly clay and lacustrine limestone, with pebble and gravels layers at the periphery of the basins. As a consequence of tectonic uplifting, rivers and valleys were considerably deepened. This phenomenon has been studied in the Allier catchment by Pastre (1987) and le Griel (1991). Alluvial terraces can be identified and dated through heavy mineral composition related to the volcanic activity of the Cantal. Data are also available in Ardennes from Voisin (1977) and, more recently, from Demoulin (1995*a*) and Pissart *et al.* (1997). From the example of the Loire Valley in Velay, south of Le Puy, Defive (1998) considers that the rate of incision was increasing from 8 Ma to the Quaternary. An estimation of 300 m during Mio-Pliocene times (8–2.8 Ma) is probable. During the Quaternary, a minimum estimation of 250 m is given (Fig. 12.7).

The rate of deepening was increasing eastward. In the Rhône catchment, the consequences of the Messinian regression (Mandier 1989) complicate the scenario. The level of the Mediterranean Sea is estimated to have lowered to −2,000 m: a narrow gorge or canyon along the Rhône Valley up to Lyons is due to regressive erosion. Acceleration of deepening affects all the tributaries of the Rhône, especially in Cévennes and Vivarais. North of Lyons, the floor of the palaeo-canyon is below the present sea level. The incision of the Rhône was considerable, 300 m at the limit Mio-Pliocene (Mandier 1989), but the valley slope back-wearing was relatively moderate. The valley shape, when preserved by lava flows, shows a good conservation of the interfluves and poor denudation on granitoid materials. A significant erosion rate is noticed only in sedimentary materials (clay and sands) in the grabens, where runoff and aeolian activity probably were the most efficient agents of denudation.

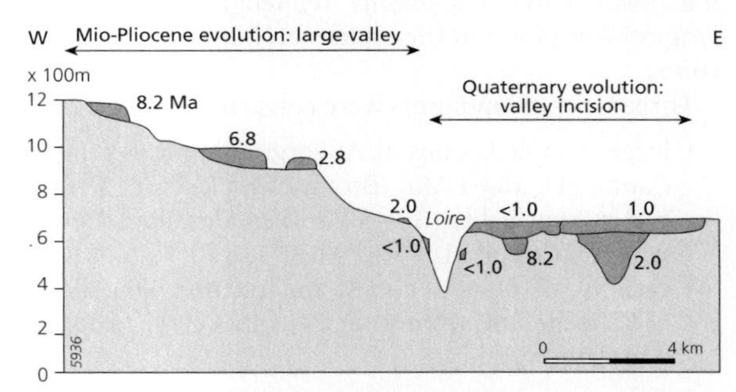

Fig. 12.7. Cross-section of the Loire Valley, 25 km south of Le Puy, showing valley incision. Approximate ages in Ma (modified after Defive 1998).

A quantitative balance of erosion rate on the interfluves is more difficult to determine. In a region where uplifting is moderate, Freytet and Morel (1987), using a morphometric method, demonstrate that the level of topographic plateaux of the Marche exactly corresponds to the Eo-Tertiary topography. In the Velay region, north of Le Puy, Chafchafi (1997) points out a moderate rate of denudation in a region where uplifting was considerable and valley incision reached down to 200 m. Nevertheless, on 80% of the area denudation is less than 25 m since the beginning of the Miocene. These observations are in complete accordance with the estimation of the volume of sediments of this epoch in the sedimentary basins around the Massif Central (le Griel 1991). Similar evolutions can be noticed in the Vosges and the Rhine Valley.

Volcanic Activity and Landforms

Volcanic activity was present from the Eocene to the Quaternary. Overall, volcanism was most marked in the Massif Central, but was also present in the vicinity of the Vosges and Schwarzwald (Kaiserstuhl in the Rhine Valley) and Ardennes-Eifel regions. Ancient volcanic events are correlated with the Eocene rifting phase: a few eroded necks, dykes, or mesa are dated from Eocene (45–60 Ma) in Forez and Limagne rift valleys. Located at the border of the graben, these lava flows are quasi-completely eroded but they form small attractive hills. This volcanism activity is mainly related to the Neogene compression phase in the alpine foreland (Michon *et al.* 1999).

Three types of landforms were constructed (Fig. 12.8):

- large stratified cones in Auvergne, Mont Dore, and Cantal (16 Ma–1 Ma) (Brousse and Lefèvre, 1966);
- major lava fields (50–80 km) in Cezallier, Devès, Aubrac, Coiron (1.5–0.7 Ma);
- systems of domes, cones, maars, and lava flows of Pliocene and Quaternary age in Velay, Ardèche, and Auvergne.

The last events are recent: during the Holocene activity occurred in Ardèche, and is dated from 12,000 BP (Defive and Berger 1991); a continuous activity is recorded in the Chaîne des Puys from 55,000 to 5600 BP. The most recent events occurred in the Alleröd in the Massif Central and Eifel, and during the Holocene in Auvergne (Gewelt and Juvigné 1988; Brousse and Lefèvre 1990; Lenselink *et al.* 1990; Michon *et al.* 1999).

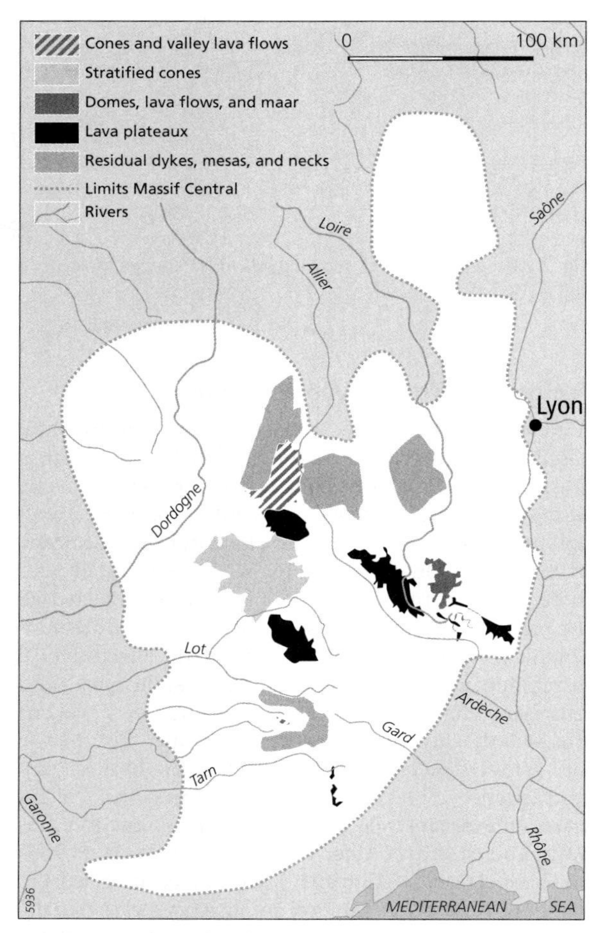

Fig. 12.8. Volcanic landform types in the Massif Central (after Hodges, in Michon *et al.* 1999).

Distribution of Surficial Deposits and Landforms

Glaciation and Glacial Deposits

The extent of Quaternary glaciation in the Vosges and the Massif Central is still the subject of much discussion. The question discussed is not the existence of a glaciation but the number of glacial events and their extent. Several sites are now major chronostratigraphical references for western Europe. According to Veyret (1981) a large ice cap extended in the main valleys of the volcanic area of Auvergne (Cantal and Mont Dore), and overflowed onto the *planèzes* (basaltic surfaces) and the piedmont region of Artense, north-west of the Cantal, and to the Dordogne valley (Fig. 12.9). Ice caps also covered Aubrac and Cezallier (Fig. 12.10). These western massifs, exposed to the wind from the Atlantic, were covered by an 80-km-long piedmont glacier at low altitude. Terminal moraines are described in the Aurillac basin, along the Dordogne Valley, and on the 'planèze' of the Cantal (Goer de Herve 1972).

The eastern massifs of Forez, Mézenc, Lozère, Margeride, and Tanargue are characterized by a small extension of high altitude surfaces, and more continental and dry conditions. Ice extent was moderate, and glacial development was limited to valley tongues and small ice caps on the highest plateaux of Forez, Margeride, and perhaps Lozère. Glacial features are of modest extension: till is recognized in a few valleys, glacial cirques are present at the upper part of few catchment basins, and some valleys have a U-shaped profile, mainly in Forez (Etlicher 1986). In any case, the alteration of landforms through glacial action was moderate. Nevertheless, the presence of these glacial deposits makes it possible to reconstruct

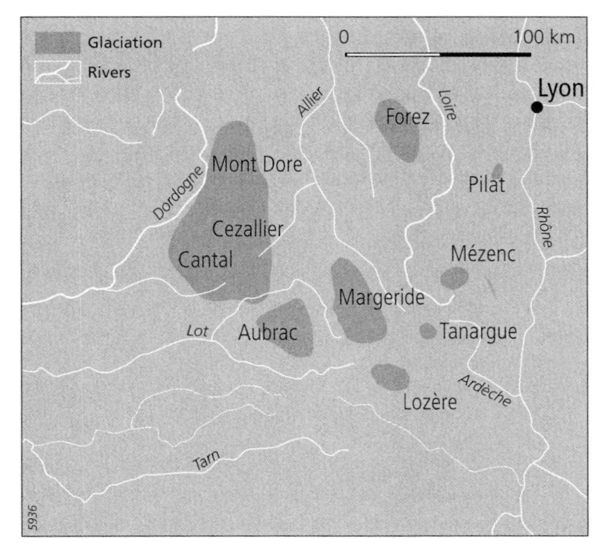

Fig. 12.9. Extent of formerly glaciated areas in the Massif Central (after Veyret 1981).

the position of the firn line, and gives information about palaeo-environmental conditions during the Quaternary glaciations.

From field observations a reconstruction of the position of the ELA (equilibrium line of alimentation) shows a drastic decline of its position from south-east to north-west summits (Fig. 12.11). In the south-eastern border of the massif the limit of the ELA was located at 1,600 m, which points to a small area of ice accumulation. In the north-west the position was near 1,200 m according to Veyret (1981) and Goer de Herve (1972). The extent of ice cover was then considerable, in accordance with the position of terminal moraines at low altitude along the Dordogne valley. The extent of ice was also considerable in the Vosges massif, especially on the west side and on the southern piedmont with its large piedmont lobe. The ELA was located at 900 m on the western slope, but probably lower around 750 m in the north of the massif (Tricart and Trautmann 1994). La Grande Pile in the southern part of the Vosges is an international reference site for glacial chronology (Woillard 1978). At this site elaborate pollen analysis of sediments reveal the oscillations of the Weichselian glaciation (Seret *et al.* 1990).

Several glacial episodes are recognized, but correlations with the main European chronostratigraphies are still uncertain, because of poor dating possibilities. Nevertheless, a group of terminal moraines has been identified, that correspond to the maximum of the last glaciation, the *moraines internes* of Etlicher and Goer de Herve (1988). Further, several morainic deposits, the *moraines externes*, are present and considered as representing a previous glacial cycle. These moraines are identified in the Aurillac basin (Veyret 1981) and Monts du Forez (Etlicher 1986). A systematic analysis of these deposits remains to be done and correlations with pre-Weichselian glaciations are still conjectural. Moreover, small landforms of morainic origin are present in a lot of valleys, e.g. in the Forez, Mont Dore, and Cantal. The features reflect a 'recurrence' phase, but the age and significance of this recurrence is debatable.

Fig. 12.10. Aubrac: a volcanic, formerly glaciated plateau; erratic blocks in the foreground and drumlins in the background (1,300 m).

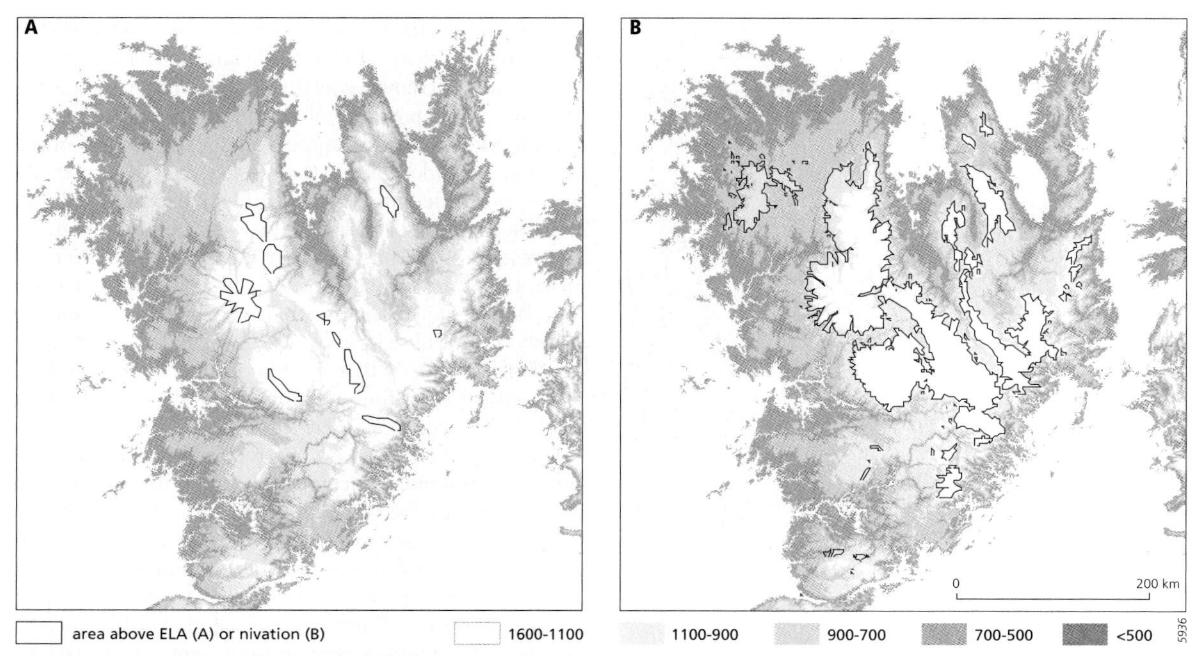

Fig. 12.11. Estimated position of the equilibrium line of alimentation (ELA) at the maximum of glaciation (A) and the area affected by nivation (B). The position of the ELA is estimated from the position of terminal moraines. Despite higher altitudes, the south-eastern part of the Massif had a small extent of ice accumulation. Minimum elevation for nivation development is calculated at 300 m below ELA (after Goer de Herve 1972; Veyret 1981; Valadas 1984; Etlicher, 1986; Etlicher and Goer de Herve 1988).

Some authors (Veyret 1981; Vergne 1989) consider it as a Late Glacial episode (Dryas); others consider it as dating to the Upper Weichselian (isotopic stadium 2) (Etlicher and Goer de Herve 1988; Macaire *et al.* 1992).

Nivation Landforms and Periglacial Deposits

In non-glaciated areas, landforms are strongly affected by Quaternary frost action. This action may lead to an especially severe denudation of regolith through frost shattering, nivation, gelifluction, and runoff. The variety of landforms depends on preglacial weathering conditions and, therefore, the altitudinal position of the Eo-Tertiary peneplain and its related weathering mantle.

Nivation

Nivation is a major phenomenon, but remodelling of valleys by this process is difficult to identify. Favourable conditions of exposure in conformity with wind direction are necessary. Field observations by Valadas (1984) in Limousin and Lozère, and by Etlicher (1986) in Forez and Cévennes, reveal a high frequency of nivation benches and nivation cirques. A model derived from the position of ELA shows the possibility of a large extension of nivation phenomena in the western part of the Massif (Fig. 12.11). The high plateaux of Limousin, Livradois, Forez, Madeleine, Margeride, and Pilat are involved. From 800 m in Limousin, the elevation of the nivation activity line rises up to 1,200 m in Cévennes. This is a major phenomenon affecting a large area in Limousin; but it is a process strictly limited to a small area around the major summits of Cévennes. The same analysis performed on all the massifs emphasizes a major climate gradient during the last glacial period; the effect on landscape morphology is frequently underestimated.

Slope Deposits

Where the regolith is thin and discontinuous, frost action develops tors and talus slope deposits (screes). Where the arenaceous cover is thick enough, frost action generates a complex sequence of displaced slope deposits with three main facies (Fig. 12.12). These three facies are superposed in a more or less complex sequence, eventually with recurrence. Several generations of these facies can be seen along the valley profiles (Godard *et al.*, 1994). *Head* (*arène remaniée à blocs*), is defined as an arenaceous sand with blocks mobilized by gelifluction during the warm season. It can be considered as a part of the active layer, thawing in summer and moving down. Displace-

Fig. 12.12. A classic sequence of bedded grus and head in gneissic material (Massif de Pilat, eastern Massif Central, 1,220 m).

ment is significant, several cm/yr. The *bedded grus* (*arène litée* or *arène fauchée*), is a regolith displaced by frost creep. Displacement is laminar and results from seasonal growing and thawing of ice lenses.

The *in situ regolith* (*arène en place*), is affected by ice lensing but the absence of displacement is proved by the continuity of diaclases or quartzite veins. The transition between gelifluction and frost creep, progressive or through an angular unconformity, depends on the thermal regime of the soil during the glacial period and local topographic conditions. Usually, the transition is progressive at the upper part of the slope, where denudation is predominant and there is no unconformity between in situ grus and head. At the foot or in the concave part of the slope, an unconformity frequently separates head and bedded grus, because of the aggradation of permafrost due to the proximity of the water table. This sequence cannot be interpreted in chronological or chronostratigraphical terms. Below the head deposit, a compacted fragic horizon, called *crassin* in the Vosges, reveals the presence of the permafrost table (van Vliet Lanoë and Valadas 1983; Godard *et al.* 1994). The origin of this fragic horizon is related to the abundance of silt produced by frost shattering as revealed by the grain-size curve. Slope wash and wind accumulation explain a higher percentage of silt in the deposits on the lower part of especially leeward slopes. A minimum of 15–20% silt is sufficient to make gelifluction active.

The head matrix is always poorly sorted. Mineralogical analysis reveals the predominance of polymineral grains and K-feldspars in coarse sand and of quartz and plagioclases in fine sand. This spectrum is interpreted as the result of differential frost shattering of minerals according to a resistance scale of minerals specific to frost action (Etlicher and Lautridou 1998). Grain-size curves always show poor sorting indexes; fine sand and silt are dominant, and this is confirmed by the results of experimental frost shattering of grus. Five different types of granulometric sorting can be distinguished (Fig. 12.13), dependent on the processes involved (Etlicher 1986). Lacotte (1982) in Limousin and Etlicher (1986) in Forez and Pilat have shown that the surface occupied by head deposits, calculated as a percentage of the total surface, increases with altitude. In Forez for example, values are 15% at 850 m, 50% at 950 m, and 90% above 1,200 m. Local conditions of lithology and climate are also noticeable factors. The sequences are thicker and more frequent in a continental situation (Vosges, Morvan, Forez, and Vivarais), whereas in Limousin and Ardennes such a sequence is rare or limited to favourable local situations. Head is present, but bedded grus is exceptional. According to Etlicher (1986) the head

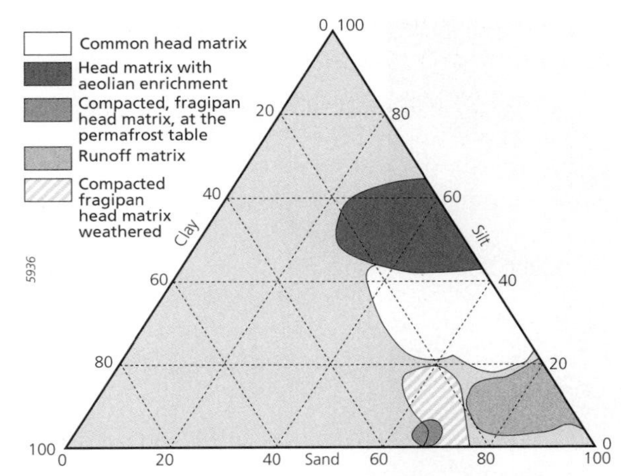

Fig. 12.13. Five types of matrix in displaced slope deposits (after Etlicher 1986).

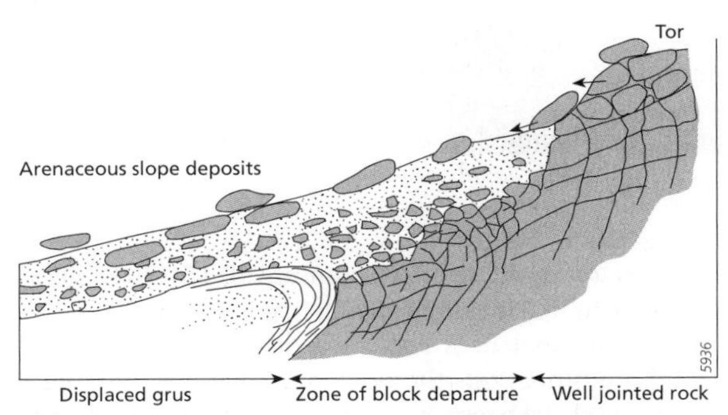

Fig. 12.14. Origin of blocks in periglacial displaced slope deposits. Notice differences in dimensions, shape, and quantity of blocks dependent on their origin (after Etlicher 1986).

blocks have two origins (Fig. 12.14). First, in the departure zone, quarrying and dislocation of rock outcropping is possible if freezing water enlarges the diaclases. The curvature of joints is due to the downslope migration of pebbles and blocks. Particles are small because joints are close together. The most frequent size of blocks is 12–35 cm, according to observations in Limousin (Lacotte 1982) and Forez (Etlicher 1986). Secondly, in case rock exposures are present in the upper part of the slope, blocks falling from the summit tors can be added to the previous processes. Boulders are broken by frost heaving, which is related to the presence of silt filling joints. Silt material results from comminution of the face of the blocks and can also be of aeolian origin. Large round blocks (1–3 m) are at the surface of the grus and migrate to the lower part of the slope through gelifluction.

Anthropogenic Erosion, Soils, and Peat Bogs

Peat bog development, and its subsequent degradation due to agricultural practice, characterizes the Holocene geomorphologic evolution. The filling up of several glacially overdeepened lakes in Artense, Cezallier, and Forez has been dated to the Alleröd or Boreal (de Beaulieu *et al.* 1982); lacustrine sediments were covered with peat bog layers. Elsewhere, peat bog development took place during the Atlantic owing to a climate deterioration or, later, to human activity. Peat degradation took place since the Neolithic period in relation to the start of agricultural development. Degradation accelerated in medieval times as a result of an increase of cultivated areas. This was notable in the upper parts of the Vosges and Massif Central. These new agriculture lands were occupied during the tenth to twelfth centuries and many rivers and creeks were diverted for irrigation or drainage of valley bottoms.

Pollen analyses of sites in the Massif Central and Vosges allow a synthesis of the Holocene vegetation history. Regression of cold steppe communities in favour of *Pinus* woods was noticeable around 9800 BP (Denèfle *et al.* 1980; Janssen and van Straten 1982; de Beaulieu *et al.* 1988; Vergne 1989). A dry, warm period favoured the development of *Corylus* and *Quercus* woodland. This woodland development occurred somewhat later in the Velay than in north-western regions. *Abies* and *Fagus* woodland appeared around 5000 BP; *Abies* was more abundant in the eastern part, *Fagus* in the western part (Fig. 12.15).

The oldest indicator of human activity is dated from 6500 BP (chaîne des Puys, Denèfle *et al.* 1980). More

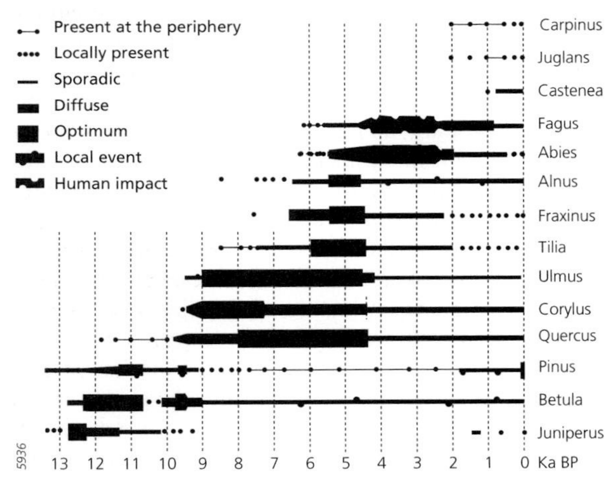

Fig. 12.15. Holocene history of vegetation in the Massif Central (after de Beaulieu *et al.* 1988).

generally, human impact on landscape seems to be predominant after 2600 BP when cereal pollens, the regression of forest taxons, the introduction of *Plantago* and *Rumex*, the expansion of *Castanea* and *Juglans*, and *Cannabis* are noted. Expansion of *Calluna* can be interpreted as a development of high altitude heathland. This taxon appears at Gallo-Roman times, when the practice of aestival grazing (cattle on the summit heathland during summer) seems to begin. The medieval activity seems to be mainly cultivation and a regression of forest is obvious (*Abies* and *Fagus*). In contrast, the nineteenth century is characterized by a new development of *Calluna*, related to the extension of land pasture on the summit plateaux.

Many peat bogs are endangered by grazing, mainly by cows in the northern part of the Massif Central and Vosges, and by sheep in the southern part of the Massif Central. More recently, since the end of World War I, reforestation affects considerable areas of uplands. Rural depletion and farm abandonment are responsible for the extension of forest, especially with coniferous species such as fir, spruce, pine, and Douglas fir. Nowadays, several peat bogs are used as a resource for urban water supply. The creation of drainage ditches in water catchments are a serious threat that new nature protection policies attempt to reduce or to stop. Runoff erosion on slopes is significant, especially on the Mediterranean border of the Massif Central (Cévennes, Vivarais, and Velay), but it is also sporadically present in the Vosges and the western part of the Massif Central. Valleys are also affected by human activity. Many valleys have an intermittent flow as drainage is often subsurface, except in winter and when the snow melts. Recent studies (Cubizolle 1997) clearly show a modification of hydrologic and geomorphologic conditions of drainage since AD 1700. During the first stage (1700–1875), flooding was frequent and local populations considered the instability of the fluvial channels to be a permanent danger. Since 1875, a stabilization of the fluvial beds, and a general tendency to vertical erosion have affected all the rivers. The reduction of fluvial activity can be interpreted in terms of change of land use. Reduction of surficial runoff and coarse bedload are due to a better land cover and an increase of forested land following rural depopulation. There has also been a decline of cultivation giving rise to an expansion of pasture. Water catchment for urban supply and irrigation, dam building, and gravel-pitting in the river channel all act to modify the geodynamic conditions, constraining rivers to vertical incision. Dams or reservoir lakes on the main fluvial routes interrupt the transit of the sediment load.

Conclusion: A Model for Altitudinal Zonation of Landforms and Deposits

The French Uplands show a vertical zonation of landscapes and landforms according to a model defined by Etlicher (1986):

- At the lower level, grus cover is thin and discontinuous. Grus is mainly in situ, and displacement affects only steep, primary valleys (gelivation valleys). Major landforms correspond to gently inclined plateaux, which were inherited from the Tertiary peneplain or pediments. The exhumation of Tertiary sediments is recent, so alteration is quite moderate and soils are thin. Tors are frequent; they are covered by small pinewoods in the eastern part of the Massif Central and oakwoods in Limousin and Vosges. Primary valleys are deeply incised with narrow gorges where rock outcrops. Alluvial ground is strictly limited to the immediate vicinity of the river: the phreatic zone is a poor supply. This level is usually called *pays coupé* (incised plateaux). Agricultural decline is marked due to poor soil quality and climatic and edaphic dryness. These areas have attracted urban development and recreational interests.
- The middle level (700–900 m) is a 'hill and basin' landscape. Concave profiles are predominant in contrast to the lower level where convex profiles of valleys are a major landscape feature. Landforms show a low, gentle relief, surficial deposits are omnipresent. Tors and blockfields are occasionally present on the top of the hills if lithological conditions are favourable. Slopes are covered with displaced arenaceous deposits. Lynchets resulting from agricultural practices interrupt the regularity of the slope. The contrast between basins and hills is often of structural origin. The hills correspond to quartz veins or small intrusive massifs. Basins are developed in heterogeneous granite, schist, or gneiss, or in faulted zones. The valley entrenchment is quite moderate, leading to a concave profile. This type of landscape was described in the Morvan and named *Morvan troué* (Beaujeu Garnier 1951; Nieuwenhuis 1971). Because of the presence of thick surficial regolith (in situ or displaced by periglacial action) farming is important, especially milk production. Isolated forests of montane level (*Fagus, Abies*) (*étage montagnard*) become more important with elevation. Around the summits forest is the main land use and cultivation ceases.
- The summit area is a landscape of peat and heather pastures. The soils are crypto-podzolic soils and peat bogs exist where local conditions of drainage are difficult. It corresponds to the summits and plateaux of the crystalline horsts and to the highest volcanic plateaux and cones. Often these elements look like a landscape peneplain of

Tertiary age. The grus cover is discontinuous and thinner than on the previous level. This is due to glacial or periglacial denudation during the Quaternary. Valleys are deeply entrenched showing local erosion by glacial tongues. U-shaped valleys, cirques, erratics, and moraines are present. Peat and heather are the main surface covers on crystalline rocks due to the local thermal conditions and the presence of podzolic soils (Legros 1972). Cultivation is locally possible on volcanic continental plateaux (Velay). These plateaux are called *Montagne* or *Hautes Chaumes* in Forez and Vosges.

Abandonment of agricultural practices has often led to reafforestation with coniferous species (fir, spruce, Douglas fir). Elsewhere abandoned pastures are colonized by several tree species such as *Sorbus* and *Betula*. Summit plateaux may become completely covered by forest except for those crests where altitude and wind prevent forest growth. The upper natural limit of the tree line is at 1,500 m in the northern Massif Central, and at 1,350 m in the Vosges. The upper community, *étage subalpin*, is treeless. In the vicinity of snowbanks or blockfields, specialized communities of high ecological interest characterize this treeless *étage subalpin* level (Thébaud *et al.* 1992).

Regional Patterns

Ardennes

The Ardennes form the western part of a larger massif, the Eifel, which extends into Luxembourg and Germany. This plateau has a south-west gentle slope in the direction of the Parisian Basin. It culminates at the 'signal of Botrange' (694 m) near the German border.

The Ardennes landscape exhibits a strong contrast between gently undulating monotonous plateaux, and steep scarps along the main rivers. Valleys are strongly incised, the maximum incision is observed in the French Ardennes along the Meuse valley and its tributaries (Semois and Ourthe), and in the Belgian part of the massif.

Eruptive rocks are not apparent, except for straight sills in metamorphic sediments. Folded sedimentary rocks, such as quartzite, shale, phyllite, and limestone of Cambrian to Devonian age constitute the basement. Anticlines and synclines are oriented SW–NE. A topographic level has truncated the sedimentary series and three structural units can be distinguished (Fig. 12.16):

- an anticline from Rocroi to Neufchateau and Bastogne;
- a syncline of Devonian age, the Dinant syncline. This unit is thrust up on the next unit;

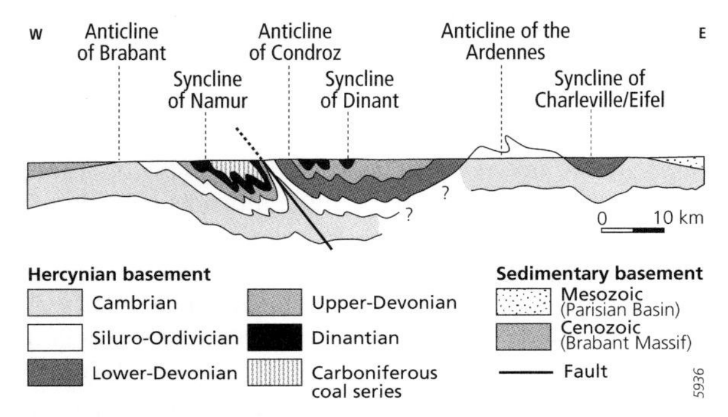

Fig. 12.16. Geological transect of the Ardennes.

- the coal syncline of Namur-Liège, limited by a major fault, the *faille du Midi*. This is linked southward with the Condroz anticline.

Several types of landscape can be distinguished related to lithological conditions (Demoulin 1995*b*). The Ardennes plateau is a monotonous schist plateau in the western part, the 'plateau de Rocroi'. It is more varied in the eastern part of the Meuse, especially around the Hautes Fagnes where several depressions and plateaux at different elevation can be recognized. The Fagnes-Famenne region is a medium-sized basin, resulting from differential weathering in schist and shale. In the Condroz region, a more varied rock lithology (sandstone, shale, quartzite, phyllade) allows the formation through differential weathering of parallel ridges and basins derived from the summit planation surface. These features are considered to be a good example of Appalachian ridges. The Condroz is also a limestone region with karst features such as the famous caves of Han sur Lesse near Dinant. For many Belgian authors, the Condroz is not included in the Ardennes *sensu stricto*. Except in the case of the Condroz, the homogenous lithology (shales, phyllades) of the Ardennes facilitates a good conservation of planation surfaces. A moderate and recent uplifting (Demoulin, 1995*a*) does not allow the elaboration of successive erosion cycles. According to Klein's (1990) definition, the relief evolution is typically acyclic in this massif. The main plateaux are part of a distinctive topographic surface, worked during a long period (Fig. 12.17). Through the analysis of weathered mantles or sedimentary deposits several erosion surfaces are identified that correspond to several phases of topographic planation during the Cretaceous and Early Tertiary. This surface is dated to pre-Senonian age on the north, Dano-Montian age on the eastern part, and pre-Tongrian and Eocene age on the western part of the massif. Thick alteration mantels

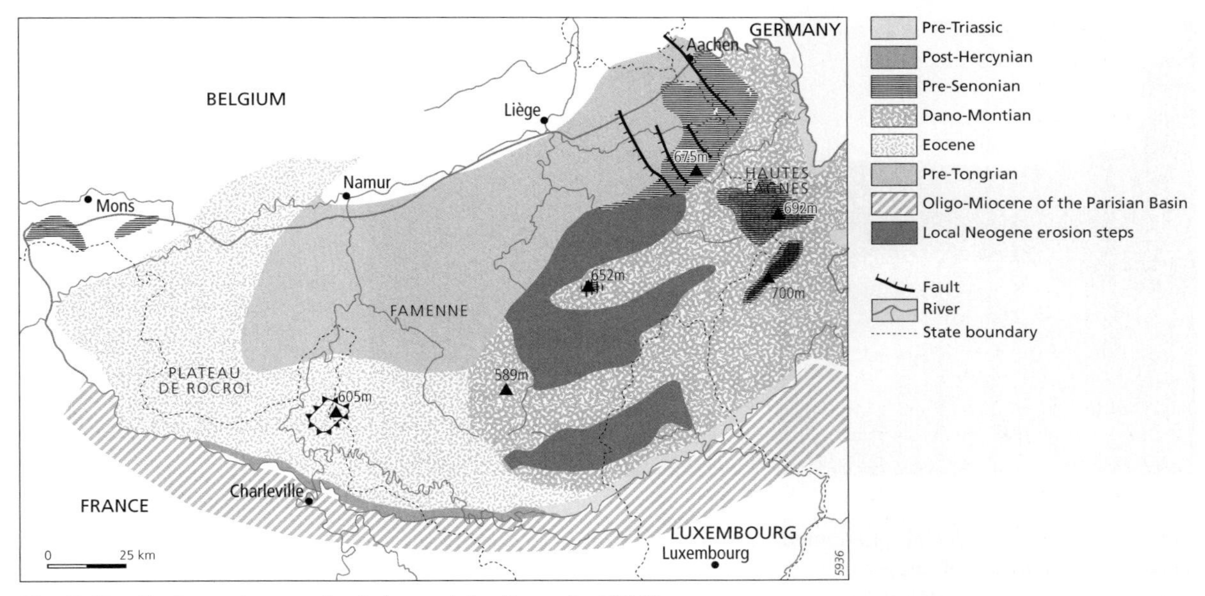

Fig. 12.17. Erosion surfaces on the Ardennes (after Demoulin 1995*b*).

(10–30 m), siliceous alterations (*Pierre de Stonne*), and gravelly fluvial deposits are proof of a long erosion phase. Cyclic levels of Neogene age are recognized only on limited areas in the eastern part. Oligocene deposits occur on lower steps, near the Hautes Fagnes, and form a summit surface. In this case a real cyclic evolution exists, according to the *Piedmont Treppen* model.

Other local erosion steps exist but are limited in extension in the Famenne, north of Givet. They are of Quaternary age, and connected with alluvial terraces of the Meuse. They are interpreted as the result of frost action and aeolian erosion working on highly susceptible material during subsequent glaciations. Probably, the same evolution is responsible for the lower steps eroded in shales at the foot of the Appalachian ridges of Condroz. The Meuse and several tributaries in the area of Liège cross these ridges from Givet to Namur.

The post-Hercynian planation is also obvious at the southern periphery of the massif (Infra-Triassic or Infra-Liassic) around Charleville-Mézières and Sedan in northern France.

Recent uplifting of Mio-Pliocene age has been demonstrated by Demoulin (1995*a*). A major problem for geomorphologists is the direction of the Meuse, flowing northwards through the western part of the massif. From Tricart (1948), it is considered as superimposed on an Oligo-Miocene surface. Belgian authors describe a complex evolution with many captures and successive courses as the uplifting of the massif increased. This uplift stimulated the regressive erosion of the tributaries of the Sambre–Meuse flowing to the 'sillon of Condroz'. A tectonic deformation of an old alluvial terrace of the Meuse, the *trainée mosane*, is obvious. Some alluvial deposits of the river are 120 m above the talweg at Givet. At present, uplifting is still being registered from 1945 to 1990 by geodetic measurements. In contrast, subsidence is important near Liège and Aachen associated with the development of the Cologne sedimentary basin.

In the upper part of the massif seven different alluvial terraces are distinct in the Meuse valley. First, there was a reduction of the flow and the energy of the Meuse after the capture of the Moselle by the Rhine and later, of the Aisne-Bar by the Seine, and this stopped the incision (Pissart *et al.* 1997). The general slope of the alluvial terraces increases from the upper to the lower terraces. This fact can be related to an effect of regressive erosion caused by the subsidence of the basin of the Rhine–Meuse alluvial plain. Abandoned loops are frequent on the Meuse and Semois which provide a good measurement of incision. Lithological differences (quartzite, limestone, and shale), cause a succession of gorges and basins. The contrast is impressive between the complexity of a folded structure, steep slopes of the incised valleys, and the monotony of plateaux (Fig. 12.18).

Although at a short distance from the front of the Scandinavian ice sheet, the massif was not high enough for its own ice cap development during the late

Fig. 12.18. The Meuse valley near Dinant (Belgium). In the background, the main Dano-Montian peneplain is shown. A folded vertical layer of Carboniferous limestone is exposed through differential erosion.

Pleistocene, but periglacial phenomena are present. Loess and aeolian silt of allogenous origin cover large areas. The abundance of zircon reveals a northern origin from the glacial till of the Scandinavian ice sheet. Aeolian silt is often mixed with local material. Therefore, the classic sequence of slope deposits is rare in the Ardennes valleys. These deposits have been reworked by creep and gelifluction. Frost action was severe on shale and schist. On quartzite material, however, scree and stone runs are present on the valley slopes, due to mudflow activity. Pissart (2000) described circular structures called *viviers* on the Hautes Fagnes and defined these forms as mineral palses or lithalsas. Poorly present in other countries, these forms are small basins surrounded by a circular wall of mineral soil. They are formed by segregation and aggregation of ice in a discontinuous permafrost environment. These features were formed during the Younger Dryas time, after the Laachersee tephra event.

Limestone and dolomite are frequent in the syncline of Dinant and the Sambre-Meuse depression of Namur. Some of these rock layers have a high ratio of calcite (80–100%) so karstic evolution is general. Dolines, caves, swallow holes, and resurgences in the Lesse valley and dry valleys and caves in the Condroz are abundant and create a typical karstic landscape, very different from the micaschiste plateau. Present evolution is very active, partly because of the high annual rainfall. Displacement of *pertes* (water holes), new dolines, and collapse events are frequent.

On the plateaux, the humid climate (1,200–1,500 mm of precipitation) and the low winter temperatures are unfavourable to cultivation. Rainfall intensities are moderate: 200 mm/10 days occurs once in ten years. Also, 54% of the annual rainfall is transferred to the hydrological network (Petit in Demoulin 1995*b*). These conditions are favourable for a widespread development of peat bogs on the highest plateaux of schist and quartzite. Reforestation with coniferous trees is widespread.

Vosges

The massif is asymmetric: the westward slope is gentle and the plateau is derived from the post-Hercynian and Eogene peneplain. Eastward, several local erosion steps, interpreted as faulted blocks (*collines sous vosgiennes*), are residual elements of a local pediment.

The contrast is also between sandstone in the northern part (*Vosges gréseuses*) and the igneous Vosges in the south. In the *Vosges gréseuses*, the major landforms are monotonous plateaux, with acid, podzolic soils under coniferous forest. These forests cover more than 80% of the land. Two resistant sedimentary layers form a cuesta: the most important is the *grès vosgien* (*Buntsandstein*), but another cuesta rises above, in the *Conglomérat principal* (*Hauptkonglomerat*). Both are facing south-east and form residual hills (monadnocks) above the exhumed post-Hercynian peneplain, which is inclined from 1,000 m at the south-east to 500 m at the north-west limit of the massif. They form the major summits of the massif. Small granite massifs are exposed on the summit and culminate at 1,100 m. The hydrological divide between the Moselle and the Rhine tributaries is not coincident with the crest line but is situated on the western slope of the massif. On the southern part of the massif the situation is more complex. Granite exposures are dominant, and the summits are higher rounded landforms called *ballons* (1,200 m). A dense network of valleys is connected by several passes at a moderate altitude (1,000 m). A main peneplain is present on granite and on sandstone. It corresponds to a Tertiary erosion level, of Oligo-Miocene age according to Tricart (1948). At the foot of the Trias cuesta, post-Hercynian peneplain remnants form local topographic levels on the granite exposures (Col de la Schlucht). This surface is mainly of pre-Triassic age. Locally, however, it can be pre-Permian in age, suggesting that the planation of the massif was largely initiated at this time. On the eastern slope, different levels interrupt a long and steep slope (the *collines sous vosgiennes*). Mainly, this scheme results from successive fault scarps related to the subsidence of the Rhine valley. Triassic and Liassic sediments appear between the igneous basement and the Tertiary sediments of the Rhine basin. Local differential erosion phenomena introduce a great variety of landforms. A pediment of Tertiary age, quite similar to the Piedmont Rhodanien, appears locally, but with a very moderate extension. It is considered as Miocene by Vogt (1992).

The Vosges and the Schwarzwald were a unique massif. The tectonic uplifting during the Pliocene, 100–300 m according to the analysis of Vogt (1992), is responsible for the present elevation of the massif. As for the French Massif Central, the main phase of uplifting is considered as Neogene. Here, a noticeable subsidence of the Rift Valley of Alsace seems to have happened during the Pliocene, in accordance with an important continental sedimentation on the piedmont. Large alluvial fan systems south of Colmar represent the ultimate phases of sedimentation.

The Quaternary glacial activity was marked. Cirques, now occupied by lakes, eroded the east flank of the Hautes Chaumes. Due to local snow accumulation, a majority of cirques are oriented eastwards, i.e. leeward, whilst the longest ice tongues flowed on the western slope of the massif. A steeper slope on the east side and probably the persistence of catabatic heating on the eastern slope led to a more rapid melting of ice tongues. In contrast, on the western side a moderate slope involved a cooler and more humid climate, so snowfall was abundant at lower elevations. As a result, the Quaternary glacial erosion was more efficient on the Lorraine (western) side than the Alsace side. Seret (1966) and André (1991) identify several systems of glacial valleys, with overdeepened profiles, moraines, and *verrous*. In the Moselle valley these features are present over a distance of 40 km (Fig. 12.19). On the northern part, glacial tongues flowed down to altitudes as low as 700 m a.s.l. (Tricart and Trautmann 1994).

Despite a modest elevation, the landscape is mountainous. The Vosges is a humid mountain range: the rainfall rises from 900 mm at the western limit of the massif to 2,000 mm/yr on the main summits. These conditions are responsible for the development of the Hautes Chaumes, a denuded, woodless area on the upper summits of the massif. The absence of a forest is still debated because of the moderate elevation of the area. Is it due to natural climate conditions or also to human action? The heather cover is used for cow and sheep grazing in an agricultural system called *estive*. As in other mountainous areas, the livestock was sent for grazing on the Hautes Chaumes during summer, while farmers use the lowlands for cultivation or forage supply. Details about vegetation and soils are available from Carbiener (1962, 1972). An increasing acidity of soil and water during recent times has been noticed but the possible explanations, modification of the thermal regime or acid rain, are still discussed.

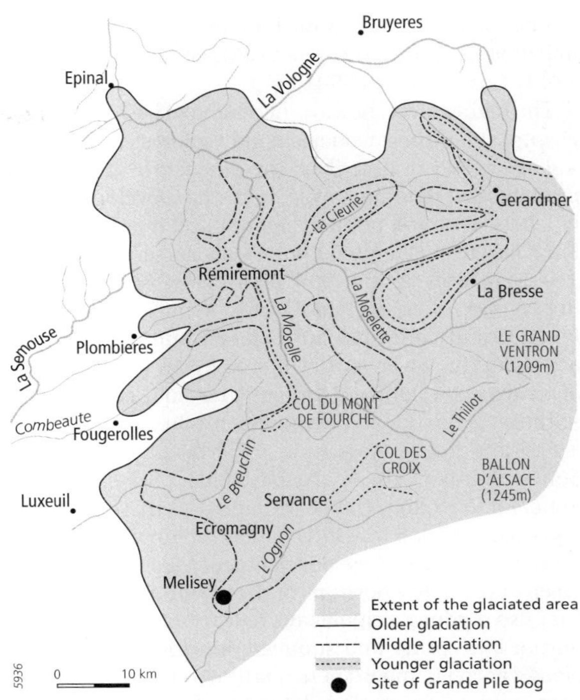

Fig. 12.19. Extent of formerly glaciated areas on the west side of the Vosges (modified after Seret 1966).

Massif Central

Limousin: Plateaux and Basins: Differential Erosion

Limousin is an area consisting of three levels of plateaux, at 950–1,050 m, 700 m, and 450 m. Explanations are cyclic (Baulig 1928), tectonic (le Griel 1991), or resulting from differential weathering where the higher plateaux are formed in more resistant rocks (alkaline granites) (Beaujeu Garnier 1951; Winckell 1971; Godard *et al.* 1994).

The higher level shows a succession of basins, used for pasture, and intermediate, mainly afforested plateaux. Valadas (1984) described the impact of cultivation practice on the landforms and the development of lynchets. From an original smooth and regular slope, a long practice of cultivation produces series of steps, 10–20 m wide and 1–5 m high called *banquette*. The upper part is eroded and forms a negative lynchet. Boulder and corestone are exposed and soil horizons are thin. The lower part is altered by an accumulation of arable soil which thickens the soil profile. This anthropogenic horizon frequently attains a thickness of 1.5 m and has homogenous properties, a dark colour, and a high per cent of organic matter. Several factors are cited to explain the development of these lynchets: pipkrakes develop in winter when the upper horizon is humid. This needle ice development can produce frost creep of the upper horizons of the soil profiles. Rill wash can also produce a significant redistribution of particles during late spring and summer rainstorms. Moreover, plough action is important for particle transfer. In the past, land cultivation was predominant for wheat, barley, and rye. Lynchets are a frequent phenomenon in Limousin, but can be seen in other crystalline regions of the French Massif Central and Vosges. Moderate slopes of 3–4° and a position on the lower part of slopes where periglacial deposits are present, are favourable conditions for their development. Radiocarbon ages obtained from charcoal layers intercalated in the slope deposits range from the seventh to the twelfth century (Valadas 1984).

Auvergne: Volcanic Landforms and Glaciation

Two major volcanic massifs occur in the Auvergne: an ancient and large system in the Cantal, and a complex system more to the north, the Mont Dore, of younger age.

In the Cantal the earliest volcanic episodes are dated to 29 Ma (Goer de Herve 1972; Nehlig *et al.* 2001). Effusive activity predominated at the beginning (Miocene) and was replaced by explosive phenomena (end of Miocene and Pliocene) with pyroclastic deposits and breccias. Activity seems to have stopped around 13 Ma with lava and pyroclastic deposits (lahar, debris flows). During this period, the Cantal was more than 3,500 m high at its maximum development. From recent interpretations (Nehlig *et al.* 2001), large debris avalanches occurred between 8.7 to 6.5 Ma. These phenomena affected 2,000 km^2. The present morphology results from the collapse of the large volcanic area. The layers of lava and breccias or pyroclastic deposits are entrenched now by a divergent valley system creating triangular plateaux (*planèzes*) around the centre of the massif (Fig. 12.20). Nowadays, the centre of the cone is a major caldera excavated by glacial erosion during the Quaternary. This erosion caused the present height of the Plomb du Cantal to be only 1,852 m (Fig. 12.21).

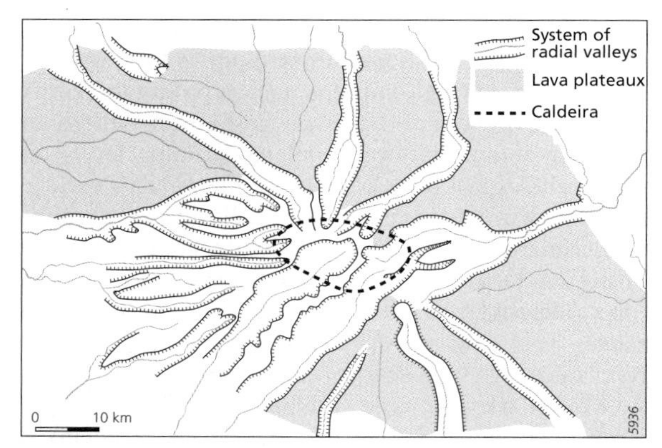

Fig. 12.20. Landforms in Cantal (modified after Valadas 1984).

In the Mont Dore area volcanic activity was more recent. The oldest lava is dated 16.2 Ma, but the major phase of construction of the complex was Villafranchian (3–2.5 Ma). The ultimate event, contemporaneous with the Quaternary glaciations, occurred around 0.5 Ma. A shorter and narrower valley system is also present. Avalanche debris occurrences are less important but contributed to the destruction of a large part of the original edifice by 1.5 Ma. The major lava fields of Cézallier and Aubrac present a monotonous landscape: plateaux at 1,000–1,300 m are widespread over the granite basement. These lava fields are 100–300 m thick and 20–30 successive lava flows are superposed. Outpourings took place along major fault zones, but at present their location cannot be deduced from topographic analysis. Variety was introduced by the presence of small, explosive 50–100-m-high cones, on the Devès and the Cézallier. Such elements are not present on the other lava fields. Glaciation has affected the two main lava fields, Cézallier and Aubrac. The major landforms result from the incision of large valleys in the main stratified cones. The plateaux were occupied by thin and stagnant ice caps. Their impact can be detected only by the presence of scarce glacial till, erratics, and, exceptionally, some drumlins. As a consequence of this moderate action, the presence of glaciers was debatable for a long time.

In contrast, the great valleys of the stratified cones, Cantal and Mont Dore, show large U-shaped valley sec-

Fig. 12.21. A major caldera (1,750 m) in Cantal. View of the central depression of the Cantal: the dotted line shows the extent of the caldera and the Puy Griou (arrowed) is considered as a part of the main volcanic pipe under the original crater.

tions, glacial basins, and morainic ridges (Fig. 12.20). The development of steep valleys was favoured by the internal structure of the cone, with superposed resistant lava flows and breccia or ash layers. Inclined plateaux are located on the periphery of the massif. To the north-west, the ice overflowed from the main valleys and spread out on the piedmont of Artense (Veyret 1981). Confluent glacial tongues from the Mont Dore constituted a powerful ice lobe 80 km wide. The ice movement strongly eroded the granite grus, and an impressive knob and basins landscape with many rock outcrops and peat bogs resulted. This area is the only one to show a glacial landscape analogous to the main ice fields of northern Europe (*fjellet* in Scandinavia).

The *chaîne des puys* consists of eighty volcanic features (domes, ash cones) on the granite plateau west of the Limagne. From 55,000 to 5600 BP volcanic eruptions created superimposed structures and small lava fields, during successive phases of activity. Strombolian cones, ash cones, domes (Puy de Dome), and maars present a great variety of spectacular landforms. Lakes now occupy some craters.

Volcanic soils are favourable for cultivation and grazing but very susceptible to erosion. These volcanic plateaux are non-forested and land use is mainly for cattle grazing. Valadas (1984) describes grassland scratch (*écorchure de pelouses*) on the Sancy and Mont Dore volcanoes. Situated on snowy slopes, this phenomenon is frequent at the upper limit of the forest near the tree line. Here snow banks are frequent and thick. This leads to the slope being wetted by snowmelt for several months in spring. Overgrazing, periglacial activity, and wind erosion in winter are the main processes responsible for degradation of the vegetation.

Cévennes, Vivarais, Lyonnais: 'Serres', Dissection, Periglacial Deposits

An altitudinal zonation of soil and vegetation can be demonstrated in this part of the massif in relation to the following conditions. Tectonic uplifting has been more vigorous in relation to the proximity of the base level of the Rhone valley. Slope angles frequently reach 20% from the summit at 1,500 m to the piedmont at 200 m. These steep slopes are especially important in several horsts separated by a complex fault system (Pilat, Tanargue, Lozère). A strong climatic contrast exists between summits and piedmont. The effect of humid Atlantic rainfalls is increased by altitude and exposure to west winds. The low temperature, wind, and snow are limiting factors for most vegetation and animal life. Piedmonts have the warm, temperate conditions of a Mediterranean climate. On their leeward side, adiabatic heating is frequent and summer drought possible. Severe rainstorms are usual in autumn when warm and humid air from the Mediterranean sea, in association with a low pressure system over the British Isles, is forced to rise. Considerable rainfall amounts results: at 1,400 m in Tanargue the mean annual precipitation is 2,500 mm and rates of 400 mm/24 hours occur once in two years. In winter, blizzards are dangerous (Staron 1993).

Altitudinal zonation of landforms, soils, vegetation, and human activities have been described by Bornand *et al.* (1997) for Margeride, by Legros (1972) for Pilat, and by

Fig. 12.22. Blockfield, or *chirat*, in the Pilat (1,300 m).

Etlicher (1986) for Forez and Pilat. In Margeride, a transect from the summit to the basal valley shows this variation. On the plateau heather and peat bogs prevail. Below, a forest zone isolates the heather from grazing and cultivation areas. Beech forest and at lower elevation pine forest occupy the steeper slopes. In some areas, mainly in the eastern part of the Massif Central, blockfields are abundant and called *clapiers* or *chirats* locally (Fig. 12.22). Blockfields occur under specific conditions, such as the presence of igneous rocks, that are not sensitive to microgelivation, and dry and cold palaeo-environmental conditions in non-glaciated areas. Blockfields are present on gneiss and quartzite, but also on phonolithes. Some of these blockfields are considered as rock glaciers or as complex talus slope deposits (Valadas 1984; Etlicher 1986). They are fed by rockfall on the upper part, and they flow down like a glacier in the lower part. *Chirats* are specific to this part of the Massif Central; they cover large areas and are similar to blockfields described in the Appalachian highlands.

The Velay region displays a complex system of volcanic dome structures of Pliocene and Quaternary age. The composition of lava is either trachytic or phonolithic or occasionally, basaltic. Volcanism began at 11 Ma. Explosive structures, now seen as maars, were frequent. Characteristic of this area is the importance of phreatomagmatic phenomena. The Velay system lies on a sedimentary graben, probably affected by a phreatic zone. Some of these maars contain sediment sequences that enable a detailed chronostratigraphical subdivision of the Weichselian and Holocene (Guiter *et al.* 2003). Similarly to the Pilat region, periglacial activity involved the development of blockfields and/or rock glaciers in phonolithic materials (Valadas 1984).

Erosion is strong on the uplands during non-snowy winters when soil and vegetation are not protected by snow cover, but is also due to the intensity of rainfall, especially in spring, and to seasonal dryness in summer. On the uplands, the summit heather surface is affected by gullying, wind erosion, and frost shattering. Pasture activity (sheep, goat, and cattle) on the uplands is in decline but was very important in the nineteenth century (Muxart *et al.* 1986, 1987). Erosion features consist of 1–3-m-deep gullies, exhuming the underlying periglacial cover of pebbles and blocks (Valadas 1984). Soils are skeletal and the regeneration of the vegetation cover is very slow. Reforestation by coniferous species is a difficult process as it is adversely affected by the nature of soils and regoliths, and the unsuitable climatic conditions (Franc and Valadas 1990). The karstic plateau of the Causses is an exception to this rule. Here, monotonous limestone plateaux with caves, dolines, and striking canyon-gorges show submediterranean vegetation despite a high elevation (1,000–1,200 m).

References

André, M. F. (1991), L'Empreinte glaciaire dans les Vosges: vallées de la Moselle et de la Moselotte. Presses Universitaires, Nancy.

Autran, A. (1984), Chevauchements synmétamorphes varisques: note de synthèse. *Programme géologie profonde de la France*, BRGM, Orleans, 81–3.

Bard, J. P., Burg, J. P., Matte, P., and Ribeiro, A. (1980), La Chaîne hercynienne d'Europe Occidentale en termes de tectonique des plaques. In: M. Slanski (ed.) *Géologie de l'Europe, 26e Congrès Géologique International*, 233–46.

Baulig, H. (1928), *Le Plateau central de la France et sa bordure méditerranéenne*. Armand Colin, Paris.

Beaujeu Garnier, J. (1951), Quelques données nouvelles à propos des Massifs Anciens. *Revue de Géomorphologie Dynamique* 4: 21–37.

Bergerat, F. (1987), Paleo-champs de contraintes tertiaires dans la plateforme européenne au front de l'orogène alpin. *Bulletin de la Société géologique de France* 3: 611–20.

Bornand, M., Lagacherie, P., and Robbez Masson, J. M. (1997), Cartographie des pédopaysages et gestion de l'espace. *Cartographie pour la gestion des espaces naturels*. Cemagref, Grenoble.

Brousse, R., and Lefevre, C. (1966), Nappes de ponces du Cantal et du Mont Dore. Aspects volcanologiques, pétrographiques, minéralogiques. *Bulletin de la Société géologique de France* 7/8: 223–45.

—— and Lefevre, C. (1990), *Le Volcanisme en France et en Europe limitrophe*. Masson, Paris.

Carbiener, R. (1962), Les Sols et la végétation des Chaumes du Champ du Feu. *Bulletin Association l'Étude Quaternaire*, 18–33.

—— (1972), Aspects de l'écologie de la grande crête des Hautes Vosges. *Congrès National de l'Association des Professeurs de Biologie-Géologie*, Strasburg, 8.

Chafchafi, A. (1997), Le Velay septentrional: morphogénèse et morphostructure. Saint-Étienne, University of Saint-Étienne Press.

Coque-Delhuille, B. (1979), Les Formations superficielles et leur signification géomorphologique dans les régions de roches cristallines: l'exemple des plateaux de la Margeride occidentale. *Revue de Géologie Dynamique et Géographie Physique* 21: 127–46.

Cubizolle, H. (1997), *La Dore et sa vallée*. Études Foréziennes, Saint-Étienne.

de Beaulieu, J. L., Pons, A., and Reille, M. (1982), Recherches pollen-analytiques sur l'histoire de la végétation de la bordure nord du Massif du Cantal (Massif Central). *Pollen et Spores* 24: 251–3.

—— —— —— (1988), Histoire de la végétation du climat et de l'action de l'homme dans le Massif Central Français depuis 15,000 ans. *Travaille Institut France de Pondichéry* 24: 37–80.

Defive, E. (1998), L'Encaissement du réseau hydrographique dans le bassin de la Loire supérieure. *Bulletin Association Géographique France* 1: 121–8.

—— and Berger, E. T. (1991), Réflexions préliminaires à l'étude des rythmes et vitesse de l'érosion aux confins du Velay oriental et du Haut Vivarais. *Geo-physio* 21: 23–32.

Demoulin, A. (1995*a*), Les Surfaces d'erosion méso-cénozoiques en Ardenne-Eifel. *Bulletin Société Geologique France* 166: 573–85.

—— (1995*b*), *L'Ardenne, Essai de geographie physique. Hommage au Professeur A. Pissart*. Dept. de Géographie physique et Quaternaire, University of Liège.

Denèfle, M., Valadas, B., Vilks, A., and Lintz, G. (1980), L'Évolution holocène de la végétation en Limousin. *Bulletin Association Étude Quaternaire* 4: 189–99.

Didier, J., and Lameyre J. (1971), Les Roches granitiques du Massif Central. *Géologie, Géomorphologie et structure profonde du Massif Central français*, Symposium Jung, Clermont Ferrand, 133–55.

Etlicher, B., (1986), *Les Massifs du Forez, du Pilat et du Vivarais: régionalisation et dynamique des héritages glaciaires et périglaciaires en moyenne montagne cristalline*. Études Foréziennes, Saint-Étienne.

—— and Goer de Herve, A. (1988), La Déglaciation würmienne dans le Massif Central français: le point des travaux récents. *Bulletin Association France l'Étude Quaternaire* 2/3: 103–10.

—— and Lautridou, J. P. (1998), Gélifraction expérimentale d'arènes de roches cristallines: bilan d'essais de longue durée. *Permafrost and Periglacial Processes* 10: 1–16.

Faure, M., Leloix, C., and Roig, J. Y. (1997), Polycyclic evolution of the Hercynian Belt. *Bulletin Société Géologique France* 168: 695–705.

Franc, A., and Valadas, B. (1990), Stations forestières et paysages: les granites du Massif Central. *Revue Forestière Française* 42: 403–16.

Freytet, P., and Morel, J. L. (1987), Réhabilitation de la morphométrie en tant qu'étude des surfaces d'aplanissement. *Revue de Géomorphologie dynamique* 26: 77–8.

Gewelt, G., and Juvigné, E. (1988), Tephrochronologie du Tardiglaciaire et de l'Holocène dans le Cantal, le Cezallier et les Monts Dore. *Bulletin Association l'Étude Quaternaire* 33: 25–34.

Godard, A., Lagasquie, J. J., and Lageat, Y. (1994), *Les Régions de socle*. Publication Faculté des lettres, Clermont Ferrand.

Goer de Herve, A. (1972), *La Planèze de Saint-Flour*. Annales Scientifique Université Clermont Ferrand, Thesis, 2 vols.

Guiter, F., Andrieu-Ponel, V., de Beaulieu, J.-L., Cheddadi, R., Calvez, M., Ponel, P., Reille, M., Keller, T., and Goeury, C. (2003), The last climatic cycles in Western Europe: a comparison between long continuous lacustrine sequences from France and other terrestrial records. *Quaternary International* 111: 9–74.

Janssen, C. R., and van Straten, R. (1982), Premiers résultats des recherches palynologiques en Forez. Plateau Central. *C.R. Académie des Sciences*, Paris, 294 (ser. 2): 155–8.

Klein, C. (1990), *L'Évolution géomorphologique de l'Europe hercynienne occidentale et centrale*. Mémoires et Documents. CNRS, Paris, 2 vols.

Lacotte, R. (1982), Formes et formations à blocailles en Limousin. *Recherches géographiques à Strasbourg* 16/17: 25–44.

Larqué, P., and Weber, F. (1978), Séquence sédimentaire et lithostratigraphie de la série paléogène du Velay. *Science Géologie Bulletin* 31: 151–8.

Ledru, P., Lardeaux, J. M., Santallier D., Autran, A., Quenardel, M., Floc'h, J. P., Lerouge, G., Maillet, N., Marchand, J., and Ploquin, A. (1989), Où sont les nappes du Massif Central français? *Bulletin de la Société géologique de France* 8: 605–18.

le Griel, A. (1991), L'Évolution géomorphologique du Massif Central Français. Thesis, Lyons.

Legros, J. P. (1972), Sols sur granite du Massif Central. In: *Les Sols dans le paysage et leurs aménagements*. Service Étude des Sols, Centre de la Recherche Agronomique, Montpellier, 175: 89–101.

Lenselink, G., Kroonenberg, S. B., and Loison, G. (1990), Pleniglacial to Holocene paleoenvironments in the Artières basin (Limagne rift valley Massif Central). *Bulletin Association l'Étude Quaternaire* 2: 136–56.

Macaire, J. J., Cocirta, C., de Luca, P., Gay, I., and Goer de Herve, A. (1992), Origines, âges et évolution d'un système lacustre polyphasé au Tardi et au Post-glaciaire: le lac Chambon, Massif Central, France. *Comptes Rendus Académie des Sciences*, ser. 2, 315: 1119–26.

Mandier, P. (1989), *Le Relief de la moyenne vallée du Rhône au Tertiaire et au Quaternaire*. Document du Bulletin Bureau de Recherches Géologiques et Minieres, Orleans, 151.

Michon, L., Merle, O., and Berger, E., (1999), Spatial and temporal distribution of tertiary volcanism in the Massif Central (France). *EUG X*, Strasburg, 321–30.

Muxart, T., Billard, A., Cohen, J., Cosandey, C., Denèfle, M., Fleury, A., and Guerrini, M. C. (1986), Dynamique physique récente de versants des hautes terres cévenoles (Espérou, Lingas) et occupation humaine. *Revue de Géographie des Pyrénées et du Sud Ouest* 57: 375–94.

Muxart, T., Cosandey, C., Billard, A., and Valadas, B. (1987), Dynamique des versants et occupation humaine dans les Cévennes (montagne du Lingas). *Bull. Assoc. Géogr. Fr.* 64: 1–40.

Nehlig P., Leyrit H., Dardon A., Freour G., Goer de Herve A., Huguet D., and Thieblemont D. (2001), Constructions et destructions du stratovolcan du Cantal. *Bulletin de la Société géologique de France* 172: 295–308.

Nieuwenhuis, J. D. (1971), Weathering and planation in the Morvan. *Revue de Géomorphologie dynamique* 3: 97–120.

Pastre, J. F. (1987), Les Formations plio-quaternaires du basin de l'Allier et le volcanisme régional. Thesis, University of Paris, 6.

Penck, W. (1925), Die Piedmontflachen des südlichen Schwartzwaldes. *Zeitschrift der Gesellschaft für Erdkunde zu Berlin*, Berlin, 81–108.

Perpillou, A. (1940), *Le Limousin: Étude de géographie physique*. Chartres, Thèse Lettres.

Pierre, G. (1990), Génération d'altérites dans le Massif Central français: implications géomorphologiques. *Physio-géo*, Paris, 20: 31–49.

—— (1999) Les Sables ferruginisés du plateau de Rocroi: contribution à l'étude de la morphogenèse tertiaire de l'Ardenne méridionale. *Géologie de la France* 3: 3–10.

Pissart A., (1962), Les Aplanissements tertiaires et les surfaces d'érosion anciennes de l'Ardenne du Sud Ouest. *Annales de la Société Géologie Belgium* 85: 71–150.

—— (2000), Remnants of lithalsas of the Hautes Fagnes, Belgium: a summary of present-day knowledge. *Permafrost and Periglacial Processes* 11: 327–55.

—— Krook, L., and Harmand, D. (1997), La Capture de l'Aisne et les minéraux denses des alluvions de la Meuse dans les Ardennes. *Comptes rendus de l'Académie des sciences. Série II, Sciences de la terre et des planètes*, 325/6: 411–17.

Seret, G. (1966), Les Systèmes glaciaires du bassin de la Moselle et leurs enseignements. Thesis, Brussels.

—— Dricot, E., and Wansard, G. (1990), Evidence for an early glacial maximum in the French Vosges during the last glacial cycle. *Nature* 346: 453–6.

Simon Coinçon, R. (1987), *Le Rôle des paléo-altérations et des paléoformes dans les socles: l'exemple du Rouergue (Massif Central français)*. Thesis, University of Paris 1. Mémoires Sciences de la Terre 2. ENSMP, Paris.

—— Schmitt, J. M., and Clément, Y. (1983), Mise au point sur les paléoaltérations tertiaires du Massif Central. *Géologie de la France* 1/1–2: 155–65.

Staron, G. (1993), *L'Hiver dans le Massif Central français*. Études Foréziennes, Saint Etienne.

Thébaud, G., Schaminee, H. J., and Hennekens, S. M. (1992), Contribution à l'étude de l'étage subalpin des moyennes montagnes ouest européennes: quelques groupements végétaux foréziens comparés à leurs homologues d'autres massifs. *Bulletin Société Botanique du Centre-Ouest* 1410: 45–63.

Tricart, J. (1948), *La Partie orientale du bassin de Paris*. Thesis, University of Paris. SEDES, Paris.

—— and Trautmann, M. (1994), Complexité du Quaternaire à Bellefosse (flanc Ouest du Champ-du-Feu, Bas-Rhin. *Revue de Geomorphologie dynamique* 43: 113–24.

Valadas, B. (1984), *Les Hautes Terres du Massif Central français. Contribution à l'étude des morphodynamiques récentes sur versants cristallins et volcaniques*. Doctoral thesis, University of Paris 1.

van Vliet-Lanoë, B. and Valadas, B. (1983), A propos des formations déplacées des versants cristallins des massifs anciens: le rôle de la glace de ségrégation dans la dynamique. *Bulletin Association l'Étude Quaternaire* 16: 153–60.

Vergne, V. (1989), L'Évolution tardiglaciaire et holocène d'un piedmont de moyenne montagne cristalline: l'Artense. Thesis, University of Paris 1.

Veyret, Y. (1981), *Modelés et formations glaciaires dans le Massif Central français: problèmes de distribution et de limites dans un milieu de moyenne montagne*. Thesis, University of Lille III.

Vogt, H. (1992), *Étude géomorphologique du rebord occidental du fossé rhénan*. Strasburg, Oberlin.

Voisin, L. (1977), *Le Modelé schisteux en Europe Occidentale: analyse morphologique d'une région-type: l'Ardenne occidentale*. Thesis. H. Champion, Paris.

Winckell, A. (1971), Rôle respectif de la tectonique cassante et de l'érosion différentielle dans l'élaboration du relief de Nord et Nord Ouest de la montagne limousine. Thesis, Clermont Ferrand.

Woillard, G. (1978), Grande Pile peat bog: a continuous pollen record for the last 140 000 years. *Quaternary Research* 9: 1–21.

13 The Parisian Basin

Yvette Dewolf and Charles Pomerol

Introduction

The Parisian basin is a geographical entity whose limits are easily defined by the Armorican massif, the Massif Central, the Vosges, the Ardennes, and the English Channel. Both Burgundy and Poitou are transitional areas. The Paris basin, a more restrictive term, corresponds according to some geologists (Cavelier and Lorenz 1987) essentially to the Tertiary 'part' of the basin: the Île de France and surroundings. The relief of the Parisian basin results from two sets of factors: tectonic and climatic. These have operated from Triassic times until the Pleistocene and have led to the development of a geographically simple whole in its gross structure and form. However, within this framework individual natural regions (or geotypes) may be recognized.

Origin and Geological Evolution

The Parisian basin is frequently considered as a model for sedimentary basins, displaying as it does, a classic framework of sedimentary formations (Pomerol 1978; Cavelier and Pomerol 1979; Cavelier *et al.* 1979; Pomerol and Feugueur 1986; Debrand-Passard 1995). This is evident from the geological map of France (Fig. 13.1), and on the related cross-section (Fig. 13.2). Indeed, the section shows the superposition of strata in a subsiding area, with a maximal thickness (3,200 m) in the Brie country. This arrangement illustrates the geometric definition of the Parisian basin, an intracratonic basin, 600 km in diameter, limited towards the west by the Armorican massif, the south by the Massif Central, the east by the Vosges, and the north-east by the Ardenno-Rhenan massif. The following geological overview is based upon the previously mentioned studies and the geological time scale (see Fig. 13.4). However, the analysis of the evolution of these sedimentary areas from Triassic to Neogene shows that the area named as the 'Parisian basin' was included in successive palaeogeographies (which were strongly influenced by adjacent seas) and overflowed across the basement regions that now act as the limits of the basin. The chronological order of the geological formations involved in the evolution of the Parisian basin according to Robin *et al.* (2000) is used in the following text.

During the Triassic, the future Parisian basin was a gulf of the German Sea. This sea transgressed westwards and reached the meridian of Paris during the Keuper. The Lower Triassic (Buntsandstein) consists of continental detrital deposits: the Vosgian Sandstones, followed by the *Bigarrés* (Mottled) Sandstones, red in colour, and reaching up to 600 m in thickness. They constitute the present-day *Vosges gréseuses* landscape (see Fig. 13.2). They are covered by a conglomerate, then by the lagoonal and marine Voltzia Sandstones, indicating the beginning of the Middle Triassic transgression. The geological divisions in order are: 1. Wellenkalk, 2. Anhydrit Gruppe, 3. Upper Muschelkalk, 4. Lettenkohle (coal in clays) made of brackish clays with carbon vegetal remains. The Upper Muschelkalk forms a cuesta in the eastern Parisian basin (Fig. 13.3). In the Upper Triassic (Keuper), the lagoons progressed westwards, and reached as far as the present lower Seine valley. It was a lagoonal-brackish transgression, while the shoreline of the German Sea moved eastward. The Keuper is a palaeogeographical paradox; it consists of a lagoonal transgression westwards and a marine regression eastwards. The Triassic terminated with the return of the German Sea and deposition of the Rhetian Avicula Sandstone.

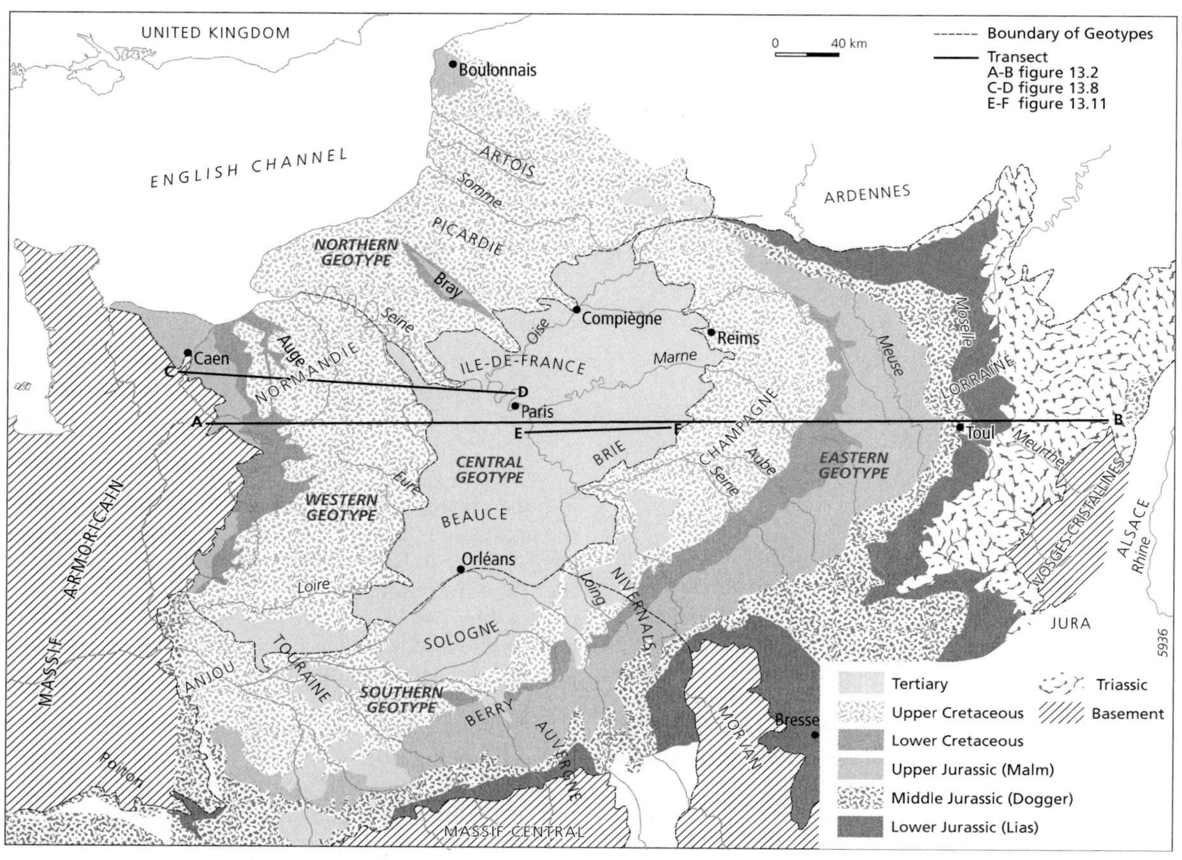

Fig. 13.1. Simplified geology of the Parisian basin, including geomorphological regions with five Geotypes as well as locations of place-names, regions, and rivers mentioned in the text (modified after Mégnien and Debrand-Passard 1980); A–B transect, Fig. 13.2; C–D transect, Fig. 13.8; E–F transect, Fig. 13.11.

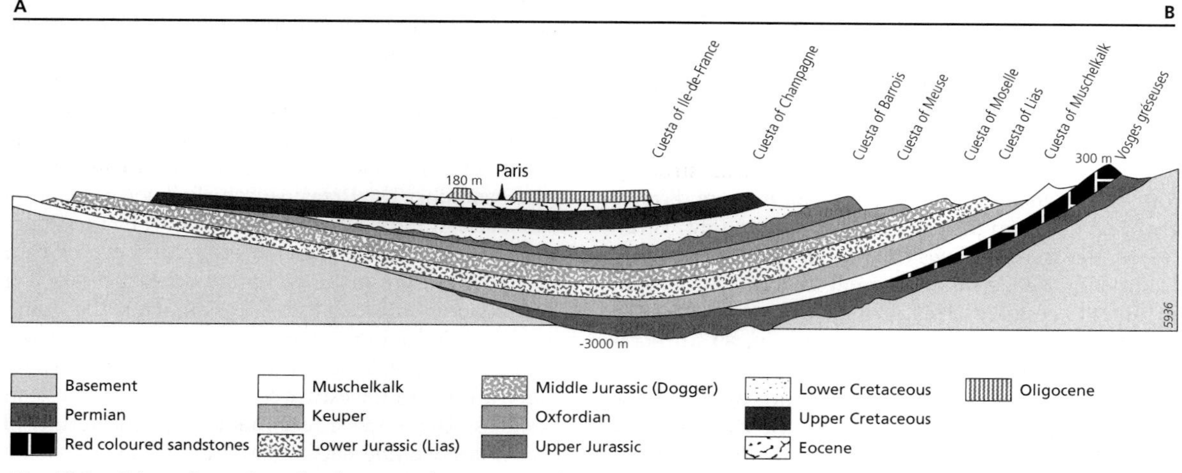

Fig. 13.2. Schematic section of sedimentary formations in the Parisian basin (after Pomerol in Debelmas 1974).

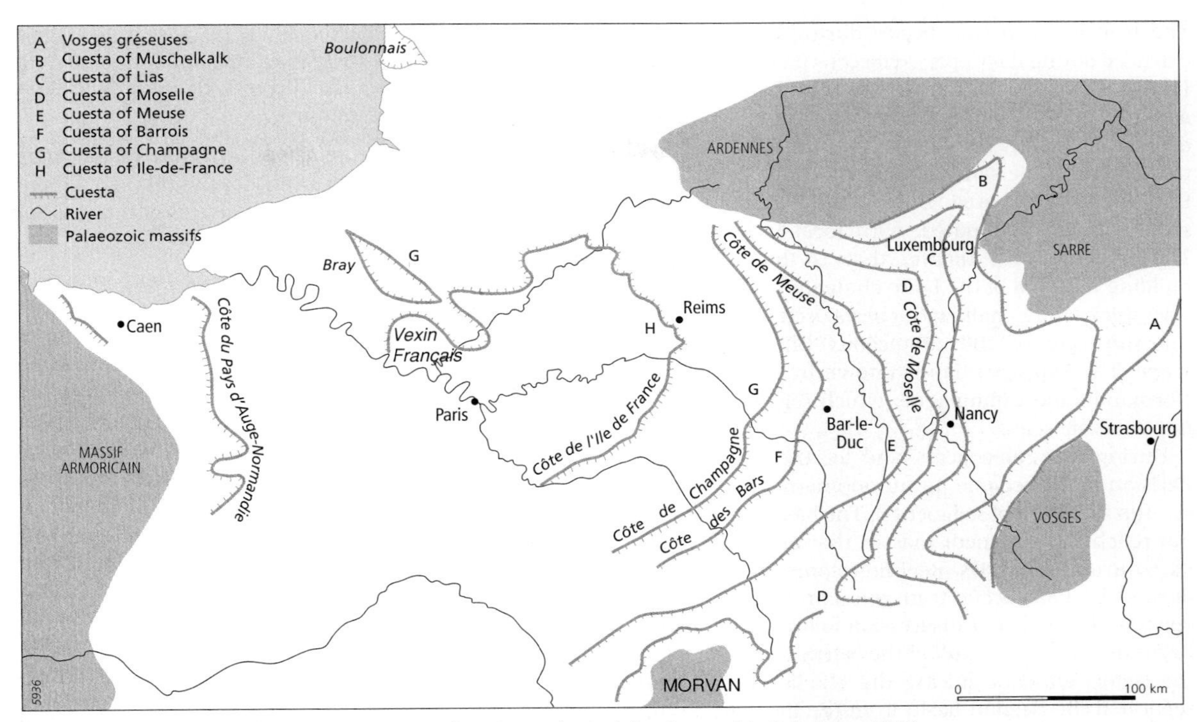

Fig. 13.3. Distribution of the main cuestas in the Parisian basin (after Pomerol in Debelmas 1974).

During the Liassic, the Parisian basin was always under the influence of the German Sea, trangressive since the Rhetian. The following depositions occurred: sandstones, limestones, schistose, and bituminous marls. In Normandy, the transgression came from England and made a junction with the German trangression. In the same way, the transgression of the pre-Atlantic Ocean crossed the Poitou Strait. So, the Parisian basin represented the junction of three seas: North, German, and Atlantic. In the Middle Jurassic, the slightly regressive Toarcian sea was present during the Aalenian and represented a wide littoral zone, where ferruginous oolites were deposited. This iron ore (*minette*) was strongly exploited in Lorraine. The new Burgundy Strait replaced the Vosgian Strait as a transgression route, and the Parisian basin showed Thetis influences. During the Bajocien, oolitic limestones, echinoderm limestones, and calcarenites were deposited on a wide platform, which now constitutes the Moselle cuesta (see Figs. 13.2 and 13.3), which dominates the Lorraine Liassic depression. The Bajocian is overlain by the oolitic and gravelly limestones of the Bathonian, named the *Pierre de Caen* (Caen Stone) in Normandy, a good building stone. After this long duration of limestones deposition, the Callovian marls were deposited, and now form the roof of the Dogger oil fields. This clay sedimentation continued during the Lower Oxfordian (*Vaches noires* cliffs: Black Cows cliffs). It is replaced during the Upper Oxfordian by a belt of corallian reefs now forming the Meuse cuesta (Figs. 13.2 and 13.3). During the Kimmeridgian, the reefs were displaced southeastwards, while new detrital deposits, including some marls, were formed. These were followed by the micritic limestones of the base of the Tithonian.

At the end of the Tithonian (Purbeckian facies) and the beginning of the Cretaceous (Wealdian facies) the Parisian basin was continental and characterized by fluvial sands and clay sedimentation with vegetal remains. The marine deposits of the Lower Cretaceous overlay the previous detrital deposits. The transgression came via the Burgundy Strait, related to the Mesogean seas. The deposition included Valanginian limestones and marls, Hauterivian marls, Barremian clays, Aptian marls. During the Albian, the Burgundy transgression merged with the Nordic gulf transgression coming from England as a new strait opened between the North Sea and the Mesogean sea. The Albian sea was widely transgressive and deposited glauconitic sands and clays.

The transgression that began during the Albian was extended during the Upper Cretaceous (Cenomanian). The Poitou Strait opened during the Liassic and was then active. Near the Armorican basement, the Cenomanian deposits are mainly detrital, and pass laterally northwards to the glauconitic sands and marly chalk, in which the Eurotunnel was bored. During the Turonian, chalky sedimentation dominated, with local facies (chalk of Touraine, or *craie tuffeau*, that is the well known building material of the Loire chateaux). On top of this, very thick white chalk with cherts occurs, which forms the substrate of the Cretaceous Champagne landscape. The Maastrichtian is known by rare remains in Normandy and Champagne, which represent traces of the Mesozoic seas.

During the Palaeogene, and for the first time, the Parisian basin became an autonomous basin, of small size (up to 250 m of sediments). The basin sediments did not reach the basement massifs that are considered as its natural boundaries. Reef limestones were deposited during the Palaeocene transgression. After a lagoonal episode, the Eocene transgression followed. By then the Parisian basin was a gulf of the North Sea. The uplift of the Artois anticline during the Middle Eocene finally separated the Belgian basin from the Parisian basin. A calcareous sediment, known as *calcaire grossier*, was deposited. This material has been used to build a variety of Parisian monuments. Subsequently, the sea advanced from the west (English Channel), and sedimentation consisted of an alternation of detrital sediments, limestones, evaporates, clays, and sands. These Eocene and Oligocene formations did not reach eastwards beyond the area of Reims. They give rise to the cuesta of the Île de France (see Figs. 13.2 and 13.3). From the Palaeocene to the Oligocene the area of maximum subsidence gradually shifted southwards from the region of Compiègne towards Orleans. During the Neogene, the evolution of the Parisian basin was wholly continental. At the margins, the seas were transgressive in the low valley of the Loire during the Miocene and Pliocene and shelly sands (*faluns*, crags) were deposited.

Using data from sequential stratigraphic analysis recorded in a great number of drillings, recent studies have provided this overview of the Parisian basin:

> an extensive basin, resulting from the thermal consequence of the Permian extension, progressively evolving towards a compressive basin. This becomes a permanent condition from the Upper Cretacious to the Present. Many tectonic and thermic events are superimposed on this evolution. Intraplate deformations during the Aalenian, Berriasian, Aptian, Upper Cretaceous, Middle Eocene–Lower Oligocene, Upper Miocene and at the Lower-Middle Pleistocene boundary. They are the consequence of Tethysian or Atlantic events, such as the opening of the Gascogne Gulf (Berriasian–Aptian), or the convergence of Africa and Eurasia (Upper Cretaceous to Present). (Guillocheau *et al.* 1999)

According to Robin *et al.* (2000), nine major transgressive–regressive stratigraphic cycles (Fig. 13.4) were registered in the sedimentary sequences of the Parisian

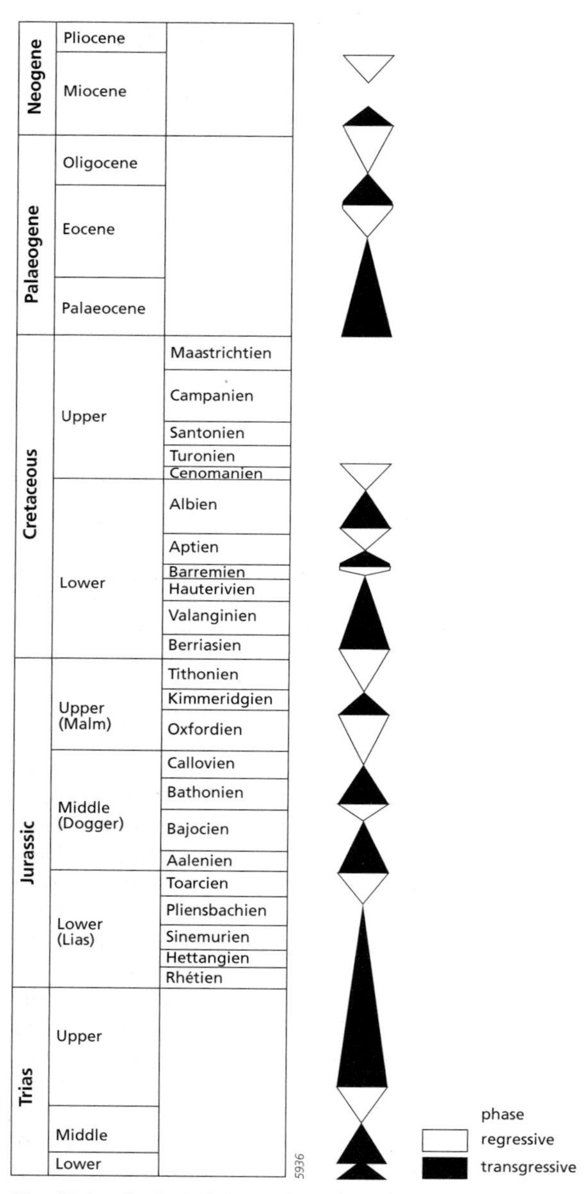

Fig. 13.4. Geological timescale and major stratigraphic cycles (simplified after Robin *et al.* 2000).

basin from Trias to Upper Cretaceous (Cenomanian), with a duration varying between 10 and 40 million years. During the Cenozoic three major cycles occurred. The measures of accommodation, i.e. available space for the sediments created by tectonic and eustatic variations, allow to recognize four main stages in the geological history of the Parisian basin:

- Triassic and Jurassic: strong accommodation (20–80 m/Ma),
- Cretaceous: accommodation becoming lower (5–20 m/Ma),
- Tertiary–Lower Quaternary: very low accommodation rate (0–2 m/Ma),
- Middle and Upper Quaternary: generalized uplift of the basin.

This decrease of the accommodation space, and of the subsidence, is interpreted as the expression of the thermal subsidence related with a Permo-Triassic extension (Brunet and le Pichon 1982). The basin remains extensive up to the Turonian, except during the Lower Berriasian and the Upper Aptian, before becoming definitively compressive. So, during its Mesozoic story, the space which would become the Parisian basin was a Germanic margin during the Triassic, a Tethysian margin during the Lower Cretaceous, or an alternation of thresholds and straits during the Jurassic, and the Middle and Upper Cretaceous. It was an autonomous basin only during the Cenozoic, but because of the very slow rate of subsidence, it hardly merits the name of sedimentary basin.

Geomorphological Evolution

The morphogenesis of the Parisian basin was controlled by tectonic and climatic factors. As shown above correlations of extensive surficial formations have provided a complex framework for the evolution of the Parisian basin. This evolution is also related to a strong subsidence, which permitted the accumulation of 3,000 m of varied sediments from the Triassic to the Lower Neogene. During this long period (220 Ma), twelve transgressive phases have been identified (Fig. 13.4), separated by phases of sea-level lowstand during which erosion surfaces were formed. Some of these surfaces are covered with alterites and characteristic deposits related to specific morphoclimatic conditions.

But if this sedimentary basin is characterized by 'plains', the tectonic activity produced important deformations associated with subsidence and also some uplifting of the surrounding massifs: the Ardennes, Vosges, Massif Central and the Armorican massif, each associated with grabens, big undulations, folds, and warpings.

Structural Geomorphology

In the Parisian basin five 'geotypes' (Fig. 13.1) are distinguished.

Eastern Geotype

This geotype occurs from the west of the Vosges to the cuesta of the Île de France. It is dominated by a succession of concentric monoclinal landforms. Their continuity is interrupted by streams, mainly converging towards the centre of the basin. This geotype is wide in its central area where seven cuestas have formed (Fig. 13.3). However, it narrows to the north as it approaches the Ardennes as well as to the south towards the Morvan.

Towards these old massifs, the cuestas show a strong influence of the tectonics that repeatedly affected these basements. For example, close to the Vosges cuestas of Triassic and Liassic are broken up and dissected by faults whilst the north-east plateaux of Triassic sandstones descends in a staircase of faults towards the Plain of Alsace. A succession of grabens marks the eastern limit of this geotype. These depressions are part of a European rifts system, related to an ENE–WSW-directed phase that functioned during the Upper Eocene and the Lower Oligocene. These are the grabens in the catchment of the Loire and its tributaries.

The high Saône basin constitutes the northern part of the Bresse depression, formed between the Jura and faulted plateaux of Bourgogne. The infilling of this depression whose thickness varies according to the regional subsidence, was completed by lacustrine and fluviatile deposits at the end of the Pliocene and in the beginning of the Pleistocene.

In the southern part of this geotype, grabens occupied by the Loire river and its tributaries disturb the regularity of the cuestas. The palaeo-course of the Loire, that can be traced for more then 100 km, is now filled with fluviatile deposits. The Bourgogne Strait, now a long corridor occupied by the river Doubs, constitutes of a karstic plateau to the east and is dislocated in blocks by faults to the north and the west. In this region alpine material testifies to the last direct fluviatile link (Plio-Pleistocene) between the Vosges and the Alps on one hand, and the Bresse depression on the other.

The lines of the cuestas perfectly reflect the amphitheatre-like structure of this eastern geotype. The highest altitudes are found in the peripheral zone: 200–400 m to the north, 400–600 m to the east and south-east whilst in the central plateaux altitudes of only 100 and 200 m are reached. The pattern of landforms is directed by both subsidence and differential erosion of the rocks. It is a simple alternation of monoclinal cuesta forms, weakly inclinated surfaces that are the back-slopes of cuestas, and depressions, located in the easily eroded marls and chalk in the Champagne region.

Southern Geotype

This geotype is fundamentally influenced by the presence of the Massif Central. Its diversity reflects the past tectonic activity of this massif. Thus, the relief of the Nivernais is bounded by faults that extend into the sedimentary cover and were part of tectonic events in the Auvergne. In the south-west part cuestas occur in the Berry, but they disappear under Tertiary detrital formations derived from the massif basement. In the Touraine, uplift activity of various ages has produced anticline domes and the surface of the plateau reaches 160 m in altitude. In summary, three developments especially characterize this southern part of the Parisian basin:

1. the formation of a series of lacustrine depressions that formed due to local subsidence during the Eocene: Berry, Touraine, Central Poitou, Brenne;
2. the succession of detrital sediments at the northern margins of the Massif Central;
3. the existence in Sologne of a region of maximum subsidence that enhanced transgressions; with the sea penetrating the regions of Poitou, Sologne, and Anjou several times.

Only the Nivernais had an independent evolution, related to the progressive uplift of the edges of the Morvan, which resulted in activating north–south faults.

Western Geotype

This geotype is especially remarkable for the plane geometry of a surface that spreads from the centre of the basin as far as the Callovo-Oxfordian cuesta (G in Fig. 13.3), that dominates the peripheral depression to the east of the Armorican massif. However, tectonic activity disturbed the regularity of this wide occidental surface during the uplift that completed the Tertiary evolution of the Armorican massif.

Northern Geotype

This geotype constitutes one of the most distinctive parts of the Parisian basin; it is the only part of the basin where anticlines and synclines give the landscape a characteristic morphological style. Its structural history has closely determined the landform and main drainage patterns. For example, downstream of Paris the orientation of the river Seine (Fig. 13.5) is controlled by the 'faults of the Seine', which are associated with a major gravimetric anomaly. The erosion of anticlines has produced the formation of the 'button-hole' ('weald') of the Bray, and the half 'button-hole' of the Boulonnais. In the eroded centre of these anticlinal structures Jurassic sediments appear under thick layers of chalk, the omnipresent rock of this area. At the margins of the region of subsidence of the Île de France high landforms rise above the chalk plains. These are related to the

Fig. 13.5. The meandering River Seine, west of Paris (Bernières with ruins of the Château Gaillard).

cuesta of the Île de France, which consists of coarse Eocene limestone. This morphological relationship is seen again to the north in the Plateau of Vexin (Fig. 13.3).

Central Geotype

This geotype is defined by a morphological style which shows a simple system of stepped plateaux, depressions, and by segments of valleys that follow preferential orientations in a particular Hercynian direction (N 110°). Tectonically, the essential feature was a tendency to subsidence, interrupted frequently during the first part of the Tertiary but persisting, although attenuated, until the Pliocene. A consequence of this subsidence pattern was that the sedimentation alternated between marine, lagoonal, lacustrine, and fluviatile deposits. This explains the origin of an extremely varied lithology. Subsequent erosion has revealed the resistant rocks and has undermined the fragile rocks, such as sands and clays. The result is a succession of stepped plateaux that characterizes this central geotype. However, tectonic activity complicates this simple pattern. In the western part of this geotype a NW–SE-directed series of domes and basins predominates due to such activity. Moreover, elsewhere in the geotype general deformations occur repeatedly. For example, tectonic movement led to lacustrine sedimentation in depressions in the Beauce and Sologne region. In the Upper Miocene or even in the Lower Pliocene uplift along the north border of the Massif Central and subsequent erosion produced extensive pediments across which detrital sediments were transported.

Palaeosurfaces and Related Surficial Formations

The genesis of these surfaces together with their correlative deposits is an important part of the history of the basin. They bear witness to phases of continental climatic conditions that developed when there was a hiatus in marine sedimentation. Some of these phases lasted several millions of years. Their study is particularly interesting for the reconstruction of the palaeoenvironmental and palaeoclimatic conditions that controlled their genesis.

The High Surface

The High Surface is also called the 'Fundamental Surface' because it was the last general surface that extended over the whole basin, and upon it the hydrographic network was inscribed. It is a typical polygenetic surface, which results from an evolution during several phases of erosion. In the western to central areas of the basin, where it is best preserved, the different segments of this surface can be linked to form a remarkably continuous surface of more or less uniform height. This testifies to 'the achievement of the morphological unit of the Parisian basin' (Cholley 1957), before the tectonic events disturbed and sometimes broke this unit. This High Surface declines gently but steadily towards the centre of the basin, from 365 m near the Armorican massif to 235 m in Pays d'Auge, to 190 m on the plateau between Risle and the Eure river, and to 170 m in the centre near Paris (Fig. 13.6). The first trace of this surface developed on the chalky layers of the Upper Cretaceous ('clays with flints surface').

During the Eocene, partial surfaces developed, marked with silicifications, alterations, and palaeosols. These surfaces were then rapidly covered with marine, lagoonal, and lacustrine sediments. During the Oligocene a transgression moved across this levelled morphology. The subsequent regression was followed by a lacustrine episode related to a new subsidence phase of the limestones of Beauce region. The following phase of continental conditions created a polygenetic surface characterized by a variety of surficial formations. These formations consist of:

Clays with flints: resulting from the decalcification in situ of the chalk and consisting of a clayey matrix with flints. They have a variable thickness from some metres to tens of metres. The sediments are non-stratified and

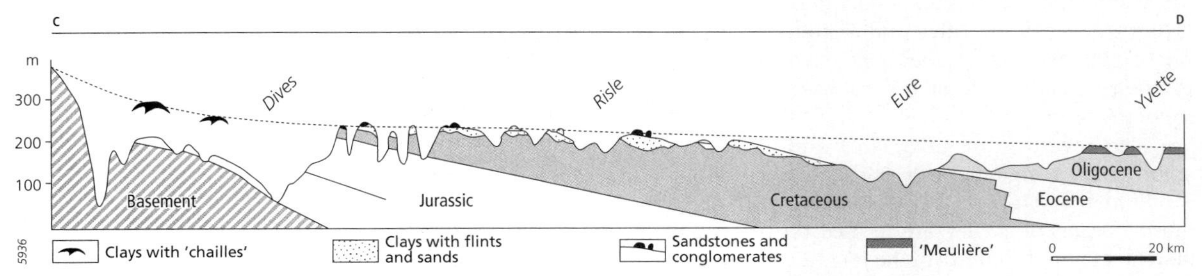

Fig. 13.6. Main polygenetic erosion surface in the west of the Parisian basin (after Thiry *et al.* 1983).

Fig. 13.7. Siliceous conglomerate (location Breteuil in Normandy).

linked to the karstic development, which resulted in a highly irregular geomorphology. Studies by Laignel (1997) and Quesnel (1997) show that at the western margins of the Parisian basin these formations have been submitted over a long time to alternating Tertiary transgression–regression events. This is reflected by several sequences of sediments with clays, sands, and loess. Palaeoclimatic changes from wet tropical in the lower Eocene, to alternating wet and dry in the upper Eocene have been inferred. In the oldest clays with flints a part of the silica is released and redistributed into the pedogenic profile. Alteration of clay minerals and of the flints have produced silcretes.

Heterochronous silicifications: on the polygenetic High Surface silicifications occur in almost all Tertiary formations. They can be grouped into three types: pedogenetic silcretes, *sarsens*, and silicified limestones:

• Pedogenic silcretes are characterized by a typical vertical columnar structure, that obliterates the primary sedimentary structure, and by illuvial silicified caps (Thiry 1981, 1983; Thiry and Simon-Coinçon 1999). These silcretes follow the polygenetic palaeosurface extending from the Massif Central basement to the centre of the basin.

• North-west of this area, slabs and blocks of sandstones and conglomerates with rounded flints and sandy matrix (Fig. 13.7) are associated with Oligocene marine sediments. They are located on the Normandy plateaux up to the cuesta of the Pays d'Auge (Fig. 13.8). The discovery of a marine fauna (Cavelier and Dewolf 1967) in some blocks of conglomerates and the perfect preservation of the shells suggest a rapid silicification, in a fluvio-marine landscape that may be concomitant with the regression of the sea. Thus, the age of this silicification could be Upper Oligocene. This silicification process could also have affected sandstones in the central part of the basin. In both cases, conglomerates and sandstones do not show pedological features, which are also lacking in the *Pierres de Stonne* in the Ardennes and in the *Quartzites des Hautes Fagnes* at the Belgian–German border. These two formations are considered Oligocene in age (Voisin 1988; Demoulin 1990), and such extensive silicifications are thought to originate in a climatic context named 'attenuated subtropical with distinct seasons' by Koeniguer and Obert (1984).

• Silicified lacustrine limestones are found in the centre of the basin. Almost all these limestones contain cherts whose size varies from centimetres to several tens of metres long. Generally, silicified facies are discordant to sedimentary structures. However, the replacement of the limestone by silica with conservation of the structure implies that calcite solution and silica precipitation occurred simultaneously. The limestone being pure implies that the silica must come from formations surrounding the lacustrine depressions, perhaps siliceous sands and clays with flints. The weathering products of these silicified lacustrine limestones form a particular type of siliceous and cavernous rocks called *meulières* (Fig. 13.9, see also Fig. 13.5). The meulières have been quarried during several centuries for millstones and

Fig. 13.8. Cuesta of the Pays d'Auge-Normandie; Upper Cretaceous/Upper Jurassic sediments (see Fig. 13.3 G).

Fig. 13.9. Meulière and *argile à meulière*.

building stone. The weathering complex that rests on limestone formations consists of a silicified mass (insoluble residues) in a matrix of the *argiles à meulières*. Meulières and argiles à meulières result from a weathering sequence with karstification of the limestones and silica redistribution (Menillet 1987, 1988). These surficial features are signs of a period of palaeo-landscape stability under alternating wet and dry climatic conditions, which preceded both a tilt movement of the High Surface and the formation of a vast pediment from the Massif Central to the English Channel, with its associated detritic sediments. This tectonic episode brought to an end the evolution of the polygenetic High Surface of the Parisian basin.

Fossilized Surfaces, Exhumed Surfaces

With the continuing uplift of the borders and the subsidence of the basin the fossilized surfaces are covered by either marine or continental sediments. Their subsequent exhumation produces a morphology of stepped

Fig. 13.10. Sandstone of Fontainebleau, south-east of Paris (orientation N 110°).

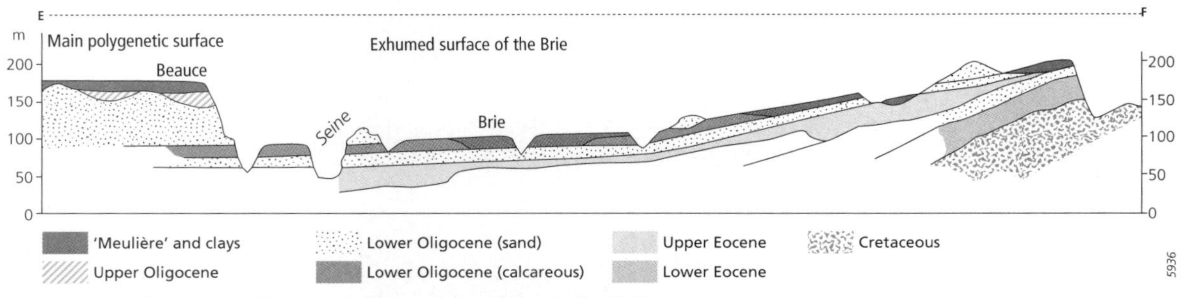

Fig. 13.11. Polygenetic surfaces, east of Paris (after Thiry *et al.* 1983).

plateaux in the basin, which have long been considered as structural surfaces. In fact, most of these are erosional surfaces which reveal karstic palaeoforms and palaeoformations, e.g. silicifications, clays, etc. (Fig. 13.10). The most significant of these surfaces developed during the first lower Oligocene tectonic phase (Cavelier and Pomerol 1979) at the end of the lacustrine episode and before a marine transgression. The duration of this phase varies from east to west from 1 to 2 Ma. Occasional sections, exposed under a sandy cover, show a karstic palaeorelief with a silicified limestone whose sedimentary structures have been preserved. This surface appears as an erosive surface with a strong chemical alteration as testified by karstification and silicification (Dewolf *et al.* 1985). The extension of the exhumed surface can be traced as far as the cuesta of the Île de France. Tilting gently towards the centre of the basin it rises strongly towards the east, from 80 to 200 metres in altitude (Fig. 13.11). The exhumed surface, called the *Surface de Brie*, shows a desilicification that is revealed by some rocks characterized by meulière ('meulière of Brie'). This polygenetic surface, that was exhumed during the Pleistocene, also shows a new phase of weathering (argilification and desilicification).

Evolution of the Hydrographic Network

Within the Parisian basin there are several fluvial basins. Major ones are those of the Loire and the Seine, and there are three to the east: the Meuse, the Moselle, and the Meurthe. In the north-west the Somme river, like the Seine, continues into the submarine river pattern in the English Channel.

The River Loire

The present course of the Loire consists of two sections: an upper, south–north-oriented course to approximately

Fig. 13.12. Campanian chalk at the cliffs of Etretat in the Pays de Caux (coast of the English Channel).

Orleans, and a course downstream from there oriented east–west until the estuary. Upstream, the course is located in a tectonic depression parallel to a major gravimetric anomaly. This depression consists of tilted blocks of Oligocene age (Debrand-Passard *et al.* 1992). Towards the north, the valley of the Loire formerly ran along the line of the Loing valley towards the river Seine. During the Pliocene the sediments of this river system were characterized by the presence of the volcanic mineral augite. This river flow, called *Fleuve à augite*, testifies again that the course of the Loire was towards the Seine (Tourenq and Pomerol 1995). The diversion of the Loire westwards occurred at the end of the Tiglian or at the beginning of the Lower Pleistocene (Rasplus 1987). The change of course is explained by tectonic movements.

The River Seine

The catchment of the river Seine occupies a major part of the Parisian basin. The subsidence of the basin explains the convergence of most drainage directions towards the central depression. Downstream of Paris the river follows the 'faults of the Seine', which are associated with a major gravimetric anomaly. The hydrographic network of the Seine is more or less adapted to the structure of the basin. However, in detail, its development is still in debate. Concerning the dating of the surface on which the current network developed, Tricart (1988) suggested that the polygenetic erosion surface formed after the regression of the Oligocene sea. It was affected by the uprising of the Vosges, and was warped irregularly. The westward tilt of the surface explains the tendency of the Seine to capture rivers with an originally north–south orientation, such as the rivers Aire and Marne.

The previously mentioned extensive pediment, which extends from the region near Orleans in the south to the English Channel in the west contains granitic gravels, indicative of an origin from the Massif Central. They date from the Middle or Upper Pliocene. Their extent probably corresponds to a Palaeo-Loire/Palaeo-Seine, before the incision of present-day valleys took place. That incision of the valleys is probably older than 1 Ma is indicated by morphostratigraphical evidence as well as by the presence of augite, which comes from volcanism (Tourenq and Pomerol 1995) dated at the Pliocene–Pleistocene transition. Along the French coastline of the English Channel steep cliffs in Upper Cretaceous chalk have been formed (Fig. 13.12).

The sequential alluvial system of the Seine is very complex (Lécolle 1989). This is particularly the case in its downstream reach. Here two palaeo-estuaries have been identified. The valley continues in the English Channel and forms troughs which are filled by alluvial gravels (Lautridou *et al.* 1999; Antoine *et al.* 2000) (Fig. 13.13). Upstream from Elbeuf a sequential system that was repeated sixteen times during 600,000 years has been identified. The large meanders of the Seine in the Parisian region are incised in soft Tertiary sediments (see Fig. 13.5). The age of floodplain sediments upstream from Paris is indicated by the presence of thick (> 5 m) Late Glacial silts in the bottom of the

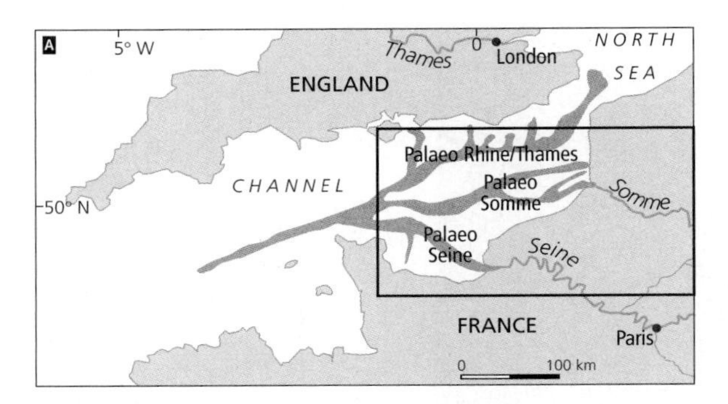

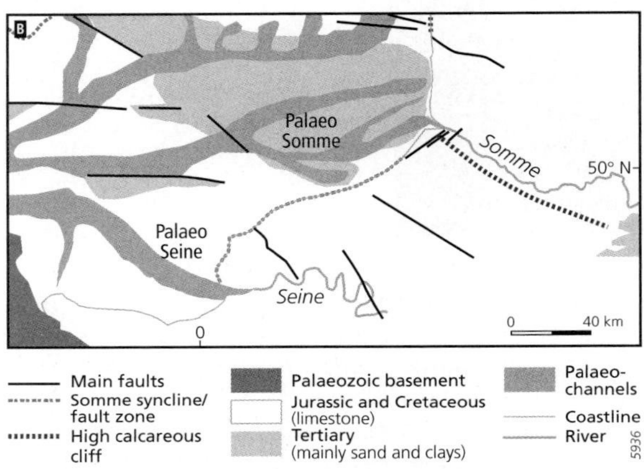

Fig. 13.13. A. Location of the Seine and Somme river valleys and their extension beneath the present-day English Channel; B. Simplified geology and structure of the Eastern Channel (after Antoine *et al.* 2000).

valley (Roblin-Jouve 1980). The 'silty floods' were fed by erosion of periglacial loess dating to 12,000 years BP (Bölling interstadial).

The question remains what the respective impact was of the tectonic and climatic variations in the evolution of the Seine network. The present hydrographic network of the basin often appears to be closely dependent on the regional structural organization. However, some anomalies cannot be explained by underlying structural features. The incision of the valleys and the formation of the terraces seem to be related to (*a*) glacio-eustatic control in the lower Seine and the English Channel sections, (*b*) climatic control in the more upstream sections, and (*c*) in general, to the uplift of the western plateaux.

Meuse, Moselle, Meurthe Rivers

The south–north orientation of the rivers Meuse, Moselle, and Meurthe in the eastern part of the basin can be explained by their initiation on the High Surface. The drainage direction of this surface was towards the coastal plain located to the north of the Ardennes (Condroz). The progressive uplift of the Ardennes massif explains the superimposition of the Meuse, resulting in repeated erosion and the creation of the cuestas. The large valleys of the eastern part of the basin were originally antecedent, and then became superimposed on this Tertiary erosion surface, leading to a well-developed system of stepped terraces that controlled the routing of valleys during the Pleistocene. These stepped terraces are preserved as remnants of high terraces up to more than 300 m in altitude; a series of middle terraces at about 250 m; and a system of low terraces in the Moselle and Meurthe rivers.

The morphological evolution of the Lorraine continental basin is characterized by a system of river captures. The best-known is the capture of the Meurthe by the Upper Moselle, as shown by the abrupt change of orientation of the course of the Moselle. A petrographic indication for capture is the presence of flinty alluvium from the Vosges in the valley of the Meuse to the north of the Moselle–Meuse palaeoconfluence, but which is absent in the valley of the Moselle downstream from Toul. The age of this capture is dated to the Middle Pleistocene (about 250,000 years, according to Harmand and Le Roux 2000).

The Inshore Rivers of the English Channel

On first examination, the inshore rivers appear to be a perfect adaptation to the regional structure. Their parallel courses run SE–NW in the direction of synclines as is seen in Picardy and in Artois. However, the relationships are more complex than this. The original slope system was a fluviomarine Pliocene plain. But a new slope system developed due to the resumption of structural changes and this obliged the rivers progressively to adapt. The River Somme has the most typical alluvial stepped system of terraces in the basin. Studies by Lautridou *et al.* (1999) have shown that there existed a palaeo-submarine valley which converged with the palaeo-valley of the Seine to form *la rivière de la Manche*, which ended far into the English Channel (Fig. 13.13). This system probably formed after the Elsterian. The sequence of terraces in the Somme valley, according to Antoine *et al.* (2000), is made up of ten alluvial bodies from 5 to 55 m relative height above the modern valley bedrock, each representing the morphosedimentary result of a glacial–interglacial climatic cycle. This terrace sequence, which is chronologically controlled by magneto-stratigraphy, ESR-, U/Th-, luminescence-, and radiocarbon-dating, serves as a reference sequence for

Pleistocene river development in northern France. The alluvial terrace bodies are overlain by a layer of loess, sandy loess, and solifluction deposits with intercalated palaeosols. These palaeosols, which are assigned ages varying from Oxygen Isotope Stages 1 to 15, are correlated with the underlying terrace bodies (Antoine *et al.* 2000).

Valley Formation and Quaternary Surficial Deposits

Pleistocene fluvial dynamics are responsible for the excavation and deposition of sediments and the shaping of the valleys. Lateral and vertical erosion alternated in correspondence with the phases of glacial–interglacial cycles. During cold periods periglacial conditions prevailed. 'Periglacial' is defined here according to French (2000) and is regarded as synonymous with cold-climate non-glacial conditions, that is, environments in which frost-related processes dominate. The Parisian basin has not been covered by Pleistocene glaciations, but it was located on the margins of the north-west European ice caps and in the neighbourhood of the Vosges glaciers.

The Shaping of Slopes

Evidence for cold-climate activity on slopes is exemplified by:

- *Asymmetric valleys* of which the slopes facing east and north are often very gentle. Apart from a thermal control on slope evolution, slopes also experienced the influence of a predominant wind direction: especially palaeowinds of west and north-west directions occurred. These palaeowinds brought snow and rain, and so encouraged gelifraction and gelifluxion, which led to the lowering of the slope angle.
- *Versants réglés*, characterized by straight and steep slopes (20–27°). This shape frequently occurs on chalky rocks and more especially on chalk slopes in Normandy.
- *Cryopediments*, which consist of long slopes with an usually rather uniform low gradient (1–4°). Their form is often attributed to the action of periglacial processes such as gelifraction and gelifluxion, and various processes of slope-wash. Such slopes have been observed in Normandy (Pays d'Auge) (Houari 1994).

Periglacial Surficial Formations

Such formations are mainly linked to the shaping of the slopes and include head deposits, *grèzes*, and *bief à silex*, all assumed to have a common origin due to frost-shattering, and also loess. The major characteristics of these periglacial deposits are:

1. *Head deposits*: this type of deposit, frequently found along the coast of northern France, in southern England, and on the Channel Islands, occurs as a heterogenous, unsorted mixture of frost-shattered angular blocks in an abundant matrix of sand, silt, or clay. These formations accumulated on the slopes by cryoturbation and solifluction, and are sometimes up to 30 m thick. Usually stratification is absent. Heads indicate cold periods that are characterized by a strong humidity. The size of material varies with the lithology of the bedrock: 'meulières de Beauce' consist of blocks that reach a size of one metre, whereas chalks and shales are reduced to thinner plates and much smaller particles. The disposition of the long axis of the blocks is mostly parallel to the slope (ploughing blocks). Head formation is known from Weichselian, Saalian, and Elsterian glacial periods.

2. *The bief à silex*: this is a particular form of head. It originates from the frost shattering of flints enclosed in a matrix of clays or sands. The material has been deposited on slopes by solifluxion and rainwash (Laignel 1997). Chalky slopes in Normandy are covered with this formation.

3. *Grèzes and grèzes litées*: although discussion continues, the cryoclastic origin of these materials is generally accepted (Dewolf 1988; Bertran *et al.* 1992; Ozouf *et al.* 1995). They are distinguished from other slope deposits by the uniform size of gelifracts that constitute the main characteristic of these formations. Their size varies with the lithology of the bedrock: from 1 to approximately 4 cm. Moreover, the grèzes are not screes. The use of the term 'screes' as a synonym for 'grèzes' should be avoided, because the grèzes accumulate by solifluction and runoff processes, not by successive rockfalls. The climatic context for the development of the grèzes is a rapid alternation of short and superficial freezing and thawing cycles plus the contribution of short-lived and often renewed snow cover or snow patches. These provide the water required for active frost shattering and frost creep. The grèzes litées show a regular superposition of beds (Fig. 13.14): thin layers with openwork structure, and thick layers with an abundant matrix. The mechanism of sheet solifluction seems to be an essential process in its formation. In the Parisian basin, the most extended and thick deposits of the grèzes are located in Lorraine (cuesta of the Meuse), and Bourgogne regions. The thickness of these deposits is variable, from 2–3 m on gentle slopes to 10–15 m on steeper valley slopes. In the Parisian basin the deposits are more frequent on east-oriented slopes. They seem to be linked to westerly winds and a long duration of snowcover. Grèzes and heads were especially formed at

Fig. 13.14. Slope deposits of the *Grèzes litées* type.

the beginning of cold periods. Their main difference concerns the lithology of the bedrock; the grèzes always derive from limestones. On the other hand, the morphology of the slopes is also different, being short and steep for heads, but long and gentle for grèzes.

4. *Loess and silty loess*: these sediments are of aeolian origin and developed in cold steppe-like climates in Europe during relative dry phases. The source of the Parisian basin loess was situated in the emerged English Channel and North Sea, as well as in the large alluvial plains of the Seine and the Somme rivers. In the north of the Parisian basin, loess often contains periglacial structures such as large frost cracks, cryoturbations, and ice-wedges. In Normandy a stratified and non-calcareous form of loess prevails, which is called *loess à doublets*. In the valley of the Seine, in the centre of the basin, and in the region of Caen a calcareous loess with humic soils is found. These loess deposits contain a record of interglacial or interstadial palaeosols. Lautridou (1985) distinguished four loess cycles from Elsterian to Weichselian age. Loess deposits are clear indications of cold and dry periods, with or without the presence of permafrost, whereas grèzes and heads are correlated with cold but humid phases. Intraloessic palaeosols and intragrèzes palaeosols demonstrate the return of more temperate climatic conditions and especially the development of a vegetation cover.

General Conclusions

The physical aspects of the Parisian basin result from the interplay of tectonic events, particularly subsidence, and climatic variations. This interplay can be traced from Triassic to Pleistocene times. The landscapes are varied and include many plateaux that represent old erosion surfaces, numerous valleys of variable form and origin, and a range of correlative surficial formations. The result is a diversified region with a complex geomorphology.

References

Antoine, P., Lautridou, J. P., and Laurent, M. (2000), Long-term fluvial archives in northwest France: response of the Seine and Somme rivers to tectonic movements, climatic variations and sea-level changes. *Geomorphology* 33: 183–207.

Bertran, P., Coutard, J. P., Francou, B., Ozouf, J. Cl., and Texier, J. P. (1992), Données nouvelles sur l'origine du litage des grèzes: implications paléoclimatiques. *Géographie physique et Quaternaire* 46: 97–112.

Brunet, M. F., and le Pichon, X. (1982), Subsidence of the Paris Basin. *Journal Geophysical Research* 87: 8547–60.

Cavelier, Cl., and Dewolf, Y. (1967), Sur une brèche marine à éléments continentaux du Stampien des environs de Damville-Eure (France) *C.R. Som. Société Géologique de France*, 274–5.

—— and Lorenz, J. (1987), Bassin Parisien ou Bassin de Paris? Le choix d'un titre. In: Aspect et évolution géologiques du Bassin Parisien. Mémoires H.S. *Bulletin d'information des Géologues du Bassin de Paris*, 9–12.

—— and Pomerol, Ch. (1979), Chronologie et interprétation des évènements tectoniques cénozoïques dans le Bassin de Paris. *Bulletin Société Géologique France* 7/21: 33–48.

—— Mégnien, C., Pomerol, Ch., and Rat, P. (1979), Le Bassin de Paris. *Bulletin d'information des Géologues du Bassin de Paris* 16/4, 1–5.

Cholley, A. (1957), *Recherches morphologiques*. Librairie A. Colin, Paris.

Debelmas, J. (1974), *Géologie de la France*. Doin, Paris.

Debrand-Passard, S. (1995), Histoire géologique résumée du sud au centre du Bassin de Paris. *Bulletin d'information des Géologues du Bassin de Paris* 32/3: 15–25.

—— Gros, Y., Lablanche, G., Menot, J. Cl., Clozier, L., and Tourenq, J. (1992), Age, genèse et évolution du fossé de la Loire: nouvelle approche stratigraphique, morphologique et structurale. *Bulletin d'information des Géologues du Bassin de Paris* 29/4: 63–74.

Demoulin, A. (1990), Les Silicifications tertiaires de la bordure Nord de l'Ardenne et du Limbourg méridional (Europe N.O). *Zeitschrift für Geomorphologie* 34: 179–97.

Dewolf, Y. (1988), Stratified slope deposits. In: M. J. Clark (ed.) *Advances in Periglacial Geomorphology*, 90–2.

—— Freytet, P., Obert, D., Plet, A., and Pomerol, Ch. (1985), Une surface polygénique: «la Surface de Brie». La Coupe de la sablière des Vieux Rayons (Fontainebleau)—France. *Bulletin d'information des Géologues du Bassin de Paris* 25/2: 21–5.

French, H. M. (2000), Does Lozinski's periglacial realm exist today? A discussion relevant to modern usage of the term 'periglacial'. *Permafrost and Periglacial Processes* 11: 35–42.

Guillocheau, F., Robin, C., Allemand, P., Bourquin, S., Dromard, G., Friedenberg, R., Garcia, J. P., Gaulier, J. M, Gaumet, F., Grosdoy, B., Hanot, F., LeStrat-Mettraux, M., Nalpas, T., Prijac, C., Rigollet, C., Serrano, O., and Granjean, G. (1999), Evolution géodynamique du Bassin de Paris: Apport d'une base de données stratigraphiques 3D. *Bulletin d'information des Géologues du Bassin de Paris* 36/4: 3–35.

Harmand, D., and le Roux, J. (2000), La Capture de la Haute Moselle. *Bulletin d'information des Géologues du Bassin de Paris* 37/33: 4–15.

Houari, A., (1994), Le Modelé marno-calcaire du bassin de la Dives à partir de la région de Livarot (pays d'Auge) Normandie. Thesis, University of Caen.

Koeniguer, J., and Obert, D. (1984), Bois silicifié de feuillus dans les sables du Massif des Trois Pignons (Fontainebleau) France. *Bulletin d'information des Géologues du Bassin de Paris* 21/2: 57–9.

Laignel, B. (1997), Les Altérites à silex de l'ouest du Bassin de Paris. Caractérisation lithologique, genèse et utilisation potentielle comme granulats. Thesis, University of Rouen.

Lautridou, J. P. (1985), Le Cycle périglaciaire Pléistocène en Europe du Nord-Ouest et plus particulièrement en Normandie. Thesis, University of Caen.

—— Auffret, J. P., Baltzer, A., Clet, M., Lecolle, F., Lefebvre, D., Lericolais, G., Roblin-Jouve, A., Balescu, S., Carpentier, G., Descombes, J. Cl., Ochietti, S., and Rousseau, D. D. (1999), Le Fleuve Seine, le fleuve Manche. *Bulletin Société Géologique de France* 170/4: 545–58.

Lécolle, F. (1989), Le Cours moyen de la Seine au Pléistocène moyen et supérieur. Thesis, University of Caen.

Mégnien, Cl., and Debrand-Passard, S. (1980), Synthèse géologique du Bassin de Paris. *Mémoires Bureau deRecherches Géologiques et Minieres*, 102 Map G. BRGM, Orleans.

Ménillet, F. (1987), Les Meulières du Bassin de Paris (France) et les faciès associés. Rôle des altérations supergènes néogènes à quaternaire ancien dans leur genèse. Thesis, University of Strasburg.

—— (1988), Les Accidents siliceux des calcaires continentaux à lacustres du Tertiaire du Bassin de Paris. *Bulletin d'information des Géologues du Bassin de Paris* 25/4: 57–70.

Ozouf, J. Cl., Coutard, J. P., and Lautridou, J. P. (1995), Grèzes, grèzes litées: historique des définitions. *Permafrost and Periglacial Processes* 6: 85–7.

Pomerol, Ch. (1974), Le Bassin de Paris. In: Debelmas (ed.) *Géologie de la France*. Doin, Paris.

—— (1978), Évolution paléogéographique et structurale du Bassin de Paris du Précambrien à l'Actuel, en relation avec les régions avoisinantes. *Geologie en Mijnbouw* 57: 533–43.

—— and Feugueur, L. (1986), *Guide géologique du Bassin de Paris*. Masson, Paris.

Quesnel, F. (1997), Cartographie numérique en géologie de surface. Application aux altérites à silex de l'ouest du Bassin de Paris. *Documents Bureau deRecherches Géologiques et Minieres* 263.

Rasplus, L. (1987), Anjou, Maine, Touraine et Brenne: la marge sud-ouest du Bassin Parisien. In: *Aspect et Evolution géologique du Bassin Parisien. Mémoires H.S. 6, Bulletin d'information des Géologues du Bassin de Paris*, 181–203.

Robin, C., Guillocheau, F., Allemand, P., Bourquin, S., Dromard, G., Gaulier, J. M. and Prijac, C. (2000), Échelles de temps et d'espace du contrôle tectonique d'un bassin flexural intracratonique: le Bassin de Paris. *Bulletin Société Géologique de France* 171/2: 181–96.

Roblin-Jouve, A. (1980), Le Paysage paléolithique de la vallée de la Seine de Corbeil à Bray-sur-Seine. Thesis, University of Paris 7.

Thiry, M. (1981), Sédimentation continentale et altérations associées: calcitisations, ferruginisations, silicifications. Les Argiles Plastiques du Sparnacien du Bassin de Paris. *Société Géologique Mémoires* 64.

—— (1983), Silicifications continentales. In: *Sédimentologie et Géochimie de la surface. Colloque à la mémoire de G. Millot*. BRGM, Orleans, 177–93.

—— Delaunay, A., Dewolf, Y., Dupuis, C., Ménillet, F., Pellerin, J., and Rasplus, L. (1983), Les Périodes de silicification au Cénozoïque dans le Bassin de Paris. *Bulletin Société Géologique de France* 7/25: 31–4.

—— and Simon-Coinçon, R. (1999), *Palaeoweathering, palaeosurfaces and related continental deposits*. International Association of Sedimentologists, Special Publication 27.

Tourenq, J., and Pomerol, Ch. (1995), Mise en évidence par la présence d'augite du Massif Central, de l'existence d'une pré-Loire, pré Seine coulant vers la Manche au Pléistocène. *Comptes Rendus Académie Scientifique Paris* 320: 1163–9.

Tricart, J. (1988), La Partie orientale du Bassin de Paris. *Mémoires et Documents CNRS* 5, 39–58.

Voisin, L. (1988), Introduction à l'étude de la Pierre de Stonne et les formations siliceuses associées au sud-ouest de l'Ardenne. *Mémoires H.S. Histoire National. Charleville-Mézières.*

14 French Alps and Alpine Forelands

Yvonne Battiau-Queney

Introduction

The French Alps are the western part of the 1,200-km-long Alpine range extending eastward to the Vienna basin (Fig. 14.1). They have the highest summits of the range, in the Mont-Blanc massif (4,807 m a.s.l.). In France, the chain has an arcuate form, convex to the north and west. It lies between Lake Geneva (46° 25′ N) and the Mediterranean coast (approximately 43° 35′ N). The Rhône valley forms a clear geological and morphological western limit. To the north (towards the Jura range) and the south-west (towards the ridges of Provence) the boundary is not so well defined. The French Alps and Alpine forelands have been thoroughly studied for over a century by many researchers from the Universities of Grenoble, Lyons, Aix-en-Provence, Nice, and Chambéry. First, it is necessary to outline the great diversity of landforms in relationship to the complex geological history, tectonics, and lithology. The importance of the Alpine karst landforms and caves must be emphasized; studies of these forms have been extended substantially in the last twenty years and they give many new insights into the Plio-Pleistocene tectonics and climates of this region. The past and present role of glaciers is another important topic in this chapter. From recent studies, we now have a much better knowledge of the transition from the last glacial period to the Holocene. It was impossible to write a chapter on the Alps and ignore the fact that the inhabitants of the Alps have to cope with many permanent natural hazards. The chapter ends with a short synthesis of the main morphogenic systems, which characterize the French Alps and forelands.

Principal Geographical Features

Climate

In the north, the climate is oceanic and precipitation is evenly distributed throughout the year (Fig. 14.2). A high relief, with landforms oriented transverse to the general western atmospheric circulation, results in a great variety of regional climates: from west to east, the continental effect is marked by a decreasing precipitation at the same altitude. Exposure and altitude combine to create contrasting local climates. Temperature inversion is frequent, especially when cold air is trapped in deep valleys. In the south, the climate is sub-Mediterranean: the average annual number of precipitation days is lower, with many more hours of sunshine and drier summers; by contrast, in winter heavy showers are frequent here and snowfall is considerable. The boundary between both climatic areas is fairly sharp, depending upon the regional topography. According to Bénevent (1926), the first geographer to study the French alpine climate, this boundary follows the southern edge of the Vercors, the north edge of the Devoluy, and the south edge of the Pelvoux massif (also called *Massif des Écrins*). Southern influences dominate the weather of the north–south-oriented upper Durance valley.

Vegetation

The alpine vegetation is well known as a result of extensive studies by Ozenda (1985). First, it is shown to be closely related to altitude, and distinct zones are clearly recognized: submontane, montane, subalpine, alpine, and glacial. However, owing to the latitudinal range and eastward continental trend, other differences are evident (Veyret and Veyret 1979; Julian and Bravard

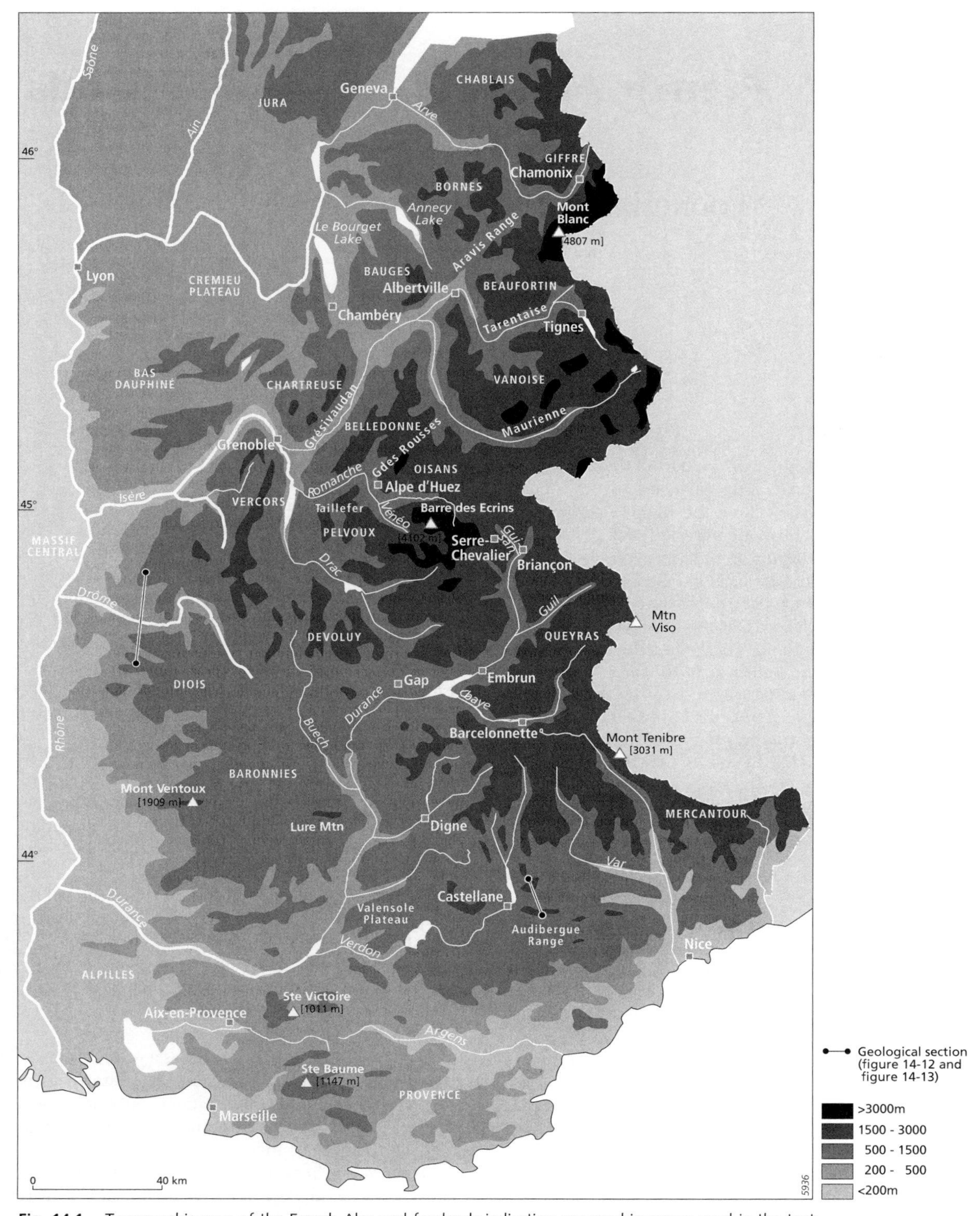

Fig. 14.1. Topographic map of the French Alps and forelands indicating geographic names used in the text.

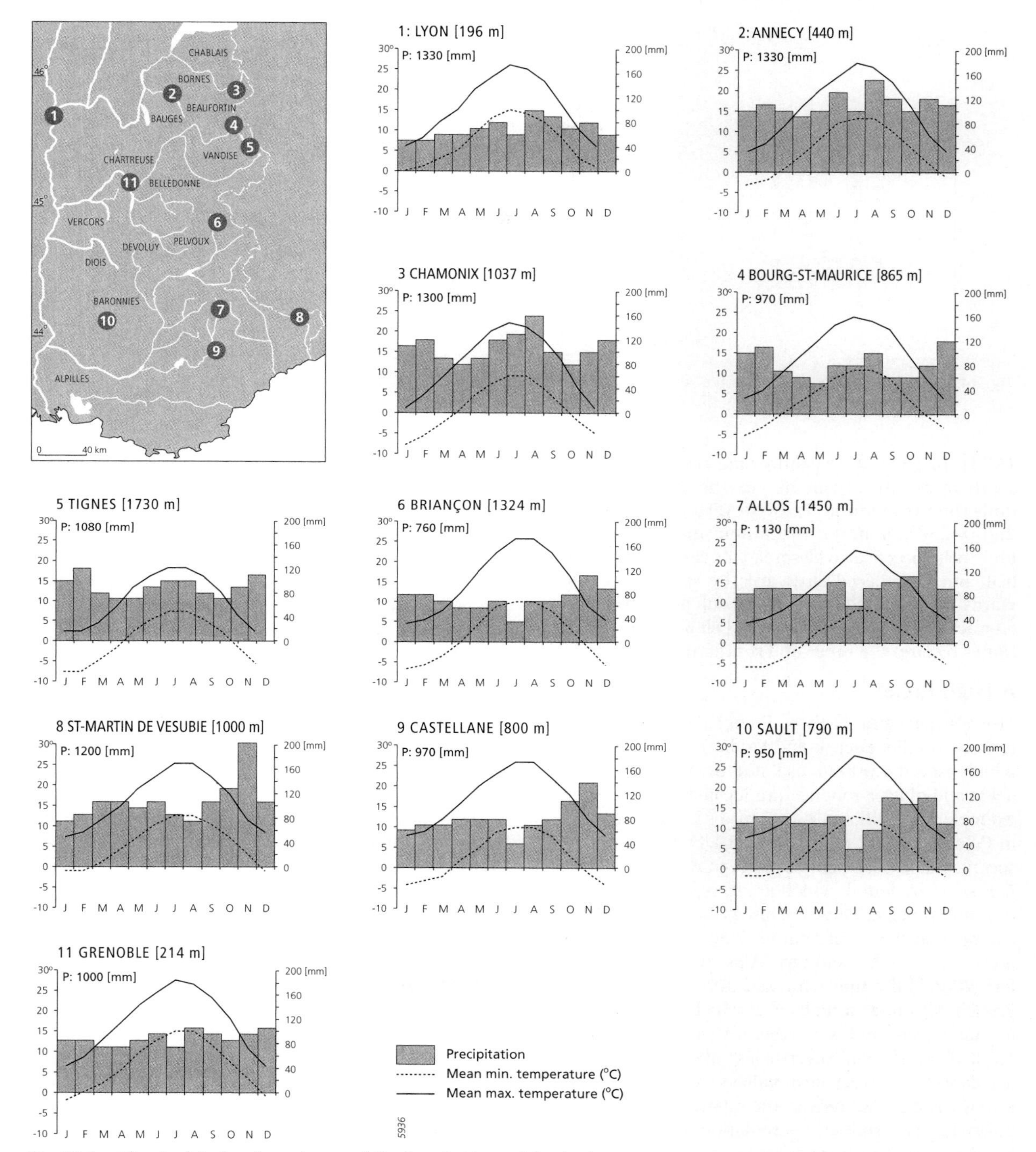

Fig. 14.2. Climate data for eleven towns of the French Alps and forelands.

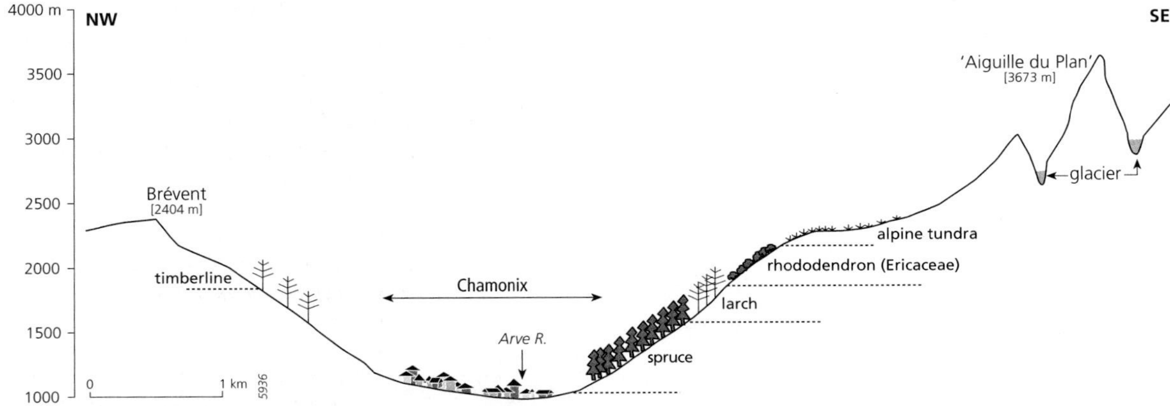

Fig. 14.3. Zonation of vegetation in the Arve valley at Chamonix.

1984). In general, the timberline rises eastwards and southwards. At Chamonix, in the Arve valley, the timberline is found at 1,900 m (Fig. 14.3). It reaches 2,200–2,400 m in the Mercantour massif. In the montane belt, beech is widespread in the northern outer belt, but is replaced eastwards by spruce and southwards by *Pinus sylvestris*. In the subalpine belt, spruce characterizes the outer northern belt and Cembran pine (*Pinus cembra*) the inner and southern French Alps.

A High Relief

The northern and southern French Alps also differ with respect to relief energy and landforms. Those summits which exceed 4,000 m a.s.l. and the most extensive ice fields and glacier tongues are located in the north: for example, Mont-Blanc, 4,807 m a.s.l., Barre-des-Écrins in Pelvoux, 4,102 m a.s.l. (Fig. 14.1). In the north the spectacular escarpments of lower Cretaceous *Urgonian Limestone* are found, which are so typical of the scenery around Grenoble. These are also present in Provence (for example on the Sainte-Baume ridge) but are absent in most areas of the southern Alps, where the strongest formation is the upper Jurassic *Tithonic Limestone*. The French Alps have a high relief which exceeds 2,000 m in many areas and is consistently over 3,000 m in the Mont-Blanc massif. Nevertheless, the mountain blocks are dissected by very deep valleys, which penetrate far into the ranges. Moreover, the relatively low cols at the valley heads permit easy circulation from one valley to another. The Alps have never been a human barrier, in contrast to the Pyrenees which, although generally lower, are less extensively dissected. Where there are high steep slopes, natural hazards are commonplace. They are frequent and widely distributed.

An Extensively Glaciated Chain during the Pleistocene

It was not until the end of the Miocene that the high relief was created. The rugged scenery was largely produced by the numerous ice masses, which developed over the Alps during the Pleistocene. The inner and highest areas were strongly carved by glaciers and the effects of the last glaciation are very distinct in the field. The outer areas and forelands received thick glacial and fluvio-glacial deposits, which belong to at least two glaciations, the Riss and Würm (Fig. 14.4).

Geological Formation

On the basis of data obtained from surface geology and ECORS-CROP (1989) deep seismic profiles, the structural form and history of the French Alps are fairly well understood (Ricou 1994; Lemoine *et al.* 2000). The western Alps are situated at the site of the Tethys ocean, which opened between Laurasia and Gondwana in late Jurassic times, about 145 Ma (Dercourt *et al.* 2000). The opening came after a rifting phase that lasted from upper Triassic to the end of middle Jurassic times and resulted in the fragmentation of Pangaea. The Ligurian segment of the Tethys ocean formed between Europe-Iberia on the northern side, and the Apulo-African microplate on the southern side. The western Alps (including the whole of the French Alps) are entirely made from European or oceanic crust, in contrast to the eastern Alps which are made of African crust. In the Pelvoux area, the rifting phase was marked by normal faults, tilted blocks, and half-grabens.

During upper Cretaceous times, the Tethys Ocean began to shrink, partly as a result of the subduction of

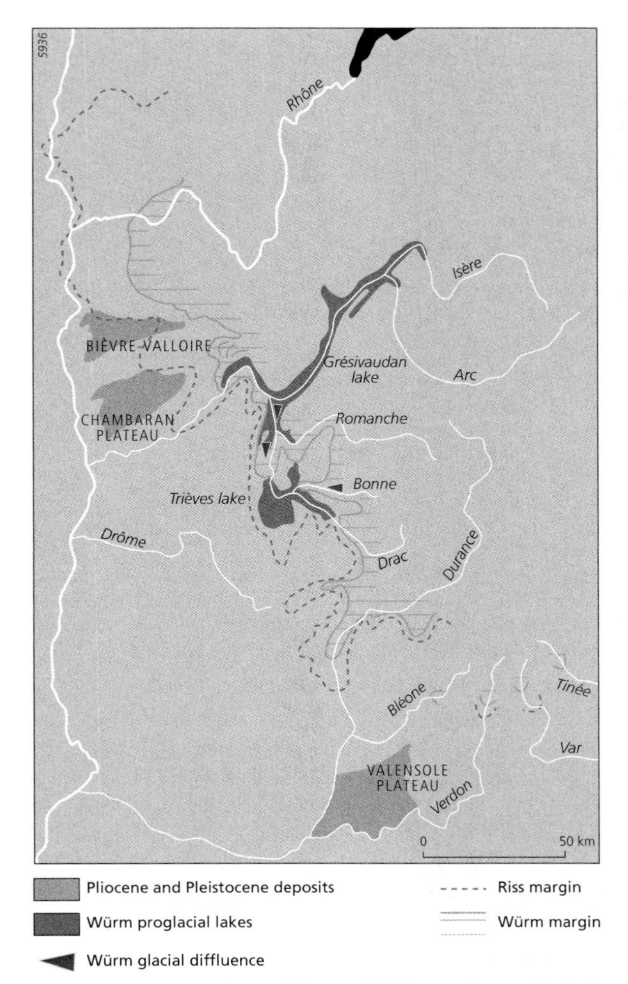

Fig. 14.4. Western limit of Riss and Würm valley- and piedmont-glaciers in the French Alps and forelands, estimated from glacial landforms and deposits (after Monjuvent, in Debelmas 1974).

oceanic crust beneath Apulo-Africa. Finally, the ocean closure was followed by continental collision. An accretion prism was formed on the leading edge of the subducting European plate. This was accompanied by extensive tectonic inversion, when normal faults were converted into thrusts. Finally, the huge continental collision and compression led to rapid uplift. However, the present high relief of the French Alps did not develop until the end of the Miocene. The elevation of the Outer Crystalline massifs and simultaneous removal of their sedimentary cover was immediately followed by the folding of the Mesozoic sediments in the Prealps.

Plate activity persists to the present, although there are few historical records of large-scale seismicity. Current deformations of the topographic surface can be measured by satellite. The Alps are presently rising at a mean rate of 1 mm per year. This elevation results from two mechanisms. First, the continuing convergence of European and African plates at an average rate of 1 cm/year. Secondly, an isostatic readjustment, following erosion of the interiors of the ranges and the removal of the load of Pleistocene ice. In France, faster rates, reaching several mm per year, are observed locally, especially in the interior. As erosion does not seem to keep pace with upheaval (in the Isère basin, this upheaval has been estimated by Chardon and Castiglioni (1984) to be around 0.2–0.25 mm/year), another mechanism has to be put forward to explain why the height of the chain does not increase faster. Late- and post-orogenic extensional tectonics and near-surface normal faulting have recently been described, and these effects have been related to subcrustal ductile faults (Tricart *et al.* 2001). Once the crustal thickness has reached a certain limit, depending upon the mechanical resistance of the crust, a gravitational mechanism tends to stretch the continental crust. This explains why the mountains cannot rise indefinitely.

Present Geological and Topographic Arrangement

An overall view of the French Alps shows a series of arcuate structural belts (Fig. 14.5). The present structural organization is inherited from the rifting and passive margin development and the lower Mesozoic palaeogeography. At that time, different blocks were separated by major normal faults, which subsequently behaved as thrust faults and lines of weakness. Inward, the main structural belts are:

The French Alpine Forelands

Strictly speaking, the alpine 'forelands' correspond to the peripheral regions that were subsiding and received erosion products from the rising mountains. Compared with the central and eastern Alps, this structural domain is poorly developed in France. From Geneva to the Chambéry region, a relatively thin upper Oligocene and Miocene 'molasse' formation rests over an irregular basement of Mesozoic rocks.

In *Bas-Dauphiné*, between the Rhône and Isère valleys, the Neogene formation is considerably thicker. Here, Oligocene fluvial, lacustrine, or brackish sediments, and Miocene marine molasse may reach several kilometres. In Oligocene times, the accumulation kept pace with

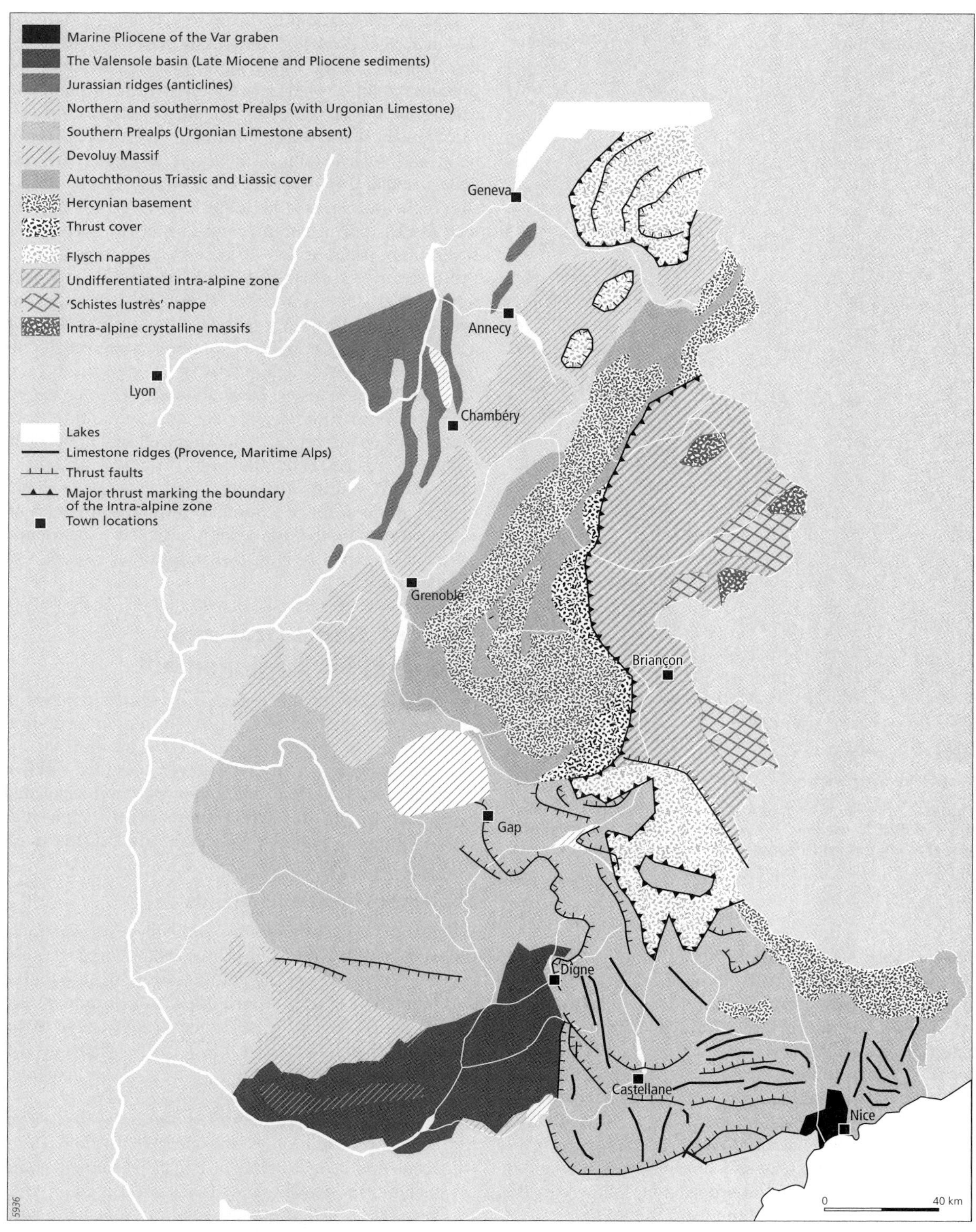

Fig. 14.5. The main structural units of the French Alps.

subsidence, the sediment supply being mainly from the west. In the Miocene, an erosion surface (the 'Rhodanian Piedmont') developed over the crystalline rocks of the Massif Central and passed eastwards to an accumulation of marine molasse in Bas-Dauphiné. It was not before the end of Miocene and the rapid elevation of the alpine area that the main source of detrital sediments became established in the Alps, when in the upper Miocene and Pliocene coarse sand and quartz gravel spread westwards. The Rhône had to migrate in the same direction and is now confined at the foot of the Massif Central. During the Messinian Mediterranean regression, it cut a deep valley that was temporarily invaded by the Pliocene Sea. Later, during the Pleistocene, the area was widely covered with glacial or fluvio-glacial deposits.

Owing to tectonics and differential erosion, the area *south of the Isère valley*, which follows a major NE–SW fault, is divided into small basins. The Neogene deposits are much thinner than the Mesozoic marine succession in this area. The Miocene molasse is found in only a few areas. Three structural lines intersect this area: the Hercynian line NE–SW-oriented, the Pyrenean line east–west oriented, and the north–south-oriented Alpine line. Most of the scarps marginal to the basins are fault-controlled. In the western Provence, the Valensole plateau unquestionably belongs to the alpine foreland. It consists of a thick succession of Mio-Pliocene alluvial deposits carried from the Alps by the ancestors of the Durance and the Verdon rivers. The Miocene peri-alpine sea has covered the western part of Provence, but not the south and south-east. Here, the alpine folded ridges are directly in contact with a platform of Jurassic and Cretaceous limestones. The Miocene rivers flowed to the north and west in a low relief landscape. The present southwards-draining river system developed during the Pliocene, concomitantly with the opening of the western Mediterranean Sea.

The Prealps

In these limestone ranges altitudes are moderate, the summits rarely exceeding 2,000–2,300 m a.s.l. except in the Devoluy (2,793 m) and the Aravis Range (2,752 m). Nevertheless, the relief is high with more than 1,000 m of altitude range between summits and valley floors. In the north, they form prominent massifs edged by impressive limestone cliffs, at least 1,000 m high above the forelands and the Alpine Trough. Owing to two successive folding events, the landscape is more complex south of Vercors.

- The *Chablais-Giffre massif* differs structurally from the others: it is made of a pile of nappes that were transported northwards from the internal zone over the Mont-Blanc massif, before the latter was uplifted. The gypsum-rich Triassic clay has facilitated thrusting over the Oligocene molasse. Resistant conglomerates (*nappe de la Brèche*) and limestones give hogbacks and homoclinal scarps that contrast sharply with the monotonous landscapes formed from shale, slate, and marl (Fig. 14.6).

Fig. 14.6. Typical landscape of Chablais-Haut Giffre, east of Geneva. The region is made up of a pile of nappes which were moved northwards and folded in wide synclines. Here, high homoclinal scarps are built from limestones and calcareous conglomerates. The forested gentle slopes in the front of the picture are associated with shales.

• The *Bornes, Bauges, Chartreuse, and Vercors massifs* are separated by deep transversal troughs, the Arve, Annecy, Chambéry, and Isère *cluses*. They are built from folded Mesozoic sediments. The fold axes here are NE–SW between the Arve and Isère valleys and north–south in Vercors. Each massif has its own style of folding and landforms, but everywhere the thick and very resistant Lower Cretaceous Urgonian limestone plays a major role in the development of these landforms.

• The Urgonian limestone is absent from the *Diois, Baronnies*, and *Devoluy massifs*. Here, marls are much thicker than the two resistant limestone formations: the Upper Jurassic Tithonic and Upper Cretaceous Turonian limestones. The strata were folded twice, along different axes: the first, 'Pyreneo-Provençal', was east–west-oriented; the second, the Alpine proper, was north–south oriented. Therefore, there is a generally compartmentalized relief with short limestone ridges.

• The *Mont-Ventoux–Lure Mountain* axis is generally considered as the southern limit of the Alps. Although this is considered still to belong to the Prealps, the form is essentially Provençal, with east–west folding axes. Mont-Ventoux is the highest point in Provence (1,909 m a.s.l.). It is a huge Urgonian limestone mass, 1,300 m high, the scarp edge of which faces north. It is renowned for its distinctive vegetation, which ranges from Mediterranean to endemic alpine species, and for its rough, windy climate.

• The *southern Prealps from Digne to Nice* form an arch convex to the south. The east–west folds are thrust to the south.

The Alpine Trough (*Sillon Alpin*)

Usually the three outer zones (forelands, Prealps, External Crystalline massifs) are considered as autochthonous, contrasting with the internal thrust zone. However, even in this outer zone the forelands and Prealps have moved significantly northwards or westwards. The Mont-Blanc massif itself and other External Crystalline massifs were also thrust to the north-west (Gourlay 1986). The crystalline rocks were intensely deformed in a ductile manner, on the outer edge of the massifs, and affected by strike-slip faulting. In response to strong uplift, they were subsequently stripped of their sedimentary cover, which slipped under gravity and folded as it did so. The contact with the Prealps is, therefore, not a simple stratigraphic contact but an important fault zone with strike-slip and vertical components. As rivers and ice easily eroded this fault zone there arose a major topographic feature, the *Sillon alpin*. It is well developed from Chamonix, in the upper Arve valley, to Grenoble along the Isère valley, then along the Drac valley.

The External Crystalline Massifs

These form the backbone of the French Alps. From the Mont-Blanc massif to the Mercantour, gneiss, micaschist, and intrusive granite are the main rocks of this high mountain belt. They belong to the old Hercynian basement. They presently outcrop because the area was strongly elevated and stripped of its sedimentary cover, from the upper Miocene onwards. They are still rising at a rate of a few mm/yr. The crystalline belt is absent between the Pelvoux and Mercantour massifs: around Embrun and Barcelonnette, nappes of flysch, which belong to the internal zone, were transported over autochthonous Mesozoic rocks and preserved in an area of low elevation.

The Crystalline massifs were intensely glaciated during the Pleistocene. Nowadays, the most beautiful glaciers are restricted to this zone, especially in the Mont-Blanc and Pelvoux massifs. In contrast to the main Alpine valleys, which skirt the massifs, they have deep and narrow valleys. The altitude range everywhere exceeds 1,500 m and reaches frequently 2,000–2,500 m (Fig. 14.7).

Curiously, despite ancient and recent glaciation, the inherited landforms are obviously well preserved in the present landscape: elements of the old pre-Triassic surface are widespread in the Grandes-Rousses, Taillefer, and Pelvoux massifs (Battiau-Queney 1997). Remnants of the Triassic sedimentary cover (mainly quartzite and dolomite) help to reconstitute this > 220 Ma-old surface. Fault-scarps inherited from the rifting period and subsequently reactivated, as well as evidence of glacial activity are illustrated in Fig. 14.8. Locally, the crystalline basement has been deformed and incorporated in a thrust/fold structure during the alpine compression. All these observations clearly prove that detailed and regional landforms are everywhere structurally controlled. This is also the case for some flat levels, developed alongside the Chamonix valley, which were previously classically interpreted as pure glacial features (Veyret and Veyret 1979).

The Intra-Alpine Zone

This zone does not have the highest summits, but it does have the highest average altitude. It is still being uplifted relatively quickly (several mm/year in the Briançon area, which is 12 cm higher relative to Gap than it was 80 years ago). It is the most heterogeneous geomorphologic belt in the French Alps for two reasons: there is a great variety of rocks and also a great tectonic complexity, the latter due to its position in the most compressed area in the plate collision zone. In spite of its limited outcrops, the Triassic gypsum played everywhere a major role in the thrusting and folding mechanisms.

Fig. 14.7. The Romanche valley, seen from Alpe d'Huez westward. The north–south flat valley floor, at 710–720 m a.s.l., is due to postglacial alluvial and lacustrine sedimentation. The Würm Romanche Glacier (flowing from left to right) has overdeepened a pre-glacial basin in soft Mesozoic shales. First deposited in a half-graben basin, shales were intensely crushed and folded during the Alpine collision period. In the foreground shales give long gentle slopes. In the background, the crystalline basement outcrops on the Taillefer massif (2,000–2,800 m) are shown.

Fig. 14.8. Areal glacial scouring (Grandes-Rousses massif, north of Alpe d'Huez). In the foreground (2,500–2,600 m), the crystalline basement has been dislocated by faults in a series of small horsts and half-grabens in which Mesozoic sediments may be found. Areal glacial scouring has slightly overdeepened the less resistant rocks, leading to a 'knob and kettle' topography. In the background, to the south, the western Pelvoux massif (3,000–3,465 m a.s.l.) presents a typical cirque landscape, in which just a few small glaciers remain.

The structure is extremely complex in the *Vanoise* massif (Debelmas and Rampnoux 1994). This is a thrust belt with a generally westwards movement except in the east part where counter-thrusts are oriented in the opposite direction. In geologically recent times the whole area has been folded and uplifted. Owing to the strong compression they suffered, rocks are generally intensely crushed and fissured. The form of the pyramidal summits is attributed to the dominance of a well-developed vertical cleavage.

Queyras and Ubaye have two distinct areas (Lemoine and Tricart 1988; Evin 1997):

1. An outer belt made of Triassic and Jurassic carbonate and quartzite, which belongs to the passive continental margin of the Tethys ocean. The strata here are vertical or inclined to the west or south-west and incorporated

Fig. 14.9. The Chatelet bridge across the Ubaye river. The Ubaye has cut a deep and narrow gorge in upper Jurassic limestone. The single-arch bridge lies 100 m above the river.

in a pile of thrusts. They give a series of steep, narrow, and short ridges that have considerably hindered the transmigration of people in the past. Vertiginous bridges (Fig. 14.9) cross the deep and narrow valleys (e.g. Guil and Ubaye).

2. An inner belt mainly made of *Schistes lustrés* which formed at the bottom of the Tethys Sea and were then metamorphosed and thrust westwards when the ocean closed. As in Vanoise, all the eastern area was later backthrust eastwards, i.e. in a direction opposing that of the regional trend. The *Schistes lustrés* are associated with a few ophiolites (pillow-basalt, gabbro, and serpentinite) which are fragments of the Tethys oceanic crust. Long and gentle schist slopes contrast with the few steep gabbro hills. Good examples are the gabbro of Mont-Pelvat (3,220 m a.s.l.) in eastern Ubaye, and the basalt of Mont-Viso, which is just beyond the French–Italian border.

The boundary between the external and internal zones is a major thrust (the 'Penninic Front', in terms of the classical nomenclature) which dips far down into the lower crust, and was active during the Oligocene (Fig. 14.5). Recently it has been shown that this major thrust, associated with a seismic boundary in the Briançon area, was subsequently reactivated as an extensional detachment from the Miocene onwards (Tricart *et al.* 2001).

Structural Landforms

Lithological and Tectonical Control

The regional physiographic organization of the French Alps is modelled on the arcuate belts of the structural organization. Each structural belt has specific landforms.

In the northern forelands, most facies of the molasse (calcareous sandstone, gravel, and sands) are soft. They give gentle slopes and hills contrasting with the higher limestone ridges that are characteristic of the Jura. Good examples are seen around Chambery and Le Bourget Lake. Between the Rhône valley and the spectacular external limestone cliffs of Vercors, Grande Chartreuse, Bauges, and Bornes, plateaux and elongated flat-topped hills are isolated by wide corridors with alluvial terraces, carved during the Pleistocene period by braided rivers or piedmont glaciers coming from the Alps.

In the northern Prealps, two formations are dominant in the landscape: the lower Cretaceous Urgonian Reef-limestone, and the upper Jurassic Tithonic limestone. These exhibit impressive rocky cliffs, above marl slopes (Fig. 14.10). The landforms are controlled by folds but are often inverted compared to the underlying structure. Perched synclines characterize the Bauges and Chartreuse massifs. The wide folds of the Vercors massif give a high plateau topography with broad synclinal depressions that are isolated from the surrounding valleys and plains by scarps up to 1,600-m high. The Bourne flows across a fold axis in a 500–700-m-deep canyon. It is a good example of antecedence: the river has cut its gorge as the Vercors massif was rapidly rising during the whole of the late Neogene (Fig. 14.11).

In the south-western Prealps (Diois, Baronnies), an inverted relief with perched synclines has been formed by differential erosion between resistant limestone beds and thick marls (Fig. 14.12). The completely different fold forms of the Prealps of the Provence give rise to a series of southwards-facing scarps. In the Castellane region, landforms are well adapted to the imbricated fold-and-thrust structure (Fig. 14.13). The deformed structure is detached from its substrate along a décollement plane,

Fig. 14.10. The northern face of Mont-Granier (Chartreuse massif, between Grenoble and Chambéry). Situated at the northern edge of the Chartreuse massif, Mont-Granier (1,933 m a.s.l.) is bordered by nearly 1,000-m-high cliffs made of Urgonian limestone overlying calcareous marl and marl. The northern face is nearly vertical and devoid of forest: it is the scar of a huge rock-slide that occurred in AD 1248. The fallen mass (approximately 500 million m^3) spread over the plain below and covered several villages. It killed at least 1,000 people. According to Goguel and Pachoud (1972), the detachment took place along a slightly inclined joint plane in the basal marls and started after an ordinary rockfall of the upper scarp. The huge land mass probably slid over an air-cushion (photo: Jean-François Billet).

where Triassic gypsum acts as a lubricant. Here the resistant formation is the Tithonic limestone through which the Verdon cuts an impressive canyon.

The External Crystalline massifs have the highest summits of the French Alps. Most of the rocks that outcrop here are resistant. However, they give rise to a range of different landforms because the density and geometry of fractures largely control the slope and cliff geometry. The dome-like Mont-Blanc is built of massive granite. The 'Aiguilles Rouges' are made of strongly deformed mica-schist and gneiss. In the intra-alpine zone, almost every type of rock is present somewhere: schist, mica-schist, gneiss, and metamorphic facies of sedimentary rocks. Limestone and dolomite, quartzite and shale have an important morphological role. Some of the highest Vanoise summits (3,600–3,800 m a.s.l.) are carved in Liassic limestone.

Karst Features

Owing to the widespread outcrop of limestone in the French Alps and forelands, it is not surprising to find karst landforms in every region except the External Crystalline massifs (Nicod 1984; Chardon 1989*b*; Maire 1990; Delannoy 1990, 1991; Lismonde and Delannoy 1990; Audra 1994). Karst landforms and cave systems are especially interesting for two reasons: (1) they give impressive landforms of great scenic beauty; (2) cave systems give unrivalled data for unravelling the Cenozoic development of the Alps. The *gouffre Berger* (–1,241 m), in Vercors, was explored in 1954 and was at that time the deepest known cave in the world. These days, the deepest known cave is the *gouffre Jean-Bernard* (Haut-Giffre massif), at a depth of 1,602 m (Lips *et al.* 1993). Generally, these very deep caves are situated close to the main transversal valleys (the alpine *cluses*), where the altitude range reaches or exceeds 1,000 m. Three factors explain the great diversity of alpine karst landforms: (1) the geological structure (thickness and nature of the limestone, folding, and faults), (2) the rate of uplift, and incision of the main valleys since the upper Miocene, and (3) the environmental conditions (past and present, including the type of vegetation, the influence of snow and ice, and the abundance of precipitation).

Three of the many areas that have been thoroughly studied are described here. First, *the Désert-de-Platé*, close to the Haut-Giffre massif and the Arve valley (Maire 1990). It is largely situated above the timberline, between 1,600 and 2,800 m a.s.l., and is characterized by a cold, wet, and snowy climate (2,400–2,800 mm of precipitation per year, 65–70% of which falls as snow). The snow-line is situated around 2,500 m a.s.l. The Désert-de-Platé forms a plateau-like area, surrounded by impressive limestone cliffs above the deep and wide Arve and Giffre valleys. The most typical and widespread landforms are bare limestone pavements with a multitude of solution pits and grooves. The steep-sided dolines were overdeepened by ice scouring so that they now trap snow and are sites of rapid solution. According to Maire (1990), the solution rate averages 104 mm/ka in this 'nival' high mountain karst.

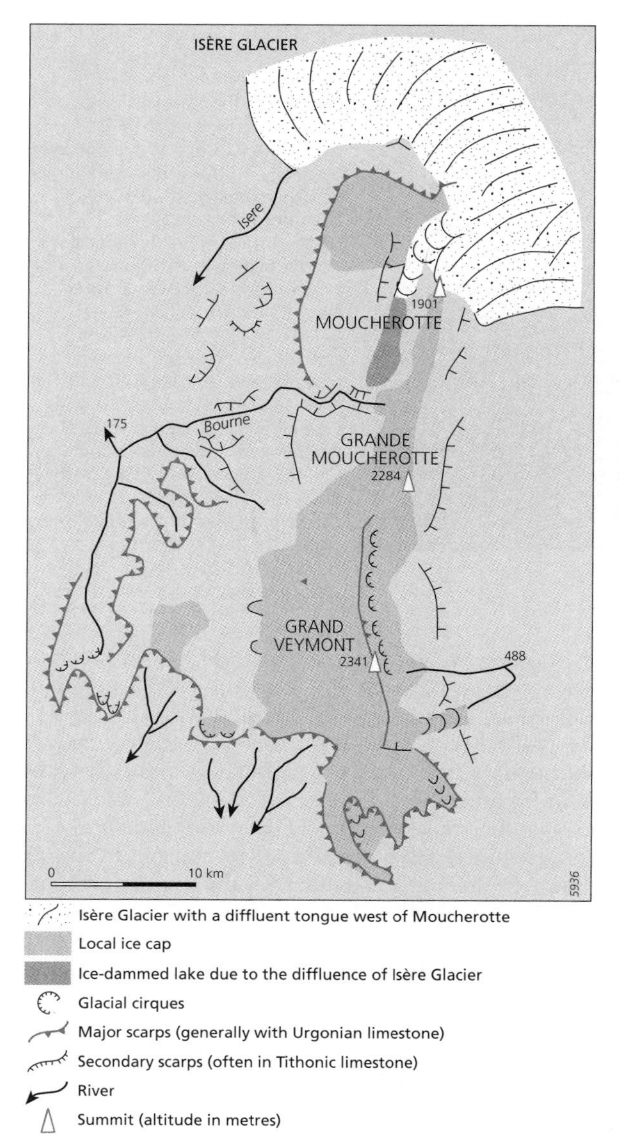

Fig. 14.11. Sketch-map of the Vercors massif (modified after Delannoy 1990).

Secondly, *the Vercors massif* (Delannoy 1991; Audra 1994; Delannoy and Caillault 1998) forms a high, extensively forested plateau with wide outcrops of Urgonian limestone. During the Würm and Riss, it supported a local ice cap adjacent to the huge Isère glacier (Fig. 14.11). Karst landforms are both numerous and diverse and cave systems are very well developed. A shallow subhorizontal system formed in upper Miocene and early Pliocene times and is now perched and inactive because of the regional Pliocene uplift and rapid incision of the main valleys during the Pleistocene glaciations. The deep subvertical cave systems are younger: their development took place during the glacial periods. The present solution rate in the Vercors massif, 120–250 mm/ka, according to Delannoy (1991) and Audra (1994), is one of the highest rates ever recorded in karst areas. This is attributed to the dense forest cover and high annual precipitation (1,200–1,500 mm).

Thirdly, *the Audibergue Range* (Julian and Nicod 1984) belongs to the Prealps of Castellane (Fig. 14.13). It reaches 1,642 m a.s.l. and lies entirely within the montane belt, with abundant snow and annual precipitation exceeding 1,300 mm. A Mediterranean influence is seen in the drier summer. Beech and fir cover the northwards-facing slopes, contrasting with shrubs and heath vegetation on the southern slopes. Dolines and karren (*lapiés*) are present everywhere A typical 'nival karst' is found on the summit ridge where the fracture pattern evidently controls both the spatial distribution and individual shape of dolines. Good examples of poljes are present in the range. Among them, the Caille polje is 3.5 km long and has a very flat floor at 1,123–1,125 m a.s.l. It is frequently inundated owing to the repeated obstruction of ponors or sinkholes. A deep cave system is also well developed in this range. The estimated solution rate here averages 62 mm/ka in the range (Julian and Nicod 1984).

Triassic *gypsum* that outcrops at the periphery of the Variscan basement and along the numerous thrust-faults of the intra-alpine zone displays many of the typical karstic landforms (Nicod 1988, 1993; Chardon 1991). Good examples of dolines (2.5–3-m-deep vertical-sided dolines and 30 m-deep, 50–100-m-wide funnel-shaped dolines) are found between 2,000 and 3,000 m a.s.l. in the upper valley of the 'Doron des Allues' (Vanoise massif). The denudation rate may reach 1,500 mm/ka, according to the ionic concentrations in spring water (Chardon 1991).

Glaciations and Glacial Landforms

Chronology

Although the French Alps must have been glaciated several times during the Pleistocene, only the last glacial period (Würm) is clearly identified in the mountains, and the last two periods (Riss and Würm) in the forelands. We have only the limited evidence of cave deposits in the limestone terrain for the pre-Riss glaciations. For example, in the Jean-Bernard cave (Haut-Giffre), U/Th dating of speleothems proves that some

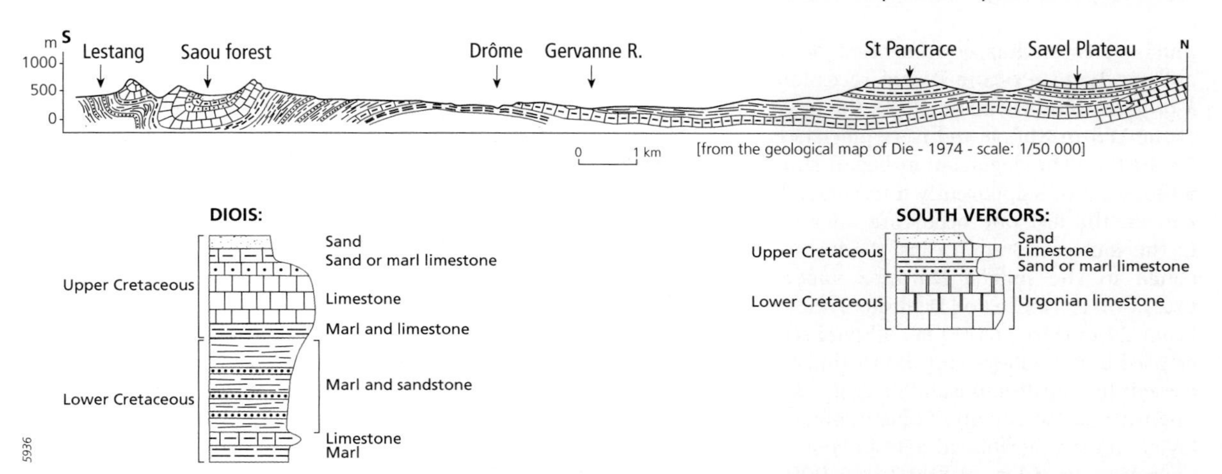

Fig. 14.12. Geological section of the Diois and southern Vercors massifs. From north to south, the section shows the changing facies of the lower Cretaceous: the Urgonian limestone is absent from Diois, where upper Cretaceous limestones give perched synclines. The Saou forest is a cradle-shaped, east–west-oriented basin; it is almost completely isolated from the peripheral regions by impressive rock cliffs (reproduced from Battiau-Queney 1993).

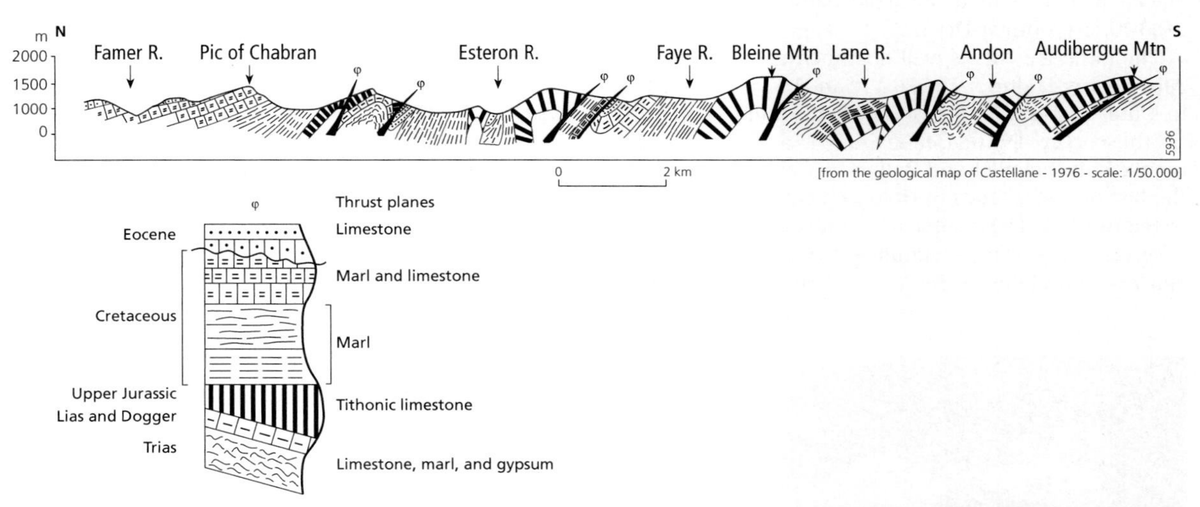

Fig. 14.13. Geological section of the Prealps near Castellane. Situated in the southern Prealps, this region is characterized by east–west ridges made of Tithonic limestone. The Mesozoic strata have been folded and thrust southwards, giving an imbricated structure. Karst landforms are widespread in the whole region (see the Audibergue area in text) (reproduced from Battiau-Queney 1993).

fluvio-glacial deposits are older than 200,000 years (Lips *et al.* 1993). Similarly, in the 'Dent de Crolles' cave system (Chartreuse), the age of stalagmites, ranging between 177 and 268 ka, proves that the underlying glacial deposits are older (isotopic stage 8) (Audra 1994). Likewise, in close proximity to the Isère valley, the *Grotte Vallier* system (Vercors), which is about 10 km long and 400 m deep, has developed since the Pliocene, and several glacial events have been recognized in the cave deposits. The earliest are older than the Brunhes/Matuyama inversion (780 ka). But outside the cave systems, there is no trace of these ancient glaciations.

From evidence found in the forelands, all researchers agree to place the widest expansion of the alpine glaciers in the Riss period (Fig. 14.4). However, in the mountain chain itself, only the Würm glaciation has left incontrovertible evidence. For the last twenty years, much research has concentrated on the chronology and mechanism of the Würm recession. According to Monjuvent and Nicoud (1988), the maximum advance of the Würm Rhône and Isère glaciers occurred before 38,400 BP. The deglaciation began shortly afterwards, although it was apparently interrupted by several readvances, the last one occurring later than 22,300 BP. In the southern French Alps, the last ice-advance happened in the middle Durance valley from 29,000 to 24,000 BP (Jorda and Delibrias 1981; Rosique 1997). From 24,000 to 18,000 BP, alluvial sedimentation decreased in the valleys, but the Durance glacier was still present just upstream from Sisteron around 18,000 BP. Then it receded rapidly. Palaeoecological, geomorphologic, and stratigraphical data indicate a rapid climatic improvement from 13,300 to 11,000 BP (Jorda and Rosique 1994). This was marked by the fast retreat of mountain glaciers, extension of forest to high altitude environments, and deepening of valleys. Then followed a cold recurrence that is indicated by renewed periglacial activity which occurred between 11,000 and 10,300 BP (younger Dryas).

The Little Ice Age is well documented in the French Alps (Leroy-Ladurie 1967; Vivian 1975). According to data acquired in Austria, Switzerland, and Italy (Röthlisberger 1986; Deline 1999), and in the Chamonix Valley (Jaillet and Ballandras 1999), it could be the last one of a series of Holocene cold recurrences. It lasted until AD 1850, after which a general retreat was observed. However, a significant readvance of some glaciers (especially in the Mont-Blanc massif) did take place between 1960 and 1984. The evolution of some ice-tongues is well documented for the last 100 years. This is the case of the Sarennes Glacier (Grandes-Rousses), close to Alpe d'Huez (Valla 1993). The first scientific study of this glacier dates back to as early as 1905, and it has been surveyed systematically since 1948. The budget was negative, except for a few years in the 1960s and the 1970s. Since 1985, the volume of ice has decreased significantly, with a simultaneous front-retreat and lowering of the ice-surface. In the Chambeyron massif (Ubaye), the southernmost French glaciers seem to have evolved in the same way. A slight retreat of the front has resulted in the denudation of rock bars and important lowering of the ice-surface, which has become progressively debris-covered near the front (Assier 1996). It is obvious that the small glaciers might entirely disappear in the near future. Owing to their much slower response to short-term climatic changes, the larger glaciers, such as the Argentière Glacier and the Mer de Glace, in the Mont-Blanc massif, are not currently threatened (Vivian 1975; Lliboutry and Reynaud 1981).

Mode of Glaciation

In the Alps, three types of glacier were present in the last Glacial period: local *ice caps* and *cirque-glaciers*, *valley-glaciers*, and wide *piedmont glaciers* in the northern forelands. This last type was fed by the huge Rhône and Isère Glaciers. The pre-glacial relief closely controlled the glacier morphology. Only the first (Fig. 14.14) and second types are still present in the Alpine regions. According to Vivian (1975), the western Alps (including

Fig. 14.14. The Girose ice-cap seen from the south (Pelvoux massif). This dome-shaped glacier has formed on a high arched ridge between 3,050 and 3,500 m a.s.l., carved in Palaeozoic gneiss. Glacial tongues flow down the northern slope (in the foreground). To the left, the jagged granite arête of the Rateau (3,809 m) is shown.

Fig. 14.15. The Rateau Glacier (Pelvoux massif). The glacier is located on the north slope of the Rateau (3,809 m a.s.l., unseen on the picture) and below sharp ridges exceeding 3,400 m. It divides into two tongues, one on each side of a secondary ridge (2,800 m). Below, in the foreground, a huge moraine of the Little Ice Age is well preserved.

western Switzerland and Italy) had a total of 568 km^2 of ice in the 1970s, 99% of which was present between Lake Geneva and the Pelvoux massif. The largest French glacier is the 'Mer de Glace', developed between 3,900 m and 1,400 m a.s.l., on the north face of the Mont-Blanc massif (Lliboutry and Reynaud 1981). It is 12 km long and has an average thickness of 400 m. It is the fourth largest Alpine glacier, after the Swiss Aletsch, Gorner, and Fiesch Glaciers. In the French Alps, the mean altitude of the glacier snouts is 2,730 m a.s.l., being around 2,660 m on north-facing slopes and 2,900 m on south-facing slopes. The equilibrium line altitude is approximately 3,000 m. When they are entrenched below steep slopes, which furnish large volumes of debris, some valley-glaciers have a more-or-less continuous debris cover.

Glacial Landscapes

When considering the effects of past glaciations, two areas that differ completely should be noted. The forelands, which received piedmont glaciers, are characterized by morainic and fluvio-glacial systems (Mandier 1988). In contrast, the mountains are mainly characterized by erosional landscapes. While, in some places, the ice was able to overdeepen large valleys, elsewhere it was unable to remove such delicate pre-glacial rock surfaces as ripple marks (Battiau-Queney 1997). Plainly, the depth of the abrasion depended to a large extent on the proximity to the main ice-flow channels. Which of the great variety of glacial landforms was produced apparently depended on both the glacier morphology and the pre-glacial relief. Ice-scouring on plateau-like areas gave a fjell morphology, which is well represented in the Pelvoux massif and on the high surfaces of the northern Prealps. In terms of their shape, depth, and wall steepness and the presence or absence of rock-bars and moraines, the corries are extremely variable. Classic U-shaped valleys are present, especially in crystalline rocks, but many ancient glaciated valleys have a V-profile. Others have a flat floor due to post-glacial fluvial or lacustrine infilling (Fig. 14.7). Remarkably fresh moraines from the Little Ice Age are frequent in the highest areas (Fig. 14.15). In some cases, the recent retreat of the glacier-front has led to the formation of proglacial lakes.

Effects of Glaciation on the Cave-System and Karst Development

In the northern French Alps, the Isère valley constitutes the regional base-level for the surface drainage and cave-system development. During each glaciation, the Isère Glacier further incised the valley which was subsequently partially filled with alluvial and lacustrine sediments during the short interglacial periods. The valley floor was probably at 500 m a.s.l. in the early Pleistocene and 200 m below sea level during the maximum extension of the Rissian glaciation. It is now around 300 m a.s.l. in the Gresivaudan (between Albertville and Grenoble). In the northern Prealps massifs at least, caves, which are presently perched above the active systems, already existed in the Pliocene. However, the glacial periods greatly favoured the development

Fig. 14.16. The western Marinet rock glacier (in the Ubaye region). This tongue-like mass of angular debris is the most spectacular rock glacier of the French Alps. It is 750 m long and from 2,750 m to 2,530 m a.s.l. It has a series of bulging lobes at the base and a steep front. It is a periglacial landform and not a glacier covered with debris. Electrical resistivity soundings have shown the presence of a thick ice core just behind the front (photo: Michèle Evin).

of caves for two reasons: firstly, glacial entrenchment has steepened groundwater hydraulic gradients; secondly, glacial meltwaters had strong abrasive effects when they flowed through the vadose zone, owing to their high velocity and coarse clastic load. In contrast, where the underground flow was dammed with ice, the caves were then infilled. Varved clays are the most typical deposits in this type of environment. At the surface, an ice cover has considerably favoured the development of limestone pavements with a stepped topography adapted to the dip of strata. This type of glacio-karstic landscape characterizes the northern Prealps, especially the Haut-Giffre and Vercors massifs.

Rock Glaciers

This landform, which was first described in Alaska a century ago, is well represented in the southern French Alps (Assier 1996). In Ubaye, Queyras, and the Mercantour National Park, they are found between 2,200 and 2,800 m a.s.l. Rock glaciers are composed of angular rock debris and interstitial ice. They have flow-like features at the surface and a very steep snout (Fig. 14.16). Although they resemble true glaciers in shape, most are essentially periglacial landforms, associated with discontinuous permafrost. Electrical resistivity soundings have shown the presence of ice-cores in most rock glaciers above 2,600 m (Evin and Fabre 1990). Nowhere is the ice purely of glacial origin. In the case of the small rock glaciers, interstitial ice, from whatever source, is simply preserved because the annual mean temperature is < -2 °C.

Natural Hazards

Natural hazards are extremely diverse in the French Alps. Chardon (1990) proposed a genetic classification. He outlined the importance of short periods, when rapid climatic and geomorphologic changes led to a general destabilization of slopes, for example when the Würm glaciers rapidly retreated from the valleys.

Earthquake Hazards

Certain natural hazards are related to tectonics forces in mountain ranges that are still affected by plate convergence. In such cases, seismic activity is recurrent, especially at the edges of the main structural belts. Along active faults (for example in Ubaye), the intensity can reach VII–VIII on the MSK scale. Historic earthquakes are known in the foreland, especially along the main alpine cluses. The most recent one occurred in Annecy in 1996. The soft Pleistocene sediments of the Isère valley trough might possibly amplify earthquakes generated along the Belledonne fault, near Grenoble, in a manner similar to the earthquakes at Kobe in 1995 or Mexico City in 1985.

Hazards Related to Glaciers

Glacier falls are rare but are often catastrophic in their effects. The most recent event happened in August 1949, when the lower part of the Tour Glacier (Mont-Blanc massif) collapsed in a few seconds, killing six persons. The brutal release of great quantities of water, due to the sudden draining of ice- or moraine-dammed lakes, creates a cataclysmic flood. It occurred for example in the Val-de-Lans near Moucherotte, which was occupied by an ice-dammed lake (Fig. 14.11). Large quantities of sub- or intraglacial meltwater can equally be released in a few seconds. This type of event occurs every third or fourth year in the Trient Glacier (Mont-Blanc massif), generally in June and August, leading to a sudden swelling of the outlet torrent. It is seen with the Chauvet Glacier (Ubaye), the last event dating from July 1997, when an estimated volume of 94,000 m^3 of water was released in a few minutes. This led to mud and blocks covering a forest of larches and temporarily damming the Ubaye River below. Nowadays, this type of hazard is relatively well understood and can be prevented by drainage and regular monitoring of the most dangerous places.

Hazards Related to Snow Avalanches

Snow avalanches are common everywhere in the French Alps (Bravard 1990). They result from instability of snow covers, which may be caused by many factors. Two main types are observed: powder avalanches, which can produce a very dangerous wind blast, and slab avalanches, which move over a well-defined sliding plane. Those channelled in narrow tracks are less dangerous than the slab avalanches, which may start anywhere after heavy snowfall, and can be triggered by a single skier. The density, wetness, and thickness of snow are important factors in the mechanism of avalanche. Much research has been carried out in France on this topic, especially by the CEMAGREF of Grenoble (Valla 1990) and ANENA (Association Nationale pour l'Étude de la Neige et des Avalanches). Catastrophic events are often those for which monitoring has not been easy. In February 1970, at Val-d'Isère, thirty-nine people were killed by a huge avalanche. In February 1999 an avalanche in the upper Arve valley, near Chamonix, killed twelve persons and destroyed several houses. This was in a place that had previously been considered to be quite safe. In the several days preceding the movement, more than 2 m of snow had accumulated. The powdered snow rushed down the north-facing slope from 2,500 m a.s.l. at 150–200 km.p.h. It crossed the valley floor at 1,380 m a.s.l. and advanced up the opposite slope, where it demolished several chalets. Despite a progressively improving knowledge of snow behaviour and the spatial distribution of past avalanches, exceptional snowfalls can lead to unexpected events in regions of high mountainous relief.

Mass-Movement Hazards

In a high-relief mountain area, with > 1,000-m-deep valleys and steep slopes, natural hazards related to gravity-induced mass movements are common. All types of movements have occurred in the French Alps. The main processes depend upon the proportions of granular solids, water, and air. The velocity ranges from slow creeping movements of the regolith to unobstructed free-air rockfalls.

Rock falls are considerably facilitated by the state of rocks where these were intensely fractured and crushed during the Alpine compression. They are commonplace in crystalline and limestone rocks. In many places, they threaten roads and villages. Such is the case at Séchilienne, in the Romanche valley, entrenched between the Belledonne and Taillefer massifs (Chardon 1987). The valley floor lies at 300–350 m and the summits exceed 2,000 m a.s.l. The valley was entirely occupied by the Würm Romanche Glacier up to 1,200 m. On the south-facing slope, crushed and intensely fractured schist crops out on a bench at 800 m a.s.l. Open fissures are present in the rock, which forms the edge of this bench. At the base, a steep (33–35°) debris slope is regularly fed by rockfalls, especially in winter and spring. Several processes are active; frost action could be the most efficient, but pipe-flow, solifluction, and debris-flows are also active. Moreover, the river undercuts the slope base. The main road below (from Grenoble to Briançon) was first protected by a 3-m-high wall. A new road had to be constructed in 1986. Several houses had to be abandoned. The place is constantly monitored, because a rock slide could dam the valley and create a lake, which would be a direct threat for the city of Grenoble below.

Slab and toppling failures, sometimes followed by rock avalanches, are less frequent but difficult to forecast. Tensile joints formed on some cliffs immediately after the main valley-glaciers disappeared. At that time, they promoted huge failures along the deepest-valley slopes (Maurienne, Tarentaise, and Romanche valleys). In many cases, these high rock cliffs have not yet stabilized. A well-documented event that affected the northern face of Mont-Granier (Chartreuse massif) in AD 1248 probably killed about 1,000 people (Goguel and Pachoud 1972; Pachoud 1991) (Fig. 14.10).

Landslides are frequent in soft or foliated rock, such as clay, marl, shale, and schist. Slumps are widespread in the Oxfordian black marls of the Barcelonnette area (Durance valley).

Mudflows and *debris-flows* are common in the Alps (Fig. 14.17). An unusual type of debris-flow, called 'torrential lava', is one of the less easily predicted and most destructive. It is locally repetitive at 20–50-year intervals. The flow is channelled in a torrent bed. Under very intense rainfall the torrent is transformed into a muddy fluid that can move enormous blocks. The discharge

Fig. 14.17. Talus slope and debris-flow in the area of the Izoard pass. The *Casse déserte* is located 1,000 m south of the pass, in the Queyras regional nature reserve. Here, at 2,200–2,600 m a.s.l., the Triassic dolomite has been strongly weathered forming red-coloured pillars. In the background, to the right, the slope is covered with loose angular clasts. In the foreground, a debris-flow with > 0.50 m clasts in the central channel bordered by lateral levees occurs.

may be 5–10 times the volume of a 100-year-long normal streamflow discharge in the same catchment (Thouret *et al.* 1995). On 24 July 1995, a downpour that lasted less than three hours gave rise to a series of debris-flows in the Guisane valley, east of the Pelvoux massif. They caused severe damage in the tourist resort of Serre-Chevalier. The total precipitation is unknown for lack of a gauge in the area but the volume of torrential lava on the alluvial fan of a right-bank torrent has been estimated at 100,000 m^3 (Lahousse and Salvador 1998).

Stream-Flood Hazards

Catastrophic floods are due to heavy and intense precipitation, either in winter, in the northern forelands and the southern Prealps, or in summer (from May to September), in the rest of the Alps. In June 1957, exceptional floods occurred in the Ubaye and Guil valleys, due to severe downpours (> 300 mm in three days) and rapid snow-melting following a föhn event (Tricart 1959). The discharge of the Guil, usually around 10 m^3/s, rose to 1,500 m^3/s in a few hours. The torrent carried huge blocks and a great quantity of tree-trunks that temporarily dammed the flow in the narrow parts of the valley and close to the bridges. Following the sudden release of this pent-up flood-water, enormous damage was caused in the valley below. The single road and several villages were destroyed. The recurrence interval of such events is difficult to estimate. It could be one century. Recent studies in the Buech watershed, a tributary of the Durance river, show that torrential activity seems to have slowed down in the last century. Severe floods, which were frequent in the nineteenth century become rare and the present active belt of the torrents is narrower than a century ago. Similar observations were made in the Ubaye and Guisane watersheds (Lahousse and Salvador 1998). Paradoxically, this evolution increases risks. For instance, in Serre-Chevalier, new houses were built on an alluvial fan in the last thirty years. In the absence of a known historical event, this area seemed out of reach of the torrent, despite a 1993 report by the public Service de Restauration des Terrains en Montagne (RTM), which considered the level of risk to be expected. In fact it was severely devastated during the exceptional event of July 1995.

Man-Induced Hazards

Natural hazards concern all parts of the Alps and after any event with death of people and severe damage, the question of human responsibility is raised. In fact, although some hazards might be triggered by human activities (road-widening, mines, deforestation, and ski activity), processes are induced by natural factors and the greatest disasters are due to exceptional meteorological events or to an unpredictable combination of circumstances. The damaging consequences of such processes could be avoided, in some cases at least. For example, in July and August 1987, the French Alps experienced a series of floods of varying amplitude, which were exceptional in the fact that they concerned the whole Alpine Range. In some places torrential floods formed: on 14 July, twenty-three persons were killed at Grand-Bornand (northern Savoy) in a campsite unfortunately located on the floodplain of a torrent. In other cases, natural hazards are not always avoidable and people have to live with them. Landslides and toppled cliffs were certainly more massive and frequent just after deep valleys were free from ice, before people began to settle in these mountains. Nowadays, we have a better knowledge of the areas liable to the most dangerous processes. Problem stems from the development of tourist activities (*sensu lato*), when these do not take into account the potential risk.

Alpine Morphogenic Systems

The French alpine morphogenic systems are closely comparable in altitude with the various vegetational thresholds (Chardon 1984, 1989*a*). Above the snow-line, at *c.*3,000 m a.s.l. glaciers can develop and glacial processes are active. The –2 °C isotherm, at *c.*2,700–2,600 m, marks the lower limit of discontinuous permafrost. So, between these two levels, periglacial processes are active. Between 2,700–2,600 m and the tree line (at *c.*2,000 m at the 45 °N latitude), pastures are more widespread than bare rock. Nival processes are associated with frost action and runoff erosion. Below the timber line, at *c.*2,000–1,800 m, the dominant processes are fluvial, though their effectiveness relates strongly to the slope steepness and valley depth. When considering the morphological environment, altitude is merely one factor among many others: wide areas in the forelands are more elevated than the deepest valleys.

According to Chardon (1989*a*), three main landscapes and environments are present in the severely glaciated French Alps:

- The main valleys are wide and flat-floored in their lower and middle parts: they were overdeepened by glaciers, then infilled by alluvial deposits. Man (by river training, marshland drainage, etc.) has considerably transformed the landscape. Huge debris cones and slump deposits are often inherited from the immediate postglacial period and the tensile joint formation, but landslides and rock falls may still occur. In their upper reaches many valleys are narrower and deeply entrenched. Glacial landforms are widespread and well preserved. Catastrophic rock-slope failures, debris-flows, floods, snow avalanches, and sudden release of glacial meltwater may be all present here.
- The medium-sized mountain environment (Prealps) was stabilized about 10,000 years ago, when these slopes became forested. Structural and inherited glacial landforms are common and largely unaffected by modern active processes. Meanwhile, even in this relatively stable environment, mass-movements may be dangerous. Here, karstic landforms reach their maximum development.
- The high mountain environments present two types: (1) the northern type (e.g. Mont-Blanc massif), which receives heavy precipitation (up to 3,000 mm p.a.) and is still widely glaciated; (2) the southern type (Queyras, Mercantour), which receives 1,000–2,000 mm of precipitation and where bare rock suffers from severe frost action. The finest rock glaciers are found here. The Pelvoux massif forms an intermediate environment: on the north face, glaciers are widespread, but elsewhere, the landscape shows some similarities with the southern Alps. Francou (1988, 1993) has studied the conditions of frost shattering and talus slope formation in the north-eastern part of the Pelvoux massif.

In any case, it must be remembered that lithology and structure strongly control a variety of landforms and have in combination with geomorphology and climate factors created extraordinarily varied landscapes. Consequently, most major landforms are structurally controlled or inherited from glacial periods.

Acknowledgements

The author gratefully acknowledges Jacqueline Domont (University of Lille 1) who drew the original figures, Peter Walsh (University College, London) who considerably improved the first English version, and also Michèle Evin and Jean-François Billet who kindly provided photographs.

References

Assier, A. (1996), *Glaciers et glaciers rocheux de l'Ubaye*. Association Sabença de la Valeia, Barcelonnette.

Audra, P. (1994), Karsts alpins; genèse de grands réseaux souterrains. *Karstologia Mémoires* 5.

Battiau-Queney, Y. (1993), *Le Relief de la France: coupes et croquis*. Collection Géographie. Masson, Paris.

—— (1997), Preservation of palaeolandforms in glaciated areas: examples from the French western Alps. In: M. Widdowson (ed.) *Palaeosurfaces: recognition, reconstruction and palaeoenvironmental interpretation*. Geological Society Special Publication, London 120: 125–32.

Bénevent, E. (1926), *Le Climat des Alpes Françaises*. Mémorial de l'Office National Météorologique de France, Ministère des Travaux Publics, Éditions Étienne Chiron, Paris.

Bravard, Y. (1990), Les Avalanches à Chamonix. *Revue de Géographie Alpine* 78: 125–43.

Chardon, M. (1984), Montagne et haute montagne alpine: critères et limites morphologiques remarquables en haute montagne. *Revue de Géographie Alpine* 72: 213–24.

—— (1987), Géomorphologie et risques naturels: l'éboulement de Séchilienne (Isère) et ses enseignements. *Revue de Géographie Alpine* 75: 249–61.

—— (1989*a*), *Montagnes et milieux montagnards; géographie physique des montagnes*. Collection Grenoble Sciences, Grenoble University 1.

—— (1989*b*), Les Karsts de l'avant-pays alpin au nord des Alpes occidentales françaises. *Karstologia* 13: 21–32.

—— (1990), Quelques réflexions sur les catastrophes naturelles en montagne. *Revue de Géographie Alpine* 78: 193–213.

—— (1991), Approche géomorphologique des karst du gypse de la Vanoise: la zone alpine et glaciaire du vallon du Fruit-Gébroulaz (Alpes–France). *Karstologia* 17: 31–42.

Chardon, M., and Castiglioni, G-B. (1984), Géomorphologie et risques naturels dans les Alpes. In: *Les Alpes, 25ème Congrès International de Géographie, Paris-Alpes, Août 1984*, 13–41.

Debelmas, J. (1974), *Géologie de la France*. 2 vols. Éditions Doin, Paris.

—— and Rampnoux, J-P. (1994), *Guide géologique du parc national de la Vanoise. Itinéraires de découverte*. Bureau de Recherches Géologiques et Minieres, Orléans, and Parc national de la Vanoise, Chambéry.

Delannoy, J-J. (1990), Le Massif de la Chartreuse, Alpes françaises du Nord: contribution à l'étude des paysages karstiques et de l'organisation des réseaux souterrains. *Karstologia* 15: 25–40.

—— (1991), *Vercors, histoire du relief. Carte géomorphologique commentée*. Parc naturel régional du Vercors.

—— and Caillault, S. (1998), Les Apports de l'endokarst dans la reconstitution morphogénique d'un karst: exemple de l'Antre de Vénus (Vercors, France). *Karstologia* 31: 27–41.

Deline, P. (1999), Les Variations holocènes récentes du glacier du Miage (Val Veny, Val d'Aoste), *Quaternaire* 10/1: 5–13.

Dercourt, J., Gaetani, M., Vrielynck, B., Barrier, E., Biju-Duval, B., Brunet, M. F., Cadet, J-P., Crasquin, S., and Sandulescu, M., (eds.) (2000), *Atlas Peri-Tethys, Palaeo-geographical maps*. CCGM/CGMW, Paris.

ECORS-CROP Deep Seismic Sounding Group (1989), Mapping the Moho of Western Alps by wide-angle reflection seismics. *Tectonophysics* 162: 193–202.

Evin, M. (1997), *Géologie de l'Ubaye*. Association Sabença de la Valeia, Barcelonnette.

—— and Fabre, D. (1990), The distribution of permafrost in rock glaciers of the Southern Alps (France). *Geomorphology* 3: 57–71.

Francou, B. (1988), *L'Éboulisation en haute montagne: Andes et Alpes*. Editec, Caen.

—— (1993), *Hautes montagnes; passion d'explorations*. Masson, Paris.

Goguel, J., and Pachoud, A. (1972), Géologie et dynamique de l'écroulement du Mont granier dans le massif de Chartreuse en novembre 1248. *Bulletin Bureau de Recherches Géologiques et Minières* (ser. 2), III. 1: 29–38.

Gourlay, P. (1986), La Déformation du socle et des couvertures delphino-helvétiques dans la région du Mont-Blanc (Alpes occidentales). *Bulletin Société Géologique France* 8: 159–69.

Jaillet, S., and Ballandras, S. (1999), La Transition Tardiglaciaire/ Holocène à travers les fluctuations du glacier du Tour (Vallée de Chamonix, Alpes du nord françaises). *Quaternaire* 10/1: 15–23.

Jorda, M., and Delibrias, G. (1981), Données nouvelles sur le Pleistocène supérieur des Alpes Françaises du Sud: le Würm récent du bassin de la Bléone (Alpes de Haute Provence). *Bulletin Association France Études Quaternaire* 3–4: 173–82.

—— and Rosique, T. (1994), Le Tardiglaciaire des Alpes Françaises du Sud: Rythme et modalités des changements bio-morphoclimatiques. *Quaternaire* 5/3–4: 141–9.

Julian, M., and Bravard, Y. (1984), Ecosystèmes alpins et aménagement. In: *Les Alpes, 25ème Congrès International de Géographie, Paris-Alpes, Août 1984*, 55–73.

—— and Nicod, J. (1984), Un karst subalpin méditerranéen: la région d'Audibergue-Mons, Alpes maritimes et Var. *Karstologia* 3: 52–8.

Lahousse, P., and Salvador, P-G. (1998), La Crue torrentielle du Bez (Hautes-Alpes, Briançonnais), 24 juillet 1995. *Geodinamica Acta* 11: 163–70.

Lemoine, M., and Tricart, P. (1988), *Queyras: un océan il y a 150 millions d'années*. Bureau de Recherches Géologiques et Minières, Orléans, et Parc naturel régional du Queyras, Guillestre.

—— de Graciansky, P-C., and Tricart, P. (2000), *De l'océan à la chaîne de montagnes tectonique des plaques dans les Alpes*. Société Géologique France, Collection Géosciences.

Leroy-Ladurie, E. (1967), *Histoire du Climat depuis l'An Mil*. Flammarion, Paris.

Lips, B., Gresse, A., Delamette, M., and Maire, R. (1993), Le Gouffre Jean-Bernard (–1602 m, Haute-Savoie, Fr.). Ecoulements souterrains et formation du réseau, *Karstologia* 21: 1–14.

Lismonde, B., and Delannoy, J-J. (1990), Le Massif de la Chartreuse, Alpes françaises du Nord: contribution à l'étude des paysages karstiques et de l'organisation des réseaux souterrains. *Karstologia* 15: 25–40.

Lliboutry, L., and Reynaud, L. (1981), Global dynamics of a temperate valley glacier, Mer de Glace, and past velocities deduced from Forbes bands. *Journal of Glaciology* 27/96: 207–26.

Maire, R. (1990), La Haute Montagne calcaire. *Karstologia Mémoires* 3.

Mandier, P. (1988), *Le Relief de la moyenne vallée du Rhône au Tertiaire et au Quaternaire*. 3 vols. Documents Bureau de Recherches Géologiques et Minières 151.

Monjuvent, G., and Nicoud, G. (1988), Modalités et chronologie de la déglaciation würmienne dans l'arc alpin occidental et les massifs français: synthèse et réflexions. *Bulletin Association France Études Quaternaire* 2/3: 147–56.

Nicod, J. (1984), Les Massifs karstiques des Alpes occidentales: trame structurale et bioclimatique. *Karstologia* 3: 3–11.

—— (1988), Le Beaufortin oriental. Carte géomorphologique au 1/25000ème. *Revue de Géographie Alpine* 76: 121–46.

Nicod, J. (1993), Recherches nouvelles sur les karsts des gypses et des évaporites associées. *Karstologia* 21: 15–30.

Ozenda, P. (1985), *La Végétation de la chaîne alpine dans l'espace montagnard européen*. Masson, Paris.

Pachoud, A. (1991), Une catastrophe naturelle majeure: l'écroulement du Mont Granier dans le massif de la Chartreuse au XIIIème siècle. *La Houille Blanche* 5: 327–32.

Ricou, L-E. (1994), Tethys reconstructed: plates, continental fragments and their boundaries since 260 Ma from Central America to South-eastern Asia. *Geodinamica Acta* 7/4: 169–218.

Rosique, T. (1997), Détritisme et morphogenèse à la fin du Würm dans les Alpes françaises méridionales (Moyenne Durance). *Quaternaire* 8/1: 39–48.

Röthlisberger, F. (1986), *10 000 Jähre Gletschergeschichte der Erde*. Sauerländer, Aarau.

Thouret, J-C., Vivian, H., and Fabre, D. (1995), Instabilité morphodynamique d'un bassin-versant alpin et simulation d'une crise érosive (l'Église-Arc 1800, Tarentaise). *Bulletin Société Géologique France* 166/5: 587–600.

Tricart, J. (1959), Mécanismes normaux et phénomènes catastrophiques dans l'évolution des versants du bassin du Guil (Hautes-Alpes, France). *Zeitschrift für Geomorphologie* 3: 227–301.

—— Schwartz, S., Sue, C., Poupeau, G., and Lardeaux, J-M. (2001), La Dénudation tectonique de la zone ultradauphinoise et l'inversion du front briançonnais au sud-est du Pelvoux (Alpes occidentales): une dynamique miocène à actuelle. *Bulletin Société Géologique France* 172: 49–58.

Valla, F. (1990), Accidents d'avalanches dans les Alpes (1975–1989). *Revue de Géographie Alpine* 78: 145–55.

—— (1993), Sarennes: un glacier sous haute surveillance (Massif de l'Oisans—France). *Revue de Géographie Alpine* 3: 33–49.

Veyret, P., and Veyret, G. (1979), *Les Alpes françaises. Atlas et géographie de la France moderne*. Flammarion, Paris.

Vivian, R. (1975), *Les Glaciers des Alpes occidentales; étude géographique*. Allier, Grenoble.

Environment and Human Impact

15 Climate: Mean State, Variability, and Change

Cor Schuurmans

Introduction

This chapter gives a description of the main characteristics of present-day climate. In describing the mean state and its variability, attention is also given to the underlying causes. For comparison, there is a short summary of early European climate, from the last glacial maximum, through the Holocene and up to the Little Ice Age (the period AD 1400–1850). The chapter finishes with a comprehensive section on climate change, with emphasis on the anthropogenic causes of recent changes.

Mean State of Present-Day Climate

The climate of western Europe has a maritime character. The weather mainly originates from the North Atlantic Ocean and its neighbouring seas. Further inland, in what is usually called central Europe, climate changes to a more continental type, but certain maritime features are still present. It is therefore called an altered maritime climate. Only in the most southern part, southern France for instance, is the Atlantic character lost and several new features are present. These features are characteristic of a Mediterranean climate.

Main Features

Climates may be called cold or warm, dry or wet, gloomy or sunny, depending on the prevailing temperatures, amount and frequency of precipitation, and the number of hours of bright sunshine. Such terms, however, are not objective unless certain, generally accepted, reference values are used. In the past different sets of reference values were proposed, each of them defining a system of climatic types. A well-known classification system was the one developed by Köppen (1936). The Köppen system distinguished eleven main climate types, based on well-defined temperature and precipitation characteristics. These were mainly referring to the response of vegetation, natural as well as cultivated, to climatic conditions. The eleven Köppen climates are indicated by the letters A–E, with some subdivision, using other letters. In the Köppen classification the whole of western Europe has a Cf climate, which means a moist, temperate climate, without a specific dry season. Cf climates occupy 22% of the globe (oceans included).

A second method to describe climate is by using the well-known definition of climate as being the average weather conditions in a certain area, over a given period of time. In practice, however, there is no direct information about weather conditions. Separate observations of temperature, precipitation, sunshine, etc. are available. For each of these quantities frequency distributions can be made, but this does not mean that a combination of prevailing temperature, precipitation, sunshine values, etc. makes sense. A climatology of weather types, based on a combination of simultaneously observed temperature, humidity, etc., is useful, but not generally available. Consequently, we are limited to the presentation of an example (Schuurmans and Krijnen 1971). In this study, for the station De Bilt (The Netherlands) weather types have been defined on the basis of a combination of daily mean temperature, amount of precipitation, and duration of sunshine. For each of the three elements, three classes are considered, resulting in twenty-seven different weather types. In Table 15.1 the climatology of warm (A), dry (D), and sunny (S) days is shown for the period 1881–1970. ADS-days are defined as: daily mean temperature in the highest $33^1/_3$% class

Table 15.1. *Mean frequency of occurrence of warm, dry and sunny days at De Bilt, The Netherlands (% of the number of seasonal days)*

	Winter (djf)	Spring (mam)	Summer (jja)	Autumn (son)
1881–1970	2	13	18	7
1991–2000	3	17	22	7

Sources: First row, Schuurmans and Krijnen (1971); second row, KNMI.

of the seasonally varying frequency distribution, no precipitation, and the duration in hours of sunshine for at least 50% of the day. From this table it is clear that ADS-days are a weather type of spring and summer, which is due to the fact that warm/mild days in autumn and winter are usually cloudy and wet. The table also contains the seasonal percentages of ADS-days for the recent decade 1991–2000. Although the numbers for the recent decade are somewhat higher than for the 1881–1970 period, indicating somewhat better weather conditions than earlier in the twentieth century, one may conclude that the ADS-climatology developed for the period 1881–1970 gives a reliable indication of the seasonal number of days of this weather type. Unfortunately, a generally accepted definition of weather types is still not available.

The most common method of climate description is the classical way of presenting maps of the geographical distributions of average temperature, precipitation, etc., separately. The period of averaging may differ, but a minimum period of 30 years (adopted by the World Meteorological Organization) is needed to account for the sometimes large differences in weather from one

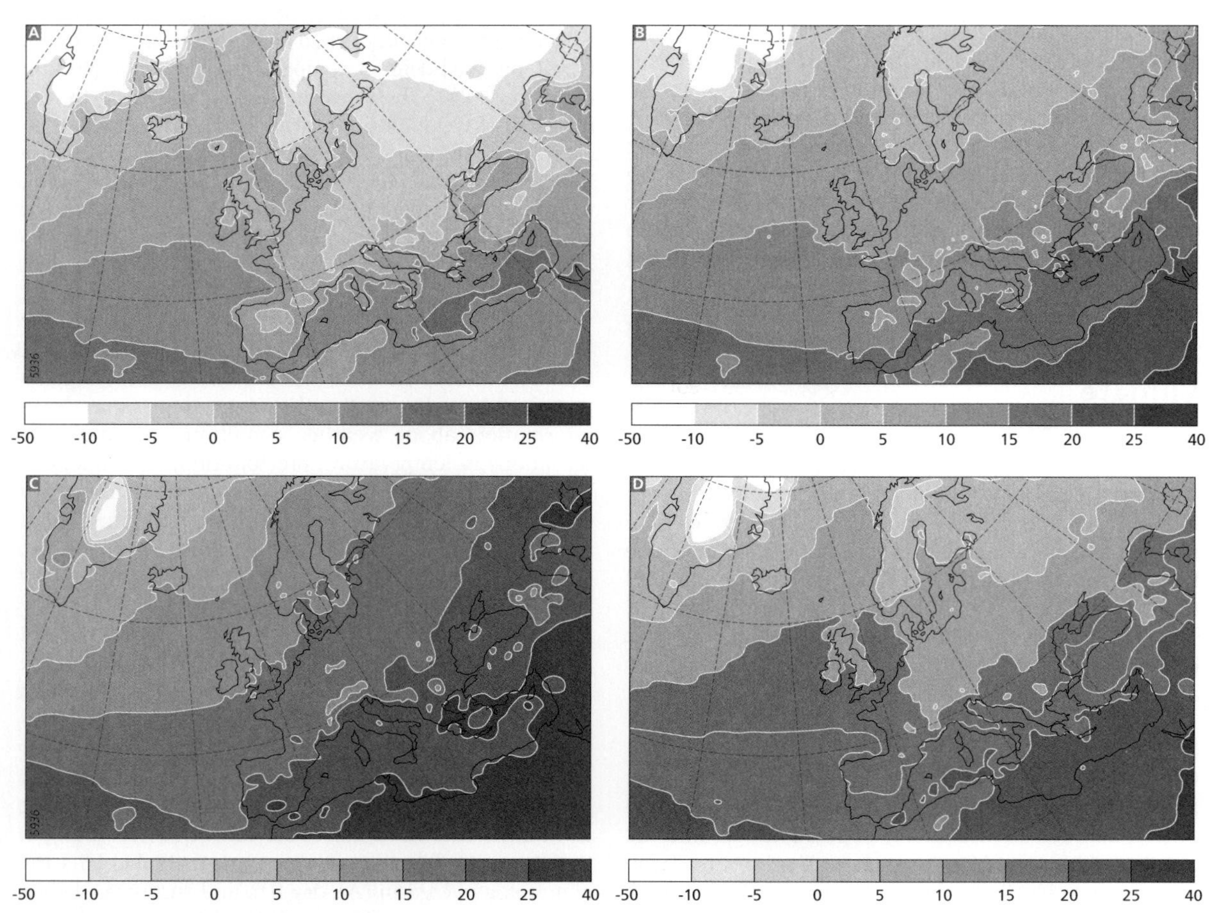

Fig. 15.1 (A–D). Mean surface air temperature in °C (1951–80) for winter (A), spring (B), summer (C), and autumn (D) (*Source*: European Climate Assessment 1995; data from Legates and Willmott 1990*a*).

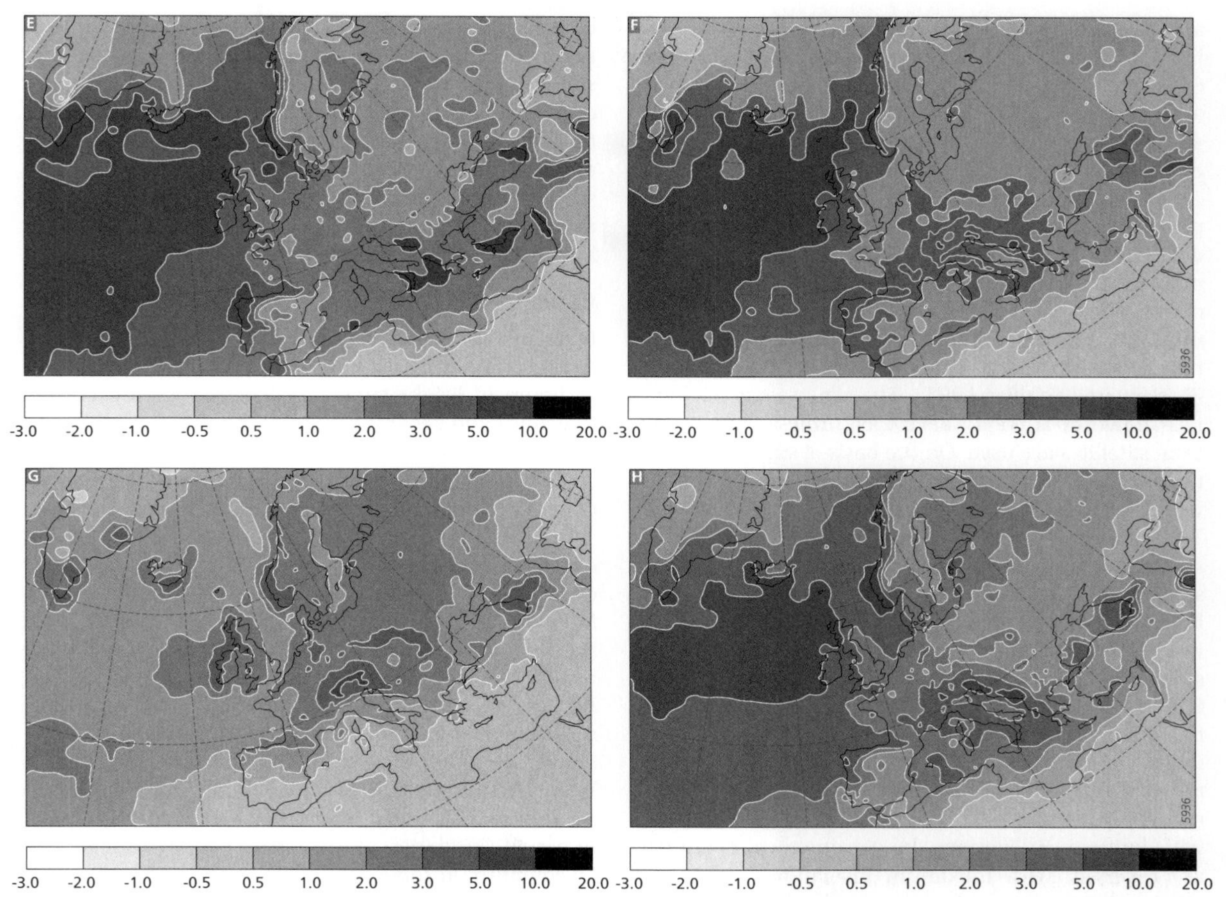

Fig. 15.1 (E–H). Mean precipitation in mm/day (1951–80) for winter (E), spring (F), summer (G) and autumn (H) (*Source*: European Climate Assessment 1995; data from Legates and Wilmott 1990*b*).

year to the next, especially in the temperate and high latitudes.

Seasonal maps of average temperature in western Europe and adjacent areas are shown in Fig. 15.1 (A–D). In winter (December, January, February) mean temperatures in western Europe are generally between 0 and 5 °C. Only along the Atlantic seaboard and the northern shores of the Mediterranean mean temperatures a few degrees above 5 °C are found. Inland areas in Germany and the French Alps have winter mean temperatures below 0 °C. In summer (June, July, August) practically the whole area of western Europe has mean temperatures between 15 and 20 °C. Mean temperatures of 20 to 25 °C, or even a little bit higher, are found only in parts of southern France, especially the areas bordering the Mediterranean. Spring (March, April, May) and autumn (September, October, November) show a quite similar temperature distribution, with mean temperatures between 5 and 10 °C in the north-eastern part and between 10 and 15 °C in the south-western part of western Europe. In autumn the borderline between these zones is shifted a little bit to the north-east, due to the higher sea temperatures in autumn as compared to spring.

Seasonal amounts of precipitation are shown in Fig. 15.1 (E–H). Areal variability over western Europe is much larger than for temperature, reflecting the strong influence of topography in relation to prevailing winds and oceanic moisture sources. This is especially the case in winter, where values differ between 1–2 and 3–5 mm/day over relatively small distances. In summer most of the area shows a mean intensity of 2–3 mm/day, with lower values of 1–2 mm/day, or even less in south-western France and particularly the Mediterranean sea shores. Higher values of 3–5 mm/day in summer are found in Alpine regions of Germany and France. As with temperature, the precipitation patterns of spring and autumn are remarkably similar, with autumn being by far the wetter of the two seasons. For a large part of western Europe spring is the driest season of the year.

Although the differences in mean temperature and precipitation in western Europe are quite pronounced, differences in sunshine and cloudiness are perhaps more striking. Western Europe, being located in the zone of westerlies is not blessed with a very sunny climate. On the contrary, mean cloudiness may exceed 70% in certain (northern) parts in summer as well as in winter (Rossow and Schiffer 1991). The climatology of cloudiness however is poorly developed, unless observations from satellites are used. On the basis of surface observations it is better to use data on sunshine duration, which of course is highly correlated with cloudiness. The pattern of annual mean sunshine duration (not shown here) exhibits an increase in the number of sunshine hours from about 1,700 in Denmark to over 2,700 hours in south-eastern France. However, in some of the more inland areas of western Europe the number of sunshine hours is even less than the 1,700 in the north. Generally, the coastal areas have more sunshine than the inland zone immediately next to it, while further inland it might be sunnier again. In winter the gradient north–south is stronger than in the annual mean, mainly because of the strong shortening of daytime with latitude. In summer the astronomical conditions are reversed, with maximum insolation at northern latitudes. However, this does not lead to a reversed gradient. Even in summer the southern part of western Europe is, with about 1,000 hours of sunshine, about twice as sunny as some of the inland areas and the countries bordering the North Sea.

In relation to cloudiness and precipitation atmospheric humidity has to be considered, which is climatologically an almost neglected element. Seasonal mean maps (1961–90) of atmospheric water vapour pressure (in hPa) for Europe are, however, given in Schönwiese *et al.* (1993). When those values and those of the average air temperature are combined the more common relative humidity (in %) can be computed. It turns out that average relative humidity in winter is between 85 and 90%, in spring between 70 and 80%, in summer between 60 and 70%, and in autumn between 75 and 90%. The lowest values for each season refer to the southern part of the area, the highest values to the northern part.

Finally, the wind climate must be considered. Wind directions in autumn and winter are most frequently between south and west in the northern half of the area and between west and north in the southern part. In spring and summer west to north winds prevail over most of the area, except south-eastern France, where winds in this season often have an easterly component. Annual mean wind velocities are generally between 3 and 7 m/sec, with the higher values in the coastal regions and the autumn/winter season, and the lowest values further inland and in the spring/summer season.

Causes

In discussing the physical causes of the present state (mean temperature, yearly amount of precipitation, number of hours of bright sunshine, average windspeed and direction) and geographical distribution of climates one should distinguish between global, regional, and local factors. Global factors are the geographical distribution of net radiation and the general circulation of the atmosphere and ocean. Regional factors are the distribution of land and sea, ocean currents, the vegetation cover on land, and the presence of large areas of snow and ice. More locally, the height and nature of the surface and the level of urbanization play a role. Of course, these factors are not acting independently of each other. Rather, it is their complex interplay that eventually results in the observed climate. As such, the climate system is too complex to explain fully the observed climate. However, our insight into the many interactive processes within the climate system is sufficiently advanced to permit simulation of the earth's climate numerically by computer in some detail.

The earth's net radiative balance is at the basis of the development of global climate. It consists of three components, namely the incoming solar radiation, the solar radiation reflected by the earth (surface and clouds), and the outgoing infrared radiation of the earth. Figure 15.2A, B, C shows the net radiative balance at the top of the atmosphere, as annual mean values and for winter and summer conditions separately, as observed by satellite.

On a global scale the net radiative balance decreases from the equator, where it is positive, to the poles, where it is negative. In winter, but also with the annual mean, the whole area of western Europe has a negative radiative balance, which means that heat from elsewhere has to be imported to keep us warm. Only in summer is the net radiative balance slightly positive (of the order of 50 W/m^2, being the difference between some 320 W/m^2 absorbed solar radiation and about 270 W/m^2 outgoing infrared). Figure 15.2A shows that in first approximation the decrease with latitude is the same at all geographical longitudes, except for the

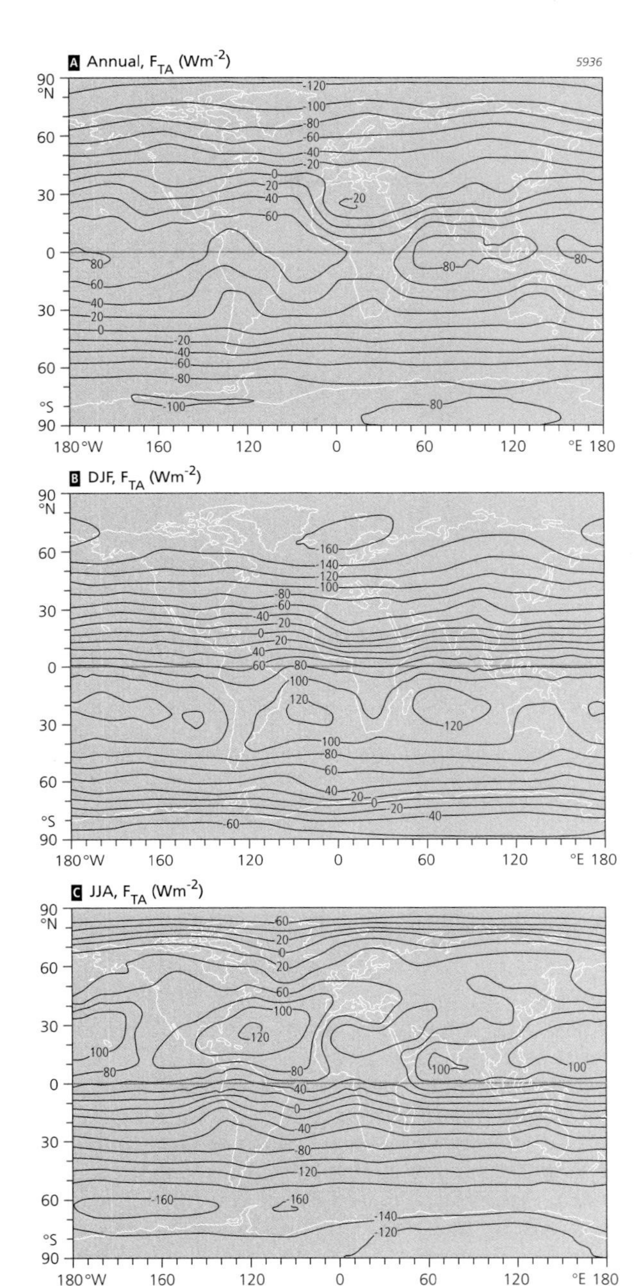

Fig. 15.2. Global distributions of the net incoming radiation F_{TA} (in W/m^2) at the top of the atmosphere, for annual (A), northern winter (B), and northern summer (C) mean conditions (after Peixoto and Oort 1992).

strange anomaly over the Sahara which has to do with the large reflection of solar radiation, combined with the large flux of outgoing long-wave radiation. The net radiative balance on the planet with an atmosphere at rest gives rise to a large temperature gradient between the tropics and the poles. This gradient causes a meridional pressure gradient in the atmosphere, which will start atmospheric motion. In principle this motion could consist of a large meridional overturning transporting warm air from the tropics to the poles and cold air from the poles in the direction of the equator. For reasons well understood, this simple solution is not what we observe on our planet. Instead of one meridional circulation cell we observe a three-cell structure for each hemisphere, in which the middle cell moves in an opposite direction to the neighbouring two. Moreover, the middle and polar cell are extremely weak compared with the tropical, or so-called Hadley cell. This means that in the extra-tropics the zonally symmetric structure of the atmospheric circulation is lost. Only in the Hadley cell, with rising air in the Intertropical Convergence Zone and descending air in the subtropics, connected at the surface by the well-known trade winds, does the meridional, quasi-zonally symmetric circulation play a role in the required poleward transport of heat.

Outside the tropics the poleward heat transport is dominated by a regime of wave-like horizontal motions, embedded in a strong westerly current, known as the jet stream. Figure 15.3 shows the zonal mean location and strength of the jet stream at our latitudes, for the annual mean and in the winter and summer seasons. Western Europe over its whole latitudinal extent is located in this so called zone of westerlies. In this zone, wave motions, consisting of areas of low and high pressure, on average move to the east. The low pressure areas often contain smaller-scale low pressure systems known as depressions or cyclones. These are vital in our weather development bringing most of the rain, especially in winter. Due to their cloudiness and strong winds they are really the 'bad weather' systems.

Weather maps of the Atlantic-European area show that these extra-tropical cyclones or storms do not move exactly west–east along latitude circles. Rather, they follow more a SW–NE course in the direction of Scandinavia. Indeed, the average storm track over the North Atlantic Ocean is from an area near Florida, then passing Iceland, before heading to the north of Scandinavia. However, very large deviations from this average track occur, in such a way that the whole of western Europe from time to time may be visited by Atlantic storms, especially in winter. The reason for the SW–NE course of the Atlantic storm track is the existence of a pattern

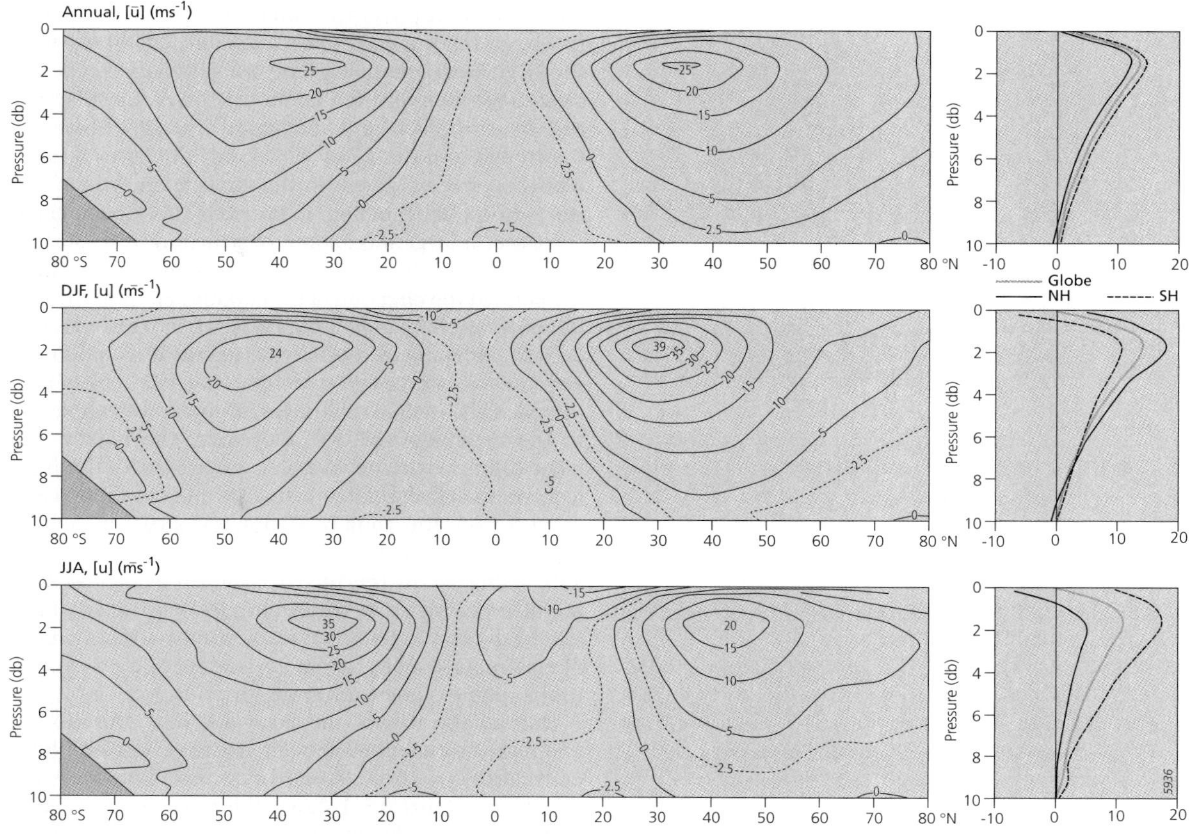

Fig. 15.3. Zonal mean cross-sections of the zonal wind component u (in m/s) for annual, northern winter (DJF) and northern summer (JJA) mean conditions. Vertical profiles of the hemispheric and global mean values are shown at the right (in m/s) (after Peixoto and Oort 1992).

of almost stationary planetary waves. These waves are of a larger scale than the eastward moving waves and are due to the large mountain barriers such as the Rocky Mountains, the Urals, and the Himalayas, but partly also to land–ocean distribution in the Northern Hemisphere. The planetary wave pattern has a 3-wave structure with troughs over Canada, East Russia, and East Asia, and with ridges in between (Fig. 15.4). Troughs are filled with relatively cold air and the ridges with relatively warm air. The eastern Atlantic and western Europe happen to be located in a ridge area of the planetary wave system, which is one of the main causes of the relatively mild climate of western Europe. The second main cause is the transport of warm water by the Gulf Stream. The Gulf Stream is part of the thermohaline circulation of the ocean, which is primarily driven by temperature and salinity gradients. Over the North Atlantic it is also driven by the westerly winds in the atmosphere. These winds have a south-westerly component, due to the northward branch of the Atlantic ridge in the planetary wave. On the other hand the long planetary wave over the Atlantic is partly shaped by the distribution of ocean surface temperatures. On the climatological timescale atmosphere and ocean may be considered as a coupled system. A third factor that enhances temperatures, at least in the winter half year, is the westerly wind itself. This wind brings air masses towards the continent which have their source area over the relatively warm oceans. In the summer half year however, these air masses originate from an ocean that is relatively cool compared to the continent, causing a moderating effect on summer temperatures. The combination of these factors is responsible for the large difference between temperatures in western Europe and those of the eastern side of the Eurasian continent, at the same latitude and height above sea level, especially in winter. For instance, Brest

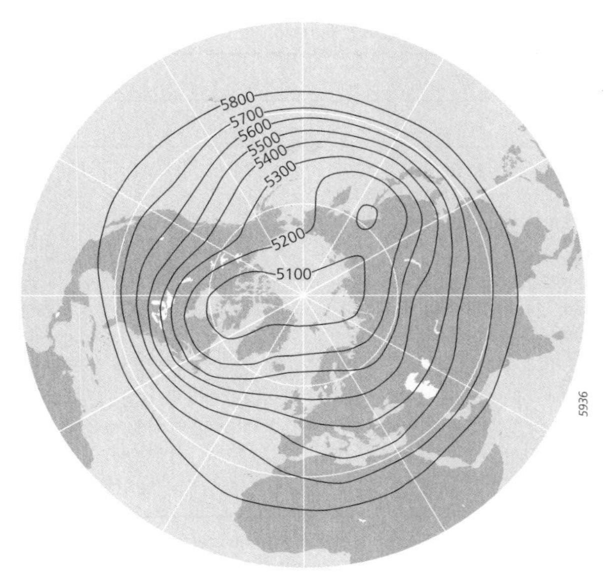

Fig. 15.4. Mean geopotential height (m) of the 500 hPa level for January, 1949–98 (after data from NCEP-Reanalysis).

(France) and Kholmsk (Russia), both at about 48 °N, have mean winter temperatures of respectively +6 °C and −8 °C.

Regional and Local Characteristics

The influence of the Atlantic ocean gradually diminishes inland. The inland climate is therefore called an altered maritime climate. Table 15.2 gives an idea of the rate of decrease of the maritime character of temperature with distance from the Atlantic coast. These data suggest that the temperature change with distance from the sea is rather slow, but this might be quite different for other climate elements. For instance, wind speed, which is higher over sea than over land, decreases very rapidly going inland. Gradients are very large over the first 10–20 km from the coast, but wind speeds do not change so much at larger distances. Also sunshine in certain parts of the year may show a strong decrease over the first tens of kilometres inland from the coast.

Herewith, we have entered the domain of the local climatic factors. Small changes in height or other surface characteristics may change the general tendency of the region. This may be noted at the scale of some tens of kilometres and continues to be of importance at much smaller scales, down to specific features of the landscape (dunes, small lakes, forests) and built-up areas. The climates at the individual stations, to be discussed later on in this subsection, should only be marginally influenced by such local factors, except for height. The height above sea level has a strong influence on temperature (on average, temperature drops with height by nearly 1 °C per 100 metres) and even small hills may have an influence on the amount and kind of precipitation. This is the reason that in climatic tables the height of stations must be given. Mountain areas, of course, in addition to local effects, also have a regional to near-global effect, depending on their scale and orientation to the airflow. The Alps for instance do have a strong regional effect on European climate, but this effect is mainly downwind, which means that western Europe is less affected by it. However, since the Alps may be considered as 'the water tower of Europe', there is in any case a strong influence through north-westward flowing rivers such as the Rhine.

To illustrate characteristic regional differences of western European climate specific data on temperature, precipitation, and sunshine for a limited number of places are collected (Table 15.3). From these data the following conclusions may be drawn.

Temperature

In addition to the seasonal averages given in Fig. 15.1A, B, C, D, which mainly show a latitudinal temperature gradient and a seasonally varying land–sea contrast, the station data, Table 15.3(*b*) gives some indication of a lowering of temperature with height. More generally the data on minimum and maximum temperatures show that in winter minimum temperatures over the area differ much more than in summer, while for maximum temperatures the opposite is true. This can be explained by differences in local radiational effects: in winter due to outgoing long-wave radiation, in summer by incoming solar radiation. In the long run, however, local effects on outgoing long-wave radiation remain, while those of incoming solar radiation partly cancel out. This is probably the reason why the absolute T_n values differ

Table 15.2. *Mean temperatures in July and January (1931–1960) at stations along 52 °N, going inland from Ireland into Poland*

	Mean T July	Mean T January
Valentia, Ireland	+15.0	+7.0
Felixstowe, England	+17.2	+3.9
De Bilt, The Netherlands	+17.0	+1.7
Hannover, Germany	+17.4	+0.1
Potsdam, Germany	+18.4	−1.1
Poznan, Poland	+18.8	−2.2
Warsaw, Poland	+19.2	−3.5

Source: Müller (1987).

TABLE 15.3(*a*). *Climate data of some selected stations in western Europe*

	January					July				
	T_x	T_n	RR	RD	SS	T_x	T_n	RR	RD	SS
Copenhagen	2.0	−2.0	49	17	36	21.8	13.6	71	14	239
Hamburg	2.3	−2.6	57	18	51	22.2	12.3	82	17	220
De Bilt	4.3	−0.8	68	21	56	22.1	12.7	77	17	199
Berlin	1.7	−3.5	44	17	56	23.8	13.3	71	14	242
Brussels	4.3	−1.2	83	23	54	22.7	12.1	97	17	198
Cologne	4.3	−2.2	51	17	47	23.4	12.4	75	15	183
Dresden	1.8	−3.6	38	16	61	23.6	13.4	109	15	227
Luxemburg	2.5	−1.4	73	19	50	22.9	13.0	66	15	225
Brest	8.6	3.7	133	22	66	19.4	12.0	62	14	210
Paris	6.0	0.9	54	17	64	24.6	14.5	55	12	231
Munich	1.1	−5.7	59	16	65	22.9	12.1	140	16	232
Bordeaux	9.2	1.7	90	16	81	25.4	13.7	56	11	262
Grenoble	5.5	−1.5	80	14	80	26.4	13.8	70	10	291
Toulouse	8.5	0.8	49	14	80	26.8	14.7	44	9	262
Marseilles	10.0	1.5	43	8	134	28.9	17.1	11	2	368

Note: T_x = mean maximum temperature in °C; T_n = mean minimum temperature in °C; RR = amount of precipitation in mm; RD = number of days per month with precipitation; SS = number of sunshine hours. Data refer to the 30-year period 1931–60; in some cases less than the maximum number of years is included.

Source: Müller (1987).

TABLE 15.3(*b*). *Climate data of some selected stations in western Europe*

	Height (m)	Absolute T_x	Absolute T_n	Highest RR in 24 hours	Dryest month	Sunniest month
Copenhagen	9	32.7	−24.2	77	March	May
Hamburg	14	35.7	−29.1	68	March	June
De Bilt	3	36.8	−24.8	74	March	June
Berlin	51	37.8	−26.0	125	April	June
Brussels	100	38.8	−18.7	75	May	May
Cologne	68	36.9	−22.9	—	March	June
Dresden	246	36.8	−27.0	114	February	June
Luxemburg	334	36.8	−19.6	67	March	July
Brest	98	35.2	−14.0	57	June	May
Paris	52	39.6	−17.0	56	March	June
Munich	527	35.2	−29.6	155	March	July
Bordeaux	47	38.6	−15.2	54	April	July
Grenoble	223	39.4	−20.0	135	March	July
Toulouse	151	40.2	−19.2	78	June	July
Marseilles	3	39.0	−16.8	86	July	July

Note: Symbols as in table 15.3(*a*). Data refer to the 30-year period 1931–60, but in some cases they are based on much longer (absolute temperatures) or shorter (24-hour precipitation) records.

Source: Müller (1987).

more strongly than the absolute values of T_x (see Table 15.3(*b*)).

Precipitation

The seasonal patterns in Fig. 15.1E, F, G, H already do show a clear height effect on precipitation. This is further illustrated by the station data. See, for example, the average precipitation amounts (Table 15.3(*a*)) and the highest 24-hour values (Table 15.3(*b*)). More detailed studies, however, reveal that upland areas do not always have more precipitation than low-lying parts. The orientation of the orographic features with respect to the prevailing winds is of major importance. Also the location of a mountain range with respect to a neighbouring range has an effect. An example of this is the southern Schwarzwald in Germany, which receives less rain than the lower northern part of the Schwarzwald. This is presumably because of the influence of the Vosges, located west of the southern Schwarzwald, causing a so-called rain shadow effect. Table 15.3(*a*) also gives the number of rainy days in January and July. In winter as well as in summer these numbers are quite high, especially in the northern half of western Europe. To a certain extent these large numbers are somewhat misleading, however. It is true that many places have a probability of a rainy day of more than 50% in all seasons, but one has to realize that nearly all days with precipitation are for the most part dry. In De Bilt (The Netherlands), for instance, where the probability of a rainy day is 52%, it only rains 7% of the time. In other words, for 93% of time it is dry weather. Finally, data on the driest month (Table 15.3(*b*)) indicate that spring is the driest season, apart from areas in southern France which are under the influence of the Mediterranean. These tendencies for dryness are mainly caused by anticyclonic developments, which in spring are strong over the North Sea and adjacent areas, and in summer over the Mediterranean.

Sunshine

The number of hours of bright sunshine in January and July (Table 15.3(*a*)) are a further illustration of the variability in the geographical distribution of sunshine. From the data on the sunniest month (Table 15.3(*b*)) an additional conclusion may be drawn, namely that the seasonal variation of cloudiness has some influence on the seasonal variation of sunshine hours, although it is clearly less important than the seasonal change in the length of the day. The role of cloudiness in determining the number of sunshine hours is in fact a complicated one. In the winter season the mornings are usually the most cloudy parts of the day, while in summer the afternoons are cloudiest. In mountain regions we find a similar effect in such a way that in winter mountain tops are in bright sunshine, while the valleys are cloud-covered. In summer the opposite is mostly true.

Not included in Table 15.3 is data on the wind and in particular on the wind speed and storminess. As far as storminess is concerned, this will be discussed in the next section. Data on wind speed are presented in much detail in the *European Wind Atlas* (Troen and Lundtang Petersen 1989). In this atlas emphasis is on the speed and power of the wind at 50 m above ground level. The major areas of high wind energy resources in Europe include: the British Isles, Denmark, and the coastal areas of Northern Germany, The Netherlands, Belgium, and north-western France. France in addition has some local wind systems of high intensity, such as the Mistral between the Alps and the Massif Central, and the Tramontana, north of the Pyrenees.

Climate Variability

Climate varies on all timescales and is also intrinsically variable. This means that if the main driving force of climate, solar radiation, would be exactly constant (except for its annual variation), climate would still show differences from year to year. In other words, the climate system is of such a complex nature that under the same stimulus it will never produce exactly the same response. Intrinsic variability is part of what is called natural variability of climate. This includes all climate variability due to changing climate factors, internal or external to the system, except anthropogenic causes. Internal causes of climate variability are numerous: changes in soil moisture content in continental areas, anomalies of sea surface temperature, differences in extent of sea ice, to name a few. External factors are mainly variations in solar radiation, either of a direct nature or due to enhanced absorption and reflection of sunlight as, for example by volcanic dust. Internal and external climate variability are strongly interlinked. In practice they are hard to separate, since their manifestations in the atmosphere–ocean system may be of exactly the same nature, taking the form of intensity changes or phase changes of the natural oscillations of the system, such as ENSO (El Niño Southern Oscillation) and NAO (North Atlantic Oscillation).

Observed Climate Variability

Everywhere we observe differences in climatic elements (temperature, rain, sunshine) from year to year without being able to attribute these differences to a specific cause. Nevertheless, for practical purposes, these differences have to be included in the definition and description of climate. Usually this is done by computing the standard deviation of the climatic element concerned. Especially when the element is almost normally (Gaussian) distributed, as is temperature, then the standard deviation is a practical measure of variability. For instance, when we know the standard deviation σ of the yearly averages (over a certain period of, say, 30 years), for N-year averages the standard deviation is equal to σ divided by $\sqrt{N}$. So, when locally the standard deviation of annual mean temperature is 1 °C, for decadal means, in first approximation, it will amount to 0.32 and for 30-year averages to 0.18 °C. Sometimes, instead of the mean and the standard deviation, the full frequency

TABLE 15.4. *Number of hours of bright sunshine at De Bilt, The Netherlands*

	Jan.	Feb.	Mar.	Apr.	May	June	July	Aug.	Sept.	Oct.	Nov.	Dec.
Maximum possible	248	280	372	420	496	510	496	434	360	310	240	217
Average 1961–90	47	73	107	152	197	192	187	185	134	103	55	43
Maximum since 1849	109	135	195	259	331	301	307	313	240	189	113	75
Minimum since 1849	19	26	46	41	104	93	106	105	68	34	18	13

Source: KNMI.

distribution is given. This makes it possible also to compute other indicators such as the median value or the value that is exceeded only in a given percentage of years. For certain applications the range and the extreme values are important. See for instance Table 15.4 for sunshine at De Bilt (The Netherlands). In this table the theoretically possible maximum number of hours of sunshine for each month is given. In combination with the observed maximum and minimum monthly values over the last 150 years and a 30-year average value for each month, a relatively complete picture can be obtained of the mean state and variability of the sunshine climate at that particular location.

Apart from the variability indicators mentioned above, there is range of phenomena, belonging to weather and climate, that is not fully covered by the usual temperature, precipitation, sunshine, and wind data. These include heat and cold waves, heavy precipitation occurrences, snowfall, lightning, droughts, severe storms, and tornados. Sometimes these phenomena are also called extremes, but they are of a different character than the extreme values belonging to the frequency distributions of temperature, precipitation, etc. Unfortunately, a climatology of these special phenomena, in the sense of average occurrences for a given period (of, say, 30 years), is in most cases not available. Such information can be derived only from large data sets of daily observations. This can be illustrated by occurrence of severe storms in The Netherlands since 1911. Severe storms can be defined as having windspeeds of Beaufort scale 10 or larger (> 89 km/h). At sea such storms cause very big waves and the water surface appears white. On land trees get uprooted and houses may be seriously damaged. The number of such storms observed since 1911 is about thirty. This means on average one severe storm every three years: fortunately, a quite rare phenomenon. Severe storms, however, are not uniformly distributed in time. Some years had more than one severe storm (1928, 1943, 1944, 1983, and 1990), while on the other hand the storm-free periods varied from zero to about ten years (1929–38). It is clear that from such a time series it is difficult, if at all possible, to say if storminess is increasing or decreasing. One aspect we understand is the occurrence of multiple storms in certain years. This invariably is a result of persistence or recurrence of some specific circulation type over the North Atlantic, which itself is probably caused by a certain configuration of the boundary conditions (distribution of sea surface temperature, limit of sea ice, etc.). It might be possible in future to predict the probability of occurrence of such circulation types, and thus the probability of occurrence of severe storms some months in advance.

Any future prediction of anomalous seasons or years most probably will be (partly) based on climate models. For a given or predicted development of the boundary conditions, such models are able to compute the accompanying development of the atmospheric circulation. In order to avoid the prescription of oceanic boundary conditions, coupled models of atmosphere and ocean must be used. Such models are already in existence and their validity is tested in simulations of observed climate variations and interannual variability. State of the art coupled models at present are able to simulate the main features of the interannual variability. These main characteristics are: (1) variability generally is higher over the continents than over the oceans; (2) at mid-latitudes (western Europe) standard deviations of annual mean temperature are of the order of 1 °C or less and are smaller in summer than in winter; (3) variability of temperature is unrelated to the level of temperature except in winter at high latitudes, where it is negatively correlated.

On this last basis one would expect that climatic warming would reduce the variability. This is in contradiction to the common belief, which is that global warming will give rise to an increase of variability. Until now statistical studies of time series of temperature variability over the foregoing century have not shown any tendency either to an increase or to a decrease.

A further aspect of variability of climate is the variability in space. In other words, how strongly do climatic elements vary with distance and, more specifically, how strongly does time-variability vary with distance. In general, one might say that the coherence (or correlation in space) of temperature and precipitation increases with latitude. This is simply explained by the increase in size of the underlying atmospheric circulation systems with latitude. Temperature, however, is more strongly correlated with distance than precipitation, which in turn is related to the size of the circulation structures. In general, monthly mean temperatures in western Europe are significantly (95% level) correlated over an areal extent of at least 1,000 km, especially in winter. In summer significant correlations do not extend beyond 500 km. From the few studies of the spatial correlation of year-to-year variability of temperature it may be concluded that statistically significant correlations are limited to some 500 km only (Schuurmans and Coops 1984).

Early European Climate

How does climate vary at timescales of decades, centuries, millennia? Without a clear definition of what we call climate change, decadal, centennial, and millennial fluctuations of climate may be considered as just a long-term manifestation of climate variability. Regular observations of the main elements (temperature, precipitation, sunshine, wind, and humidity) are available only for recent decades or, as far as temperature and precipitation are concerned, perhaps a century. For a few places in western Europe, however, observational records of temperature and precipitation go back to about AD 1700. For the time before 1700 only descriptions of weather and climate are available. Such historic sources of climate information, supplemented by early observational records, have resulted in a more or less reliable climatic history of western Europe for a period starting around AD 800. In this sense, (western) Europe is unique in the world (see e.g. Flohn and Fantechi 1984; Glaser 2001; Jones *et al.* 2001). Concentration on observed climate variability as revealed by instrumental measurements means in practice that only the variability at the decadal timescale and the available trends over the period since about 1850 can be considered. Discussion of climate variations at the 100- or 1,000-year timescale is only possible when historic and other proxy data are included in the discussion. A comprehensive treatment of the subject is found in the work of Lamb (1977).

Part of the western European landscape is shaped by ice masses of the British and Scandinavian ice caps and the concomitant changes of sea level during the ice age epoch. The last glacial maximum is more than 10,000 years ago and the climate of the geological period since then, the Holocene, was not very much unlike the climate that we have now. It is sometimes even stressed that the Holocene climate has been relatively constant, but according to recent studies this might not have been the case. A classical subdivision of the Holocene (Roberts 1998), mainly based upon vegetation characteristics, considers the following phases: the Preboreal (a warming period between 10,500 and 9000 BP), the Boreal (a mild period between 9000 and 8000 BP), the Atlantic (a somewhat warmer period from 8000–5000 BP), the Subboreal (mild times between 5000 and 2500 BP) and the Subatlantic (a period of more variable climate from 2500 BP to present). The long Atlantic period is also called the Postglacial Climatic Optimum. This was a period in which the climate was a few degrees warmer than it is now, due to a benificial increase of solar radiation in the summer season. This was caused by a change in the parameters (precession, obliquity) of the Earth's orbit around the sun (Hartmann 1994). Geological evidence shows that during this period of time the climatological belts were shifted northwards, in such a way that part of North Africa was wetter than it is now (green Sahara) and southern Europe probably dryer. In any case, also in western Europe the climate during the Holocene Optimum was most probably quite different from the present. In addition, recent climate studies show that also other climate changes occurred during the Holocene. For instance, a well-documented case of a rapid deterioration of climate occurred at about 2700 BP (van Geel *et al.* 2000). The evidence originates from observed changes in peat species (van Geel *et al.* 1996). On the basis of a close relation with certain indicators of solar activity, a periodicity of 2,400 years of such climate cooling events has been proposed. Anyhow, about 400 years ago, in and around the seventeenth century, western Europe (and also some other parts of the world) experienced a very cold period, generally known as the Little Ice Age (LIA). For this period our evidence is on a solid basis of historic descriptions (including pictorial information such as the well-known paintings of winter scenes from the Dutch School) and early instrumental observations. From the available data it has been derived that the frequency of severe winters in Holland in the LIA was about three per decade, against less than one per decade at present (van den Dool *et al.* 1978). From historical evidence only, we know that some centuries before the LIA climate went through an opposite extreme, the Medieval Optimum. The nature and timing of this event has long been unclear, but research over the last twenty-five years has shown that the twelfth and thirteenth centuries had the warmest episodes, but that the whole of the Medieval Optimum is characterized by large fluctuations between warmer and cooler decades. This large variability over

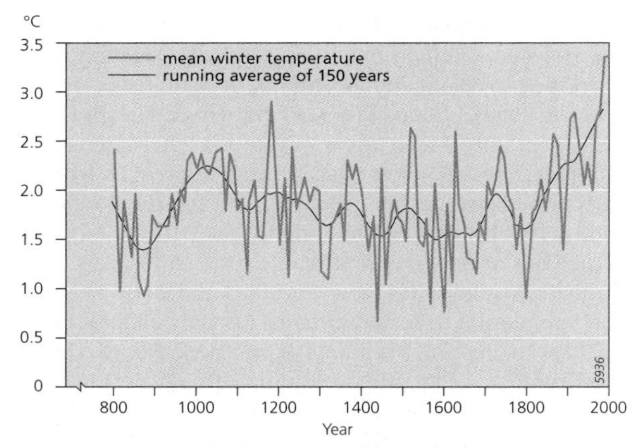

Fig. 15.5. Decadal mean winter temperatures in °C of the Low Countries for the period 800–2000. The smooth curve is a running average of 150 years (modified after van Engelen *et al.* 2001).

a decadal to centennial timescale is also an apparent feature in the LIA, as is clearly shown by a reconstruction of temperatures in Holland (see Fig. 15.5). The reasons for this variability can be internal, but external causes such as solar activity and volcanic eruptions are also possible. Crowley *et al.* (2000) suggest that over the last 1,000 years on average nearly 25% of the decadal scale variability may be attributed to volcanic eruptions. The contribution in the interval 1400–1850, including the LIA, could even be as high as 45%.

Concomitant with the temperature variations, variations of precipitation and storminess also occurred, but a reconstruction is more difficult than for temperature. Moreover, there is no easy relation between temperature and precipitation and/or storminess. As we will see in the next section, very much depends on the nature and types of the atmospheric circulation over the Atlantic and our area.

Information about the general circulation in earlier times is even more sparse than information on temperature and precipitation. It is only for the last 100 years that some insight is obtained about changes of prevailing winds and pressure patterns, for the most part only as far as surface data are concerned. A rather complete record of atmospheric observations, including data on the location and strength of the jetstreams is available only from the end of the 1950s onwards. From this more complete data it has been learned that apart from interannual differences, variations on a decadal scale are also present. Of course, from the long records of temperature and precipitation, it was already known that decadal and centennial changes of weather exist, but on the basis of the more complete data it may be concluded that such changes involve also the whole of the atmospheric (and oceanic) circulation. As pointed out already by Sutcliffe (1966) in his lucidly written book *Weather and Climate*, it is merely an illusion that significant changes of climate in the past seem to be so rare. Short-term fluctuations, which are now experienced as dominant landmarks in the recent record are completely lost in the distant perspective of the climate, thousands, or even hundreds of years ago.

However, as long as there is no clear-cut explanation of the decadal to centennial variations of western European climate, apart from the suggested relation with solar activity and volcanic eruptions, it will be impossible to say how these variations will affect us in future. Since they took place before the present era of man-made influences on climate (see next section), they have also to be taken into account in all considerations about future developments of climate. This in particular involves the variability of the thermohaline circulation of the ocean. Through its influence upon the temperature pattern of the North Atlantic Ocean and location and intensity of the Gulf Stream, major effects upon the climate of western Europe have been recorded and most probably will continue to take place (Rahmstorf 2000).

North Atlantic Oscillation

The North Atlantic Oscillation (NAO) explains a relatively large part of the natural variability of the climate. There is some relation between variations in the strength of the westerlies and variations of temperatures, precipitation, and storminess. The point, however, is how to characterize the strength of the westerlies. As early as the beginning of the twentieth century, meteorologists discovered that both the low pressure area near Iceland and the high pressure area near the Azores played a role in the strength of westerlies in the Atlantic–European area. The Icelandic Low and the Azores High were called centres of action and the pressure difference between these centres was adopted as a measure of the intensity of the westerly circulation. Variations of this intensity were studied and the quasi-periodic changes in the intensity were called the North Atlantic Oscillation. While this development took place in the first half of the twentieth century, it was only in the 1990s that the North Atlantic Oscillation really was recognized as a fundamental mode of variation of the atmospheric circulation. A simple index, defined as the mean difference between the normalized sea-level pressure at Lisbon (Portugal) and Stykkisholmer (Iceland) was adopted to show the variation in time. Figure 15.6 shows this variation for winter (December–March) for the period 1864–2000.

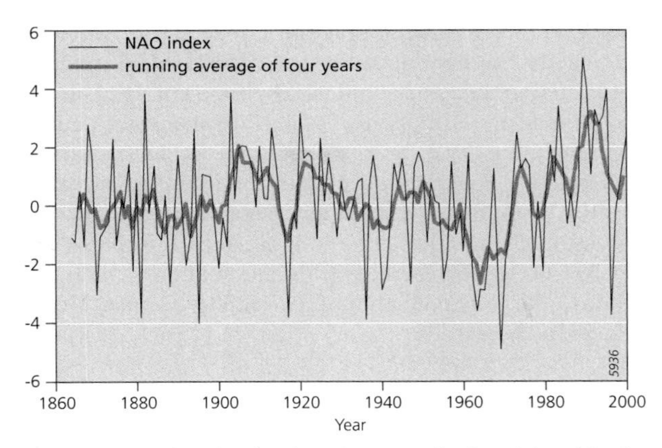

Fig. 15.6. NAO index for the winter months December–March (updated after Hurrell and van Loon 1997). The smooth curve is a running average of 4 years.

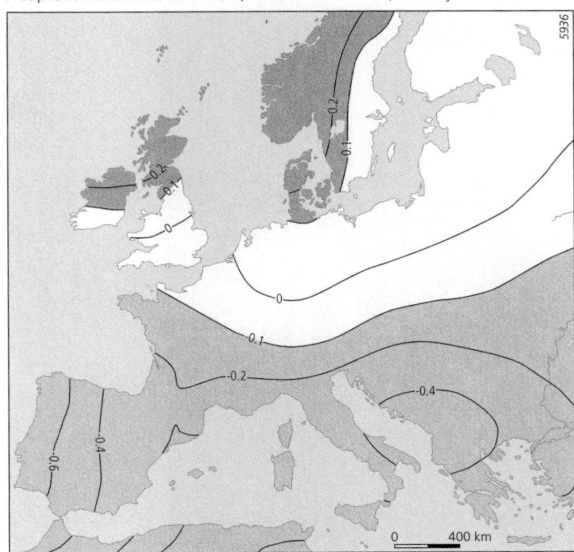

Fig. 15.7. Precipitation anomalies associated with the NAO. The figure shows the changes in precipitation (in mm/day) corresponding to a unit change of the NAO index (after Hurrell and van Loon 1997).

The NAO index clearly varies on many timescales from one year to perhaps half a century. Of course, it is the variability on these relatively long, decadal to centennial, timescales that is most interesting in relation to climate variation in Europe.

In the positive NAO-phase case, the westerlies are relatively strong and the circulation over the North Atlantic is more zonally directed, transporting cyclones directly towards western Europe, with a possibility of having an influence quite deep into the continent. In the negative NAO-phase, the westerlies are weak and the storm track over the North Atlantic is usually split into a northern branch with cyclones moving even to the north of Scandinavia, and a southern branch which brings Atlantic cyclones into the Mediterranean area and North Africa. In the latter case the circulation over the North Atlantic is also known as a blocking, since in fact the normal westerly air current is blocked by a huge and persistent anticyclone at about 60 °N, between Iceland and Scandinavia.

It follows that the NAO has a strong impact on European weather and climate. Atlantic storms, for instance, responsible for most of our precipitation in the winter half-year, will deposit this precipitation at more northern latitudes in the NAO-plus phase than in the NAO-minus phase. Figure 15.7, which shows the difference in precipitation between winters with a high and winters with a low NAO index, amply demonstrates this effect. The correlation between the amount of precipitation in winter and the NAO index amounts to +0.77 in Bergen (Norway) and to −0.69 in Madrid. Going north the correlations gradually change from negative to positive. This means that, for example, in southern Sweden the period 1980–2000 had 50% more precipitation than the foregoing twenty-year period 1960–1980. Also the relation with temperature is very strong as is shown in Fig. 15.8: winters with a high NAO index are mild, while winters with a low NAO index are relatively cold. The NAO index alone explains on average about 40% of the temperature variance in winter. Or, to put it in other terms, the correlation coefficient between winter temperature and the NAO index in The Netherlands is of the order of 0.6, while the difference in winter temperature between the five winters with the highest and the five with the lowest NAO index amounts to about 5 °C, with the maximum temperature difference being about 8 °C. In the summer half-year the NAO is much more weakly developed than in winter. Generally speaking, in summer the climate of western Europe is much less affected by the large-scale circulation over the North Atlantic. Blocking is nevertheless of strong importance as is also the shape and precise location of the Icelandic Low and Azores High as centres of action.

Changes in the large-scale atmospheric circulation and pressure distribution over the North Atlantic area are a direct cause of climate fluctuations in western Europe. The causes of these (multi-year to multi-decadal) changes in the circulation might be quite diverse,

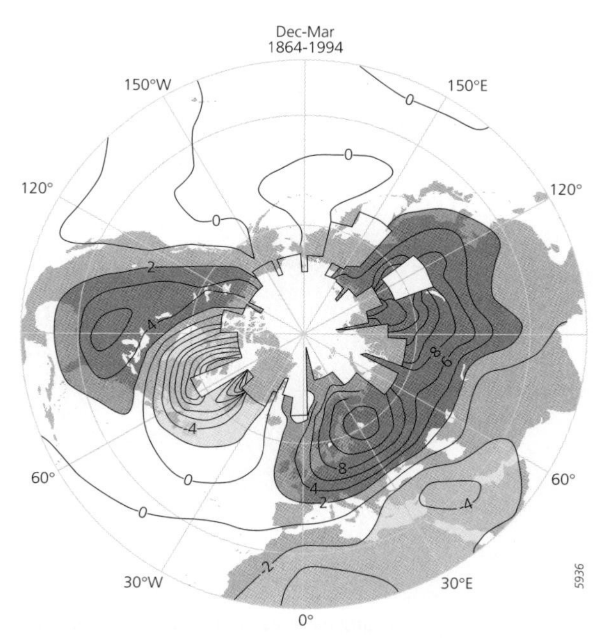

Fig. 15.8. Temperature anomalies associated with the NAO. The figure shows the change in winter temperature (in 10^{-1} °C) corresponding to a unit change of the NAO index (after Hurrell and van Loon 1997).

ranging from internal atmospheric variability or coupled atmosphere–ocean modes, to external factors, such as solar activity. Rodwell *et al.* (1999) have shown that the variation of the NAO index, as shown in Fig. 15.6, can be simulated by an atmospheric general circulation model (GCM), forced from below by observed sea surface temperatures. This is an indication that processes in the ocean, influencing the sea surface temperature (SST), may be responsible for atmospheric circulation changes. Here we could think of internal ocean waves or changes in the thermohaline ocean circulation. However, SST anomalies also may be the result of air–sea interaction processes, dynamical as well as thermodynamical.

An alternative explanation of the variations of the NAO index over the past fifty years is based upon stratospheric processes, related to solar activity, volcanic dust, ozone depletion, and enhanced greenhouse cooling of the stratosphere (Shindell *et al.* 2001). At the present stage of research no definite conclusions can be drawn. It may be expected, however, that full understanding of the various causes will open possibilities of short-term climatic predictions, as in the case of the tropical atmosphere–ocean phenomenon El Niño (Philander 1990). This quasi-periodic warming event in the eastern tropical Pacific causes large anomalies of weather all over the tropics, but also elsewhere. Extensive studies, however, have shown that the impact of El Niño, as far as the higher latitudes are concerned, is mainly restricted to the Pacific and its neighbouring land areas. Some very minor influences of El Niño on western European climate have been discovered, as, for instance, a tendency for more cyclonic conditions (Fraedrich 1990). Though not completely negligible, in general it is assumed that the climate of western Europe is not really affected by the El Niño phenomenon, which explains a large part of climate variability in many other places in the world.

Man-Made Climate Change

Man's activities always have caused changes in the environment, but the spatial scale of these changes was usually too small to induce extensive changes of climate. According to certain studies (Budyko 1974), even massive deforestation in earlier centuries will have caused only small changes of climate over an area like Europe, since the forest was generally replaced by other vegetation. Only when complete destruction of vegetation takes place, as occurred for instance in mountain regions, is precipitation affected substantially. A more recent source of man-made climate change in Europe is city growth. Large cities to a certain extent create their own climates, which in many aspects differ from the climates of the surrounding rural areas. The climate of a large city is warmer and dryer than the climate at that location would otherwise have been. Differences may amount to 1 °C or more. The reasons for the temperatures to be higher is heat release, related to energy consumption, and increased absorption of solar radiation during the day, which significantly reduces night-time cooling. The dryness of the air in large cities is due to the rapid run-off of precipitation, preventing evaporation of water. Rainfall itself is not necessarily changed over the city, but the large built-up area influences the air motion over and around cities, in such a way that in some cases a precipitation maximum is created some distance downstream of the city (Henderson-Sellers and Robinson 1986). Man-made changes in the course of rivers, or the construction of large water reservoirs may influence local climate at a scale and magnitude comparable to the effect of city growth.

However, since the twentieth century, man-made climate changes no longer seem to be restricted to the local and regional scale. The notion of (inadvertent) global-scale anthropogenic climate change gradually became more realistic and by the end of the twentieth century it became generally accepted and commonly known as 'global warming'. Since 1988 the subject has been under scientific review by the UN Intergovernmental Panel on Climate Change (IPCC).

Physical Aspects of Global Warming

All through the twentieth century individual researchers have presented theoretical evidence of the possibility that anthropogenic changes in the composition of the atmosphere could have major effects on climate. The most important constituents discussed in this respect were carbon dioxide, sulfuric aerosols, and ozone. As evidence from observations and from numerical experiments with climate models became available, the scientific community of climatologists started to react. Soon also society as a whole got alarmed and the subject of man-made climate change became a major issue in the field of environmental problems. What is the physical nature of the issue?

Physical climatology shows that planetary climate strongly depends on the composition of the planet's atmosphere. Planet Earth, for instance, would have a colder climate by about 30 °C, if the atmosphere contained only nitrogen and oxygen, which make up about 99% of the present atmosphere. It is the remaining less than 1% of atmospheric trace gases (water vapour, carbon dioxide, etc.), that keeps the planet at a habitable average temperature of 15 °C. Increases of these trace gas concentrations by the same theory are shown to cause warming. The mechanism behind the warming is the so-called 'greenhouse effect'. In short this means that the added trace gases have little effect on the incoming solar radiation, but they hamper the outgoing infrared radiation, emitted by the earth's surface and atmosphere. The thus enhanced infrared back radiation from the sky causes a warming at the surface and the lower atmosphere up to a certain height. At the new equilibrium temperature the outgoing infrared radiation at the top of the atmosphere again equals the net incoming solar radiation. On the basis of this required radiative balance at the top of the atmosphere, by simple computation it can be estimated what the temperature increase for a given increase of atmospheric trace gases should be. Following the example of CO_2, one easily computes that the temperature increase needed to re-establish radiative balance at the top of the atmosphere, at the doubling of the CO_2-concentration, amounts to 1.2 °C (see Appendix). However, in the complex climate system any temperature change, due to whatever cause, will be followed by a number of after-effects, also called feedback effects (e.g. changes in water vapour content of the air, changes in cloudiness, changes in the albedo of the earth's surface). Theoretical model experiments including these feedbacks have made clear that an initial warming, due to enhanced greenhouse back radiation, will be amplified, i.e. the feedback effects generally turn out to be positive. The computed warming of 1.2 °C at $2 \times CO_2$ could in the end become two to four times as large, so that the result could be a warming of 2–5 °C on a global scale.

This basic theory of the greenhouse effect has been known for quite some time. What happened during the last twenty-five to thirty years is that: (1) the climate models became fully three-dimensional and physically much more complete, including even chemical processes of the atmosphere; (2) observations of atmospheric composition and climate convincingly showed that man-made influences really were taking place. The symptoms of anthropogenic change were first of all clear in measurements of the concentration of atmospheric carbon dioxide. Routine measurements started in 1958 and they show a continuous increase of some 0.5% per year, in parallel to the increase in the consumption of fossil fuels (coal, oil, gas). However, not all released CO_2 accumulates in the atmosphere. Some 45% is taken up by the oceans and by vegetation. On the other hand, these sinks of CO_2 may also act as a source, which makes the estimation of future changes of atmospheric CO_2, given the fossil fuel consumption, quite difficult. Apart from CO_2, also other trace gases are becoming more abundant in the atmosphere, such as methane, nitrous oxide, chloro-fluorocarbons and tropospheric ozone. Reasons for their increase vary from direct emission (methane and chloro-fluorocarbons) to production in complex chemical reactions between gases due to various human activities (energy production, traffic, agriculture).

A new symptom of anthropogenic change was added in 1984 when the strong depletion of stratospheric ozone in spring over Antarctica was discovered. In a very short time it was convincingly shown that chloro-fluorocarbons were the main cause of the creation of the ozone hole. And again in the course of the 1990s it became clear that still another source of atmospheric pollution was affecting our climate. In industrial areas and heavily populated parts of the globe with a large fossil fuel consumption sulfuric acid aerosols are formed in the atmosphere. In large concentration these particles have the direct effect of reflecting solar radiation, which causes a cooling that counteracts the increasing greenhouse warming. However, they may also have a secondary or indirect effect, which is much less clear, in which they influence cloudiness and precipitation. In the 1990s it was also ascertained that the rate of global warming over the past twenty-five years can be explained only if, besides the increasing greenhouse effect, the tropospheric aerosol effect is also taken into account.

Recent Changes of Climate

On a global scale mean temperatures started to increase after 1975, and especially the 1990s have shown a number of very warm years (see Fig. 15.9A). Statistically, as well as physically, the conclusion has been drawn that the global warming since 1975 cannot be explained without taking the increasing greenhouse effect into account. This is nicely shown by van Ulden and van Dorland (2000). On the basis of the best available data on solar irradiance variations and amounts of volcanic dust over the period they estimated the contribution of these external forcings to the variability of global mean temperature. The same was done for the source of internal variability related to ENSO. Figure 15.9B shows the model-computed contributions, while Fig. 15.9C gives the residual variation of the global mean temperature. The shape and magnitude of the residual variation are in good agreement with the hypothesis that this change is caused by the anthropogenic forcing since 1850. A similar result was also obtained in a GCM study by Tett *et al.* (1999), but in this GCM simulation the contribution of internal variability was not allowed to vary in time. Already in IPCC-1995 it was concluded that: 'The balance of evidence suggests a discernible human influence on global climate', but this conclusion was based upon a much more general analysis, taking detailed comparisons between predicted and observed temperature changes as a function of latitude and height into account.

Average temperature of Europe over the past 150 years runs almost parallel to the global average, but its variability is much larger, due to the smaller area included. Like global mean temperature, temperatures in western Europe also increased strongly in the first half of the twentieth century. This increase was largest in the northern parts of Europe. A relative maximum was reached around 1940 for winter and around 1950 for summer temperatures. After some twenty-five years of a relatively small cooling, further warming started around 1975, in winter as well as in summer. Due to the larger interannual variability of European mean temperatures, it is still impossible to draw the same conclusion about an anthropogenic cause, as for the global mean. Nevertheless, we may say that the increasing temperatures in Europe over the last decades are not in contradiction with the expected warming due to the increasing greenhouse effect. This was concluded also in the First European Climate Assessment (ECA 1995). In this publication the geographical distribution of the

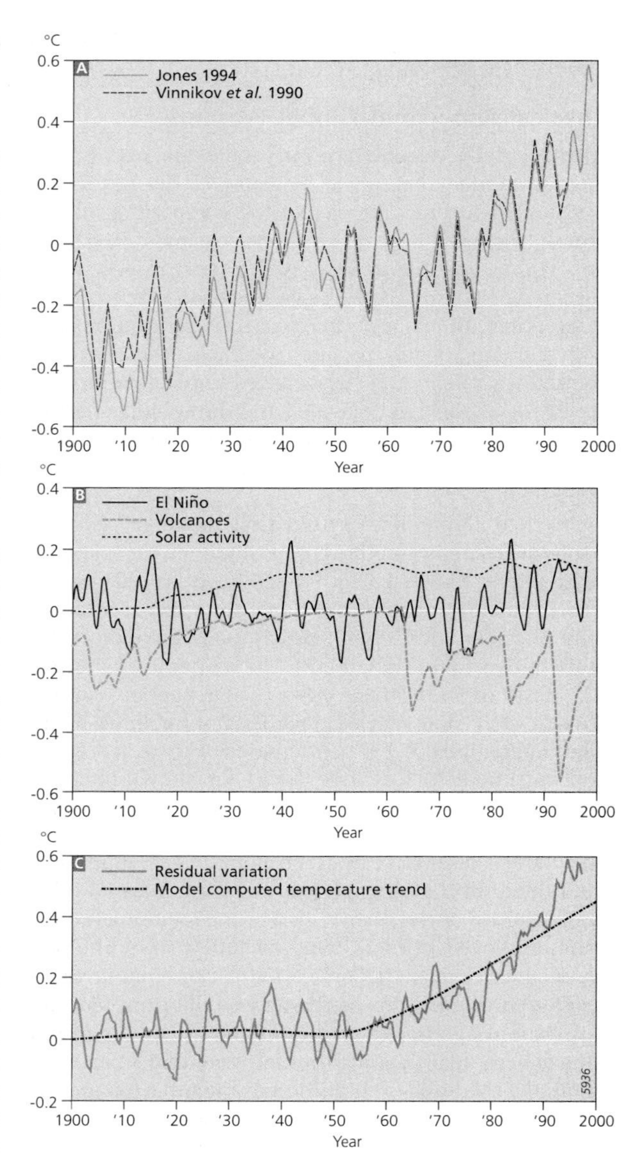

Fig. 15.9. Variability of global mean temperature in °C since 1900 and its causes. (A) Two approximations of the trend of global mean surface air temperature for the period 1900–2000, with respect to the normal 1961–90. (B) Three estimates of natural contributions to the variability. (C) Residual variation of the global mean temperature after subtraction of the contributions by the three natural forcings, and model computed temperature trend associated with the anthropogenic increases of the atmospheric concentration of greenhouse gases and aerosols. (Data from van Ulden and van Dorland 2000.)

TABLE 15.5. *Average differences of temperature (ΔT) and precipitation (ΔRR) in 1991–2000, with respect to mean temperature (T) and precipitation amount (RR), for the period 1961–1990 at De Bilt, The Netherlands*

	Jan.	Feb.	Mar.	Apr.	May	June	July	Aug.	Sept.	Oct.	Nov.	Dec.
T	2.2	2.5	5.0	8.0	12.3	15.2	16.8	16.7	14.0	10.5	5.9	3.2
ΔT	+1.2	+1.2	+1.7	+1.2	+1.0	+0.2	+1.2	+1.1	+0.6	−0.3	+0.4	+0.4
RR	69	49	65	52	61	70	76	71	67	75	81	83
ΔRR	−1	+5	+1	−3	+5	+19	−4	+3	+20	+19	+15	+10

Note: Temperatures in °C, precipitation in mm.

Source: KNMI.

warming was shown for all four seasons, while also the changes in the amount of precipitation were shown. The main conclusion was that the whole of western Europe was becoming warmer in all seasons and, as far as the northern part is concerned, also wetter. In the first approximation the patterns of temperature and precipitation change resembled those given in Figs. 15.8 and 15.7, showing the responses to an enhanced NAO index. As shown in Fig. 15.6 the NAO index indeed has been at a relatively high level since 1975. The question therefore arises if perhaps the increasing greenhouse effect could stimulate the intensity and persistence of the NAO. Up till now a convincing answer to this question is not available.

In the years since the publication of the First European Climate Assessment (ECA 1995), temperatures remained above the level of the 1961–90 norm. On the broader scale of the Northern Hemisphere the warming also continued, although some larger areas, for example around Greenland and over the Sahara, did show a small cooling in the decade 1991–2000, compared to 1961–90. Global warming therefore is not as uniformly distributed as the term suggests. As an illustration temperature and precipitation changes at De Bilt (The Netherlands) averaged over 1991–2000 are presented in Table 15.5. Temperature changes are greatest in late winter and late summer. Deviations of 10-year averages of more than 1.5 °C in winter months or more than 1 °C in summer months are exceptional (95% significance level). Precipitation has most strongly increased in the months August to December. Here deviations of 10-year averages of about 20 mm may be called statistically significant at the 95% level. The deviations shown in the table are probably representative for a large part of western Europe. In particular the recent increase of precipitation amounts has been reported for other areas in western Europe too. The latter have given rise to severe floods in various countries.

The impact of the recorded changes in the last decade on various sectors of society and life is very difficult to assess, but it has certainly not been negligible. According to public press and other media many people feel that the frequency of occurrence of extremes has increased. However, this is very difficult to ascertain. Reasons are the rareness of these events, by definition, and the lack of reliable observations over a sufficiently long period. Nevertheless, an assessment of observed daily temperature and precipitation extremes in Europe has been attempted (ECA 2002). Main conclusions are that the warming trend at 45% of all stations is associated with a significant decrease in the number of frost days and generally accompanied by a slight increase in wet extremes.

Future Climate

Prediction of future climate is not yet possible. Nevertheless, many publications exist in which outlooks of climate up to the year 2100 are given. Such outlooks are based upon scenarios of future levels of man-made climate forcing, which themselves are based upon scenarios of future use of fossil fuels and other activities influencing the composition of the atmosphere and/or the nature of the earth's surface. How do these outlooks of climate for the next 50–100 years look like and how reliable are they? All figures given are based upon or in line with the recently issued Third Assessment Report of the Intergovernmental Panel of Climate Change (IPCC 2001). A distinction should be made between projections of average global climate and projections of regional climate. Based on different scenarios global average temperature up to 2100 is assumed to increase by 1.4–5.8 °C. The large range of uncertainty is due to the spread of the scenarios, but also to different results of climate models using the same scenario and even intra-model variability, i.e. the same model using the same forcing scenario may produce a different degree of warming depending on initial conditions and spatial resolution. Atmospheric aerosols, also related to man's activities, will delay the warming, but the magnitude of their cooling effect will not very much increase in future. Global warming will be accompanied by an increase in the intensity of the hydrological cycle by 3–10% per degree C warming, depending on location and season. At higher temperatures the largest increase is to be expected. The increase in intensity of the hydrological cycle involves both precipitation and evaporation to the same extent. In addition, the warming will cause a rise of sea level of 9–88 mm by 2100. Its average value, 47 mm, is more than twice the observed sea-level rise in the foregoing century. Most of this sea-level rise will be due to expanding warmer oceans and only a small part to the melting of glaciers and ice caps. The Antarctic ice cap, however, could even increase in volume, due to an increasing mass balance (more accumulation than melting and production of icebergs), according to Gregory and Oerlemans (1998).

Outlooks of regional climate are highly uncertain, even if they are produced by comprehensive climate models and based on the same scenarios as used for the global warming projections. The different scenarios by themselves introduce a large degree of uncertainty, but even for the same scenario models differ in their solutions, from almost no change in atmospheric circulation to major changes, as for instance a strong increase in the NAO index. Since climate strongly depends on the type and intensity of the circulation in a region, it is obvious that model results also differ as far as the pattern and magnitude of changes in temperature and precipitation are concerned. The joint use of different model results on the basis of regionalization or downscaling methods provides probably the most reliable outlook. In the recent IPCC report (IPCC 2001) regional outlooks for temperature and precipitation changes in 2071–100 (with respect to 1961–90) are given for various regions, including northern Europe and the Mediterranean area. For temperature a greater or much greater warming than in the global case seems to be indicated for the whole of Europe, for winter and summer. For the precipitation outlook the results are mixed. In winter a strong increase of precipitation in northern Europe is expected, while for the Mediterranean area results are inconsistent (for greenhouse warming only) or no change (for greenhouse warming combined with aerosol cooling). In summer, the inconsistent results and a no-change forecast apply to northern Europe, while for the Mediterranean area dryer conditions are expected. This increase in precipitation in northern Europe, especially in winter, and a drying of southern Europe, especially in summer, have been a systematic result of several models over a number of years. On the other hand, Giorgi and Francisco (2000) show that the dominant uncertainty in the simulation of regional climate change is due to inter-model variability, with inter-scenario and internal model variability playing secondary roles. The fact that equally sophisticated models still show different solutions with the same forcing, could be an indication of the lack of predictability of regional climate on the timescale of decades.

Finally, on the important issue of possible changes in storminess in our area, model results are even less reliable and less consistent than for precipitation. The latter could be partly due to the lack of comparable analysis techniques. The main reason, however, is the limitations of the models used. The frequency and intensity of extra-tropical storms strongly depend on details of the vertical distribution of the changes in the horizontal temperature gradient over the North Atlantic. For the lower layers of the atmosphere, near the surface, models generally compute a decrease of the meridional temperature gradient, because polar areas warm more than the tropics. This would lead to a tendency of less storminess. In the upper troposphere, however, the horizontal temperature gradient, according to most models, strongly increases, which would mean an increase of storminess. The net result is very difficult to establish. This also applies to the prediction of a possible shift in the storm tracks.

In any case anthropogenic forcing is not the only cause of decadal change of regional climate. Natural causes, atmosphere–ocean interaction, and solar-volcanic changes of the stratosphere also play a role. Results of present climate research, taking the various causes into account, are very promising. Therefore, it is not unrealistic to expect that in the course of the present century, climate forecasts of useful skill and reliability become available.

Appendix

The equation expressing the radiative balance at the top of the atmosphere reads

$$S\,(1-\alpha) = I \qquad (15.1)$$

in which S = global average of incoming solar radiation, α = reflected amount of solar radiation (by clouds and the earth's surface) and I = outgoing long-wave radiation. Note that the geographic distribution of the radiation balance is given by Fig. 15.2.

What happens when the CO_2-concentration of the atmosphere increases? First of all, this will lead to a decrease of the outgoing long-wave radiation. We call this

$- \Delta I$. This so-called greenhouse trapping will give rise to an increase of temperature at the earth's surface by ΔT. In order to re-establish the radiative balance, temperature will have to increase by such an amount that the greenhouse trapping is exactly compensated for by an enhanced long-wave emission $+ \Delta I$. In formula this means that

$$dI/dT \cdot \Delta T = + \Delta I \qquad (15.2)$$

or, after rearrangement

$$\Delta T = 1/\ dI/dT \cdot \Delta I \qquad (15.3)$$

The outgoing long-wave radiation may be written as

$$I = \varepsilon\, \sigma\, T^4 \qquad (15.4)$$

σ = Stefan-Boltzmann constant ($5.67 \cdot 10^{-8}$ W/m^{2}K^4) and ε = constant of proportionality, with an empirical value of 0.6.

This means that

$$dI/dT = 4\varepsilon\, \sigma\, T^3 \qquad (15.5)$$

Substitution in (15.3) and using (15.4) leads to

$$\Delta T/T = {}^1/_4 \cdot \Delta I/I \qquad (15.6)$$

By substitution of (15.1) we can even eliminate ε and arrive at the final formula

$$\Delta T = T \cdot \Delta I\ /\ 4\ S\ (1 - \alpha) \qquad (15.7)$$

On the right-hand side of the formula the present values of the four parameters are known as ΔI (at $2 \times CO_2$) = 4.2 W/m^2, T = 288 K, S = 342 W/m^2, $\alpha = 0.3$.

Inserting these values leads to $\Delta T = 288 \times 4.2 / 4 \times 342 \times 0.7 = 1.26$ K.

References

Budyko, M. I. (1974), *Climate and Life*. International Geophysics Series, 18. Academic Press, New York.

Crowley, T. J., *et al.* (2000), Causes of climate change over the past 1000 years. *Science* 289: 270–7.

European Climate Assessment (ECA) (1995), *Climate of Europe: recent variations, present state and future developments*. KNMI, De Bilt.

—— (ECA), (2002), *Climate of Europe: assessment of observed daily temperature and precipitation extremes*. KNMI, De Bilt.

Flohn, H., and Fantechi, R. (eds.) (1984), *The Climate of Europe: Past, Present and Future*. Academic Sciences Library. Reidel, Dordrecht.

Fraedrich, K. (1990), European Grosswetter during the warm and cold extremes of the El Niño/Southern Oscillation. *International Journal of Climatology* 10: 21–31.

Giorgi, F., and Francisco, R. (2000), Evaluating uncertainties in the prediction of regional climatic change. *Geophysical Research Letters* 27: 1295–8.

Glaser, R. (2001), *Klimageschichte Mitteleuropas: 1000 Jahre Wetter, Klima, Katastrophen*. Wissenschaftliches Buchgesellschaft, Darmstadt.

Gregory, J. M., and Oerlemans, J. (1998), Simulated future sea-level rise due to glacier melt based on regionally and seasonally resolved temperature changes. *Nature* 391: 474–6.

Hartmann, D. L. (1994), *Global Physical Climatology*. Academic Press, New York, 316–17.

Henderson-Sellers, A., and Robinson, P. J. (1986), *Contemporary Climatology*. Longman, Harlow, 284–95.

Hurrell, J. W., and Loon, H. van (1997), Decadal variations in climate associated with the North Atlantic Oscillation. *Climatic Change* 36: 301–26.

IPCC-1995 (1996), *Climate Change 1995: The Science of Climate Change*. Cambridge University Press, Cambridge.

IPCC-2001 (2001), *Climate Change 2001: The Science of Climate Change*. Cambridge University Press, Cambridge.

Jones, P. D. (1994), Hemispheric surface air temperatures: a reanalysis and update to 1993. *Journal of Climatology* 7: 1794–802.

—— Osborn, T. J., and Briffa, K. R. (2001), The evolution of climate over the last millennium. *Science* 292: 662–7.

Köppen, W. (1936), Das Geographische System der Klimate. In: W. Köppen and R. Geiger (eds.) *Handbuch der Klimatologie*, i. p. C. Borntrâger, Berlin.

Lamb, H. H. (1977), *Climate: Present, Past and Future*. Methuen, London.

Legates, D. R., and Willmott, C. J. (1990*a*), Mean seasonal and spatial variability in global surface air temperature. *Theory Application Climatology* 41: 11–21.

—— —— (1990*b*), Mean seasonal and spatial variability in gauge-corrected, global precipitation. *International Journal of Climatology* 10: 111–27.

Müller, M. J. (1987), *Handbuch Ausgewählter Klimastationen der Erde*. Gerold Richter, University of Trier.

Peixoto, J. P., and Oort, A. H. (1992), *Physics of Climate*. American Institute of Physics, New York.

Philander, S. G. H. (1990), *El Niño and the Southern Oscillation*. Academic Press, New York.

Rahmstorf, S. (2000), The thermohaline ocean circulation: a system with dangerous thresholds? *Climatic Change* 46: 247–56.

Roberts, N. (1998), *The Holocene: an environmental history*. Blackwell, Oxford.

Rodwell, M. J., Rowell, D. P., and Folland, C. K. (1999), Oceanic forcing of the wintertime North Atlantic Oscillation and European climate. *Nature* 398: 320–3.

Rossow, W. B., and Schiffer, R. A. (1991), ISCCP cloud data products. *Bulletin of American Meteorological Society* 72: 2–20.

Schönwiese, C-D., Rapp, J., Fuchs, T., and Denhard, M. (1993), *Klimatrend-Atlas, Europa 1891–1990*. Johann Wolfgang Goethe-Universität, Frankfurt am Main, Berichte der Zentrums für Umweltforchung, 20. ZUM, Frankfurt.

Schuurmans, C. J. E., and Coops, A. J. (1984), Seasonal mean temperatures in Europe and their interannual variability. *Monthly Weather Review* 112: 1218–25.

—— and Krijnen, H. J. (1971), Weertypeklimatologie voor De Bilt, 1881–1970. *KNMI Wetenschappelijk Rapport*, 71–6.

Shindell, D., Schmidt, G. A., Miller, R. L., and Rind, D. (2001), Northern Hemisphere winter climate response to greenhouse gas, ozone, solar and volcanic forcing. *Journal of Geophysical Research* 106: 7193–210.

Sutcliffe, R. C. (1966), *Weather and Climate*. Weidenfeld and Nicolson, London.

Tett, S. F. B., Stott, P. A., Ingram, W. J., and Mitchell, J. F. B. (1999), Causes of twentieth-century temperature change near the Earth's surface. *Nature* 399: 569–72.

Troen, I., and Lundtang Petersen, E. (1989), *European Wind Atlas*. Risø National Laboratory, Roskilde, Denmark.

van den Dool, H. M., Krijnen, H. J., and Schuurmans, C. J. E. (1978), Average winter temperatures at De Bilt (The Netherlands), 1634–1977. *Climatic Change* 1: 319–31.

van Engelen, A. F. V., Buisman, J., and IJnsen, F. (2001), A millennium of weather, winds and water in the Low Countries. In: P. D. Jones, A. D. Davies, A. E. J. Ogilvie, and Briffa, K. R. (eds.) *History and Climate*. Climatic Research Unit, University of East Anglia, Norwich, 101–25.

van Geel, B., Buurman, J., and Waterbolk, H. T. (1996), Archeological and palaeoecological indications of an abrupt climate change in The Netherlands, and evidence for climatological teleconnections around 2650 BP. *Journal of Quaternary Science* 11: 451–60.

—— Heusser, C. J., Renssen, H., and Schuurmans, C. J. E. (2000), Climate change in Chile at around 2700 BP and global evidence for solar forcing: a hypothesis. *The Holocene* 10: 659–64.

van Ulden, A. P., and Dorland, R. van (2000), Natural variability of global mean temperatures: contributions from solar irradiance changes, volcanic eruptions and El Niño. In: *Proceedings 1st Solar and Space Weather Euroconference*, Santa Cruz de Tenerife, Tenerife, Spain, 25–9 September 2000, ESA SP-463, 213–18.

Vinnikov, K. Ya., Groisman, P. Ya., and Lugina, K. M. (1990), Empirical data on contemporary global climate changes (temperature and precipitation). *Journal of Climatology* 3: 662–77.

16 Weathering and (Holocene) Soil Formation

Jan Sevink and Otto Spaargaren

Introduction

The soils of western Europe are marked by superficial accumulation of organic matter as a result of its climate being temperate, rather wet, and leading to leaching and soil acidification. Soils are largely of Holocene age, many properties still being determined by their highly varied parent material. Even more prominent are the impacts of man in this densely populated and highly industrialized part of Europe, causing many soils to be partly or completely of anthropogenic origin. In this chapter, these main soil traits, their origin and distribution will be discussed, attention also being paid to the terminology used and its backgrounds.

General Trends in Soil Genesis in Western Europe

In the temperate, humid climate of western Europe, organic matter tends to decompose slowly, giving rise to accumulation of above-ground litter and to dark-coloured, humus-rich mineral topsoils. The retarded decomposition is associated with a low activity of soil biota and rather massive production of soluble organic acids, and particularly brought about by the prevailing site conditions such as relatively poor drainage, high precipitation, low temperature, and acid parent material. These site conditions are found over large tracts of western Europe, which therefore have acid, nutrient-poor soils with prominent accumulation of organic matter, such as *Podzols* and *Histosols* of the North European Lowlands, and *Umbrisols* of the middle and high altitude mountains. Bioturbation in these soils is generally weak and therefore soils have distinct horizons and a sharp boundary between the organic surface and mineral subsurface horizons, and thus exhibit large contrasts in soil properties with depth. Where parent materials are more basic and capable of neutralizing acids (e.g. limestone and marl) or climatic conditions are more favourable (e.g. in southern France), litter decomposition proceeds faster and soil biota are more active. In such soils, finely divided organic matter is intimately mixed with mineral material, litter layers are thin or absent and soil reaction tends to be neutral to slightly acid. Moreover, soil horizon differentiation is less prominent as a result of bioturbation, particularly by earthworms.

The variation in topsoil properties is so prominent that it was immediately recognized in the early days of soil science and led to a still existing terminology for the description of topsoil organic matter. These classic terms are *mor, moder, and mull* (see e.g. Kubiena 1953), reflecting increasing decomposition and biological activity, and are still extensively used in national classification systems (Fig. 16.1).

In temperate climates physical weathering is rather weak. However, particularly at higher altitudes and towards the more continental central Europe, it is more prominent because of frequent alternation of frost and thaw leading to rapid fracturing and disintegration of otherwise massive rocks. Net precipitation is generally high enough to leach soluble components such as carbonates and gypsum, but soil temperature conditions are unsuited for rapid chemical weathering of the most common rock-forming minerals (feldspars, micas, and quartz). In the acid, nutrient-poor soils mentioned above, these mobile elements include iron and aluminium, which are leached from the topsoil and may accumulate at some depth, giving rise to *Podzols*. Where weathering occurs under more favourable conditions and soils are less acid and well drained, iron and aluminium released

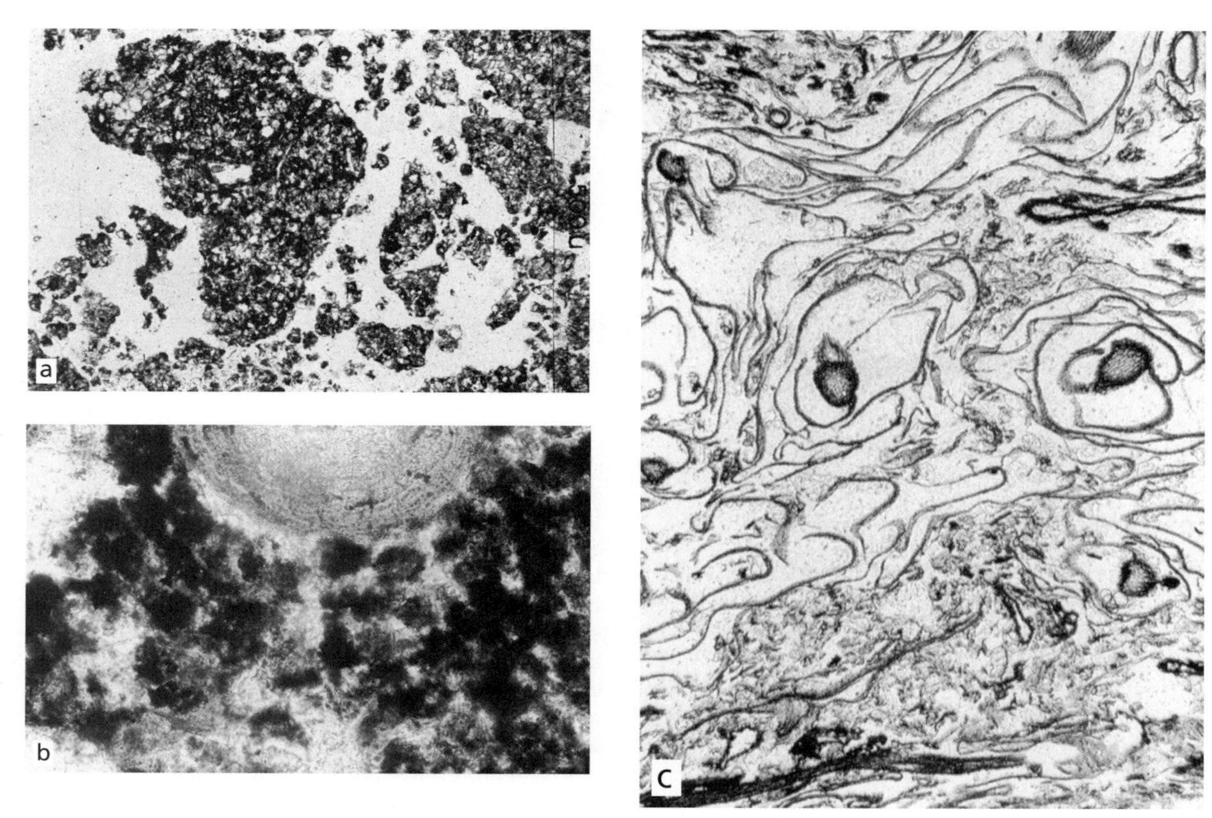

Fig. 16.1. Microphotos of thin sections of different types of organic soil material: (*a*) mull (photo: International Soil Reference and Information Centre, ISRIC), (*b*) moder, and (*c*) mor.

by weathering are immobilized. Iron combines with clay and humus into iron-clay-humus complexes, giving the soil a characteristic brown colour. This formed the basis for the classification of soils as 'Brown soils' in older central and north-west European classification systems (e.g. German *Braunerde* and *Braunlehme*; French *Sols bruns*; Dutch *Bruine bosgronden*) and for the description of their genesis as resulting from 'brunification' (Duchaufour 1998). Where the climate becomes more Mediterranean, the soils become more reddish in colour, due to a gradual change in the mineralogy of the iron compounds formed. This gave rise to the classic distinction between brunification and rubefaction as major soil-forming processes. Clay translocation is a common feature in these soils and gives rise to textural contrast between the topsoil that has been washed out ('eluviated') and the subsoil with illuviated clay, which often induces seasonal water stagnation and associated pseudogley or even stagnogley. The effects of clay translocation are especially evident in homogeneous medium to fine textured parent materials such as loess and were also recognised very early, leading to the use of terms in France such as *sol lessive* and *lessivage*.

Taking into account the geological history of western Europe, it is not surprising that most soils are of relatively recent, 'post-glacial' age. In the north and in the Alps, glacial and periglacial deposits dominate, dating back to the Last Glacial Period or, in the Alps, even of recent age. In many hilly or mountainous areas that remained outside the reach of the Pleistocene ice caps, pre-Holocene soils have been removed by periglacial processes, e.g. in the Eifel-Ardennes, the French Massif Central, and Britanny. Furthermore, during the Late Pleistocene (aeolian) cover sand and loess deposits accumulated in a broad zone south of the border of the Fennoscandian ice cap, covering any earlier soil. In areas with steep relief such as the Alps, Holocene soil formation has been limited by erosion, while in the Rhine–Meuse delta and further north along the coasts of the North Sea and Baltic Sea fluvial and marine deposits of recent age abound. Thus, pre-Holocene soils are largely restricted to the rare stable older land surfaces where periglacial processes were hardly active and which remained outside the reach of the Würmian/Weichselian aeolian deposition, such as the coastal terraces of the Languedoc and the Gironde terraces in southern France. On such surfaces, polygenetic soils survived which often reflect earlier, warmer periods by their reddish colour and strong weathering. The geological structure of western Europe is complex, giving rise to a large variety of parent materials, occurring in an intricate mosaic. Since most soils are of relatively recent age, soil properties such as colour, texture, soil depth, and carbonate content still strongly depend on the characteristics of the parent material. As a consequence, soil patterns commonly reflect the geological structure of the area concerned.

In parts of western Europe, such as the loess belt and southern France, agriculture had already started by about 4000 BC and since then vegetation has become increasingly affected by man, being gradually transformed from natural forests into secondary open forests and shrub lands. Typical examples are the transformation of birch-oak forests into heathlands on the Pleistocene deposits of the North European Lowlands, and the development of maquis at the expense of oak forests in southern France. In the Middle Ages the pressure on the land was already such that major parts of western Europe were completely deforested and cultivated, causing extensive erosion of soils on slopes, accumulation of eroded soil material on foot slopes, and silting of river valleys. At many places man-induced aeolian activity became rampant leading to large-scale soil degradation, while peat areas were extensively exploited for fuel. At the same time, man increasingly interfered with natural drainage and started to reclaim low-lying, poorly drained coastal areas. In the south and centre of western Europe slopes were extensively terraced. As a result of these human activities, even before the introduction of modern mechanized agriculture, natural non-ploughed soils were relatively rare and restricted to more remote and poorly accessible non-productive areas. In the past century, human impacts on soils have sharply increased, not only through mechanization and intensification of agriculture but also through large-scale air, water, and soil pollution, amongst which, in particular, acid atmospheric deposition. The result is that current properties of west European soils are largely anthropogenic in origin.

In the various western European countries, terminologies and classification systems developed rather independently, hampering the communication between soil scientists at the international level and causing misunderstanding and even geopolitically induced strife. To overcome these problems, international classification systems and terminologies are increasingly used, though even nowadays uniformity has not been reached. In this chapter, we will employ the widely accepted terminology and classification system of the FAO. Its correlation with national systems is given in Table 16.1.

Dominant Soil-Forming Processes

As stated above, the land surfaces and associated regoliths in which western European soils have formed are mainly of Late-Pleistocene to Holocene age and thus soil-forming processes often have only been active over periods of less than about 10,000 years. The transformation of rock into soil therefore is not very advanced and soil-forming processes involved are largely limited to those active under the present-day temperate, humid climatic conditions.

Weathering

Physical weathering was intense and widespread during glacial periods, giving rise to phenomena such as frost wedges, solifluction deposits, and debris slopes. On slopes, extensive debris mantles were formed, which nowadays serve as parent material for soils (Fig. 16.2). Under present-day conditions physical weathering is weak, except for high-altitude soils where frost action and rock fracturing are prominent. Fragmentation of rock material through root activity is an additional process during the Holocene, which is particularly active at lower altitudes.

In the temperate humid climate, chemical weathering comprises the gradual dissolution of easily soluble soil components such as calciumcarbonate, a partial or complete dissolution of non-phyllosilicates (e.g. feldspars and amphiboles) often accompanied by residual accumulation of sesquioxides (Al- and Fe-(hydr)oxides) and gradual degradation of micas into clay-sized phyllosilicates. Mobile compounds released by chemical weathering such as basic cations (e.g. Na, K, Mg, and Ca) are leached, leading to residual accumulation of non-mobile elements and resistant minerals, while neoformation of secondary minerals, i.e. by precipitation from the soil solution, plays a very minor role if occurring at all. The latter is due to the combination of slow weathering and concurrent limited release of elements such as Si and Al, and massive leaching. In the weathering of phyllosilicates, a specific sequence can be distinguished (Jackson 1964): from mica through illite to vermiculite and finally smectite with a concurrent loss of magnesium and potassium (Fig. 16.3). The original structure

TABLE 16.1. *Correlation of the FAO soil groups with soil groups or orders in other north-west European soil classification systems*

FAO-ISRIC-ISSS, 1998	Characteristics	Belgium	Denmark	France	Germany	Netherlands
Acrisols	Acid soils with accumulation of clay in the subsoil	Abd	Lessive	Luvisol dégradé	Versauerter Parabraunerde	Brikgrond
Albeluvisols	Deep tonguing of the clay-depleted eluvial horizon into a clay-enriched illuvial horizon	Abc(m), Lbc(m)	Stagnogley	Luvisol dernique	Stagnogley	Brikgrond
Andosols	Weak to moderate soil development in volcanic ashes	—	—	Andosol, Vitrisol	Andosol	—
Anthrosols	Soils profoundly modified by centuries-long agricultural activity	Zem, Zdm	Kolluvial jord	Anthroposol	Anthropogene Boden	Eerdgrond
Arenosols	Coarse-textured materials with little soil development	Zap	Blegsol, Brunsol	Arénosol	Regosol	Vaaggrond
Calcisols	Soils with accumulation of secondary carbonates	—	—	Calcosols	—	—
Cambisols	Weak soil development	Gbb	Brunjord	Brunisol	Braunerde, Terra fusca	Vaaggrond
Chernozems	Thick, humus-rich, base-saturated surface horizon; presence of secondary carbonates	Abe, Lbe	—	Chernosol	Tschernosem	—
Fluvisols	Recently deposited alluvial material	Aep, Afp	Gley, Blegsol Stagnogley	Fluviosol, Thalassosol	Auenboden, Gleye, Marschen	Vaaggrond
Gleysols	Influence of groundwater	Ahx, Aix	Gley, Stagnogley	Réductisol, Rédoxisol	Gleye, Marschen	Vaaggrond
Histosols	Organic soils with accumulated peat	V	Histosol	Histosol	Moore	Veengrond
Leptosols	Shallow or very gravelly soils	Gbb97	Ranker, Rendzina	Lithosol	Rohboden, (Para)Rendzina	—
Luvisols	Neutral soils with accumulation of clay in the subsoil	Aba, Lba	Lessive	Luvisol	Parabraunerde, Terra rossa	Brikgrond
Nitisols	Soils with deep accumulation of clay in the subsoil	Aba, Uba	—	Fersialsol	Plastosole, Latosole	—
Phaeozems	Soils with a thick, humus-rich, base-saturated surface horizon	Abek, Lbek	—	Phaeosol	Brauner Steppenboden	Eerdgrond
Podzols	Soils with illuvial, cheluviated alumino-organic complexes	Zag, Sag	Podzol	Podzosol	Podsol	Podzol
Regosols	Non-developed soils in medium- to heavy-textured materials	Acp, Lcp	Rajord	Régosol	Locker Syrosem, Regosol, (Para)Rendzina	Vaaggrond
Solonchaks	Saline soils	—	—	Salisol, Sodisol	Brackmarsch	—
Umbrisols	Soils with a thick, humus-rich, base-desaturated surface horizon	Zag(o)	Ranker	Rankosol	Ranker-Syrosem	Eerdgrond
Vertisols	Soils that shrink and swell upon drying and wetting	Ubb	—	Vertisol	Pelosol	—

Fig. 16.2. *Hakenwerfen* in schist in the hills north of Wiesbaden, central Germany, resulting from physical weathering under past periglacial conditions causing fragmentation and down-slope movement of the angular schist fragments formed.

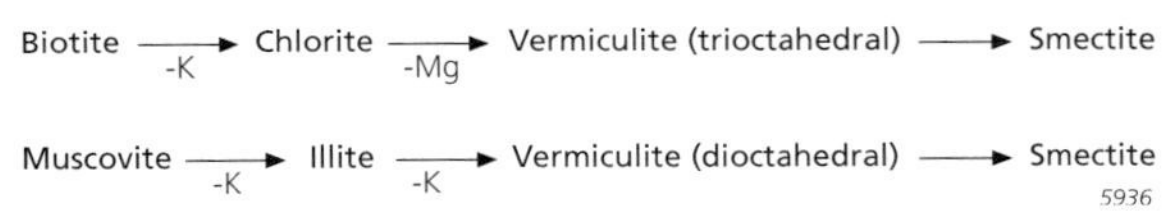

Fig. 16.3. Weathering sequence of mica in temperate humid climates.

(2:1 layer) of the phyllosilicate is largely preserved. The processes involved have been extensively investigated (McBride 1994).

New formation of other clay minerals such as kaolinite and smectite is rare because concentrations of the relevant elements (Al, Si) in the soil solution remain too low (Bolt and Bruggenwert 1976). Any significant occurrence of these minerals is an indicator for their inheritance from the parent rock or from earlier weathering stages, which may have survived. Typical examples of the inheritance of kaolinite from earlier weathering stages are the *arène*, i.e. the deeply weathered sandy regolith over granites in France, which often has been attributed to deep weathering in the Tertiary (Duchaufour 1998), and the occurrence of laterites in the Mainz region of Germany, which have formed under humid tropical conditions prevailing during the Tertiary Era (Doebl 1973).

In acid soils such as *Podzols*, aluminium and iron released by weathering are complexed by organic acids and leached into the soils. Aluminium may precipitate and eventually give rise to the formation of Al-hydroxy interlayered vermiculite and smectite, often referred to as soil chlorite. Iron is either completely leached from the soil, particularly if poorly drained, or precipitates as poorly crystalline, yellowish-brown hydroxides.

Under less acid conditions, where Al is released by weathering but not mobilized in the form of organic complexes, Al is often incorporated in the clay minerals to form soil chlorite. The behaviour of iron is somewhat more complex: if weathering occurs under reducing conditions iron released is mobile and will be leached, whereas under oxidative conditions it is more or less immobile (see e.g. Bolt and Bruggenwert 1976; McBride 1994). In such aerobic, oxidative environments, with less complexing organic acids but still in the presence of organic matter, the iron released precipitates as ferrihydrite or goethite in combination with organic matter (iron-humus-clay complexes), producing the characteristic brown colour of the temperate soils of western Europe. Where organic matter is scarce or absent, iron precipitates as goethite or hematite in iron-clay complexes, producing redder colours as observed in the Mediterranean soils of southern France and in palaeosols formed during periods of warmer climate (Segalen 1964; Cornell and Schwertmann 1996).

Organic Matter Accumulation

As described above, site conditions are generally conducive for retarded decomposition of litter. In the litter layer, described as O or H horizon, depending on wetness conditions (O for dry conditions and H for wet), and in the underlying mineral A horizon, soluble organic acids are produced, which in permeable soils are washed into deeper soil horizons. This enhances the progressive

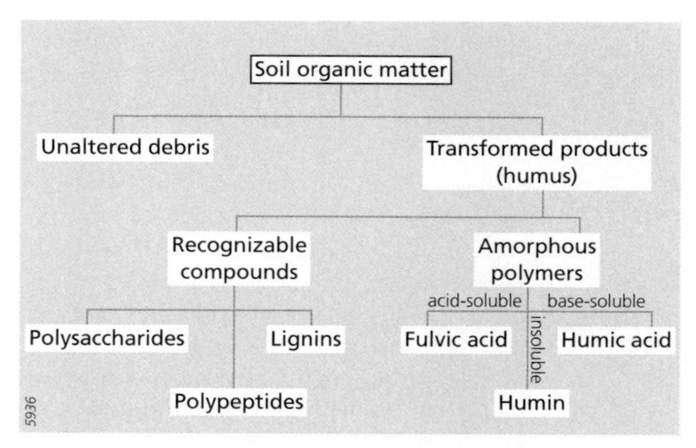

Fig. 16.4. Soil organic matter components and pathways (after McBride 1994).

leaching and soil acidification caused by percolation of infiltrated water.

Particularly French and German soil scientists extensively studied the factors controlling decomposition of organic matter and the pathways and organic compounds involved, emphasizing the prominent role of organic matter dynamics in western European soil genesis. Results have been summarized in various handbooks, including Kubiena (1953), Duchaufour (1998), and Scheffer and Schachtschabel (1998). Distinction is made between unaltered vegetation debris, which dominates the upper part of the litter layer, and transformed products or humus. This humus comprises recognizable compounds, such as microbial cell structures and decomposing roots and leaves, and amorphous products of which the origin could not be readily identified (see Fig. 16.1). The latter products dominate the lower O and the A horizon. Processes and products involved are highly complex (Stevenson 1994; Swift and Spark 1998). In Fig. 16.4, generalized pathways in litter decomposition and products involved are indicated. In soil geography, however, generally an even more simple distinction is made between groups of humus compounds, based on their solubility and relevance for soil-forming processes and properties. These are fulvic acids, humic acids, and humine.

A retarded decomposition, as in the O-layer of *Podzols* with mor-type humus, is associated with the production of particularly fulvic acids and dominance of humine, which is inherited from the organic input. A more rapid decomposition, as in many *Luvisols* and *Cambisols* with mull-type humus, is associated with humic acids and humine of microbial origin. However, the specific effects of a number of soil components complicate this simple pattern. In soils that contain abundant finely divided calcium carbonate, chemical decomposition is impeded. Nevertheless, bioturbation is prominent, giving rise to calcareous soils with dark-coloured humic topsoils, known as *Rendzinas*, but currently classified as *Rendzic Leptosols* (Calcareous mull, see Fig. 16.5).

Upon weathering of volcanic materials containing glass and other highly weatherable minerals, poorly or non-crystalline compounds are formed, such as allophane, imogolite, and aluminium hydroxides. These compounds immobilize the fulvic and humic acids produced by chemical decomposition of litter, favouring rapid accumulation of organic matter. These so-called *Andosols* have specific chemical and physical properties, such as a very low volume weight and very high water-holding capacity and phosphate fixation. Moreover, even if well drained these soils are very high in organic matter (Figs. 16.5 and 16.6).

Lastly, iron also combines with fulvic and humic acids into less decomposable iron-humus complexes (Acid mull), though its impact on decomposition is considerably less and fair quantities of finely divided iron (hydr)oxides are required for a significant impact. Differences in composition of the organic matter types are illustrated by Fig. 16.5, in which the size of the half-circles reflects the amount of C present. Three types of humin are distinguished on the basis of their presumed reaction pathway: inherited = inherited from the litter; insolubilized = combined with Al and/or Fe into insoluble complexes; humin bound to iron and clay = bound in clay-iron-humus compounds. The terminology for various humus types, used throughout western Europe, is given in Table 16.2.

Bioturbation

Bioturbation stands for mixing of soil material by larger soil fauna elements, such as earthworms and moles, and is the major process counteracting soil horizon differentiation brought about by, for example, leaching and clay translocation. It is responsible for the development of the Ah horizon and plays an important role in soil aggregation. Bioturbation is closely connected to the cycle of organic matter, being most prominent in mull soils, i.e. well-drained, nutrient-rich, neutral to slightly acid soils, while in mor or peat soils bioturbation is negligible. Additionally, it strongly depends on land use, for example being strongly enhanced by transformation of forests into meadows because of the concurrent changes in input (relative increase of root litter and more easily decomposable litter) and soil climate, while being reduced by intensive soil tillage. A clear example is the development of a plough pan as a result of the mechanical stress executed by ploughs, leading to a loss

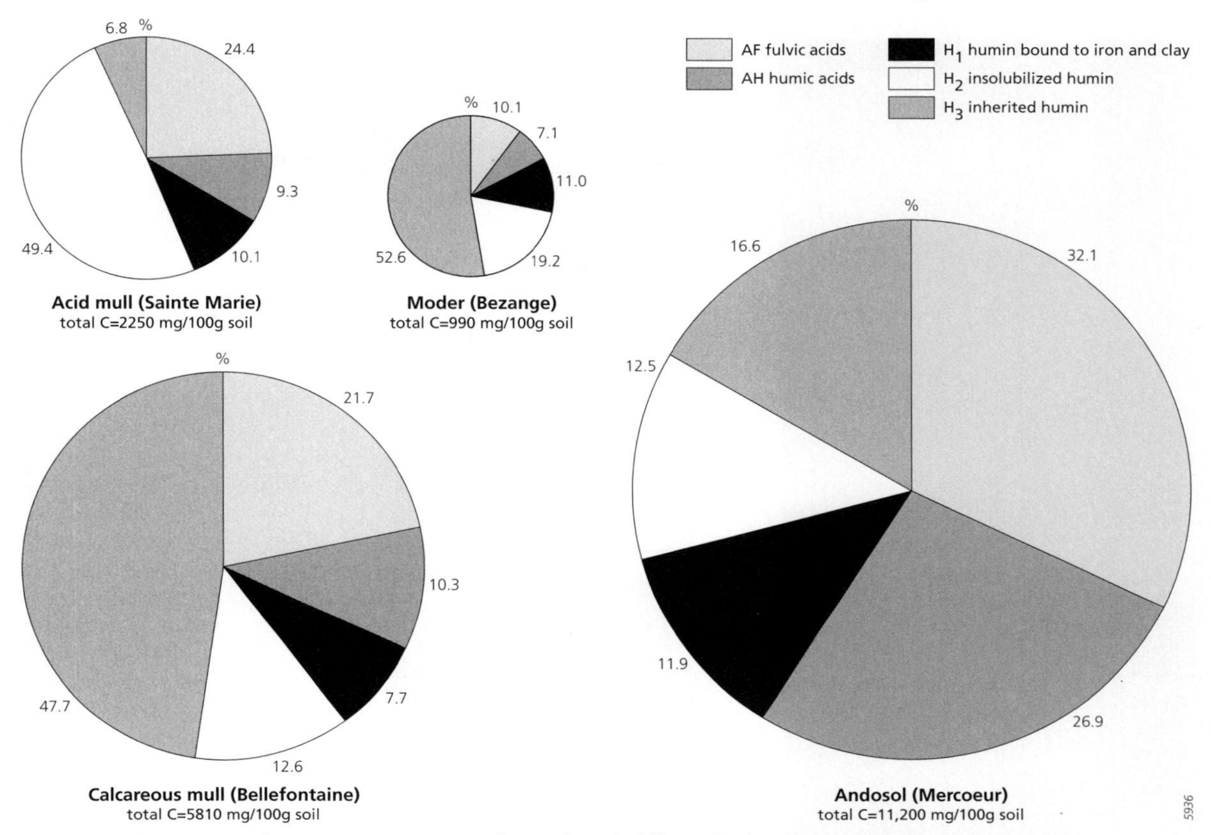

Fig. 16.5. Composition of organic matter in mineral topsoils with different kinds of humus (after Duchaufour 1982).

of biopores and damage to the macro-fauna, inducing a diminished bioturbation and eventually the development of a pan. Bioturbation is thus one of the soil-forming processes, which is most easily affected by man and may lead to rapid changes in soil properties. The role of bioturbation is particularly visible in the recently reclaimed lake bottoms of the central Netherlands, where soil fauna was completely absent. It was only upon the introduction of moles that significant bioturbation took place; a process that could easily be monitored in the field and was associated with major changes in soil properties (Kooistra *et al.* 1985; Jongmans *et al.* 2001). Other soils in which bioturbation plays a prominent role are *Chernozems* and related steppe soils with deep A horizons largely composed of faunal excrements (Fig. 16.7).

Hydromorphism: Gley, Pseudogley, and Stagnogley

Many soils in western Europe are affected by either groundwater, prolonged wetness caused by excess rainfall, or temporary wetness induced by a slowly permeable subsoil. The first category occurs particularly in the coastal areas of Denmark, Germany, The Netherlands, and Belgium, as well as in the estuarine and deltaic areas in France. Also soils in inland valleys frequently show influences of groundwater. The last category abounds on tablelands and flat uplands with finer-textured parent materials and texturally contrasted soils, such as on the Tertiary and Mesozoic sedimentary rocks of the Paris basin and central and southern Germany. In such soils anaerobic and reductive conditions occur, in which iron and manganese are reduced and become mobile. In all cases, both elements are subject to redistribution, a phenomenon described as hydromorphism. In soils with high groundwater tables, the anaerobic and reductive conditions occur not far below the surface. As a result subsoils often have bluish or greyish colours. The topsoil may be oxidized, in which case mull-type organic matter prevails, while, if waterlogged, the topsoil may become peaty. Fluctuating groundwater tables are responsible for alternating reduction/oxidation regimes just above the permanently reduced part of the soil. Here, iron and

Fig. 16.6. Dark-coloured *Andosol* in stratified volcanic ash layers in the French Massif Central near Clermont-Ferrand. The Ah horizon is strongly rooted and very porous and has a low bulk density (volume weight of around 0.7 t/m^3).

Fig. 16.7. Mole burrows and *krotovinas* (filled-in mole burrows) in a *Chernozem* from Germany (photo: ISRIC).

manganese are mobilized during the reduction phase and transported to zones where air enters the soil. This transport is driven by concentration gradients, which are maintained by the continuous oxidation of Fe and Mn in the oxidized zone. Transport and oxidation result in accumulation of yellowish or brownish iron and purplish-black manganese hydroxides along pores, voids, and structural elements. The resulting hydromorphic properties are known as *gley*.

Pseudogley formation is the result of a surplus of rainwater, which remains in the soil for a sufficient long period

TABLE 16.2. *Humus types in relation to drainage (horizontal) and decreasing biological activity (vertical)*

	Aerated conditions	Humid conditions more or less aerated	Temporary waterlogging (fluctuating water table)	Permanently waterlogged (surface water table)
Incorporated humus (stable clay-humus complex)	mull	hydromull		
Incompletely mixed (clay-humus complex often unstable)	moder	hydromoder	anmoor	
Superposed humus (thick A_0 horizon)	mor	hydromor	hydromor	peat

Source: After Duchaufour (1977).

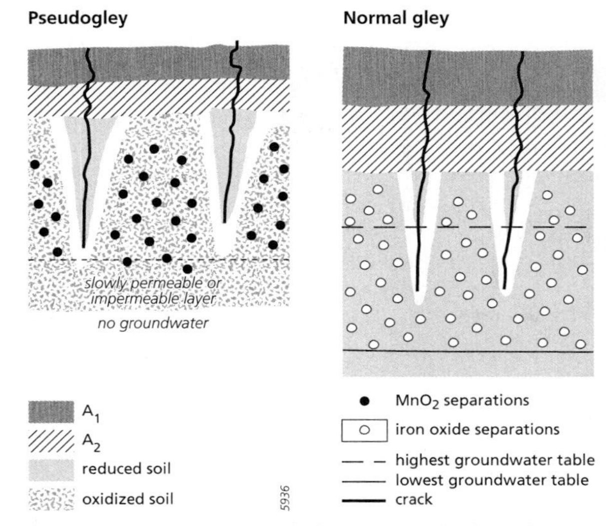

Fig. 16.8. Characteristics of gley and pseudogley (after van Schuylenborgh, 1973).

Fig. 16.9. Pronounced stagnogley features in the lower part of the whitish eluvial horizon and the upper part of the slowly permeable grey subsoil in a Dystric Planosol from Ratum, eastern Netherlands (photo: ISRIC).

to produce a reductive environment. This can be due to a temporary high precipitation, a low hydraulic conductivity, or a combination of both (Segalen 1964; Schlichting and Schwertmann 1973; Begon and Jamagne 1973; Blume and Schlichting 1985; Duchaufour 1998; Scheffer 1998; Scheffer and Schachtschabel 1998). Morphological phenomena associated with pseudogley are opposite to those of gley (Fig. 16.8). Water entering the soil will reduce the iron and manganese along voids and pores, which are then transported to still oxidized parts inside the structural elements. This process results in grey or white colours on the outside of peds, and yellowish, brownish, or reddish colours inside the peds.

Stagnogley is found where water stagnates on an impermeable layer in the soil. Here a pseudo-groundwater level is created. The result of stagnating water in the soil is prolonged wetness above and just below the stagnating layer, which causes iron and manganese to be mobilized and removed from the topsoil, generally through lateral flow. The mobilization of these elements has a number of effects. First, iron does not function anymore as a cementing agent in structural elements, which then tend to destabilize. Secondly, the Fe^{2+} in the soil solution may play a role in the disintegration of clay minerals ('ferrolysis'; Brinkman 1979). In any case, stagnogley tends to enhance the effect of the impermeable layer as clay is removed from the overlying horizon (Fig. 16.9). In time, pseudogley may develop into stagnogley as a result of increasing stagnation of water on the subsoil brought about by combinations of clay translocation, ferrolysis, and gradual loss of soil structure in the eluvial horizon as a result of declining organic matter, clay, and iron content. Examples of such development have been described by Duchaufour (1998). The transition is gradual and the differentiation between pseudogley and stagnogley is often problematic.

Clay Translocation or Lessivage

Clay transport is a common phenomenon in soils of the more continental parts of western Europe, where soils dry out during summertime. Upon wetting at the onset of the autumn rains, the ionic concentration of the soil solution is low, which enhances the dispersion of clay particles. Translocation of clay and associated finely divided iron will then occur by water seeping through the topsoil. In the lower part of the soil, which is still dry, water is drawn into the structural elements by capillary forces and the clay-sized particles are filtered out at the surface of the peds or along pores as clay-iron cutans or

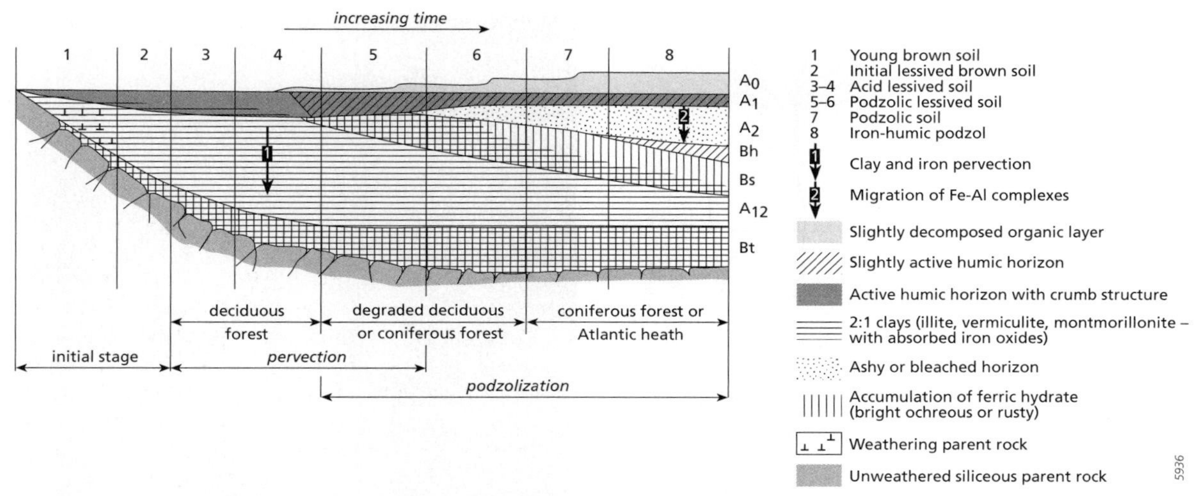

Fig. 16.10. Chronosequence of soils with clay translocation towards degraded soil with secondary podzol, France (after Duchaufour 1977). Soil horizon designation: A_0 = H, $A_1 - A_{12}$ = Ah, A_2 = E.

films. Thus, the upper part of the soil is slowly depleted of clay and iron over time, whereas the subsoil is enriched in these soil components. The process leads to development of the 'argic' horizon (FAO-ISRIC-ISSS 1998), which is characteristic for such soils as *Luvisols* and *Albeluvisols*, as well as occasionally the albic horizon. The removal of dispersed clay has been described as 'eluviation' or 'impoverishment' and the accumulation in the subsoil as 'clay-illuviation'. The process of eluviation and illuviation together is also known as 'lessivage'. In time, prolonged clay translocation will reduce the permeability of the subsoil and thus enhance water stagnation. Eventually, this may lead to pseudogley and even stagnogley, and the topsoil may become so seriously acidified that peptization is no longer possible. In such soils, in which the exchange complex may become largely saturated with Al^{3+} and H^+, clay translocation is no longer an active process and chemical breakdown of clay described as ferrolysis (Brinkman 1979) and even podzolization may occur (Brahy *et al.* 1996). In these soils, the argic horizon can be considered a fossil phenomenon, which under agriculture and concurrent liming and fertilization can be reactivated due to the reduced acidity and higher base saturation of the topsoil. These processes have been described for agricultural soils on loess in the southern part of The Netherlands and in Belgium. Clay translocation is promoted by a pronounced seasonal contrast in precipitation. Such contrast particularly occurs in the somewhat more continental climates of central Germany and France, while under more oceanic climatic conditions contrasts may be weak. Consequently, textural differences between E and Bt horizons are larger, the abruptness of this change is more prominent, and pseudogley more common in the more continental areas (Fig. 16.10).

Podzolization

In acid soils with poor litter decomposition, water seeping through the organic litter layer and the humus-enriched surface horizon takes up organic acids. It enters the acid, often coarse-textured mineral soil below forming chelates with aluminium (with or without iron). They pass through the soil with the percolating water and precipitate deeper in the profile as alumino-organic compounds. Whether iron-organic compounds precipitate depends on the drainage, since under wet conditions most of the iron remains mobile and is removed from the system with the drainage water. The process is known as podzolization, which combines *cheluviation* (eluviation of chelates) and *chilluviation* and is responsible for the formation of *Podzols* in the northern part of western Europe. It gives rise to the ash-coloured albic horizon after which the Podzols have been named (*pod* is 'ash' in Polish) and the illuvial brown- and/or black-coloured spodic horizon. In weaker expressed *Podzols*, such as the *podzosols ocriques* or *sols ocre podzoliques* (CPCS 1967), the albic horizon may be faint or absent, and the occurrence of cheluviation is established by chemical criteria (such as oxalate extractable Fe, Al, and organic matter). Under well-drained conditions and in iron-

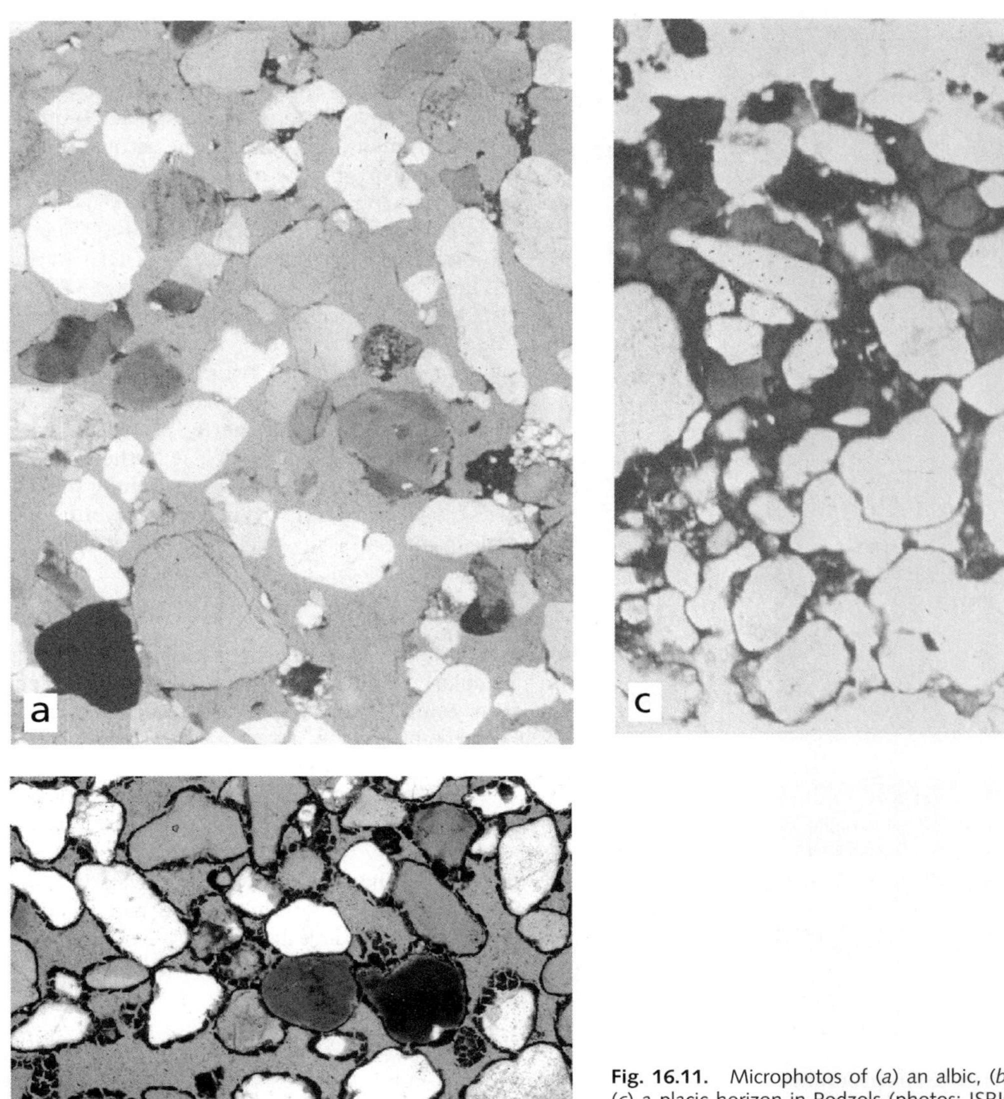

Fig. 16.11. Microphotos of (*a*) an albic, (*b*) a 'humus B' and (*c*) a placic horizon in Podzols (photos: ISRIC).

containing parent materials, the spodic B may differentiate into an upper 'humus B' (Bh) and a lower 'iron B' (Bs) horizon, the latter often consisting of lamellae or iron bands. Various forms of spodic B horizons are recognized such as the strongly indurated *Ortstein* to the completely cemented thin 'Placic' horizon (Figs. 16.11 and 16.12). The dense variants have in common that they often lead to water stagnation, a pseudo-groundwater level, and formation of small lakes or *vennen*.

Less Frequent Processes

Salinization

Saline soils are rare. They may be found on salt-containing sedimentary rocks or sites affected by saline groundwater originating from such rocks, which are relatively common in the Mesozoic sedimentary complexes of, for example, the Elzas and the Paris basin, where Permian and Triassic strata are often high in rock salt.

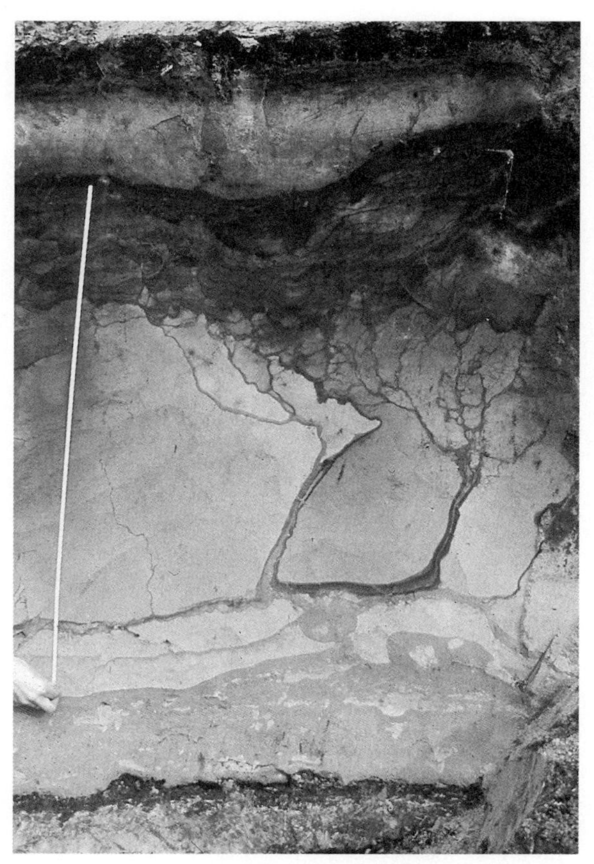

Fig. 16.12. *Podzol* with strongly developed humic spodic B (Bh) and humus fibres from the Lutterzand in the eastern Netherlands, developed in Late Weichselian cover sand, covering an Allerød palaeosol (upper part of the ruler = top of the Bh horizon).

Furthermore, soils with saline groundwater are relatively widespread in the low-lying coastal areas, where inundation with or seepage of seawater is common. The latter holds in particular for the polders of The Netherlands.

Soil Ripening

Soil ripening is the process, which occurs when recent marine or alluvial sediments are drained, either through their reclamation or by natural causes, such as abandonment of river channels and silting up of tidal flats. The main processes involved are loss of excess water, oxidation of primary minerals such as sulphides and of organic matter, and subsidence due to settlement of the solid particles under the influence of gravity. Ripening has been extensively studied in The Netherlands (e.g. Pons and Zonneveld 1965) being highly relevant for land use after reclamation of former sea and lake beds. Of particular concern are the processes that occur upon ripening of sediments containing iron sulphides (notably pyrite). Upon drainage, the soils become extremely acid. This acidity is caused by the oxidation of pyrite in the sediments, producing sulphuric acid in the soils. Characteristically, such soils show orange-coloured depositions of meta-stable jarosite ($KFe(SO_4)_2(OH)_6$) in the subsoil. These 'acid sulphate soils' or 'cat clays' (*Thionic Fluvisols/Gleysols*) have been reported from Jutland, Denmark (Madsen *et al.* 1984), along the North Sea coast in Germany (*Maibolt* or *Schwefelsaure Organomarsch*; Mückenhausen 1977), and from polders in the western part of The Netherlands on old marine clays (de Bakker 1979). In the western provinces of The Netherlands an entire generation of the first farmers to occupy some of the polders established during the nineteenth century had to struggle to survive because of the very low yields they achieved on such soils.

Melanization

In central Germany, in the region around Braunschweig, and particularly in the former East Germany, soils occur with deep, humus-rich, almost base-saturated surface horizons, which strongly resemble the steppe soils—*Chernozems*—occurring in the Ukraine and Russia (Fig. 16.13). These soils are formed under a warmer and more continental climate than that of the present-day (Mückenhausen 1977). Their occurrence can therefore be considered as relic, and many show nowadays forms of degradation such as a decrease of organic matter, depletion of nutrients, and discolouring. The process is described as 'melanization' and has a major bioturbation component.

Soil-Forming Processes in High-Altitude Mountains

The main soil-forming processes are physical weathering, both by freezing/thawing and differences in insulation, snowmelt-induced accumulation of organic matter, and redox-processes. The physical weathering produces shallow and loose, gravelly and stony soils with little fine earth material to fill the cavities. The gravel, stones, and boulders have, characteristically, sharp edges. Frost shattering is the most common phenomena; however, the initial process of differential swell-and-shrink of individual minerals in the rock under the influence of large temperature variations, especially during the summer and on south-facing slopes, together with the establishment of *lichen*, should not be underestimated. This process paves the way for water to enter the rock and, by freezing, to further open small cracks,

Fig. 16.13. (*a*) *Chernozem* near Hildesheim, Germany, and (*b*) earthworms (*Lumbricus terrestris*), responsible for the strong bioturbation in this soil.

finally breaking it up into large boulders and smaller fragments. Water stagnation of snowmelt on still-frozen subsoil in springtime can cause wetness for a significant period, especially in (almost) level areas. Here organic matter can accumulate to give rise to shallow or moderately thick peaty layers. Additionally, the stagnation may lead to reduction of iron and weak cheluviation, giving rise to 'micro-podzols' (e.g. Kubiena 1953), which are quite widespread in the mountains.

Soil Classifications in Western Europe

National Soil Classification Systems

The soil classification systems of France and Germany are mainly based on pedogenetical features and processes, whereas the systems of Belgium, Denmark, and The Netherlands have been devised with a strong linkage to conditions important to local soil use and management. This makes the latter three less widely applicable and less easy to correlate than the former two systems. Table 16.1 shows the correlation between the groups as used in FAO-ISRIC-ISSS (1998) and the corresponding orders or groups used in national classification systems. This table clearly demonstrates the wide variety of criteria used as well as the need for a common international classification system.

France

Several soil classification systems have been used in France during the past forty years. The most well known is the *Classification des sols* (CPCS 1967), which has also been used widely in soil mapping in francophone Africa and Asia. The major units of the system are *Classe* (class), *Sous-classe* (subclass), *Groupe* (group), and *Sous-groupe* (subgroup). The basis of class definition is the degree of profile development, mineral alteration, composition of organic matter, and predominant soil-forming factors. Subclasses are defined by the pedoclimate, groups according to the morphological features, and subgroups by the intensity and expression of the characteristic soil morphology. Recently, pedologists in France developed the *Référentiel Pédologique* (AFES 1995). The system is morphogenetic, based on the notion that the soil cover can be characterized at various organizational levels, the smallest one being the arrangement of elementary soil particles. At higher levels horizons are recognized, which are considered homogenous bodies in the soil cover. The soil system then comprises of three-dimensional associations of horizons, which grade vertically and laterally into each other.

The system as such is not considered a hierarchical classification but a pedological reference system. In 1995, 102 *Références* were identified, but their number is still growing.

Germany

The system of soil classification in Germany, published in 1977, recognizes six levels of classification (Mückenhausen 1977): four main groups or *Abteilungen* (namely, terrestrial soils, hydromorphic soils, sub-hydric soils, and peat), 18 classes, 56 types, and 210 subtypes. The system is based on four main properties of soils: (*a*) direction and degree of leaching, (*b*) build-up of the soil profile (inclusive of the litter layer but exclusive of geological strata), (*c*) permeability of the parent material, and (*d*) the soil dynamics resulting from (*a*), (*b*), and (*c*). Emphasis in the system is on differences in soil-forming processes, which are poorly quantified, leaving many possibilities for differences in interpretations. At the same time this renders the system rather unsuited for practical applications in for example land use planning.

Belgium

The Belgian soil classification is a parametric system. It uses the parent material in terms of textural classes, the drainage class from very dry to extremely wet, and the type of profile development, e.g. chernozemic A-horizon, textural B-horizon, or no profile development, to distinguish between the various soils. Soils are indicated by a combination of letters, e.g. Aba: dry loamy soil with a textural B-horizon.

Denmark

Danish soils are classified according to parent material, pedological development and soil properties related to cultivation (Madsen 1983). The system is hierarchical with four levels: order, group, series, and phase. Soil names stem from other (inter)national systems or are of Danish origin, e.g. *Rendzina* (FAO-UNESCO 1974), *Histosol* (FAO-UNESCO 1974; Soil Survey Staff 1975), *Stagnogley* (Mückenhausen 1977), *Lessivejord* (Duchaufour 1977), and *Brunjord* (Danish).

The Netherlands

The Dutch system of soil classification (de Bakker and Schelling 1989) is based mainly on the nature of the parent material, soil development as expressed by the presence of illuvial humus and/or clay, and type and thickness of the surface horizon. At the highest level of subdivision five main groups are recognized: organic or *peat* soils, mineral soils with little or no pedogenetical features called *vague* soils, mineral soils with a thick, dark-coloured surface horizon called *earth* soils, mineral soils with illuvial humus or *podzol* soils, and mineral soils with illuvial clay termed *brick* soils. At lower levels, human influences, wetness conditions, textural differentiation and degree of ripening of the sediments are introduced. The system uses local, Dutch names to differentiate between soils at the lower levels, which renders correlation with other systems more difficult.

Soil Distribution in Western Europe

For reasons of uniformity and correlation, further use is made of the World Reference Base for Soil Resources (WRB) (FAO-ISRIC-ISSS 1998) to illustrate the soil distribution and variability in western Europe. The WRB has evolved from the subsequent Legend and Revised Legend of the Soil Map of the World (FAO-UNESCO 1974; FAO-UNESCO-ISRIC 1988), and has been accepted by the International Union of Soil Sciences as the international language for soil classification and correlation. Table 16.1 shows the correlation with the national nomenclatures of Belgium, Denmark, France, Germany, and The Netherlands.

From north to south, increasingly more developed soils are found. This is caused by the glacial history of north-western Europe, in particularly Denmark and Schleswig-Holstein in Germany. Here, soils have developed only over some 10,000–15,000 years, when the ice of the Weichselian period retreated. Soils found in the upland parts of these regions are mainly: (*a*) *Regosols*, non-developed soils in medium- to heavy-textured materials and *Arenosols*, non-developed soils in coarse-textured materials, e.g. dunes; (*b*) *Cambisols*, weakly developed soils; and (*c*) *Podzols*, soils with illuvial, cheluviated alumino-organic complexes. In places, liming practices in Denmark have influenced the latter soils so strongly that the process of cheluviation has been reduced considerably or stopped altogether, turning them into 'modern-day palaeosols' or fossil *Podzols*. In poorly drained positions, *Gleysols*, wet soils influenced by groundwater, and *Histosols*, organic soils with accumulated peat, dominate. *Fluvisols*, young alluvial soils, frequently occur in valleys. More to the south, in the sandy North German Plain and the adjacent sandy districts of The Netherlands, *Podzols* prevail, with *Gleysols* and *Fluvisols* occupying the valleys. In Ost Frisia (the north-western part of this region), as well as along the coast of Schleswig-Holstein and in the north-eastern part of The Netherlands, large areas have been empoldered starting some 1,000 years ago (Schlichting 1960; de Bakker 1979). Here *Fluvisols* in marine deposits prevail as well as *Histosols* on marine deposits. To the east (Mecklenburg, Vorpommern), most soils are

Fig. 16.14. Plaggen soil near Brecht, Belgium, with a thick surface horizon overlying a disturbed horizon showing spade marks (photo: Roger Langohr).

Fig. 16.15. Active shifting sand from the central Netherlands (photo: Jan van Mourik).

more or less hydromorphic, either as a result of shallow groundwater or by stagnation on impervious layers such as glacial till, marine deposits, and placic horizons.

Many parts of this region have been strongly altered by man. Century-long agricultural activity has resulted in profoundly modified soils (e.g. the *Plaggenesch* in Germany, the *plaggen bodem* in The Netherlands and Belgium), particularly in those areas where natural fertility is low. Such soils are considered *Anthrosols* (Fig. 16.14). More recently, industrial activity and urbanization have led to highly disturbed areas with often unnatural, man-made soils. These soils classify as *Anthropic Regosols*. The practice of *plaggen* fertilization has led in the past to the development of shifting sands (Koster *et al.* 1993), which covered large areas, particularly in The Netherlands (Fig. 16.15). They have been mostly stabilized by afforestation with pines, but recent environmental policies encourage reactivation of drift-sand regions by removing vegetation and surficial organic horizons. The lowlands of Belgium and The Netherlands comprise mainly of *Fluvisols*, *Gleysols*, and *Histosols*, with *Arenosols* occupying the dune areas along their coast. Here also large man-made areas occur, in particular near the ports of Rotterdam and Antwerp, where *Anthropic Regosols* are found. Also peat excavation for fuel consumption has largely altered the original soils.

The belt of loess deposits, stretching from Belgium and northern France through the southern tip of The Netherlands to the Harz Mountains in Germany and onwards, consists of originally calcareous loess, deposited during the Saalian and Weichselian Ice Age. These soils are mostly decalcified depending on their age and show accumulation of clay in the subsoil, translocated from the upper part of the soil (*Luvisols*). Because of their fertile nature, these soils have been used for agriculture for a long time. This has resulted in widespread truncation of the soil caused by water erosion and filling of valleys with colluvial deposits derived from the loess soils, giving rise to *Regosols*. Some loess soils show clear evidence of their polygenetic nature. Having been subjected to periglacial conditions during the Weichselian, fossil frost wedges and polygonal structural patterns can be found where these soils have not been extensively used. These soils are considered *Albeluvisols*.

The low mountain ranges in Germany (e.g. Harz, Rothaar, Eifel, Schwarzwald), France (e.g. Vosges, Massif Central), and Belgium/Luxembourg (Ardennes) are characterized by shallow (*Leptosols*) and non-developed (*Regosols*) or weakly developed soils (*Cambisols*), and locally *Umbrisols* (soils with a thick, humus-rich, base-desaturated surface horizon, Fig. 16.16). The limited soil development is due to the combination of periglacial denudation, cool climate, and man-induced erosion. Where volcanism has been active (e.g. in the Eifel and the Massif Central), *Andosols* (weakly to moderately developed soils in volcanic ashes) occur.

Fig. 16.16. *Umbrisol* in deeply weathered granite (*arène*) near Langogne, Massif Central, France.

In the large sedimentary basins of western Europe (e.g. the Paris basin, the Rhine graben, the basin of Augsburg in Germany), soils in the low-lying parts are dominantly *Fluvisols* and *Gleysols*. The surrounding areas are occupied with *Cambisols*, *Luvisols*, *Albeluvisols*, and occasionally *Regosols/Leptosols*, of which in particular the *Rendzic Leptosol* (soil with a thick, humus-rich, base-saturated surface horizon over marl) of the northern and eastern parts of the Paris basin has to be mentioned. Soils in these basins are strongly dependent on the parent material, ranging from *Vertisols* in the more continental basins on heavy clays, to *Chromic Luvisols* and *Acrisols* on deeply weathered marl and greywacke. Incidentally, on quartzitic sandstone, as for example near Fontainebleau (France), even *Podzols* occur.

Western France comprises the level, low-lying terrain of the Garonne–Dordogne, the Loire, and the Seine estuaries with the adjacent, gently undulating hills of Normandy and Brittany. The river valleys are occupied by *Fluvisols*, with *Arenosols* along the coast and in places even *Podzols* (Les Landes), whereas in the uplands mainly *Luvisols* and *Cambisols* are found. Occasionally, saline soils (*Solonchaks*) occur in the estuaries where marine influences and evaporation are strong. In the south, the Rhone valley and the Camargue (the Rhone Delta) are occupied by *Fluvisols* with, particularly in the Camargue, important stretches of *Solonchaks*. Along the south coast of France *Luvisols* prevail, with inclusions of *Vertisols* (soils that shrink and swell upon drying and wetting) and *Calcisols* (soils with secondary accumulation of carbonates). It is in this southern area, notably on the marine and river terraces, that older strongly developed soils such as *Chromic Luvisols*, *Nitisols*, and *Acrisols* are rather common, marked by their reddish colour and deep weathering, evidencing earlier warmer climatic conditions.

Fig. 16.17. Scree slope with *Leptosols* and *Regosols* in the French Alps (photo: ISRIC).

The foothills and mountains of the Alps and Pyrenees in Germany and France are characterized by very young soils (*Regosols, Leptosols* in the mountains, *Cambisols, Regosols*, and *Skeletic Leptosols* (extremely gravelly soils) in the foothills) (Fig. 16.17). In the French Jura, mainly *Rendzic Leptosols* and *Phaeozems* (soils with a thick, humus-rich, base-saturated surface horizon) are found. In the south-east of Germany climate becomes more continental with an almost equal annual rainfall and evapotranspiration, and with warm summers and cold winters. Here *Chernozem* soils with a thick, humus-rich, base-saturated surface horizon and presence of secondary carbonates occur, which form the utmost western tip of the belt of *Chernozems* occupying parts of the Czech Republic, Slovakia, Hungary, Romania, Ukraine, and the Russian Federation through into Asia.

An overview of western European soils is given in Sheet V of the *Soil Map of the World* (FAO-UNESCO 1981), the *Soil Map of the European Communities* (CEC 1985), and the forthcoming Soil and Terrain (SOTER) database for Europe, currently under construction at INRA, Toulouse.

Anthropogenic Impacts

Pre-Industrial Anthropogenic Impacts

Widespread deforestation, which accompanied the gradual introduction of animal husbandry, later on followed by agriculture, took place throughout north-western Europe since the beginning of the Neolithic period. Another more direct and major impact that merits attention is the terracing of slopes, sometimes accompanied by irrigation, that was widely practised and had already started in the Roman Period (Fig. 16.18). In the North European Lowlands a complex land use system based on export of nutrients and soil material from mostly heath lands with a low nutrient status to arable lands near farms existed. By continuing this system for many centuries the plaggen soils originated (Fig. 16.19). At the same time, in the 'export areas' soils degraded and became sensitive to wind erosion, giving rise to major drift sand complexes (Koster *et al.* 1993). In specific coastal areas the human impact caused by land reclamation, peat exploitation, and construction of mounds determined the soil landscape at a regional scale.

Recent Anthropogenic Impacts

Since the nineteenth century, when industrialization took off in Europe, man has modified large tracts of land. Not only agricultural land has been affected, but large-scale industries, mining, urbanization, and land reclamation, to name a few activities, have greatly altered the European landscape. Manuring of agricultural soils has shifted from application of organic to inorganic fertilizers and liming. This greatly altered the chemical status of the soils because soil acidity decreased and more optimal doses of individual elements, mainly Na, K, and P, could be administered. This has eventually led to 'precision agriculture', which is practised in some

Fig. 16.18. Formerly intensively cultivated terraces in the source area of the Loire, central France, at an altitude of about 1,200 m a.s.l.

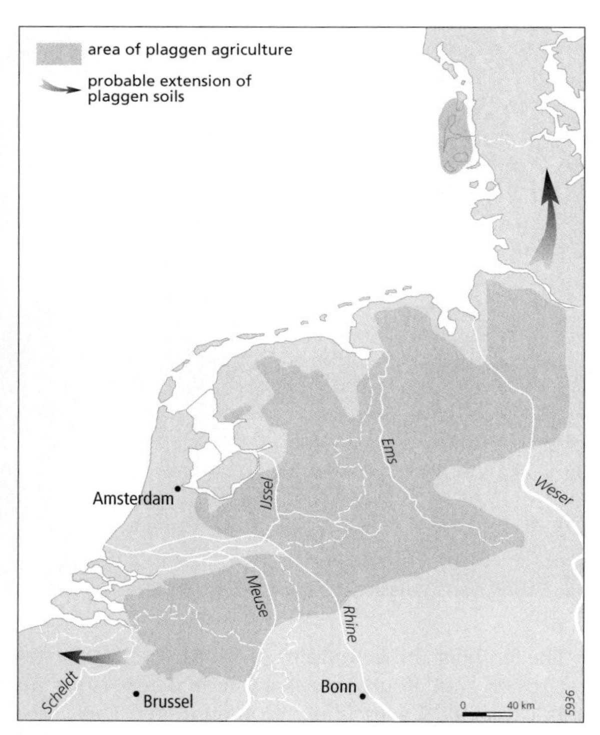

Fig. 16.19. Distribution of plaggen agriculture in north-western Europe, based on the map of Pape (1972), and additional data indicating an extension towards Schleswig-Holstein and Flanders.

technologically highly developed farms (Bouma *et al.* 1999). In precision agriculture, soil nutrient status and crop nutrient requirements are carefully matched by using detailed soil information and GPS (Global Positioning Systems) for on-the-spot application of fertilizer. The aim is to apply optimum amounts of fertilizer in terms of the soil nutrient status and the prevention of losses by leaching of nutrients. The latter is particularly important to prevent groundwater pollution, for example by nitrogen fertilizer, but also contributes to the reduction of costs of fertilizer applications. In addition, controlling groundwater levels, especially in the Low Countries, opened up opportunities to utilize permanently wet soils, e.g. the backswamp soils and peats of the Rhine–Meuse delta, for a wider range of use. A negative side-effect of the lowering of the groundwater level has been the rapid oxidation and subsequent subsidence of peat areas.

Large-scale industrial plants and harbour facilities require solid foundation and, as a consequence, soils have been completely changed to fulfil the requirements. Large parts near the harbours of Antwerp, Hamburg, and Rotterdam have been raised with sand-in-water slurry in order to provide the building grounds for the necessary structures. Such soils are often mixed with building rubble and industrial by-products such as fly ash, sometimes giving rise to very peculiar soils (Zevenbergen *et al.* 1998) (Fig. 16.20). Mining activities, both opencast and underground, have not only changed the landscape by often leaving behind hills of waste, but created, as a result of subsequent restoration, areas with entirely new, man-made soils. Large parts of the Borinage in Belgium, the region around Lille in northern France, the extreme south of the province of Limburg in The Netherlands, the Ruhr area in Germany, and the brown coal-mine areas of former East Germany show many examples of such reshaped lands. Moreover, the industrial activities in these regions have led to strong urbanization creating city agglomerates dominated by urban soils.

Urbi-Anthropic Regosol — Strande, Schleswig-Holstein, Germany

	pH $CaCl_2$	% Org. C	BD g/cm^3	mg/kg Zn	DTPA Cu	extractable Cd	Pb
0– 30 cm	7.6	6.5	0.71	83	52	3	2
30– 60 cm	7.9	13.5	0.45	232	30	13	4
60–120 cm	9.6	13.9	0.50	105	44	6	5

5936

Photo: ISRIC
Data: Institut für Pflanzenernährung und Bodenkunde, Universität Kiel

Fig. 16.20. Chemical composition of a recent man-made soil consisting of urban sludge in northern Germany (photo: ISRIC).

Fig. 16.21. Glass city in the western part of The Netherlands (photo: Stiboka collection).

Horticultural activities, especially in Belgium, Germany, and The Netherlands have resulted in so-called 'glass cities', large areas completely covered with glass-houses (Fig. 16.21). In case of extreme rainfall excess precipitation can be discharged only with great difficulty and flooding occurs during the extreme events. Empolderment of large areas in the former Zuiderzee in the central part of The Netherlands, especially during the years before and after World War II, has created extensive new land. These soils are representative of the former sea-bottom, and are fine-layered and often fine-textured with an initial high water content. Reclamation of these soils has been carried out by initial dredging of drainage canals and by planting large areas with reed in order to extract the excess of water. Land consolidation and landscaping, particularly levelling of existing terraces in order to make large-scale mechanized farming feasible, has altered the soils of the European loess belt considerably, with additional negative side-effects as accelerated soil erosion.

On the other hand, soils on the steep slopes of the Mosel and Ahr rivers in Germany have been strongly modified by terracing for vineyards. Similar trends, i.e. both levelling of terraces and creation of new terraces, can be observed throughout France.

An indirect impact of modern industry all over western Europe has been a massive input of atmospheric fallout, often not originating from the region itself, and being composed of contaminants and nutrients such as nitrogen. At the start, this input was dominated by sulphur dioxide from industrial exhausts and power plants, causing serious acidification of precipitation and consequently of soils and waters. Later on, the contribution of sulphur dioxide declined because of environmental protective measures, whereas deposition of nitrogen compounds, largely from bioindustry, increased and in places reached very high levels, particularly in the Low Countries. This not only led to forest dieback and further soil acidification but also to eutrophication and major changes in vegetation as well as (ground)water pollution (Heij and Schneider 1991).

References

AFES (Association française d'étude des sols) (1995), *Référentiel Pédologique*. INRA, Paris.

Begon, J. C., and Jamagne, J. (1973), Sur la génèse des sols limoneux hydromorphes en France. In: E. Schlichting and U. Schertmann (eds.) *Pseudogley & Gley*. Chemie, Weinheim, 307–18.

Blume, H.-P., and Schlichting, E. (1985), Morphology of wetland soils. *Proceedings. Workshop Wetland Soils IRRI*, 161–76.

Bolt, G. H., and Bruggenwert, M. G. M. (1976), *Soil chemistry, A. Basic Elements*. Developments in Soil Science 5A. Elsevier, Amsterdam.

Bouma, J., Stoorvogel, J., van Alphen, B. J., and Booltink, H. W. G. (1999), Pedology, Precision Agriculture, and the Changing Paradigm of Agricultural Research. *Soil Science Society of America Proceedings* 63: 1763–77.

Brahy, V., Delvaux, B., and Deckers, J. (1996), Undisturbed soils developed in loess: Toposequence in the Bertem Forest. *Pédologie Themata* 2: 42–60.

Brinkman, R. (1979), *Ferrolysis, a soil-forming process in hydromorphic conditions*. Thesis, Agricultural University. Pudoc, Wageningen.

CEC (Commission of the European Communities) (1985), *Soil Map of the European Communities* 1 : 1 000 000. Directorate-General for Agriculture. Luxembourg.

Cornell, R. M., and Schwertmann, U. (1996), *The Iron Oxides. Structure, Properties, Reactions, Occurrence and Uses*. VCH, Weinheim.

CPCS (Commission de Pédologie et de Cartographie des Sols) (1967), *Classification des sols*. INRA, Thivernal—Grignon.

de Bakker, H. (1979), *Major soils and soil regions in the Netherlands*. Junk, The Hague, and Pudoc, Wageningen.

—— and Schelling, J. (1989), *Systeem van bodemclassificatie voor Nederland, de hogere niveaus*. 2nd (rev.) edn. With English summary. Pudoc, Wageningen.

Doebl, F. (1973), Ein 'Aquitan'-Profil von Mainz-Weissenau (Tertiär, Mainzer Becken). Mikrofaunische, sedimentpetrographische und geochemische Untersuchungen zu seiner Gliederung. *Geologisches Jahrbuch* A/A5.

Duchaufour, P. (1977), *Pédologie*, i. *Pédogenèse et Classification*. Masson, Paris.

—— (1982), *Pedology: pedogenesis and classification*. George Allen & Unwin, London.

—— (1998), *Handbook of Pedology. Soils, Vegetation, Environment*. Balkema, Rotterdam.

FAO-UNESCO (1974), *Soil Map of the World 1 : 5 000 000*, i. *Legend*. UNESCO, Paris.

—— (1981), *Soil Map of the World 1 : 5 000 000*, v. *Europe*. UNESCO, Paris.

FAO-UNESCO-ISRIC (1988), FAO-UNESCO *Soil Map of the World. Revised Legend*. World Soil Resources Report 60. Rome.

FAO-ISRIC-ISSS (1998), *World Reference Base for Soil Resources*. World Soil Resources Report 84. FAO, Rome.

Heij, G. J., and Schneider, T. (1991), *Acidification Research in the Netherlands*. Studies in Environmental Science 46. Elsevier, Amsterdam.

Jackson, M. L. (1964), Chemical composition of soils. In: F. E. Bear (ed.) *Chemistry of the Soil*. Reinhold, New York, 71–141.

Jongmans, A. G., Pulleman, M. M., and Marinissen, J. C. Y. (2001), Soil structure and earthworm activity in a marine silt loam under pasture versus arable land. *Biological Fertilized Soils* 33: 279–85.

Kooistra, M. J., Bouma, J., Boersma, O. H., and Jager, A. (1985), Soil structure differences and associated physical properties of some loamy typic Fluvaquents in The Netherlands. *Geoderma* 36: 215–28.

Koster, E. A., Castel, I. I. Y. and Nap, R. L. (1993), Genesis and sedimentary structures of late Holocene aeolian drift sands in northwest Europe. In: K. Pye (ed.) *The dynamics and environmental context of aeolian sedimentary systems*. Geological Society Special Publication 72: 247–67.

Kubiëna, W. L. (1953), *Bestimmungsbuch und Systematik der Böden Europas*. Enke, Stuttgart.

McBride, M. B. (1994), *Environmental chemistry of soils*. Oxford University Press, New York.

Madsen, H. B. (1983), *A Pedological Soil Classification System for Danish Soils*. Ministry of Agriculture, Bureau of Land Data (ADK), Vejle.

—— Jensen, N. H., Jacobsen, B. H., and Platou, S. W. (1984), *Potentielt svovlsure jorder i Jylland*. Dansk Landbruksministeriet/ Miljøministeriet.

Mückenhausen, E. (1977), *Entstehung, Eigenschaften und Systematik der Böden der Bundesrepublik Deutschland*. DLG-Verlag, Frankfurt am Main.

Pape, J. C. (1972), Oude landbouwgronden in Nederland. *Boor en Spade* 18: 85–115.

Pons, L. J., and Zonneveld, I. S., (1965), *Soil ripening and soil classification. Initial soil formation in alluvial deposits and a classification of the resulting soils*. ILRI Publication 13, Wageningen.

Scheffer, F., and Schachtschebel, P. (1998), *Lehrbuch der Bodenkunde*. Enke, Stuttgart.

Schlichting, E. (1960), *Typische Böden Schleswig-Holsteins*. Schriftenreihe der Landwirtschaftlichen Fakultät der Universität Kiel 26.

—— and Schwertmann, U. (1973), *Pseudogley & Gley*. Verlag Chemie, Weinheim.

Segalen P. (1964), *Le Fer dans les sols*. ORSTOM, Paris.

Soil Survey Staff (1975), *Soil Taxonomy. A basic system of soil classification for making and interpreting soil surveys*. Agricultural Handbook 436. Soil Conservation Service, US Dept. of Agriculture, Washington, DC.

Stevenson, F. J. (1994), *Humus chemistry. Genesis, Composition, Reactions*. Wiley, New York.

Swift, R. S., and Spark, K. M. (eds.) (1998), Understanding & managing organic matter in soils, sediment and waters. *Proceedings of the 9th International Conference of the International Humic Substances Society*. University of Adelaide, Adelaide, Australia, 21–25 September 1998.

van Schuylenborgh, J. (1973), Sesquioxide formation and transformation. In: E. Schlichting und U. Schwertmann (eds.) *Pseudogley & Gley*. Verlag Chemie, Weinheim, 93–102.

Zevenbergen C., van Reeuwijk, L. P., Bradley, J. P., Coomans, R. N. J., and Schuiling, R. D. (1998), Weathering of MSWI bottom ash with emphasis on glassy constituents. *Journal Geochemical Exploration* 62: 293–8.

17 Forests and Forest Environments

Josef Fanta

Introduction

North-western Europe has on various counts a very heterogeneous character. Crystalline and metamorphic bedrocks of various ages and Tertiary and Quaternary deposits define its geology and geomorphological features. The area belongs to several climatic zones and parts of it went through quite different processes during their Quaternary development. All these aspects were of essential importance for forests—their origin, development, species composition, structural features, and the character of their environments. During the postglacial period favourable climatic conditions enabled trees to migrate from the refuges in the south and south-east of Europe to the north and north-west. With the exception of extreme conditions all the dry land of north-western Europe was covered with forests whose species composition varied, depending on local conditions of the physical environment. Natural woods and forests, both closed and open and continuously changing in time, contributed greatly to natural landscape diversity. Since the Neolithic and especially in the Middle Ages, human influence becomes the crucial factor of forest development, the impact being superimposed on natural conditions and evolutionary processes. Man not only drastically reduced the forested area in Europe, but the use of forests over several millennia also strongly changed the conditions for the functioning of forests as natural ecosystems. As a result, the man-made forests of today often have little in common with natural forest communities, which once covered the European continent. Nevertheless, even these man-made forests have important functions: they greatly influence the local climate and the hydrological regime of the landscape; they protect steep slopes against erosion and are an important source of biodiversity; and they contribute strongly to the variety of landscape structure as well as to the protection of the environment.

This chapter provides a general survey of the phytogeographical, palaeoecological, and environmental aspects of forests in north-western Europe. For a proper insight the following components are taken into consideration:

- the abiotic component (the physical environment: topography, climate);
- the phytogeographical component (horizontal distribution and altitudinal zonation);
- the historical component (postglacial development, early impact of humans on forests);
- the ecological component (distribution and ecological properties of trees, main forest types);
- the forest use component (organized forestry and its development and the present situation of forests and forestry.

Topography and Climate

The significance of topographical and climatic features for forests and their environment is shortly summarized. The main topographical feature of Europe is its west–east running mountain ranges giving rise to a distinct NW–SE altitudinal gradient. This involves extended lowlands in north-western regions, uplands (*Mittelgebirge*) in the central part of the area, and a chain of high mountains in the south and south-east. This specific topography has a threefold effect for forests (Jahn 1991):

- the influence of the warm Mediterranean climate is obstructed by high mountains in the south;

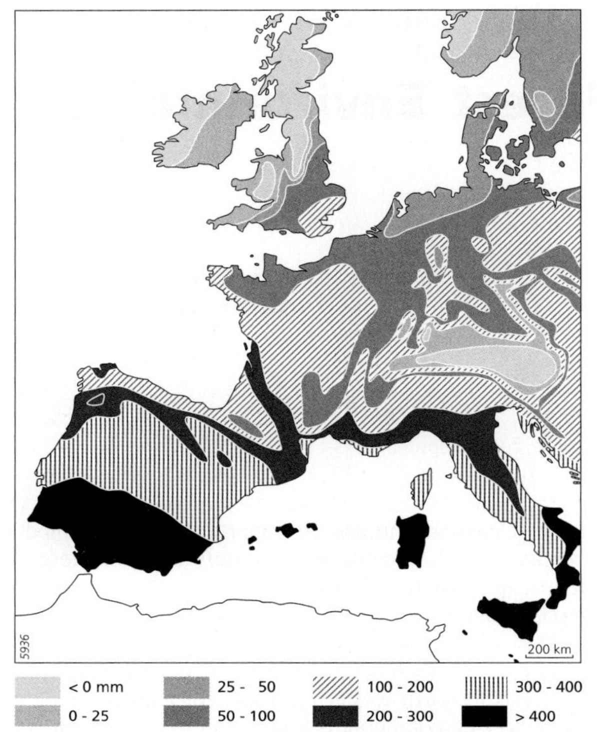

Fig. 17.1. Climatic precipitation deficit in mm, expressed as potential evapotranspiration less total precipitation, during June through August (after Blüthgen and Weischet 1980).

- the migration of southern floral elements to the north is impeded;
- with the prevailing westerly winds the moist air masses born above the Atlantic Ocean and the North Sea are able to penetrate deep into the continent.

Due to the last feature, the whole area is under the strong influence of the Atlantic climate. A distinct west–east climatic gradient with transition from Atlantic to Subatlantic and central European climate exists. Specific climatic conditions have a prominent influence on composition, distribution, and altitudinal zonation of forests (e.g. the typical asymmetry of vegetation and forest altitudinal zonation in mountain ranges due to rain shadow).

A rather small part of the area (south France) has distinctive climatic features of the Mediterranean, different from the Atlantic and Subatlantic climate. Southern France has a very mild climate with warm winters and sunny, cloudless summers with maximum sunshine. Rainy periods are in autumn, winter, and spring. Specific vegetation and evergreen forests are characteristic of this area.

Regional differences in climatic conditions and their impact on vegetation and forests are best illustrated by the precipitation deficit, expressed as the difference between the potential evapotranspiration and total precipitation. Figure 17.1 shows the pattern of this deficit during the summer months June to August. The deficit rises from < 0 in the Alps and higher uplands in south Germany to about 300 mm in southern France and even to > 400 mm in the western Mediterranean.

Phytogeographical Division

The phytogeographical division of the area stems from the west–east climatic and NW–SE altitudinal gradients and their influence on vegetation. The division of Europe into phytogeographic regions and provinces by Meusel *et al.* (1965/84), further adapted by Jahn (1991), is used to demonstrate the horizontal pattern and vertical zonation of vegetation and forests (Fig. 17.2). The horizontal division into phytogeographical provinces reflects the decreasing influence of the Atlantic climate along the west–east gradient as well as the influence of the Mediterranean Sea and the Alps. Moreover, the altitudinal zonation strongly contributes to the spatial diversity of climatic features and, consequently, to the diversity of forests.

Horizontal Division

(See Figs. 17.3 and 17.4 for topographic information.)

The Atlantic Province

Characteristic climatic features of the coastal regions from France to Denmark are a high variability of weather conditions, rather high precipitation, relatively small temperature differences between summer and winter, high air humidity, and strong winds. Lasting snow cover is exceptional. In the northern part, along the North Sea coast, forests grow on medium to poor, silicate-rich soils of mostly glacial origin. Beech (*Fagus sylvatica*) and oaks (*Quercus robur* and *Q. petraea*) dominate in forest communities accompanied by birches (*Betula pendula* on dry and *B. pubescens* on wet soils). The understorey is poor in species, with *Ilex aquifolium* and *Lonicera periclymenum* as characteristic shrubs, and *Vaccinium myrtillus* and *Deschampsia flexuosa* as typical herbal species. Deforested landscapes covered with heathlands were mostly converted into arable land and later partly afforested with Scots pine (*Pinus sylvestris*) and to a lesser extent with Douglas fir (*Pseudotsuga menziesii*) during the nineteenth and twentieth centur-

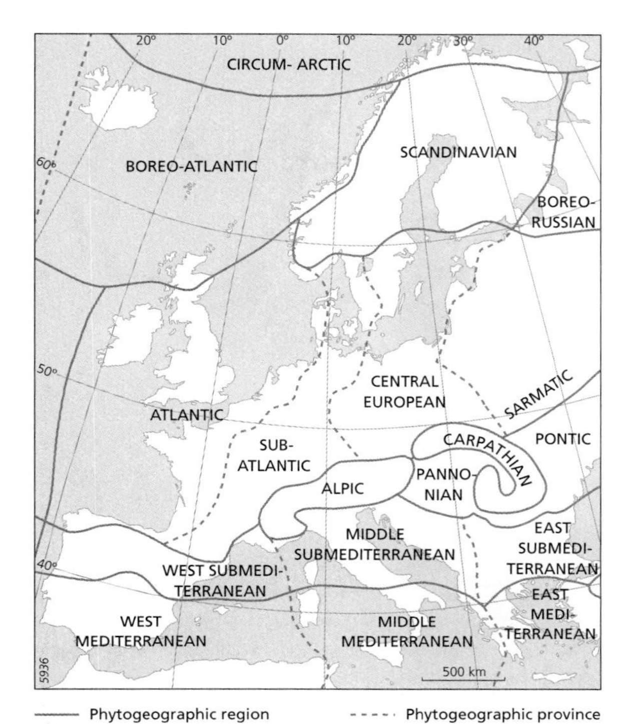

Fig. 17.2. Phytogeographic regions and provinces (after Meusel *et al.* 1965/84).

ies. The north-western part of the province along the Atlantic coast (from Bretagne to Belgium) bears poor soils originated from crystalline rocks and sandstone. Beech (*Fagus sylvatica*) is the dominant species in the tree layer; oaks (*Quercus robur* and *Q. petraea*) are admixed. The understorey of evergreen species gives these Atlantic forests a specific physiognomy formed by *Taxus baccata, Buxus sempervirens*, and *Hedera helix*. The field layer is rich in ferns. Due to intensive use and misuse of these forests in the past (deforestation, coppice management), only small remnants have survived, mostly in the form of coppice woodlots. In contrast to the more inland situation, the beech has been able to survive here due to vegetative resprouting. Extensive deforested areas today are covered with Atlantic heathlands. Dry heathlands are dominated by *Calluna vulgaris, Erica cinerea*, and *Ulex europaeus*. Under wet conditions *Erica ciliaris, E. tetralix*, and *Ulex minor* prevail. Extensive areas of these degraded landscapes have been afforested in the nineteenth and twentieth centuries using exotic coniferous tree species such as *Pinus pinaster* and Douglas fir (*Pseudotsuga menziesii*).

The extensive lowlands of the Parisian basin were originally covered by beechwoods. In the western part of the area oakwoods were dominant: *Quercus robur* along the coast, *Quercus petraea* inland, and *Quercus pubescens* in the warmer valleys of the river Loire and its tributaries. Hornbeam (*Carpinus betulus*) and silver birch (*Betula pendula*) are also present. Today, beech is absent over most of the area, its occurrence being limited to several small isolated islands along the Loire river. Due to the ancient colonization and intensive agriculture, forest cover has been strongly reduced. In the broad area along the French Atlantic coast (Les Landes, La Garonne) beech (*Fagus sylvatica*) is absent and oaks dominate the tree layer (*Quercus robur* and *Q. petraea*, with *Q. pyrenaica* and *Q. pubescens* admixed). Human colonization over several millennia has completely changed the landscapes: original woodlands have been replaced by heathlands and blown sands. Forest restoration activities in the past two centuries were dominated mostly by commercial intentions, using mainly *Pinus pinaster* and *P. maritima* for afforestation.

The Subatlantic Province

This phytogeographical unit takes in a broad north–south running belt from south-western Sweden to southern France (Fig. 17.2). The strongly varying topography and hydrology, as well as a distinct north–south gradient (relief and temperature), are at the roots of a variety of site conditions and a corresponding richness in forest types. The climate is temperate with prevailing Atlantic features at higher altitudes in the uplands. Snow cover is present in the winter and the sequencing of seasons is clearly developed as a result of higher temperature differences between the warm and cold parts of the year. The extensive plains in the northern part of this belt bear deciduous forests dominated by beech (*Fagus sylvatica*). The main forest types are:

- acidophilous beech forests on acid sandy and sandy-loam soils;
- oak-beech forests on medium-rich sites with groundwater influence (with *Quercus petraea*);
- birch-oak forests on very dry and poor soils (with *Quercus robur*);
- hornbeam-oak woodlands on gley and stagnogley soils (with *Quercus robur*);
- alder woods on sites with stagnating water.

Forests in the area were already destroyed on a large scale in prehistoric times and then survived only in remnants on sites not suitable for agricultural production. During the past two centuries mainly conifers were used for afforestation, with Scots pine (*Pinus sylvestris*) prevailing. The uplands (*Mittelgebirge*) cover an extensive area with very different site and forest growth conditions.

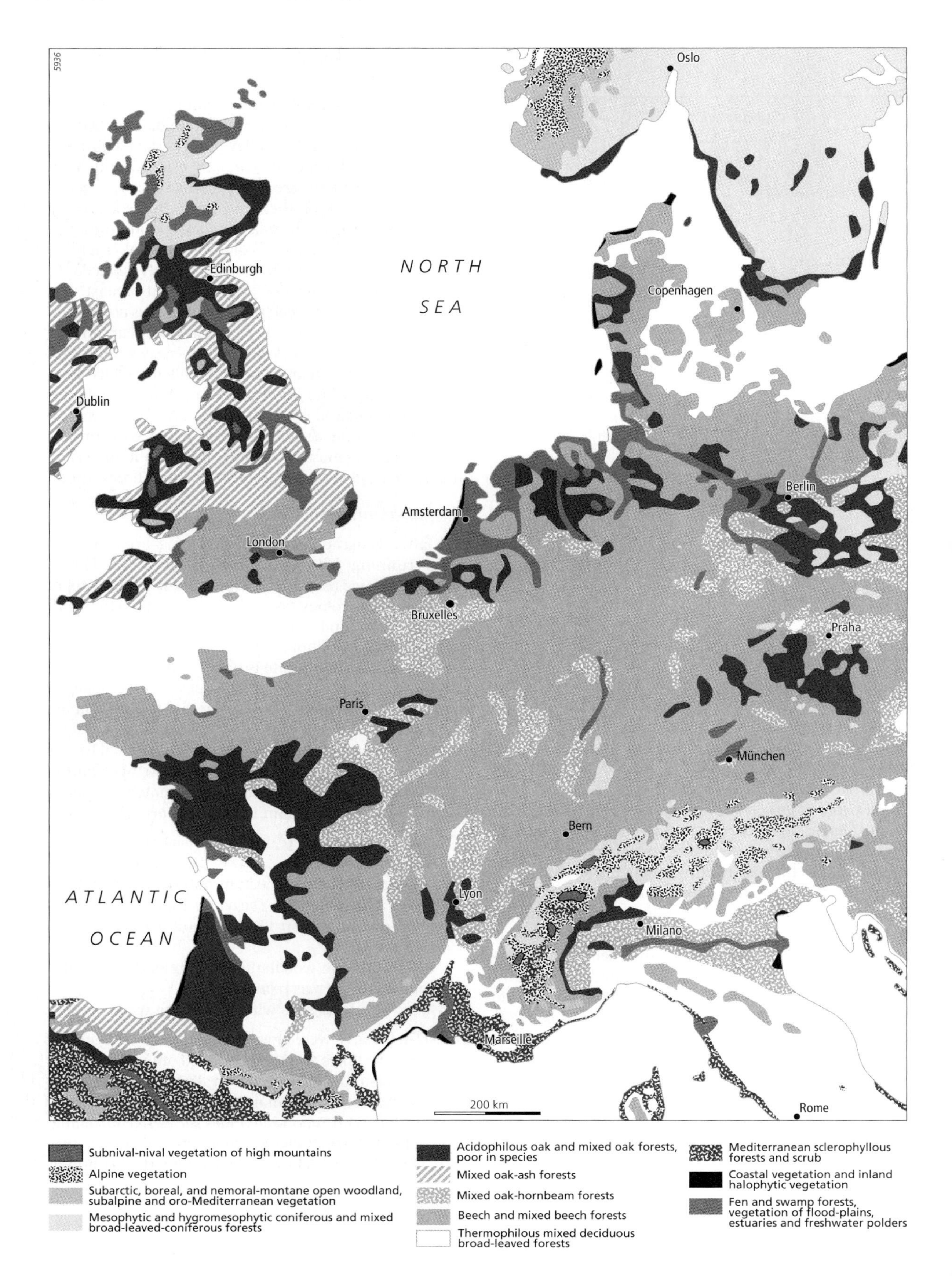

5936
Oslo
NORTH SEA
Edinburgh
Copenhagen
Dublin
Berlin
Amsterdam
London
Bruxelles
Praha
Paris
München
Bern
ATLANTIC OCEAN
Lyon
Milano
Marseille
200 km
Rome
Subnival-nival vegetation of high mountains
Alpine vegetation
Subarctic, boreal, and nemoral-montane open woodland, subalpine and oro-Mediterranean vegetation
Mesophytic and hygromesophytic coniferous and mixed broad-leaved-coniferous forests
Acidophilous oak and mixed oak forests, poor in species
Mixed oak-ash forests
Mixed oak-hornbeam forests
Beech and mixed beech forests
Thermophilous mixed deciduous broad-leaved forests
Mediterranean sclerophyllous forests and scrub
Coastal vegetation and inland halophytic vegetation
Fen and swamp forests, vegetation of flood-plains, estuaries and freshwater polders

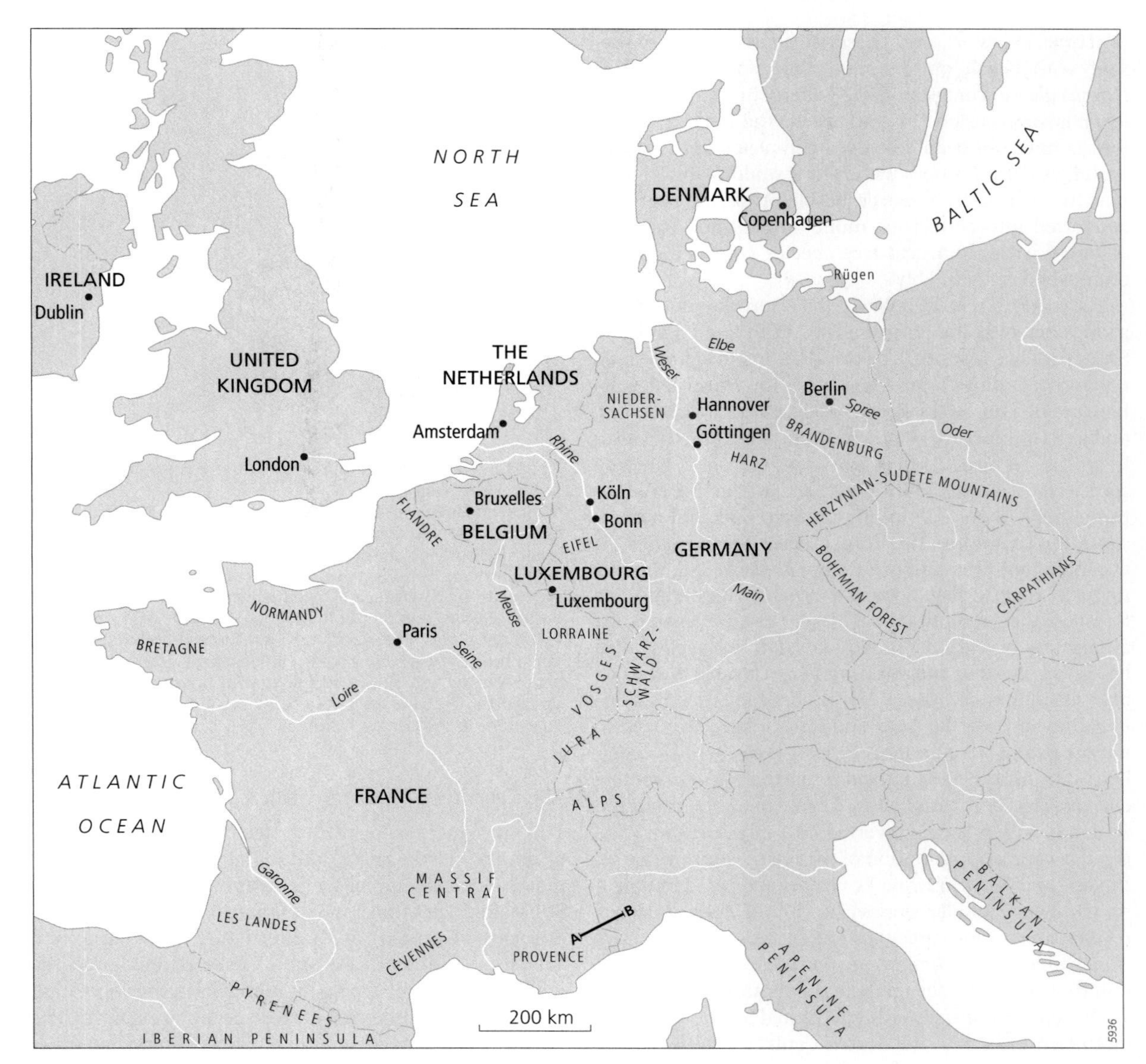

Fig. 17.4. Location map.

The altitudinal vegetation zonation in the higher mountain ranges is strongly developed, but the position of particular mountain groups in the north–south gradient leads to distinct differences among particular mountain ranges. With the exception of specific sites (e.g. extremely dry or wet, alluvial carrs) the Subatlantic uplands are the domain of beech (*Fagus sylvatica*). Jahn (1991) divides this area into three latitudinal sections:

The northern uplands involve mountain groups as far south as the river Main and westwards of the river Rhine. In this area, beech (*Fagus sylvatica*) is dominant in forest communities, especially on acid soils. At lower elevations sessile oak (*Quercus petraea*) is admixed. On mesotrophic sites sycamore (*Acer pseudoplatanus*), ash

Fig. 17.3. (*opposite*) Map of the natural vegetation (original scale 1 : 2.5 mill.) (simplified after Bohn and Neuhäusl 2001).

(*Fraxinus excelsior*), and cherry (*Prunus avium*) are present with beech on the mountain slopes while elm (*Ulmus glabra*) comes in along watercourses. The herbal layer is species-rich. In contrast to the plains, upland forests have not been destroyed completely and survive on hilly sites and slopes with rather shallow soils. In the past two centuries, however, beech forests were broadly converted into coniferous monocultures and the area of forest with a natural tree species composition has diminished considerably.

The middle part of the upland complex involves mountain groups between the Lorraine Hills in the north and the Swiss Jura in the south. The area has favourable climatic and soil conditions for forest growth. Natural forests here are rich in species. Beech (*Fagus sylvatica*) dominates, and a rich admixture of the European silver fir (*Abies alba*) is typical. With the exception of the Vosges, Norway spruce (*Picea abies*) is naturally present in the higher montane (oreal) zone and, locally, in deep and cold depressions and ravines. In the Vosges, beech (*Fagus sylvatica*) accompanied by sycamore (*Acer pseudoplatanus*) grows up to the highest elevations and forms the timberline. For forests of the submontane and lower montane zones, an admixture of various broad-leaved tree species with beech is characteristic, forming very variable and valuable forest stands. These mountains have been colonized since the Middle Ages and forests have survived to a large extent only on inaccessible slopes, while accessible areas have been changed to pastures. For commercial reasons, beech and silver fir forests in the montane zone have often been converted into spruce stands. The regular vegetation zonation of the north–south running Vosges and Schwarzwald is remarkably asymmetrical on windward and lee slopes (Fig. 17.5). Western slopes receive much more precipitation; eastern slopes lie in a rain shadow with yearly precipitation equalling only half that of the western part of the mountains.

The southern uplands are represented by the extensive mountain groups of the Massif Central, the Cévennes, and the eastern slopes of the Pyrenees. The influence of Atlantic (in west and north-west) and (sub)Mediterranean (in south and south-east) climatic features is prominent and mirrored in the vegetation zonation in particular mountain ranges. The woodlands of the submontane and montane zones are formed by down oak (*Quercus pubescens*) and *Buxus sempervirens*, with mixtures of *Acer* ssp., *Fraxinus excelsior*, *Tilia platyphylos*, and *Sorbus aria*. At their upper limits, which may lie between 1,000 and 1,500 m a.s.l., *Quercus pubescens* is replaced by *Quercus petraea*. Beech (*Fagus sylvatica*) is present to some extent on cool, north-facing slopes and at the highest elevations.

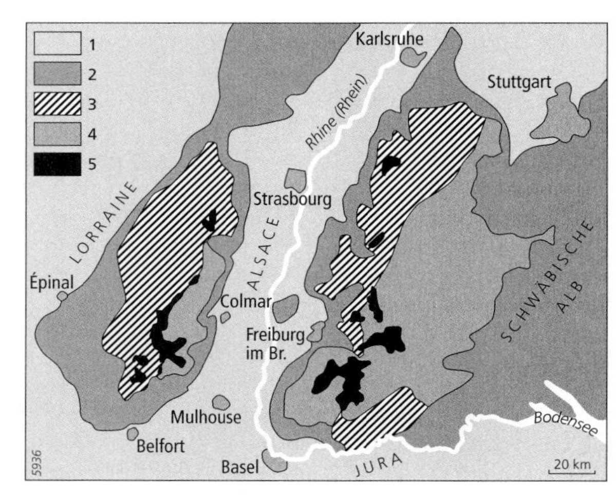

Fig. 17.5. Zonation of vegetation in the Vosges (left) and the Schwarzwald (right).
1. Plain and lower foothill zone with riparian woodlands or plantations in the alluvium of the Rhine, and *Quercus pubescens* woodlands or vineyards on the adjacent slopes. 2. Foothill zone with *Carpinus betulus* and *Quercus petraea*, and submontane zone with *Fagus sylvatica*. 3. Montane and upper montane zone on the western slopes of both mountain massifs, with *Fagus sylvatica* and *Abies alba* woodlands. 4. Montane and upper montane zone on the eastern slopes, with dominant *Abies alba* (locally with *Pinus sylvestris*). 5. Oreal zone with *Fagus sylvatica* in the Vosges and *Picea abies* in the Schwarzwald (after Jahn 1991, based on Ozenda *et al.* 1979).

The Central European Province

This province is a rather heterogenous area with a similar north–south altitudinal gradient and geomorphological and phytogeographic structure to the Subatlantic province. The climate is typically central European, transient, with alternating Subatlantic and subcontinental features. Snow cover is regular in the winter months; most precipitation falls in the vegetation period. Atlantic tree and herbal species are absent. The northern plains and lowlands of glacial (loamy moraine) deposits were originally covered with mesotrophic beechwoods. Most of these woodlands were cleared or converted into Scots pine monocultures, but remnants are still present throughout the area and along the Baltic coast (e.g. Rügen). The central hilly part between the rivers Weser, Elbe, and Oder (south of Berlin) has a very heterogeneous character, which is reflected in the composition of the original woodlands. Mixed oak, birch, and hornbeam forests alternated with beech-dominated woodlands on drier sites. An extensive complex of floodplain forests is characteristic of lowlands along the rivers Elbe and Spree. A large part of this area was deforested and converted into agricultural land. But large remnants of semi-natural floodplain forests still exist there. These show a typical zonation with willow and poplar groves on sites undergoing yearly flooding, oak, ash, and elm forests on higher embankments, and alder carrs on sites with stagnating water. Through the area runs the western border of hemiboreal and nemoral Scots pine forests, which occupy dry and nutrient-poor sites, often mixed with deciduous trees (oak, birch). The central and west European uplands as far as the foothills of the Alps are the domain of upland mesotrophic beech woodlands and mixed beech–fir forests (Hartmann 1974; Hartmann and Jahn 1967). At higher elevations, Norway spruce is also found, and in the upper montane zone it dominates. But most original deciduous and mixed woodlands had been cleared and the remaining forests converted into Norway spruce monocultures.

The West Sub-Mediterranean Province

This province coincides with southern France between the eastern Pyrenees and the western Alps (Provence). Subatlantic oakwoods with *Quercus pubescens*, *Q. ilex*, and *Buxus sempervirens* represent the native woodlands in this area. Extensive areas of degraded land, resulting from several thousand years of colonization, have now been afforested with *Pinus halepensis* and, to a lesser extent, with *Pinus maritima*. Towards the foothills of the Massif Central and to the north, sweet-chestnut forests (*Castanea sativa*) represent the millennia-long tradition of forest culture. The steep slopes of the Mediterranean coast are unable now to bear forest growth as they are fully degraded and eroded due to past deforestation, forest grazing, and other human impacts. As substitute communities, various types of shrubby thorny vegetation formations (*garrigues*, *maquis*) have developed.

The Alpic/Alpine Region

The Alps, with highest elevations exceeding 4,000 m a.s.l., exhibit a harsh climate with low temperatures and short summers, high precipitations, and long winters with a lasting and thick snow cover. At high elevations and on north-exposed slopes snow cover may persist throughout the year. According to the orographic structure, the Alps can be divided into the Pre-Alps, Intermediate Alps, and Inner Alps. Due to the strong climatic influence, the western Alps reveal a specific zonation of forest communities (Fig. 17.6) with thermophilous broad-leaved tree species prevailing in the sub-Mediterranean south-west and coniferous trees (*Abies alba*, *Picea abies*, *Larix decidua*) on northern slopes and at high altitudes. The height of the alpine timberline strongly depends on orography, climatic features, and

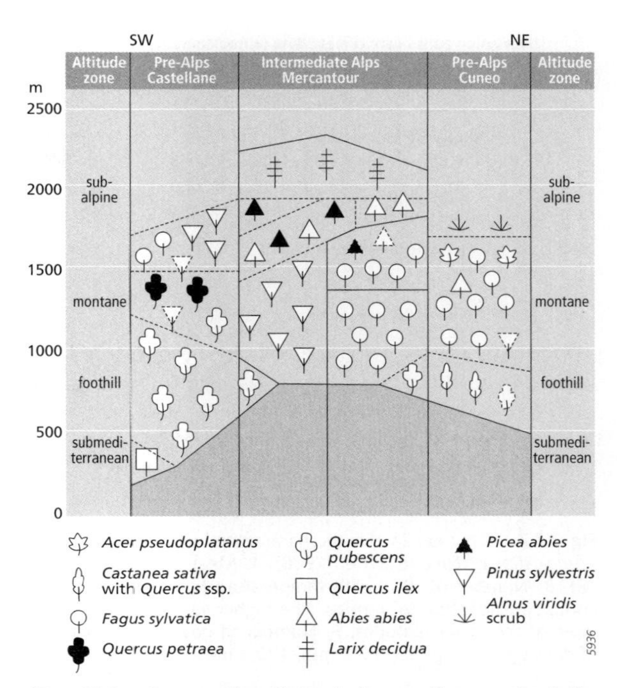

Fig. 17.6. Cross-section through the southern part of the western Alps (Castellane/France to Cuneo/Italy, see profile A–B in Fig. 17.4) (modified after Mayer 1984).

soil. Green alder (*Alnus viridis*) forms thick shrubby growth along torrents and on wet slopes (alder torrents) not suitable for high forest growth.

The Altitudinal Zonation

Uplands in north-western Europe do not form an extensive orographic complex. They are separated into rather small upland units of various altitudes—in fact, mountain islands surrounded by lowlands. In this situation, the prevailing climatic features in particular phytogeographic provinces have a strong influence on the local mesoclimatic features in each of the mountain ranges. Haeupler (1970) derives the altitudinal zonation in European mountains from the climatic zones, the geographical location, and the extent and height of the mountains. With regard to these aspects the altitudinal and vegetation zonation differs strongly in particular west-European mountain ranges (Fig. 17.7). Only the Alps with their huge mass and extent create their own climate with clearly continental features. This is reflected in the absence of deciduous tree species in the Inner Alps; their place is taken by Norway spruce (*Picea abies*). With increasing altitude precipitation increases and the average annual temperature decreases by about 0.6 °C per

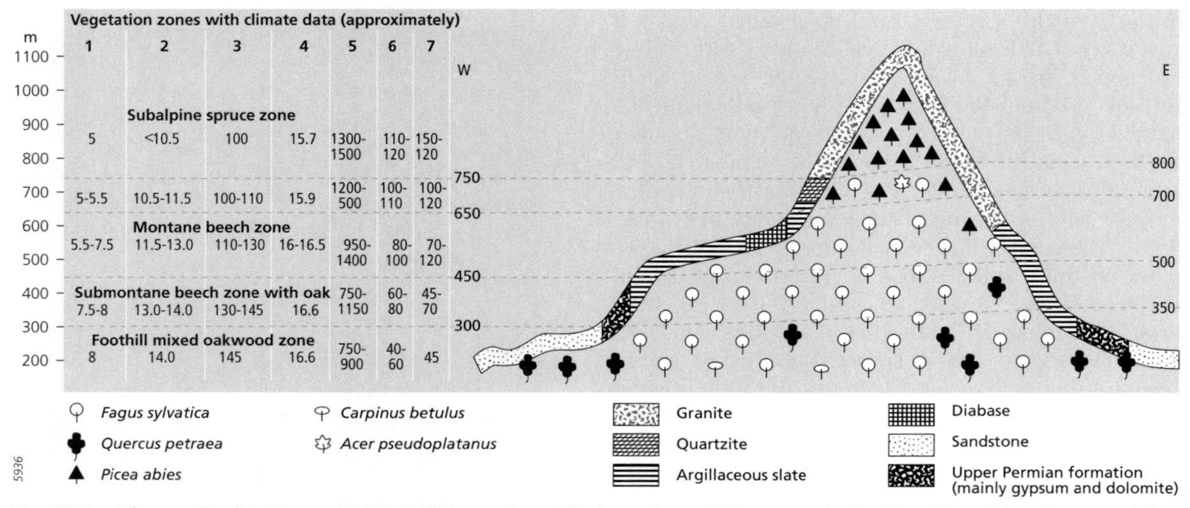

Fig. 17.7. Changes in climatic parameters with increasing altitude, and vegetation zones in the Harz Mountains, Germany (after *Klima-Atlas of Niedersachsen* 1960). 1. Mean annual temperature (*t*, °C). 2. Mean temperature growing period (May–September, °C). 3. Number of days with temperature higher than or equal to 10 °C. 4. Mean annual range of temperature (°C). 5. Mean annual precipitation (*p*, mm/a). The higher values apply to the windward sides or the higher altitudes, the smaller values to the lee sides or the lower altitudes. 6. Number of days with snow cover > 0 cm. 7. Index of dryness $p/t + 10 \times k/120$ (k = mean number of days with precipitation > 1 mm; 120 = mean number of days with precipitation > 1 mm within the area of former West Germany).

100 m altitude. As a result, the length of the vegetation period shortens and, consequently, the growing conditions for forest trees change considerably. In mountains strongly influenced by the Atlantic climate deciduous forests rise to the highest altitudes and form the timberline (Massif Central, Vosges). Mountain ranges influenced by the subcontinental climate (Harz, Bohemian Forest, Schwarzwald, and especially the Alps) bear, at their highest altitudes, a distinct belt of coniferous tree species, which form the timberline. Deciduous trees generally do not grow higher than the higher montane (oreal) zone.

In Europe the following altitudinal zones are distinguished:

- *The plains and foothill zone* (planar and colline zone) in the lowlands. These zones today bear only remnants of former deciduous forests. The most important species originally were oaks (*Quercus robur* and *Q. petraea*), hornbeam (*Carpinus betulus*), lime (*Tilia cordata*). In areas strongly influenced by the atlantic climate, beech (*Fagus sylvatica*) can play a role in these lowland forests.
- *The submontane zone* in the foothills of mountain ranges. This zone was originally covered by mixed deciduous forests with beech and oaks as the main tree species. Admixture of other broadleaf trees, e.g. ash (*Fraxinus excelsior*), elms (*Ulmus* ssp.), and others strongly depends on local site conditions.
- *The montane zone* covers altitudes between about 400–1,000 m a.s.l. In the west European mountains, beech (*Fagus sylvatica*) is dominant. European silver fir (*Abies alba*), a shade-tolerant species, accompanies beech in most west European mountain ranges. In the higher montane zone, sycamore (*Acer pseudoplatanus*) forms the natural admixture to beech. In mountain ranges influenced by subcontinental climate Norway spruce (*Picea abies*) is present.
- *The subalpine zone* forms a (usually small) belt between the closed montane forests and the timberline. The climate is harsh and makes the existence of a closed forest impossible. In particular west European mountain ranges, a distinct difference between the atlantic and central European/subcontinental mountains can be observed in vegetation and forest composition. In the former ones beech is the most important constituent of forests, accompanied by sycamore; in the latter ones subalpine forests are dominated by Norway spruce (*Picea abies*) with a small admixture of beech (*Fagus sylvatica*), rowan (*Sorbus aucuparia*), and downy birch (*Betula pubescens*). On specific sites, however, beech can extend uphill through the spruce belt to make contact with shrub communities (*Pinus mugo*) above the timberline, and as a network of vegetatively reproduced plant shoots (or *polycormon*), it can form the timberline (Fanta 1981).

Postglacial Forest History

Our knowledge of the Tertiary tree flora of western Europe is far from complete. Palynological research confirms that the Tertiary tree flora was much richer than that of today. In the face of the continental ice sheet advancing from the north the European Tertiary flora retreated to the south. This southwards migration, however, was hindered by the west–east European mountain ranges (central European uplands—the Hercynian-Sudete mountains, and the Alps and the Carpathians) which, in the Quaternary period, developed persisting snow and/or ice sheets and mountain glaciers. As a result, some Tertiary tree species, not able to overcome these physical barriers, became extinct in Europe, but survived in North America and Asia. Consequently the European tree flora became considerably impoverished. Frenzel (1968) and Campbell (1982) state that at least thirteen deciduous and seven coniferous tree species disappeared from the European flora during the Quaternary (amongst these were *Carya, Liquidambar, Liriodendron, Magnolia, Chamaecyparis, Pseudotsuga, Sequoia, and Taxodium*).

During the Weichselian/Würm period tree species retreated to the utmost south (Iberian, Apennine, and Balkan peninsulas). They survived in wooded steppe and tundra and/or in islands of mixed woodlands along the Mediterranean coast (Beug 1967; Frenzel 1968). The vast area of western Europe was covered with subarctic tundra and cold steppe with low vegetation (grasses, sedges, arctic herbal species, the dwarf birch *Betula nana*, and the dwarf willow *Salix herbacea*, etc.), while Denmark and north-western Germany lay under ice (Fig. 17.8). The northward migration of tree species from their refuges in the south started some 14,000–10,000 years ago. The sequence of postglacial vegetation and forest development involves periods with distinctly different climates and, consequently, with strongly different forest composition (Table 17.1).

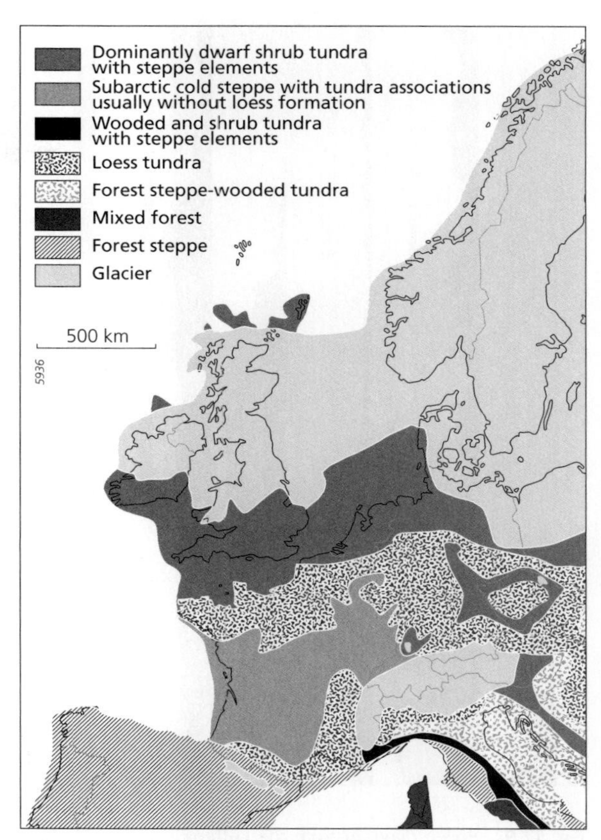

Fig. 17.8. Vegetation of western Europe during the last glacial period (after Frenzel 1968).

Pollen diagrams document both the general trend and local differences in migration of trees from the south to the north and west of Europe. Figure 17.9 shows a typical sequence of the postglacial tree species succession from its beginning to the present. The early successional forests were formed by *Salix, Betula*, and *Pinus*—pioneer

TABLE 17.1. *Periods of Holocene vegetation and forest development in western Europe*

Period	Approx. years BP	Forest vegetation	Archeological age
Preboreal	until *c.*9,500	tundra—treeless or poor in trees; *Salix, Betula, Pinus*	Palaeolithic
Boreal	9,500–7,500	*Corylus, Pinus*	Mesolithic
Atlantic	7,500–5,000	*Corylus, Quercus, Ulmus*	Early Neolithic
		Tilia, Fraxinus, Alnus	Middle Neolithic
Subboreal	5,000–2,500	Mixed *Quercus* woods	Late Neolithic
		Fagus increasing	Bronze Age
Subatlantic	2,500–1,000	Mixed woods rich in *Fagus, Abies*; *Picea* in the mountains	Iron Age
Recent	since 1,000	strong impact of humans; deforestation, plantations	Historical Age

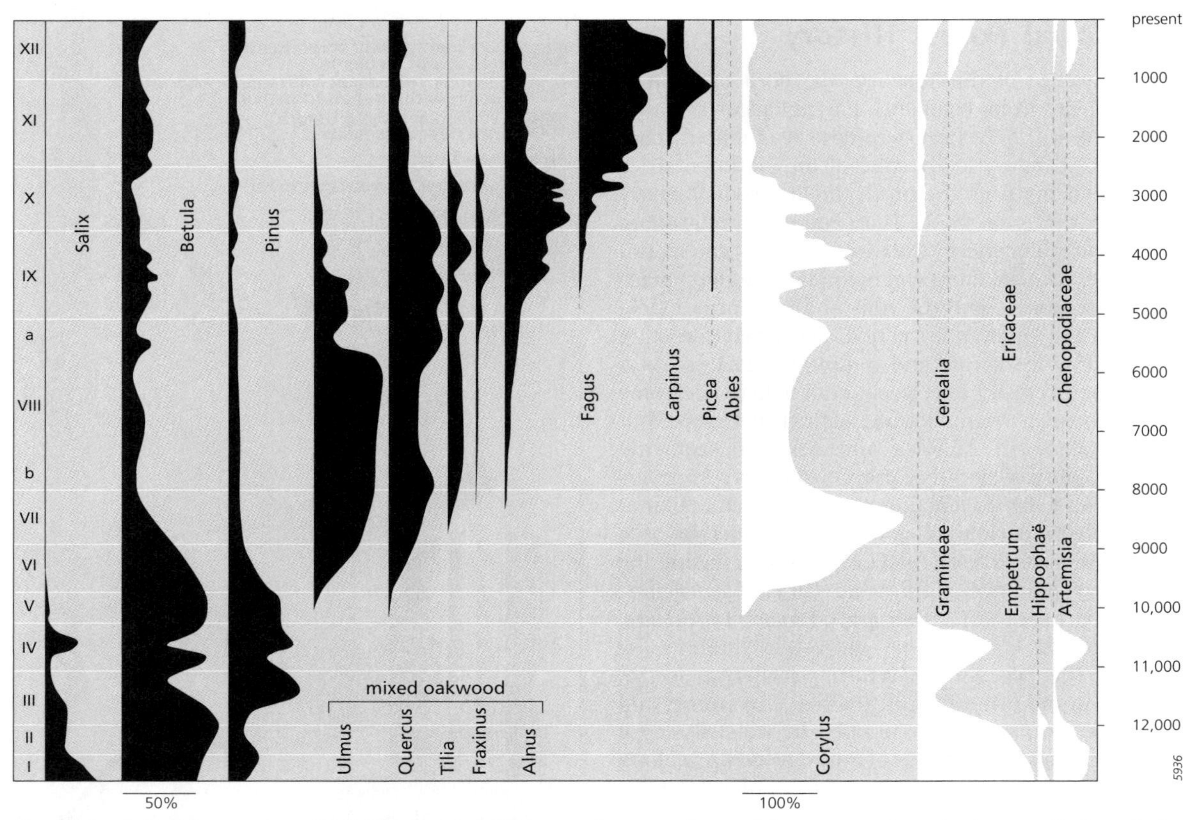

Fig. 17.9. Pollen diagram from the Luttersee near Göttingen, Germany. A complete sequence from the Ice Age (I) through the Late-Glacial period (II–IV) and middle Holocene warm period (VIII) up to the present (XII) is shown. The pollen zone numbering follows Overbeck (1975) (after Walter and Straka 1970).

tree species characterized by an early and abundant production of light seeds spread by wind (anemochory) over great distances. Extension of forest cover in this period was accompanied by decreasing cover of low vegetation. In the Boreal, Atlantic, and Subboreal periods mixed oakwoods fully developed (*Quercus, Ulmus, Tilia, Fraxinus; Corylus* in the shrub layer), replacing *Betula* and *Pinus*. A rather high number of tree species, their different ecological properties and seed dispersal (e.g. *Ulmus*—anemochory; *Quercus*—zoochory), and varying soil conditions enabled differentiation of the forest cover into various forest types. In the Subatlantic period the shade-tolerant *Fagus* and *Carpinus* spread, replacing shade-intolerant *Quercus* and *Corylus*. *Ulmus* and *Tilia* nearly disappeared but survived in reduced amounts on specific sites. The migration routes were species-specific, depending on the location of the refuge in the south, the physical barriers in the form of mountain ranges, seed production and the spreading ability of particular species, and the competition of the already present vegetation. A detailed picture of the migration routes of the most important tree species is shown in Fig. 17.10 (Taberlet *et al.* 1998), based on the chloroplast analysis (cp DNA) of the refugial populations in the Iberian and Apennine peninsulas and in the Balkans. In the last two millennia, humankind has become the main factor disturbing the migration process, hindering the natural spread of tree species by deforestation, by changing forest composition, and more recently by planting conifers.

Species Ecology

Quercus robur and *Quercus petraea*

Oaks survived in scattered and isolated islands along the Mediterranean coast and in larger populations in the Iberian, Apennine, and Balkan peninsulas (Fig. 17.10). The migration to the north took place in two steps: first,

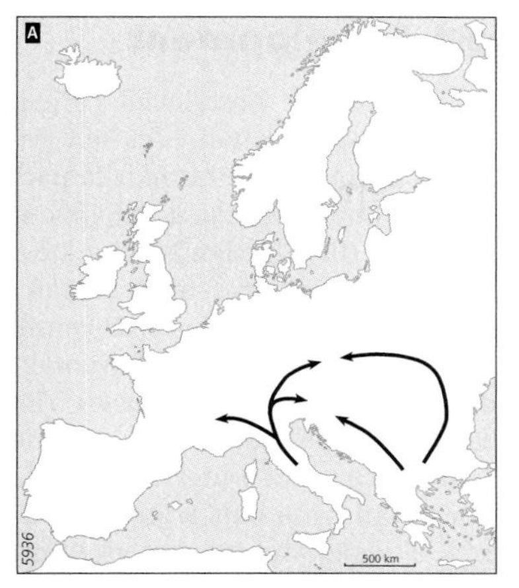

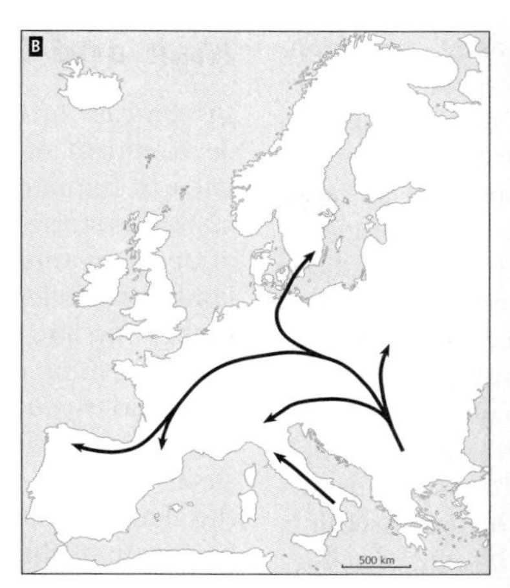

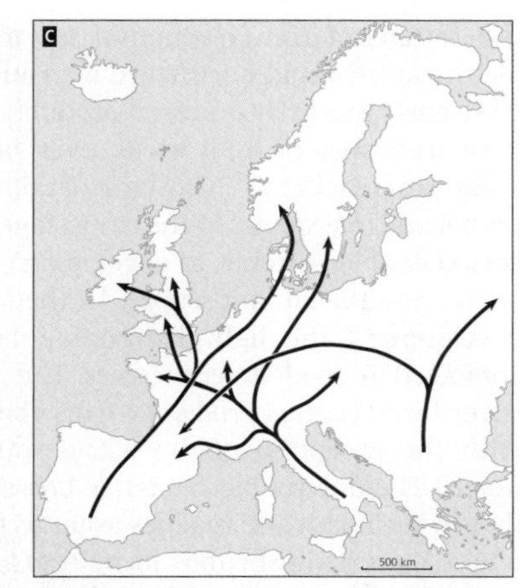

Fig. 17.10. Postglacial colonization routes of (*a*) silver fir, *Abies alba*; (*b*) common beech, *Fagus sylvatica*; and (*c*) white oak, *Quercus* ssp. (after Taberlet *et al.* 1998).

in the late Weichselian (some 13,000–11,000 BP) when the Alps and the Carpathians formed barriers that delayed the migration northwards until 10,000 BP, while the Pyrenees were passed 1,000 years earlier (Huntley and Birks 1983). The second phase started around 10,000 BP. Within the following 3,000 years, up to around 7000 BP, the whole of western and central Europe was colonized. From the genetic point of view, most of the recent oak populations in western Europe are a mixture of progenies originated in refuges in the Iberian and the Apennine peninsulas. In the last 3,000 years a southward withdrawal of oak has been observed in northern regions and in higher elevations in central Europe. In contrast, at the same time, an increase in oak pollen values has been established in pollen records for the west of France. In their distribution, oaks prefer lowlands and submontane sites. The ecological amplitude (wet–dry) of the pedunculate oak (*Quercus robur*) is broader than that of the sessile oak (*Quercus petraea*), which prefers intermediate sites.

Fagus sylvatica

The main refuges of beech were situated on the Apennine peninsula and in the Balkans (Fig. 17.10). The northward migration from the first refuge was obstructed by the geographical barrier of the Alps, while the migration from the Balkans could take place via the lowland corridor between the eastern Alps and the western Carpathians. As a result, the whole of Europe had been colonized by beech from the Balkans (Huntley and Birks 1983; Taberlet *et al.* 1998; Pott 2000). First traces of the beech expansion in the south date from 9000 BP. Around 7000 BP, beech reached the south-east foothills of the eastern Alps, moved through the lowlands between them and the western Carpathians, and started its migration into north-western Europe (4000 BP). At 2000 BP the whole region as far as the Atlantic and Baltic coasts was colonized with the exception of Normandy and Denmark. In the south-west, the Pyrenees form its limit. To the north the distribution area is limited by low temperature (spring frosts, which prevent regeneration). To the east the continental climate gives the limit.

In its natural area beech occupies hilly, preferably submontane and montane sites, irrespective of the mineral richness of the soil substratum. Both lack and surplus of soil water determine its occurrence along the dry–wet gradient. Due to drought it fails to colonize dry lowlands, but it is rather aggressive in the wet Atlantic climate on poor and dry sandy soils. The last phases of beech migration throughout north-western Europe have been heavily hindered by Man and his colonization of the lowlands, which led to deforestation, forest fragmentation, and grazing. The recent expansion of beech in the north-western European lowlands (Sissingh 1970), even in inland dune fields, might be seen as a natural continuation of beech migration after a break of 2,000 years caused by human influences. This expansion leads to the recent forming of new climax beech communities under contemporary conditions (Fanta 1995).

Betula pendula and *Betula pubescens*

Birches survived the glaciation in the shrub and wooded tundra and mixed woodlands in a broad belt in the south of Europe. As cold-tolerant pioneer species they started the northward migration much earlier than other trees and colonized nearly the whole continent after the retreat of the continental ice masses. Important factors in the rapid northward migration of both species were early maturity and seed production in trees and an easy transport of light seeds over large distances by wind (anemochory). Also lack of competition by other herbal and tree species in the early stages of the postglacial period enabled birches to develop first-forest formations in the postglacial landscapes. In the later stages of this development the light-demanding birches were outcompeted by shade-tolerant trees. The natural area of the silver birch (*Betula pendula*) covers the whole continent with the exception of the warm south and the high north. The downy birch (*Betula pubescens*) extends further to the north and reaches as far as the polar tree line, where it grows in shrubby formations. Its southern limit runs along the southern foothills of the Alps.

Both species grow on poor to very poor soils. *Betula pendula* occurs on drier sites while *B. pubescens* prefers moist to wet sites and even acid peaty sites and bogs, both in the high north and high in the mountains. The birches are short-lived trees. Commercial European forestry has not seen birches as valuable. In man-made forests birches mostly survive on forest margins and have been tolerated on extreme sites. However, their pioneer character and ecological properties make them very valuable species in forest restoration activities.

Abies alba

This species survived the glacial period in several refuges in the Balkans, the Apennine peninsula, south-eastern France, and the Pyrennees (Huntley and Birks 1983; Taberlet *et al.* 1998). Most of the actual populations originate from refuges in the Balkans and Central Italy, and their mixture in the contact zone ('suture zone'; Taberlet *et al.* 1998). The refuge in south-eastern France probably contributed to the genetic polymorphism of present populations, too. The remaining refuge in the Pyrenees did not expand in the postglacial period and remained isolated. European fir colonized mountain regions in the southern part of Europe, it does not occur in lowlands. As a shade-tolerant species, fir often forms a natural admixture in beech forests. Remnants of primeval forests show structured stands in which fir trees can reach the age of more than 600 years and great dimensions (height 50–60 m, thickness 2 m). Its ecological optimum is on rather heavy and moist (gley) soils. The last two hundred years have shown a retreat of fir from mountain forests due to the combined impact of insect plagues (*Dreyfusia*), air pollution, improper forest management (extensive clearings, conversion to Norway spruce), and deer browsing, which made the natural regeneration of fir impossible.

Man and Forest Development

In contrast to the first half of the postglacial period, development of forests in its second half is closely related to human activities. Pronounced human impact on forests started in the Neolithic, with the development of agrarian settlements in (fertile) lowlands. The loess landscapes offered the best conditions: mild climate, fertile soils, and open woodlands suitable for year-round use for shifting cultivation or cattle grazing. The conditions also encouraged the resprouting of tree species. The first agricultural settlements were established on gentle slopes along streams, in the reach of water but outside the flooded area and with surrounding woodlands as an inexhaustible source of wood and grazing grounds. The first activities of Neolithic people, small in extent, probably led to 'enrichment' of Neolithic landscapes by making openings in closed forests, creating forest margins, and allowing various plant species of open places to establish. For example, the first farmers of the Early Neolithic Linear Bandkeramik culture reached the loess regions of Germany about 5500 cal yr BP (see e.g. Kalis *et al.* 2003). These people were full-time farmers and they cultivated cereals such as einkorn (*Triticum monococcum*), emmer (*T. dicoccum*), and barley (*Hordeum vulgare*), and other crops. However, they formed only a sparse population and their impact on the landscape will have been limited and of local character. This impact is also not easily detectable in pollen diagrams, as most cultivated plants are cleistogamic (self-pollinating) and so do not readily emit pollen into the atmosphere. In the Bronze Age (Subboreal, *c.*3500–2500 BP) shifting cultivation had already been abandoned. Consequently, the permanent use of grounds led to deforestation of broader areas and the population growth led to colonization of submontane areas. Vast woodlands were opened up and specific forms of human impact on woodlands, including their exploitation and use of forest products, developed with various consequences for forests (Küster 1998, 1999).

Deforestation and Forest Fragmentation

Throughout the ages, reclaiming land for agriculture took place at the expense of forests and woodlands. In forest openings settlements were established and shifting cultivation applied, enabling temporary use and, afterwards, restoration of exhausted fields. By the Bronze Age this form of exploitation led, on poor soils in western Europe, to development of heathland communities in place of the forest. This process of forest fragmentation continued, in fact without change throughout millennia, until the eighteenth century AD.

Fig. 17.11. Largely deforested hilly landscape in the surroundings of Alpirsbach, South Germany, beginning nineteenth century. (*Source*: comm. postcard, lithography Th. Dipold, 1839.)

On poor soils sheep-grazing on heathland was the main form of agrarian use of the landscape. Grazing, sod-cutting, and heath-burning prevented forest regeneration, while soil exhaustion and acid heath litter supported podsolization. Combined impact of increasing population growth and climatic warming in the Middle Ages led to development of blown sand areas in Belgium, The Netherlands, Germany, and Denmark. On more fertile soils in the submontane zone, the agrarian exploitation was based on alternating arable and fallow land. Mountain areas had been colonized latest, in the Middle Ages. Here, the main form of agricultural use was mountain pasture. Over the centuries all exploitable land had been taken and remaining forests were limited to poor and inaccessible sites not suitable for agricultural use (Fig. 17.11).

Forest fragmentation and landscape deforestation did not go on continuously at the same pace. During its prehistorical and historical development, western Europe experienced periods of population explosions. Warmer climates led to active colonization, e.g. during the Roman, Franconian, and Carolingian periods, and the thirteenth century AD. The Great Migration, warfare, and plague outbreaks, especially in the fourteenth and fifteenth centuries, led to population declines. Deforestation and forest fragmentation reached its maximum and the wooded area of western Europe was at its lowest level at the end of the Middle Ages and at the beginning of the New Era (seventeenth–eighteenth centuries). The real change in this situation came only in the nineteenth century when governmental afforestation actions were started in most countries.

Woodland Pasture and Pannage

Grazing of (mixed) herds of cattle, sheep, goats, horses, and pigs in the surroundings of agrarian settlements happened throughout the year in warmer areas. Light oak-dominated forests with their all-year green understorey of shrubs, herbs, and grasses were especially suitable for this use. Grazing generally prevented forest regeneration and led to development of light park forests dominated by old trees. Under a heavy grazing pressure these ultimately changed to open pastures. A lower pressure enabled development of (thorny) shrubs and/or regeneration of some tree species in a vegetative way (oaks, hornbeam, lime, ash) that could survive better than others (beech). Oak woodlands were vital for pannage and were promoted as coppice in forest regeneration since early times. In contrast, yew (*Taxus baccata*) was actively eliminated, being poisonous to farm animals. Elm, ash, and hornbeam were lopped and shredded, and twigs with leaves were dried to be used as fodder in the winter. Goats are considered to be the most destructive grazers (browsing on trees and shrubs), while cows and sheep are predominantly grass-eaters. Natural regeneration of oakwoods by seed was practically impossible due to the eating of acorns by pigs (*pannage*). In warmer areas with mild, snow-less winters, forest grazing took place throughout the year. In colder areas, litter raking was widely used to supply organic material for bedding in cowsheds in winter and to produce manure for arable land.

As early as in prehistoric times woodland pasture had led to changes in forest structure and composition. Applied in an intensive way, it caused heavy forest degradation and, finally, full elimination of forest growth followed by development of grasslands or heathlands. But because of its vital importance for agrarian communities, some forest was maintained for millennia and in the Middle Ages such forest often became an issue of social and political contention and litigation among feudal landowners and municipalities. Due to the devastating effects on forests, woodland pasture was ultimately abolished in the nineteenth century in most west European countries. However, remnants of this form of woodland use have still been maintained in several areas (see e.g. Pott and Hüppe 1991). The practice of cattle grazing in high mountains (*Almwirtschaft*) led, on accessible slopes, to a lowering of the alpine timberline. Heavy erosion (both surface and torrential) and/or development of avalanches and landslides often accompanied its practice and sometimes caused irreversible changes in mountain environments.

Fuelwood, Charcoal, Timber, and Other Wood Products

Specific properties of wood and timber species were known already in early times, and selection of wood species for various purposes was applied since time immemorial with, as a result, a successive change in species composition of existing woodlands in the surroundings of the settlements. Construction timber was transported over land (oak) or by water (fir, spruce) to distant places. In the eleventh century, the forests of Normandy were largely destroyed, being exploited for shipbuilding to conquer England. In the seventeenth–eighteenth centuries, full-length fir logs from the Schwarzwald and Bohemian Forest were transported to Holland to be used as shipmasts in the then flourishing Dutch shipbuilding industry. With the increasing population pressure in the Middle Ages, wood consumption increased. The development of cities, metal processing, glass furnaces, salt-works, and lime-burners consumed huge quantities of wood. For direct use as fuelwood, hardwood species were valued (oak, beech). By the early Middle Ages direct use of fuelwood was replaced by charcoal burned in piles on site in the woods (Fig. 17.12). Due to charcoal production, deforestation proceeded at high speed, though coppicing was introduced as the first form of forest management. The speed of resulting deforestation slowed down only in the industrial period when hardcoal and lignite replaced wood as a source of energy.

Until the Late Middle Ages, not only fuelwood and construction timber were important. Wicker walls, tools, baskets, various vessels, fencing, wagons, etc., were also produced from selected wood species. The elastic timber of the yew was used to produce bows. Lime was valued as source of bast for matting, etc. Willow supplied flexible twigs for baskets. Hard *Buxus* (box) timber was used

Fig. 17.12. Forest exploitation in France in the eighteenth century. Left: Felling of old trees; young growth remains intact. Middle: Full harvesting of wood and timber to bring forward various products. Right: Various phases of charcoal burning in piles. More than half the fuelwood was burned in piles to produce charcoal. The volume of charcoal was four times less than fuelwood, and this has implications for the transport economy.

(*Source*: Duhamel du Monceau: *De l'exploitation des bois*, 1764.)

to produce spindles. Resinous pine was valued as kindling wood. Sycamore and beech wood was burned and the ash containing potassium used to produce glass. The early European civilization grew up with the forest and depended strongly upon its products.

Coppice Management

Man not only utilized and exploited the forests by using its sources and raw materials, he also actively affected it by adapting its structure and composition to produce materials, tools, and other products to supply his requirements. Coppicing might be seen as an example of such an approach. This management form had been practised probably since the Roman conquest (Burschel and Huss 1997). It probably developed from incidental use in prehistoric times and via Roman sweet chestnut culture into a planned management executed according to specific rules. It provided products in the form of offshoots of various sizes, which could be harvested in short rotation of 10–30 years. Oaks (esp. *Quercus robur*), ash, birch, alder, hazel, hornbeam, and lime are suitable species to be managed in this way due to their ability to form vegetative offsprings. In contrast, beech is a tree species that generally does not resprout as abundantly although beech coppice management was applied as well in the area, especially along the Atlantic coast. As a result beech was eliminated and replaced by oak, hornbeam, and other resprouting deciduous species. Coppice management attained much attention in the Middle Ages, with an increased demand for charcoal production. Later on, production of tanning bark became important. In some areas (e.g. Rhineland, The Netherlands) it was the primary goal of coppice management. On poor, sandy soils a maximum of four oak coppice rotations ensured a profitable production of wood and bark. Thereafter, the rootstocks had to be pulled up, the soil manured and ploughed and acorns sown to establish new stands.

Combination of coppice and high forest management methods gave rise to the coppice-with-standards forest (*Mittelwald*). This coppicing was practised in regular 10–30-year rotations on trees forming the understorey. The upperstorey was formed by valuable tree species producing first-class thick logs in a rotation of 100 or more years. This management form is still broadly used in Europe, especially in France. Another form of coppice management applied in former times was pollarding—coppicing of trees at a height of *c.*1.5 m. In this way twigs and offshoots were produced on ash, elm, and lime. Offshoots were harvested with leaves in the summer and stored, to be used as fodder in the winter. Pollarding also avoided browsing of offshoots by livestock grazing in the forest. Pollarding of willow trees ensured high production of long, thin twigs for wickerwork. Coppice management had a strong influence on forest composition. It was the earliest form of monoculture as tree species not suitable for this management form (e.g. beech) were removed. On poor soils repeated coppicing led to soil degradation, causing centuries-long and even irreversible changes in forest soils, forest composition, and the diversity of the herbal layer (Pott 1981).

Deforestation and combination of the above methods of forest use have completely changed not only forests but also whole landscapes. Not only their physiognomy has changed. The water balance and water supply over large areas have also altered completely, as have the surface processes (erosion, deflation) and the soil chemistry in terms of the carbon, nitrogen, potassium, and phosphorus turnover. Surface flow and erosion increased dramatically in deforested landscapes. As a result, local relief changes occurred and huge amounts of fertile soil were transported by rivers (e.g. Bork *et al.* 1998). Disastrous situations (development of inland sand dunes, flooding, gully erosion in hilly areas) have been documented over time in various parts of north-western Europe as a result of extensive deforestation. The forested area in some west European landscapes has been reduced to 1% (e.g. The Netherlands: 1.2% in 1850), with small remnants of forests surviving mostly as fully degraded coppice. In the Mediterranean regions of Europe forest destruction and degradation set in much earlier and with still greater consequences than in its Subatlantic and central parts. Long settlement history, high population pressure, dry climate, extensive goat grazing on steep slopes, soil erosion, and resulting lack of forest natural regeneration produced a severe landscape degradation. Over large areas forests disappeared completely and were replaced by xerophytic shrubs (*garrigues, maquis*) and low vegetation (Pignatti 1978; DiCastri *et al.* 1981).

Ecological Characteristics of Main Forest Types

Three interrelated conditions play a role in the developments of (regional and local) forest communities:

- possibility of immigration of trees during the post-glacial period;
- relation between ecological properties of particular tree species and environmental (site) conditions;
- competitive ability of particular tree species.

North-western Europe is for the greater part an area of broad-leaved forests. The natural occurrence of native coniferous tree species is limited to the montane and subalpine zone of the south-east European mountains—European silver fir (*Abies alba*), Norway spruce (*Picea abies*), stone pine (*Pinus cembra*), and larch (*Larix decidua*). Scots pine (*Pinus sylvestris*), one of the postglacial pioneers, largely disappeared in the course of the post-glacial development, being replaced by the deciduous tree species. It survived only on very marginal sites—poor sandy soils in north-west European lowlands (e.g. van Zeist 1955*a*, *b*) and on dry, rocky slopes in the mountains in the south. The occurrence of yew (*Taxus baccata*) is difficult to document palynologically due to a lack of microfossils in pollen diagrams (Huntley and Birks 1983), but its distribution throughout Europe is well known from botanical evidence.

Distribution and Ecological Properties of Trees

The distribution of particular tree species, and their ecological properties and site requirements can be generally derived from distribution maps (e.g. Rubner 1953; Meusel *et al.* 1965/84; Bohn and Neuhäusl 2001; see Fig. 17.3). Important ecological characteristics of trees are related to light requirements, temperature, climate (preference of maritime or continental features), and soil conditions (especially moisture, acidity, and nutrient supply). These characteristics are described in the dendrological, geobotanical, and forestry literature (e.g. Rubner 1953; Bonnemann and Röhrig 1971; Ellenberg 1978, 1988; Ellenberg 1991). The ecological characteristics of trees must be always seen in relation to site properties. They may alter under extreme conditions. Soil requirements may be compensated by mild climate; soil moisture by high air humidity under an Atlantic climate; mineral-rich soil may balance low temperature at higher elevations. This mutual substitution of site factors is known as 'the law of relative environmental constancy', formulated by Walter and Straka (1970).

Most Important Woodlands: Beechwoods and Mixed Oakwoods

Availability of nutrients and water is important for the growth conditions of, and competition among, trees. As a result, the distribution of trees follows in general these gradients of the environmental conditions, with their prevailing horizontal continuity and altitudinal zonality. However, sometimes there is also remarkable discontinuity depending on specific soil or site conditions. With regard to this, two main groups of forest communities are distinguished:

- zonal communities, the distribution of which follows the ecological gradient situation of the broader region,
- azonal communities, depending on specific site or soil conditions (e.g. excessive water supply) and not related to the regional gradient situation.

Beech and mixed oakwoods are typical representatives of the zonal forest communities. Riparian woodlands or alder carrs are typical azonal communities.

Beechwoods

Beech (*Fagus sylvatica*) is the tree with the most widespread natural distribution in western Europe. It migrated to north-western Europe only in the last 4,000 years and, consequently, its distribution throughout the continent has been strongly affected by humans. The combination of shade tolerance, medium temperature, and soil moisture requirements and indifference to soil chemical properties (pH, Ca, N) gives the species a broad ecological amplitude. It can grow and survive under very different conditions and outcompete, via its shade tolerance, other more sensitive and/or specialized species. It forms rather dark stands with limited conditions for a luxurious herbal layer (with the exception of spring-aspect herbs) and often has a thick layer of slowly decaying litter (Fig. 17.13). Under normal conditions the species regenerates abundantly naturally from seeds (generative reproduction). Seed-setting is mostly massive and periodical, in intervals of 4–8 years. At the maritime and subalpine timberline sites the species shows a strong vegetative reproduction ability. Under natural conditions beech can reach a physical age of 250–350 years and mighty dimensions. Due to its opportunistic character, the beech can be outcompeted by other tree species only on extreme sites (e.g. dry, wet, stony, or compacted soils). Under natural conditions beech is a species dominating terminal stages of forest succession. In the course of the succession it easily regenerates under trees of the pioneer or transient succession stages (e.g. birch and oaks, respectively). Natural communities reveal a rather high spatial and temporal diversity (differences between young and old growth; differences among the initial, transient, and terminal stages of succession). In managed beech forests the diversity is mostly reduced (closed one-layer canopy). Beech forests grow in both lowlands and in mountains, forming woodlands of similar physiognomy. Lowland beechwoods have been heavily affected and reduced in extent by humans and survive today only as isolated fragments. Mountain beechwoods (Fig. 17.14) have survived except where converted into coniferous monocultures by commercial forestry.

Oakwoods

Several European oak species survived the glaciation in southern refuges and are growing now in north-western

Fig. 17.13. Native beech forests originally covering extensive areas of European lowlands and uplands.

Fig. 17.14. On steep slopes in high mountains beech trees suffer from creeping snow and often display polycormonal growth forms due to vegetative offsprings.

Europe under very different ecological conditions. Jahn (1991) distinguishes cool-temperate oakwoods and warm-temperate oakwoods.

Cool-temperate oakwoods occupy sites in the Atlantic and Subatlantic provinces, from lowlands to the submontane zone of the south-east European mountains, outside the sites taken up by beech. The main tree species are *Quercus robur*, *Quercus petraea*, and *Carpinus betulus*. Oaks can reach a physical age of several centuries and grow to large dimensions. The pedunculate oak (*Quercus robur*) has a broader ecological amplitude than the sessile oak (*Quercus petraea*). Both species are rather indifferent to soil nutrient conditions. They form rather light stands with favourable conditions for the development of the second tree layer (*Carpinus betulus*), shrub, and herbal layers. Seed-setting is abundant in mast years. Both oaks regenerate naturally in gaps and/or under the canopy of other shade-intolerant tree species such as birch or pine. Oak woodlands suffered heavily under human impact from the very beginning of the colonization. Due to their abundant herbal layer, they could be grazed throughout the year. As a result, extensive deforestation and/or degradation into dwarf shrub heathlands occurred. Age-long coppice management also affected these woodlands in a negative way (Pott 1981). Especially on nutrient-richer soils a number of other deciduous trees can coexist with both oaks and hornbeam, forming a range of forest communities. The most important of these are species-poor mixed birch–oak woods and beech–oak woods, and species-rich mixed hornbeam–oak woods. The latter growing on richer soils. A prominent feature of natural mixed oak forests was a rich structure and diversity, and a high amount of dead wood, both standing and lying on the ground (Fig. 17.15). Research shows that 30–40% of species diversity—bacteria, fungi, insects, birds, mammals—depends on dead wood. Warm-temperate oakwoods are

Fig. 17.15. Structure and species-rich natural mixed oak forests with dead wood.

concentrated in the south of the area. The dominating species is the downy oak (*Quercus pubescens*). Along the west Mediterranean coasts evergreen holm oakwoods (*Quercus ilex*) occur. These oakwoods have been exposed to human activities by agriculture, vine production, and sweet chestnut culture since the very beginning of the colonization of the Mediterranean area. On steep slopes, deforestation and erosion led to irreversible degradation of sites and the development of thorny shrub formations. In dry summer periods forests in this area are strongly endangered by fire.

Amongst the above zonal communities various azonal forest communities occur on specific sites. Their species composition depends strongly on soil conditions (very dry or wet; steep slopes; nutrient-rich or poor; and combination of these soil properties). Jahn (1991) distinguishes four main groups of these azonal communities:

1. Mixed maple–ash communities with lime and elm, on slopes in the submontane and montane zones. Due to very different site conditions, these forests reveal a great variability in composition, spatial structure, and growth. To this group the following forests can be added:

- forests on steep slopes in ravines and deep valleys;
- forests on nutrient-rich sites (sycamore–ash woods on shady slopes; maple–lime woods on sunny slopes);
- forest on nutrient-poor and dry, stony slopes (*Tilia cordata*, *Quercus petraea*).

2. Riparian woodlands with a clearly developed zonation depending on the flooding regime. Willow and poplar groves tolerant to repeated flooding are found along the low riverbanks, and hard broadleaf species (pedunculate oak, ash, elm) on higher sites. Due to thousands of years of land use and river embanking, these forests have largely disappeared, being converted into grasslands in the north and vineyards in the south. Only very small remnants are found today showing species richness and a complicated vertical structure with climbing plants such as *Clematis vitalba*, *Humulus lupulus*, *Lonicera periclymenum*, and others.

3. Alder and ash woods on rich swamps. Today these have been mostly drained and converted to meadows or the other (drier) forest types of which only small remnants survive.

4. Birch and pine woods on poor bogs. A surplus of stagnating water (unpermeable grounds) of a low pH forms a very specific niche for the development of species-poor forest communities with dominating *Betula pubescens* and/or *Pinus sylvestris*.

Organized Forestry

By the Middle Ages, deforestation and degradation of forests was recognized as a dangerous phenomenon and the first royal and manorial prescriptions and ordinances were issued to protect forests. The *Corps des Maîtres des Eaux et Forests* was established in 1291 in France. A French forest ordinance was issued in 1318 and the *Majestas Carolina* in 1355 (Badré 1983). First management measures were taken to restore degraded forests. The first sowing of pine seeds was carried out in the Nürnberger Reichswald in 1368. Around 1514 the first afforestations took place in The Netherlands. The lowest point for forest area in west Europe was reached between the end of the seventeenth and the beginning of the nineteenth century. By then the lowlands were fully deforested, cover sand areas changed into inland sand dunes and heathlands, and the remaining small coppices and grazed woods were fully degraded. Submontane landscapes were for the most part cleared and mountain woodlands subjected to a heavy pressure (clearings, mountain pasture). These circumstances led to serious and well-considered attempts to improve the situation.

Commercial Forestry

In France, between 1661 and 1669, *La Grande Reformation des Forêts du Royaume* was carried out (Minister Colbert's Ordinance from 1669). In 1713 the German forester H. C. von Carlowitz presented the basic ideas on forestry as a rational, fully organized activity in his work *Sylvicultura Oeconomica*. Von Carlowitz defined, for the first time, the basic principle of forestry: the unity of sustainable existence of the forest and sustainable yield of timber and wood, to be achieved in a continual process of planning and management. Similar development took place in France (Duhamel du Monceau: *De l'exploitation des bois*, 1764). At this time, and hand in hand with the changed forest ownership (state, municipalities, manorial forest owners, small private owners), forestry as a rational activity was born. This development was also supported by social and economic changes which took place during the eighteenth century. These included the introduction of hard coal as a source of energy, the planting of potatoes, the stable feeding of animals which was more efficient than pannage, and the invention of fertilizers enabling higher livestock yields than could be achieved from woodland pasture, etc. The main tasks of the emerging forestry in this restoration period (*Waldaufbauphase*) were reafforestation of the huge areas of derelict land that was not suitable any more for agricultural production, and improvement, by applying management measures, of heavily degraded forests. In its 300 year-long history European forestry developed into a rational commercial activity oriented towards timber production, with its own consolidated organization, which also involved forestry education and the establishment of professional rules and skills.

Due to afforestation the forested area has grown slowly but steadily in the past 300 years. The sandy landscapes in the wide belt along the Atlantic and North Sea coasts (from Les Landes in France to the northernmost point of Denmark) experienced a profound change by afforestation of blown sands and dry heathlands. For example, in The Netherlands the forested area has grown from 1.2% in 1850 to 9% at the end of the twentieth century. Due to extreme environmental conditions on the one hand and commercial considerations on the other, pines have been mostly used in these afforestations: *Pinus maritima* and *Pinus pinaster* in the south, and *Pinus sylvestris* in the north. These man-made coniferous plantations today dominate the young forests in these areas (Fig. 17.16). Further inland, *Pseudotsuga menziesii* has often been used, and to a lesser extent *Larix kaempferi*. In forests of the European uplands today *Picea abies* is dominating.

Improvement of degraded and overused existing forests was made possible by the introduction of high forest management with its silvicultural systematics. This led to artificial afforestation of cleared areas, and the use of thinnings and felling according to established regeneration rules and procedures. Coppicing as a cheap method of fuelwood production has been gradually abandoned. Coppice woodlands have been converted into coppice-with-standards stands or into high forests, enabling high-quality timber production. Also in this case, the commercial imperative has often led to introduction of fast-growing conifer plantations or to mixtures of exotic trees with indigenous deciduous species. Thus, European uplands and mountain regions today bear a high percentage of non-indigenous Norway spruce, while Douglas fir has been widely planted in the warm southern mountain ranges. Sweet chestnut (*Castanea sativa*), introduced by the Romans, has been cultivated in plantations on lower southern slopes of the Massif Central and the western Alps.

As a result of the systematic silvicultural treatment and legal protection, the status and production capacity of European forests has improved considerably. In the past 300 years of the forest restoration period huge and ever-increasing amounts of wood and timber have been produced and timber markets have been supplied with this valuable raw material (Table 17.2). Further increase of forested areas can be expected in the future, related to the recent change in the European agricultural policy, where a surplus of agricultural production has led to set-aside agricultural land. Also the need to improve the ecological infrastructure of European landscapes (nature development areas, corridors, stepping-stones) and to meet recreational need, especially in the surroundings of big cities and industrial agglomerations, has played a role in planning and establishing new forests.

Problems of Contemporary Forests and Forestry

Against these achievements should be set the serious ecological misachievements and setbacks of the mainly

Fig. 17.16. Man-made coniferous plantation forests mostly have a very low structural and biological diversity.

TABLE 17.2. *Forest extension in north-west European countries*

	Belgium	Denmark	France	Germany	Luxembourg	The Netherlands
Total forest area (ha)	680,000	484,000	15.1 mill.	9.69 mill.	260,000	355,000
Land area %	19.2	11.0	25.4	27.7	33.0	8.7
Forest area per inhabitant (ha)	0.06	0.09	0.26	0.12	0.23	0.02
Coniferous forests %	41.6	70.5	36.5	69.3	63.0	71.0
Deciduous forests %	58.4	29.5	63.5	30.7	37.0	29.0
Growing stock m^3/ha	122	115	127	159	—	78
Forest harvesting mill. m^3	2.6	1.9	32.9	46.9	0.28	1.16
Private ownership %	58.1	60.9	70.8	35.0	55.0	55.0
Public ownership %	41.9	39.1	29.2	65.0	45.0	45.0

Source: Beckel (1995).

commercially and technically oriented west European forestry. These include:

- physical instability of single-species coniferous monocultures;
- insect plagues;
- impact of acid litter on soil processes (acidification, podsolization, soil compaction);
- reduced biological diversity in all ecosystem compartments;
- reduction and even elimination of vital ecosystem processes;
- high sensitivity of man-made forests to environmental stresses, both anthropogenic (atmospheric pollution, climatic change) and natural (climatic events and fluctuations).

With these ecological shortcomings and imperfections many contemporary forests do not fulfil their stabilizing functions in the landscape properly. Moreover, neglecting the above-mentioned ecological aspects in forestry practice has ever-increasing economic consequences. Even under 'normal' conditions, casual fellings are higher in man-made coniferous forests than in woodlands with more or less natural tree species composition. In extreme cases, such as during the severe storms of 1972–3 (Belgium, The Netherlands, Niedersachsen), 1990 and 1999 (southern Germany), or 2002 (eastern France), forest management could be fully disrupted over large areas. Wind-blown timber is usually devalued, timber markets disrupted, forest planning distorted, and afforestation measures hampered by high costs and a lack of proper planting stock. In the last three decades of the twentieth century awareness has grown in several European countries that fundamental changes in forest management are necessary. Whilst economic damage must be avoided as much as possible it is equally important to improve the ecological functioning of forests and woodlands as irreplaceable components of European landscapes (Fanta 1997). Also various aspects of nature protection and biodiversity conservation must play a more prominent role in this new development. Forests in western Europe have a huge potential in this direction (Scherzinger 1996) and they can and must play an important role in creating national and international ecological networks (EECONET; Bennett 1991; NATURA 2000). With the development of these ideas a new period of forest reconstruction and rehabilitation of more natural forests has begun.

Multifunctional Forestry

The need for change arises from the increasing knowledge that forests fulfil various functions. Social and ecological functions in the densely populated and highly diversified Europe appear to be as important to society as the timber production function. As a result, the concept of multifunctionality of forests has been defined during the last decades of the twentieth century (Angelidis *et al.* 1997) so that:

- the production function includes food and raw materials (primarily wood) supplies;
- the ecological function results from biodiversity preservation, protection of landscapes, and regulation of interactions with the abiotic environment;
- the social function includes use of forests as areas for leisure activities, and the heritage, historical, and cultural aspects of forests.

The integration of principles of multifunctionality and sustainability indicates the future development of European forestry. Recent development shows that the above principles can be best realized in forestry practice by applying the methods of close-to-nature forestry. This form of forestry:

- is flexible enough to be adopted under various environmental and forest conditions;
- enables integration of ecological and economic goals;
- enables maximum use of natural ecosystem processes (especially natural selection and regeneration) to achieve management goals;
- gives rise, in the long term, to development of self-sustaining forest ecosystems of a high biodiversity and ecological stability that can, better than commercial plantations, withstand risks arising from both persisting anthropogenically induced environmental stresses, climate change (Puhe and Ulrich 2001) and fluctuations of natural environmental factors;
- makes it possible to co-ordinate and integrate interests of forestry with those of nature protection, landscape preservation, and conservation of biodiversity.

In forest management terms this means developing and applying a new forest paradigm, establishing criteria and indicators to fulfil various functions of forests simultaneously, and, when necessary, sequencing them according to the state of the forest, the management goals, and the site conditions. All this enables maximum use of natural processes instead of expensive, continuous forest manipulation (see e.g. Burschel and Huss 1997). To facilitate this development, networks of protected forests have been established in all west European countries. Long-term research programmes have been developed to obtain scientific information on natural processes in forest ecosystems. Conversion of coniferous monocultures growing on inappropriate sites aims at the rehabilitation of a more natural forest structure and composition. This will increase their ecological and physical stability.

References

Angelidis, A., Rey, G., and Hermeline, M. (1997), *Europe and the Forest*, iii. European Parliament, Luxembourg.

Badré, L. (1983), *Histoire de la Forêt Française*. Arthaud, Paris.

Beckel, J. (1995), *Satellite Remote Sensing Forest Atlas of Europe*. Perthes, Gotha.

Bennett, G. (ed.) (1991), *Towards a European Ecological Network*. Institute for European Ecological Policy, Arnhem.

Beug, H.-J. (1967), Probleme der Vegetationsgeschichte in Südeuropa. *Berichte Deutsch. Botan. Gesselschaft* 80: 682–9.

Blüthgen, J., and Weischet, W. (1980), *Allgemeine Klimageographie*, 3. De Gruyter, Berlin.

Bohn, U., and Neuhäusl, R. (eds.) (2001), *Map of the Natural Vegetation of Europe*, Scale 1 : 2.5 mill. Landwirtschaftsverlag, Münster-Hiltrup.

Bonnemann, A., and Röhrig, E. (1971), *Waldbau auf ökologischer Grundlage*, i. Parey, Hamburg.

Bork, H.-R., Bork, H., Dalchow, C., Faust B., Piorr, H. R., and Schatz, Th. (1998), *Landschaftsentwicklung in Mitteleuropa*. Klett-Perthes, Gotha.

Burschel, P., and Huss, J. (1997), *Grundriss des Waldbaus*, 2. Parey, Berlin.

Campbell, J. J. N. (1982), Pears and persimmons. A comparison of temperate forests in Europe and eastern North America. *Vegetatio* 49: 85–101.

DiCastri, F., Goodall, D. W., and Specht, R. L. (eds.) (1981), *Mediterranean-type shrublands*. Ecosystems of the World 11. Elsevier, Amsterdam.

Ellenberg, H. (1978), *Vegetation Mitteleuropas mit den Alpen in ökologischer Sicht*. Ulmer, Stuttgart.

Ellenberg, H. (1988), *Vegetation ecology of Central Europe*. Cambridge University Press, Cambridge.
—— (1991), Zeigerwerte von Pflanzen in Mitteleuropa. *Scripta Geobotanica* 18.
Fanta, J. (1981), *Fagus silvatica* L. und das Aceri-Fagetum an der alpinen Waldgrenze in mitteleuropäischen Gebirgen. *Vegetatio* 44: 13–24.
—— (1995), European beech (*Fagus sylvatica* L.) in the Dutch part of the NW-European diluvium. *Ned. Bosbouw Tijdschrift* 67: 225–34.
—— (1997), Rehabilitating degraded forests in Central Europe into self-sustaining forest ecosystems. *Ecological Engineering* 8: 289–97.
Frenzel, B. (1968), *Grundzüge der pleistozänen Vegetationsgeschichte Nord-Euroasiens*. Steiner, Wiesbaden.
Haeupler, H. (1970), Vorschläge zur Abgrenzung der Höhenstufen der Vegetation im Rahmen der Mitteleuropakartierung. *Göttinger Flor. Randbriefe* 4: 3–15.
Hartmann, F. K. (1974), *Mitteleuropäische Wälder*. Fischer, Stuttgart.
—— and Jahn, G. (1967), *Waldgesellschaften des mitteleuropäischen Gebirgsraumes nördlich der Alpen*. Fischer, Stuttgart.
Huntley, B., and Birks, H. J. B. (1983), *An atlas of past and present pollen maps for Europe: 0–13,000 years ago*. Cambridge University Press, Cambridge.
Jahn, G. (1991), Temperate deciduous forests of Europe. In: E. Röhrig and B. Ulrich (eds.) *Temperate Deciduous Forests*. Ecosystems of the World 7: 377–502.
Kalis, A. J., Merkt, J., and Wunderlich, J. (2003), Environmental changes during the Holocene climatic optimum in central Europe —human impact and natural causes. *Quaternary Science Reviews* 22: 33–79.
Küster, H.-J. (1998), *Geschichte des Waldes. Von der Urzeit bis zur Gegenwart*. Beck, Munich.
—— (1999), *Geschichte der Landschaft in Mitteleuropa*. Beck, Munich.
Mayer, H. (1984), *Wälder Europas*. Fischer, Stuttgart.
Meusel, H., Jäger, E., and Weinert, E. (1965/84), *Vergleichende Chorologie der zentraleuropäischen Flora*. 3 vols. Fischer, Jena.
Overbeck, F. (1975), Botanisch-Geologische Moorkunde. Wachholtz, Neunmünster.
Ozenda, P. (ed.) (1979), *Vegetation map (scale 1: 3mil.) of the Council of Europe member states*. Coll. Sauvegarde de la Nature 16. Council of Europe, Strasburg.
Pignatti, S. (1978), Evolutionary trends in mediterranean flora and vegetation. *Vegetatio* 37: 175–85.
Pott, R. (1981), Der Einfluss der Niederholzwirtschaft auf die Physiognomie und die floristisch-soziologische Struktur von Kalkbuchenwäldern. *Tüxenia* NS 1: 233–42.
—— (2000), Palaeoclimate and vegetation—long-term vegetation dynamics in central Europe with particular reference to beech. *Phytocoenologia* 30: 285–333.
—— and Hüppe, J. (1991), *Die Hudelandschaften Nordwestdeutschlands, Westfäll*. Museum für Naturkunde, Münster.
Puhe, J., and Ulrich, B. (2001), *Global Climate Change and Human Impacts on Forest Ecosystems*. Springer, Berlin.
Rubner, K. (1953), *Die pflanzengeographischen Grundlagen des Waldbaues*. 3rd edn. Neumann, Radebeul-Berlin.
Scherzinger, W. (1996), *Naturschutz im Wald*. Ulmer, Stuttgart.
Sissingh, G. (1970), Dänische Buchenwälder. *Vegetatio* 21: 4–6.
Taberlet, P., Fumagalli, L., Wust-Saucy, A.-G., and Cosson, J.-F. (1998), Comparative phylogeography and postglacial colonization routes in Europe. *Molecular Ecology* 7: 453–64.
van Zeist, W. (1955*a*), Some radio-carbon dates from the raised bog near Emmen (The Netherlands). *Palaeohistoria* 4: 113–18.
—— (1955*b*), Pollen analytical investigations in the northern Netherlands. *Acta Botanica Neerlandica* 4: 1–81.
Walter, H., and Straka, H. (1970), *Arealkunde*, 2nd edn. Ulmer, Stuttgart.

18 Geomorphic Hazards and Natural Risks

Olivier Maquaire

Introduction

Western European countries are subject to natural phenomena that can cause disasters. Their origins are various: geophysical (earthquakes), hydrometeorological (sea storms, floods, and avalanches), or geomorphologic (landslides). They are fairly widespread but less frequent and of relatively low intensity compared with other regions of the world; for example, an earthquake in France or Belgium is not likely to be as violent as in Greece or Japan. Some of the countries concerned, such as France and Germany, are subject to all the hazards mentioned above, while Denmark and The Netherlands are seldom exposed to earthquakes and never to avalanches because they have no mountains. Man is not responsible for phenomena such as earthquakes, but contributes significantly to the onset and aggravation of other hazards, and is sometimes largely responsible for the direct and indirect consequences, having built and maintained installations in 'risk' sectors. The number of victims and the cost of the damage may be high, depending on the circumstances, the intensity, and the duration of the phenomenon.

Western European countries have experienced real natural disasters in the distant or recent past. Floods following a storm wave in The Netherlands in 1953 were responsible for some 2,000 deaths and damage amounting to over 3 billion Euros. Two hundred people died in the most destructive flood ever known in France in 1930 in the Tarn (Ledoux 1995). Natural phenomena such as these can recur with at least the same intensity but may entail much greater damage because of increased human occupation in the sectors concerned: the flooding submerges zones which are much more urbanized than they were in the nineteenth century. Whether prevention measures are taken depends on the level of risk which the populations concerned are prepared to accept. These measures should be associated with spatial and temporal forecasts and preceded by an analysis of the processes for these phenomena to be fully understood.

In order to remove the ambiguities and the inaccuracies of terminology that are observed all too often, it is necessary in the first instance to define 'geomorphic hazards and natural risks', particularly in terms of the notions of risk, hazards, and vulnerability. Secondly, a brief description, definition, and typology of each of the four hazards concerned—floods, earthquakes, landslides, avalanches—follows. Examples of past and recent events illustrate their consequences for the environment and their human impact. Thirdly, the steps taken by societies to combat these risks—indemnity and prevention—will be considered.

Further information may be found in the works of Blaikie *et al.* (1994); Horlick-Jones *et al.* (1995); Casale (1996). The collective work of C. and C. Embleton (1997) on the geomorphologic hazards of Europe was used extensively in drafting this chapter. All regions and towns cited in the text are indicated in Figs. 18.1 and 18.6.

General Concepts

The terminology used in this chapter conforms to the definitions proposed by Varnes (1984) or is based on the international multilingual glossary of terms concerning disaster management that was drawn up as part of the International Decade for Natural Disaster Reduction (IDNDR) (Leroi 1996; Finlay and Fell 1997).

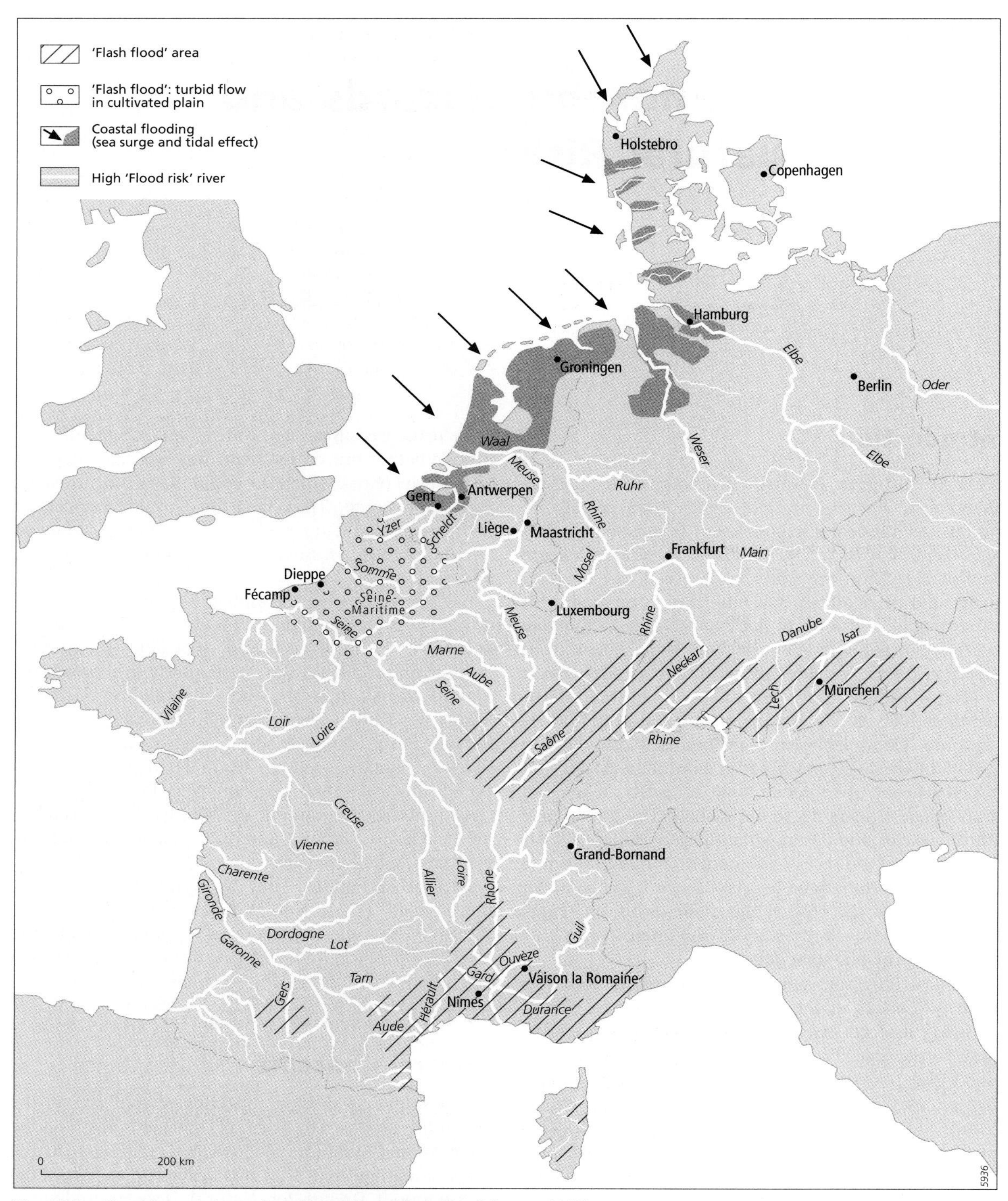

Fig. 18.1. Flood hazards in western Europe (modified after Salomon 1997).

Natural Disaster and Natural Risk

A natural disaster is an event which, when it occurs, is irrespective of the capacity of the community concerned to cope. A natural event is defined as a disaster a posteriori, after it has occurred in a given territory. Depending on the capacity of societies to react and to protect themselves, the event becomes a disaster or not. The management of the event depends on the level of development of the community affected. *The risk*, on the other hand, is only *potential*. It measures the possible event a priori. Where the disaster cannot be prevented research must concentrate largely on forecasting, and therefore on the risk, in order to anticipate disasters, 'even if relief and compensation are often easier than prevention' (Ledoux 1995). The concept of risk is a theoretical model that is the result of two other concepts: *hazard* and the *vulnerability* associated with human presence (people, houses, economic activities, infrastructures, property). A risk exists when there is a probability of a sudden onset of a destructive natural phenomenon and where it may cause damage to people, property, and activities. Risk is a relative notion that depends on the vulnerability of the social group and its perception of the hazard. The level of risk must be established by physical values for each of the hazards concerned. In some cases account is taken of interaction between the various hazards in a multi-risk measure. These values may be qualitative or quantitative (Leroi 1996). This depends largely on the knowledge of the phenomenon, its complexity, the technical and financial resources involved, and the existence of powerful and effective tools, such as geographic information systems (GIS), physical models, etc.

Hazard

Hazard is a complex notion that depends on the diversity of the natural phenomena. It meets the criteria of brevity, suddenness, and of a random nature. These natural phenomena or *hazards* must be defined in terms of intensity and spatial and temporal occurrence, in other words, where, when, how, and why does the phenomenon occur? The *spatial occurrence probability* (spatial extension or location) is governed by 'permanent' factors of predisposition or susceptibility such as slope, geology, and land use. The spatial extent will be fairly precise, with a departure zone and a dissipation (current, exceptional). The *probability of supervention or temporal occurrence* (return period) depends on the combination or conjunction, complex or otherwise, of natural or human onset factors, climatic conditions (rainfall, freezing, thaw, sun), earth tremors, and human activity. For floods this may be the conjunction of snowmelt and heavy spring rainfall. For a landslide it might be preceding climatic conditions (Flageollet *et al.* 1999), such as a rise over several years in underground water, associated with a succession of several very wet years involving interstitial suppressions (Maquaire 1990). This probability is quantitative (return time of 10 years, 30 years, 100 years) or qualitative (negligible, low, high). The *intensity* expresses the significance of the phenomenon. It is either real, i.e. measured (the magnitude of an earthquake on the Richter scale, the height of the water for a flood), or estimated (the intensity of an earthquake on the MSK scale, the duration of the submersion). A hazard level (or hazard degree) may be assessed by combining the elementary cause, for example, for a landslide, with a combination of the intensity and the temporal occurrence. For a flood, depending on its type, we may combine two of the most penalizing parameters—the height or duration of submersion and the speed of the floodwaters.

Vulnerability

Vulnerability is difficult to define, as there is no intrinsic vulnerability, but a vulnerability for each hazard. Vulnerability is therefore defined as the level of potential damage, ranging from 0 to 1, to a given exposed site, which is subjected to a possible or real phenomenon of a given intensity (Varnes 1984; Flageollet 1988; Leone *et al.* 1996; Leroi 1996). Vulnerability assessment thus poses problems of understanding the interaction between this phenomenon and the exposed site. This interaction can be expressed by so-called damage functions, or by extension, vulnerability functions. These enable a structuring of the various components of the vulnerability concept, which is the first step towards vulnerability assessment. After Leone *et al.* (1996), three main groups of exposed sites susceptible to damage are defined. These are property or land, including any structures on it, but also whole areas or types of land use, means, people, and the various activities and functions. Each group has its corresponding specific type of damage function:

- a structural-damage function for material assets;
- a corporal-damage function for people;
- an operational-damage function for the various activities and functions.

Inundations: Flood Hazards and Their Management

Every year catastrophic floods occur in countries throughout north-western Europe, causing considerable damage, inconvenience, and, yet more important, many deaths (on average a dozen a year in France, where they affect about 10% of the territory). They are caused either by flooding following winter or spring rainfall, or by flash floods over steeply sloping watercourses with a strong spate (summer storms in Mediterranean areas), or by melting snows in spring, or finally in coastal zones by a rise in sea level caused by cyclonic storms. Major floods are historic milestones: their exact dates and the damage they caused has often been known for hundreds of years. Flooding is even more devastating in our own time, because of an increasing number of human installations along river banks which increases vulnerability. Also, the risks are increased by changes in the characteristics of the river flood level, due to human intervention. The cost of flood damage and protection is now so high that flood warning systems must be developed and installed in order to reduce the risks.

The natural causes of flooding relate to the nature of the rainfall determined by duration and intensity, which affects the saturation of the ground and its retention capacity, and the spatial extent of the rainfall. In addition to the rainfall characteristics there is also a rise in river levels in spring, when ice break-up may occur and large masses of snow meltwater are released. In low coastal areas deep cyclonic depressions may cause a rise in sea level and therefore a risk of flooding. In addition to these natural causes, human intervention may change, provoke, or reduce flows. When the concentration and infiltration time of the waters is reduced their mobility and their destructive potential increase. Such activities include:

- human occupation and construction (factories and housing developments, roads and railways), increasing the extent of impermeable surfaces, causing an increase of urban rainfall runoff;
- land consolidation measures, such as the removal of hedges, the formation of larger land parcels and the introduction of drainage networks, which concentrate flows and reduce the response time of runoff from the catchments;
- other agricultural practices, such as deforestation and rapid crop-rotation over large cultivated areas, which reduce the soil's capacity to store water, thereby also increasing runoff;
- the drainage of marshes and other wetlands, especially in areas bordering rivers, which reduces their capacity to store water, to regularize the flow rate, and to reduce flood levels;
- the building of embankments along watercourses preventing the spread of the flow and aggravating the risk downstream in case of extreme events.

Flooding of low areas may result in damage to or complete destruction of buildings and other infrastructure, damage to cropland, and disruption of public transport and services; negatively affect water supply systems; and even endanger sensitive installations, such as, for example, in France, the nuclear power station in the Gironde estuary in 1999.

A Typology and Distribution of Flood Hazards

Fattorelli *et al.* (1999) offer a classification of the main types of flood in European countries, mainly relating to the meteorological situations: winter rainfall floods, summer convectional storm-induced floods, convective frontal storm-induced floods, snowmelt floods, urban sewer flooding, sea surge and tidal flood threat, and dam-break flood risk. Simplifying this, we may distinguish four major types of flooding in north-western Europe (Fig. 18.1).

Extensive Floods, More or Less Cyclic

These occur in winter or spring, due to depressions from the west associated with a well-developed warm front. Heavy, continuous, and prolonged precipitation can lead to soil saturation and consequent high volumes of runoff. The rise in the waters is progressive and slow. The flooding may last from several days to several weeks and flow volumes are often considerable (Fig. 18.2). Generally, there is relatively low danger to the population in these cases as they can be evacuated in time, but they often cause considerable material damage when they occur in densely populated areas. There are numerous examples of large floods but only a few memorable cases are quoted here. The most devastating floods in France include the Tarn in 1930, where there were 200 fatalities, 10,000 disaster victims, thousands of domestic animals killed, 3,000 houses destroyed, 500 factories out of commission, and losses amounting to billions of Euros (Ledoux 1995). Examples in Belgium include the flooding in the Meuse basin on 10 January 1926, which affected most of the city of Liège causing considerable damage and disruption. Plans for protection by means of dykes were drawn up after this flood (Heyse 1997). In the winter of 1994/5, in The Netherlands, some 300,000 people and millions of cattle had to be evacuated from the floodplains of the Rhine, Waal, and Maas (van Dorsser 1997). The alarming pictures of the disastrous flooding of the River Oder in the summer of 1997 recalled the last great high-water events on the Rhine in 1993 and 1995. The dykes at various locations along the River Oder were unable to withstand the enormous water pressure. Large areas were flooded in Czechia, Poland, and to a lesser extent in Germany. Over

Fig. 18.2. Aerial photograph of the Ill river flood of 3 February 1976, 30 km south of Strasburg (France) (photo: Jean Trautmann).

a hundred people lost their lives and the flood damage amounted to billions of Euros. The winter floods of 1999, 2000, and 2001 caused spectacular flooding in France and in several other European basins. We should also mention floods arising from cyclonic summer rainfall or associated with summer thunderstorms, such as the flooding of 21 July 1980 in the Meuse basin and of 6 June 1985 in the Schelde basin, south of Ghent in Belgium (Heyse 1997).

Snowmelt Floods

Rapid snowmelt can sometimes cause flooding, especially in the spring when warm southern airstreams can influence Alpine or upland areas and create sudden snowmelt, frequently accompanied by heavy rainfall. This phenomenon is usually very localized and quite slow, affecting urban areas at the bottom of valleys. In Denmark, in 1970, for example, the stream through Holstebro was flooded: in this case the damage was extensive because the road embankments, normally retaining the water, collapsed due to faulty construction (Moller 1997). Very steep watersheds may sometimes produce flash floods, since floodwater velocity can be high (Fattorelli *et al.* 1999).

Flash Floods

Flash floods, characterized by very rapid and high flow rises of sometimes only a few hours duration, are caused by high intensity storms in spring, autumn, or winter. These sudden, and usually localized torrential rains can be very destructive, and affect mainly Mediterranean regions, the upstream regions of large basins, and mountain watercourses. The flood wave causes the most damage (both material and human), largely because of the gravitational force of the moving water, but also because they carry moving objects such as trees and cars, which increase their destructive power. South-eastern France has suffered several major destructive floods in recent years (Joly *et al.* 1997). For example, a flooding in the Vaucluse in the autumn of 1992 killed forty-six people, including thirty-two who were the victims of a flash flood in the river Ouvèze in the small town of Vaison-la-Romaine on 22 September. Damages totalled 45 million Euros for the five departments and 718 municipalities affected. These rapid floods are aggravated by strong urban rainfall streaming due to large impermeable surfaces and to inadequate drainage channels (Salomon 1997).

Flash flooding is traditionally associated with mountainous regions and a Mediterranean climate; however, it also occurs in the widely cultivated plains of north-western Europe. The discharges, which are mistakenly called mudflows, are actually turbid flows that gather on cultivated land in the alluvial plains and may then empty into urban areas through a network of dry valleys. The storms usually occur in spring but they can also happen in the winter. In Haute-Normandie (France) over 180 floods of this type were recorded between 1960 and 1998 (Delahaye and Hauchard 1998; Delahaye *et al.* 1999). Flood events have increased consistently since the beginning of the 1990s, particularly in the Seine-Maritime region of France (Figs. 18.3, 18.4),

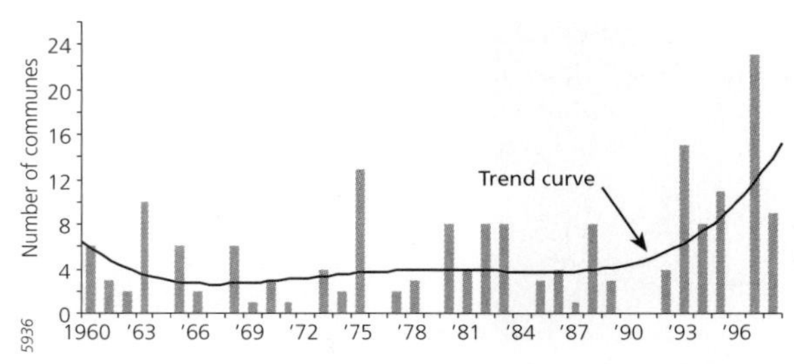

Fig. 18.3. Number of communes affected by flash floods (turbid flow) in the Seine-Maritime region of France between 1960 and 1998 (Delahaye, unpublished and 1999).

where several people were killed in 1997 (Delahaye 1999).

Flooding Associated with Cyclonic Storms

One of the major problems of flooding affecting many coastal areas is related to the sea surge and tidal effects. Moreover, associated with this problem is the phenomenon of coastal erosion which may lead to flooding (Fattorelli *et al.* 1999). In north-western Europe extensive coastal areas are low-lying and the combination of high tides, low atmospheric pressure, and strong onshore winds producing tidal surges are a major threat (van Dorsser 1997). Coastal flooding can also be caused by seepage of seawater through natural or artificial embankments, breaches of these embankments, or in exceptional cases through ocean swells causing waves to overtop the embankments. The problem has worsened because of land subsidence. The flood of 1953 in The Netherlands is a good example. One-third of the land in The Netherlands is below sea level at high tide. In the lowlands of The Netherlands and Belgium, around the Zeeland and Schelde estuaries the incoming tide extends some tens of kilometres inland and into several tributaries where the risk of flooding by storm surge is great. During the storm surge of 1 February 1953 all sea defences in Belgium, The Netherlands, and north-west Germany were severely tested, with extensive damage in places and hundreds of breached dykes (Fig. 18.5). The north-westerly gales forced water into the funnel-like estuary of the Schelde, increasing the level by 2.5 m above normal high tide and there was flooding when the protective dykes overflowed or were breached (Heyse 1997). In the Schelde basin 24,000 ha were flooded and another 16,000 ha in The Netherlands; 1,860 people died. The cost of the damage was estimated at more than 3 billion Euros (Ledoux 1995). A short time after the 1953 disaster the Dutch government decided to repair the damage and increase the height and strength of the complete coastal dyke system in the country, the so-called 'Delta-Plan'. It should be noted that these sea storms are responsible for coastal recessions of several metres a year, particularly of sandy beaches. Good examples are seen in Germany (Garleff *et al.* 1997), France (Joly *et al.* 1997), Belgium (Heyse 1997), and Denmark (Moller 1997).

Flood Risk Assessment and Reduction

Flooding hazards are typified by the height, speed, and duration of submersion. Flows that rise slowly in broad

Fig. 18.4. Gully that appeared in twelve hours on 13 May 1998 after a 80-mm storm in a small dry valley south of the town of Fécamp (Seine-Maritime, France). Of the catchment basin, 80% (3 km^2) is mainly cultivated. The incision reached a maximal depth of 8 m in the surficial deposits (silt, clay to flintstone) and in the chalk (photo: Daniel Delahaye).

Fig. 18.5. Devastation of villages and arable land in the province of Zeeland, The Netherlands, after the extreme spring tide and storm flood of 1 February 1953 (Photo: De ramp. Nationaal Rampenfonds, Amsterdam).

floodplains of large river basins and for which a flood warning can be issued do not have the same consequences as the sudden flooding of small basins due to heavy local storms which are difficult to predict. Progress is made in the techniques of spatial forecasting of storms, but in many instances there is still not sufficient time to alert local populations when rapid flood rises occur due to extreme rainfall events. Flood risk assessment for complex settings generally requires the use of hydrological simulation models, which simulate river flow and which are based on an assumed temporal and spatial distribution of precipitation. These models enable the evaluation of the effectiveness of various flood mitigation techniques (Fattorelli *et al.* 1999). At present, in most of the major European river basins (Fig. 18.1) flood warning systems have been installed and some, such as the Rhine and the Donau have systems that link the various countries from source to mouth of the water course. The Donau Accident Emergency Warning System, a co-operative programme linking the eleven countries along its course, was designed to monitor flood and pollution hazards. Moreover, some rivers are regulated by the construction of upstream reservoirs retaining flood overflows as well as supporting low water levels.

Rivers can also be embanked. In Belgium the SIGMA plan aims to reinforce and raise the existing 480 km of embankments and to build flood-control polders and a storm-surge barrier in the Schelde estuary. In The Netherlands, the Delta Plan was executed in order to close all the Dutch estuaries south of the Rotterdam Waterway with the exception of the Western Schelde, which form the waterway to the harbour of Antwerp. Already between 1701 and 1707 the Pannerdens Kanaal was excavated to regulate the water distribution between the rivers Waal, Rhine, and IJssel to reduce the risk of river flooding. Moreover, several storm surge and river barrages have been constructed to further regulate water levels in the rivers Meuse and Rhine. After the 1995 flood event a new Delta Plan for the rivers has now been started to raise and strengthen the dykes. In addition the governments of Germany, France, Belgium, and The Netherlands are engaged in consultations on the regulation of the water supply of the rivers. In all these countries the closure of water storage-basins and the effects on runoff caused by deforestation and the expansion of urban areas exacerbated the situation (van Dorsser 1997). In some cases, installations such as the hydroelectric station built on the Rhine in the Alsace aggravated the situation by raising the water levels and accelerating current velocities. The hydro-electric station can also accelerate flood waves in such a way that the flow of the upper Rhine comes into concordance with those of its tributaries Neckar, Main, and Moselle.

Though the benefit of large reservoirs or so-called retention basins is beyond doubt, it is evident that they cannot provide an answer to the problems in the catchment areas today. Control of the flows downstream is a remedy but not a preventive measure. Current practices tend towards managing river discharge as far upstream as possible in the catchment basins in an attempt to reduce the frequency and height of high water levels. Such programmes are promising but difficult to implement as they require intercommunal and even international structures and concerted land control measures, as well as the acceptation of the practices involved by a variety of stakeholders. In short, integrated land use management is required and this is still embryonic in many regions.

After the high flow events in 1993, 1995, and 1998 in The Netherlands, those living further down the Rhine are very sensitive about their safety. Flood levels must be lowered without raising the dykes. A partial restoration of the buffering function of the original river floodplain is envisaged by lowering and locally widening the embanked floodplains, and even by locally lowering or removing dykes and thereby bringing 'retention polders' into service again (Salomon 1997). In February 1995, the International Commission for the Protection of the Rhine (ICPR) commissioned the project group Action Plan on Flood Defence to draft an action plan for the complete Rhine drainage basin. The plan must also integrate and continue the ecological improvement of the Rhine and its floodplains. Between now and 2020 the Rhine will be 'renatured' from the Bodensee in Switzerland to the North Sea. The planning covers the renaturation and reforestation of 1,200 km^2 of old alluvial zones, 300 km^2 of which will once more be liable to inundation. This huge hydro-ecological project is estimated to cost around 5 billion Euros in the next one or two decades of its execution. At the same time, parallel action was started for the rivers Sarre/Moselle (International Commission for the Protection of the Sarre/Moselle—ICPSM) and the river Meuse (International Commission for the Protection of the Meuse—ICPM) (Corbonnois and Humbert 2000).

Earthquakes in Western Europe—a Moderate Seismic Zone

General Characteristics

The seismicity of western European countries is regarded as moderate when compared to Mediterranean countries such as Greece, Turkey, and Italy. Though several hundred earthquakes are recorded every year, only a dozen or so have a magnitude of more than 3.5. However, destructive earthquakes have occurred in the past and this should be taken into consideration, given the unpredictability of the phenomenon. Most earthquakes occur because of the rapid propagation of fractures in the contact zones between continental plates where the deformations are strongest. In Europe, earthquakes arise from the slow collision of the African and Eurasian plates and they are distributed along fault lines and old fold zones such as the Pyrenees, the southern Alps, or the Rhine graben (Fig. 18.6). An earthquake is measured in terms of magnitude and intensity. Its epicentre corresponds to its position on the earth's surface, vertical to the seismic focus where the intensity is highest. The frequency and duration of the tremors have a significant bearing on the effects at the surface. The magnitude is a measure of the energy freed by an earthquake using the recording of seismic waves and is classified on the Richter scale. In theory this scale has no limit, but the highest known magnitudes so far do not exceed 9.5 (Chili 1960). In Europe an earthquake with a magnitude approaching 9 was recorded in Lisbon in 1975. The intensity is a reflection of the effects and the damage caused by the earthquake at a given location observed at the earth's surface. It is at its maximum at the epicentre and decreases concentrically with distance. Several scales of intensity have been defined, which makes it difficult to compare information from widely differing sources and dates. The most frequently used are the Mercalli scale, dating from 1902 and modified in 1956, and the MSK scale, named after Medvedev, Sponheuer, and Karnik, three European seismologists. These two scales comprise twelve degrees written in roman numerals from I to XII. Since January 1997 most European countries have used a new scale, EMS 92 (European Macroseismic Scale 1992).

As is the case for most hazards it is impossible to forecast the time and exact place of an event. Earthquakes can occur anywhere along fault lines and at variable depths. In western Europe they are mostly superficial. The focus is situated less than 25 km in the earth's crust and neither the surface movement nor the damage is usually very great. The distribution of seismic zones is well known from the establishment of several monitoring networks, and a data bank, including data on historical seismicity. The SIRENE–SISFRANCE bank, for example, contains the characteristics of 5,000 historic and contemporary earthquakes felt in Europe. Several national earthquake catalogues are also available, for example in France (Lambert *et al.* 1996), and in Germany (Leydecker 1986; Grünthal 1988). Several monitoring networks equipped with numeric seismic stations send data, either deferred or close to real-time, to processing centres such as the Centre Sismologique Euro-Mediterranéen—CSEM, and the Bureau Central Sismologique Français—BCSF, in Strasburg. These centres can establish the characteristics of earthquakes very quickly and can inform the authorities. One example is the ORFEUS (Observatories and Research Facilities for European Seismology) network, which combines several networks of Digital broadboard seismograph stations in the European Mediterranean area.

Seismic Activity in Historical Times and Geographic Distribution

A region or location that has experienced seismic shocks in the past may sooner or later experience shocks of the same or greater intensity again. The main seismic zones have been located (Fig. 18.7) as a result of the many seismic monitoring networks and the research on historic seismicity. France is without doubt the country most

Fig. 18.6. Geomorphological hazards, and location of the main regions, towns, and rivers cited in the text.

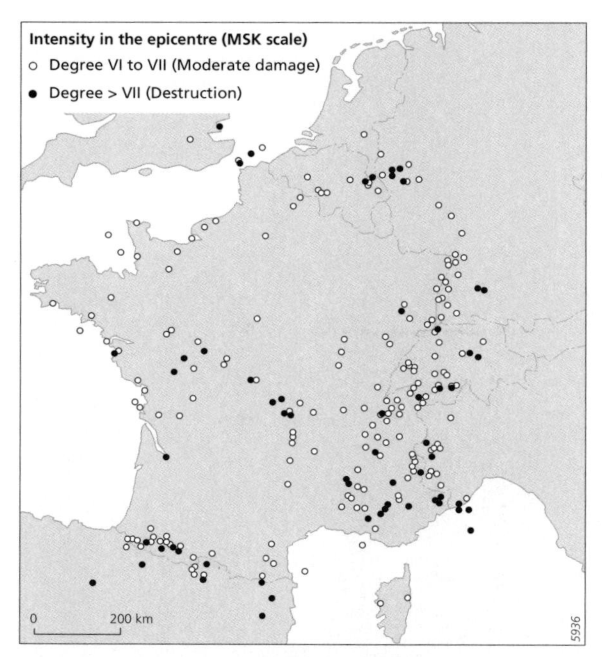

Fig. 18.7. Epicentres of major earthquakes in France and neighbouring countries during the period 1200–1994 (after Lambert *et al.* 1996).

affected by seismic activity in western Europe. The sedimentary Parisian and Aquitaine basins are regarded as almost aseismic and the large Hercynian region extending from the Massif Central to Bretagne has a low weak seismicity, but in the other regions the seismic activity, both past and present, is far from negligible. A dozen historic earthquakes prior to 1900 showed a macroseismic intensity in excess of VIII (on the MSK-scale) and since 1900 there have been twenty or more earthquakes with a local magnitude equal to or greater than 5, with M = 6.2 in the so-called Lambesc earthquake in the Provence on 11 June 1909, in which forty people died (Lambert 1997). For the earthquake at Basle in Switzerland on 18 October 1356 (intensity X), the damage extended between 15 and 30 km into the three neighbouring countries (France, Switzerland, and Germany) and several hundred people died. Two intensity VIII earthquakes occurred in 1477 and 1490 near Clermont-Ferrand and one at Remiremont in the Vosges on 12 May 1682. In the Pyrenees, at Arette, in August 1967, an earthquake (M = 5.3) caused severe damage to the town (Fig. 18.8). The strongest earthquake (M = 5.6) recorded in France for the last forty years occurred on 18 February 1996 (Madariaga and Perrier 1991). In Belgium, recorded earthquakes do not exceed 6.5 on the Richter scale and a maximum intensity of VII at the epicentre. The earthquake at Liège on 8 November 1893 (M = 4.9) was the most severe in Belgium until the one near Oudenaarde in 1938 (M = 5.6). In both cases there was extensive damage to houses and other property (François *et al.* 1987; Heyse 1997). The major earthquake at Roermond (province of Limburg, south-eastern Netherlands) on 13 April 1992 involved both Dutch and Belgian territory, and caused considerable damage in the town of Roermond and in several nearby villages (van Dorsser 1997). Two major earthquakes occurred in the Limburg area in the past with epicentres at Maastricht (April 4, 1761, MSK = 5) and Eindhoven (6 April 1843, MSK = 6.5). Apart from these few cases The Netherlands may be regarded as aseismic, like Denmark and Luxembourg. This is also true for northern and central Germany (Garleff *et al.* 1997). Seismic activity occurs mainly in the Rhine graben and spreads in a NW–SE direction along the Rhine valley as far as the Roer graben. Likewise, seismic activity is recorded in the Alps and Alpine foreland. For several earthquakes recorded in the past (Leydecker 1986), the intensity varied from V to a maximum of VIII, as for example, in the Ulm region in 1911 and 1978 (Lambert 1997).

Evaluation and Reduction of Seismic Hazard

Seismic hazard is defined as the probable level of ground shaking associated with the recurrence of earthquakes. The assessment of seismic hazard is the first step in the evaluation of seismic risk, obtained by combining the seismic hazard with vulnerability factors (type, value, and age of the buildings and infrastructures, population density, land use, date, and time of the day). Frequent, large earthquakes in remote areas result in high seismic hazard but pose no risk. On the contrary, moderate earthquakes in densely populated areas entail small hazard but high risk (Grünthal *et al.* 1999*a*, *b*). Unlike other hazards such as landslides or floods, it is impossible to prevent the occurrence of an earthquake when the state of the constraints of the earth's crust has reached the rocks' threshold of resistance to cracking. Consequently, in order to reduce seismic risk we can only reduce the vulnerability of people and property to earthquakes. In theory, therefore, these precautions are necessary (Lambert 1997):

- Buildings, networks and infrastructures must be resistant;
- Constructions must be moved away from dangerous sites: the proximity of active faults, unstable slopes, liquefied land, and a topographical or geological context which favours the amplification of seismic vibrations;

Fig. 18.8. Severe damage of the church belltower and destruction of several houses during the Arette earthquake of 13 August 1967: magnitude 5.3 on the Richter-scale and intensity VII–VIII on the MSK-scale (Pyrenees, France) (photos: Mairie d'Arette).

- Emergency services and communications must remain operational;
- The population must be informed and prepared for the possibility of such a phenomenon—instructions regarding behaviour, evacuation, etc.

After Giardini *et al.* (1999), the basic elements of modern probabilistic seismic hazard assessment are grouped into four main categories:

- Earthquake catalogues and databases: the compilation of a uniform database and catalogue of seismicity for the historical (pre-1900), early-instrumental (1900–64), and instrumental periods (1964–today);
- Earthquake source characterization: the creation of a master seismic source model to describe the spatial-temporal distribution of earthquakes, using evidence from palaeoseismic events, mapping, geodetic measurements, and geodynamic modelling;
- Strong seismic ground motion: the evaluation of ground shaking as a function of earthquake size and distance, taking into account propagation effects in different tectonic and structural environments and using direct measures of ground motions and their effects;
- Computation of seismic hazard: assessment of the probability of occurrence of ground shaking in a given time period, to produce maps of seismic hazard and related uncertainties at appropriate scales.

The Global Seismic Hazard Assessment Program (GSHAP) was launched in 1992 by the International Lithosphere Program (ILP). The GSHAP project published in 1999 a seismic hazard map of the world depicting the levels of chosen ground motions that probably will, or will not, be exceeded in specified exposure times. This Global Seismic Hazard Map depicts Peak Ground Acceleration (PGA) with a 10% chance of exceedance within an exposure time of 50 years (Fig. 18.9). PGA, a short-period ground motion parameter, which is proportional to force, is the most commonly mapped ground motion parameter because the

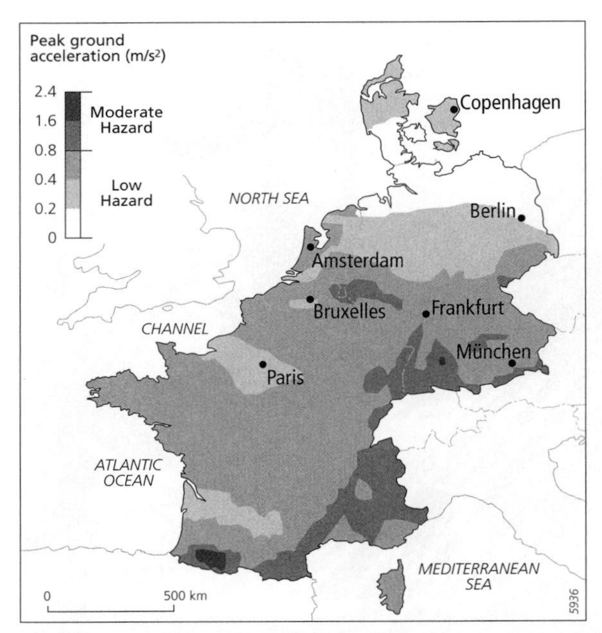

Fig. 18.9. Map of horizontal peak ground acceleration seismic hazard in western Europe, representing stiff site conditions for an exceedance or occurrence rate of 10% within 50 years (modified after Grünthal *et al.* 1999*a*, *b* fig. 4).

seismic provisions in current building codes specify the horizontal force a building should be able to withstand during an earthquake. As indicated in Fig. 18.9 and as mentioned earlier, the great majority of west European countries are in the low hazard zone (equivalent to 0–8% g, where g equals the acceleration of gravity). Moderate hazard zones (8–25% g) are found mainly in the Alpine massifs, in the Limburg region, and in the western Pyrenees, which exhibit maximum values for western Europe. It should be noted that the scale of seismic hazard on the world map also includes high- and very high-risk zones (25–40% g and > 40% g).

General mapping is useful in visualizing and assessing relative hazard levels over vast regions, but it must be complemented by research into seismic microzonage, generally carried out on a scale of 1/5,000 or 1/10,000. Microzonage takes account of changes in vibratory movement in the ground that may be due to differences in relief and geology. It also takes account of unstable slopes and liquefied soil which could provoke seismic tremors. Microzoning is useful in drafting recommendations regarding installations and land use, in order to avoid building in the most dangerous zones, and in identifying the most sensitive structures in a given zone. Regional maps are essential, given the increased vulnerability and the increase in population density in certain areas. For example, one study showed that if an earthquake identical to that of 1909 (the Lambesc earthquake in the Provence, in which forty people died) were to occur now, the death toll would be from 400 to 1,000 and the damage would be in the order of 1 billion Euros (Madariaga and Perrier 1991).

Mass Movements

There is a wide variety of mass movements. Mostly they are occasional phenomena of low amplitude with limited effects. However, there are also mass movements that cause extensive and costly damage because of their diversity, their frequency, and their wide geographic distribution. These phenomena are usually found in mountain regions owing to the instability of slopes and cliffs. They can also occur on plains and plateaux where they are associated with the use or dissolution of the subsoil. The wide variety of these phenomena, in terms of form, the volumes mobilized and the speed of movement, the frequency of occurrence and their spatial distribution in the countries concerned, and prevention and forecasting methods will be reviewed. Further details are found in general works or specialist articles such as Flageollet (1988, 1995, 1999); Almeida-Teixeira *et al.* (1991); Dikau *et al.* (1996*a*, *b*); Soldati (1996); Matthews *et al.* (1997); Dikau (1999); Schrott and Pasuto (1999); and in the proceedings of the 'Landslide' symposiums which take place every four years (Senneset 1996; Bromhead *et al.* 2000); the latest was held in Rio de Janeiro in June–July 2004 (Lacerda *et al.* 2004).

Types of Mass Movement

The form and scope of mass movements are very diverse because of: (*a*) the multitude of initiating mechanisms (erosion, deformation, dissolution, and rupture under static or dynamic load); (*b*) initiating mechanisms combined with landforms (height and gradient of the slope etc.); (*c*) the lithology (characteristics and sensitivity of materials—solid, plastic, viscous, liquid); (*d*) the rock or sediment structure (overhang, fracturing, superimposed layers); (*e*) the characteristics of the water table; and (*f*) the relative proportions of water and solid materials. Consequently, classifications exist either based on landform, type of process, speed, volume, displacement agent, cause of movement, age, or the risks involved. Often the speed of movement is chosen as the classification criterion as it indirectly expresses the risks involved. Two major types are distinguished, depending on the speed of movement (Fig. 18.10). The deformation of slow movements is progressive and may be accompanied

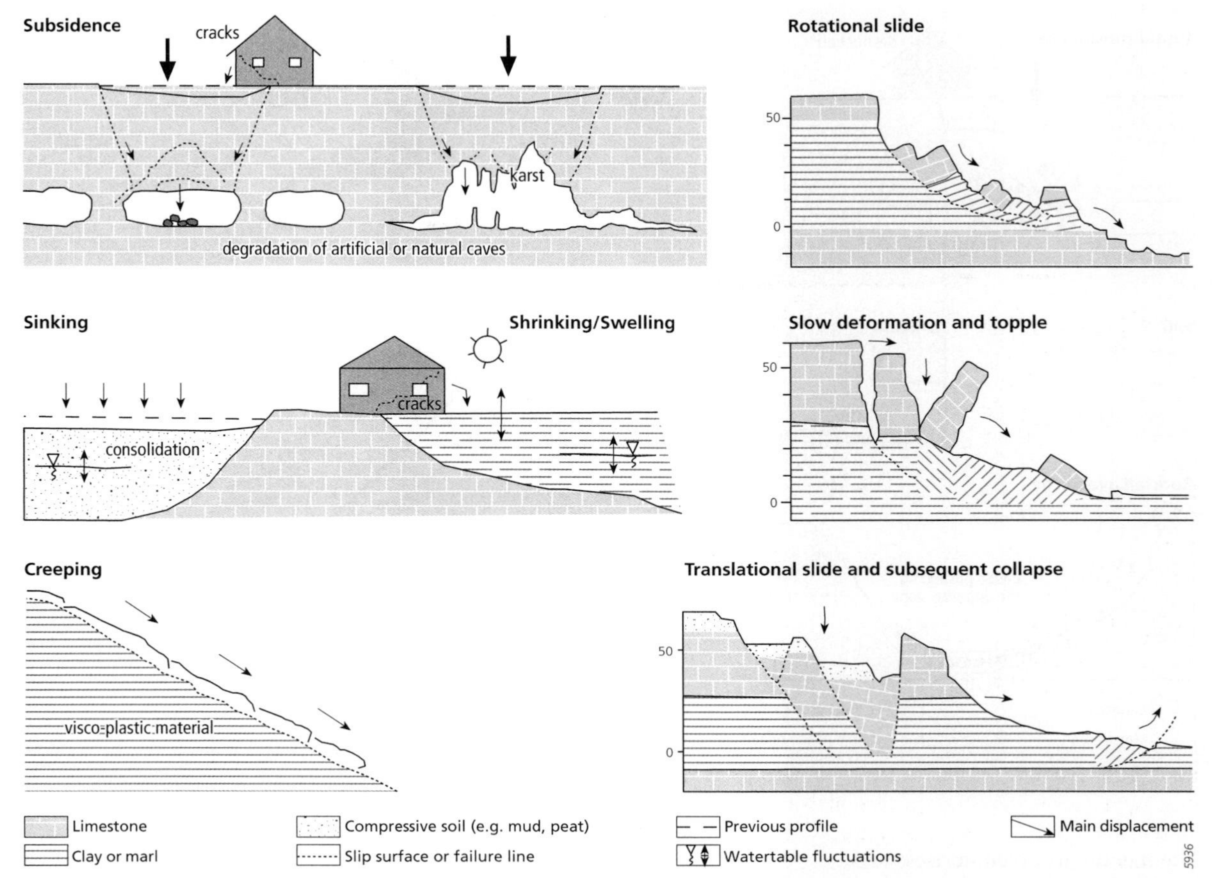

Fig. 18.10. Typology of mass movements based on the speed of movement: A. slow movements, and B. rapid movements with materials propagated in a mass or reworked.

by rupture, but shows no sudden acceleration. These movements can be monitored and controlled and are usually not a direct threat to personal safety.

The following *slow movements* are distinguished (Fig. 18.10A):

- 'subsidence' of material with slow surface deformation, with or without fractures, due to the evolution of natural or artificial caves (subterranean quarry or mines for iron, salt, coal, etc.), forming circular or oval cavities or surficial depressions;
- 'sinking' of material by the consolidation of compressible soil (mud, peat);
- 'shrinking-swelling' of clayey soils depending on water content;
- 'creeping' of visco-plastic material on a shallow slope;
- 'landslide' (e.g. slump, rotational slide, translational slide) corresponding to a mass displacement of coherent soil (clay or marl) along a circular, curvilinear, flat, or complex rupture surface.

Two groups of *rapid movements* are distinguished (Fig. 18.10B), depending on whether the materials are propagated in a mass or reworked: The first comprises:

- 'brutal subsidence' by rapid spontaneous collapse producing sinkholes or shafts caused by the rapid rupture of natural or artificial subterranean cave vaults;
- 'falls' (stone fall, pebble fall, boulder fall, debris fall, soil fall) arising from the mechanical development of cliffs or extensively fractured rocky escarpments;
- 'rock fall, rockslide, rock fall avalanche' of slabs from cliffs or rocky escarpments, depending on pre-existing discontinuity slabs.

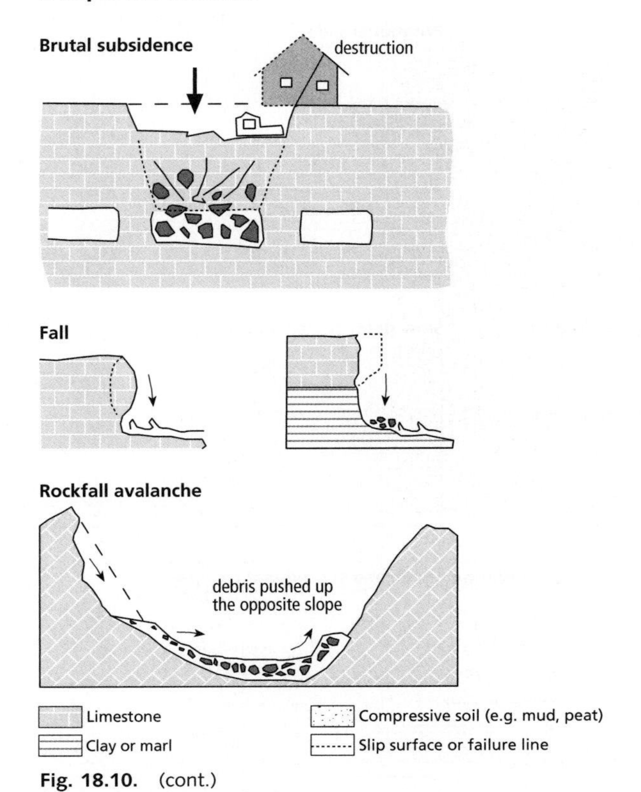

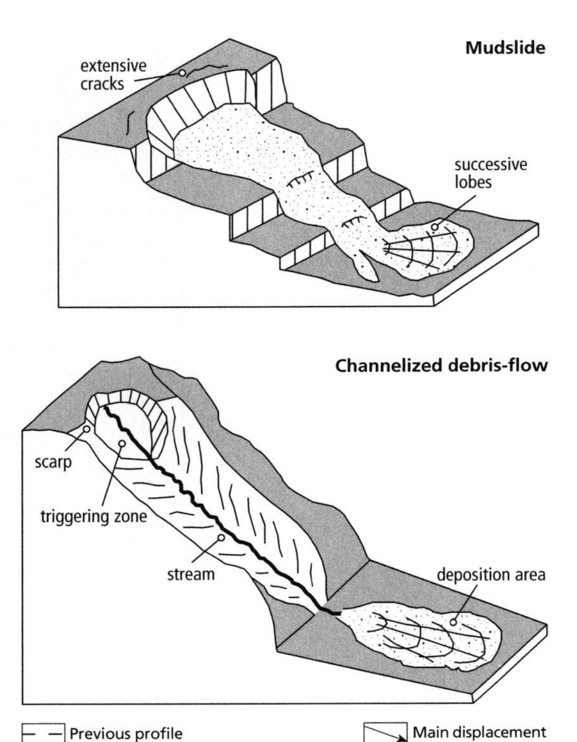

Fig. 18.10. (cont.)

The second group comprises:

- 'debris flow' arising from the transport of material in viscous or fluid flows in mountain stream beds;
- 'soil flow, mudflow, mudslide' (Fig. 18.11), generally due to the evolution of the front of landslips; their propagation mode is somewhere between mass displacement and viscous or fluid transport.

Slow movements may accelerate and end in a sudden rupture after a phase of warning signs (cracks, deformations, subsidence); this is true for landslides. Rapid movements may be a danger to life because of their scale and the sudden onset. It should be noted that the site and the composition of the movement (vertical or lateral) should be taken into account to distinguish between these phenomena. Movements may be limited in time and space, but can also be very extensive. Large movements affecting valley or mountain slopes may travel fairly long distances, burying people and destroying property (Flageollet 1988). Moreover, it should be stressed that many landslides are both composite and complex and they can change over time from one type to another, for example, a slide may develop into a flow. They can also be superficial (a few metres thick) or deep (several tens or hundreds of metres thick). According to Flageollet (1996) it is possible to distinguish mass movements on the basis of several criteria (e.g. morphology, speed) in 'stabilized' landslides (of Holocene, Pleistocene, or pre-Quaternary age), 'dormant episodic infrequent' landslides (recent or ancient history) and 'active intermittent or continuous' landslides (recent history).

Causes of Triggering or Reactivation

Mass movements are associated with a gravitational displacement of masses of destabilized earth under the effect of natural processes, such as snowmelt, exceptionally heavy rainfall, earthquake, action of the sea, etc., or to human activity such as terracing, deforestation, mining activities, and groundwater extraction. The causes are generally multiple, due to a conjunction or superimposition of several factors, including predisposition by slope or nature of the land and triggering factors.

Fig. 18.11. Mudslide that occurred during the spring of 2001 near the town of Corps (Trièves, southern French Alps).

The role of water is significant but very varied. First, it acts on the state or consistency of materials, mainly the clays, and may lead to an increase in water retention capacity. This then may produce a change from a solid to a plastic and then to a liquid state. Second, it acts on the underground water by means of the Archimedes thrust (Terzaghi's law) which lifts or lightens the soil (Terzaghi 1950). The consequence for the slopes is a decrease in the friction forces along the surfaces of the rupture, which promotes movement.

Figure 18.12 demonstrates the various phases in the development of the Villerville-Cricqueboeuf landslide at the Normandy coast (France), in relation to the piezometric and rainfall data (Maquaire 1990, 2000). It should be noted that the values for the high water levels, and to a lesser extent the low water levels, rise

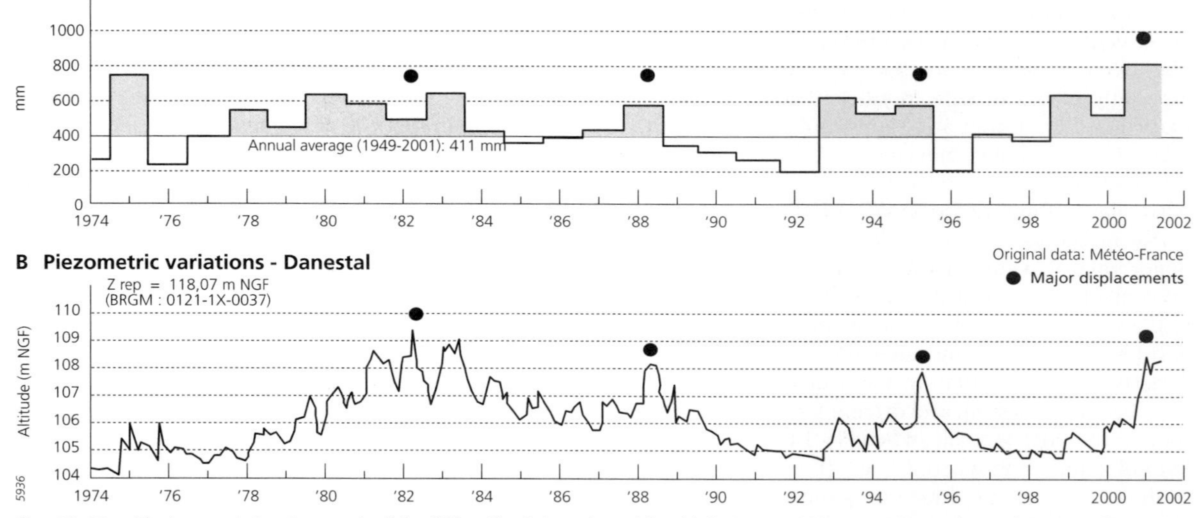

Fig. 18.12. Pluri-annual development of the Villerville-Cricqueboeuf landslide (Normandy coast, France) in relation to the piezometric and climatic data between 1974 and 2001 (modified after Maquaire 2000 fig. 2).

Fig. 18.13. House destroyed after the sudden major Villerville landslide of 10–11 January 1982 (Normandy coast, France). The house fell down about 3 m vertically. At the rear, the main scarp is about 3–4 m high. In the background, above the landslide scarp, other buildings remained intact.

progressively from one year to another. This pluriannual rise of the average water level reflects the cumulative effect of rainfall from previous years. The rise which began in 1978 continued until March 1982 and April 1983 (Fig. 18.12B). Then there was a period of several years in which the average levels decreased. The 1978 increase corresponds perfectly with the non-cyclical annual pluviometry fluctuations (Fig. 18.12A), which correspond to a period of effective rainfall considerably higher than the average (1949–98). The high level of the groundwater observed in 1982 is in phase with the onset of the major movement of 10–11 January 1982 (Fig. 18.13). In the same way, but to a lesser degree, the reactivation of February 1988 occurs after two hydrologic years that were above average, but especially after excessive rainfall in previous months had caused an abrupt and significant rise in the water table. For the major movement of January–February 1995, excessive rains in the two previous years, and in particular in the three previous months, contributed to a rise in the water table. This analysis confirms the major role of water in the onset and reactivation of movements. It is possible to quantify the respective role of the various factors for a slope, and in particular the part played by water, by calculating a safety factor (F). It derives from calculating for a given geometry the ratio between resisting and driving forces. If F is greater than 1, the slope is stable; when F = 1 the slope is at the limit of failure, and when F is less than 1 it is unstable. In the case cited here, concerning the Normandy coast, the slope stability analysis shows that a rise of 1 metre in the water table diminishes the general safety factor by approximately 6% (Fig. 18.14) (Maquaire 2000). But we should not forget the role of the sea, which prevents the creation of a balanced slope because of its erosive action at the toe of the slope. A recession of approximately ten metres in width will decrease the overall safety factor by 2% for this slope.

Distribution of Mass Movements

France is most affected by landslides, both in the number recorded and the variety; all types are represented. Denmark and, to a lesser extent Luxembourg, are the least affected. In Denmark, there are only a few mudslides along coastal cliffs in morainic deposits. In Luxembourg, there are soil collapses caused by natural dissolution and the mining of gypsum as well as occasional slope failures. The Netherlands has experienced subsidence caused by coal mining from 1900 to the mid-1970s, and by marl excavation near Maastricht since the seventeenth century. Presently, it is still occurring in regions where oil and gas are extracted on a large scale, especially in the northern part of the country, in the Groningen basin. The regions involved are fairly extensive, but the subsidence values are small, up to several decimetres. Germany experiences mass movements of several types, including rockfalls, slides, and debris flows, particularly in the Alps and Alpine foreland, in the central and southern Uplands (*Mittelgebirge*), and rarely, in the North German Lowlands and coasts. Thus we may observe landslides on the fronts of the secondary chalk cuestas in northern Bayern, extensively all along the Isar valley, south of Munich in Miocene marls, and both rockfalls and landslides in the Sarre-Nahe region. In karst regions we

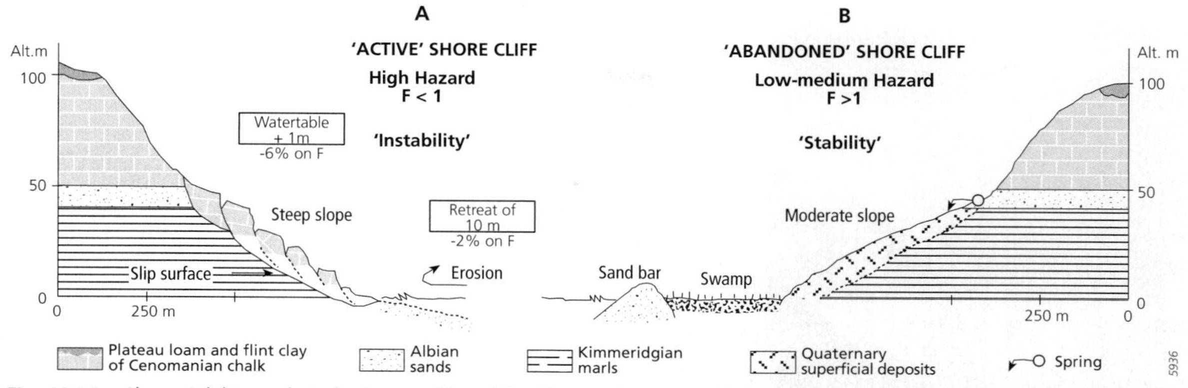

Fig. 18.14. Slope stability analysis for two profiles of the Normandy coast, including global safety factors (F) and hazard levels.

may also find that soluble formations in subterranean caves have collapsed (sulphate and chloritic rocks, but less so with calcareous rocks). This is seen in the Harz mountains and along the fringes of other central German Uplands and also at a few places in the North German Lowlands (Grunert and Hardenbicker 1997). At Lüneburg (Niedersachsen), 169 buildings were demolished between 1949 and 1973 because of subsidence caused by salt mining and the karstification of gypsum (Flageollet 1988). In Belgium, there are mass movements and subsidence associated with karst processes in limestone and marl excavations and related to coal mining (Heyse 1997). Rockfalls and landslides are fairly common in the hilly regions of Belgium, e.g. along the Meuse valley and its tributaries. Landslides and debris flows are most commonly encountered in the Ardennes.

In France, mass movements are mainly concentrated in the Alps and to a lesser extent in the Pyrenees. All types, both large and small, can be observed in the French Alps (Flageollet 1988; Ledoux 1995). Several dozen mass movements have been recorded in the Barcelonnette basin (southern French Alps) in the black marls topping the flysch of the thrust sheets (Fig. 18.15). The oldest are fossilized or dormant (Weber 2001). The most recent triggering or reactivations are associated with climatic conditions (Flageollet *et al.* 1999). In one example, three main earthflows have fossilized a gullied topography (Malet *et al.* 2000, 2001; Maquaire *et al.* 2001, 2003) (Fig. 18.16) and the streams are now subject to frequent channelized debris flows. In the Pyrenees the variety of earth movements is almost as wide as in the Alps, although instances are fewer in number. Landslides occur in the Jura on the valley slopes and the edges of plateaux in Lias marls. Falls occur in the chalky cliffs of the Ain valley and in the high chains of the Jura. In Lorraine, landslides occur mainly in the Toarcien marls on the slopes of the Moselle between Nancy and Metz (the Corny landslide). In Normandy, the movements are distributed as follows: rockfalls and slides on the coastal cliffs, extensive landslides (Fig. 18.17) on the coast at Bessin, in particular the Bouffay landslide in August 1981 (Maquaire 1990), and landslides in the Pays d'Auge on the valley slopes. Subsidence, caused by human activity, has been observed in the coal basins, the Lorraine iron and salt mines, and above underground quarries (e.g. marl, gypsum, chalk) in the Paris region. The dissolution of karst also entails subsidence as in the Paris region in the gypsum and in the Orleans region in the chalk. Several thousand detached houses have been damaged by the shrinkage of clay soil, followed by swelling. This was due to a pluviometric deficit that prevailed for almost ten years from 1989 onwards. Regions most affected by these processes were located in the Paris region, northern France, and Aquitaine.

Characterization of Hazard: Prevention, Prevision, and Protection

The prevention of disasters due to mass movements involves defining where and explaining why a given type of movement will occur. By analysing geologic, geomorphologic, and geotechnical data, including historical inventories of mass movement events, and statistical probability methods, the respective weight of the various instability factors can be determined (Soeters and van Westen 1996; van Westen 2000). The answer to the question 'when?' is obtained by analysing climatic conditions that lead to instability, and by establishing a relationship with the recorded earth movements (Maquaire 1990; Matthews *et al.* 1997; van Asch and Buma 1997; Flageollet *et al.* 1999). Flageollet *et al.* (1999) showed that the triggering or reactivation of many landslides may be explained by specific meteorological conditions.

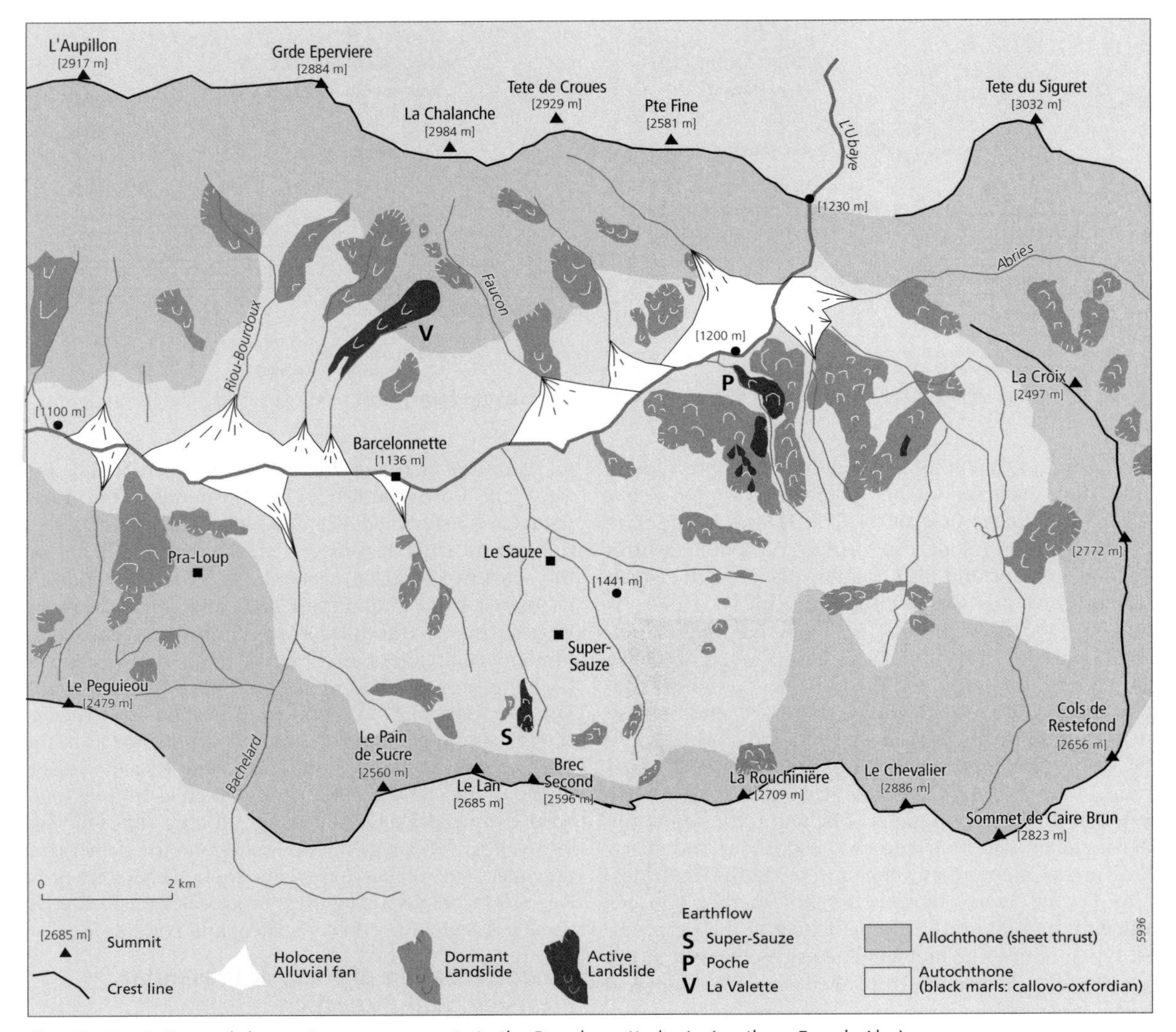

Fig. 18.15. Active and dormant mass movements in the Barcelonnette basin (southern French Alps).

However, some landslides bear no obvious relation to meteorological conditions, be they recent or more distant. The great diversity of the phenomena indicates that the conditions triggering them cannot be uniform. For deep slides, they involve the analysis of rainfall over previous weeks or months and taking into account the resulting piezometric variations. For shallow mudslides, however, it is the intensity and quantity of rainfall over previous days and hours that are most relevant. Nevertheless, rainfall is only one element of the system that accelerates or triggers landslides and is combined with other factors, such as land use. Instability can even occur following relatively dry months, whether or not they are preceded by heavy annual rainfall with high water tables. Conversely, heavy rainfall in earlier years or decades can be enough to trigger instability even if little rain has fallen over the immediately preceding months. These complex relations between the type of mass movement, actual and previous meteorological conditions, their generation (triggering or reactivation), and the initial degree of slope stability makes it difficult to define 'universal laws'.

The risk from mass movements can be reduced and protection can be assured by preventive or curative works of two types. Preventive defence tackles the causes of

Fig. 18.16. In the Barcelonnette basin, France (location in Fig. 18.15), the La Valette earthflow occurred in the spring of 1982 at the contact between the flysch of the thrust sheets and the autochthonous black marls in the upper part of the slope (altitude *c.*1,800 m). Then, the flow progressed and fossilized a gullied topography up to an altitude of 1,200 m, over a distance of more than 1,000 m. In the foreground, the forested in-situ crest remained above the flow (photo: Jean-Philippe Malet, 1999).

the active or potential mass movement and usually involves deep drainage or reforestation. Passive defence restrains the material, which is or may be in movement. This generally involves terracing, reprofiling, and the building of retaining structures. For example, in natural or artificial underground areas in urban zones any space that develops that could initiate mass movements is frequently filled by injecting concrete, sand, or bentonite. In high-risk cases where there is a high level of vulnerability and no way of reducing the hazard level, the slopes must be monitored. Generally surface movements and variations in the water level are recorded continually and an alert is sounded automatically when a fixed threshold is exceeded. In France, for example, two main sites in the Alps are fitted with instruments—the La Clapière landslide at Saint-Étienne-de-Tinée (Alpes-Maritimes) and the rockfall at the 'Ruines de Séchilienne' near Vizille (Isère). Some sites are also fitted with instruments to serve as scientific observatories in order to provide a better understanding of the various processes and to develop and validate predictive models. In the experimental site of the Super-Sauze earthflow (Fig. 18.18) in the Barcelonnette basin (Alpes-de-Haute-Provence, France) mass movements have been monitored and modelled since 1995 (Flageollet *et al.* 2000; Malet *et al.* 2000, 2001; Maquaire *et al.* 2001).

Avalanches

A brief reference must be made to avalanches, which affect only the mountain massifs in the French and German Alps, the Pyrenees, and to a lesser extent the Vosges, Schwarzwald, Massif Central, and Jura. For detailed description of the conditions under which the avalanches are triggered and spread the reader should consult Salm (1982) and Jamieson and Johnston (1993). Every year numerous small and occasionally large, destructive avalanches occur in these regions, many victims are claimed, and property is destroyed on a large scale (Fig. 18.19). In the French Alps, the most deadly avalanche of the twentieth century was at Val-d'Isère (Savoie) in 1970, with thirty-nine victims. In the Pyrenees, in 1601, the villages of Chèze and Saint-Martin (Hautes-Pyrénées) were destroyed with 107 victims. The most recent disastrous avalanche in the French Alps was at the base of the valley of Chamonix, on 9 February 1999. It destroyed seventeen chalets in the hamlet of Montroc, killing twelve people. It was an avalanche of powdery snow, more or less as an aerosol, which crossed the valley and extended up the opposite slope (Fig. 18.20).

Characteristics

An avalanche is a mass of snow that descends a slope at a speed of more than 1 m/sec. It is therefore a rapid gravitational movement. Depending on the nature of the flow several types of avalanche are distinguished:

- the 'aerosol avalanche' is a very rapid flow of a cloud comprising a mixture of air and snow particles, which can reach speeds of up to 400 km/h. It can cross the valley bottom and advance up an opposing slope because of its low density. The destruction is due to a significant blowing effect;
- the 'dense avalanche' is a flow of snow, which moves over the soil surface following the depressions of the relief (corridors, ravines, talwegs). The speed

Fig. 18.17. Aerial view of the Bouffay landslide that occurred on 5 August 1981 on the coast in Bessin (Normandy, France). This complex slide (translational slide and subsequent collapse, see Fig. 18.10) is 350 m long and 200 m wide.

rarely exceeds 80–100 km/h. The destruction is due to the enormous thrust developed by the mass of snow on obstacles;

- the 'mixed avalanche', in which both types of flow are present.

An avalanche is triggered by instability in the snow mantle, which usually comprises a number of stratified layers with distinct mechanical properties. Its heterogeneity is the result of variations in the relief and is also influenced by meteorological conditions. Many factors may lead to the triggering of an avalanche; some are related to the site, and are of a fixed nature, whilst others are variable. The fixed factors, which influence the stability of the snow mantle, are the relief, landform, slope angle, exposure, and vegetation. The variable factors are height and intensity of recent snowfall, rain, wind, thermic factors, and the structure of the snow cover. The onset may be natural or artificial, i.e. deliberate in the case of a preventive triggering or involuntary for example in the case of an overload by a skier.

Forecasting and Prevention

Methods for defining the avalanche hazard and of forecasting its approach are essentially the same as for landslides. In terms of prevention and protection strategies four categories can be distinguished:

- 'permanent and active' defence works, to contain the snow mantle by installing small wood banquettes, nets, racks, grids, fences, etc.;
- 'permanent and passive' defence works, by installing paravalanche galleries, stem posts, dykes, and avalanche detectors with a warning system on roads;
- 'active temporary measures', which are of limited duration and comprise, for example, preventive triggering by explosives;
- 'passive temporary measures' such as the temporary evacuation of the threatened population.

Temporal forecasting requires niveo-meteorological knowledge, which enables interpretation of the development of (in)stability of the snow. Spatial forecasting comprises the mapping of avalanche zones (e.g. corridors, spread zones). Since the disaster at Val-d'Isère in France in 1970 'Cartes de Localisation Probable des Avalanches' (CLPA: maps of the probable location of avalanches) have been constructed on a scale of 1 : 25,000 using the history of known avalanches and an analysis of the relief, lithology, and vegetation cover. These maps are historical documents and as such cannot be used as direct risk forecasting instruments, but they contribute to regulation mapping as basic documents. That these maps do not always depict the hazard zones correctly and precisely is shown, for example, by the Montroc avalanche in 1999 (Fig. 18.21). It moved considerably beyond the CLPA limits plotted in 1971 and slightly amended in 1991, which were based on witnesses' accounts of an avalanche in 1908.

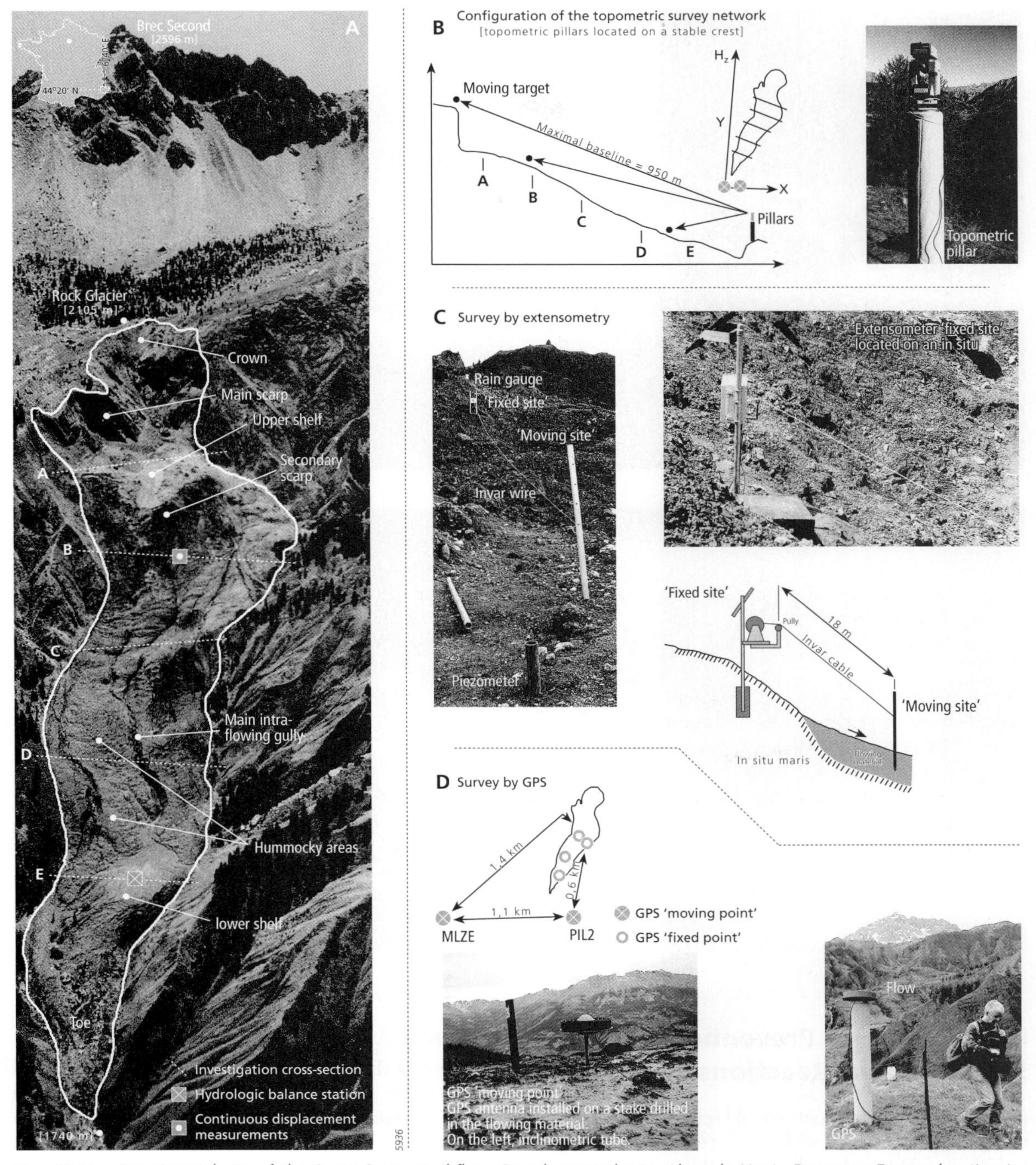

Fig. 18.18. Experimental site of the Super-Sauze earthflow, Barcelonnette basin, Alpes-de-Haute-Provence, France; location in Fig. 18.15 (after Malet *et al.* 2001).

Fig. 18.19. Rescue operations after the Saint-Colomban-les-Villards avalanche of 20 January 1981 in the Savoie, France (photo: François Valla, Cemagref, 21 February 1981).

Fig. 18.20. Aerial view of the deposit zone of the Péclerey avalanche of 9 February 1999 in the hamlet of Montroc (near Chamonix, France) (photo: François Rapin, Cemagref, 12 February 1999).

Indemnity and Prevention: Society's Two Reactions to Risks

It is clear that governments should reach a consensus on acceptable levels of risk, which in the end involves a political choice. This should be compatible with economic development and with social demand for protection against risks. There is no such thing as 'zero risk' and governments may favour either assistance and indemnity or prevention as a means of dealing with the risk.

Indemnity

Usually it is the State that declares a state of natural disaster after consulting technical reports regarding the exceptional nature of the phenomenon. Once the situation is recognized as a natural disaster the victims are compensated in some countries by the State automatically. This is the case in Belgium, where the money comes from Disaster Funds financed by the Treasury. In France since July 1982, on the other hand, it is compulsory to take private insurence against natural

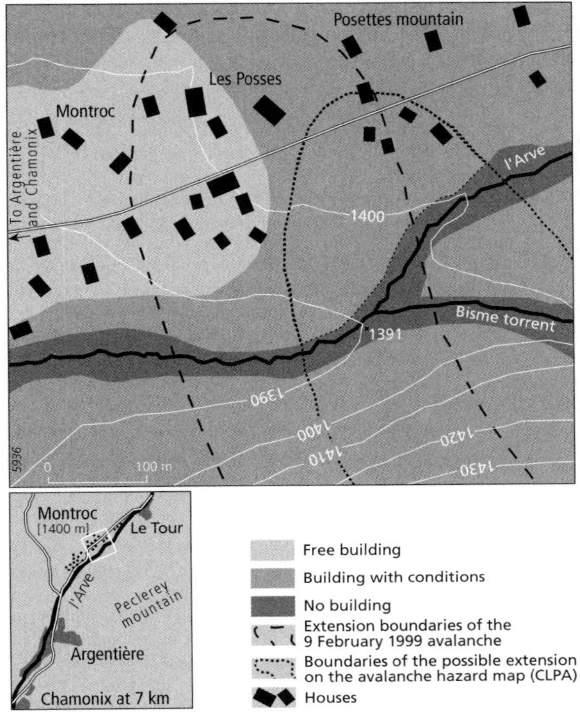

Fig. 18.21. Extension of the avalanche of 9 February 1999 in the village of Montroc, valley of Chamonix, France; zones of building restrictions are indicated.

disasters even in areas where the risk is low or nonexistent, based on the principle of solidarity. The list of communes declared threatened in one way or another is published by the State in the *Official Journal*, and those who have suffered damage must declare it to their insurers within ten days. In Germany insurance is optional. There is no law, but action may be taken case by case, depending on the gravity of the situation. A similar situation exists in The Netherlands. In recent years Belgium has been examining the possibility of a compulsory private insurance scheme like the French system because of the handicaps of the Disaster Fund, which has proved to be very inefficient. It is now specified that fire insurance contracts should in future cover flood damage, earthquakes, tidal surges, and landslides.

The Prevention of Natural Risks

This can also be ensured by regulations and in particular by mapping risk zones (Pissart and Closson 1999). Prevention may be implicitly included in local building regulations and may be mapped using historic data on hazards. France has been a pioneer in the field, having been responsible for preventive cartography for over twenty years. The State drafts prevention plans for natural risks (*Plan de Prévention des Risques naturels prévisibles—PPR*), particularly for regions of high risks. The PPR comprises graphic documents on a scale of 1 : 10,000 to 1 : 5,000. The plans include the information map for natural phenomena, the hazard map, the vulnerability map, and the PPR zoning plan showing three zones in three colors: prohibition (red), low restriction (blue), and zones with no specific restriction (white). Additional regulations are provided that specify measures for prohibition, protection, and safety, the management and use of buildings, public works, open spaces, and cultivated areas. Similar preventive measures are now written in parts of Belgium. In Germany, preventive cartography is not yet systematic, but is included indirectly in building laws. For example, landslides, debris flows, and avalanches in the German Alps have been mapped in detail and inventoried by the Geological Information Survey system. Obviously, these detailed mapping procedures are very time-consuming. Of the 17,000 French communes threatened by one or more natural risk only 1,700 were covered by a risk prevention plan in 1999. And even then errors in assessment of risks are made. This is exemplified by the Montroc avalanche (Fig. 18.21): of the twenty chalets damaged twelve were in the white zone (building permitted) and eight in the blue zone (building permitted under certain conditions). In this case the preventive cartography should therefore be revised and the zone affected will be converted to a red zone (building prohibited).

Conclusion

The technologically advanced nations of western Europe face numerous risks for relative frequency and intensity of hazards as compared with other parts of the world. Each of these countries defines its own acceptable risk thresholds, and there are large differences between countries as to the way to deal with these natural phenomena. For indemnity there is either the adoption of a solidarity principle with compulsory insurance for everyone, or state intervention case by case. For prevention there is compulsory mapping in some cases. Hazards vary little at the human level, while the vulnerable spaces are changing constantly as man changes his use of the land. It is mainly vulnerability, or rather its perception and management, that changes the risk level. This is particularly true for the risk of earthquakes, where man cannot influence the hazard in any way. For other hazards humans may be partly responsible for the onset or aggravation of the hazard, but may also be able to reduce the hazard by taking appropriate steps (e.g. land drainage, reinforcement, sustaining walls, protective dykes, deflecting walls). In recent years the European Union has strongly promoted research on environmental hazards and risks and their effects on society.

Acknowledgements

I would like to thank all the people who have advised me or who have given me help regarding the collection of information or the edition of this chapter: C. Ancey, F. Rapin, and F. Valla, Cemagref, Grenoble; M. Casabonne, Mayor of Arette; D. Delahaye, University of Rouen, for his share in editing the section on flooding and for Fig. 18.1; M. Granet, Institut de Physique du Globe, Strasburg; G. Grünthal, GeoForschungs-Zentrum (GFZ), Potsdam; J. Lambert, BRGM, Orleans; A.V. Poschinger, University of Munich; C. Puissant, Mons (Belgium); J.-C. Salomon, University Michel-de-Montaigne of Bordeaux; A. Souriau and M. Sylvander, Observatoire Midi-Pyrénées; J. Trautmann, Faculty of Geography, Strasburg; Margaret Nelson for the translation; and a very special thank you to my Ph.D. researchers, J. Ph. Malet and A. Remaitre, who drew all the original figures.

References

Almeida-Teixeira, M. E., Fantechi, R., Oliviera, R., and Gomes Coelho, A. (1991), Natural hazards and engineering geology. Prevention and control of landslides and other mass movements. *Proceedings of the European School of Climatology and Natural Hazards Course*, Lisbon, 28 March to 5 April 1990. Office for Official Publications of the European Communities.

Blaikie, P., Cannon, T., Davis, I., and Wisner, B. (1994), *At risk: natural hazards, people's vulnerability and disasters*. Routledge, London.

Bromhead, E., Dixon, N., and Ibsen, M. L. (eds.) (2000), Landslides, in research, theory and practice. *Proceedings of the VIIIth International Symposium on Landslides*, 26–30 June 2000, Cardiff. T. Telford, London.

Casale, R., (ed.) (1996), Hydrological and hydrogeological risks. *Proceedings of the First Review Meeting*, Brussels, 30–1 January 1995.

Corbonnois, J., and Humbert, J. (2000), Ressources et gestion de l'eau dans les bassins français de la Meuse, de la Moselle et du Rhin. In: J. P. Bravard (ed.) *Les Régions françaises face aux extrêmes hydrologiques. Gestion des excès et de la pénurie*. SEDES, Paris, 119–49.

Delahaye, D. (1999), Originalité des risques hydrologiques en Seine-Maritime: la catastrophe de Saint-Martin-de-Boscherville en juin 1997. *Études Normandes* 1–2: 157–70.

—— and Hauchard, E. (1998), Analyse spatiale des processus de ruissellement en Pays de Caux au travers de quelques épisodes critiques. *Bulletin de l'Association Géographique Français* 3: 306–16.

—— Gaillard, D., and Hauchard, E. (1999), Analyse des processus de ruissellement et d'inondation dans le Pays de Caux (France), intérêt d'une approche géomorphologique. In: S. Wicherek (ed.) *Paysages agraires et environnement*. CNRS, Paris, 209–19.

Dikau, R. (1999), The recognition of landslides. In: R. Casale and C. Margottini (eds.) *Floods and landslides*. Springer, Berlin, 39–44.

—— Brunsden, D., Schrott, L., and Ibsen, M. (eds.) (1996*a*), *Landslide recognition: identification, movement and causes*, Wiley, Chichester.

—— Cavallin, A., and Jäger, S. (1996*b*), Databases and GIS for landslide research in Europe. *Geomorphology Special Issue* 15: 227–39.

Embleton, C., and Embleton C. (eds.) (1997), *Geomorphological hazards of Europe*. Developments in Earth Surface Processes 5. Elsevier, Amsterdam.

Fattorelli, S., Dalla Fontana, G., and Da Ros, D. (1999), Flood hazard assessment and mitigation. In: R. Casale and C. Margottini (eds.) *Floods and landslides*. Springer, Berlin, 19–38.

Finlay, P. J., and Fell, R. (1997), Landslides: risk perception and acceptance. *Canadian Geotechnical Journal* 34: 169–88.

Flageollet, J. C. (1988), *Les Mouvements de terrain et leur prévention*. Masson, Paris.

—— (1995), Knowledge of landsliding for prevention and rescue. In: T. Horlick-Jones, A. Amendola, and R. Casale (eds.) *Natural risk and civil protection*. Commission of the EC, London, 255–67.

—— (1996), The time dimension in the study of mass movements. *Geomorphology* 15: 185–90.

—— (1999), Landslide hazard—a conceptual approach in risk viewpoint. In: R. Casale and C. Margottini (eds.) *Floods and landslides*. Springer, Berlin, 3–18.

—— Maquaire, O., Martin, B., and Weber, D. (1999), Landslides and climatic conditions in the Barcelonnette and Vars basins (Southern French Alps, France). *Geomorphology Special Issue* 30: 65–78.

—— Malet, J. P., and Maquaire, O. (2000), The 3-D structure of the Super-Sauze earthflow: a first stage toward modelling its behaviour. *Physics and Chemistry of the Earth*, Part B 25/9: 785–91.

François, M., Pissart, A., and Donnay, J. P. (1987), Analyse macroseismique du tremblement de terre survenu à Liège le 8 novembre 1983. *Annales Société Géologique Belge* 109: 529–38.

Garleff, K., Kugler, H., Poschinger, A. V., Sterr, H., Strunk, H., and VillWock, G. (1997), Germany. In: C. Embleton and C. Embleton (eds.) *Geomorphological hazards of Europe*. Developments in Earth Surface Processes 5. Elsevier, Amsterdam, 147–77.

Giardini, D., Grünthal, G., Shedlock, K., and Zhang, P. (1999), Global seismic hazard map produced by the Global Seismic Hazard Assessment Program (GSHAP). Map and legend.

Grunert, J., and Hardenbicker, U. (1997), The frequency of landsliding in the north Rhine area and possible climatic implications. In: J. A. Matthews, D. Brunsden, B. Frenzel, B. Gläser, and M. M. Weib (eds.) *Rapid mass movement as a source of climatic evidence for the Holocene*. Gustav Fischer, Stuttgart, 159–70.

Grünthal, G. (1988), Erbebenkatalog des Territoriums der Deutschen Demokratischen Republik und angrenzender Gebiete von 823 bis 1984. *Zentralinst. Physik der Erde* 99: 177 incl. Updates up to 1991.

—— and the GSHAP Region 3 Working Group (1999*a*), Seismic hazard assessment for Central, North and Northwest Europe: GSHAP Region 3. *Annali Di Geofisica* 42/6: 999–1011.

—— Bosse, C., Sellami, S., Mayer-Rosa, D., and Giardini, D. (1999*b*), Compilation of the GSHAP regional seismic hazard for Europe, Africa and the Middle East. *Annali di Geofisica* 42/6: 1215–23.

Heyse, I. (1997), Belgium. In: C. Embleton and C. Embleton (eds.) *Geomorphological hazards of Europe*. Developments in Earth Surface Processes 5. Elsevier, Amsterdam, 31–59.

Horlick-Jones, T., Amendola, A., and Casale, R. (eds.) (1995), *Natural risk and civil protection*. Commission of the European Communities. London.

Jamieson, J. B., and Johnston, C. D. (1993), Shear frame stability parameters for large-scale avalanche forecasting. *Annals of Glaciology* 18: 268–73.

Joly, F., Bocquet, G., Bravard, J. P., Chardon, M., Kieffer, G., Paskoff, R., and Vogt, J. (1997), France. In: C. Embleton and C. Embleton (eds.) *Geomorphological hazards of Europe*. Developments in Earth Surface Processes 5. Elsevier, Amsterdam, 115–45.

Lacerda, W., Ehrlich, M., Fontoura, S. A. B., Sayão, A. S. F. (eds.) (2004), Landslides: Evaluation and stabilization. *Proceedings of the IXth International Symposium on Landslides*, 28 June–2 July, Rio de Janeiro. Balkema, Rotterdam.

Lambert, J. (1997), *Les Tremblements de terre en France, hier, aujourd'hui, demain*. BRGM, Orléans.

—— Levret-Albaret, A., Cushing M., and Durouchoux, Ch. (1996), *Mille ans de séismes en France. Catalogue d'Epicentres Paramètres et Références*. Ouest Éditions, Presses Academiques, Nantes.

Ledoux, B. (1995), *Les Catastrophes naturelles en France*. Payot et Rivages, Paris.

Leone, F., Asté, J. P., and Leroi, E. (1996), Vulnerability assessment of elements exposed to mass-movement: Working toward a better risk perception. In: K. Senneset (ed.) *Landslides*. Balkema, Rotterdam, 263–9.

Leroi, E. (1996), Landslide hazard-risk maps at different scales: Objectives, tools and developments. In: K. Senneset, (ed.) *Landslides*. Balkema, Rotterdam, 35–51.

Leydecker, G. (1986), Erdbebenkatalog für die Bundesrepublik Deutschland mit Randgebieten für die Jarhe 1000–1981. *Geolog. Jahrbuch*, ser. E, 36.

Madariaga, R., and Perrier, G. (1991), *Les Tremblements de terre*. Presses de CNRS, Paris.

Malet, J. P., Maquaire, O., and Klotz, S. (2000), The Super-Sauze flowslide (Alpes-de-Haute-Provence, France). Triggering mechanisms and behaviour. In: E. Bromhead, N. Dixon, and M. L. Ibsen (eds.) *Landslides, in research, theory and practice*. T. Telford, London, 999–1004.

—— —— and Calais, E. (2001), The use of Global Positioning System techniques for the continuous monitoring of landslides. Application to the Super-Sauze earthflow (Alpes-de-Haute-Provence, France). *Geomorphology* 43/1–2: 33–54.

Maquaire, O. (1990), *Les Mouvements de terrain de la côte du Calvados*. Documents du BRGM 197. Éditions du BRGM, Orléans.

—— (2000), Effects of groundwater on the Villerville-Cricqueboeuf landslides. Sixteen years of survey (Calvados, France). In: E. Bromhead, N. Dixon, and M. L. Ibsen (eds.) *Landslides, in research, theory and practice*. T. Telford, London, 1005–10.

—— Flageollet, J. C., Malet, J. P., Schmutz, M., Weber, D., Klotz, S., Albouy, Y., Descloîtres, M., Dietrich, M., Guérin, R., and Schott, J. J. (2001), A multidisciplinary study for the knowledge of the Super Sauze earthflow in Callovian-Oxfordian black marl (Super Sauze, Alpes-de-Haute-Provence, France). *Revue Française de Géotechnique* 95/96: 15–32.

—— Malet, J.-P., Remaître, A., Locat, J., Klotz, S., and Guillon, J. (2003), Instability conditions of marly hillslopes: towards landsliding or gullying? The case of the Barcelonnette Basin, South East France. *Engineering Geology* 70: 109–30.

Matthews, J. A., Brunsden, D., Frenzel, B., Gläser, B., and Weib, M. M. (eds.) (1997), *Rapid mass movement as a source of climatic evidence for the Holocene*. Gustav Fischer, Stuttgart.

Moller, J. T. (1997), Denmark. In: C. Embleton and C. Embleton (eds.) *Geomorphological hazards of Europe*. Developments in Earth Surface Processes 5. Elsevier, Amsterdam, 91–8.

Pissart, A., and Closson, D. (1999), The importance and the problems of cartography, an example: the cartography of natural constraints on a territory of 74 km^2 in Belgium (the Sprimont territory). In: R. Casale and C. Margottini (eds.) *Floods and landslides*. Springer, Berlin, 125–32.

Salm, B. (1982), Mechanical properties of snow. *Review of Geophysics and Space Physics* 20: 1–19.

Salomon, J. N. (1997), *L'Homme face aux crues et aux inondation*. Presses Universitaires de Bordeaux, Bordeaux.

Schrott, L., and Pasuto, A. (eds.) (1999), Temporal stability and activity of landslides in Europe with respect to climatic change (Teslec). *Geomorphology Special Issue* 30: 1–211.

Senneset, K. (ed.) (1996), Landslides, *Proceedings of VIIth International Symposium on Landslides*, 17–21 June, Trondheim.

Soeters, R., and van Westen, C. J. (1996), Slope instability. Recognition, analysis and zonation. In: Turner A. K. and R. L. Schuster (eds.) *Landslides: investigation and mitigation*. Special Report 247, Transportation Research Board, National Research Council. National Academy Press, Washington, DC, 129–77.

Soldati, M. (ed.) (1996), Landslides in the European Union. *Geomorphology Special Issue* 15: 183–364.

Terzaghi, K. (1950), Mechanism of landslides. In: S. Paige (ed.) Application of geology to engineering practice. *Review of Geological Society of America*, 83–123.

Turner, A. K., and Schuster, R. L. (eds.) (1996), *Landslides: investigation and mitigation*, Special Report 247. Transportation Research Board, National Research Council. National Academy Press, Washington, DC.

van Asch, T. W. J., and Buma, J. T. (1997), Modelling groundwater fluctuations and the frequency of movement of a landslide in the Terres Noires region of Barcelonnette (France). *Earth Surface Processes and Landforms* 22: 131–41.

van Dorsser, H. J. (1997), The Netherlands. In: C. Embleton and C. Embleton (eds.) *Geomorphological hazards of Europe*. Developments in Earth Surface Processes 5. Elsevier, Amsterdam, 325–42.

van Westen, C. J. (2000), The modelling of landslide hazards using GIS. *Surveys in Geophysics* 21: 241–55.

Varnes, D. J. (1984), Landslides hazard zonation: a review of principles and practice. *Natural Hazard* 3: 1–63.

Weber, D. (2001), Contribution de la géomorphologie à la connaissance des mouvements de terrains dans les 'Terres noires' alpines: le glissement-coulée de Super Sauze (Alpes de Haute Provence, France). Ph.D. Thesis, Louis Pasteur University, Strasburg.

19 Air, Water, and Soil Pollution

Andrew Farmer

Introduction

The physical environment of western Europe (its air, water, and soil) has been affected by a wide range of pollutants for centuries. Localized pollution of water from anthropogenic sources has been observed since the time of the Roman Empire and by the medieval period cities already experienced air pollution problems. As will be seen, proposals to tackle pollution in the Rhine stretch back to the fifteenth century. However, extensive pollution of the environment was a characteristic of the industrial revolution and major and widespread impacts have been observed throughout the nineteenth and twentieth centuries. Only in the last few decades have the emissions (and, therefore, impacts) of many of these pollutants declined due to measures taken by the countries of the region, both collectively and individually (Farmer 1997).

This chapter presents an overview of trends in air, water, and soil pollution. In each case the pollutants of most concern will be discussed, indicating their sources and impacts; locations are indicated in Fig. 19.1. In each case the measures that have been adopted to reduce these pollutants will be described, not least to suggest trends for the future. The monitoring of pollutant emissions, concentrations in the environment, and their specific impacts have generated enormous quantities of data over many years. Basic 'state of environment' information is produced at the municipal, regional, national, and international level. The latter includes reports produced by EU institutions, especially the European Commission and the European Environment Agency, as well as other multilateral co-operative institutions such as the Rhine Commission.

Air Pollution

Introduction

Severe air pollution sources are concentrated, among other regions, in the traditional heavy industry complexes in north-eastern France (Fig. 19.2), Luxembourg (Fig. 19.3), the Meuse valley in Belgium, and in the huge Ruhr industrial complex in western Germany. The range of air pollutants produced by human activity, as well as the impacts that they cause, are extensive. This section will focus on the following pollutants: ammonia, nitrogen oxides, ozone, particulates, and sulphur dioxide. These result in a range of impacts from direct effects on human health and on vegetation to damage to buildings and materials and acidification and eutrophication of soils and water. Many other air pollutants are produced by a variety of processes. These include heavy metals (such as cadmium and lead) and persistent organic pollutants. These pollutants can cause localized impacts (around sources) as well as transboundary impacts.

Ammonia

Ammonia is produced from a few industrial processes. However, by far the largest source is agriculture. The intensification of agriculture (e.g. animal housing) and the increased use of fertilizers have resulted in a major increase in emissions across much of Europe. Ammonia, in high concentrations, can cause damage to vegetation near sources. However, the major concern from this pollutant results from its deposition as ammonium, which causes eutrophication of soils and, through the nitrogen cycle processes in soils and waters, also causes acidification. Emissions from animals occur from the

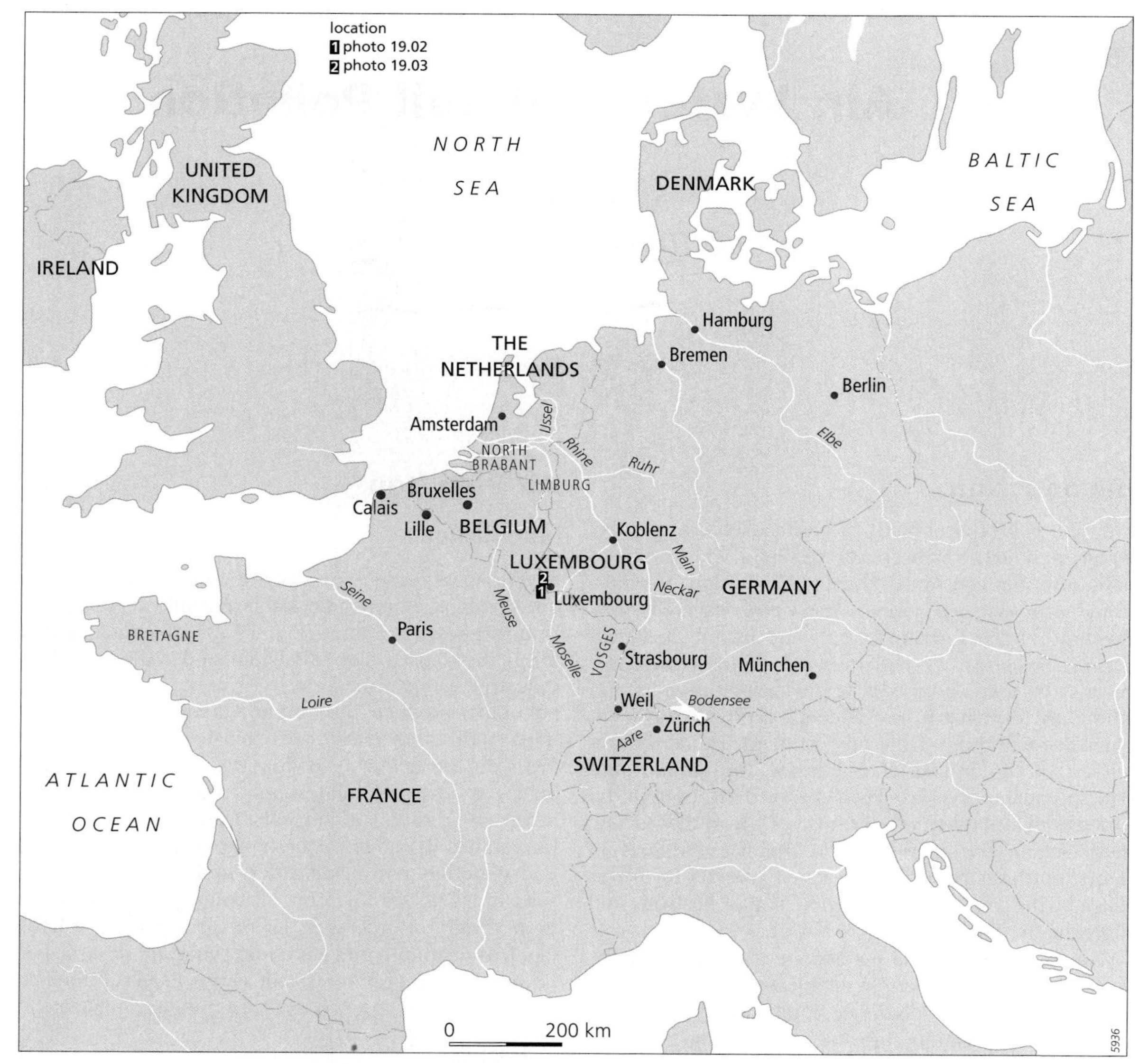

Fig. 19.1. Locations mentioned in the text.

unutilized nitrogen excreted in their waste. These emissions, that occur as waste, are produced, stored, or spread on fields. The highly intensive animal husbandry in Belgium, The Netherlands, Denmark, north-west Germany, and Bretagne all show elevated ammonia emissions.

Nitrogen Oxides

Nitrogen oxides are produced by the combustion of fossil fuels, ranging from power stations to motor vehicles. Most of the nitrogen comes from the reaction of atmospheric nitrogen and oxygen within the combustion chamber. There are two forms of nitrogen oxides—nitrogen monoxide (NO) and nitrogen dioxide (NO_2). Nitrogen oxides cause respiratory health problems and direct damage to vegetation. They also form a major component within acid rain. This not only causes acidification, but also eutrophication of soils and waters due to the nitrogen that they contain. The most important sources of nitrogen oxides are industry (especially any

Fig. 19.2. Severe pollution in the 1970s by steelworks and blast-furnaces in the valley of the 'Hauts-Fourneaux' in the old industrial complex of Longwy, north-eastern France (photo: Pim Beukenkamp).

Fig. 19.3. Air pollution in the 1970s by steel industry (ARBED) in the town of Esch sur Alzette, Luxemburg (photo: Pim Beukenkamp).

major combustion process, such as energy generation) and motor vehicles. In both cases policies have been implemented to reduce emissions and, in western Europe, total emissions are now declining. The highest concentrations of nitrogen oxides are found in the larger municipal areas across western Europe.

Ozone

Ozone is a secondary pollutant, i.e. it is not emitted by a specific source, but generated in the atmosphere by photochemical reactions on other pollutants. The primary reactants include nitrogen dioxide and volatile organic compounds (VOCs), which react in the presence of sunlight to produce ozone. VOCs do not generally have direct impacts on the environment themselves, although there are some specific toxic VOCs. They are most commonly emitted either through the use of solvents (e.g. in paint spraying) or through incomplete combustion of fuels such as petrol in vehicles. Ozone causes respiratory health problems for susceptible populations as well as damage to vegetation and materials. The production of ozone can be limited either by the quantity of VOCs available or the quantity of nitrogen oxides. The limiting pollutant varies across Europe. VOCs, nitrogen oxides, and ozone itself are also all readily transported across great distances, thus making effective control of ozone pollution more problematic than for the other individual pollutants. Ozone concentrations tend to be higher in the summer in sunnier environments and are highest in rural areas. However, regular exceedence of ozone standards occurs across much of western Europe.

Particulates

Particulates include any solid components suspended in the atmosphere. These can have a relatively inert chemistry in themselves or consist of toxic materials, such as metals. Particulates arise from many sources. Combustion processes usually result in the emission of incompletely combusted particles (smoke) and particles can be ejected into the air from processes such as mining or the action of vehicle movements on dusty roads. Localized impacts of dusts on soils and vegetation may occur around sources such as quarries. However, the primary concern with particulates is with those with a diameter of less than 10 μm (called PM10) and particularly those with a diameter of less than 2.5 μm (called PM2.5). These are able to penetrate deep into the respiratory tracts of humans and can result in health effects.

Sulphur Dioxide

Sulphur dioxide pollution is primarily caused through the combustion of fuels containing sulphur, i.e. coal and oil. The principal sources of pollution are energy generation, although some transport emissions can also be important. Sulphur dioxide can cause impacts on health and damage vegetation. However, it is best known as the major contributor to acid rain. Although emissions have declined dramatically in recent years the damage caused by acidification may take decades to recover.

Trends in Emissions

Table 19.1 shows the total emissions of sulphur dioxide, nitrogen oxides, VOCs, and ammonia in six western European countries in 1999. The differences between countries largely reflect their size. However, it is important to note that emissions are not uniformly distributed. Concentrations of heavy industry (e.g. the Ruhr industrial area in Germany) have higher emission levels, as do those areas with high concentrations of intensive animal units (e.g. the provinces of Noord-Brabant and Limburg in The Netherlands). The trends in emissions can be understood by reference to Table 19.2. This places the emissions in a policy context. For each pollutant the second column gives the percentage reduction in emissions between 1990 and 1999. This demonstrates significant differences between pollutants and countries (EMEP 2001). The greatest reductions are seen for sulphur dioxide. This high-profile pollutant (the focus of the pan-European acid rain debate of the 1980s and early 1990s) was the subject of a variety of United Nations and European Union (EU) emission reduction policies. All countries have achieved major reductions, but it should be noted that the high performance of Germany reflects the inclusion of the sources from the former German Democratic Republic following reunification which had not been subject to the previous extensive pollution control legislation of the Federal Republic during the 1980s. Less progress has been made on nitrogen oxide and VOC emissions, partly due to the slower introduction of controls on vehicle emissions compared to industry. The smallest reductions have been with ammonia. The greatest progress on ammonia is seen in Denmark and The Netherlands, which historically recognized

TABLE 19.1. *Total emissions of sulphur dioxides, nitrogen oxides, non-methane volatile organic compounds (NMVOCs), and ammonia, expressed in kT/year*

Country	Sulphur dioxide	Nitrogen oxides	NMVOCs	Ammonia
Belgium	186	292	271	103
Denmark	56	210	128	96
France	682	1,530	1,784	805
Germany	831	1,637	1,653	624
Luxembourg	4	16	15	7
Netherlands	100	408	282	175

TABLE 19.2. *Progress towards meeting air pollution emission reduction targets in western Europe. For each of sulphur dioxides, nitrogen oxides, non-methane volatile organic compounds (NMVOCs), and ammonia, the target from the 2001 EU National Emission Ceilings Directive (NECD) is given, together with progress made in reductions by 1999*

Country	Sulphur dioxide NECD target for 2010 (%)	Sulphur dioxide change 1990–99 (%)	Nitrogen oxides NECD target for 2010 (%)	Nitrogen oxides change 1990–99 (%)	NMVOCs NECD target for 2010 (%)	NMVOCs change 1990–99 (%)	Ammonia NECD target for 2010 (%)	Ammonia change 1990–99 (%)
Belgium	–73	–50	–48	–14	–59	–21	–31	–4
Denmark	–70	–69	–53	–23	–50	–24	–46	–25
France	–71	–47	–57	–18	–57	–27	–1	+2
Germany	–90	–84	–61	–39	–69	–49	–28	–18
Luxembourg	–73	–74	–53	–31	–52	–30	–7	–3
Netherlands	–75	–51	–55	–27	–63	–44	–43	–25

ammonia as a pollution issue and took domestic steps to reduce pollution prior to any international agreement.

Table 19.2 also indicates how far these emission reductions contribute towards the longer-term policy goal required by the 2001 EU National Emission Ceilings Directive. By 1999 all countries were well on the way to achieving the 2010 goals for sulphur dioxide. However, much remains to be done to achieve the targets for the remaining pollutants (sulphur oxides, nitrogen oxides, and ammonium). Emissions of fine particulates have also declined significantly during the 1990s. The percentage change in emissions of primary and secondary fine particles 1990–9 varies from about −23 for France and Belgium to −60 for Germany, with Denmark, The Netherlands, and Luxembourg in intermediate positions. However, the impact of these particulates is highly localized (e.g. on populations close to emission sources such as urban roads) and thus general emission data provide limited information.

Air Pollution Impacts

Ambient Air Quality

Ambient air quality concerns the concentrations of pollutants in the air itself, as opposed to their impacts following deposition onto soils, water, etc. Ambient air quality has changed considerably over recent decades in western Europe. Before the 'clean-up' of coal- and oil-fired power generation, the principle concerns arose from concentrations of sulphur dioxide and total suspended particulates. High concentrations of both had dramatic effects on health and the wider environment. The impacts of ambient air pollution on health are difficult to determine. The different air pollutants will have varying impacts on different susceptible population groups (e.g. those with pre-existing respiratory or cardiac problems). Current EU legislation requires authorities in member states to warn populations if air pollutant concentrations are likely to exceed 'alert' thresholds, so that individuals can reduce exposure. The simplest means of assessing air quality changes is to examine exceedence of statutory quality standards. For example, the average number of days with exceedence of 100 μg/m³ of ozone (8-hour mean) for urban stations only (after EEA 2001) for the period 1990–9 varies from 5 to 40 days. Relatively low values are found for Denmark, The Netherlands, and France, whereas higher values are found for Belgium and Germany. However, the interannual variation is also very large. Table 19.3 clearly shows the reduction in concentrations of specific air pollutants in selected western European cities (μg/m³) for the period 1990–1995/6. It is clear that good comparative information is not complete. However, air quality problems occur across the region for a range of pollutants (EEA 1999; EEA 2002). There is wide year-to-year variation, as climatic conditions are important in allowing the build-up (or dissipation) of air pollutants in urban areas.

An important biomarker of air pollution impacts are lichens. These organisms are highly sensitive to a range of air pollutants and, for many years, the presence and absence of different species have been used to map air quality across much of Europe (Gilbert 1992). Urban environments showed dramatic impacts on lichen flora, as was well documented in Paris as far back as the mid-nineteenth century (Nylander 1866, 1896). However, air

TABLE 19.3. *Concentrations of air pollutants in selected western European cities (μg/m³)*

City	Year	Sulphur dioxide		Nitrogen dioxide		Particulates	
		Annual average	Maximum 24 hr	Annual average	Maximum 24 hr	Annual average	Maximum 24 hr
Berlin	1990	57	348	39	98	78	311
	1995	20	169	33	64	52	165
Bremen	1990	18		38		32	
	1995	10	100	33	76	28	152
Brussels	1990	19	100	41	96	25	96
	1996	14	60	37	93	22	104
Hamburg	1990	26	177	47	97	50	267
	1995	14	111	37	80	41	180
Lille	1990	26					
	1995					18	75
	1996	28	106				
Zürich	1990	15	68	44	98	38	118
	1995	11	43	36	72	31	90

quality has improved markedly, so that urban lichen flora have begun to recover in Paris (Seaward and Letrouit-Galinou 1991) and in other cities, such as Munich (Kandler 1987). However, increasing emissions of 'new' pollutants, such as ammonia, have begun to affect lichens in more rural areas (de Bakker 1989).

Deposited Air Pollutants

Air pollutants may be deposited in different forms. They may dry deposit onto surfaces, or accumulate in rain, snow, or mist and be wet deposited. The quantities deposited will depend upon the concentrations of pollutants in the atmosphere. However, high concentrations will also form where the quantities of deposited precipitation are low (e.g. cloud formations at higher elevations). The three principle concerns over deposited pollutants are:

- Acidification: gases such as sulphur dioxide and nitrogen oxides dissolving in rain to form sulphuric and nitric acids;
- Eutrophication: gases such as nitrogen oxides and ammonia adding nitrogen to ecosystems; and
- Direct toxicity: pollutants such as heavy metals and persistent organic compounds being deposited in dry or wet form.

While deposited acid pollutants may cause some direct effects on vegetation, it is via soil processes that most impacts occur. Soils have a natural buffering capacity against acidity, their slow weathering releasing base cations that neutralize acid. The rate at which this happens depends upon the minerals present. Soils on limestone, for example, have a high capacity to buffer deposited acidity, while those derived from granite have very little buffering capacity. Thus the same quantities of acidity deposited in different regions of western Europe can have very different impacts. Soil acidification has a number of consequences. First the buffering cations are lost. Cations such as calcium and magnesium are plant nutrients and so their loss may affect plant growth. In acid conditions aluminium will take over as the buffering cation. This is toxic to both plants and animals and a number of effects of acidification on both soils and waters result from the presence of aluminium rather than the direct effect of pH. Examples of the detailed changes that can occur in soil chemistry in The Netherlands are given by de Vries and Leeters (1996). The deposition of nitrogen has major consequences for natural ecosystems. Many such ecosystems are naturally poor in nutrients and the addition of nitrogen may stimulate the growth of aggressive species which may replace those of greater biodiversity interest. The most extensive studies of nitrogen deposition effects in western Europe have been undertaken in The Netherlands, including chalk grassland (Bobbink 1991) and heathlands (van der Eerden *et al.* 1991).

While individual studies of changes in soil chemistry or effects on ecosystems provide an understanding of the impacts of deposited air pollutants, they do not provide information on the extent of these impacts. To achieve this the critical load concept was developed, that represents the quantity of deposited pollutant (acidity or nitrogen) that a soil can accommodate before adverse changes occur (Bull 1992). Critical loads depend upon soil type and vegetation type and these are easily mapped. By combining information on critical loads with levels of deposition, areas can be identified where such loads are exceeded and, therefore, adverse effects can be predicted to be taking place. Various countries have undertaken detailed assessments of critical loads and their exceedence. However, Europe-wide assessments are at a larger scale. Table 19.4 provides an example of the degree of exceedence of critical loads in western Europe for acidity and nitrogen in 1990 and how these are predicted to change by 2010. Exceedence for both types of deposition in 1990 was widespread. By 2010

TABLE 19.4. *Ecosystems where critical loads for acidification and eutrophication are exceeded in 1990, and following implementation of the 2001 EU National Emission Ceilings Directive (NECD) (%)*

Country	Acidification: area unprotected (1990)	Acidification: area unprotected after NECD	Eutrophication: area unprotected (1990)	Eutrophication: area unprotected after NECD
Belgium	58.4	7.4	100.0	83.4
Denmark	13.8	1.5	63.0	28.9
France	25.8	4.2	92.0	70.9
Germany	79.5	7.1	99.0	72.9
Luxembourg	66.7	0.9	100.0	75.1
Netherlands	89.3	23.7	98.0	87.0

TABLE 19.5. *Results from national forest-damage surveys 1997–2000: Percentage of trees with defoliation greater than 25%*

Country	1997	1998	1999	2000
Belgium	17	17	18	19
Denmark	21	22	13	11
France	25	23	20	18
Germany	20	21	22	23
Luxembourg	30	25	19	23
Netherlands	35	31	13	22

Note: In 1999 The Netherlands reduced its annual monitoring from 200 plots to 11 plots.

remaining problems with acidification will be much reduced and, as we saw above, significant progress towards this goal has already been achieved. However, exceedence of the critical loads for nitrogen will be improved only to a limited degree and this issue will require further action in the future.

Probably the most publicized impacts of deposited air pollution (together with ambient air pollution) in western Europe have been the effects on the health of forests. This was particularly highlighted in Germany, where a new disease phenomenon was identified, i.e. *Waldsterben* (Schutt and Cowling 1985). For many years surveys of tree health have been undertaken across Europe and recent data for western Europe are provided in Table 19.5. One of the most careful analyses of forest 'decline' has been undertaken in the Vosges Mountains of France. Landemann (1995) concluded that the large-scale studies did not provide any clear indication of a substantial deterioration of the forest condition in France. A recent review of forest condition in Europe and North America (Innes and Skelly 2002) concluded that while forest health in many areas is being adversely affected by air pollution, this is limited to specific regions with important pollutant sources. Furthermore, they argue that the concept of widespread regional forest decline (*Waldsterben*) should be discarded.

Policy Responses

The countries of western Europe, through the European Union, now largely have a common policy framework for tackling air pollution problems. The need for common, concerted action is driven by the transboundary nature of many air pollutants. Traditionally the European Union has attempted to tackle the problem of air pollution through a range of different measures, including:

- Ambient air quality standards: the EU has, for many years, set obligatory air quality standards to protect human health, especially in urban areas, thus requiring member states to adopt whatever measures are needed to achieve these. The 1996 Air Framework Directive is now the main instrument establishing limit values for individual air pollutants.
- Industrial emissions: the EU also sets requirements (including some specific emission limits) for air pollutants from industrial sources. The 1996 IPPC Directive requires regulators to determine 'best available techniques' for a range of processes, whereas the 1999 Solvents Directive adopts emission limits for VOCs (or National Emission Ceilings) for specific industrial processes as a means of reducing tropospheric ozone levels. Other limits are established for processes such as large combustion plants and incinerators.
- Vehicle emissions: there is a long history of EU legislation in this area, now being developed within the Auto-Oil Programme. Particular attention has been focused on lead, sulphur, NOx, and particulate emissions from different vehicle types.
- National Emission Ceilings: there is a history of EU legislation which caps emissions from individual member states, e.g. of sulphur from large combustion plants. The 2001 National Emissions Ceilings Directive is 'effects-based', capping member states' emissions based on their relative impacts in the environment.

The European Commission published the Clean Air for Europe (CAFE) Programme (European Commission 2001; Farmer 2001*a*), which should lead to thematic strategies under the Sixth Environmental Action Programme. Areas that are proposed to be incorporated into this ambitious new framework are air quality standards including: (1) emissions from stationary sources, (2) fuel standards, (3) vehicle emissions standards, and (4) acidification and ozone strategies. The CAFE programme highlights a number of issues that need to be addressed. It considers that the most pressing of these are the problems of particulates and tropospheric ozone. The former has multiple sources and our understanding of its health impacts is far from clear. Ozone is a secondary pollutant with multiple primary pollutant sources. Both are transboundary in nature and both are difficult to control to a level which should protect health and vegetation. The programme also states that other issues still remain to be fully addressed, including nitrogen oxides, acid deposition, and eutrophication. In conclusion, it must be noted that the dramatic decline in sulphur dioxide emissions in recent years will result in less acidification of soils and water. However, deposited nitrogen remains a problem and the problems of ambient air quality from particulates and ozone still require major efforts before they are tackled.

Water Pollution

Introduction

Freshwater surface, groundwater, and coastal waters are affected by a wide range of pollutants. These may range from large industrial discharges to runoff from roads and diffusion of agricultural fertilizers through soils. The following, currently particularly relevant elements will be addressed:

- The problem of diffuse agricultural pollution affecting surface waters and groundwaters;
- Point source pollution from urban waste water discharges; and
- Coastal pollution, particularly trends in the discharge of hazardous substances.

Water pollution is highly site-specific and the problems of individual rivers, for example, are often of high public profile. As a result this section takes the River Rhine as a case study examining changes in different pollutant pressures and how international collaboration has addressed these. Sources of water pollution are usually classified as 'point' or 'diffuse' sources. A point source usually takes the form of discharge from specific outfalls, etc., such as from an industrial installation or a sewage treatment works. A diffuse source does not have a single point of entry into a water body and may represent the slow ingress of a pollutant into, for example, a river over a wide area (D'Arcy *et al.* 2000). Although these pollution sources are often quite different, overlaps between them may occur. For example, runoff from urban areas is an important source of diffuse pollution, but combined storm water/sewage overflows, containing urban runoff, are a major point source problem. Point source pollution has been progressively regulated for many years. The single outflow to receiving waters has made such sources amenable to 'end-of-pipe' technical solutions and to be subject to 'traditional' regulation—i.e. the issuing of a permit with specified discharge limits. Diffuse pollution is less amenable to traditional technical or regulatory environments. Figure 19.4 provides overview data for western Europe on river concentrations of two pollutants. It can be seen that while there are significant improvements in phosphorus concentrations in most countries, nitrate pollution remains more problematic. This reflects the types of measures adopted, with point sources of phosphorus being tackled in many cases, while many nitrogen sources are yet to be addressed, though some measures have been achieved in Denmark.

Nitrogen and phosphorus pollution lead to eutrophication—the excess growth of plants (including algae) which can result in the decline of other aquatic life. This is a serious problem in many rivers, but also in parts of the coast. Eutrophication of coastal waters can lead to the production of phytoplanktonic blooms. Certain phytoplankton produce toxins that can lead to the death of marine organisms and can accumulate in bivalve molluscs and be transported into the human food chain. In 1995 the blooms of the toxin-producing *Gymnodinium* on the south coast of Bretagne and the coastal waters of the Pays-de-la-Loire resulted in heavy losses of fish,

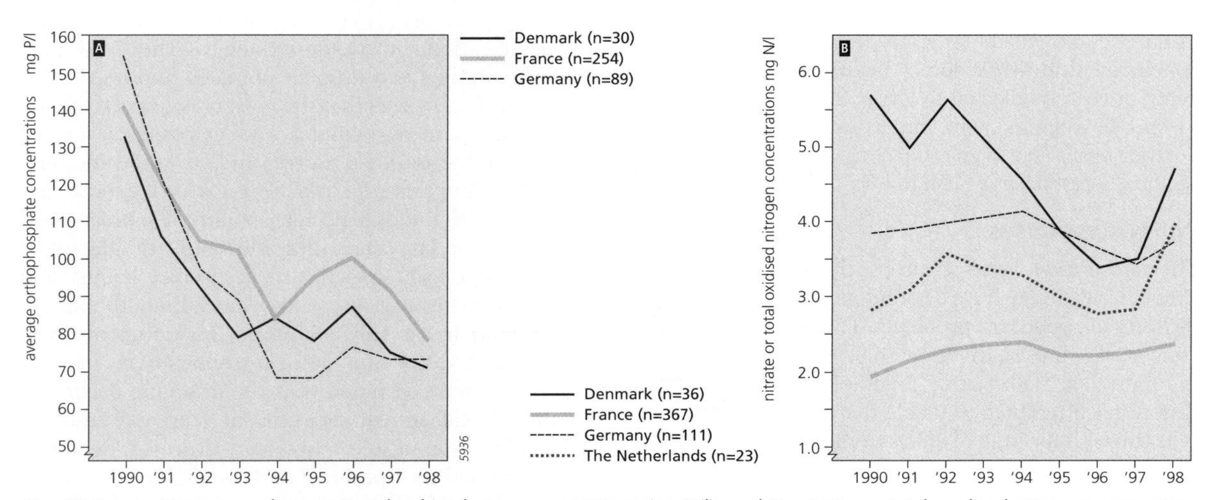

Fig. 19.4. A. Mean annual average orthophosphate concentrations (μg P/l), and B. nitrate or total oxidized nitrogen concentrations (mg N/l) at aggregated river stations (number given in brackets) between 1990 and 1998 in selected western European countries.

shellfish, and other invertebrates. Losses of farmed fish were linked to blooms of *Hetersigma akishiwo* in western Bretagne in 1994. Toxin accumulation in shellfish for human consumption (diarrheic shellfish poisoning) has regularly affected the waters of the northern part of the Seine estuary during the 1990s, as well as areas of western Bretagne and the Pays-de-la-Loire. More dangerous has been the increasing occurrence of paralytic shellfish poisoning first found in France in Bretagne, but now occurring in other coastal waters (IFEN 1999).

Point Source Pollution

Point sources of pollution to waters include industrial discharges and domestic wastewater treatment works. Both types of point source can result in the input of chemicals or organic matter that reduces the oxygen concentration of waters and, therefore, can reduce aquatic biota. Industry is also a significant source of a range of toxic substances, such as heavy metals, while domestic wastewater represents an important source of nutrients that lead to eutrophication of waters.

Industrial Sources

The discharges of toxic substances by industry were a major cause of the extreme pollution of a number of rivers in western Europe. However, major improvements have been achieved in this area. Table 19.6 provides data on the total discharge of a range of hazardous substances to the north-east Atlantic ocean. It is important to note that these figures include industrial sources as well as some diffuse sources. It can be seen that significant improvements have occurred in the discharge of most hazardous substances between 1990 and 1998. It is important to note that the figures include all riverine inputs, so that downstream countries naturally have higher discharge levels.

Domestic Wastewater

Sewage treatment works collect wastewater and may provide various levels of treatment, from simple removal of solids to complex treatment removing nutrients, etc. The changes in pollution from these sources is best viewed in relation to the objectives of the EU's Urban Wastewater Treatment Directive (European Commission 1998; Farmer 2001*b*). For nutrients this, *inter alia*, required nutrient removal for discharges to waters that are eutrophic or that may become eutrophic by the end of 1998. Member states had two options. They could either designate individual waters as 'sensitive', in which case wastewater discharged from sewage treatment works serving a population equivalent (pe—a standard measure of organic input) of more than 10,000 pe would require either 80% phosphorus removal and/or 70% nitrogen removal (depending on the potential impact) or a combined phosphorus and nitrogen removal of 75%. Alternatively, member states could designate their entire territories as sensitive and meet a 75% reduction for both parameters for all wastewater treatment plants. The main findings in 2001 of the European Commission with respect to compliance with these requirements were:

- *Belgium*: It was found that especially in Wallonia many of the treatment plants lack adequate nutrient removal. Much greater progress has been made in Flanders.
- *France*: For the whole of France, at the end of 1998, 130 of 158 agglomerations in sensitive areas did not meet the requirements for nutrient removal (Farmer *et al.* 2000). Moreover, data published in 1996 shows that 58% of phosphorus was removed in the Rhine–Meuse catchment and only 30 of the 62 agglomerations above 10,000 pe removed 70% or more of the nitrogen input.
- *Germany*: Much of the territory of Germany has extensive phosphorus removal, but additional treatment in many wastewater treatment works is required to remove nitrogen. Additional parts of Germany in catchments of the Baltic and North Seas should also be designated as sensitive.

TABLE 19.6. *Total direct and riverine inputs of hazardous substances to the north-east Atlantic ocean from western Europe (tonnes/yr) for 1990 and 1998*

Country	Cadmium		Mercury		Lead		Zinc		Lindane		PCB7	
	1990	1998	1990	1998	1990	1998	1990	1998	1990	1998	1990	1998
Belgium	4.30	2.45	4.05	0.61	27.50	51.50	285.0	498.50	102.0	110.5	30.5	135.0
Denmark	0.56	0.0	0.13	0.0	7.22	0.0	83.36	0.0	26.0	0.0	30.3	0.0
France	4.70	0.0	5.70	0.0	150.0	0.0	227.50	0.0	175.0	0.0	100.0	0.0
Germany	8.54	6.25	10.54	2.15	212.75	181.50	1762.2	1280.5	342.0	261.0	143.5	88.0
Netherlands	10.30	9.70	3.18	2.15	346.50	257.50	1993.0	1397.0	16.7	259.5	150.0	181.0

- *Luxembourg*: The entire territory has been designated as sensitive. While treatment removes more than 75% of phosphorus prior to discharge, additional investment is needed before the requirement for 75% removal of nitrogen has been met.
- *The Netherlands*: The entire territory has been designated as sensitive. By the end of 1998 the requirement for 75% removal of phosphorus had been met, but nitrogen removal at that date was only 60%. It is expected that full conformity with the Directive will be achieved by 2005.

Agricultural Diffuse Pollution

Agricultural diffuse pollution includes a range of different pollutant types, such as nutrients (from fertilizers and manure), pesticides, and sediments (e.g. from ploughed fields). Figure 19.5A provides data on total pesticide consumption in western Europe in the 1990s. While pesticide reduction has been achieved in countries such as Denmark and The Netherlands, in most other countries little change has occurred. Total fertilizer inputs are also given in Fig. 19.5B. Again there is significant variability between countries in their performance, although improvement is more widely distributed than with pesticides. The specific problems and measures adopted are best considered on a country-by-country basis.

Denmark

Of the total area of Denmark, 62% is agricultural land. There are sizeable areas of cereals, particularly barley grown for livestock, a substantial dairy herd, and a highly successful pig sector. Denmark faces one of the worst aquatic nitrate pollution problems in Europe, with parts of the country exceeding 100 mg/l of nitrate in freshwaters. However, there has been some decline in agricultural pressures with a decline in total fertilizer consumption of 36% and in potash and phosphorous by 62% since the early 1980s. The trend has been facilitated by a reduction in the agricultural area of 13.5% between 1983/4 and 1998/9. Total pesticide use, in terms of active ingredients, has also fallen by 21%. The first nitrate (environment) action plan was introduced in Denmark in 1987 and achieved a reduction of 14% in the volume of nitrate leached into watercourses. At a similar time the national pesticide reduction plan was initiated and achieved its aim of reducing pesticide consumption by half. Other instruments have also been used to reduce the environmental burden of diffuse pollution, including charges, taxes, and duties, for example, there is nitrogen tax, which farmers must be registered for by law, and a retail tax on pesticides of 37%.

France

France is the EU's largest agricultural producer, responsible for one-quarter of the EU's total farm output; agriculture occupies approximately 55% of the land area. One-third of farmland is given over to grazing (chiefly for cattle) and the other two-thirds to arable crops. Intensification and modernization have environmental impacts, which vary depending on the region and the type of farming operation. For example, livestock production has become a major source of water pollution affecting many areas of France, but is best epitomized

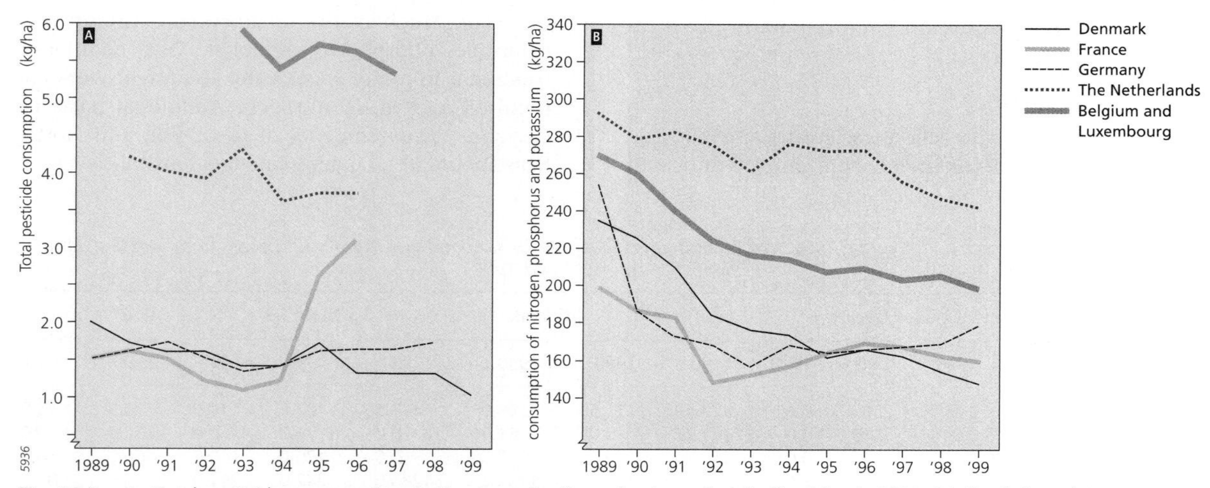

Fig. 19.5. A. Total pesticide consumption (active ingredient) per hectare of agricultural land, 1989–99 (kg/ha), and B. consumption of nitrogen, phosphorus, and potassium fertilizer per unit agricultural land 1989–99 (kg/ha).

in Bretagne. In this region, nitrate concentrations are often higher than 50 mg/l and in spring higher than 100 mg/l, due to the intensive production of cattle, pigs, and poultry. Pesticide residues in groundwater are found over large areas. The government response to these concerns has not been as proactive as in many other countries.

Germany

Agriculture accounts for about 50% of land use in Germany. Arable and permanent crops cover two-thirds of the total farmed area with the remainder dedicated to grassland. Specialized dairy farming is very important in grassland areas of the North German Lowlands, in the central and southern Uplands, and in the Alps. Most of the eastern *Länder* are dominated by large-scale arable farming. The most fertile regions in the western *Länder* (Niedersachsen and Schleswig-Holstein) also tend to specialize in arable farming. Very intensive, specialized livestock production is found in coastal areas of the north-west. Mixed farms, however, are still an important part of the farming sector, especially in central and southern Germany. Around 2% of Germany's agricultural land is farmed organically, which makes it one of the largest organic sectors in the EU. German agriculture is not as intensive or industrialized as in, for example, The Netherlands, but there are regions with very high input levels and livestock densities, particularly in the north. Reports from the German Joint Commission on Water Research indicate that nitrate and phosphate pollution of many freshwater bodies, in particular coastal waters, has historically been at a relatively high level. The Federal Environment Agency (Schulz 1999) records that between 1993 and 1997 diffuse nitrate agricultural runoff and groundwater contamination accounted for 72% of total nitrate point and diffuse pollution levels. Similarly, agricultural runoff and groundwater phosphate emissions into surface waters accounted for 66% of total point and diffuse phosphorous emissions. There is a substantial body of national legislation concerned with input use, good agricultural practice, and pollution control but it is notably less stringent than in Denmark.

The Netherlands

The Netherlands is one of the most intensively farmed areas in Europe and despite its small land size, is ranked the third largest agricultural exporter in the world, after the US and France. The main areas of production are horticulture, pigs, poultry, and dairy farming. Livestock farming is concentrated in the south and east of the country and accounts for about 60% of The Netherlands' agricultural production. Arable farming is concentrated in the north and south-eastern regions of the country and is mostly given to the production of cereals, animal fodder, sugar beet, vegetables, and flowers. Due to the intensiveness of agricultural practice, and particularly livestock production, The Netherlands has suffered some of the worst pollution levels in Europe. The Dutch government has historically failed to take action in proportion to the scale of the problem and is under considerable pressure to curb the environmental impacts of intensive farming. The most significant concern has been the high nitrate and phosphate concentrations in freshwaters that have been held attributable to the vast quantities of manure produced from intensive livestock farming. It has been estimated that The Netherlands produces more manure per hectare than in any other country in Europe, an amount equivalent to about 340 kg of nitrogen per hectare, nearly five times the EU average. This manure and slurry is spread on the fields and leaches into freshwater systems. Dutch soils are thought to be heavily contaminated with nitrate and some are saturated with phosphates. It has been recorded that on average, use of nitrogen on grassland was 679 kg/N/ha in 1993. It is also broadly estimated that agriculture's overall contribution to surface-water pollution (from the application of organic and inorganic mineral inputs) amounts to 40% of total phosphate pollution and 56% of total nitrogen pollution (Smit 2000). Other pollution sources are just as much of a concern, for example the adverse effects of pesticide pollution on both surface and groundwater. Dutch maximum permissible concentrations of agricultural pesticides in surface water are exceeded in more than 50% of all surface waters and sometimes by factors of up to 100–1,000; this is mainly because The Netherlands has extremely high levels of pesticide use, four times the EU average. A national pesticide action plan is in place to combat this problem. A family of measures to reduce water pollution include an increasingly strict manure policy, including stringent mineral accounting systems, and a 20% cut to the country's pig population.

Case Study: The River Rhine

It is not possible to provide extensive detail for the many different water bodies affected by pollution. However, it is worth examining probably the highest profile case for this region—the Rhine, which has a long history of multiple pollutant problems and for which solutions have been developed by the different countries of the region. The Rhine is one of Europe's most important and famous rivers. Originating in Switzerland it is 1,320 km long and flows via France, Germany, and The

Netherlands to the North Sea. With its tributary rivers it has a catchment of around 200,000 km^2. Centuries of human settlement and manipulation of the river have, not surprisingly, resulted in the degradation of the environment. It is also clear that conflicts can arise in the role of the Rhine for human purposes, for example, between its role as a drinking-water source and its use for wastewater disposal. Degradation started early. One of the earliest records of problems in the environment was noted in 1449. The effects of overfishing and of pollution had led to significant declines in the fish population. As a result the Strasburg Regulations were adopted. These formed the first international agreement to protect the Rhine. However, many more changes to the environment were to take place, before improvements would be seen. Even though pollution problems had been highlighted for many years, it was not until the 1960s that the issue became of high enough profile for action to be considered. However, by this stage water quality had deteriorated to the point that little aquatic life remained along much of the river. Discharges of organic wastes (e.g. from sewage) had reduced oxygen levels to below 2 mg/l, industrial discharges included large quantities of heavy metals, hydrocarbons, organochlorine compounds, pesticides, etc. Sediments became highly contaminated. As a result, those utilities providing drinking water had to employ increasingly more complex and expensive treatment methods.

Solving the problems of the Rhine was only possible through international co-operation. The Rhine catchment states (Switzerland, France, Germany, Luxembourg, and The Netherlands), therefore, established the International Commission for the Protection of the Rhine in Basle on 11 July 1950. Co-operation was strengthened particularly in 1963 with the signing of the Berne Convention on the Protection of the Rhine. The accident at the Sandoz plant near Basle resulted in the release of large quantities of toxic substances into the Rhine. These caused the death of almost all aquatic life for many kilometres downstream as far as the Lorelei, just upstream of Koblenz. The effect was equally dramatic politically, with popular concern heightened in all the Rhine states. In a short time three ministerial conferences had taken place addressing pollution to the river. These culminated in the 1987 Rhine Action Programme, in which a series of broad and challenging goals were set out for the period up to the year 2000. The specific goals were:

- Ecosystems of the Rhine should be improved sufficiently for species such as salmon and sea trout to re-establish breeding populations,
- The production of drinking water from the Rhine should be guaranteed for the future,
- The pollution of river sediments should be reduced to levels compatible with the use of sludge for landfill or sea dumping.

The Rhine Action Programme also established more specific objectives to underpin these broad goals, such as that of reducing the input of dangerous substances to the river by 50% between 1985 and 1995. Concerns were again heightened in 1988 when enormous algal blooms occurred in the North Sea related to nutrient discharges from the mouth of the Rhine. As a result the Rhine Action Programme added a further broad objective for 2000:

- The improvement of the ecological state of the North Sea.

Accident Management and Industrial Pollution Reduction

The management of accident prevention and containment is highly site-specific. The International Commission for the Hydrology of the Rhine in co-operation with several universities developed a model to enable rapid and reliable prediction of how pollutants would travel along the river in case of accidents. This model can predict the timing of maximum pollutant concentrations following an accident, thereby allowing for measures to be taken to reduce or avoid its effects. The model includes the Rhine from its outflow at the Bodensee (Lake Constance) to The Netherlands, including the estuarine rivers IJssel, Nederrijn, and Waal, the tributaries of the Aare, Neckar, Main, and Moselle, and the influence of standing waters. Moreover, a comprehensive inventory took place of all of the industrial plants along the Rhine which, in the event of an accident, could release significant pollution into the river. To tackle industrial discharges, especially of toxic substances, the promotion of Best Available Technology (BAT) in pollution prevention has been important. BAT concerns improvements to industrial processes themselves (to reduce the likelihood of pollution production) as well as effective 'end-of-pipe' technologies to treat wastewater prior to discharge. These aspects of the Rhine Action Programme have been very successful. By 1994 the Rhine Commission reported that the 50% target for 1995 for the reduction in discharge of dangerous substances had been met in most cases and 90% reduction had been achieved for many. However, some problems remained, particularly for diffuse sources such as pesticides and nutrients from agriculture. However, the improvement has been dramatic in a short time.

Reviving Salmon Populations

The Rhine Action Programme contained the objective of reviving sustainable salmon populations in the river by

2000. Salmon are an important indicator species for the general health of the river, as they require good quality water, spawning grounds, and unhindered migratory access. They are also of importance to the public, which view the species as a symbol of the state of the river. The reduction in discharges of dangerous substances was a necessary prerequisite, ensuring water quality was adequate for fish survival. A particularly important parameter in this regard has been that to increase dissolved oxygen levels. Measures to control organic discharges (especially from wastewater treatment works) were introduced and dissolved oxygen levels have increased significantly. Additionally, the large number of engineering works undertaken since the early nineteenth century posed many barriers to fish migration. Fish passages at the necessary barriers have been constructed at many places. The results have been very encouraging. Since 1990 salmon and sea trout have returned to the Rhine and its tributaries from the sea, and natural reproduction has been recorded since 1992. In 1995 nine salmon were caught at the Iffezheim barrier, just downstream of Strasburg, proving migration had occurred more than 700 km upriver.

Nutrient Discharges into the Rhine

The two nutrients of most interest are phosphorus and nitrogen. For the Rhine discharges of both are of concern, given the need to protect both the river ecosystem and that of the area of the North Sea to which the river discharges. Other factors are also important. One of these is light. In areas of high suspended solids light penetration in the water column can be severely limited and, therefore, elevated nutrient concentrations would have less impact. The Rhine Commission has undertaken two full surveys of phosphorus and nitrogen inputs in the Rhine catchment in 1985 and 1996. The sources are divided into two general types—point sources (such as a wastewater treatment works, industrial discharges) and diffuse sources (such as agricultural and other land surface runoff and drainage).

Phosphorus Sources

The data (Fig. 19.6A) show that, in 1985, point sources accounted for about 75% of the total phosphorus input to the Rhine, with urban (i.e. sewage) discharges being about twice that of industrial inputs. By 1996 the relative importance of urban and industrial sources to each other remained similar. However, by this time the relative contribution of point and diffuse sources was roughly equal. Between 1985 and 1996 the total input of phosphorus from human activity reduced from 72,400 t P/a to about 25,400 t P/a. This is a reduction of about 65% and was well above the target in the Rhine Action Programme of a 50% reduction by 1995. The decline was driven overwhelmingly by a 77% reduction from urban point sources and a 76% reduction from industrial point sources. Diffuse source inputs were reduced by 59%. These changes are illustrated in Fig. 19.7. In 1985 more than a third of all phosphorus input to the Rhine arose from urban pollution from Germany. The 81% decline in this source by 1996 is highly important in driving improvement in the river.

Nitrogen Sources

In 1985 the contribution of point sources (284,000 t N/a) was slightly more than that from diffuse sources (249,000 t N/a) (see Fig. 19.6B). By 1996 the diffuse source contribution had declined only slightly to 230,000 t N/a, while that from point sources was reduced to 162,000 t N/a. Overall this represents a reduction of 26%. This is only about half of the target of 50% for 1995 in the Rhine Action Programme. The largest decline derived from the industrial sector, which accounted for 15% of point source inputs in 1985, but only 5% in 1996 (or a 77% reduction in absolute emissions). This contrasts with a 27% reduction in discharges from sewage treatment works (Fig. 19.7). The largest single point source is, as with phosphorus, urban pollution from Germany. While some improvement has been made, this has not been so striking as for phosphorus. The large quantities of diffuse nitrogen pollution also remain of concern.

Nutrient Concentrations in the Rhine

Water quality has been monitored for many years at different locations along the length of the Rhine. Phosphorus concentrations have shown a dramatic decline (Fig. 19.8A), so that the target concentration has been met along much of the river. This trend has occurred since the mid-1970s in the mid/lower monitoring stations and since the early 1980s in the upper monitoring station (where phosphorus levels were already nearly at the target concentration). Phosphorus concentrations in the river have responded relatively well to changes in inputs and the significant decline in discharges (see above) is reflected in the improved river water quality. The data at each location show similar trends but these are very different to the trends for nitrate concentrations (Fig. 19.8B). Concentrations increase the further downstream sampling is undertaken. At all locations concentrations rose in the 1970s, reaching peaks in the mid- to late 1980s or even early 1990s. At Koblenz concentrations of nitrate more than doubled, but smaller increases were found at the other

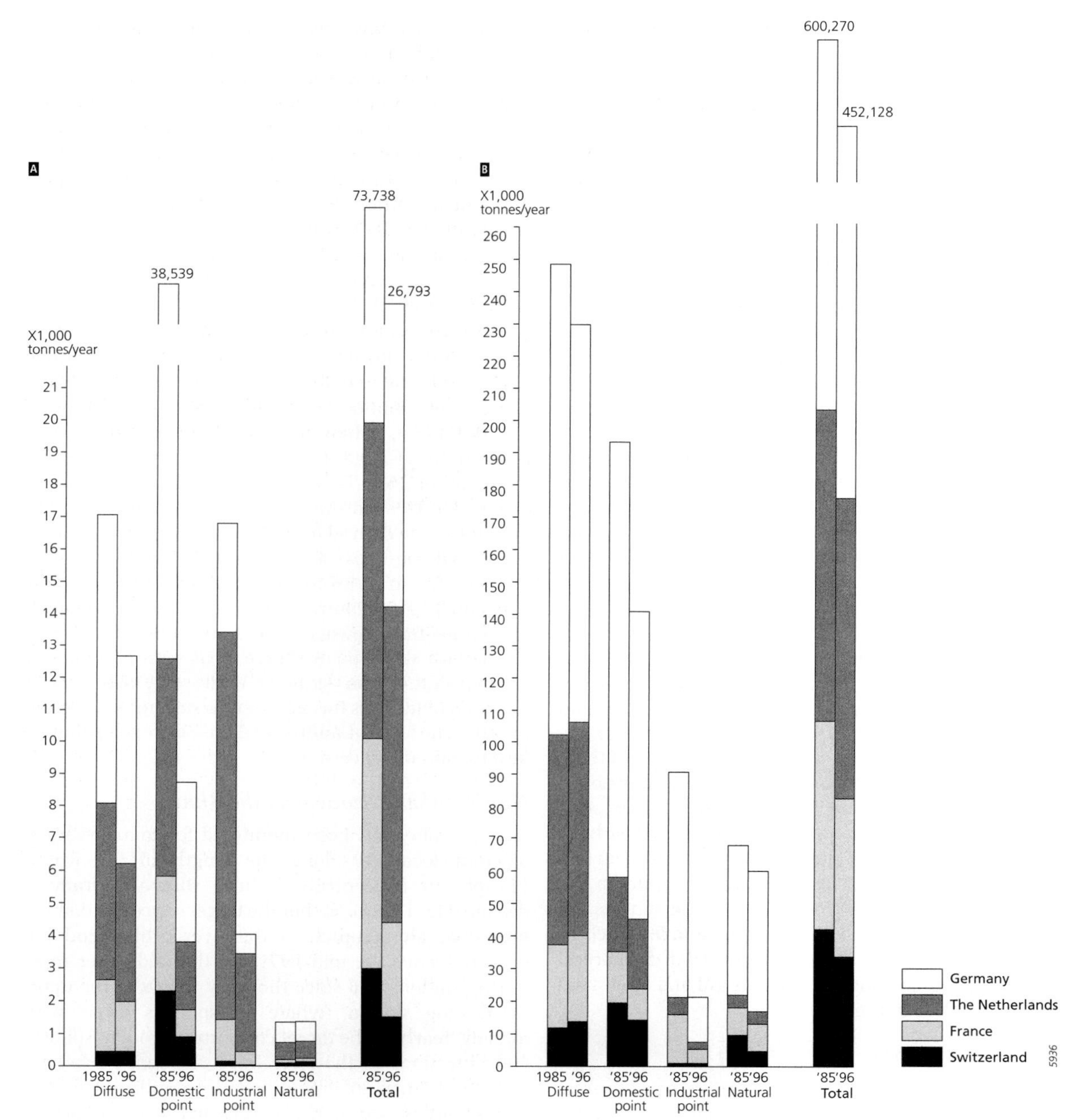

Fig. 19.6. Sources of total A. phosphorus, and B. nitrogen in 1985 and 1996 to the Rhine. All values are in tonnes per year.

two sites. In the late 1990s concentrations have begun to decline, but at no location have concentrations returned to those shown when monitoring started. The decline in nitrate concentrations is certainly less marked than the relative decline in discharges (see above). Part of the reason for this may be the large contribution of diffuse sources, especially that entering via groundwater. There is a significant time lag (of many years) between nitrogen entering groundwater and its influence on the river. Thus it may be some time before improvements in river water quality fully reflect changes in pollution sources.

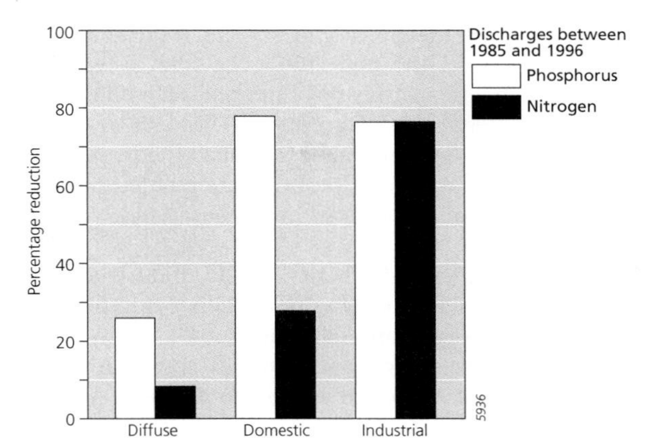

Fig. 19.7. Percentage reduction in phosphorus and nitrogen discharges from anthropogenic sources to the Rhine by source between 1985 and 1996.

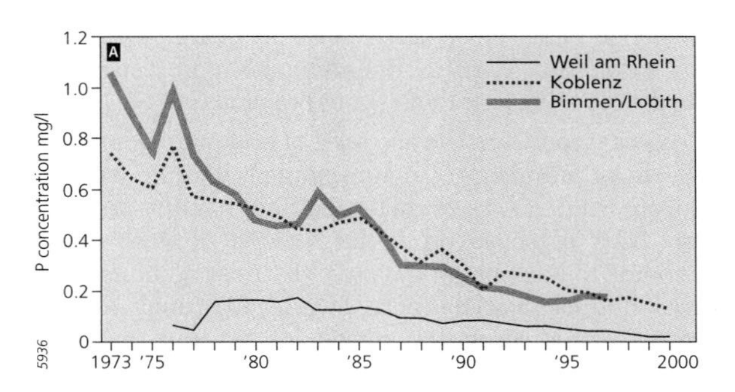

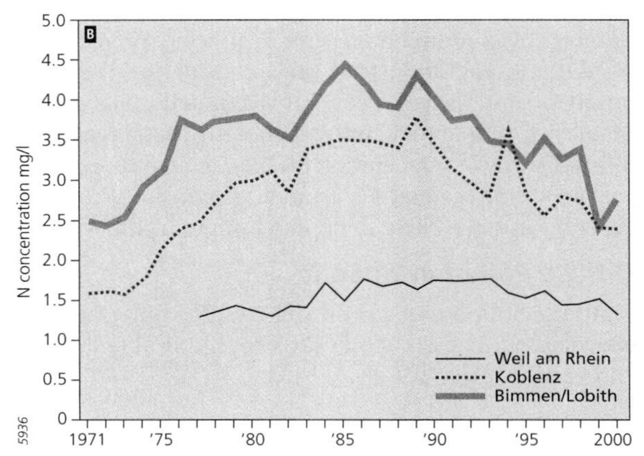

Fig. 19.8. A. Phosphorus, and B. nitrate concentrations at three locations along the Rhine, 1971–2000.

Achievements

The fifty-year history of the Rhine Commission is a success story. Water quality has improved for many parameters, the physical structure of the river has been upgraded and salmon have returned. Total discharges of phosphorus have fallen dramatically, largely due to extensive investment in phosphorus removal in domestic and industrial wastewater treatment works. Water quality monitoring indicates that the target value for a sustainable ecosystem has been met along much of the river. This success has not, however, been repeated for nitrogen. While there has been an important reduction in discharges since 1985, those that remain are highly significant, particularly given that they are carried by the Rhine into the North Sea coastal area where nitrogen loads are considered to be the key factor for eutrophication. The role of diffuse nitrogen pollution is of considerable concern. There is pressure at an EU level to invest in additional nitrogen removal on wastewater treatment works and to reduce nitrogen applications in agriculture. The future management of the Rhine will be more comprehensive in nature than anything previously.

Conclusions

It is evident that some of the extreme occurrences of water pollution of the twentieth century in western Europe have been tackled. However, the elimination of specific point sources only serves to highlight the remaining diffuse pollution problem. Some countries have been slow to take decisive action. This is not least due to the politically sensitive nature of regulating agriculture, especially in countries such as France. The main objective of the EU Water Framework Directive (2000/60/EC), adopted in 2000 (Farmer 2001*c*), is to establish a framework for the management of surface water and groundwater on the basis of the river basin. Member states have to draw up river basin management plans which contain information on the measures to be adopted to achieve 'good' water status for all surface and groundwaters to be achieved by the end of 2015.

Soil Pollution

Soils are intimately related to ecological systems, containing or supporting ecosystems specific to different soil types. Soils are finite resources. They are produced through weathering and biological processes and pollution from anthropogenic sources can alter their character in such a way as to result in major changes to ecosystems as well as knock-on impacts on ground and surface waters. There is a wide range of pressures on soils. This section will focus upon pollution impacts as a result of direct contamination of soil (e.g. through waste landfill, leaching from industrial facilities, etc.).

Contamination

Contamination of soils occurs mostly as the result of industrial activities and waste disposal. However, other activities, such as spills from transport, are also important. The following sectors have been highlighted as providing important sources of contamination:

- Industry—chemicals industry, petrochemical/oil industry, steel industry;
- Energy—gas works, petrochemical/oil industry;
- Transport—accidents, maintenance of transport vehicles, inadequate storage of hazardous chemicals, leaking from petrol storage at petrol stations;
- Households—waste disposal;
- Military—military bases, shooting ranges, warfare agent stocks, airstrips, etc.

Across the EU as a whole, the largest concentration of contaminated sites is found in the area stretching from Nord-Pas de Calais in France, across Belgium and The Netherlands to the Rhien–Ruhr region in Germany. Soil contamination, which locally or regionally poses very severe problems, is one area of environmental policy that has received little attention at the EU level and, therefore, significant divergence in approach remains between the different countries (e.g. on defining which sites are contaminated). Some basic information is as follows:

• In Belgium, the Wallonia administration identified almost 9,000 illegal waste dumps in 1980 that required remedial action, and most of these have either been cleaned up or it has been determined that no action is required. However, over 700 sites still require attention. In Flanders 8,000 contaminated sites were registered by the late 1990s and plans are in place to take remedial action where required (OECD 1998).

• In Denmark over 30,000 sites have been identified as potentially contaminated, with an estimate that the final number may be about 40,000. Of these 4,900 sites require cleaning up either because they are located in positions that can affect water quality or because of land use requirements. Old landfills account for 42% of these sites (OECD 1999).

• The first inventory in France in 1978 identified 120 contaminated sites, but further surveys were undertaken after comparable surveys for other countries often identified inventories of several thousand sites (OECD 1997). In the mid-1990s the total listed was 669. In each case ownership of the site is identified and clean-up requirements determined (IFEN 1999).

• Surveys in Germany have identified the inconceivable number of around 300,000 contaminated sites, of which nearly half are located in the new *Länder* (OECD 2001).

• In The Netherlands about 110,000 industrial sites have been identified as being potentially contaminated and there are also many old landfill sites. Clean-up of about 1,000 severely contaminated sites has been undertaken in a concerted way since the early 1980s. The Netherlands also has a major problem with contamination of sediments in the beds of Dutch watercourses. There are an estimated total of 90 million m^3 of sediments requiring remediation. This concerns not only the enhancement of water quality, as about 24 million tonnes of polluted sediments are dredged each year for navigation channel maintenance, and disposal of these sediments poses significant problems (OECD 1995).

Progress reports on the assessment and management of the huge numbers of contaminated sites in western Europe (EEA 2001) reveal that although major progress has been achieved on initial surveys of sites, much remains to be done on detailed site investigations. Progress has been variable across western Europe and even where major efforts have been funded and sustained over many years (e.g. in The Netherlands), the number of remediated sites is still small compared to the overall number of contaminated sites. Anyway, remediation of contaminated industrial sites, considering the huge numbers and large areas involved, will take many decades to implement fully. The countries invest significant sums in remediation (Table 19.7), although the per capita investment varies enormously. Projected expenditures are likely to be much more in the future.

Diffuse Pollution of Soils

Diffuse pollution impacts on soils result from a number of processes. These include the deposition of pollutants

TABLE 19.7. *Annual expenditures for contaminated sites remediation in selected western European countries*

Country	Total expenditures (€m.)	Population (000,000)	€ per capita	Total area (km^2)
Belgium (Flanders)	78.6	5.8	13.6	13,511
Denmark	90	5.3	17	43,090
Germany	57	79.4	0.7	356,970
Netherlands	550	14.9	36.9	37,330

Note: Data from Germany is projected from expenditures from some of the *Länder* (after EEA 2001).

from the atmosphere which result in the acidification and eutrophication of soils. Atmospheric deposition of pollutants can also result in localized accumulation of metals and organic compounds, such as dioxins. Processes related to agriculture also cause diffuse soil contamination. The spreading of sewage waste, the application of fertilizers, and the use of pesticides all may contaminate soils. Mineral phosphate fertilizers, for example, are usually contaminated with cadmium and continued use results in higher soil cadmium levels. Sewage sludge application is of increasing concern. The range of contaminants it contains is large—including a variety of heavy metals, trace organic compounds, and micro-organisms. Soil processes may break some of these down, but others accumulate and pose a risk for soil organisms and humans. The use of sewage sludge has also beneficial effects, due to its organic content and as a fertilizer. The implementation of the EU urban Wastewater Treatment Directive (see above) will result in additional sewage sludge generation and, therefore, more demand for disposal. The EU Directive on sewage sludge establishes quality standards for sludge applied to land and this will provide some protection from some diffuse pollution.

References

Bobbink, R. (1991), Effects of nutrient enrichment in Dutch chalk grassland. *Journal of Applied Ecology* 28: 28–41.

Bull, K. R. (1992), An introduction to critical loads. *Environmental Pollution* 7: 173–6.

D'Arcy, B. J., Ellis, J. B., Ferrier, R. C., Jenkins, A., and Dils, R. (2000), *Diffuse Pollution Impacts*. CIWEM, London.

de Bakker, A. J. (1989), Effects of ammonia emissions on epiphytic lichen vegetation. *Acta Botanica Neerlandica* 38: 337–42.

de Vries, W., and Leeters, E. E. J. M. (1996), *Effects of acid deposition on 150 forest stands in The Netherlands*. i. *Chemical composition of humus layer, mineral soil and soil solution*. DLO Winand Staring Centre for Integrated Land, Soil and Water Research. Report 69.1, Wageningen, The Netherlands.

EEA (1999), *Environment in the European Union at the Turn of the Century*. European Environment Agency, Copenhagen.

—— (2001), Indicator Fact Sheet Signals 2001. European Environment Agency, Copenhagen.

—— (2002), *Environmental Signals 2002: Benchmarking the Millennium*. European Environment Agency, Copenhagen.

EMEP (2001), *Transboundary Acidification and Eutrophication and Ground Level Ozone in Europe*. EMEP Summary Report.

European Commission (1998), *Implementation of Council Directive 91/271/EEC of 21 May 1991 Concerning Urban Waste Water Treatment, as amended by Commission Directive 98/15/EC of 27 February 1998: Summary of the Measures Implemented by the Member States and Assessment of the Information Received Pursuant to Articles 17 and 13 of the Directive*. COM (1998) 775. Commission of the European Communities, Brussels.

—— (2001), *The Clean Air for Europe (CAFE) Programme: Towards a Thematic Strategy for Air Quality*. COM (2001) 245. Commission of the European Communities, Brussels.

Farmer, A. M. (1997), *Managing Environmental Pollution*. Routledge, London.

—— (2001*a*), Clean Air for Europe: the EU's new programme for improving air quality. *Automotive Environment Analyst* 77: 20–1.

—— (2001*b*), Reducing phosphate discharges: the role of the 1991 EU Urban Waste Water Treatment Directive. *Water Science and Technology* 44: 41–8.

—— (2001*c*), The EC Water Framework Directive: an introduction. *Water Law* 12: 40–6.

—— Precioso, B. L., Latorre, F., Tuddenham, M., and Thornton, C. (2000), The role of the 1991 EU Urban Waste Water Treatment Directive in reducing phosphorus discharges in France and Spain. *European Water Management* 3: 35–43.

Gilbert, O. L. (1992), Lichen reinvasion with declining air pollution. In: J. W. Bates and A. M. Farmer (eds.) *Bryophytes and Lichens in a Changing Environment*. Clarendon, Oxford, 159–77.

IFEN (1999), *L'Environment en France*. IFEN, La Découverte, Paris.

Innes, J. L., and Kelly, J. M. (2002), Forest decline and air pollution: an assessment of 'forest health' in the forests of Europe, the Northeastern United States, and Southeastern Canada. In: J. N. B. Bell and M. Treshow (eds.) *Air Pollution and Plant Life*. Wiley, Chichester, 273–93.

Kandler, O. (1987), Lichen and conifer recolonisation in Munich's cleaner air. In: P. Mathy (ed.) *Air Pollution and Ecosystems. Proceedings of an International Symposium held in Grenoble, France, 18–24 May 1987*, 784–90.

Landemann, G. (1995), Forest decline and air pollution effects in the French mountains: a synthesis. In: G. Landemann and M. Bonneau (eds.) *Forest Decline and Atmospheric Deposition Effects in the French Mountains*. Springer, Berlin, 3–40.

Nylander, W. (1866), Les Lichens du Jardin du Luxembourg. *Bulletin Society Botanique de France* 13: 364–72.

—— (1896), *Les Lichens des Environs de Paris*. Paul Schmidt, Paris.

OECD (1995), *Environmental Performance Reviews: Netherlands*. Organization for Economic Co-operation and Development, Paris.

—— (1997), *Environmental Performance Reviews: France*. Organization for Economic Co-operation and Development, Paris.

—— (1998), *Environmental Performance Reviews: Belgium*. Organization for Economic Co-operation and Development, Paris.

—— (1999), *Environmental Performance Reviews: Denmark*. Organization for Economic Co-operation and Development, Paris.

—— (2001), *Environmental Performance Reviews: Germany*. Organization for Economic Co-operation and Development, Paris.

Schulz, D. (1999), *Potential to decrease nitrogen emissions from agriculture into the environment—the Nitrogen Reduction Programme in the Federal Republic of Germany*. Federal Environment Agency, Bismarkplatz, Berlin.

Schutt, P., and Cowling, E. B. (1985), 'Waldsterben', a general decline of forests in Central Europe: symptoms, development and possible causes. *Plant Disease* 69: 548–58.

Seaward, M. R., and Letrouit-Galinou, M. A. (1991), Lichens return to the Jardin du Luxembourg after an absence of almost a century. *Lichenologist* 23: 116–18.

Smit, H. (2000), *Manure and the Environment*. Ministry of Agriculture Nature Management and Fisheries, The Netherlands.

van der Eerden, L. J., Dueck, T. A., Berdowski, J. J. M., Greven, H., and van Dobben, H. F. (1991), Influence of ammonia and ammonium sulphate on heathland vegetation. *Acta Botanica Neerlandica* 40: 281–96.

20 Urbanization, Industrialization, and Mining

Ed de Mulder and Chris Bremmer

Introduction

Urbanization, industrialization, and mining are three different responses by mankind to survive, to develop, to create wealth and prosperity, and to organize life. Of the three, mining is the oldest, accessing specific materials provided by nature. Urbanization originally was a community's means of seeking shelter in a hostile natural environment. The co-operation involved in urbanization also provides benefit to all. Industrialization is the latest of these three processes and aims to concentrate production activities at one spot, which allows for increased scale and higher outputs. Industrialization is often linked with mining and more so with urbanization. Mining is obviously found in confined areas primarily determined by the availability of the required earth materials. Urbanization is determined by a mix of geographical (infrastructure), geological (firm underground and stable conditions), and strategic conditions. Moreover, economic, social, and political factors are increasingly important. The same can be said for the processes leading to the establishment of industrial sites.

Urbanization, industrialization, and mining have in common that they not only profit from the environment in which they operate, but also affect the natural balances of that environment. Consequently, these activities generate some response in the subsurface, either small or more significant. In the course of time humankind has faced many of these responses, but they still may cause surprises. This chapter briefly describes the impact of urbanization, industrialization, and mining on the natural environment of north-western Europe, both in terms of assets and threats. Attention is given to monitoring the Earth's response to these activities through the geological processes involved. For monitoring and prediction substantial information and knowledge of the subsurface is necessary and sources of such data are outlined. The chapter starts with some facts and figures, and is mainly based on *Urban Geoscience* by McCall *et al.* (1996). Information is also derived from the State of the Environment reports as published by the European Union. Much more attention is given to urbanization than to either of the other activities. However, since modern, urban, industrialized societies consume large amounts of primary resources, mining, industrialization, and urbanization are closely connected.

Facts and Figures

Urbanization

One of the most spectacular features on Earth in the second half of the twentieth century was the rate of expansion of cities. Until the end of World War II a minority of the world's population lived in cities, even in the USA, where urbanization had already begun in the late nineteenth century. Fifty years later, this picture has changed dramatically: by the year 2005 about 50% of the world's 7 billion people will be living in cities. Sixty of these cities will each have more than 5 million inhabitants. The strongest growth in urbanization is witnessed in developing countries, especially in Africa and Asia. This has been particularly true in the last two decades when great masses of the traditionally rural population of these regions moved into urban centres. In contrast to Africa, where the urban percentage increased by almost 20% over the last twenty years, the increase in urbanization in Europe has been very modest with only a 0.6% increase over the period 1990–5. Europe is already one of the most urbanized continents

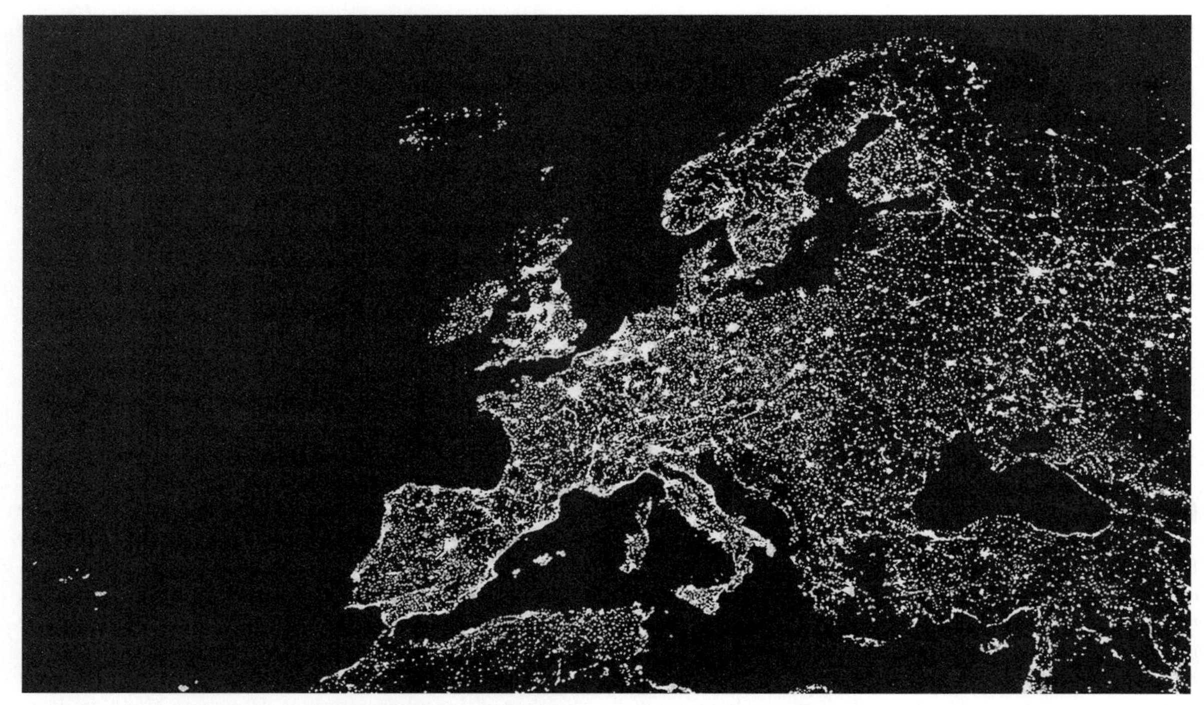

Fig. 20.1. Europe by night (after EEA 1999*a*).

(Fig. 20.1) and in 1999 70% of Europeans were urban-based (EEA 1999*a*). The European Environmental Agency expects an additional growth in the European Union of more than 4% to this figure in the period 1995–2010. In 1999 urban areas, with a population density of > 100/km^2, covered some 25% of the EU's territory. Of all European countries, Belgium has the highest percentage of people living in cities and Portugal the lowest (EEA 1999*b*). Several definitions on urban agglomerations exist. In this chapter the definition by the European Environmental Agency (EEA 1999*a*) is followed. It defines urban agglomerations as 'territorial units characterised by the presence of buildings, transport infrastructure and public amenities, and by a minimum threshold size' (p. 314). This definition does not give the minimum population density of such units, but tentatively a number of 100 inhabitants per km^2 is indicated.

Industrialization

If we do not take into consideration the small-scale industrial sites in Roman and medieval times, industrialization started in Europe in the late eighteenth century and expanded rapidly in the nineteenth and twentieth centuries. Together with (overseas) trade it has generated very substantial economic wealth for the old continent. Industrialization has also led to an intensive use of primary materials and to a growing production of waste, wastewater, and emissions of gases and aerosols. This naturally leads to an increase in environmental stress. From an environmental point of view the most important emissions are the greenhouse gases, ozone-depleting gases, and acidifying substances. In Europe alone, the annual production of CO_2 has been estimated to be about 3.3 billion tonnes in 2000 (EEA 1999*a*). Considering the actual trends and despite the political pressures it is not expected that this will decrease very much on a short timescale. In spite of numerous and considerable efforts by industry and the European governments, CO_2-production is still expected to be around 3.5 billion tonnes annually in 2010 (EEA 1999*a*).

Production of ozone-depleting substances has decreased sharply since the end of the 1980s due to strict environmental measures. The potential chloride/bromide-concentration reached a peak in 1994 and has decreased since. Soil contamination is another consequence of industrialization. An inventory of nine EU countries (amongst them: France, Germany, Benelux, and Denmark) yields the incredible number of a total of about 1,250,000 contaminated sites. Risk analysis has been performed for only a fraction of them. In the past two decades many efforts have been made by the industry and mining sectors to combat the negative impact of industrial and mining activities on the environment. In co-operation with the European Union and the United Nations Environmental Programme, several international initiatives have been launched as a positive response to the principles of Sustainable Development (Shields and Solar 2000).

Mining

In the late eighteenth, the nineteenth and the first half of the twentieth century mining of energy materials, heavy metals, and industrial minerals has been very significant in western Europe and shown generally upward volume trends. Due to (*a*) the decreased expansion possibilities caused by conflicting claims on land use and environmental constraints, (*b*) the reduced commodity prices, (*c*) the depletion of resources, and (*d*) improved long-distance transport facilities, almost all of the non-aggregate mining operations in western Europe were terminated during the past five decades in favour of giant new sites in other continents. Non-aggregate mining operations in western Europe are now confined to Finland and Portugal.

Economic growth and increase in production capacity has lead to an almost equal growth in the exploitation of primary resources. Some correcting trends are to be found in the increased use of alternative or recycled materials. With respect to the exploitation of metalliferous resources, most minerals show an increase in production statistics at least until the beginning of the 1980s. Primary aluminium production increased worldwide from 2 Mtonnes/yr in 1950 to about 15 Mtonnes/yr, and silver from 5,000 tonnes/yr to 15,000 tonnes/yr. Other commodities, such as antimony and mercury, did not see an increase in mining production (BGS 1988). Metalliferous mineral production is very much limited to a restricted number of locations since high ore grades are the results of specific geological conditions. France and Germany both still have a number of operational mines for the production of metalliferous minerals though production rates are low on a worldwide standard. The relatively small contribution of western Europe to these figures is illustrated by the fact that only 0.7% of the world aluminium production and 0.1% of the silver production is coming from this part of Europe, notably France (Lumsden 1992). The production of zinc, lead, copper, and associated minor metals is confined to two main types of deposits: (1) zinc and lead disseminations in mainly carboniferous deposits, and (2) copper, lead, zinc, and massive sulphide deposits in volcano-genic environments, for example, Hercynian areas of central France (Chessy) and Brittany. Small antimony operations also exist in Hercynian lode-type concentrations in France. Amongst the industrial minerals mined in western Europe are such common raw materials as bentonite (for cosmetics, boring mud, or pet litter), potash for fertilizers, and salt. Large salt deposits of Permian age are now mined in The Netherlands and Germany by a solution type of production technique by which brine water is pumped from a depth of more than 1,000 metres. About 6% of the world production of potash comes from France and about 8% from Germany, both mined in potassic salts.

Aggregate resources and building stone are mined all over the world for construction of buildings or infrastructure purposes. Sand, gravel, and clay are found in deltaic and alluvial plains or where large glaciers discharged their rock debris at the end of glacial periods, as in zones south of the Scandinavian countries (Denmark and Germany). Production of aggregates and building stone, in a quantitative sense, is much larger than that of minerals. The Netherlands alone produces more than 20 million tonnes/year of gravel, mostly for its own use (this amounts to 6 m^3 per person p.a.). A substantial part of this production is extracted from the river floodplains of the Rhine and Meuse as well as from (fluvio)glacial deposits in the northern half of the country. Large amounts of sands are also extracted from submarine localities in the North Sea. In the UK, the produced amount of building sand for mortar was estimated to be about 22 million tonnes/yr in 1990 (Lumsden 1992). Both Germany and Belgium are important producers of ornamental stone. In these countries ornamental stone quarries are mainly confined to specific limestone and sandstone formations (such as the Nivelsteiner, Gildehauser, and Bentheimer sandstone and Baumberger limestone in Germany; the Gobertanger and Doornik limestone, and Leede sandstone in Belgium), as well as crystalline and volcanic rocks (such as the Drachenfelser trachyte and tuffstone from Germany). Also large volumes of basaltic rocks have been and are still being used for dyke building and maintenance, e.g. in The Netherlands.

Impact on the Natural Environment

Such significant shifts in population and in land use cannot occur without consequences. When 3.5 billion people live on a small percentage of the total land area of the Earth, with all their demands for food, water, and construction materials, a tremendous impact on the

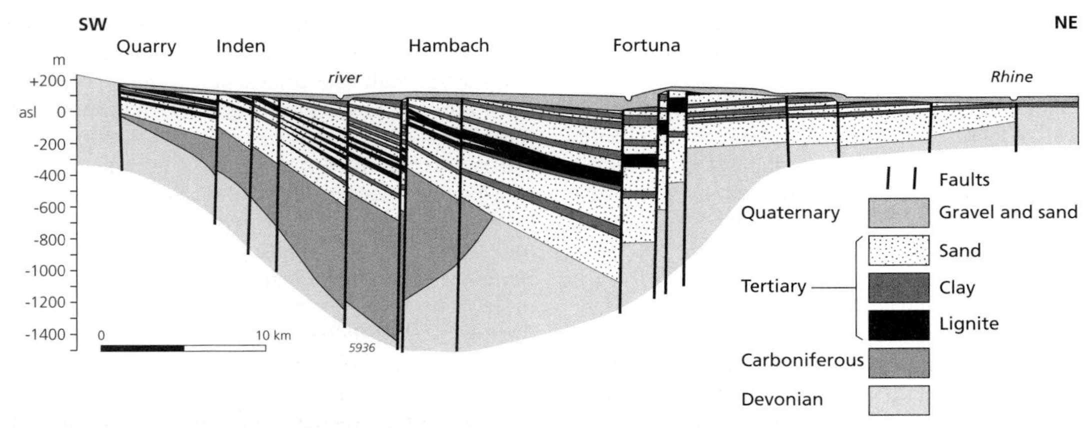

Fig. 20.2. Geological cross-section of the brown coal region near the German–Dutch border to the west of the city of Cologne (after Rheinbraun A.G. in Berendsen 1996).

Fig. 20.3. Brown coal excavation (photo: Pim Beukenkamp).

physical environment is inevitable. As a result, the balance between human occupation and the natural environment locally becomes severely disrupted. Since prices of land in cities rise irregularly but relentlessly, buildings become taller and the number of underground constructions increases. Greater and greater loads are placed on rock foundations, especially geologically less suitable sediments, since the best sites are often occupied already. This requires sophisticated engineering works to enhance their bearing capacity. Huge volumes of earth materials are being removed and redistributed every year for levelling, reclamation, mining, etc.

An extreme example of large-scale opencast mining and subsequent land reclamation is found in Germany, near the German–Dutch border, where huge amounts of brown coal are still being exploited for electricity supply. To reach the thick brown coal layers enormous volumes of the topsoil must be removed (Figs. 20.2, 20.3). Moreover, regional groundwater levels must be artificially lowered, as long as the excavation continues, by many tens to hundreds of metres, having a great impact on an extensive region surrounding the quarry sites. Since the reunification of West and East Germany the majority of the large brown coal quarries in the former German Democratic Republic (East Germany) are closed or will be closed in the near future. Consequently, in a large region around the towns of Leipzig and Halle in eastern Germany the landscape is being completely remodelled as former excavated areas are being restored to arable fields again, and to large recreational lakes and other facilities (Eissmann 2000; Eissmann and Rudolph 2002).

For every person living in the USA 85 tonnes of earth materials are moved annually, whereas for The Netherlands this figure is 70 tonnes annually (EEA 1999*a*). These values exceed annual volumes of such materials removed by geological processes due to sedimentation, erosion, and plate boundary transportation. Humankind has become a geological agent. Groundwater extraction and past mining activities in now urbanized areas may cause land subsidence and sinkholes. In other places the urban subsurface may become waterlogged as a result of leaking water mains and broken sewerage systems. In arid areas constructions may collapse due to unstable soils. These are just a few examples of geo-related problems in the subsurface of cities that will be discussed in the next sections.

There are no indications so far that this widespread urbanization trend will end soon. The Population Division of the United Nations predicts that it will continue to at least 2025 (United Nations 1995). However, it is obvious that this trend cannot continue forever. Food production, supply, and transport will pose ever-increasing problems in mega-cities. Increased urbanization may lead to social stresses in society, which may eventually result in civil disturbances. Counter-measures, for example suburbanization, may help to improve the quality of urban life, but may exert pressures on nearby green spaces. In addition, disastrous earthquakes are anticipated in the near future in cities such as San Francisco and Tokyo, similar to the one that struck the Kobe-Osaka conurbation recently.

Assets and Threats of the Subsurface to the Urban Society

Today, the subsurface is generally not a real issue when it comes to the decision on where to develop a housing or an industrial site. The only factor in this respect, which in the last few decades has often played a role, is the subsurface quality in terms of contamination. Contaminated sites may hamper further development, not because of technical problems related to decontamination but particularly because of the lengthy and costly procedures that have to be undergone before any development may take place. Technically, anything can be constructed at most sites, even on (deep) water. However, taking the subsurface quality in all its aspects into consideration at the planning stage may avoid time-consuming procedural issues as stated above, and may also generate substantial revenues. At least considerable financial losses may be prevented. This section deals with both the assets and the potential threats generated by the subsurface in relation to human development, in cities and at industrial sites. With respect to assets, specific attention is paid to urban soils, urban groundwater, mineral resources, and subsurface space. Concerning threats, the occurrence of both natural and man-induced types are discussed with special emphasis on European cities and industrial zones.

Asset: Urban Soils

One of the most obvious aspects of urbanization is the irreversible loss of agricultural soils. The European Environmental Agency (EEA 1999*a, b*) estimates that 2% of Europe's agricultural land is lost to urbanization every 10 years. This is due to construction activities and to contamination and fragmentation (enclosure by built-up areas). Urban soils play a significant role in the sustainability of urban life. They provide open spaces ('greenspace') for recreation, reduce urban noise levels, enhance air circulation, balance levels of humidity, and capture dust and gases. One hectare of urban park with trees, shrubs, and grass, can remove 900 kg of CO_2 from the air and can deliver 600 kg of O_2, both within a 12-hour period. Consequently, urban soils are also important as sinks and sources in the carbon cycle. Urban soil has many important functions for urban society and ecology, but the scientific knowledge of urban soils found in urban greenspaces is still rather poor, both in terms of processes and properties. The greenspace area in larger European cities varies greatly: from some 5% in Vienna and Budapest to over 60% in Oslo and Gothenburg (EEA 1999*b*). Large areas of the open spaces in some cities are derelict because of former mining activity. They are so damaged that they are unsuited for any use without treatment. The costs of such a treatment are generally very high and this leads urban developers to search for alternative uses for derelict land, such as isolation of the contaminated soil and/or conversion into nature reserves or woodlands.

Natural soil formation takes place on a minimum timescale of decades to a few centuries. This depends on the physical and chemical properties of the substrate. Urbanization can affect soils in many ways, not all of which are negative (e.g. garden soils). First, soil loss can be caused by burial, by water or wind erosion, and by physical or chemical damage. Water erosion is particularly effective on sloping ground (> 3 degrees). During large-scale construction activities up to 100–220 tons of soil/hectare/year may be lost, whereas only 1 ton soil/hectare/year can be developed from parent material under natural conditions. Erosion from one site often leads to unwanted sedimentation at another. This second impact of urbanization on soils often causes blocked drainage systems, decreased agricultural potential, and eutrophication of surface waters. Thirdly, urbanization may modify local groundwater tables. Vertical walls perpendicular to the topographic slope interrupt natural runoff and may cause local waterlogging. Next, compaction due to construction activities may also cause waterlogging. These developments can influence urban soil ecology by reducing oxygen supply to plant roots, leading to deteriorating vegetation conditions. De-icing and irrigation with Ca-rich waters has an impact on soil acidity. Extreme pH values cause deterioration of urban vegetation, whilst modifications in carbon content or nutrient availability affect soil structure and soil moisture. These, in turn, have an impact on the capacity of urban soils to immobilize soil contaminants or to support plant life. Construction waste may also reduce cation exchange capacity of urban soils (de Mulder *et al.*, 2001). Industrial waste will change the chemical composition and microbial population of urban soils, causing declining vegetation and exposing soils to wind or water erosion. The potential of microbiological activity to decontaminate some organic pollutants in urban soils is an asset, however.

Asset: Urban Groundwater

Potable water is one of the most important resources for the urban population. Many western European cities largely depend on groundwater. Fortunately, the amount of fresh water available is far greater than the demand in many cases. In the European Union, the average internal fresh water supply, i.e. originating from precipitation in the EU itself, amounts to approximately 1190 km^3/yr. With a total fresh water abstraction of 240 km^3/yr this is some 21% of its renewable resources (EEA 1999*a*). The Organization for European Co-operation and Development (OECD 1998) regards this a sustainable position. Urbanization affects the groundwater systems in wide areas in and around cities. Groundwater extraction on the one hand and leakage or seepage of water-carrying pipes on the other, have a strong impact on the yield and flow patterns of shallow and deeper aquifers that underlie a city. Remediation of polluted groundwater is costly and may not be, or be only partly, possible. This puts a strong emphasis on serious urban groundwater management, including protection. Another effect of urbanization on urban groundwater resources is that it results in extensive impermeable land surfaces, which reduces direct infiltration of excess rainfall, and thus increases and accelerates surface runoff. This process negatively affects recharge potentials for urban aquifers. Surface impermeabilization includes the construction of roofs and of paved areas. It may also result in higher peak discharges during rainfall and thus contribute to river discharge problems in downstream areas. It is assumed that this is one of the contributing factors to the changed discharge patterns in the Rhine–Meuse river system.

Pressurized piped water supply systems always leak and provide an important contribution to aquifer recharge. The proportion of water put into distribution that does not reach the customer can vary between 20% and > 50% (Lawrence and Cheney 1996). This significant recharge alone may compensate the reduced recharge due to impermeabilization. It can result in a net higher recharge in urbanized areas as compared to non-urbanized areas. Increased leakage due to pipe fracturing may also be anticipated in areas subject to earthquakes or land subsidence. In addition, unsewered sanitation greatly increases the rate of urban groundwater recharge. More than 90% of the water supply on domestic premises is not used for consumption and will thus, via unsewered sanitation units, eventually be added to the groundwater resources. Unlike water mains, sewerage systems are not pressurized and will leak much less water (only a small percentage). This percentage significantly rises in earthquake or subsidence-prone areas, or in urban areas where very old sewerage systems are still in place. Another possibility for recharge of suburban aquifers is through irrigation. Leaking sewerage systems also receive infiltration from high-level groundwaters and may actually drain the city. Land surface storage of (often polluted) industrial process water in unlined storage basins contribute to shallow aquifer recharge on the one hand, and present serious long-term threats to groundwater quality on the other.

A major problem with groundwater extraction is the potential risk for land subsidence, particularly in cities in wet and soft coastal and river plains (de Mulder *et al.* 1994). The ground level in the city of Venice (Italy), for example, dropped several metres due to pumping out large amounts of groundwater, and this caused periodic flooding of the city centre. Furthermore, reversal of groundwater gradients allows sewer intrusion into the aquifer. Extensive pumping in coastal areas often triggers saline groundwater intrusion from the sea and from deeper, saline groundwater resources. As the governing processes are known, successful measures to reduce such risks can now be taken. Termination of groundwater extraction at industrial scales often causes a rapid rise in groundwater levels, approaching the original levels observed before industrialization. This frequently results in flooding of basements and tunnels. Such floodings are known from many European cities, including Berlin, Hamburg, Barcelona, Paris, London, and Birmingham. Flooding can also bring the water table in contact with contaminated land where toxic material may be mobilized from the previously unsaturated zone.

Asset: Mineral Resources

Construction materials such as sand, gravel, clay, limestone, and building stone are essential for the building of cities. Apart from these, cities also need energy, mostly provided by the use of energy minerals (coal, oil, gas, radioactive minerals). Cities in developed countries consume far more construction minerals than those in developing countries. An average city of 250,000 inhabitants in western Europe consumes 1,650,000 tons of gravel and sand per year, and 500,000 tons of hard rock to be crushed for use as aggregate (EEA 1999*a*). Sand and gravel are recovered mainly from relatively young superficial deposits in alluvial and coastal plains. Crushed rocks are collected from bedrock quarries. These minerals are 'quality related to use', for example road surfacing aggregates must be of high quality (sandstone, dolerite, and some limestones). For many other applications and fill, aggregates of much lower quality can be used. The specifications for various uses are strictly set out in building codes and regulations, which differ from country to country. For instance, the grading limits for sand used for concrete differ between codes BS 882, the British Standard, and code NEN 5905, the Netherlands Standard. As an example, the chloride content for sand used for prestressed concrete should be lower than 0.01% by mass of combined aggregate according to BS 882, whereas according to NEN 5905 it should contain not more than 0.015% chloride by mass of combined aggregate. Though aggregates are easily produced from shallow open pits, the sea floor, lake bottoms, or river beds, the testing procedures as described in the building codes and regulations are often complex. Samples generally need to be tested for physical, mechanical, and chemical properties. Aggregate minerals require little processing other than size reduction by crushing, grinding, and screening, and the removal of any unwanted components such as shell, coal, or wood fragments. Pricing strongly depends on transport distance. Concrete is produced from a mixture of gravel and cement, consisting of limestone chips and sand. Its durability is controlled by its chemical composition and related resistance to weathering.

Clay is mainly used for producing bricks, pipes, and roofing tiles. Considerable volumes of clay are used in earth walls, both as barriers for noise, for light infrastructure, and for water barriers such as dykes. Building stone or dimension stone is mostly obtained from igneous and sedimentary rocks. It is less often obtained from metamorphic rocks, apart from slates, which are used as tiles. The present use of dimension stone is mainly through thin-cut slabs as ornamental facing to new buildings. Not so much in western European countries, but particularly in Italy (Carrara) and Finland, large industries have developed to excavate and process rocks. Iron and steel are frequently used as framework materials or to reinforce concrete. Gypsum is used as plaster material, asphalt for paving city roads. Recycling of demolition debris and other secondary materials has become an increasingly significant feature in the construction industry for western European cities. For example, in The Netherlands the use of recycled material rose from 13.6% in 1994 to 17.3% in 1998 (Ministerie van Verkeer en Waterstaat 2001).

Almost all energy materials such as oil, gas, coal, peat, and radioactive minerals used in cities are non-renewable. Many of the other energy sources such as hydroelectricity, direct water power, tidal and wave energy, wind power, geothermal energy, solar energy, and wood and waste burning are renewable but account for only a small portion of the total energy supply. Apart from the Reykjavik urban area in Iceland, geothermal energy is produced only on limited and local scales in European countries. Although the demand for energy materials is high, the price is still relatively low since new resources are identified quite frequently. Considerable rise of energy prices would have a significant impact on industries based in and near western European cities, greatly affecting the latter's socio-economic performances.

Asset: Underground Space

With increasing population density and accumulation of capital in urban centres more and more attention is paid to the use of underground space. All major western European cities have developed certain parts of their underground space for the construction of parking lots, infrastructure (metro system), shopping malls, and theatres. It is reasonable to expect that this tendency will continue and even accelerate as both ground prices and the expertise on underground space construction are expected to increase strongly (de Mulder *et al.* 1997; de Mulder and Bremmer 2000). Although almost any construction can be made in any place, even in the subsurface, it is evident that the subsurface conditions determine the costs involved and the environmental impact to a large extent. For example, the presence of a natural hydrological barrier at a reasonable depth will favour the construction of an underground waste disposal site, while the occurrence of thick, coarse-grained sand beds at or near the surface will save the cost of deep foundations. Ideally, this would imply that a match should be found between what the user wants and what the subsurface conditions offer. However, as more and more claims are put on specific parts of the subsurface, there is an increased risk that such claims are in conflict with each other. Sound decisions on the future destination of parts of the subsurface can only be made if national and regional authorities collect all available knowledge of the subsurface together with the relevant economic, planning, and environmental conditions involved. For example, the Dutch National Planning Authority (RPD) has recently developed a policy agenda for multiple use of space in general, and of the use of the subsurface in particular (Rijks Planologische Dienst 2001).

Impressive plans, e.g. for road and railway tunnels, are being or have already been developed for many sub-urban expansions. In the western Netherlands a 7.5-km-long train tunnel is being constructed under the 'Green Hart' in the Randstad conurbation (Amsterdam-The Hague-Rotterdam-Utrecht) for the TGV (*Train à Grande Vitesse*). In the Paris outer ring road (A86) two tunnels of 7.5 and 10 km are under construction. In 1998, the Swiss electorate agreed to the construction of the 34-km-long Lötschberg tunnel, in the south of Switzerland. A detailed blueprint exists for a 52-km train tunnel through the Alps between Turin and Lyons. Finally, plans have been developed for a 170-km tunnel connecting Stuttgart and Brescia. Although much of such infrastructure is in non-urbanized areas, nevertheless the connected cities will gain enormously in socio-economic relevance.

Threat: Natural and Man-Induced Hazards

Though natural hazards will always happen, urbanization contributes to their impact on human society. Increased population densities in vulnerable areas, coupled with a growth of industrial acitivities in those areas, have contributed to human casualties. Urbanization goes together with increased capital investments, resulting in higher economic losses once urbanized areas are affected by natural hazards. Urban expansion often also takes in areas that are less suitable to such purposes because the best sites have been occupied already. The recorded annual number of great natural catastrophes shows large variations but has increased worldwide (Fig. 20.4). Although most of these events occur in developing countries a similar trend is apparent in Europe. Storms and floods are the most common hazards and cause most damage in north-western Europe. Surprisingly, there was only one great natural catastrophe in 2002, the summer floods of the rivers Elbe and Danube in central Europe. A *great* catastrophe is defined as 'a catastrophe which distinctly overtaxed the ability

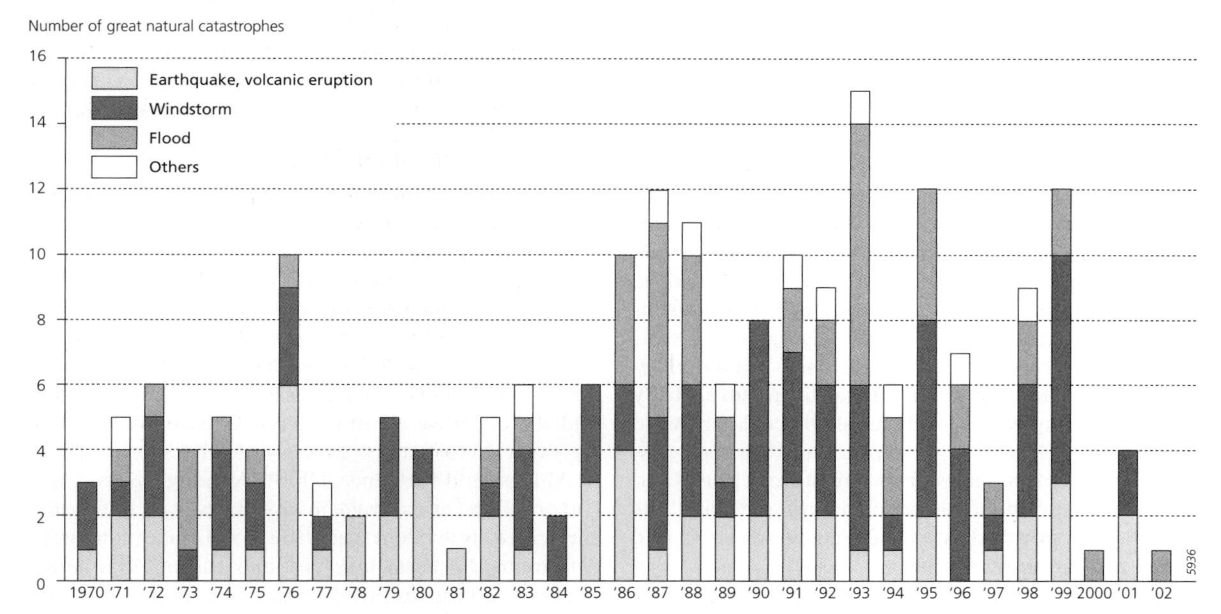

Fig. 20.4. Number of great natural catastrophes (1970–2002) according to Münchener Rück (2003).

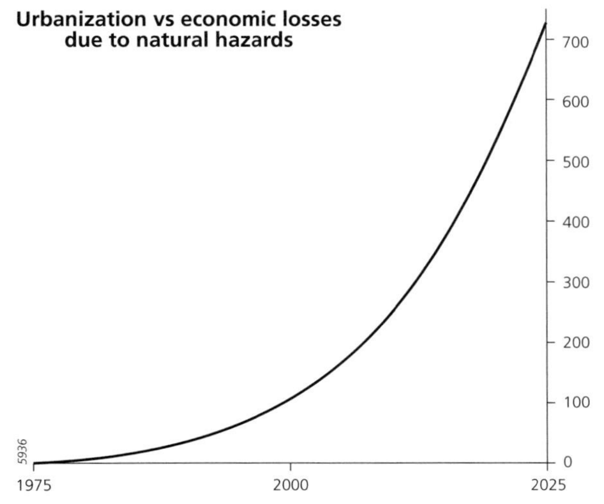

Fig. 20.5. Economic losses due to major natural disasters, period 1975–2025.

of the region to help itself, making interregional or international assistance necessary'. Münchener Rück (2003) estimates the losses in the three countries most severely affected (Germany, Czech Republic, and Austria) during the 2002 summer floods to be about €15 billion of which about €3 billion was insured. On a global level it is foreseen that the amount of insured losses will rise from €10 billion in 1990 to some €280 billion in 2025. The real losses are much higher, however. The minimum real losses are estimated to increase from €25 to almost €700 billion over the same time span (Fig. 20.5).

Natural land subsidence caused by karst and subterranean erosion may lead to sinkholes and creates substantial damage in urban areas. Such types of natural land subsidence occurring in limestones, gypsum, chalk, and rock salt are sometimes difficult to predict. Belgian, French, and German cities reported to be prone to natural land subsidence originating from karst and other dissolution phenomena include: Brussels, Paris, Strasburg, Nice, Stuttgart, and Würzburg.

Naturally degraded soils due to shrinking, swelling, or collapsing, and soils vulnerable to liquefaction, constitute a major problem in urban construction works. This is usually caused by the specific mineralogical and organic composition of the clay and peat beds or is due to the loose structure of fine-grained sands that may result in liquefaction. Special chemical or physical treatment of such soils may prevent these problems. Naturally occurring hazardous gas emissions can also harmfully affect the urban population. Radon is a highly radioactive gaseous element and is, together with its radioactive decay products, emitted from uranium-bearing formations, mainly sediments or igneous rocks, beneath dwellings. It is also derived from uranium-bearing construction materials and mine waste. Radon causes lung cancer and the use of such construction materials should be avoided in the first place. Moreover, the ventilation of cellars and the sealing of floors can prevent this risk. There are several smaller towns in Europe prone to this hazard in Belgium, Italy, the United Kingdom, and the Nordic countries.

Man-Induced Subsidence

Apart from loading the surface, the two principal causes of man-induced subsidence are fluid and gas extraction and mineral extraction. Extraction of groundwater as a resource for potable or industrial water is one of the most common causes leading to land subsidence. This in turn, can result in foundation failures, disrupted water and gas mains, and flooding. Groundwater extraction for water management or land reclamation is another possible cause of land subsidence (de Mulder *et al.* 1994). Examples of European cities suffering from groundwater extraction are Amsterdam and Venice. Extensive extraction of natural gas may lead to land subsidence as well. The second largest gas field on earth, the Groningen field, situated several kilometres deep in the north-eastern part of The Netherlands, will eventually generate a land subsidence of about 40 cm in the central part of the gas field according to recent estimates (Gussinklo *et al.* 2001) (Fig. 20.6). As the region is intensely cultivated and as water table levels must be controlled accurately the subsidence in this area has necessitated major reconstruction of hydraulic constructions, dykes, and quays. Moreover, minor earthquakes may also result from gas production in this region. Oil extraction has been reported to result in land subsidence as well. For example, the Ekofisk oilfield in the North Sea caused subsidence of the seabed of between 30 and 70 cm annually in the 1980s as a result of reservoir compaction (Lumsden 1992).

Surface collapse occurs at shallow, often abandoned mines, where voids are left after mining activities in the past. This is a rather unpredictable process in time. There is a strong correlation with the joint density in the rock and the size of the block volume: the higher the joint density and the smaller the block volume, the more likely the roof of a void is to collapse. Surface collapse is most predominant in old mining areas and occurs in numerous cities, among which are Mons, Liège, and Brussels (Belgium), Halle and Lüneburg (Germany), and Lille and Metz (France). Large areas in Normandy are

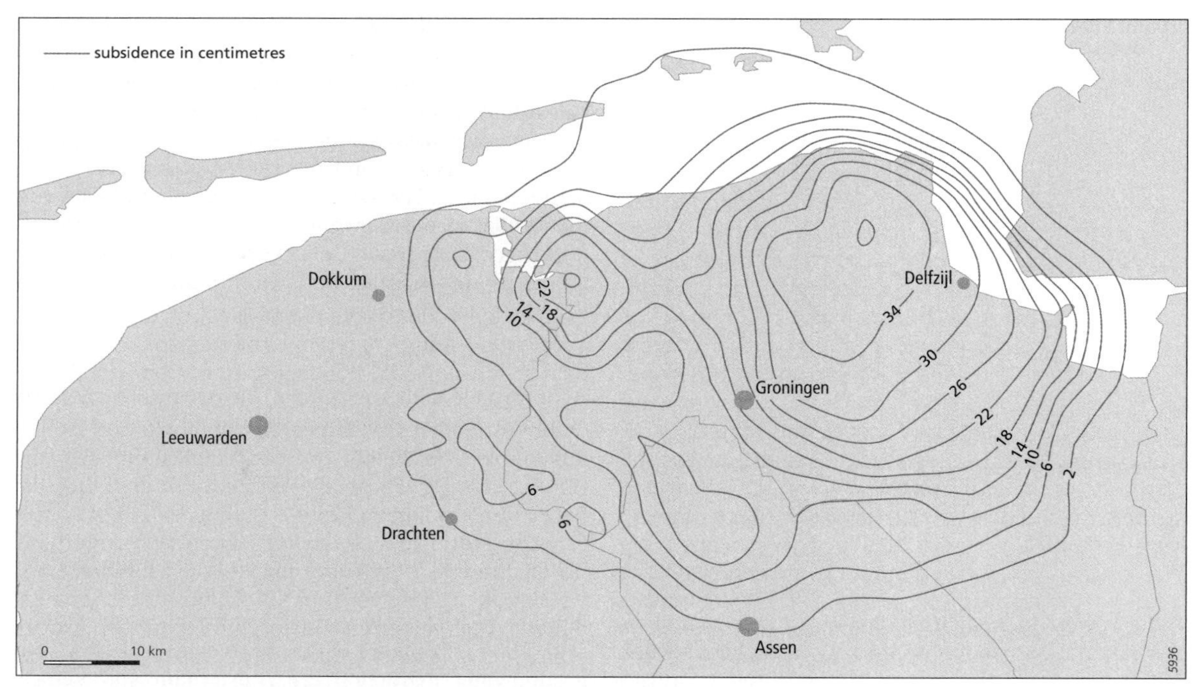

Fig. 20.6. Predicted land subsidence over the Groningen gas field (northern Netherlands) due to gas extraction (after Nederlandse Aardolie Maatschappij 1995).

affected by small, abandoned private mines that lie dormant under the surface and threaten buildings.

Contaminated Land

In general, the urban subsurface is contaminated by non-natural components. Over the centuries cities have built on their own waste. The level of contamination differs from site to site and is strongly controlled by the history of the various parts of the city. Most of the contamination is immobilized and stored in the unsaturated zone. Part of it may reach the saturated zone and become mobile, particularly while entering an aquifer. Presently immobile contaminants can become mobile when subsurface conditions change. The type of contamination depends on the type of industry active on that spot. Typical contaminants in the city subsurface are: heavy metals, arsenic, cyanides, oil products, chlorinated hydrocarbons, and polyaromatic hydrocarbons. Cleaning contaminated land costs a great deal of money and involves long time periods over which remedial measures must last. Although remediation procedures differ from country to country it can be seen that after an initial period of rather rigid long-term clean-up policy, a more liberal and decentralized remediation policy has been applied since the 1990s in most western European countries.

Rising Groundwater Tables

Termination or reduction of extensive groundwater extraction for industry, mining, and drinking water production results in a rise of the groundwater table. Leakage of water mains also adds to rising groundwater tables. Rising groundwater may mobilize contaminations and salt stored in the unsaturated zone, adding to the pollution problem. Clays, consolidated by the pumping process in earlier days, will swell again. As a result, rising groundwater will result in basement flooding, damage to foundations, and leakage and flooding in tunnels. Examples of cities suffering from rising groundwater tables are Berlin and Hamburg (Germany), Mons, Lille and Paris (France), and The Hague (The Netherlands).

Waste Disposal

Urban waste products generally come from households, from industry, or from demolition of constructions. On average in Europe most of the waste (72%) ends up in landfills, 17% is incinerated, 5% composted, and 4%

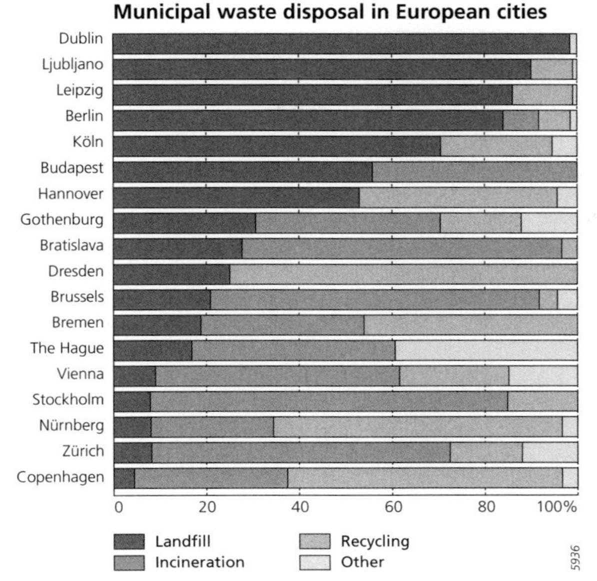

Fig. 20.7. Municipal waste disposal in European cities (after EEA 1999*b*).

recycled (EEA 1999*b*). Fig. 20.7 shows the types of municipal waste disposal in eighteen European cities. Controlled landfills usually have a stratified structure, with waste layers of 1–2 metres alternated by thin layers of inert (often soil) material. The degeneration process in waste dumps produces leachate and gas. Although increasing emphasis has been placed on alternatives to landfilling of wastes, including reduction of waste, recycling, and incineration, large quantities of landfill will still be created in the near future. Using natural geological barriers, such as buried clay beds and consolidated peat beds in combination with geotextiles in the design of waste disposal sites, will prevent leached material from entering the aquifers.

Urban Environmental Geo-indicators

Environmental indicators are essential tools in assessing the state of health of the environment and in defining the effectiveness of urban environmental policy measures. By far the majority of the commonly applied environmental indicators are in the biospheric and atmospheric domains. Indicators on the performance of the geosphere are generally lacking as instruments for environmental policy-making. However, the subsurface of cities is not perfectly stable and urban subsurface conditions change, both due to entirely natural and to man-induced causes. This indicates the need for the development of environmental geo-indicators (Berger and Iams 1996). For the urban area, McCall (1996) distinguishes intensive, rapid-onset geohazards, such as earthquakes, volcanic eruptions, tsunamis, landslides, and flooding, from pervasive, slow-onset geohazards and other non-intensive processes. This second group includes foundation problems, land subsidence, groundwater and surface water pollution, rising groundwater and sea-water levels, saltwater encroachment in coastal cities, depleting groundwater resources, contaminated and saline soils. The list of urban geohazards is based on an analysis of such hazards in 140 cities spread over all continents. From this list, nine environmental urban geo-indicators have been selected (McCall 1996):

- Destabilization due to seismicity or volcanic activity
- Flooding due to marine incursion, excessive precipitation, river overflow or rupture of reservoirs and dams
- Tsunamis
- Landslides
- Subsidence due to gravitational collapse or settling
- Contamination and pollution
- Rising water levels due to cessation of pumping activities, or to irrigation returns
- Falling water levels
- Depletion of resources such as water, building materials, fertile lands, energy materials, forests.

For each of these indicators definitions have been given and relevant processes have been assessed. Finally, measuring tools for registration of changes in the urban environment for each of these indicators were assessed, such as groundwater monitoring using benchmarks for falling water tables. Application of these geo-indicators will enable a better warning of both likely catastrophic natural hazards and of slower natural and man-induced events, which in the long run may be equally disastrous.

Subsurface Geo-information

The fact that the prosperity of cities also relies on the way in which they incorporate a sensible management of the subsurface, makes availibility of information on the subsurface highly relevant. Only by considering the pros and cons, which in terms of the subsurface translate to assets and threats, can a sensible planning policy become possible. Therefore subsurface geo-information is more and more regarded to be on an equal footing with other geo-information in city planning and management procedures. Particularly in cities where problems

TABLE 20.1 *Indication of the main geo-related environmental problems in urban areas in western Europe*

Key issues	Belgium	Denmark	France	Germany	Luxembourg	The Netherlands
Geological and topographical						
Failure of foundations and underground infrastructure	3	1	3	1	1	2
Ground collapse	3	1	3	1	2	2
Landslides and rockfalls	2	1	2	1	3	0
Land loss, erosion, and siltation	2	2	1	1	1	2
Earthquakes and volcanic eruptions	1	0	2	1	0	0
Natural radiation	2	2	1	1	2	0
Flooding and coastal inundation	2	2	3	3	2	3
Human interference						
Contaminated land	2	2	3	2	2	3
Derelict land	2	0	3	2	0	0
Groundwater pollution	2	3	3	2	3	3
Surface water pollution	2	3	3	2	3	2
Urban waste disposal	2	3	3	2	2	1
Natural resources						
Shortage of natural construction materials	0	1	2	2	3	2
Shortage of water resources	1	2	2	1	2	1
Depletion of green and arable land	1	0	2	2	2	3

Note: Relevance of risk: 0 = almost none, 1 = low, 2 = medium, 3 = high.

Source: EuroGeoSurveys, International Geological Congress, Rio de Janeiro, 9 August 2000.

concerning the subsurface have occurred in the past, substantial soil investigation programmes have been conducted for new developments (see e.g. Ellison *et al.* 1993). In The Netherlands, a small country, over 3 million boreholes and Cone Penetration Tests have been performed in the past twenty-five years, resulting in a subsurface information density of about seventy data points per square kilometre. In this case data are stored in a DINO database. This will eventually comprise over 200 different datasets on various subsurface-related topics, and already holds over 400,000 borehole descriptions (http://dinoloket.nitg.tno.nl) (de Mulder and Kooijman 2003). Usually, subsurface data are stored in the numerous files of municipal organizations, project developing companies, infrastructure bodies, architects, civil engineering companies, subsurface contractors, university departments, and geological surveys. The last-mentioned body is often charged with systematic and nation-wide subsurface data collection. Most geological surveys, as well as several major cities, have stored (parts of) their data already in digital files (de Mulder and Bremmer 2000).

On a European level a meta database system is presently being developed jointly by the geological surveys in the EU and Norway, the EuroGeoSurveys. This system (GEIXS) brings together information on specific data-items stored in geo-databases. The project is a forerunner of a larger net-based information system (www.eurogeosurveys.org.). EuroGeoSurveys has checked the problems and potentials in the subsurface compartment of urban centres in sixteen European countries. To that end an Urban Topic Network (GEURBAN) was established. This network has determined fifteen key issues, or geo-problems, related to (*a*) primary geological and topographical conditions, (*b*) secondary pollution and contamination events, and (*c*) depletion of natural resources, such as groundwater, arable land, and building materials in and around our cities (de Mulder 1998). A summary is given in Table 20.1, which clearly illustrates that high risks for underground structures and mass movements mainly apply to Belgium and France, but that flooding forms a particular risk in The Netherlands and Germany. Problems related to human interference are more or less equally spread over all countries; the same applies to natural resources.

Concluding Remarks

Western Europe has a long tradition in urbanization, industrialization, and mining. With 70% of its over 800 million people (population of entire continent) living in cities, urbanization is still slowly increasing. In the past decades mining has substantially decreased. Industrialization is coupled with economic growth. Although less dramatic when compared to other continents, urbanization and industrialization are significant phenomena in western Europe, affecting the demand and supply of natural resources such as water, space, and minerals. Through roof collapses and the rise of groundwater in abandoned mines, the negative impact of mining on cities will be noticed for many years to come. Urbanization accelerates and aggravates the risk for natural and man-induced hazards by enhancing geological instability at densely populated and capital-intensive, hazard-prone sites, such as coastal zones and river basins. On the positive side, the subsurface conditions of urban areas and their surroundings are rather well known in western European countries. This knowledge can be applied to predict, prevent, and mitigate geological hazards and can be used for optimal site selection, thus saving substantial costs to the urban society. Urban environmental geo-indicators are useful instruments to monitor changes in the geo-environmental conditions. Without question the use of underground space will increase significantly in the coming decades due to space limitations at or above the surface.

References

Berendsen, H. J. A. (1996). *De vorming van het land. Inleiding in de geologie en de geomorfologie* (in Dutch). Fysische Geografie van Nederland. Van Gorcum, Assen.

Berger A. R., and Iams, W. J. (1996). *Geoindicators, Assessing rapid environmental changes in earth systems*. Balkema, Rotterdam.

BGS (British Geological Survey) (1988). *World Mineral Statistics*. Nottingham.

de Mulder, E. F. J. (1998). Geoproblems in Urban Centres in EU countries and Norway. *Geologia dell'Ambiente, SIGEA* 4: 2–3.

—— and Bremmer, C. (2000). The downside of cities, managing urban subsurface data. *Proceedings Urban Data Management Symposium* (UDMS), Delft.

—— and J. Kooijman (2003). Dissemination of Geoscience data: the societal implications. In: M. S. Rosenbaum and A. K. Turner (eds.) *New Paradigms in subsurface prediction. Characterization of the shallow subsurface. Implications for urban infrastructure and environmental assessment*. Springer-Verlag, Berlin, 191–9.

—— Baardman, B. A. M., and ten Kate, A. M. (1997). The Underground Municipal Information System (UMIS). *Proceedings International Association Engineering Geology 2, Athens.*

—— McCall, G. J. H., and Marker, B. R. (2001). Geosciences for urban planning and management. In: P. Marinos (ed.) *Proceedings International Symposium on Engineering Geology and the Environment, 1997, Athens.*

—— Claessen, F. A. M., Satijn, H. M. C., Hannink, G., van Bruchem, A. J., and Hulsbergen, J. G. (1994). Effects of reclamation of a new polder on its environment. *Engineering Geology* 37: 15–24.

EEA (European Environmental Agency) (1999*a*). *Environment in the European Union at the turn of the century*. Environmental Assessment Report no 2, Copenhagen.

—— (1999*b*). *Europe's Environment: the Second Assessment*. Elsevier Science, Amsterdam.

Eissmann, L. (2000). *Die Erde hat Gedächtnis. 50 Millionen Jahre im Spiegel mitteldeutscher Tagebau*. Sax-Verlag, Beucha.

—— and Rudolph, A. (2002). *Metamorphose einer Landschaft. Die aufgehenden Seen um Markkleeberg*. Sax-Verlag, Beucha.

Ellison, R. A., Booth, S. J., and Strange, P. J. (1993). Geological mapping in urban areas, the BGS experience in London. *Episodes* 16.

Gussinklo, H. J., Haak, H. W., Quadvlieg, R. C. H., Schutjens, P. M. F. M., and Vogelaar, L. (2001). Subsidence, tremors and society. *Geologie en Mijnbouw / Netherlands Journal of Geosciences* 80: 121–36.

Lawrence, A. R., and Cheney, C. (1996). Urban Groundwater. In: G. J. H. McCall, E. F. J. de Mulder, and B. R. Marker (eds.) *Urban Geoscience*. Balkema, Rotterdam.

Lumsden, G. I. (1992). *Geology and the Environment in Western Europe*. Clarendon, Oxford.

McCall, G. J. H. (1996). Geoindicators of rapid environmental change: The urban setting. In: A. R. Berger and W. J. Iams (eds.) *Geoindicators, Assessing rapid environmental changes in earth systems*. Balkema, Rotterdam.

—— de Mulder, E. F. J., and Marker, B. R. (1996). *Urban Geoscience*. Balkema, Rotterdam.

Ministerie van Verkeer en Waterstaat (2001). *Tweede Structuurschema Oppervlaktedelfstoffen; Landelijk beleid voor de bouwgrondstoffenvoorziening*, Pt. 1 (in Dutch).

Münchener Rück (2003). *Topics. Annual review: natural catastrophes 2002*. Münchener Rückversicherungs Gesellschaft, Munich.

Nederlandse Aardolie Maatschappij (1995). Bodemdaling door aardgaswinning. Groningen veld en randvelden in Groningen, Noord-Drenthe en het oosten van Friesland. Statusrapport 1995, prognose tot het jaar 2050. NAM rapport 27600. Assen.

OECD (1997). Organization for European Co-operation and Development, *Environmental Data Compendium 1997*, Paris.

OECD (1998). Organization for European Co-operation and Development, *Towards sustainable development—environmental indicators*, Paris.

Rijks Planologische Dienst (2000). *Ruimtelijke Verkenningen 2000* (in Dutch). Ministry of Housing, Planning, and Environment (VROM), The Netherlands.

Shields, D. J., and Solar, S. V. (2000). Challenges to sustainable development in the mining sector. *Industry and Environment* 23, UNEP.

United Nations (1995). *Urban and Rural Areas 1994*. UN Publication ST/ESA/SER.A./147 and 148. Department for Economic and Social Information and Policy Analysis. Population Division, United Nations, New York.

21 Geoconservation

Gerard Gonggrijp†

Introduction

The detailed descriptions of the physical geography in the previous chapters show the rich geodiversity of north-western Europe, reflected in its many geological landscapes (landscapes without the biological and cultural 'furnishing'). The various geological forces, acting in time and space have created the foundation for this richness. The landscape's framework has mainly been designed by such endogenic processes as tectonics, orogenesis, and volcanism, while its details have been sculptured by such exogenic processes as weathering, gravity, and glacial-, fluvial-, aeolian-, and marine activities. These modelling processes resulted in a very diverse geology, geomorphology, and pedology. The long scientific tradition and the rich geodiversity made north-western Europe one of the classical areas for geological research. It therefore includes many of the international case studies in earth sciences and became the cradle of numerous international reference localities such as Emsian (Rheinland-Pfalz, Germany), Dinantian (Ardennes, Belgium), Aptian (Provence, France), Danian—Dane is Latin for Denmark (Stevens Klint), Tiglian (Middle Limburg, The Netherlands), Eemian (river in western Netherlands), etc. The chronological division of glacial and fluvioglacial features is primarily based on type localities (villages, rivers, etc.) in Denmark, northern and southern Germany, and The Netherlands. Moreover, a multitude of Tertiary and Pre-Tertiary stages of the standard geological timetable have been named after type localities of geological and prehistoric sites in France. Geological landscapes such as the Maare system of the Eifel (Germany, Fig. 21.1), the volcanoes on the Massif Central (France), the Saalian and Weichselian ice-pushed ridges of Germany, The Netherlands (Fig. 21.2), and Denmark as well as the impressive dunes along the coast from France to the northernmost tip of Denmark have been subjects of detailed research. These geological landscapes form a unique geological patchwork (locations: see Fig. 21.3).

The activities of humans, especially in the last century, have damaged or destroyed many of these landscapes and sites of geological interest. However, selected sites and areas representing the geogenesis of the earth should be preserved for the benefit of science, education, and human welfare. In all European countries attention is given to landscape preservation; however, policy and practice have mainly been based on specific biological, historical-cultural, and visual landscape qualities. Although the geological aspects are the essential foundation for these values, they seldom form the basis for the preservation policy. To a large extent this is due to ignorance of nature conservationists and politicians, but also to the lack of interest of the earth scientists themselves. Consequently, geology often remains 'invisible' in nature and landscape conservation plans and documents. Therefore many important geological sites (*sensu lato*) have been lost or are still being threatened. Since the 1980s this situation is changing little by little owing to the efforts of involved earth scientists. The establishment in 1988 of the European Association for the Conservation of the Geological Heritage (ProGEO) has been of great importance for this development

† To our great sorrow Gerard Gonggrijp died shortly after he had finished a draft of this chapter; Gerard devoted his scientific career to the conservation, preservation, and rehabilitation of geological, geomorphological, and pedological important sites in The Netherlands as well as in a European context; he was the first president of ProGEO (the European Association for the Conservation of the Geological Heritage).

Fig. 21.1. The Schalkenmehren Maar is a typical Maar in the Eifel (western Germany), an area that can be considered as a 'type locality' for explosive craters.

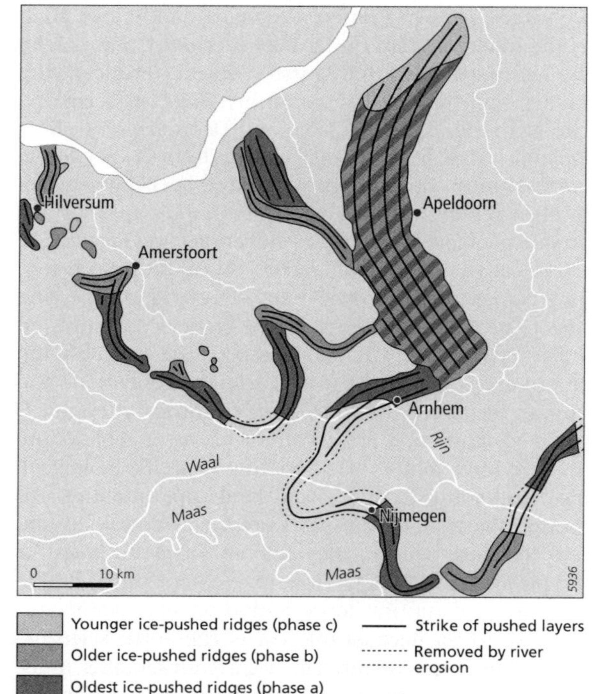

Fig. 21.2. Large (up to > 100 m high) ice-pushed ridge systems of Saalian age in the central Netherlands.

(Black and Gonggrijp 1990; Johansson and Wimbledon 1998).

Geoconservation implies the conservation and preservation of geological, geomorphological, and pedological important sites. The word 'geosite' (or *geotope* in German) has been adopted for such sites. Most attention has been focused on the first two disciplines. Geosites include a wide range of important scientific or educational earth phenomena such as: sites of palaeontological, palynological, stratigraphical, petrological, mineralogical, sedimentological, tectonical, geomorphological, or pedological interest, etc. These sites can be of a fossil (ancient) nature or can exhibit still active geomorphic processes. Stürm (1994) described these geosites (*geotopes*) as 'distinct parts of the geosphere of outstanding geological and geomorphological interest'. Geosites vary enormously in dimension: from small exposures such as a very thin and limited sediment layer or marker horizon in a particular quarry, or an individual small lake of periglacial origin, to an esker system of 100 km in length, to an area containing a system of ice-pushed ridges of several hundreds of km^2, or a volcanic area with an extension of hundreds to thousands of km^2.

Conservation in the Past

Striking natural phenomena have always had their special social and religious meaning in society. Specific phenomena in the 'geo-world', feared and admired, have played a special role in human cultural history. Geological processes and phenomena were integrated in daily and religious life. Often legends in their various versions are related to these natural phenomena. An example is the story of the origin of two isolated pingo remnants situated just 500 m from each other in the Dutch Veluwe region. The story goes that Thunar, the

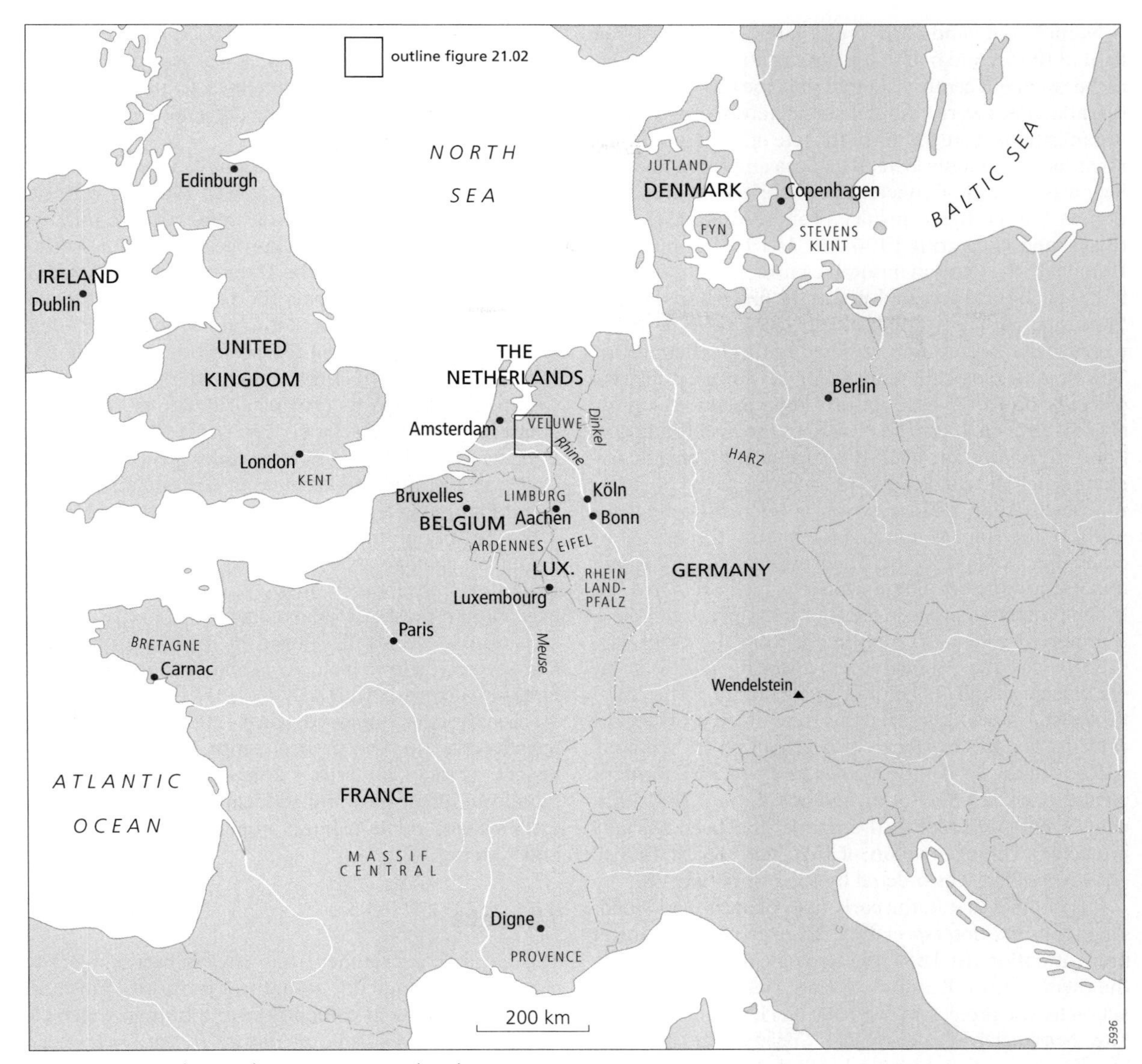

Fig. 21.3. Map showing locations mentioned in the text.

thunder god, hurled his golden hammer at a dangerous snake. The hammer smashed the head of the monster and penetrated seven miles into the trembling earth. However, Thunar, intoxicated by the ascending gases, crashed his wagon, and a second hole was formed. After the accident the people discovered two lakes as deep as the world. Likewise, the origin of big, erratic boulders has often been described in fascinating stories, in which giants, the devil, and the Flood have been held responsible for their presence. Lorelei, a German water nymph, lived on a treacherous rock in the Rhine luring harmless sailors to their death by her singing. The impressive and mysterious prehistoric parallel lines of granite standing stones near Carnac in Bretagne (France) had a ceremonial function, probably of a religious and astronomic nature. In Scandinavia big boulders were erected and engraved with runes by the Vikings to record special events. The preservation of those phenomena was based on unwritten laws of society. They were part of the cultural and natural heritage of those societies.

Nature and landscape preservation became a real item in the second half of the nineteenth and beginning of the twentieth century. At that time the first nature conservation laws were formulated and nature conservation organizations were founded. In spite of good intentions, attention was mainly focused on flora and fauna and geology was a marginal concern. Nevertheless, geoconservation in Europe has a history of more than 150 years (Black and Gonggrijp 1990). In 1819 Alexander von Humboldt, the German natural scientist, globetrotter, and father of physical geography, was the first to use the term 'natural monument' (Wiedenbein 1994). This term was reserved for scenic and striking natural phenomena. The first (geological) monument in Europe that was officially protected as a natural monument was probably the Dragon Rock (*Drachenfels*) by the river Rhine near Bonn (Germany). In 1836 this spectacular volcanic trachyte dome, threatened by mining, was saved. In 1852 the Devil's Wall (*Teufelsmauer*), a spectacular hogback north of the Harz Mountains (Germany) threatened by quarrying, was protected. Their impressive appearance and natural historical importance played a decisive role in the protection procedure. Another typical example of a protected geological monument from the early days (1840) was the Agazzis Rock, showing glacial striae on Blackford Hill in Edinburgh (Scotland). The Swiss geologist Agazzis discovered this rock which provided the evidence for the existence of a glaciation in Scotland. Already in the 1840s the *Hesselagerstenen*, an enormous erratic block 13.8 × 9.4 m and weighing 1,000 tons, near Nyborg on East Fyn (Denmark), had been officially protected. The excavation of this, the biggest Danish erratic boulder, was ordered by King Christian VIII.

It is notable that in the early days of nature and landscape conservation especially striking geological features drew attention. In 1887 the Geological Commission of the Swiss Nature Research Society started a successful action for the protection of erratic blocks (Hantke 1986). The Norwegian Geographical Society put an even broader range of sites forward. At a meeting in 1909 the society recommended the protection of all scientifically and historically important geological and mineralogical sites under the first Nature Conservation Act in 1910 (Erikstad 1984). At a meeting in 1905 of the Geological Society of Stockholm, de Geer drew attention to the need for conservation of natural monuments. This led to the first official specific inventory of geosites in Sweden by Munthe (1920). The aim of the Dutch Society for the Preservation of Nature Reserves in The Netherlands (*Natuurmonumenten*), as formulated in 1905, was similar. It included the conservation of the natural flora and fauna, and of important geological features. The Dutch earth scientist van Baren, an expert on Dutch geology, as early as 1908 warned against the destruction of scientifically important geomorphologic features, although these landforms were not spectacular at all compared with examples from abroad.

However, geoconservation began its more systematic but still modest way up only slowly, together with the revival of nature and landscape conservation, initiated by the organization of the European Year for Nature Conservation in 1970. The Danish Nature Conservation Act of 1969 mentions the preservation of geological formations as well as flora, fauna, and scenic areas as one of the purposes of the protection of nature and landscape assets. In 1969 the Netherlands Working Group GEA started its provincial inventory of geosites (Gonggrijp and Boekschoten 1981). The French Nature Conservation Act of 1976 includes geoconservation (Billet 1994), and this resulted in the establishment of eight national geological reserves in the 1980s (Martini 1994). Although there obviously have been some good examples, geology explicitly mentioned and acted upon in nature conservation plans and projects is still an exception in most European countries. Clearly, geoconservation is weakly developed in most countries and lags severely behind biological conservation. It is therefore not surprising that only in 2004 the first textbook on Geodiversity appeared (Gray 2004). Gray defines 'geodiversity' as 'the natural range (diversity) of geological (rocks, minerals, fossils), geomorphological (landform, processes) and soil features, including their assemblages, relationships, properties, interpretations and systems'.

Threats

The natural environment has always been subject to continuous change due to natural processes, but most changes are slow in comparison to a human timetable (Westbroek 1991). Defining 'nature' (*sensu stricto*) as an environment untouched by any form of human impact shows that it has almost completely disappeared in Europe. This does not mean there are no natural elements left, but present nature reserves are far from natural. Most of the north-western European reserves, such as old grazing pastures and planted woods, are more cultural than typical natural areas. Obviously, the term 'nature' inevitably suffers from inflation. Small-scale farming, exploitation of natural resources, and other activities such as building and infrastructural works had limited effects until the end of the nineteenth century when due to a strong population growth and rapid technical advancements the pressure on nature increased tremendously. Activities related to modernizing and intensifying agriculture and forestry, water management and flood control, exploitation of natural resources, urbanization, and recreation all have had or still have an irreversible impact on the geological landscape, on fossil phenomena, and on still-active geomorphic processes.

Agriculture, Cattle Breeding, and Forestry

Although the development of farming radically changed flora and fauna, the relief remained more or less untouched. Only the upper soil suffered some superficial cultivation. Real changes in the geomorphology have occurred on the poor sandy soils in north-western Europe since the Bronze and Iron Ages (Waterbolk 1962), but especially during the Middle Ages over-exploitation of these areas led to extensive sand drifts in most west European countries (Castel *et al.* 1989; Koster *et al.* 1993). At the beginning of the twentieth century many of these migrating dune fields, which in some cases threatened villages and arable fields, were stabilized by forest plantations. The largest remnants of these still actively moving drift sands are situated in the central Netherlands (Fig. 21.4). Nowadays, these regions essentially resulting from man-made wind erosion are considered as typical examples of nature development. Consequently, some of these regions are preserved even to the extent of artificially removing any vegetation that invades the dune fields. However, during the nineteenth and twentieth centuries most agricultural activities such as deep ploughing and levelling of 'waste lands', reclamation of wetlands, and extensive reallocation programmes of farmlands had drastic effects on geomorphology. Especially in the low-lying coastal regions of north-western Europe these activities destroyed many of the minor relief forms, such as patterns of (fossil) marine creeks and small river channels, partly submerged river dunes, aeolian cover sand ridges, and forms of periglacial origin, which typify these landscapes. Man-made changes in more hilly landscapes should also not be overlooked. Viniculture, for example, remodelled many slopes in Germany and France. Deforestation resulted in drastically increased soil erosion on many places.

Water Management and Coastal Protection

During the Middle Ages, a period of increasing population, the influence of the sea and rivers became a real problem. From around AD 1000 onwards dykes were built all along the Atlantic coasts and rivers to protect the population from floods. This had an enormous influence on the natural development of coastlines and river courses. Later in history the capture of river meanders, which occurred on a very extensive scale all over Europe, to provide better conditions for navigation and more rapid water drainage, completely changed development of water systems (Fig. 21.5). This process continued until about the 1980s. For example, only a fraction of the original smaller river courses in The Netherlands is still untouched; larger rivers in all west European countries are all strongly remodelled by man.

At many localities low cliffs along coasts have been smoothed and armoured to prevent erosion. The

Fig. 21.4. Migrating inland dune sands encroaching on a pine plantation, at Hulshorster Zand, North Veluwe region, central Netherlands.

Fig. 21.5. Regulation activities in the Dutch part of the River Dinkel in 1978. The German part of the Dinkel had been completely regulated in the past. A few years ago renaturation projects started in the Dutch section.

lowering of groundwater levels for agricultural and infrastructural purposes affected soils, lakes, and large areas of wetland and peatland. Soil erosion and oxidation of peat, resulting in the destruction of geological and biological archives, occurred on a wide scale.

Extraction of Mineral Resources

In the past the impact of the exploitation of mineral resources was relatively small. Regionally, the need for many kinds of natural resource had an enormous impact on landscape. Pits and quarries of many square kilometres are not exceptional. Although peat winning is a superficial activity, its impact on the landscape has been considerable. A striking example of disastrous excavation activity is the opencast mining of brown coal in Germany. Removing the covering formations and the extraction of the brown coal layers underneath has completely stripped the landscape. Between Aachen and Cologne an area of over 200 km^2 is turned into a completely anthropomorphic landscape many metres below the original surface. Apart from the complete alteration of the landscape these huge and deep opencast mines cause serious hydrological problems with national cross-border consequences. An unintentional but positive result of these activities is the creation of outcrops for geological research, which contributes to the understanding of the geological development of the area (Eissmann 2000; Eissmann and Rudolph 2002). Especially in the (former) East German region of Halle-Leipzig huge brown coal quarries are presently remodelled into forests, arable fields, and recreational areas (Fig. 21.6).

Urban and Infrastructural Expansion

The rapid extension of cities and infrastructural facilities, especially after World War II, required land and huge quantities of natural resources, such as sand and gravel. As a result many areas with important sites for geoconservation became inaccessible, hidden, or even destroyed (Gonggrijp and Boekschoten 1981; Gonggrijp 1993*b*).

Recreation and Recreational Facilities

A typical example is the creation of an increasing number of golf courses, which have an evident impact on geomorphology, especially so in relatively flat areas. Other effects are related to a strong recreational pressure in certain areas causing vulnerable vegetation to be destroyed, leading to (gully) erosion and ablation and thus damage to the original relief.

Motivation for Protection

Nature conservation is based on ethical, aesthetic, scientific, and educational motives. Nature itself does not evaluate; in nature there is no good or bad, beautiful or ugly. In fact nature and environmental conservation are human activities. The world will turn with or without orchids, pandas, birds of prey, eskers, a Danian-type locality, or even human beings. There is no natural law that prohibits or prevents the extinction of these phenomena. It is proved that big disasters in the geological past, such as the meteor impact that killed, for example the dinosaurs, give room for other species, for instance

Fig. 21.6. The Störmthaler Lake in the summer of 2001, illustrating large-scale landscape restructuring of former brown coal quarries south of Leipzig in the (former) East German Democratic Republic; groundwater rise will transform the quarries into large lakes, mainly for recreational purposes (photo: courtesy of Lothar Eissmann and Armin Rudolph 2002).

the mammals, to develop. However, the question is whether we human beings want the world to turn without these phenomena as a result of our activities, knowing that at the end the destruction of the natural environment may lead to our own termination as a species.

Recently nature conservation has been defined in terms of functions that nature fulfils for humanity (van der Maarel and Dauvellier 1978; de Groot 1992). These functions have been specified as regulation, carrying capacity, production, and information functions. Of course those functions do often conflict with each other. For example, the interests of the extraction of minerals (production) and the preservation of nature (information) are often incompatible. Therefore, a balance must be found between what has to be preserved and what should be sacrificed for the sake of 'human development'. Natural geosites are important as a basis for study, education, and orientation in space and time and therefore they fulfil the earth-scientific aspects of the information function of nature (Fig. 21.7). Besides, geology is important for its carrying capacity for the biotic and human world, and for its function of production with respect to mining (Wolfert 1995). However, the preservation of geosites is not yet generally accepted as of equal value in nature conservation, although more and wider understanding of its importance is growing (Duff 1994; Gonggrijp 1997*b*, 1999*b*; Hofmann 1998; Erikstad 1999; Gray 2004).

The most evident geological examples of geoconservation sites are the (inter)nationally defined type localities with their type sections (Wimbledon 1990; Gray 2004). These sites have been determined by (inter)national stratigraphical commissions and they form the keystones for the stratigraphical system on which the geological history of the earth has been based. Therefore, undoubtedly their protection is of highest importance. Apart from formally defined type localities there are also geoconservation sites that refer to 'classic' examples of famous geological features, such as Siccar Point in Scotland where James Hutton (1726–97), one of the founders of modern geology, first inferred and later discovered unconformity, which contributed to his ideas of uniformitarianism or actualism (Holmes 1965). Agazzis Rock is another famous example mentioned earlier. In the field of geomorphology and pedology type localities have not been defined. However, famous sites function as such, for example, the calcareous cliffs of Møns Klint in southern Denmark or the volcanic Maare system in the Eifel region of Germany. Likewise, specific

Fig. 21.7. Vertically stacked, ice-pushed sediments exposed in a former sandpit in Jutland, Denmark (photo: Eduard Koster).

sites of which the soils have been extensively described as a reference in soil classifications can be considered as pedological type localities.

Scientific Motives

Geoconservationists want to preserve sites that represent important aspects, events, moments, and phenomena of the geological history. Preservation has to be realized, because a site once investigated is not investigated forever! Even if a site has been described and sampled thoroughly, it sometimes will be necessary to repeat the research because new methods offer new possibilities for scientific interpretation. For example, in a period of twenty-five years, the type section of the Usselo layer of Weichselian Late-Glacial, Allerød age has been investigated three times for just that reason (Fig. 21.8). The internationally recognized Dutch stratotype of the Eemian is another example. In 1997 at the request of the Stratigraphical Commission of the International Geological Union this stratotype (borehole section) has been reinvestigated in detail. Evidently, any listing of geosites will never be complete and final: scientific opinions change, research methods develop, blind spots are filled. This leads to a continuous exchange of already protected sites for 'better' ones.

Educational Motives

The preservation of geosites for training and education of students, amateurs, or the general public is another reason for protection (Harley 1994). These sites should

Fig. 21.8. A re-excavation of the 'stratotype' of the Usselo layer with an Allerød peat layer and palaeosoil horizon of late Weichselian age (Veluwe area, central Netherlands). After its third excavation the trench has been filled up again to safeguard this vulnerable site of scientific interest.

Fig. 21.9. The footprints of two different dinosaurs in Upper Jurassic, steeply tilted shales, exposed in an excavation in Barkhausen near Osnabrück, Germany. This geological monument is provided with instructive displays.

be instructive and have good accessibility (Fig. 21.9), thereby enabling the demonstration of geological processes and phenomena that relate to a variety of aspects of daily life, such as the occurrence of natural hazards, floods, earthquakes, the exploitation of natural resources, or the use of the subsoil for buildings, infrastructure, etc. The recent GRECEL project (Geological Heritage: Research in Environmental Education and Co-operation on a European Level) financed by the European Commission (Socrates/Comenius programme) is a contribution of earth scientists from various countries to school education (Drandaki *et al.* 1999). This project includes five proposals for geoconservation education integrated in the environmental school programme for the secondary level.

Site Selection

At the beginning of the twentieth century the protection of sites was promoted for their uniqueness, their striking appearance, their beauty, or because they were vulnerable or threatened (Gonggrijp and Boekschoten 1981). At that time there was hardly any systematic approach. In a few cases regional or thematic geological inventories for protection purposes were executed (Geinitz 1909; Munthe 1920; Hantke 1986). Great Britain was the first country with a special earth-science conservation branch (1950) in the official nature conservation organization: the Nature Conservation Council (1949), which carried out nation-wide geoconservation inventories. And it still is the only country with geological heritage sections. Gray (2004) discusses the status of the Sites of Special Scientific Interest (SSSIs) in England and Wales; over 2,000 SSSIs of geological interest are currently being designated by the UK nature conservation agencies. These represent about one-third of all SSSIs, the remainder being of biological interest. The interest in nature and environment specially promoted by the event of the European Year for Nature Conservation 1970 had a positive effect on the interest of earth scientists in conservation too. A first provisional national overview of 115 geosites in The Netherlands, based on expert judgement, was published by de Soet (1974). This was followed by a more complete inventory, including almost 1,000 sites, as part of the national overview of the Dutch Nature Policy Plan (Gonggrijp 1993*a*). The first Danish overview of 197 sites, based on expert judgement, appeared in 1984 (Fredningsstyrelsen 1984). The German situation is different because of the federal system. Nature conservation is the responsibility of the various *Bundesländer* and each uses its own criteria. At the beginning of the 1980s all Bundesländer developed geosite overviews (Meiburg 1993). Presently, a total number of 12,000 selected sites exists of which about 5,000 have been protected in some way or another (Grube 1999). Gray (2004) mentions the so-called 'geotope protection areas', defined as 'distinct parts of the geosphere of outstanding geological and geomorphological interest', which have been identified in Germany and Austria. The status of these *Geotopes* is more or less similar to that of the SSSIs in the UK. Until now there are no completed overviews of geosites in Belgium (pers. comm.), Luxembourg, and France. But this does not mean that there are no protected geological sites. In France twelve national geological reserves, designated on palaeontological, stratigraphic, mineralogical, and stratotype grounds, have been established in the 1980s varying in size from 61 ares to

Fig. 21.10. The volcanic dome of the Puy de Dome in the Massif Central (France).

Fig. 21.11. One of the displays explaining the geological origin of the strongly eroded calcareous rocks in the massif of the Dentelles de Montmirail in the northern Provence (France).

145,000 hectares (Martini 1994; Gray 2004; Fig. 21.10, 21.11, and 21.12). The inventory of all sites of national importance has been started recently with a try-out in Bretagne (pers. comm. G. Martini). Most of the early geosite selections were based on a qualitative evaluation by expert judgement. However, recent environmental studies for conservation and physical planning purposes tend to an approach based on quantitative data with arithmetic evaluation models. Presently, classification schemes are set up, selection criteria are defined, and evaluation schemes are developed in several countries.

Classification

Classifications in relation to earth science mean an arrangement of phenomena in time, process, rock type, landform, species (fossil), mineral content, etc. The standard of the classification largely depends on the required spatial scale and application of the information; obviously, classifications vary from country to country. For example, German systems are usually based on a geological timescale in combination with lithology and geologic processes, as well as erosional or depositional landforms (Grube and Ross 1982). In The Netherlands a geogenetic or morphogenetic classification system has been developed (Gonggrijp 1999*a*, *b*), that must be suitable for supporting landscape conservation policy and must also enable digital processing for practical purposes. In order to make the classification also suitable for international use, the geomorphological regions of Europe distinguished by Embleton (1984) have been used as a framework for this classification. These regions are based on geo(morpho)logical megastructures,

Fig. 21.12. View of the Mont Ventoux (1,909 m a.s.l.), an isolated mountain consisting of calcareous rocks also called the 'Fuji-Yama' of France. This region forms part of the Mont Ventoux National Park that is classified as a UNESCO Biosphere Reserve (photo: Eduard Koster).

called *morpho-provinces*, and large-scale geological processes, called *morpho-regions*, with international dimensions. The next step in the hierarchic division was the selection of a national set of (twenty) geomorphologically and chronologically distinct landscape units, called *morpho-complexes*. These morpho-complexes form the starting point for the classification and evaluation system. The morpho-complexes in their turn are divided into *morpho-patterns* and individual *morpho-elements*, the smallest units in the system (simple landforms). On this lowest classification level geological outcrops, erratics, and pedological sites can be added to the system.

Criteria

For the proper selection of geosites a set of evaluation criteria is needed. Several authors have described such sets of criteria (e.g. Soyez 1971; Gonggrijp and Boekschoten 1981; Wimbledon 1990; Gordon 1992; Bauer and Seibel 1993). Erikstad (1994) made an inventory of criteria applied for geosite selection and distinguished primary criteria, such as rarity, representativeness, diversity, coherence, and soundness, and several secondary criteria, such as accessibility, key area, etc. The primary criteria provide information on the geological identity, the secondary ones are related to scientific, educational, and practical values. Gonggrijp (1996) discussed the various criteria and arranged them into the following seven categories:

- General selection criteria for geotopes: 'geogenetic and/or form representativity', rarity, naturalness;
- Scientific selection criteria for specific geotopes: established or potential (natural) historical documentation value, key or reference site, and chronology or age information;
- Educational selection criteria for specific geotopes: instructive site value, visibility, accessibility, and diversity;
- Selection criteria for recreational/touristic geotopes: quality/beauty, visibility, accessibility, and diversity;
- Selection criteria for the protection of geotopes: scientific value, vulnerability, and specific threats;
- Selection criteria in relation to nature development of geotope areas: actual or potential for specific processes, diversity;
- Selection on ethical principles: intrinsic value.

In the classification and selection system of Gonggrijp (1996) the first and most important criterion is *'geogenetic' representativeness*. For example low (only a few metres high), periglacial dunes of Weichselian age cover a considerable part of the Low Countries and thereby determine the image of that landscape. It is evident that the best examples of these particular landforms should be selected. Fluvio-glacial landforms, like eskers of Saalian age, are *rare* phenomena in northwestern Europe. Likewise, living peatlands have become a *rare* feature as the overwhelming majority of peatlands in Germany, Denmark, and The Netherlands have been drained, excavated, or otherwise changed by man. The criterion *'form' representativeness* is used for the selection of instructive examples suitable for education. This may be a well-developed volcanic dome or remnant of an explosive crater or a large caldera of volcanic origin. It may also be a particular sequence of a series of Weichselian cover sands on top of Saalian boulder clay on ice-pushed deposits, as shown in outcrops. Such a sequence of sediments is typical for large parts of Germany, Denmark, and The Netherlands. The criterion *naturalness* is being used to separate undisturbed landforms from affected and therefore less valuable ones. Finally, periglacial phenomena such as frost mound remnants or pingos may be a frequent phenomenon in northern regions, but such landforms are very rare in more southern regions, such as Belgium and France. Consequently, such features in the Belgian Ardennes are selected as geosites based upon the criterion of *rarity*. This classification and evaluation system can be linked to geomorphological GIS databases, such as GEOMORF in The Netherlands, resulting in map series covering whole countries.

Geoconservation Policy, Strategy, and Protection

Sound geoconservation is a complicated process, involving several groups of actors often having different interests (Gonggrijp 1999*b*). It demands an effective strategy, a sound legal framework, and a clear and transparent procedure from basic geological data to protected, managed, and monitored geosites. The awareness of the importance of the abiotic world for society, including the understanding of its economic role, is growing. Moreover, the specific role geodiversity plays in nature and landscape conservation in general and in relation to biodiversity and cultural history in particular is becoming fully recognized (Duff 1994; Hofmann 1998, 1999). Although instruments for protective measures for geosites are usually available, they are often not explicitly mentioned in spatial planning policies, strategies, and environmental assessments plans. There is still a need for a better integration of geoconservation in all conservation planning procedures. In this respect Great Britain is far ahead compared to the rest of Europe (Nature Conservation Council 1990). The indisputable relation between geodiversity and the actual or potential biodiversity could be the basis for more understanding by all actors. In this process the public's support is strongly needed. Popularization of geology can be an effective instrument, as unknown is unloved (Gonggrijp 1997*a*; Hose 1997; Meyer 1997). Since 1992 the Ministry of Environment in Denmark has published a series of popular and well-illustrated geoconservation books on Denmark. The series contains an overview of the regional geology and an extended description of the various geosites (Andersen and Sjörring 1992). In several countries geoconservation excursions are being organized regularly under the umbrella of ProGEO (Gonggrijp 1997*a*). In France several geological reserves are being provided with excellent popular explanations of the local geology. In this respect the geological reserve of the Haute-Provence covering an area of 145,000 hectares, should be mentioned (Fig. 21.13). Besides various displays explaining geology in the area, a museum supports the educational functions of this reserve (Martini 1994). The Geo-Park Volcano-Eifel in Gerolstein (Frey 1997) and the Geo-Park in

Fig. 21.13. An illustrative example of active erosional processes ('badlands formation') in the Geological Reserve Haute-Provence (France).

Fig. 21.14. The construction of an erratic boulder park (at Maarn, Province of Utrecht, The Netherlands), with help of a local nature conservation organization.

Wendelstein in southern Germany are similar good examples. In northern Germany and The Netherlands more than thirty erratic boulder gardens have been set up (Fig. 21.14), informing the public on the source areas and type of boulders and the glacial history (Schulz 1997). Depending on the national legal and planning system, there are several ways to safeguard geosites. Nowadays, the following instruments are used to reach the goals: acquisition by a state or private nature conservation organization, or allocation as a Nature Monument, Nature Reserve, or National Park by a Nature Conservation Act. The legal basis for protection can be arranged through relevant laws such as a physical planning act, mining act, environmental management act (including environmental impact assessments), forestry act, etc.

Management and Renaturation

Legal protection is not enough to safeguard geosites in the future. After realization of the protection of geosites plans for management and monitoring have to be developed and to become operative. In the past this has often been neglected.

Design and Execution

Optimal visibility and sustainability of the sites, especially in the case of educational sites, are important goals. In case of exposures a management design, including additional arrangements such as water shields, drainage provisions, shelters, removal of trees and waste, educational provisions, etc. have to be worked out (e.g. McKirdy 1990; Bennet 1992). A calculation is also an essential part of the scheme, as future management costs are often neglected (Gonggrijp 1994). Geomorphological sites may need special arrangements, for example removing the vegetation to increase the visibility of the site. In the case of active geological processes, such as aeolian or fluvial transport and sedimentation, the continuation of the process should be guaranteed. Likewise exposures will degrade and therefore have to be cleaned every now and then. This also implies that sites should be monitored.

Management and Monitoring

Partly due to the fact that arable lands of marginal economic value are taken out of production and changed into more natural areas in many west European countries the idea of 'making (new) nature' has become popular in the 1980s and 1990s. Renaturation, rehabilitation, or nature development have become modern issues. At the same time people have begun to realize that natural geomorphic processes, like sand drifting in dune fields, river meandering, or clay and silt sedimentation in Wadden areas, could be reintroduced in specific places. In The Netherlands these renaturation ideas have been incorporated in a national Nature Policy Plan in 1990, which intends the realization of a national 'Main Ecological Network' connecting nature reserves, national parks, and the countryside in between.

In other countries, such as Denmark (Brookes 1987) and Germany (Kufeld 1988) renaturation programs have also been launched. These policy plans of renaturation focus on the development of 'new nature' areas and features of high ecological value, and the keyword is biodiversity! However, renaturation is not always positive for geoconservation. Measures taken to improve the ecological situation sometimes affect or even damage geodiversity as well as archaeological and cultural historical values. Examples of activities that might affect the geological situation and thereby geoconservation's interest are: removal of polluted upper soil layers; removal of peat and/or gyttja polluted by nutrients; restoration of the former relief in case of levelling in the past; reintroduction of meandering in canalized river systems; rising of groundwater level to stimulate wetland development; removal of vegetation to stimulate aeolian activity in dune areas; removal of coastal dunes to restore marine inlets; rearrangement of former quarries. In relation to renaturation plans and their effects on geological values the following landscape types are distinguished:

Fossil Geological Landscapes

These are the landscapes developed in former geological periods under different climatic and geological conditions. They cover most of the north-western European region. Illustrative examples are old river terraces, glacial and periglacial landforms from the Saalian and Weichselian glaciations, volcanic landscapes and peniplanes. In these cases renaturation projects should be executed with great care, as any change in these particular landscapes is irreversible. For example, the removal of soil material or surficial sediments should be avoided in those cases where pedological and/or geological (i.e. palynological) archives may be destroyed.

Fossilized Geological Landscapes

Fossilized landscapes are those where recent geological processes have been fixed by human activities, such as construction of dykes, afforestation, etc. In time these fossilized landscapes will have developed their own cultural-historical and morphological elements. Ecological improvement of these areas should be weighed against the losses with respect to geology and cultural history. Reactivation of aeolian, fluvial, and marine processes in certain areas should be considered favourably provided the measures take into account the former extent and location of particular processes.

Active Geological Landscapes

With the exception of Wadden regions along the coasts of north-western Europe, some river areas, heavily degraded areas due to accelerated soil erosion, and high mountain regions, active geological landscapes hardly exist any more. From a geoconservation point of view these regions deserve particular interest.

'Exploitation Landscapes'

Exploitation landscapes are areas that have been used for extraction of mineral resources, such as pits and quarries. Many of these excavation sites have been abandoned or filled up with waste. Former excavation sites can be made suitable for recreational activities or serve renaturation purposes (Eissmann and Rudolph 2002). Moreover, some sites are very interesting from a geological point of view. In these cases consultation in the planning phase with all the interested parties including the geoconservationists is of great importance (Gonggrijp 1997*c*; Hiekel 1997).

Final Remarks

In the last thirty years geoconservation has made considerable progress. The European Association for the Conservation of the Geological Heritage (ProGEO), established in 1988, has played an important role in stimulating the formation of geoconservation working groups and national committees for geoconservation. Co-operation with international organizations such as UNESCO (World Heritage) and IUCN (International Nature Conservation) is essential to integrate geoconservation into international and national protection schemes. A fine example is the GEOSITE/GEOPARK project initiated in 1995 by the IUGS (International Geological Union) and now also under auspices of UNESCO (Wimbledon *et al.* 1999). Until 2002 twelve such GEOPARKS have been created in Europe.

References

Andersen, S., and Sjörring, S. (ed.) (1992), *Det nordlige Jylland, en beskrivelse af områder af national geologisk interesse* (in Danish). Geografforlaget, Brenderup.

Bauer, J., and Seibel, S. (1993), Geomorphologisch orientierter Naturschutz im Saarland—Anspruch und Wirklichkeit. In: D. Förster, A. Grube, and F. W. Wiedenbein (eds.) *Geotopschutz: Materialien 1, Mitwitz, D*, University of Erlangen-Nürnberg, Erlangen, 103–8.

Bennet, M. (1992), The management of Quaternary SSSI in England, with specific reference to disused pits and quarries. Conservation of geomorphology and Quaternary sites in Great Britain: an overview of site assessment. In: C. Stevens, J. E. Gordon, C. P. Green, and M. G. Macklin (eds.) *Conserving our landscape*. Proceedings of the Conference Conserving our landscape: Evolving landforms and Ice-age heritage, May 1992, Crewe, UK, 36–40.

Billet, P. (1994), L'Émergence d'un droit du patrimoine géologique en France. In: G. Martini (ed.) *Proceedings of the 1st International*

Symposium on the Conservation of our Geological Heritage, Mémoires de la Société Géologique de France 165: 17–19.

Black, G. P., and Gonggrijp, G. P. (1990), Space and time: a new approach. *Naturopa* 65: 10–13.

Brookes, A. (1987), Restoring the sinuosity of artificial straightened stream channels. *Environmental Geological Sciences* 10/1: 33–41.

Castel, I. I. Y., Koster, E. A., and Slotboom, R. (1989), Morphogenetic aspects and age of Late Holocene eolian drift sands in Northwest Europe. *Zeitschrift für Geomorphologie* 33: 1–26.

de Groot, R. S. (1992), *Functions of nature. Evaluation of nature in environmental planning, management and decision making.* Wolters-Noordhoff, Groningen.

de Soet, F. (1974), Het landschap als schakel in het systeem aardemens. *Tijdschrift voor Economische en Sociale Geografie* 65: 305–13 (in Dutch).

Drandaki, I., Diakantoni, A., Eder, W., Fermeli, G., Galanakis, D., Gonggrijp, G. P., Hlad, B., Koutsouveli, A., Martini, G., Page, K., and Patzak, M. (1999), GRECEL, Geological Heritage: Research in Environmental Education and Co-operation at European Level. In: D. Barettino, M. Vallejo, and E. Gallego (eds.) *Proceedings of the Third International Symposium on the Conservation of the Geological Heritage/Towards the Balanced Management and Conservation of the Geological Heritage in the New Millennium*, Madrid, 324–9.

Duff, K. (1994), Natural areas: a holistic approach to conservation based on geology. In: D. O'Halloran, C. Green, Stanley M. Harley, and J. Knill (eds.) *Geological and Landscape Conservation*. The Geological Society, London, 121–6.

Eissmann, L. (2000), *Die Erde hat Gedächtnis: 50 Millionen Jahre im Spiegel mitteldeutscher Tagebau*. Sax-Verlag, Beucha.

—— and Rudolph, A. (2002), *Metamorphose einer Landschaft—Die aufgehenden Seen um Markleeberg*. Sax-Verlag, Beucha.

Embleton, C. (1984), *Geomorphology of Europe*. Macmillan, London.

Erikstad, L. (1984), Registration of sites and areas with geological significance in Norway. *Norsk Geografysk Tidsskrift* 38: 199–204.

—— (1994), Quaternary geology conservation in Norway, inventory program, criteria and results. In: G. Martini (ed.) *Proceedings of the 1st International Symposium on the Conservation of our Geological Heritage, Mémoires de la Société Géologique de France* 165: 213–15.

—— (1999), A holistic approach to secure geoconservation in local physical planning. In: D. Barettino, M. Vallejo, and E. Gallego (eds.) *Proceedings of the Third International Symposium on the Conservation of the Geological Heritage/Towards the Balanced Management and Conservation of the Geological Heritage in the New Millennium*, Madrid, 69–72.

Fredningsstyrelsen (1984), *Fredningsplanlægning og Geologi. Nationale geologiske interesseområder*. København, no. 4 (in Danish).

Frey, M-L. (1997), Künstlerische Darstellungsmöglichkeiten geologischer Sachverhalte am Beispiel des Geo-Parks der VG Gerolstein. In: L. Feldmann (ed.) *Geotopschutz und seine rechtlichen Grundlagen. Schriftenreihe der Deutsche Geologischen Gesellschaft* 12: 100–9.

Geinitz, E. (1909), Unsere großen Findlinge. *Mecklenburg* 4: 83–94.

Gonggrijp, G. P. (1993*a*), Nature Policy Plan, new developments in the Netherlands. In: L. Erikstad (ed.) *Earth Science Conservation in Europe*. Norsk Institutt for Naturforskning, AAS: Nina, 5–16.

—— (1993b), River meanders or houses: a case study. In: L. Erikstad (ed.) *Earth Science Conservation in Europe*. Norsk Institutt for Naturforskning, AAS: Nina, 35–8.

—— (1994), Two geological monuments in The Netherlands: De Zândkoele and Wolterholten. In: D. O'Halloran, C. Green, M. Harley, M. Stanley, and J. Knill (eds.) *Geological and Landscape Conservation*. The Geological Society, London, 323–8.

—— (1996), *Indelings- en waarderingsmethode voor aardkundige waarden*. Wageningen, The Netherlands, IBN report 218 (in Dutch).

—— (1997*a*), Unknown, unloved. Education, the basis for protection. In: P. G. Marinos, G. C. Koukis, and G. C. Tsiambaos (eds.) *Engineering geology and the environment*. Balkema, Rotterdam, iii. 2945–8.

—— (1997*b*), Geotope motivation and selection: A way of objectifying the subjective. In: P. G. Marinos, G. C. Koukis, and G. C. Tsiambaos (eds.) *Engineering geology and the environment*. Balkema, Rotterdam, iii. 2949–54.

—— (1997*c*), Nature development: Biologist's experimental garden! Geologist's future sand box? In: P. G. Marinos, G. C. Koukis, and G. C. Tsiambaos (eds.) *Engineering geology and the environment*. Balkema, Rotterdam, iii. 2939–44.

—— (1999*a*), A classification and evaluation system for geotopes. *Proceedings of the Second International Symposium of our Geological Heritage/World Heritage: Geotope Conservation world-wide, European and Italian Experiences, Memorie descrittivedella Carta Geologica D'Italia*, Rome, LIV. 323–8.

—— (1999*b*), Geodiversity: the key to a holistic approach in renaturation. In: D. Barettino, M. Vallejo, and E. Gallego (eds.) *Proceedings of the Third International Symposium on the Conservation of the Geological Heritage/Towards the Balanced Management and Conservation of the Geological Heritage in the New Millennium*, Madrid, 77–80.

—— and Boekschoten, G. J. (1981), Earth-science conservation, no science without conservation. *Geologie en Mijnbouw* 60: 433–45.

Gordon, J. E. (1992), Conservation of geomorphology and Quaternary sites in Great Britain: an overview of site assessment. In: C. Stevens, J. E. Gordon, C. P. Green, and M. G. Macklin (eds.) *Conserving our Landscape*. Proceedings of the Conference Conserving our landscape: Evolving landforms and Ice-age heritage, May 1992, Crewe, UK, 11–21.

Gray, M. (2004), *Geodiversity. Valuing and conserving abiotic nature*. J. Wiley and Sons, Chichester.

Grube, A. (1999), Geoconservation in Germany—1996. *Proceedings of the Second International Symposium of our Geological Heritage/World Heritage: Geotope Conservation world-wide, European and Italian Experiences, Memorie descrittive della Carta Geologica D'Italia*, Rome, LIV. 265–72.

—— and Ross, P. H. (1982), Schutz geologischer Naturdenkmale. *Zeitschrift für Natur- und Landeskunde van Schleswig-Holstein und Hamburg* 89/2–3, 37–48.

Hantke, R. (1986), Erdgeschichtliche Naturdenkmäler. In: H. Wildermuth (ed.) *Natur als Aufgabe*. Schweizerischer Bund für Naturschutz, Basle.

Harley, M. J. (1994), 'Rigs'. A nationwide site conservation initiative based on local voluntary groups. In: M. Martini (ed.) *Proceedings of the 1st International Symposium on the Conservation of our Geological Heritage, Mémoires de la Société Géologique de France*, Paris, 259–60.

Hiekel, W. (1997), Biotopschutz/Geotopschutz—gegensätzliche Naturschutzrichtungen? *1 Internationale Jahrestag Deutsche Geologische Gesellschaft (Fachsektion Geotopschutz)* 5: 77–9.

Hofmann, T. (1998), Nature is more than geo(topes) and bio(topes) —some holistic considerations. In: A. Miidel (ed.) *Proceedings ProGEO '97 Scientific Conference in Estonia*, 15–17.

—— (1999), Geotope in Österreich: Heutige Situation and Chancen für die Zukunft. In: M. Felber (ed.) *Geologia Insubrica* 4/1: 87–90.

Holmes, A. (1965), *Principles of Physical Geology*. Workingham, Van Nostrand Reinhold.

Hose, T. A. (1997), Geotourism—Selling the earth to Europe. In: P. G. Marinos, G. C. Koukis, G. C. Tsiambaos, and G. C. Stournaras (eds.) *Engineering geology and the environment*. iii. 2955–60.

Johansson, C. E., and Wimbledon, W. A. P. (1998), ProGEA—The European Association for the Conservation of the Geological

Heritage. In: A. Miidel (ed.) *Proceedings ProGEO '97 Scientific Conference in Estonia*, 29–30.

Koster, E. A., Castel, I. I. Y., and Nap, R. L. (1993), Genesis and sedimentary structures of late Holocene aeolian drift sands in Northwest Europe. In: K. Pye (ed.) *The dynamics and environmental context of aeolian sedimentary systems. Geological Society Special Publication* 72: 247–67.

Kufeld, W. (1988), Geographisch-planungsrelevante Untersuchungen am Aubachsystem (südlich von Regensburg) als eine Grundlage eines Bachsanierungskonzeptes. *Berichte der Akademie für Naturschutz and Landschaftspflege* 12: 259–302.

McKirdy, A. P. (1990), A handbook of earth science conservation techniques. Appendices to *Earth science conservation in Great Britain: A strategy*. Nature Conservancy Council/English Nature, Peterborough.

Martini, G. (1994), Bilan général de la Protection du Patrimoine Géologique en France. In: G. Martini (ed.) *Proceedings of the 1st International Symposium on the Conservation of our Geological Heritage*, Mémoires de la Société Géologique de France, Paris, 111–18.

Meiburg, P. (1993), Geotopschutz und Geowissenschaftlicher Naturschutz in Hessen. In: D. Förster, A. Grube, and F. W. Wiedenbein (eds.) *Geotopschutz: Materialien 1, Mitwitz, D*, 121–37.

Meyer, D. E. (1997), Zur Bedeutung geologischer Aufschlüsse, Naturdenkmale und Lehrpfade für die Gesellschaft und den Schutz der Natur. *1 Internationale Jahrestag Deutsche Geologische Gesellschaft (Fachsektion Geotopschutz)*, 5: 28–36.

Munthe, H. (1920), Strandgrottor och närstående geologiska fenomen i Sverige. *Naturskyddsutredning*, SGU Ser. C 302, 1–67 (in Swedish).

Nature Conservancy Council (1990), *Earth-science conservation in Great Britain: A strategy*. Nature Conservancy Council/English Nature, Peterborough.

Schulz, W. (1997), Findlingsgärten in Norddeutschland. *1 Internationale Jahrestag Deutsche Geologische Gesellschaft (Fachsektion Geotopschutz)*, 5: 83–7.

Soyez, D. (1971), Erfahrungen bei der Erfassung des glazialmorphologischen Formenschatzes zur Ausweisung von Schutzgebieten in Dalarna/Schweden. *Schriftenreihe für Landschaftspflege und Naturschutz* 6: 221–37.

Stürm, B. (1994), The geotope concept: Geological nature conservation by town and country planning. In: D. O'Halloran, C. Green, M. Harley, M. Stanley, and J. Knill (eds.) *Geological and Landscape Conservation*, 27–31.

van der Maarel, E., and Dauvellier, P. L. (1978), *Naar een globaal ecologische model voor de ruimtelijke ontwikkeling van Nederland*, Rijks Planologische Dienst, Studierapport 9, The Hague, Parts 1 and 2 (in Dutch).

Waterbolk, H. T. (1962), Hauptzüge der eisenzeitlichen Besiedlung der nördlichen Niederlände. *Offa* 19: 9–46.

Westbroek, P. (1991), *Life as a geological force*. Norton, New York.

Wiedenbein, F. W. (1994), German development in earth science conservation. In: G. Martini (ed.) *Proceedings of the 1st International Symposium on the Conservation of our Geological Heritage*, Mémoires de la Société Géologique de France 165, Paris, 119–27.

Wimbledon, W. A. P. (1990), European Heritage Sites and Type Site Inventories. Earth-Science Conservation: An absolute need for science and education. *Jahrbuch Geologische Bundes-Anstalt*, 133/4: 657–8.

—— Andersen, S., Cleal, C. J., Cowie, J. W., Erikstad, L., Gonggrijp, G. P., Johansson, C. E., Karis, L., and Suominen, V. (1999), Geological world heritage: GEOSITES—a global comparative site inventory to enable prioritisation for conservation. *Proceedings of the Second International Symposium of our Geological Heritage/World Heritage: Geotope Conservation world-wide, European and Italian Experiences, Memorie descrittive della Carta Geologica D'Italia*, Rome, LIV. 45–60.

Wolfert, H. P. (1995), Use of the catena principle in geomorphological impact assessment: a functional approach. *Zeitschrift für Geomorphologie* 39: 17–31.

Subject Index

Geographic references have been omitted from the index because the extensive contents and the many location maps in most chapters provide sufficient information to locate toponyms and other geographic entities. Bold numbers denote reference to figures and tables.